Regression Analysis of Count Data, Second Edition

Students in both social and natural sciences often seek regression methods to explain the frequency of events, such as visits to a doctor, auto accidents, or new patents awarded. This book provides the most comprehensive and up-to-date account of models and methods to interpret such data. The authors have conducted research in the field for more than 25 years. In this book, they combine theory and practice to make sophisticated methods of analysis accessible to researchers and practitioners working with widely different types of data and software in areas such as applied statistics, econometrics, marketing, operations research, actuarial studies, demography, biostatistics, and quantitative social sciences. The book may be used as a reference work on count models or by students seeking an authoritative overview. Complementary material in the form of data sets, template programs, and bibliographic resources can be accessed on the Internet through the authors' homepages. This second edition is an expanded and updated version of the first, with new empirical examples and more than two hundred new references added. The new material includes new theoretical topics, an updated and expanded treatment of cross-section models, coverage of bootstrap-based and simulation-based inference, expanded treatment of time series, multivariate and panel data, expanded treatment of endogenous regressors, coverage of quantile count regression, and a new chapter on Bayesian methods.

A. Colin Cameron is Professor of Economics at the University of California, Davis. His research and teaching interests span a range of topics in microeconometrics. He is a past director of the Center on Quantitative Social Science at UC Davis and is currently an associate editor of the *Stata Journal* and *Journal of Econometric Methods.*

Pravin K. Trivedi is Professor of Economics, University of Queensland, and Distinguished Professor Emeritus and J. H. Rudy Professor Emeritus of Economics at Indiana University, Bloomington. During his academic career, he has taught undergraduate and graduate-level econometrics in the United States, Europe, and Australia. His research interests are in microeconometrics and health economics. He served as coeditor of the *Econometrics Journal* from 2000 to 2007 and has been on the board of *Journal of Applied Econometrics* since 1988.

Professors Cameron and Trivedi are coauthors of the first edition of *Regression Analysis of Count Data* (Cambridge University Press, 1998), *Microeconometrics: Methods and Applications* (Cambridge University Press, 2005), and *Microeconomics Using Stata Revised Edition* (2010).

Econometric Society Monographs

Editors:

The Econometric Society is an international society for the advancement of economic theory in relation to statistics and mathematics. The Econometric Society Monograph series is designed to promote the publication of original research contributions of high quality in mathematical economics and theoretical and applied econometrics.

Other titles in the series:

G. S. Maddala, *Limited dependent and qualitative variables in econometrics*, 9780521241434, 9780521338257

Gerard Debreu, *Mathematical economics: Twenty papers of Gerard Debreu*, 9780521237369, 9780521335614

Jean-Michel Grandmont, *Money and value: A reconsideration of classical and neoclassical monetary economics*, 9780521251419, 9780521313643

Franklin M. Fisher, *Disequilibrium foundations of equilibrium economics*, 9780521378567

Andreu Mas-Colell, *The theory of general equilibrium: A differentiable approach*, 9780521265140, 9780521388702

Truman F. Bewley, Editor, *Advances in econometrics – Fifth World Congress (Volume I)*, 9780521467261

Truman F. Bewley, Editor, *Advances in econometrics – Fifth World Congress (Volume II)*, 9780521467254

Hervé Moulin, *Axioms of cooperative decision making*, 9780521360555, 9780521424585

L. G. Godfrey, *Misspecification tests in econometrics: The Lagrange multiplier principle and other approaches*, 9780521424592

Tony Lancaster, *The econometric analysis of transition data*, 9780521437899

Alvin E. Roth and Marilda A. Oliviera Sotomayor, Editors, *Two-sided matching: A study in game-theoretic modeling and analysis*, 9780521437882

Wolfgang Härdle, *Applied nonparametric regression*, 9780521429504

Jean-Jacques Laffont, Editor, *Advances in economic theory – Sixth World Congress (Volume I)*, 9780521484596

Jean-Jacques Laffont, Editor, *Advances in economic theory – Sixth World Congress (Volume II)*, 9780521484602

Halbert White, *Estimation, inference and specification*, 9780521252805, 9780521574464

Christopher Sims, Editor, *Advances in econometrics – Sixth World Congress (Volume I)*, 9780521444590, 9780521566100

Christopher Sims, Editor, *Advances in econometrics – Sixth World Congress (Volume II)*, 9780521444606, 9780521566094

Roger Guesnerie, *A contribution to the pure theory of taxation*, 9780521629560

David M. Kreps and Kenneth F. Wallis, Editors, *Advances in economics and econometrics – Seventh World Congress (Volume I)*, 9780521589833

David M. Kreps and Kenneth F. Wallis, Editors, *Advances in economics and econometrics – Seventh World Congress (Volume II)*, 9780521589826

David M. Kreps and Kenneth F. Wallis, Editors, *Advances in economics and econometrics – Seventh World Congress (Volume III)*, 9780521580137, 9780521589819

Donald P. Jacobs, Ehud Kalai, and Morton I. Kamien, Editors, *Frontiers of research in economic theory: The Nancy L. Schwartz Memorial Lectures, 1983–1997*, 9780521632225, 9780521635387

Steinar Strom, Editor, *Econometrics and economic theory in the 20th century: The Ragnar Frisch Centennial Symposium*, 9780521633239, 9780521633659

Continued on page following the index

Regression Analysis of Count Data

Second Edition

A. Colin Cameron
University of California, Davis

Pravin K. Trivedi
University of Queensland
Indiana University, Bloomington, Emeritus

CAMBRIDGE
UNIVERSITY PRESS

CAMBRIDGE UNIVERSITY PRESS
Cambridge, New York, Melbourne, Madrid, Cape Town,
Singapore, São Paulo, Delhi, Mexico City

Cambridge University Press
32 Avenue of the Americas, New York, NY 10013-2473, USA

www.cambridge.org
Information on this title: www.cambridge.org/9781107667273

First edition published 1998
Second edition published 2013

Printed in the United States of America

A catalog record for this publication is available from the British Library.

Library of Congress Cataloging in Publication Data

Cameron, Adrian Colin.
Regression analysis of count data / A. Colin Cameron, University of California, Davis,
Pravin K. Trivedi, Indiana University, Bloomington. – Second edition.
pages cm. – (Econometric society monographs)
Includes bibliographical references and index.
ISBN 978-1-107-01416-9 (hardback) – ISBN 978-1-107-66727-3 (paperback)
1. Regression analysis. 2. Econometrics. I. Trivedi, P. K. II. Title.
QA278.2.C36 2013
519.5′36–dc23 2012043350

ISBN 978-1-107-01416-9 Hardback
ISBN 978-1-107-66727-3 Paperback

To Michelle and Bhavna

Contents

List of Figures

List of Tables

Preface

Since *Regression Analysis of Count Data* was published in 1998, significant new research has contributed to the range and scope of count data models. This growth is reflected in many new journal articles, fuller coverage in textbooks, and wide interest in and availability of software for handling count data models. These developments (to which we have also contributed) have motivated us to revise and expand the first edition. Like the first edition, this volume reflects an orientation toward practical data analysis.

The revisions in this edition have affected all chapters. First, we have corrected the typographical and other errors in the first edition, improved the graphics throughout, and where appropriate we have provided a cleaner and simpler exposition. Second, we have revised and relocated material that seemed better placed in a different location, mostly within the same chapter though occasionally in a different chapter. For example, material in Chapter 4 (generalized count models), Chapter 8 (multivariate counts), and Chapter 13 (measurement errors) has been pruned and rearranged so the more mainstream topics appear earlier and the more marginal topics have disappeared altogether. For similar reasons bootstrap inference has moved to Chapter 2 from Chapter 5. Our goal here has been to improve quality of synthesis and accessibility of material to the reader. Third, the final few chapters have been reordered. Chapter 10 (endogeneity and selection) has moved up from Chapter 11. It replaces the measurement error chapter that now appears as Chapter 13. Chapter 11 now covers flexible parametric models (previously Chapter 12). And the current Chapter 12, which covers Bayesian methods, is a new addition. Fourth, we have removed material that was of marginal interest and replaced it with material of potentially greater interest, especially to practitioners. For example, as barriers to implementation of more computer-intensive methods have come down, we have liberally sprinkled illustrations of simulation-based methods throughout the book. Fifth, bibliographic notes at the end of every chapter have been refreshed to include newer references and topics. Sixth, we have developed an almost complete set of computer code for the examples in this book.

The first edition has been expanded by about 35%. This expansion reflects the addition of a new Chapter 12 on Bayesian methods as well as significant additions to most other chapters. Chapter 2 has new sections on robust inference and empirical likelihood and includes material on the bootstrap and generalized estimating equations. In Chapter 3 and throughout the book, the term "pseudo-ML" has been changed to "quasi-ML" and robust standard errors are computed using the robust sandwich form. Chapter 4 improves the coverage and discussion of how many alternative count models relate to each other. Censored, truncated, hurdle, zero-inflated, and especially finite mixture models are now covered in greater depth, with a more uniform notation, and hierarchical count models and models with cross-sectional and spatial dependence have been newly added. Chapter 5 moves up presentation of methods for discrimination among nonnested models. Chapter 6 adds a new empirical example of fertility data that poses a fresh challenge to count data modelers. The time series coverage in Chapter 7 has been expanded to include more recently developed models, and there is some rearrangement so that the most often used models appear first. The coverage of multivariate count models in Chapter 8 uses a broader and more modern range of dependence concepts and provides a lengthy treatment of parametric copula-based models. The survey of count data panel models in Chapter 9 gives greater emphasis to moment-based approaches and has a more comprehensive coverage of dynamic panels, the role of initial conditions, conditionally correlated random effects, flexible functional forms, and specification tests. Chapter 10 provides an improved exposition of models with endogeneity and selection, including consideration of latent factor and two-part models as well as simulation-based inference and control function estimators. A major new topic in Chapter 11 is quantile regression models for count data, and the coverage of semiparametric and nonparametric methods has been expanded and updated. As previously mentioned, the new Chapter 12 covers Bayesian analysis of count models, providing an entry to the world of Markov chain Monte Carlo analysis of count models. Finally, Chapter 13 provides a comprehensive survey of measurement error models for count data. As a result of the expanded coverage of old topics and appearance of new ones, the bibliography is now significantly larger and includes more than two hundred additional new references.

To emphasize its empirical orientation, the book has added many new examples based on real data. These examples are scattered throughout the book, especially in Chapters 6–12. In addition, we have a number of examples based on simulated data. Researchers, instructors, and students interested in replicating our results can obtain all the data and computer programs used to produce the results given in this book via Internet from our respective personal web sites.

This revised and expanded second edition draws extensively from our jointly authored research undertaken with Partha Deb, Jie Qun Guo, Judex Hyppolite, Tong Li, Doug Miller, Murat Munkin, and David Zimmer. We thank them all. We also thank Joao Santos Silva for detailed comments on Chapter 10 and Jeff Racine for detailed comments on Chapter 11. The series editor Rosa Matzkin and an anonymous reviewer provided helpful guidance and suggestions for improvements for which we are grateful. As for the first edition, it is a pleasure to acknowledge the overall editorial direction and encouragement of Scott Parris of Cambridge University Press throughout the multiyear process of bringing the project to completion.

A. Colin Cameron
Davis, CA

Pravin K. Trivedi
Bloomington, IN
August 2012

Preface to the First Edition

This book describes regression methods for count data, where the response variable is a non-negative integer. The methods are relevant for analysis of counts that arise in both social and natural sciences.

Despite their relatively recent origin, count data regression methods build on an impressive body of statistical research on univariate discrete distributions. Many of these methods have now found their way into major statistical packages, which has encouraged their application in a variety of contexts. Such widespread use has itself thrown up numerous interesting research issues and themes, which we explore in this book.

The objective of the book is threefold. First, we wish to provide a synthesis and integrative survey of the literature on count data regressions, covering both the statistical and econometric strands. The former has emphasized the framework of generalized linear models, exponential families of distributions, and generalized estimating equations, while the latter has emphasized nonlinear regression and generalized method of moment frameworks. Yet between them there are numerous points of contact which can be fruitfully exploited. Our second objective is to make sophisticated methods of data analysis more accessible to practitioners with different interests and backgrounds. To this end we consider models and methods suitable for cross-section, time series, and longitudinal data. Detailed analyses of several data sets as well as shorter illustrations, implemented from a variety of viewpoints, are scattered throughout the book to put empirical flesh on theoretical or methodological discussion. We draw on examples from, and give references to, works in many applied areas. Our third objective is to highlight the potential for further research by discussion of issues and problems that need more analysis. We do so by embedding count data models in a larger body of econometric and statistical work on discrete variables and, more generally, on nonlinear regression.

The book can be divided into four parts. The first two chapters contain introductory material on count data and a comprehensive review of statistical methods for nonlinear regression models. Chapters 3, 4, 5, and 6 present models and applications for cross-section count data. Chapters 7, 8, and 9 present methods for data other than cross-section data, namely time series, multivariate and longitudinal or panel data. Chapters 10, 11, and 12 present methods for common complications, including measurement error, sample selection and simultaneity, and semiparametric methods. Thus the coverage of the book is qualitatively similar to that in a complete single book on linear regression models.

The book is directed toward researchers, graduate students, and other practitioners in a wide range of fields. Because of our own background in econometrics, the book emphasizes issues arising in econometric applications. Our training and background also influence the organizational structure of the book. But areas outside econometrics are also considered. The essential prerequisite for this book is familiarity with the linear regression model using matrix algebra. The material in the book should be accessible to people with a background in regression and statistical methods up to the level of a standard first-year graduate econometrics text such as Greene's *Econometric Analysis*. While basic count data methods are included in major statistical packages, more advanced analysis can require programming in languages such as Splus, Gauss, or Matlab.

Our own entry into the field of count data models dates back to the early 1980s when we embarked on an empirical study of the demand for health insurance and health care services at the Australian National University. Since then we have been involved in many empirical investigations that have influenced our perceptions of this field. We have included numerous data analytic discussions in this volume to reflect our own interest and those of readers interested in real data applications. The data sets, computer programs, and related materials used in this book will be available through Internet access to our individual web sites. These materials supplement and complement this book and will help new entrants to the field, especially graduate students, to make a relatively easy start.

We have learned much on modeling count data through collaborations with coauthors, notably Partha Deb, Shiferaw Gurmu, Per Johansson, Kajal Mukhopadhyay, and Frank Windmeijer. The burden of writing this book has been eased by help from many colleagues, coauthors, and graduate students. In particular, we thank the following for their generous attention, encouragement, help, and comments on earlier drafts of various chapters: Kurt Brännäs, David Hendry, Primula Kennedy, Tony Lancaster, Scott Long, Grayham Mizon, Neil Shephard, and Bob Shumway, in addition to the coauthors already mentioned. We especially thank David Hendry and Scott Long for their detailed advice on manuscript preparation using Latex software and Scientific Workplace. The manuscript has also benefited from the comments of a referee and the series editor, Alberto Holly, and from the guidance of Scott Parris of Cambridge University Press.

Work on the book was facilitated by periods spent at various institutions. The first author thanks the Department of Statistics and the Research School of Social Sciences at the Australian National University, the Department of Economics at Indiana University in Bloomington, and the University of California in Davis for support during extended leaves to these institutions in 1995 and

1996. The second author thanks Indiana University and European University Institute, Florence, for support during his tenure as Jean Monnet Fellow in 1996, which permitted a period away from regular duties. For shorter periods of stay that allowed us to work jointly, we thank the Department of Economics at Indiana University, SELAPO at University of Munich, and the European University Institute.

Finally we would both like to thank our families for their patience and forbearance, especially during the periods of intensive work on the book. This work would not have been possible at all without their constant support.

A. Colin Cameron
Davis, CA

Pravin K. Trivedi
Bloomington, IN

CHAPTER 1

Introduction

God made the integers, all the rest is the work of man.
– Kronecker

This book is concerned with models of event counts. An event count refers to the number of times an event occurs, for example, the number of airline accidents or earthquakes. It is the realization of a nonnegative integer-valued random variable. A *univariate* statistical model of event counts usually specifies a probability distribution of the number of occurrences of the event known up to some parameters. Estimation and inference in such models are concerned with the unknown parameters, given the probability distribution and the count data. Such a specification involves no other variables, and the number of events is assumed to be independently identically distributed (iid). Much early theoretical and applied work on event counts was carried out in the univariate framework. The main focus of this book, however, is on *regression analysis* of event counts.

The statistical analysis of counts within the framework of discrete parametric distributions for univariate iid random variables has a long and rich history (Johnson, Kemp, and Kotz, 2005). The Poisson distribution was derived as a limiting case of the binomial by Poisson (1837). Early applications include the classic study of Bortkiewicz (1898) of the annual number of deaths in the Prussian army from being kicked by mules. A standard generalization of the Poisson is the negative binomial distribution. It was derived by Greenwood and Yule (1920), as a consequence of apparent contagion due to unobserved heterogeneity, and by Eggenberger and Polya (1923) as a result of true contagion. The biostatistics literature of the 1930s and 1940s, although predominantly univariate, refined and brought to the forefront seminal issues that have since permeated regression analysis of both counts and durations. The development of the counting process approach unified the treatment of counts and durations. Much of the vast literature on iid counts, which addresses issues such as heterogeneity and overdispersion, true versus apparent contagion, and identifiability of Poisson mixtures, retains its relevance in the context of count data regressions. This leads to models such as the negative binomial regression model.

Significant early developments in count models took place in actuarial science, biostatistics, and demography. In recent years these models have also been used extensively in economics, political science, and sociology. The special features of data in their respective fields of application have fueled developments that have enlarged the scope of these models. An important milestone in the development of count data models for regression was the emergence of the "generalized linear models," of which the Poisson regression is a special case, first described by Nelder and Wedderburn (1972) and detailed in McCullagh and Nelder (1983, 1989). Building on these contributions, the papers by Gourieroux, Monfort, and Trognon (1984a, b) and the work on longitudinal or panel count data models by Hausman, Hall, and Griliches (1984) have also been very influential in stimulating applied work in the econometric literature.

Regression analysis of counts is motivated by the observation that in many, if not most, real-life contexts, the iid assumption is too strong. For example, the mean rate of occurrence of an event may vary from case to case and may depend on some observable variables. The investigator's main interest therefore may lie in the role of covariates (regressors) that are thought to affect the parameters of the conditional distribution of events, given the covariates. This is usually accomplished by a regression model for event count. At the simplest level we may think of this task in the conventional regression framework in which the dependent variable, y, is restricted to be a nonnegative random variable whose conditional mean depends on some vector of regressors, $\mathbf{x}$.

At a different level of abstraction, an event may be thought of as the realization of a point process governed by some specified *rate of occurrence* of the event. The number of events may be characterized as the total number of such realizations over some unit of time. The dual of the event count is the *interarrival time*, defined as the length of the period between events. Count data regression is useful in studying the occurrence rate per unit of time conditional on some covariates. One could instead study the distribution of interarrival times conditional on covariates. This leads to regression models of *waiting times* or *durations*. The type of data available – cross-sectional, time series, or longitudinal – will affect the choice of the statistical framework.

An obvious first question is whether "special methods" are required to handle count data or whether the standard Gaussian linear regression may suffice. More common regression estimators and models, such as the ordinary least squares in the linear regression model, ignore the restricted support for the dependent variable. This leads to significant deficiencies unless the mean of the counts is high, in which case normal approximation and related regression methods may be satisfactory.

The Poisson (log-linear) regression not only is motivated by the usual considerations for regression analysis but also seeks to preserve and exploit as much as possible the nonnegative and integer-valued aspect of the outcome. At one level one might simply regard it as a special type of *nonlinear* regression that respects the discreteness of the count variable. In some analyses this specific distributional assumption may be given up, while preserving nonnegativity.

In econometrics the interest in count data models is a reflection of the general interest in modeling discrete aspects of individual economic behavior. For example, Pudney (1989) characterizes a large body of microeconometrics as "econometrics of corners, kinks and holes." Count data models are specific types of discrete data regressions. Discrete and limited dependent variable models have attracted a great deal of attention in econometrics and have found a rich set of applications in microeconometrics (Maddala, 1983), especially as econometricians have attempted to develop models for the many alternative types of sample data and sampling frames. Although the Poisson regression provides a starting point for many analyses, attempts to accommodate numerous real-life conditions governing observation and data collection lead to additional elaborations and complications.

This introductory chapter presents the Poisson distribution and some of its characterizations in section 1.1, the Poisson regression model in section 1.2, and some leading examples of count data in section 1.3. An outline of the book is provided in section 1.4.

The scope of count data models is very wide. This monograph addresses issues that arise in the regression models for counts, with a particular focus on features of economic data. In many cases, however, the material covered can be easily adapted for use in other social sciences and in natural sciences, which do not always share the peculiarities of economic data.

1.1 POISSON DISTRIBUTION AND ITS CHARACTERIZATIONS

The benchmark parametric model for count data is the Poisson distribution. It is useful at the outset to review some fundamental properties and characterization results of the Poisson distribution (for derivations, see Taylor and Karlin, 1998).

If the discrete random variable Y is Poisson distributed with *intensity* or rate parameter μ, $\mu > 0$, and t is the *exposure,* defined as the length of time during which the events are recorded, then Y has density

$$\Pr[Y = y] = \frac{e^{-\mu t}(\mu t)^y}{y!}, \qquad y = 0, 1, 2, \ldots, \tag{1.1}$$

where $\mathsf{E}[Y] = \mathsf{V}[Y] = \mu t$.

If we set the length of the exposure period t equal to unity, then

$$\Pr[Y = y] = \frac{e^{-\mu}\mu^y}{y!}, \qquad y = 0, 1, 2, \ldots. \tag{1.2}$$

The probabilities satisfy the recurrence relation

$$\Pr[Y = y + 1](y + 1) = \Pr[Y = y]\mu. \tag{1.3}$$

The Poisson distribution has a single parameter μ and we refer to it as $\mathsf{P}[\mu]$. Its kth raw moment, $\mathsf{E}[Y^k]$, may be derived by differentiating the moment generating function (mgf) k times

$$\mathsf{M(t)} \equiv \mathsf{E}[e^{tY}] = \exp\{\mu(e^t - 1)\},$$

with respect to t and evaluating at $t = 0$. This yields the following four raw moments:

$$\begin{aligned}
\mu_1' &= \mu \\
\mu_2' &= \mu + \mu^2 \\
\mu_3' &= \mu + 3\mu^2 + \mu^3 \\
\mu_4' &= \mu + 7\mu^2 + 6\mu^3 + \mu^4.
\end{aligned}$$

Following convention, raw moments are denoted by primes, and central moments without primes. The central moments around μ can be derived from the raw moments in the standard way. Note that the first two central moments, denoted μ_1 and μ_2, respectively, are equal to μ. The central moments satisfy the recurrence relation

$$\mu_{r+1} = r\mu\mu_{r-1} + \mu\frac{\partial\mu_r}{\partial\mu}, \qquad r = 1, 2, \ldots, \tag{1.4}$$

where $\mu_0 = 0$.

Equality of the mean and variance is referred to as the *equidispersion* property of the Poisson. This property is frequently violated in real-life data. *Overdispersion* (*underdispersion*) means that the variance exceeds (is less than) the mean.

A key property of the Poisson distribution is additivity. This is stated by the following *Countable Additivity Theorem* (for a mathematically precise statement, see Kingman, 1993).

Theorem. *If* $Y_i \sim \mathsf{P}[\mu_i]$, $i = 1, 2, \ldots$ *are independent random variables, and if* $\sum \mu_i < \infty$, *then* $S_Y = \sum Y_i \sim \mathsf{P}[\sum \mu_i]$.

The binomial and the multinomial can be derived from the Poisson by appropriate conditioning. Under the already stated conditions,

$$\begin{aligned}
&\Pr[Y_1 = y_1, Y_2 = y_2, \ldots, Y_n = y_n \mid S_Y = s] \\
&= \left[\prod_{j=1}^{n} \frac{e^{-\mu_j}\mu_j^{y_j}}{y_j!}\right] \Big/ \left[\frac{\left(\sum \mu_i\right)^s e^{-\sum \mu_i}}{s!}\right] \\
&= \frac{s!}{y_1!y_2!\ldots y_n!}\left(\frac{\mu_1}{\sum \mu_i}\right)^{y_1}\left(\frac{\mu_2}{\sum \mu_i}\right)^{y_2}\cdots\left(\frac{\mu_n}{\sum \mu_i}\right)^{y_n} \\
&= \frac{s!}{y_1!y_2!\ldots y_n!}\pi_1^{y_1}\pi_2^{y_2}\ldots\pi_n^{y_n},
\end{aligned} \tag{1.5}$$

where $s = \sum Y_i$ and $\pi_j = \mu_j / \sum \mu_i$. This is the multinomial distribution $\mathsf{m}[s\,; \pi_1, \ldots, \pi_n]$. The binomial is the case $n = 2$.

There are many characterizations of the Poisson distribution. Here we consider four. The first, the law of rare events, is a common motivation for the Poisson. The second, the Poisson counting process, is very commonly encountered in an introduction to stochastic processes. The third is simply the dual of the second, with waiting times between events replacing the count. The fourth characterization, the Poisson-stopped binomial, treats the number of events as repetitions of a binomial outcome, with the number of repetitions taken as Poisson distributed.

1.1.1 Poisson as the Law of Rare Events

The law of rare events states that the total number of events will follow, approximately, the Poisson distribution if an event may occur in any of a large number of trials, but the probability of occurrence in any given trial is small.

More formally, let $Y_{n,\pi}$ denote the total number of successes in a large number n of independent Bernoulli trials, with the success probability π of each trial being small. Then the distribution is binomial with

$$\Pr\left[Y_{n,\pi} = k\right] = \binom{n}{k}\pi^k(1-\pi)^{n-k}, \quad k = 0, 1, \ldots, n.$$

In the limiting case where $n \to \infty$, $\pi \to 0$, and $n\pi = \mu > 0$ – that is, the average μ is held constant while $n \to \infty$ – we have

$$\lim_{n\to\infty}\left[\binom{n}{k}\left(\frac{\mu}{n}\right)^k\left(1-\frac{\mu}{n}\right)^{n-k}\right] = \frac{\mu^k e^{-\mu}}{k!},$$

the Poisson probability distribution with parameter μ, denoted as $\mathsf{P}[\mu]$.

1.1.2 Poisson Counting Process

The Poisson distribution has been described as characterizing "complete randomness" (Kingman, 1993). To elaborate this feature, the connection between the Poisson distribution and the *Poisson process* needs to be made explicit. Such an exposition begins with the definition of a *counting process*.

A stochastic process $\{N(t), t \geq 0\}$ is defined to be a counting process if $N(t)$ denotes an event count up to time t. $N(t)$ is nonnegative and integer valued and must satisfy the property that $N(s) \leq N(t)$ if $s < t$, and $N(t) - N(s)$ is the number of events in the interval $(s, t]$. If the event counts in disjoint time intervals are independent, the counting process is said to have independent increments. It is said to be stationary if the distribution of the number of events depends only on the length of the interval.

The Poisson process can be represented in one dimension as a set of points on the time axis representing a random series of events occurring at points of time. The Poisson process is based on notions of independence and the Poisson distribution in the following sense.

Define μ to be the constant rate of occurrence of the event of interest, and $N(s, s+h)$ to be the number of occurrences of the event in the time interval $(s, s+h]$. A (pure) Poisson process of rate μ occurs if events occur independently with constant probability equal to μ times the length of the interval. The numbers of events in disjoint time intervals are independent, and the distribution of events in each interval of unit length is $\mathsf{P}[\mu]$. Formally, as the length of the interval $h \to 0$,

$$\begin{aligned} \Pr[N(s, s+h) = 0] &= 1 - \mu h + o(h) \\ \Pr[N(s, s+h) = 1] &= \mu h + o(h), \end{aligned} \tag{1.6}$$

where $o(h)$ denotes a remainder term with the property $o(h)/h \to 0$ as $h \to 0$. $N(h, s+h)$ is statistically independent of the number and position of events in $(s, s+h]$. Note that in the limit the probability of two or more events occurring is zero, whereas 0 and 1 events occur with probabilities of, respectively, $(1 - \mu h)$ and μh. For this process it can be shown (Taylor and Karlin, 1998) that the number of events occurring in the interval $(s, s+h]$, for nonlimit h, is Poisson distributed with mean μh and probability

$$\Pr[N(s, s+h) = r] = \frac{e^{-\mu h}(\mu h)^r}{r!} \qquad r = 0, 1, 2, \ldots. \tag{1.7}$$

Normalizing the length of the exposure time interval to be unity, $h = 1$, leads to the Poisson density given previously. In summary, the counting process $N(t)$ with stationary and independent increments and $N(0) = 0$, which satisfies (1.6), generates events that follow the Poisson distribution.

1.1.3 Waiting-Time Distributions

We now consider a characterization of the Poisson that is the flip side of that given in the preceding paragraph. Let W_1 denote the time of the first event, and $W_r, r \geq 1$ the time between the $(r-1)^{th}$ and r^{th} event. The nonnegative random sequence $\{W_r, r \geq 1\}$ is called the sequence of *interarrival times*, *waiting times*, *durations*, or *sojourn times*. In addition to, or instead of, analyzing the number of events occurring in the interval of length h, one can analyze the duration of time between successive occurrences of the event, or the time of occurrence of the rth event, W_r. This requires the distribution of W_r, which can be determined by exploiting the duality between event counts and waiting times. This is easily done for the Poisson process.

The outcome $\{W_1 > t\}$ occurs only if no events occur in the interval $[0, t]$. That is,

$$\Pr[W_1 > t] = \Pr[N(t) = 0] = e^{-\mu t}, \tag{1.8}$$

which implies that W_1 has exponential distribution with mean $1/\mu$. The waiting time to the first event, W_1, is exponentially distributed with density

$f_{W_1}(t) = \mu e^{-\mu t}, t \geqslant 0$. Also,

$$\begin{aligned}\Pr[W_2 > t | W_1 = s] &= \Pr[N(s, s+t) = 0 | W_1 = s] \\ &= \Pr[N(s, s+t) = 0] \\ &= e^{-\mu t},\end{aligned}$$

using the properties of independent stationary increments. This argument can be repeated for W_r to yield the result that $W_r, r = 1, 2, \ldots$, are iid exponential random variables with mean $1/\mu$. This result reflects the property that the Poisson process has no memory.

In principle, the duality between the number of occurrences and time between occurrences suggests that count and duration data should be covered in the same framework. Consider the arrival time of the rth event, denoted S_r,

$$S_r = \sum_{i=1}^{r} W_i, \quad r \geq 1. \tag{1.9}$$

It can be shown using results on sums of random variables of exponential random variables that S_r has one-parameter gamma distribution with density

$$f_{S_r}(t) = \frac{\mu^r t^{r-1}}{(r-1)!} e^{-\mu t}, \quad t \geq 0. \tag{1.10}$$

This is a special case of a more general result derived in Chapter 4.10. The density (1.10) can also be derived by observing that

$$N(t) \geq r \Leftrightarrow S_r \leq t. \tag{1.11}$$

Hence

$$\begin{aligned}\Pr[N(t) \geq r] &= \Pr[S_r \leq t] \\ &= \sum_{j=r}^{\infty} e^{-\mu t} \frac{(\mu t)^j}{j!}.\end{aligned} \tag{1.12}$$

To obtain the density (1.10) of S_r, the cumulative distribution function (cdf) given in (1.12) is differentiated with respect to t. Thus, the Poisson process may be characterized in terms of the implied properties of the waiting times.

Suppose one's main interest is in the role of the covariates that determine the Poisson process rate parameter μ. For example, let $\mu = \exp(\mathbf{x}'\boldsymbol{\beta})$. Then from (1.8) the mean waiting time is given by $1/\mu = \exp(-\mathbf{x}'\boldsymbol{\beta})$, confirming the intuition that the covariates affect the mean number of events and the waiting times in opposite directions. This illustrates that, from the viewpoint of studying the role of covariates, analyzing the frequency of events is the dual complement of analyzing the waiting times between events.

The Poisson process is often too restrictive in practice. Mathematically tractable and computationally feasible common links between more general count and duration models are hard to find (see Chapter 4).

In the waiting time literature, emphasis is on estimating the *hazard rate*, the conditional instantaneous probability of the event occurring given that it has not yet occurred, controlling for censoring due to not always observing occurrence of the event. Andersen, Borgan, Gill and Keiding (1993), Kalbfleisch and Prentice (2002), and Fleming and Harrington (2005) present, in great detail, models for censored duration data based on the application of martingale theory to counting processes.

We focus on counts. Even when duration is the more natural entity for analysis, it may not be observed. If only event counts are available, count regressions still provide an opportunity for studying the role of covariates in explaining the mean rate of event occurrence. However, count analysis leads in general to a loss of efficiency (Dean and Balshaw, 1997).

1.1.4 Binomial Stopped by the Poisson

Yet another characterization of the Poisson involves mixtures of the Poisson and the binomial. Let n be the actual (or true) count process taking nonnegative integer values with $\mathsf{E}[n] = \mu$, and $\mathsf{V}[n] = \sigma^2$. Let $B_1, B_2, \ldots, B_n$ be a sequence of n independent and identically distributed Bernoulli trials, where each B_i takes only two values, 1 or 0, with probabilities π and $1 - \pi$, respectively. Define the count variable $Y = \sum_{i=1}^{n} B_i$. For n given, Y follows a binomial distribution with parameters n and π. Hence,

$$\begin{aligned} \mathsf{E}[Y] &= \mathsf{E}[\mathsf{E}[Y|n]] = \mathsf{E}[n\pi] = \pi\mathsf{E}[n] = \mu\pi \\ \mathsf{V}[Y] &= \mathsf{V}[\mathsf{E}[Y|n]] + \mathsf{E}[\mathsf{V}[Y|n]] = (\sigma^2 - \mu)\pi^2 + \mu\pi. \end{aligned} \tag{1.13}$$

The actual distribution of Y depends on the distribution of n. For Poisson-distributed n it can be found using the following lemma.

Lemma. *If π is the probability that $B_i = 1$, $i = 1, \ldots, n$, and $1 - \pi$ the probability that $B_i = 0$, and $n \sim \mathsf{P}[\mu]$, then $Y \sim \mathsf{P}[\mu\pi]$.*

To derive this result, begin with the probability generating function (pgf), defined as $g(s) = \mathsf{E}[s^B]$, of the Bernoulli random variable B,

$$g(s) = (1 - \pi) + \pi s, \tag{1.14}$$

for any real s. Let $f(s)$ denote the pgf of the Poisson variable n, $\mathsf{E}[s^n]$; that is,

$$f(s) = \exp(-\mu + \mu s). \tag{1.15}$$

Then the pgf of Y is obtained as

$$\begin{aligned} f(g(s)) &= \exp[-\mu + \mu g(s)] \\ &= \exp[-\mu\pi + \mu\pi s], \end{aligned} \tag{1.16}$$

which is the pgf of Poisson-distributed n with parameter $\mu\pi$. This characterization of the Poisson has been called the Poisson-*stopped* binomial. This characterization is useful if the count is generated by a random number of repetitions of a binary outcome.

1.2 POISSON REGRESSION

The approach taken to the analysis of count data, especially the choice of the regression framework, sometimes depends on how the counts are assumed to arise. There are two common formulations. In the first, counts arise from a direct observation of a point process. In the second, counts arise from discretization (ordinalization) of continuous latent data. Other less used formulations appeal, for example, to the law of rare events or the binomial stopped by Poisson.

1.2.1 Counts Derived from a Point Process

Directly observed counts arise in many situations. Examples are the number of telephone calls arriving at a central telephone exchange, the number of monthly absences at the place of work, the number of airline accidents, the number of hospital admissions, and so forth. The data may also consist of interarrival times for events. In the simplest case, the underlying process is assumed to be stationary and homogeneous, with iid arrival times for events and other properties stated in the previous section.

1.2.2 Counts Derived from Continuous Data

Count-type variables sometimes arise from categorization of a latent continuous variable as the following example indicates. Credit rating of agencies may be stated as "AAA," "AAB," "AA," "A," "BBB," "B," and so forth, where "AAA" indicates the greatest credit worthiness. Suppose we code these as $y = 0, 1, \ldots, m$. These are pseudocounts that can be analyzed using a count regression. But one may also regard this categorization as an ordinal ranking that can be modeled using a suitable latent variable model such as ordered probit. Section 3.6 provides a more detailed exposition.

1.2.3 Regression Specification

The standard model for count data is the *Poisson regression model,* which is a nonlinear regression model. This regression model is derived from the Poisson distribution by allowing the *intensity parameter* μ to depend on covariates (regressors). If the dependence is parametrically exact and involves exogenous covariates but no other source of stochastic variation, we obtain the standard Poisson regression. If the function relating μ and the covariates is stochastic, possibly because it involves unobserved random variables, then one obtains a *mixed Poisson regression,* the precise form of which depends on the assumptions about the random term. Chapter 4 deals with several examples of this type.

A standard application of Poisson regression is to cross-section data. Typical cross-section data for applied work consist of n independent observations, the ith of which is $(y_i, \mathbf{x}_i)$. The scalar dependent variable y_i is the number of occurrences of the event of interest, and $\mathbf{x}_i$ is the vector of linearly independent regressors that are thought to determine y_i. A regression model based on this distribution follows by conditioning the distribution of y_i on a k-dimensional vector of covariates, $\mathbf{x}_i' = [x_{1i}, \ldots, x_{ki}]$, and parameters $\boldsymbol{\beta}$, through a continuous function $\mu(\mathbf{x}_i, \boldsymbol{\beta})$, such that $\mathsf{E}[y_i|\mathbf{x}_i] = \mu(\mathbf{x}_i, \boldsymbol{\beta})$.

That is, y_i given $\mathbf{x}_i$ is Poisson distributed with density

$$f(y_i|\mathbf{x}_i) = \frac{e^{-\mu_i}\mu_i^{y_i}}{y_i!}, \qquad y_i = 0, 1, 2, \ldots \tag{1.17}$$

In the *log-linear* version of the model the mean parameter is parameterized as

$$\mu_i = \exp(\mathbf{x}_i'\boldsymbol{\beta}), \tag{1.18}$$

to ensure $\mu > 0$. Equations (1.17) and (1.18) jointly define the Poisson (log-linear) regression model. If one does not wish to impose any distributional assumptions, (1.18) by itself may be used for (nonlinear) regression analysis.

For notational economy we write $f(y_i|\mathbf{x}_i)$ in place of the more formal $f(Y_i = y_i|\mathbf{x}_i)$, which distinguishes between the random variable Y and its realization y. Throughout this book we refer to $f(\cdot)$ as a density even though more formally it is a probability mass function.

By the property of the Poisson, $\mathsf{V}[y_i|\mathbf{x}_i] = \mathsf{E}[y_i|\mathbf{x}_i]$, implying that the conditional variance is not a constant, and hence the regression is intrinsically heteroskedastic. In the log-linear version of the model the mean parameter is parameterized as (1.18), which implies that the conditional mean has a multiplicative form given by

$$\begin{aligned}\mathsf{E}[y_i|\mathbf{x}_i] &= \exp(\mathbf{x}_i'\boldsymbol{\beta}) \\ &= \exp(x_{1i}\beta_1)\exp(x_{2i}\beta_2)\cdots\exp(x_{ki}\beta_k),\end{aligned}$$

with interest often lying in changes in this conditional mean due to changes in the regressors. The additive specification, $\mathsf{E}[y_i|\mathbf{x}_i] = \mathbf{x}_i'\boldsymbol{\beta} = \sum_{j=1}^{k} x_{ji}\beta_j$, is likely to be unsatisfactory because certain combinations of $\boldsymbol{\beta}$ and $\mathbf{x}_i$ will violate the nonnegativity restriction on μ_i.

The Poisson model is closely related to the models for analyzing counted data in the form of proportions or ratios of counts sometimes obtained by grouping data. In some situations – for example when the population "at risk" is changing over time in a known way – it is helpful to reparameterize the model as follows. Let y be the observed number of events (e.g., accidents), N the known total exposure to risk (i.e., number at risk), and $\mathbf{x}$ the known set of k explanatory variables. The mean number of events μ may be expressed as the product of N and π, the probability of the occurrence of event, sometimes also called the hazard rate. That is, $\mu(\mathbf{x}) = N(\mathbf{x})\pi(\mathbf{x}, \boldsymbol{\beta})$. In this case the probability π is parameterized in terms of covariates. For example, $\pi = \exp(\mathbf{x}'\boldsymbol{\beta})$. This leads to a *rate* form of the Poisson model with the density

$$\Pr[Y = y|N(\mathbf{x}), \mathbf{x}] = \frac{e^{-\mu(\mathbf{x})}\mu(\mathbf{x})^y}{y!}, \qquad y = 0, 1, 2, \ldots. \qquad (1.19)$$

Variants of the Poisson regression arise in a number of ways. As was mentioned previously, the presence of an unobserved random error term in the conditional mean function, denoted ν_i, implies that we specify it as $\mathsf{E}[y_i|\mathbf{x}_i, \nu_i]$. The marginal distribution of y_i, $f(y_i|\mathbf{x}_i)$, will involve the moments of the distribution of ν_i. This is one way in which mixed Poisson distributions may arise.

1.3 EXAMPLES

Patil (1970), in earlier work, gives numerous applications of count data analysis in the physical and natural sciences, but usually not in the regression context.

For count data regression there are now many examples of models that use cross-sectional, time-series, or longitudinal data. Economics examples include models of counts of doctor visits and other types of health care utilization; occupational injuries and illnesses; absenteeism in the workplace; recreational or shopping trips; automobile insurance rate making; labor mobility; entry and exits from industry; takeover activity in business; mortgage prepayments and loan defaults; bank failures; patent registration in connection with industrial research and development; and frequency of airline accidents. Other social science applications include those in demographic economics, crime victimology, marketing, political science and government, sociology, and so forth.

The data used in many of these applications have certain commonalities. Events considered are often rare. Zero event counts are often dominant, leading to a skewed distribution. Also there may be a great deal of unobserved heterogeneity in the individual experiences of the event in question. This unobserved heterogeneity leads to overdispersion; that is, the actual variance of the process exceeds the nominal Poisson variance even after regressors are introduced.

Several examples are described in the remainder of this section. The first few examples are leading examples used for illustrative purposes throughout this book.

1.3.1 Health Services

Health economics research is often concerned with the link between health service utilization and economic variables such as income and price, especially the latter, which can be lowered considerably by holding a health insurance policy. Ideally one would measure utilization by expenditures, but if data come from surveys of individuals it is more common to have data on the number of times that health services are consumed, such as the number of visits to a doctor in the past month and number of days in hospital in the past year, because individuals can better answer such questions than those on expenditure.

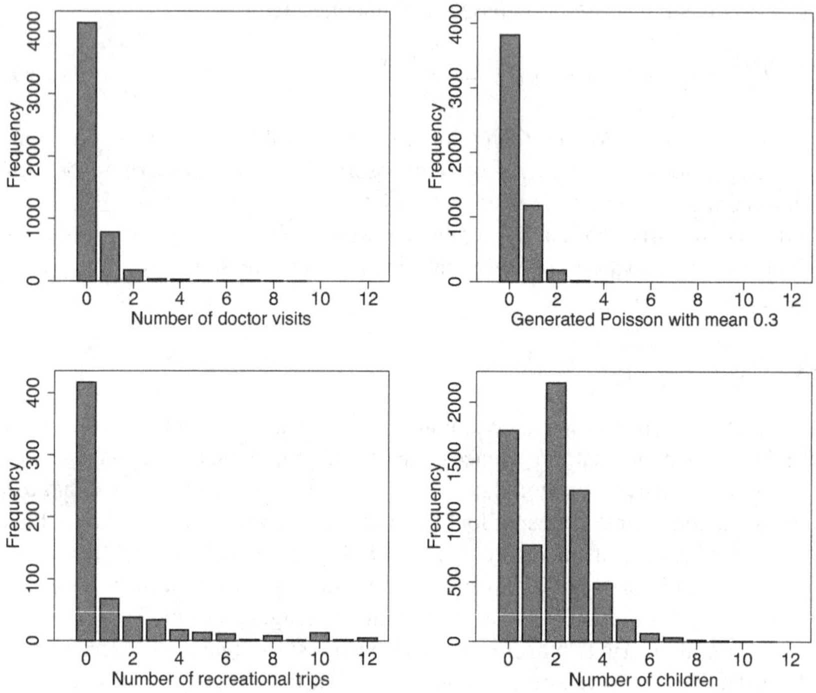

Figure 1.1. Frequency distributions of counts for four types of events: Doctor visits, generated Poisson data, recreational trips, and number of children.

Data sets with health care utilization measured in counts include the National Health Interview Surveys and the Surveys on Income and Program Participation in the United States, the German Socioeconomic Panel, and the Australian Health Surveys.

The upper left panel of Figure 1.1 presents a histogram for data on the number of doctor consultations in the past two weeks from the 1977–78 Australian Health Survey. There is some overdispersion in the data, with the sample variance of 0.64 approximately twice the sample mean of 0.30. This modest overdispersion leads to somewhat more counts of zero (excess zeros) and fewer counts of one or more than if the data were generated from the Poisson with mean 0.30; see the histogram in the upper right panel of Figure 1.1. These data from Cameron and Trivedi (1988) are analyzed in detail in Chapter 3, and similar health use data are analyzed in Chapter 6.3.

There are numerous studies of the impact of insurance on health care use as measured by the number of services. For example, Winkelmann (2004, 2006) studies the impact of health reform on doctor visits in Germany, using data from the German Socioeconomic Panel, whereas Deb and Trivedi (2002) analyze count data from the Rand Health Insurance Experiment.

1.3.2 Recreational Demand

In environmental economics, one is often interested in alternative uses of a natural resource such as forest or parkland. To analyze the valuation placed on such a resource by recreational users, economists often model the frequency of the visits to a particular site as a function of the cost of usage and the economic and demographic characteristics of the users.

The lower left panel of Figure 1.1 presents a histogram for survey data on the number of recreational boating trips to Lake Somerville in East Texas. In this case there is considerable overdispersion, with the sample variance of 39.59 much larger than the sample mean of 2.24. There is a very large spike at zero, as well as a very long tail (the printed histogram was truncated at 12 trips). Various models for these data from Ozuna and Gomaz (1995) are implemented in Chapter 6.4.

1.3.3 Completed Fertility

Completed fertility refers to the total number of children born to a woman who has completed childbearing. A popular application of the Poisson regression is to model this outcome.

The lower right panel of Figure 1.1 presents a histogram for fertility data from the British Household Panel Survey. In this case there is little overdispersion, with sample variance of 2.28 compared to the sample mean of 1.86; the range of the data is quite narrow, with 98% of observations in the range 0–5 and the highest count being 11. Most notably, the data are bimodal, whereas the Poisson is unimodal. Various models for these data are implemented in Chapter 6.5.

1.3.4 Time Series of Asthma Counts

The preceding three examples were of cross-sectional counts. By contrast, Davis, Dunsmuir, and Street (2003) and Davis, Dunsmuir, and Wang (1999) analyze time series counts of daily admissions for asthma to a single hospital at Campbelltown in the Sydney metropolitan area over the four years, 1990 to 1993. Figure 1.2 presents the frequency distribution of the count (upper left panel), a time-series plot of the counts (upper right panel), and autocorrelation coefficients of the counts (lower left panel) and of the squared counts (lower right panel). The mean of the series is 1.94 and its variance is 2.71, so there is some overdispersion. There is strong positive autocorrelation that is slow to disappear. Furthermore, the time series exhibits some periodicity. The study by Davis et al. (2003) confirms the presence of a Sunday effect, a Monday effect, a positive linear time trend, and a seasonal effect. Thus the standard time series data characteristics of continuous data also appear in this discrete data example. Time series models are presented in Chapter 7.

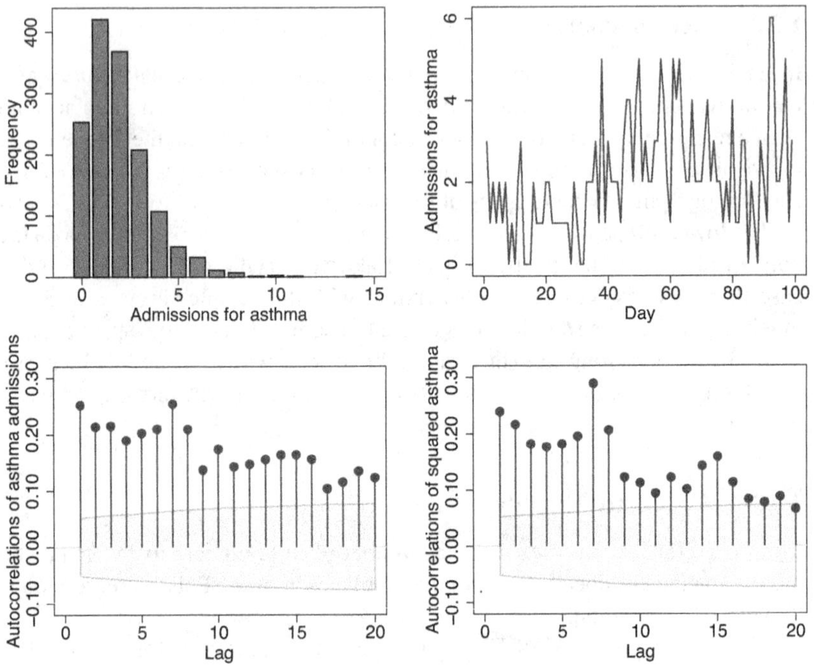

Figure 1.2. Daily data on the number of hospital admissions for asthma.

1.3.5 Panel Data on Patents

The link between research and development and product innovation is an important issue in empirical industrial organization. Product innovation is difficult to measure, but the number of patents is one indicator of it that is commonly analyzed. Panel data on the number of patents received annually by firms in the United States are analyzed by Hausman, Hall, and Griliches (1984), as well as in many subsequent studies. Most panel studies estimate static models, but some additionally introduce lagged counts as regressors. Panel data on patents are studied in detail in Chapter 9.

1.3.6 Takeover Bids

In empirical finance the bidding process in a takeover is sometimes studied either by using the probability of any additional takeover bids after the first using a binary outcome model or by using the number of bids as a dependent variable in a count regression. Jaggia and Thosar (1993) use cross-section data on the number of bids received by 126 U.S. firms that were targets of tender offers and actually taken over within one year of the initial offer. The dependent count variable is the number of bids after the initial bid received by the target firm. Interest centers on the role of management actions to discourage takeover, the role of government regulators, the size of the firm, and the extent to which the firm was undervalued at the time of the initial bid. These data are used in Chapter 5.

1.3.7 Bank Failures

In insurance and finance, the frequency of the failure of a financial institution and the time to failure of the institution are variables of interest. Davutyan (1989) estimates a Poisson model for annual data on U.S. bank failures from 1947 to 1981. The focus is on the relation between bank failures and overall bank profitability, corporate profitability, and bank borrowings from the Federal Reserve Bank. The sample mean and variance of bank failures are, respectively, 6.343 and 11.820, suggesting some overdispersion. These time series are obtained at much longer frequency than the daily asthma data, so there are only 35 observations.

1.3.8 Accident Insurance

In the insurance literature the frequency of accidents and the cost of insurance claims are often the variables of interest because they have an important impact on insurance premiums. Dionne and Vanasse (1992) use data for 19,013 drivers in Quebec on the number of accidents with damage in excess of $250 reported to police over a one-year period. Most drivers had no accidents. The frequencies are very low, with a sample mean of 0.070, though there are many observations. The sample variance of 0.078 is close to the mean. Their paper uses cross-section estimates of the regression to derive predicted claims frequencies, and hence insurance premiums, from data on different individuals with different characteristics and records.

1.3.9 Credit Rating

How frequently mortgagees or credit card holders fail to meet their financial obligations is a subject of interest in credit ratings. Often the number of defaulted payments is studied as a count variable. Greene (1994) analyzes the number of major derogatory reports made after a delinquency of 60 days or longer on a credit account, for 1,319 individual applicants for a major credit card. Major derogatory reports are found to decrease with increases in the expenditure-income ratio (average monthly expenditure divided by yearly income). Age, income, average monthly credit-card expenditures, and whether the individual holds another credit card are statistically insignificant.

1.3.10 Presidential Appointments

Univariate probability models and time series Poisson regressions have been used to model the frequency with which U.S. presidents were able to appoint U.S. Supreme Court Justices (King, 1987a). King's regression model uses the exponential conditional mean function, with the number of presidential appointments per year as the dependent variable. Explanatory variables are the number of previous appointments, the percentage of population that was in the military on active duty, the percentage of freshmen in the House of Representatives, and the log of the number of seats in the Court. It is argued that the presence of lagged appointments in the mean function permits a test for serial independence. King's results suggest negative dependence. However, it is an interesting issue whether the lagged variable should enter multiplicatively or additively. Chapter 7 considers this issue.

1.3.11 Criminal Careers

Nagin and Land (1993) use longitudinal data on 411 men for 20 years to study the number of recorded criminal offenses as a function of observable traits of criminals. These observable traits include psychological variables (e.g., IQ, risk preference, neuroticism), socialization variables (e.g., parental supervision or attachments), and family background variables. The authors model an individual's mean rate of offending in a period as a function of time-varying and time-invariant characteristics, allowing for unobserved heterogeneity among the subjects. Furthermore, they also model the probability that the individual may be criminally "inactive" in the given period. Finally, the authors adopt a nonparametric treatment of unobserved interindividual differences (see Chapter 4 for details). This sophisticated modeling exercise allows the authors to classify criminals into different groups according to their propensity to commit crime.

1.3.12 Doctoral Publications

Using a sample of about 900 doctoral candidates, Long (1997) analyzes the relation between the number of doctoral publications in the final three years of Ph.D. studies and gender, marital status, number of young children, prestige of Ph.D. department, and number of articles by mentor in the preceding three years. He finds evidence that scientists fall into two well-defined groups: "publishers" and "nonpublishers." The observed nonpublishers are drawn from both groups because some potential publishers may not have published just by chance, swelling the numbers who will "never" publish. The author argues that the results are plausible as "there are scientists who, for structural reasons, will not publish" (Long, 1997, p. 249).

1.3.13 Manufacturing Defects

The number of defects per area in a manufacturing process is studied by Lambert (1992) using data from a soldering experiment at AT&T Bell Laboratories. In this application components are mounted on printed wiring boards by soldering their leads onto pads on the board. The covariates in the study are qualitative,

Table 1.1. *Joint frequency distribution of emergency room visits and hospitalizations*

ER\HOSP	0	1	2 or more
0	72.4	7.3	1.5
1	6.4	5.0	1.4
2 or more	1.2	1.0	3.8

being five types of board surface, two types of solder, nine types of pads, and three types of panel. A high proportion of soldered areas had no defects, leading the author to generalize the Poisson regression model to account for excess zeros, meaning more zeros than are consistent with the Poisson formulation. The probability of zero defects is modeled using a logit regression, jointly with the Poisson regression for the mean number of defects (see Chapter 4 for details).

1.3.14 Bivariate Data on Health Services

Munkin and Trivedi (1999) consider the joint modeling of two count variables suspected of being strongly associated. Using a sample of 4,406 observations obtained from the National Medical Expenditure Survey conducted in 1987 and 1988, the authors consider joint modeling of the number of emergency room visits (ER) and the number of hospitalizations (HOSP). These variables should be positively correlated across individuals because a stay in a hospital often follows an emergency room visit. Table 1.1 presents an extract of their joint frequency distribution. A feature of the data is that much of the probability mass is concentrated at a relatively few pairs of realized outcomes.

1.4 OVERVIEW OF MAJOR ISSUES

We have introduced a number of important terms, phrases, and ideas. We now indicate where in this book they are further developed.

Chapter 2 covers issues of estimation and inference that not only are relevant to the rest of the monograph but also arise more generally. One issue concerns the two leading frameworks for parametric estimation and inference, namely, maximum likelihood and the moment-based methods. The former requires specification of the joint density of the observations, and hence implicitly of all population moments; the latter requires a limited specification of a certain number of moments, usually the first one or two only, and no further information about higher moments. The discussion in this chapter has focused on distributional models. The moment-based approach makes weaker assumptions, but generally provides less information about the data distribution. Important theoretical issues concern the relative properties of these two broad approaches to estimation and inference. There are also issues of the ease of computation involved. Chapter 2 addresses these general issues and reviews some of the major results that are used in later chapters. This makes the monograph relatively self-contained.

Chapter 3 is concerned with the Poisson and negative binomial models for count regression. These models have been the most widely used starting points for empirical work. The Poisson density is only a one-parameter density and is generally found to be too restrictive. A first step is to consider less-restrictive models for count data, such as the negative binomial and generalized linear models, which permit additional flexibility by introducing an additional parameter or parameters and breaking the equality restriction between the conditional mean and the variance. Chapter 3 provides a reasonably self-contained treatment of estimation and inference for these two basic models, which can be easily implemented using widely available packaged software. This chapter also includes issues of practical importance such as interpretation of coefficients and comparison and evaluation of goodness of fit of estimated models. For many readers this material will provide an adequate introduction to single equation count regressions.

Chapter 4 deals with mixed Poisson and related parametric models that are particularly helpful when dealing with overdispersed data. One interpretation of such processes is as *doubly stochastic Poisson processes* with the Poisson parameter treated as stochastic (see Kingman, 1993, chapter 6). Chapter 4 deals with a leading case of the mixed Poisson, the negative binomial. Overdispersion is closely related to the presence of unobserved interindividual heterogeneity, but it can also arise from *occurrence dependence* between events. Using cross-section data it may be practically impossible to identify the underlying source of overdispersion. These issues are tackled in Chapter 4, which deals with models, especially overdispersed models, that are motivated by non-Poisson features of data that can occur separately or jointly with overdispersion, for example, an excess of zero observations relative to either the Poisson or the negative binomial, or the presence of censoring or truncation.

Chapter 5 deals with statistical inference and model evaluation for single equation count regressions estimated using the methods of earlier chapters. The objective is to provide the user with specification tests and model evaluation procedures that are useful in empirical work based on cross-section data. As in Chapter 2, the issues considered in this chapter have relevance beyond count data models.

Chapter 6 provides detailed analysis of three empirical examples to illustrate the single equation modeling approaches of earlier chapters, especially the interplay of estimation and model evaluation that dominate empirical modeling.

Chapter 7 deals with *time-series* analysis of event counts. A time-series count regression is relevant if data are T observations, the tth of which is $(y_t, \mathbf{x}_t)$, $t = 1, \ldots, T$. If $\mathbf{x}_t$ includes past values of y_t, we refer to it as a *dynamic* count regression. This involves modeling the serial correlation in the count process.

Both the static time-series count regression and dynamic regression models are studied. This topic is still relatively underdeveloped.

Multivariate count models are considered in Chapter 8. An m-dimensional multivariate count model is based on data on $(\mathbf{y}_i, \mathbf{x}_i)$, where $\mathbf{y}_i$ is an $(m \times 1)$ vector of variables that may all be counts or may include counts as well as other discrete or continuous variables. Unlike the familiar case of the multivariate Gaussian distribution, the term "multivariate" in the case of count models has several different definitions. Hence, Chapter 8 deals more with a number of special cases and provides relatively few results of general applicability.

Another class of multivariate models uses *longitudinal* or *panel data*, which are analyzed in Chapter 9. Longitudinal count models have attracted much attention, following the work of Hausman, Hall, and Griliches (1984). Such models are relevant if the regression analysis is based on $(y_{it}, \mathbf{x}_{it}), i = 1, \ldots, n,$ $t = 1, \ldots, T$, where i and t are individual and time subscripts, respectively. Dynamic panel data models also include lagged y variables as regressors. Unobserved random terms may also appear in multivariate and panel data models. The usefulness of longitudinal data is that without such data it is extremely difficult to distinguish between true contagion and apparent contagion.

Chapters 10 through 13 contain material based on more recent developments and areas of current research activity. Some of these issues are actively under investigation; their inclusion is motivated by our desire to inform the reader about the state of the literature and to stimulate further effort. Chapter 10 deals with models with simultaneity and nonrandom sampling, including sample selection. Such models are often estimated with either less parametric nonlinear instrumental variable type estimators or fully parametric maximum likelihood methods. In Chapter 11, we review several flexible modeling approaches to count data, including quantile methods adapted to counts and other methods that are based on series expansion of densities. These methods permit considerable flexibility in the variance-mean relationships and in the estimation of probability of events and may be described as "semiparametric." Chapter 12 analyzes a selected number of count models from a Bayesian perspective. Finally, Chapter 13 deals with the effects of measurement errors in either exposure or covariates, and with the problem of underrecorded counts.

We have attempted to structure this monograph, keeping in mind the interests of researchers, practitioners, and new entrants to the field. The last group may wish to gain a relatively quick understanding of the standard models and methods, practitioners may be interested in the robustness and practicality of methods, and researchers wishing to contribute to the field presumably want an up-to-date and detailed account of the different models and methods. Wherever possible we have included illustrations based on real data. Inevitably, in places we have compromised, keeping in mind the constraints on the length of the monograph. We hope that the bibliographic notes and exercises at the ends of chapters will provide useful complementary material for all users.

In cross-referencing sections we use the convention that, for example, Chapter 3.4 refers to section 4 in Chapter 3. Vectors are defined as column vectors, with transposition giving a row vector, and are printed in bold lowercase. Matrices are printed in bold uppercase.

Appendix A lists the acronyms that are used in the book. Some basic mathematical functions and distributions and their properties are summarized in Appendix B. Appendix C provides some software information.

1.5 BIBLIOGRAPHIC NOTES

Johnson, Kemp, and Kotz (2005) is an excellent reference for statistical properties of the univariate Poisson and related models. A lucid introduction to Poisson processes is provided in earlier sections of Kingman (1993). A good introductory textbook treatment of Poisson processes and renewal theory is Taylor and Karlin (1998). A more advanced treatment is in Feller (1971). Another comprehensive reference on Poisson and related distributions with full bibliographic details until 1967 is Haight (1967). Early applications of the Poisson regression include Cochrane (1940) and Jorgenson (1961). Count data analysis in the sciences is surveyed in a comprehensive three-volume collective work edited by Patil (1970). Although volume 1 is theoretical, the remaining two volumes cover many applications in the natural sciences.

In both biostatistics and econometrics, especially in health, labor, and environmental economics, there are many regression applications of count models, which are mentioned throughout this book and especially in Chapter 6. Early examples of applications in criminology, sociology, political science, and international relations are, respectively, Grogger (1990), Nagin and Land (1993), Hannan and Freeman (1987), and King (1987a, 1987b). Basic references for count regression are provided in Chapter 3.9.

CHAPTER 2

Model Specification and Estimation

2.1 INTRODUCTION

This chapter presents the general modeling approaches most often used in count data analysis – likelihood-based, generalized linear models, and moment-based. Statistical inference for these nonlinear regression models is based on asymptotic theory, which is also summarized.

The models and results vary according to the strength of the distributional assumptions made. Likelihood-based models and the associated maximum likelihood estimator require complete specification of the distribution. Statistical inference is usually performed under the assumption that the distribution is correctly specified.

A less parametric analysis assumes that some aspects of the distribution of the dependent variable are correctly specified, whereas others are not specified or, if they are specified, are potentially misspecified. For count data models, considerable emphasis has been placed on analysis based on the assumption of correct specification of the conditional mean or of correct specification of both the conditional mean and the conditional variance. This is a nonlinear generalization of the linear regression model, in which consistency requires correct specification of the mean and efficient estimation requires correct specification of the mean and variance. It is a special case of the class of generalized linear models that is widely used in the statistics literature. Estimators for generalized linear models coincide with maximum likelihood estimators if the specified density is in the linear exponential family. But even then the analytical distribution of the same estimator can differ across the two approaches if different second moment assumptions are made. The term "quasi- (or pseudo-) maximum likelihood estimation" is used to describe the situation in which the assumption of correct specification of the density is relaxed. Here the first moment of the specified linear exponential family density is assumed to be correctly specified, whereas the second and other moments are permitted to be incorrectly specified.

An even more general framework, which permits estimation based on any specified moment conditions, is that of moment-based models. In the statistics literature this approach is known as estimating equations. In the econometrics literature this approach leads to generalized method of moments estimation, which is particularly useful if some regressors are endogenous rather than exogenous.

Results for hypothesis testing also depend on the strength of the distributional assumptions. The classical statistical tests – Wald, likelihood ratio, and Lagrange multiplier (or score) – are based on the likelihood approach. Analogs of these hypothesis tests exist for the generalized method of moments approach. Finally, the moments approach introduces a new class of tests of model specification, not just of parameter restrictions, called *conditional moment tests*.

Section 2.2 presents the simplest count model and estimator, the maximum likelihood estimator of the Poisson regression model. It also explains the notation used throughout the book. The three main approaches – maximum likelihood, generalized linear models, and moment-based models – and associated estimation theory are presented in, respectively, sections 2.3 through 2.5. Testing using these approaches is summarized in section 2.6. Section 2.7 considers robust inference that adjusts standard errors for nonindependent observations due to within-cluster correlation, time-series correlation, or spatial correlation. Throughout, statistical inference based on first-order asymptotic theory is given, with results derived in section 2.8, except for the bootstrap with asymptotic refinement, which is presented in section 2.6.5.

Basic count data analysis uses maximum likelihood extensively and also generalized linear models. For more complicated data situations presented in the second half of the book, generalized linear models and moment-based models are increasingly used, as well as more complex likelihood-based models that are estimated using simulation methods.

This chapter is a self-contained source, to be referred to as needed when reading later chapters. It is also intended to provide a bridge between the statistics and econometrics literatures. Of necessity, the presentation is relatively condensed and may be challenging to read in isolation, although motivation for results is given. It assumes a background at the level of Davidson and MacKinnon (2004) or Greene (2011).

2.2 EXAMPLE AND DEFINITIONS

2.2.1 Example

The starting point for cross-section count data analysis is the Poisson regression model. This assumes that y_i given the vector of regressors $\mathbf{x}_i$ is independently Poisson distributed with density

$$f(y_i|\mathbf{x}_i) = \frac{e^{-\mu_i}\mu_i^{y_i}}{y_i!}, \qquad y_i = 0, 1, 2, \ldots, \tag{2.1}$$

and mean parameter

$$\mu_i = \exp(\mathbf{x}_i'\boldsymbol{\beta}), \tag{2.2}$$

where $\boldsymbol{\beta}$ is a $k \times 1$ parameter vector. Counting process theory provides a motivation for choosing the Poisson distribution; taking the exponential of $\mathbf{x}_i'\boldsymbol{\beta}$ in (2.2) ensures that the parameter μ_i is nonnegative. This model implies that the conditional mean is given by

$$\mathsf{E}[y_i|\mathbf{x}_i] = \exp(\mathbf{x}_i'\boldsymbol{\beta}), \tag{2.3}$$

with interest often lying in changes in this conditional mean due to changes in the regressors. It also implies a particular form of heteroskedasticity, due to equidispersion or equality of the conditional variance and conditional mean,

$$\mathsf{V}[y_i|\mathbf{x}_i] = \exp(\mathbf{x}_i'\boldsymbol{\beta}). \tag{2.4}$$

The standard estimator for this model is the maximum likelihood estimator (MLE). Given independent observations, the log-likelihood function is

$$\mathcal{L}(\boldsymbol{\beta}) = \sum_{i=1}^{n}\{y_i\mathbf{x}_i'\boldsymbol{\beta} - \exp(\mathbf{x}_i'\boldsymbol{\beta}) - \ln y_i!\}. \tag{2.5}$$

Differentiating (2.5) with respect to $\boldsymbol{\beta}$ yields the Poisson MLE $\hat{\boldsymbol{\beta}}$ as the solution to the first-order conditions

$$\sum_{i=1}^{n}(y_i - \exp(\mathbf{x}_i'\boldsymbol{\beta}))\mathbf{x}_i = \mathbf{0}. \tag{2.6}$$

These k equations are nonlinear in the k unknowns $\boldsymbol{\beta}$, and there is no analytical solution for $\hat{\boldsymbol{\beta}}$. Iterative methods, usually gradient methods such as Newton-Raphson, are needed to compute $\hat{\boldsymbol{\beta}}$. Such methods are given in standard texts.

Another consequence of there being no analytical solution for $\hat{\boldsymbol{\beta}}$ is that exact distributional results for $\hat{\boldsymbol{\beta}}$ are difficult to obtain. Inference is accordingly based on asymptotic results, presented in the remainder of this chapter. There are several ways to proceed. First, we can view $\hat{\boldsymbol{\beta}}$ as the estimator maximizing (2.5) and apply maximum likelihood theory. Second, we can view $\hat{\boldsymbol{\beta}}$ as being defined by (2.6). These equations have a similar interpretation to those for the ordinary least squares (OLS) estimator. That is, the unweighted residual $(y_i - \mu_i)$ is orthogonal to the regressors. It is therefore possible that, as for OLS, inference can be performed under assumptions about just the mean and possibly variance. This is the generalized linear models approach. Third, because (2.3) implies $\mathsf{E}[(y_i - \exp(\mathbf{x}_i'\boldsymbol{\beta}))\mathbf{x}_i] = \mathbf{0}$, we can define an estimator that is the solution to the corresponding moment condition in the sample, that is, the solution to (2.6). This is the moment-based models approach.

2.2.2 Definitions

We use the generic notation $\boldsymbol{\theta} \in R^q$ to denote the $q \times 1$ parameter vector to be estimated. In the Poisson regression example, the only parameters are the regression parameters, so $\boldsymbol{\theta} = \boldsymbol{\beta}$ and $q = k$. In the simplest extensions, an additional scalar dispersion parameter α is introduced, so $\boldsymbol{\theta}' = (\boldsymbol{\beta}'\ \alpha)'$ and $q = k + 1$.

We consider random variables $\hat{\boldsymbol{\theta}}$ that *converge in probability* to a value $\boldsymbol{\theta}_*$,

$$\hat{\boldsymbol{\theta}} \xrightarrow{p} \boldsymbol{\theta}_*,$$

or equivalently the *probability limit* (plim) of $\hat{\boldsymbol{\theta}}$ equals $\boldsymbol{\theta}_*$,

$$\text{plim}\, \hat{\boldsymbol{\theta}} = \boldsymbol{\theta}_*.$$

The probability limit $\boldsymbol{\theta}_*$ is called the *pseudotrue value*. If the *data generating process* (dgp) is a model with $\boldsymbol{\theta} = \boldsymbol{\theta}_0$, and the pseudotrue value actually equals $\boldsymbol{\theta}_0$, so $\boldsymbol{\theta}_0 = \boldsymbol{\theta}_*$, then $\hat{\boldsymbol{\theta}}$ is said to be *consistent* for $\boldsymbol{\theta}_0$.

Estimators $\hat{\boldsymbol{\theta}}$ used are usually root-n consistent for $\boldsymbol{\theta}_*$ and asymptotically normally distributed. Then the random variable $\sqrt{n}(\hat{\boldsymbol{\theta}} - \boldsymbol{\theta}_*)$ *converges in distribution* to the multivariate normal distribution with mean $\mathbf{0}$ and variance $\mathbf{C}$,

$$\sqrt{n}(\hat{\boldsymbol{\theta}} - \boldsymbol{\theta}_*) \xrightarrow{d} \mathsf{N}[\mathbf{0}, \mathbf{C}], \tag{2.7}$$

where $\mathbf{C}$ is a finite positive definite matrix. It is sometimes notationally convenient to express (2.7) in the simpler form

$$\hat{\boldsymbol{\theta}} \overset{a}{\sim} \mathsf{N}[\boldsymbol{\theta}_*, \mathbf{D}], \tag{2.8}$$

where $\mathbf{D} = \frac{1}{n}\mathbf{C}$. That is, $\hat{\boldsymbol{\theta}}$ is asymptotically normal distributed with mean $\boldsymbol{\theta}_*$ and variance $\mathbf{D} = \frac{1}{n}\mathbf{C}$. The division of the finite matrix $\mathbf{C}$ by the sample size makes it clear that as the sample size goes to infinity the variance matrix $\frac{1}{n}\mathbf{C}$ goes to zero, which is to be expected because $\hat{\boldsymbol{\theta}} \xrightarrow{p} \boldsymbol{\theta}_*$.

The variance matrix $\mathbf{C}$ may depend on unknown parameters, and the result (2.8) is operationalized by replacing $\mathbf{C}$ by a consistent estimator $\hat{\mathbf{C}}$. In many cases $\mathbf{C} = \mathbf{A}^{-1}\mathbf{B}\mathbf{A}'^{-1}$, where $\mathbf{A}$ and $\mathbf{B}$ are finite positive definite matrices. Then $\hat{\mathbf{C}} = \hat{\mathbf{A}}^{-1}\hat{\mathbf{B}}\hat{\mathbf{A}}'^{-1}$, where $\hat{\mathbf{A}}$ and $\hat{\mathbf{B}}$ are consistent estimators of $\mathbf{A}$ and $\mathbf{B}$. This is called the *sandwich* form, because $\mathbf{B}$ is sandwiched between $\mathbf{A}^{-1}$ and $\mathbf{A}'^{-1}$ transposed. A more detailed discussion is given in section 2.5.1.

Results are expressed using matrix calculus. In general, the derivative $\partial g(\boldsymbol{\theta})/\partial \boldsymbol{\theta}$ of a scalar function $g(\boldsymbol{\theta})$ with respect to the $q \times 1$ vector $\boldsymbol{\theta}$ is a $q \times 1$ vector with j^{th} entry $\partial g(\boldsymbol{\theta})/\partial \theta_j$. The derivative $\partial \mathbf{h}(\boldsymbol{\theta})/\partial \boldsymbol{\theta}'$ of an $r \times 1$ vector function $\mathbf{h}(\boldsymbol{\theta})$ with respect to the $1 \times q$ vector $\boldsymbol{\theta}'$ is an $r \times q$ matrix with jk^{th} entry $\partial h_j(\boldsymbol{\theta})/\partial \theta_k$.

2.3 LIKELIHOOD-BASED MODELS

Likelihood-based models are models in which the joint density of the dependent variables is specified. For completeness a review is presented here, along with results for the less standard case of maximum likelihood with the density function misspecified.

We assume that the scalar random variable y_i given the vector of regressors $\mathbf{x}_i$ and parameter vector $\boldsymbol{\theta}$ is distributed with density $f(y_i|\mathbf{x}_i, \boldsymbol{\theta})$. The likelihood principle chooses as estimator of $\boldsymbol{\theta}$ the value that maximizes the joint probability of observing the sample values $y_1, \ldots, y_n$. This probability, viewed as a function of parameters conditional on the data, is called the *likelihood function* and is denoted

$$\mathsf{L}(\boldsymbol{\theta}) = \prod_{i=1}^{n} f(y_i|\mathbf{x}_i, \boldsymbol{\theta}), \tag{2.9}$$

where we suppress the dependence of $\mathsf{L}(\boldsymbol{\theta})$ on the data and have assumed independence over i. This definition implicitly assumes cross-section data but can easily accommodate time-series data by extending $\mathbf{x}_i$ to include lagged dependent and independent variables.

Maximizing the likelihood function is equivalent to maximizing the *log-likelihood function*

$$\mathcal{L}(\boldsymbol{\theta}) = \ln \mathsf{L}(\boldsymbol{\theta}) = \sum_{i=1}^{n} \ln f(y_i|\mathbf{x}_i, \boldsymbol{\theta}). \tag{2.10}$$

In the following analysis we consider the local maximum, which we assume to also be the global maximum.

2.3.1 Regularity Conditions

The standard results on consistency and asymptotic normality of the MLE hold if the so-called *regularity conditions* are satisfied. Furthermore, the MLE then has the desirable property that it attains the Cramer-Rao lower bound and is fully efficient. The following regularity conditions, given in Crowder (1976), are used in many studies.

1. The pdf $f(y, \mathbf{x}, \boldsymbol{\theta})$ is globally identified and $f(y, \mathbf{x}, \boldsymbol{\theta}^{(1)}) \neq f(y, \mathbf{x}, \boldsymbol{\theta}^{(2)})$, for all $\boldsymbol{\theta}^{(1)} \neq \boldsymbol{\theta}^{(2)}$.
2. $\boldsymbol{\theta} \in \boldsymbol{\Theta}$, where $\boldsymbol{\Theta}$ is finite dimensional, closed, and compact.
3. Continuous and bounded derivatives of $\mathcal{L}(\boldsymbol{\theta})$ exist up to order three.
4. The order of differentiation and integration of the likelihood may be reversed.
5. The regressor vector $\mathbf{x}_i$ satisfies
   (a) $\mathbf{x}_i'\mathbf{x}_i < \infty$
   (b) $\dfrac{\mathsf{E}[w_i^2]}{\sum_i \mathsf{E}[w_i^2]} = 0$ for all i, where $w_i \equiv \mathbf{x}_i' \dfrac{\partial \ln f(y_i|\mathbf{x}_i, \boldsymbol{\theta})}{\partial \boldsymbol{\theta}}$
   (c) $\lim_{n\to\infty} \dfrac{\sum_{i=1}^n \mathsf{E}[w_i^2 \mid \boldsymbol{\Omega}_{i-1}]}{\sum_{i=1}^n \mathsf{E}[w_i^2]} = 1$, where $\boldsymbol{\Omega}_{i-1} = (\mathbf{x}_1, \mathbf{x}_2, \ldots, \mathbf{x}_{i-1})$.

The first condition is an obvious *identification* condition, which ensures that the limit of $\frac{1}{n}\mathcal{L}$ has a unique maximum. The second condition rules out possible problems at the boundary of $\boldsymbol{\Theta}$, and it can be relaxed if, for example,

$\mathcal{L}$ is globally concave. The third condition can often be relaxed to existence up to second order. The fourth condition is a key condition that rules out densities for which the range of y_i depends on $\boldsymbol{\theta}$. The final condition rules out any observation making too large a contribution to the likelihood. For further details on regularity conditions for commonly used estimators, not just the MLE, see Newey and McFadden (1994).

2.3.2 Maximum Likelihood

We consider only the case in which the limit of $\frac{1}{n}\mathcal{L}$ is maximized at an interior point of $\boldsymbol{\Theta}$. The MLE $\hat{\boldsymbol{\theta}}_{\mathsf{ML}}$ is then the solution to the first-order conditions,

$$\frac{\partial \mathcal{L}}{\partial \boldsymbol{\theta}} = \sum_{i=1}^{n} \frac{\partial \ln f_i}{\partial \boldsymbol{\theta}} = \mathbf{0}, \tag{2.11}$$

where $f_i = f(y_i|\mathbf{x}_i, \boldsymbol{\theta})$ and $\partial\mathcal{L}/\partial\boldsymbol{\theta}$ is a $q \times 1$ vector.

The asymptotic distribution of the MLE is usually obtained under the assumption that the density is correctly specified. That is, the data generating process (dgp) for y_i has density $f(y_i|\mathbf{x}_i, \boldsymbol{\theta}_0)$, where $\boldsymbol{\theta}_0$ is the true parameter value. Then, under the regularity conditions, $\hat{\boldsymbol{\theta}} \xrightarrow{p} \boldsymbol{\theta}_0$, so the MLE is consistent for $\boldsymbol{\theta}_0$. Also,

$$\sqrt{n}(\hat{\boldsymbol{\theta}}_{\mathsf{ML}} - \boldsymbol{\theta}_0) \xrightarrow{d} \mathsf{N}[\mathbf{0}, \mathbf{A}^{-1}], \tag{2.12}$$

where the $q \times q$ matrix $\mathbf{A}$ is defined as

$$\mathbf{A} = -\lim_{n\to\infty} \frac{1}{n} \mathsf{E}\left[\sum_{i=1}^{n} \left.\frac{\partial^2 \ln f_i}{\partial\boldsymbol{\theta}\partial\boldsymbol{\theta}'}\right|_{\boldsymbol{\theta}_0}\right]. \tag{2.13}$$

A consequence of regularity conditions three and four is the *information matrix equality*

$$\mathsf{E}\left[\frac{\partial^2\mathcal{L}}{\partial\boldsymbol{\theta}\partial\boldsymbol{\theta}'}\right] = -\mathsf{E}\left[\frac{\partial\mathcal{L}}{\partial\boldsymbol{\theta}}\frac{\partial\mathcal{L}}{\partial\boldsymbol{\theta}'}\right], \tag{2.14}$$

for all values of $\boldsymbol{\theta} \in \Theta$. This is derived in section 2.8. Assuming independence over i and defining

$$\mathbf{B} = \lim_{n\to\infty} \frac{1}{n} \mathsf{E}\left[\sum_{i=1}^{n} \left.\frac{\partial \ln f_i}{\partial\boldsymbol{\theta}}\frac{\partial \ln f_i}{\partial\boldsymbol{\theta}'}\right|_{\boldsymbol{\theta}_0}\right], \tag{2.15}$$

the information equality implies $\mathbf{A} = \mathbf{B}$.

To operationalize these results, one needs a consistent estimator of the variance matrix in (2.12). There are many possibilities. The expected *Fisher information estimator* takes the expectation (2.13) under the assumed density and evaluates at $\hat{\boldsymbol{\theta}}$. The *Hessian estimator* simply evaluates (2.13) at $\hat{\boldsymbol{\theta}}$ without taking the expectation. The *outer product* (OP) *estimator* evaluates (2.15) at $\hat{\boldsymbol{\theta}}$ without taking the expectation. It was proposed by Berndt, Hall, Hall, and

Hausman (1974) and is also called the BHHH estimator. A more general form of the variance matrix of $\hat{\boldsymbol{\theta}}_{\mathsf{ML}}$, the sandwich form, is used if the assumption of correct specification of the density is relaxed (see section 2.3.4).

As an example, consider the MLE for the Poisson regression model presented in section 2.2. In this case $\partial\mathcal{L}/\partial\boldsymbol{\beta} = \sum_i (y_i - \exp(\mathbf{x}_i'\boldsymbol{\beta}))\mathbf{x}_i$ and

$$\partial^2\mathcal{L}/\partial\boldsymbol{\beta}\partial\boldsymbol{\beta}' = -\sum_i \exp(\mathbf{x}_i'\boldsymbol{\beta})\mathbf{x}_i\mathbf{x}_i'.$$

It follows that we do not even need to take the expectation in (2.13) to obtain

$$\mathbf{A} = \lim_{n\to\infty}\frac{1}{n}\left[\sum_i \exp(\mathbf{x}_i'\boldsymbol{\beta}_0)\mathbf{x}_i\mathbf{x}_i'\right].$$

Assuming $\mathsf{E}[(y_i - \exp(\mathbf{x}_i'\boldsymbol{\beta}_0))^2|\mathbf{x}_i] = \exp(\mathbf{x}_i'\boldsymbol{\beta}_0)$ – that is, correct specification of the variance – leads to the same expression for $\mathbf{B}$. The result is most conveniently expressed as

$$\hat{\boldsymbol{\beta}}_{\mathsf{ML}} \overset{a}{\sim} \mathsf{N}\left(\boldsymbol{\beta}_0, \left(\sum_{i=1}^{n} \exp(\mathbf{x}_i'\boldsymbol{\beta}_0)\mathbf{x}_i\mathbf{x}_i'\right)^{-1}\right). \tag{2.16}$$

In this case the Fisher information and Hessian estimators of $\mathsf{V}[\hat{\boldsymbol{\theta}}_{\mathsf{ML}}]$ coincide and differ from the outer-product estimator.

2.3.3 Profile Likelihood

Suppose the likelihood depends on a parameter vector $\boldsymbol{\lambda}$ in addition to $\boldsymbol{\theta}$, so the likelihood function is $\mathcal{L}(\boldsymbol{\theta}, \boldsymbol{\lambda})$. The *profile likelihood* or *concentrated likelihood* eliminates $\boldsymbol{\lambda}$ by obtaining the restricted MLE $\hat{\boldsymbol{\lambda}}(\boldsymbol{\theta})$ for fixed $\boldsymbol{\theta}$. Then

$$\mathcal{L}_{pro}(\boldsymbol{\theta}) = \mathcal{L}(\boldsymbol{\theta}, \hat{\boldsymbol{\lambda}}(\boldsymbol{\theta})). \tag{2.17}$$

This approach can be used in all situations, but it is important to note that $\mathcal{L}_{pro}(\boldsymbol{\theta})$ is not strictly a log-likelihood function and that the usual results need to be adjusted in this case (see, for example, Davidson and MacKinnon, 1993, chapter 8).

The profile likelihood is useful if $\boldsymbol{\lambda}$ is a nuisance parameter. For example, interest may lie in modeling the conditional mean that is parameterized by $\boldsymbol{\theta}$, with variance parameters $\boldsymbol{\lambda}$ not intrinsically of interest. In such circumstances, there is an advantage to attempting to estimate $\boldsymbol{\theta}$ alone, especially if $\boldsymbol{\lambda}$ is of high dimension.

For scalar θ the profile likelihood can be used to form a likelihood ratio $100(1-\alpha)\%$ confidence region for θ, $\{\theta : \mathcal{L}_{pro}(\theta) > \mathcal{L}(\hat{\theta}, \hat{\boldsymbol{\lambda}}) - \chi_1^2(\alpha)\}$, where $\mathcal{L}(\hat{\theta}, \hat{\boldsymbol{\lambda}})$ is the unconstrained maximum likelihood. This need not be symmetric around $\hat{\theta}$, unlike the usual confidence interval $\hat{\theta} \pm z_{\alpha/2}\mathsf{se}[\hat{\theta}]$, where $\mathsf{se}[\hat{\theta}]$ is the standard error of $\hat{\theta}$.

Other variants of the likelihood approach can be used to eliminate nuisance parameters in some special circumstances. The *conditional likelihood* is the joint density of $\boldsymbol{\theta}$ conditional on a sufficient statistic for the nuisance parameters and is used, for example, to estimate the fixed effects Poisson model for panel data. The *marginal likelihood* is the likelihood for a subset of the data that may depend on $\boldsymbol{\theta}$ alone. For further discussion, see McCullagh and Nelder (1989, chapter 7).

2.3.4 Misspecified Density and Quasi-MLE

A weakness of the ML approach is that it assumes correct specification of the complete density. To see the role of this assumption, it is helpful to begin with an informal proof of consistency of the MLE in the usual case of correctly specified density.

The MLE solves the sample moment condition $\partial\mathcal{L}/\partial\boldsymbol{\theta} = \mathbf{0}$, so an intuitive necessary condition for consistency is that the same moment condition holds in the population or $\mathsf{E}[\partial\mathcal{L}/\partial\boldsymbol{\theta}] = \mathbf{0}$, which holds if $\mathsf{E}[\partial \ln f_i/\partial\boldsymbol{\theta}] = \mathbf{0}$. Any density $f(y|\mathbf{x}, \boldsymbol{\theta})$ satisfying the regularity conditions has the property that

$$\int \frac{\partial \ln f(y|\boldsymbol{\theta})}{\partial\boldsymbol{\theta}} f(y|\boldsymbol{\theta})dy = \mathbf{0}, \tag{2.18}$$

see section 2.8 for a derivation. The consistency condition $\mathsf{E}[\partial \ln f(y|\mathbf{x}, \boldsymbol{\theta})/\partial\boldsymbol{\theta}] = \mathbf{0}$ is implied by (2.18), if the expectations operator is taken using the assumed density $f(y|\mathbf{x}, \boldsymbol{\theta})$.

Suppose instead that the dgp is the density $f^*(y_i|\mathbf{z}_i, \boldsymbol{\gamma})$ rather than the assumed density $f(y_i|\mathbf{x}_i, \boldsymbol{\theta})$. Then (2.18) no longer implies the consistency condition $\mathsf{E}[\partial \ln f(y|\mathbf{x}, \boldsymbol{\theta})/\partial\boldsymbol{\theta}] = \mathbf{0}$, which now becomes

$$\int \frac{\partial \ln f(y|\mathbf{x}, \boldsymbol{\theta})}{\partial\boldsymbol{\theta}} f^*(y|\mathbf{z}, \boldsymbol{\gamma})dy = \mathbf{0},$$

because the expectation should be with respect to the dgp density, not the assumed density. Thus misspecification of the density may lead to inconsistency.

White (1982), following Huber (1967), obtained the distribution of the MLE if the density function is incorrectly specified. Then the ML estimator is called the quasi-MLE or pseudo-MLE, which we denote $\hat{\boldsymbol{\theta}}_{\mathsf{QML}}$. In general $\hat{\boldsymbol{\theta}}_{\mathsf{QML}} \xrightarrow{p} \boldsymbol{\theta}_*$, where the pseudotrue value $\boldsymbol{\theta}_*$ is the value of $\boldsymbol{\theta}$ that maximizes $\text{plim}\, \frac{1}{n}\sum_{i=1}^{n} \ln f(y_i|\mathbf{x}_i, \boldsymbol{\theta})$ and the probability limit is obtained under the dgp $f^*(y_i|\mathbf{z}_i, \boldsymbol{\gamma})$. Under suitable assumptions,

$$\sqrt{n}(\hat{\boldsymbol{\theta}}_{\mathsf{QML}} - \boldsymbol{\theta}_*) \xrightarrow{d} \mathsf{N}[\mathbf{0}, \mathbf{A}_*^{-1}\mathbf{B}_*\mathbf{A}_*^{-1}], \tag{2.19}$$

where

$$\mathbf{A}_* = -\lim_{n\to\infty}\frac{1}{n}\mathsf{E}_*\left[\sum_{i=1}^{n} \left.\frac{\partial^2 \ln f_i}{\partial\boldsymbol{\theta}\partial\boldsymbol{\theta}'}\right|_{\boldsymbol{\theta}_*}\right] \tag{2.20}$$

and

$$\mathbf{B}_* = \lim_{n\to\infty} \frac{1}{n} \mathsf{E}_* \left[\sum_{i=1}^{n} \left. \frac{\partial \ln f_i}{\partial \boldsymbol{\theta}} \frac{\partial \ln f_i}{\partial \boldsymbol{\theta}'} \right|_{\boldsymbol{\theta}_*} \right], \tag{2.21}$$

where $f_i = f(y_i|\mathbf{x}_i, \boldsymbol{\theta})$ and the expectations E_* are with respect to the dgp $f^*(y_i|\mathbf{z}_i, \boldsymbol{\gamma})$.

The essential point is that if the density is misspecified the MLE is in general inconsistent. Generalized linear models, presented next, are the notable exception. These models include the Poisson.

A second point is that the MLE has the more complicated variance function with sandwich form (2.19), because the information matrix equality (2.14) no longer holds. The *sandwich estimator* of the variance matrix of $\hat{\boldsymbol{\theta}}_{\mathsf{ML}}$ is $\hat{\mathbf{A}}^{-1}\hat{\mathbf{B}}\hat{\mathbf{A}}^{-1}$. Different estimates of $\hat{\mathbf{A}}$ and $\hat{\mathbf{B}}$ can be used (see section 2.3.2), depending on whether or not the expectations in $\mathbf{A}_*$ and $\mathbf{B}_*$ are taken before evaluation at $\hat{\boldsymbol{\theta}}_{\mathsf{ML}}$. The *robust sandwich estimator* does not take expectations.

2.4 GENERALIZED LINEAR MODELS

Although the MLE is in general inconsistent if the density is incorrectly specified, for some specified densities consistency may be maintained even given partial misspecification of the model. A leading example is maximum likelihood estimation in the linear regression model under the assumption that y_i is independently $\mathsf{N}[\mathbf{x}_i'\boldsymbol{\beta}_0, \sigma^2]$ distributed. Then $\hat{\boldsymbol{\beta}}_{\mathsf{ML}}$ equals the OLS estimator, which may be consistent even given nonnormality and heteroskedasticity, because the essential requirement for consistency is correct specification of the conditional mean: $\mathsf{E}[y_i|\mathbf{x}_i] = \mathbf{x}_i'\boldsymbol{\beta}_0$.

A similar situation arises for the Poisson regression model. Consistency essentially requires that the population analog of (2.6) holds:

$$\mathsf{E}[(y_i - \exp(\mathbf{x}_i'\boldsymbol{\beta}_0))\mathbf{x}_i] = \mathbf{0}.$$

This is satisfied if $\mathsf{E}[y_i|\mathbf{x}_i] = \exp(\mathbf{x}_i'\boldsymbol{\beta}_0)$. More generally such results hold for ML estimation of models with specified density that is a member of the linear exponential family, and estimation of the closely related class of generalized linear models.

Although consistency in these models requires only correct specification of the mean, misspecification of the variance leads to invalid statistical inference due to incorrect reported *t*-statistics and standard errors. For example, in the linear regression model the usual reported standard errors for OLS are incorrect if the error is heteroskedastic rather than homoskedastic. Adjustment to the usual computer output needs to be made to ensure correct standard errors. Furthermore, more efficient generalized least squares (GLS) estimation is possible.

We begin with a presentation of results for weighted least squares for linear models, results that carry over to generalized linear models. We then introduce generalized linear models by considering linear exponential family models. These models are based on a one-parameter distribution, which in practice is too restrictive. Extensions to two-parameter models have been made in two ways – the linear exponential family with nuisance parameter and generalized linear models. Further results for generalized linear models are then presented before concluding with extended generalized linear models.

2.4.1 Weighted Linear Least Squares

Consider the linear regression model for y_i with nonstochastic regressors $\mathbf{x}_i$. In the linear model the regression function is $\mathbf{x}_i'\boldsymbol{\beta}_0$. Suppose it is believed that heteroskedasticity exists and as a starting point can be approximated by $\mathsf{V}[y_i] = v_i$, where v_i is known. Then one uses the *weighted least squares* (WLS) estimator with weights $1/v_i$, which solves the first-order conditions

$$\sum_{i=1}^{n} \frac{1}{v_i}(y_i - \mathbf{x}_i'\boldsymbol{\beta})\mathbf{x}_i = \mathbf{0}. \tag{2.22}$$

We consider the distribution of this estimator if the mean is correctly specified, but the variance is not necessarily v_i. It is helpful to express the first-order conditions in matrix notation as

$$\mathbf{X}'\mathbf{V}^{-1}(\mathbf{y} - \mathbf{X}\boldsymbol{\beta}) = \mathbf{0}, \tag{2.23}$$

where $\mathbf{y}$ is the $n \times 1$ vector with i^{th} entry y_i, $\mathbf{X}$ is the $n \times k$ matrix with i^{th} row $\mathbf{x}_i'$, and $\mathbf{V}$ is the $n \times n$ weighting matrix with i^{th} diagonal entry v_i. These equations have the analytical solution

$$\hat{\boldsymbol{\beta}}_{\mathsf{WLS}} = (\mathbf{X}'\mathbf{V}^{-1}\mathbf{X})^{-1}\mathbf{X}'\mathbf{V}^{-1}\mathbf{y}.$$

To obtain the mean and variance of $\hat{\boldsymbol{\beta}}_{\mathsf{WLS}}$, we assume $\mathbf{y}$ has mean $\mathbf{X}\boldsymbol{\beta}_0$ and variance matrix $\boldsymbol{\Omega}$. Then it is a standard result that $\mathsf{E}[\hat{\boldsymbol{\beta}}_{\mathsf{WLS}}] = \boldsymbol{\beta}_0$ and

$$\mathsf{V}[\hat{\boldsymbol{\beta}}_{\mathsf{WLS}}] = (\mathbf{X}'\mathbf{V}^{-1}\mathbf{X})^{-1}\mathbf{X}'\mathbf{V}^{-1}\boldsymbol{\Omega}\mathbf{V}^{-1}\mathbf{X}(\mathbf{X}'\mathbf{V}^{-1}\mathbf{X})^{-1}. \tag{2.24}$$

One familiar example of this is OLS. Then $\mathbf{V} = \sigma^2\mathbf{I}$ and $\mathsf{V}[\hat{\boldsymbol{\beta}}_{\mathsf{WLS}}] = (\mathbf{X}'\mathbf{X})^{-1}\mathbf{X}'\boldsymbol{\Omega}\mathbf{X}(\mathbf{X}'\mathbf{X})^{-1}$, which simplifies to $\sigma^2(\mathbf{X}'\mathbf{X})^{-1}$ if $\boldsymbol{\Omega} = \sigma^2\mathbf{I}$. A second familiar example is GLS, with $\mathbf{V} = \boldsymbol{\Omega}$, in which case $\mathsf{V}[\hat{\boldsymbol{\beta}}_{\mathsf{WLS}}] = (\mathbf{X}'\boldsymbol{\Omega}^{-1}\mathbf{X})^{-1}$.

It is important to note that the general result (2.24) represents a different situation from that in standard textbook treatments. One begins with a working hypothesis about the form of the heteroskedasticity, say v_i, leading to a *working variance matrix* $\mathbf{V}$. If $\mathbf{V}$ is misspecified then $\hat{\boldsymbol{\beta}}_{\mathsf{WLS}}$ is still unbiased, but it is inefficient and most importantly has a variance matrix of the general form (2.24), which does not impose $\mathbf{V} = \boldsymbol{\Omega}$. One way to estimate $\mathsf{V}[\hat{\boldsymbol{\beta}}_{\mathsf{WLS}}]$ is to specify $\boldsymbol{\Omega}$ to be a particular function of regressors and parameters and obtain consistent estimates of these parameters and hence a consistent estimator $\hat{\boldsymbol{\Omega}}$ of $\boldsymbol{\Omega}$. Alternatively, White (1980) and Eicker (1967) gave conditions under which one need not specify the functional form for the variance matrix $\boldsymbol{\Omega}$, but can instead use the matrix $\widetilde{\boldsymbol{\Omega}} = \mathsf{Diag}[(y_i - \mathbf{x}_i'\hat{\boldsymbol{\beta}}_{\mathsf{WLS}})^2]$, where $\mathsf{Diag}[a_i]$ denotes the diagonal matrix with i^{th} diagonal entry a_i. The justification is that even though $\widetilde{\boldsymbol{\Omega}}$ is not consistent for $\boldsymbol{\Omega}$, the difference between the $k \times k$ matrices $\frac{1}{n}\mathbf{X}'\mathbf{V}^{-1}\widetilde{\boldsymbol{\Omega}}\mathbf{V}^{-1}X$ and $\frac{1}{n}\mathbf{X}'\mathbf{V}^{-1}\boldsymbol{\Omega}\mathbf{V}^{-1}\mathbf{X}$ has probability limit zero.

These same points carry over to generalized linear models. One specifies a working variance assumption such as variance-mean equality for the Poisson. The dependence of v_i (and hence $\mathbf{V}$) on μ_i does not change the result (2.23). The estimator retains consistency if the mean is correctly specified, but variance-mean inequality will lead to incorrect inference if one does not use the general form (2.24). Generalized linear models have an additional complication, because the mean function is nonlinear. This can be accommodated by generalizing the first-order conditions (2.22). Because in the linear case $\mu_i = \mathbf{x}_i'\boldsymbol{\beta}$, (2.22) can be reexpressed as

$$\sum_{i=1}^{n} \frac{1}{v_i}(y_i - \mu_i)\frac{\partial \mu_i}{\partial \boldsymbol{\beta}} = \mathbf{0}. \tag{2.25}$$

The first-order conditions for generalized linear models such as the Poisson are of the form (2.25). It can be shown that the preceding discussion still holds, with the important change that the matrix $\mathbf{X}$ in (2.23) and (2.24) is now defined to have i^{th} row $\partial \mu_i / \partial \boldsymbol{\beta}'$.

There is considerable overlap in the next three subsections, which cover different representations and variations of essentially the same model. In sections 2.4.2 and 2.4.3, the density is parameterized in terms of the mean parameter; in section 2.4.4, the density is parameterized in terms of the so-called canonical parameter. The latter formulation is used in the generalized linear models literature. To those unfamiliar with this literature, the mean parameter formulation may be more natural. For completeness both presentations are given here. Other variations, such as different ways in which nuisance scale parameters are introduced, are discussed at the end of section 2.4.4.

2.4.2 Linear Exponential Family Models

This presentation follows Cameron and Trivedi (1986), whose work was based on Gourieroux, Monfort, and Trognon (1984a). A density $f_{\mathsf{LEF}}(y|\mu)$ is a member of a *linear exponential family* (LEF) with *mean parameterization* if

$$f_{\mathsf{LEF}}(y|\mu) = \exp\{a(\mu) + b(y) + c(\mu)y\}, \tag{2.26}$$

where $\mu = \mathsf{E}[y]$, and the functions $a(\cdot)$ and $c(\cdot)$ are such that

$$\mathsf{E}[y] = -\left[c'(\mu)\right]^{-1} a'(\mu), \tag{2.27}$$

where $a'(\mu) = \partial a(\mu)/\partial \mu$ and $c'(\mu) = \partial c(\mu)/\partial \mu$, and

$$\mathsf{V}[y] = \left[c'(\mu)\right]^{-1}. \tag{2.28}$$

The function $b(\cdot)$ is a normalizing constant. Different functional forms for $a(\cdot)$ and $c(\cdot)$ lead to different LEF models. Special cases of the LEF include the normal (with σ^2 known), Poisson, geometric, binomial (with number of trials fixed), exponential, and one-parameter gamma.

For example, the Poisson density can be written as $\exp\{-\mu + y \ln \mu - \ln y!\}$, which is an LEF model with $a(\mu) = -\mu$, $c(\mu) = \ln \mu$ and $b(y) = -\ln y!$ Then $a'(\mu) = -1$ and $c'(\mu) = 1/\mu$, so $\mathsf{E}[y] = \mu$ from (2.27) and $\mathsf{V}[y] = \mu$ from (2.28).

Members of the exponential family have density $f(y|\lambda) = \exp\{a(\lambda) + b(y) + c(\lambda)t(y)\}$. The LEF is the special case $t(y) = y$, hence the qualifier linear. The natural exponential family has density $f(y|\lambda) = \exp\{a(\lambda) + \lambda y\}$. Other exponential families come from the natural exponential family by one-to-one transformations $x = t(y)$ of y.

A regression model is formed by specifying the density to be $f_{\mathsf{LEF}}(y_i|\mu_i)$ where

$$\mu_i = \mu(\mathbf{x}_i, \boldsymbol{\beta}), \tag{2.29}$$

for some specified mean function $\mu(\cdot)$. The MLE based on an LEF, $\hat{\boldsymbol{\beta}}_{\mathsf{LEF}}$, maximizes

$$\mathcal{L}_{\mathsf{LEF}} = \sum_{i=1}^{n} \{a(\mu_i) + b(y_i) + c(\mu_i)y_i\}. \tag{2.30}$$

The first-order conditions using (2.27) and (2.28) can be rewritten as

$$\sum_{i=1}^{n} \frac{1}{v_i}(y_i - \mu_i)\frac{\partial \mu_i}{\partial \boldsymbol{\beta}} = \mathbf{0}, \tag{2.31}$$

where

$$v_i = \left[c'(\mu_i)\right]^{-1} \tag{2.32}$$

is the specified variance function that is a function of μ_i and hence $\boldsymbol{\beta}$. These first-order conditions are of the form (2.25), and as seen subsequently we obtain results similar to (2.24).

It is helpful at times to rewrite (2.31) as

$$\sum_{i=1}^{n} \frac{(y_i - \mu_i)}{\sqrt{v_i}} \frac{1}{\sqrt{v_i}} \frac{\partial \mu_i}{\partial \boldsymbol{\beta}} = \mathbf{0}, \tag{2.33}$$

which shows that the standardized residual is orthogonal to the standardized regressor.

Under the standard assumption that the density is correctly specified, so the dgp is $f_{\mathsf{LEF}}(y_i|\mu(\mathbf{x}_i, \boldsymbol{\beta}_0))$, application of (2.12) and (2.13) yields

$$\sqrt{n}(\hat{\boldsymbol{\beta}}_{\mathsf{LEF}} - \boldsymbol{\beta}_0) \overset{d}{\rightarrow} \mathsf{N}[\mathbf{0}, \mathbf{A}^{-1}], \tag{2.34}$$

where

$$\mathbf{A} = \lim_{n\to\infty} \frac{1}{n} \sum_{i=1}^{n} \frac{1}{v_i} \frac{\partial \mu_i}{\partial \boldsymbol{\beta}} \frac{\partial \mu_i}{\partial \boldsymbol{\beta}'}\bigg|_{\boldsymbol{\beta}_0}. \tag{2.35}$$

Now consider estimation when the density is misspecified. Gourieroux, Monfort, and Trognon. (1984a) call the estimator in this case the *pseudo-maximum likelihood* (PML) *estimator*. Other authors call such an estimator a *quasimaximum likelihood estimator*. Throughout this book we use the term "quasi-ML," because this has become the common terminology. Note that this is a departure from the first edition of this book.

Assume $(y_i, \mathbf{x}_i)$ is independent over i, the conditional mean of y_i is correctly specified as $\mathsf{E}[y_i|\mathbf{x}_i] = \mu(\mathbf{x}_i, \boldsymbol{\beta}_0)$, and

$$\mathsf{V}[y_i|\mathbf{x}_i] = \omega_i \tag{2.36}$$

is finite but $\omega_i \neq v_i$ necessarily. Thus the mean is correctly specified, but other features of the distribution such as the variance and density are potentially misspecified. Then Gourieroux et al. (1984a) show that $\hat{\boldsymbol{\beta}}_{\mathsf{LEF}} \xrightarrow{p} \boldsymbol{\beta}_0$, so the MLE is still consistent for $\boldsymbol{\beta}_0$. The intuition is that consistency of $\hat{\boldsymbol{\beta}}_{\mathsf{LEF}}$ essentially requires that

$$\mathsf{E}\left[\frac{1}{v_i}(y_i - \mu_i)\frac{\partial \mu_i}{\partial \boldsymbol{\beta}}\bigg|_{\boldsymbol{\beta}_0}\right] = \mathbf{0},$$

so that the conditions (2.31) hold in the population. This is the case if the conditional mean is correctly specified because then $\mathsf{E}[y_i - \mu(\mathbf{x}_i, \boldsymbol{\beta}_0)|\mathbf{x}_i] = 0$. Also

$$\sqrt{n}(\hat{\boldsymbol{\beta}}_{\mathsf{LEF}} - \boldsymbol{\beta}_0) \xrightarrow{d} \mathsf{N}[\mathbf{0}, \mathbf{A}^{-1}\mathbf{B}\mathbf{A}^{-1}], \tag{2.37}$$

where $\mathbf{A}$ is defined in (2.35) and

$$\mathbf{B} = \lim_{n\to\infty} \frac{1}{n} \sum_{i=1}^{n} \frac{\omega_i}{v_i^2} \frac{\partial \mu_i}{\partial \boldsymbol{\beta}} \frac{\partial \mu_i}{\partial \boldsymbol{\beta}'}\bigg|_{\boldsymbol{\beta}_0}. \tag{2.38}$$

Note that v_i is the working variance, the variance in the specified LEF density for y_i, whereas ω_i is the variance for the true dgp. Note also that (2.37) equals (2.24), where $\mathbf{X}$ is the $n \times k$ matrix with i^{th} row $\partial \mu_i/\partial \boldsymbol{\beta}'$, $\mathbf{V} = \mathsf{Diag}[v_i]$, and $\boldsymbol{\Omega} = \mathsf{Diag}[\omega_i]$, confirming the link with results for weighted linear least squares.

There are three important results. First, regardless of other properties of the true dgp for y, the QMLE based on an assumed LEF is consistent provided the conditional mean is correctly specified. This result is sometimes misinterpreted. It should be clear that it is the assumed density that must be LEF, whereas the true dgp need not be LEF.

Second, correct specification of the mean and variance is sufficient for the usual maximum likelihood output to give the correct variance matrix for the

QMLE based on an assumed LEF density. Other moments of the distribution may be misspecified.

Third, if the only part of the model that is correctly specified is the conditional mean, the MLE is consistent, but the usual ML output gives an inconsistent estimate of the variance matrix, because it uses $\mathbf{A}^{-1}$ rather than the sandwich form $\mathbf{A}^{-1}\mathbf{B}\mathbf{A}^{-1}$. This is correct only if $\mathbf{B} = \mathbf{A}$, which requires $\omega_i = \upsilon_i$, that is, correct specification of the conditional variance of the dependent variable.

Correct standard errors are obtained by using as variance matrix $\frac{1}{n}\hat{\mathbf{A}}^{-1}\hat{\mathbf{B}}\hat{\mathbf{A}}^{-1}$ where $\hat{\mathbf{A}}$ and $\hat{\mathbf{B}}$ equal $\mathbf{A}$ and $\mathbf{B}$ defined in, respectively, (2.35) and (2.37), evaluated at $\hat{\mu}_i = \mu(\mathbf{x}_i, \hat{\boldsymbol{\beta}}_{\text{LEF}})$, $\hat{\upsilon}_i^2 = \left[c'(\hat{\mu}_i)\right]^{-1}$, and $\hat{\omega}_i$. The estimate of the true variance $\hat{\omega}_i$ can be obtained in two ways. First, if no assumptions are made about the variance one can use $\hat{\omega}_i = (y_i - \hat{\mu}_i)^2$ by extension of results for the OLS estimator given by White (1980). Even though $\hat{\omega}_i$ does not converge to ω_i, under suitable assumptions $\hat{\mathbf{B}} \xrightarrow{p} \mathbf{B}$. Second, a structural model for the variance can be specified, say as $\omega_i = \omega(\mathbf{x}_i, \boldsymbol{\beta}, \boldsymbol{\alpha})$, where $\boldsymbol{\alpha}$ is a finite dimensional nuisance parameter. Then we can use $\hat{\omega}_i = \omega(\mathbf{x}_i, \hat{\boldsymbol{\beta}}_{\text{LEF}}, \hat{\boldsymbol{\alpha}})$, where $\hat{\boldsymbol{\alpha}}$ is a consistent estimate of $\boldsymbol{\alpha}$. Particularly convenient is $\omega_i = \alpha\upsilon_i$, because then $\mathbf{B} = \alpha\mathbf{A}$ so that $\mathbf{A}^{-1}\mathbf{B}\mathbf{A}^{-1} = \alpha\mathbf{A}^{-1}$.

As an example, consider the Poisson regression model with exponential mean function $\mu_i = \exp(\mathbf{x}_i'\boldsymbol{\beta})$. Then $\partial\mu_i/\partial\boldsymbol{\beta} = \mu_i\mathbf{x}_i$. The Poisson specifies the variance to equal the mean, so $\upsilon_i = \omega_i$. Substituting in (2.35) yields

$$\mathbf{A} = \lim \frac{1}{n}\sum_i \mu_i \mathbf{x}_i \mathbf{x}_i',$$

and similarly substituting in (2.38) yields

$$\mathbf{B} = \lim \frac{1}{n}\sum_i \omega_i \mathbf{x}_i \mathbf{x}_i'.$$

In general for the correctly specified conditional mean, the Poisson QML estimator is asymptotically normal with mean $\boldsymbol{\beta}_0$ and variance $\frac{1}{n}\mathbf{A}^{-1}\mathbf{B}\mathbf{A}^{-1}$. If additionally $\omega_i = \upsilon_i = \mu_i$ so that the conditional variance of y_i is correctly specified, then $\mathbf{A} = \mathbf{B}$ and the variance of the estimator simplifies to $\frac{1}{n}\mathbf{A}^{-1}$.

2.4.3 LEF with Nuisance Parameter

Given specification of a true variance function, so $\omega_i = \omega(\cdot)$, one can potentially obtain a more efficient estimator, in the same way that specification of the functional form of heteroskedasticity in the linear regression model leads to the more efficient GLS estimator.

Gourieroux et al. (1984a) introduced the more general variance function

$$\omega_i = \omega(\mu(\mathbf{x}_i, \boldsymbol{\beta}), \alpha) \tag{2.39}$$

by defining the LEF *with nuisance parameter* (LEFN)

$$f_{\text{LEFN}}(y|\mu, \alpha) = \exp\{a(\mu, \alpha) + b(y, \alpha) + c(\mu, \alpha)y\}, \tag{2.40}$$

where $\mu = \mathsf{E}[y]$, $\omega(\mu, \alpha) = \mathsf{V}[y]$, and $\alpha = \psi(\mu, \omega)$, where $\psi(\cdot)$ is a differentiable function of α and ω and $\psi(\cdot)$ defines for any given μ a one-to-one relationship between α and ω. For a given α this is an LEF density, so the functions $a(\cdot)$ and $c(\cdot)$ satisfy (2.27) and (2.28), with $c(\mu, \alpha)$ and $a(\mu, \alpha)$ replacing $c(\mu)$ and $a(\mu)$.

Gourieroux et al. (1984a) proposed the *quasigeneralized pseudomaximum likelihood* (QGPML) estimator $\hat{\boldsymbol{\beta}}_{\mathsf{LEFN}}$ based on LEFN, which maximizes with respect to $\boldsymbol{\beta}$

$$\begin{aligned}\mathcal{L}_{\mathsf{LEFN}} = \textstyle\sum_{i=1}^{n}\{&a(\mu_i, \omega(\widetilde{\mu}_i, \widetilde{\alpha})) + b(y_i, \omega(\widetilde{\mu}_i, \widetilde{\alpha}))\\ &+ c(\mu_i, \omega(\widetilde{\mu}_i, \widetilde{\alpha}))y_i\}\end{aligned} \tag{2.41}$$

where $\widetilde{\mu}_i = \mu(\mathbf{x}_i, \widetilde{\boldsymbol{\beta}})$ and $\widetilde{\boldsymbol{\beta}}$ and $\widetilde{\alpha}$ are root-n consistent estimates of $\boldsymbol{\beta}$ and α. The first-order conditions can be reexpressed as

$$\sum_{i=1}^{n} \frac{1}{\widetilde{\omega}_i}(y_i - \mu_i)\frac{\partial \mu_i}{\partial \boldsymbol{\beta}} = \mathbf{0}. \tag{2.42}$$

Assume $(y_i, \mathbf{x}_i)$ is independent over i, and the conditional mean and variance of y_i are correctly specified, so $\mathsf{E}[y_i|\mathbf{x}_i] = \mu(\mathbf{x}_i, \boldsymbol{\beta}_0)$ and $\mathsf{V}[y_i|\mathbf{x}_i] = \omega(\mu(\mathbf{x}_i, \boldsymbol{\beta}_0), \alpha_0)$. Then $\hat{\boldsymbol{\beta}}_{\mathsf{LEFN}} \overset{p}{\to} \boldsymbol{\beta}_0$, so the QGPMLE is consistent for $\boldsymbol{\beta}_0$. Also

$$\sqrt{n}(\hat{\boldsymbol{\beta}}_{\mathsf{LEFN}} - \boldsymbol{\beta}_0) \overset{d}{\to} \mathsf{N}[\mathbf{0}, \mathbf{A}^{-1}], \tag{2.43}$$

where

$$\mathbf{A} = \lim_{n\to\infty} \frac{1}{n}\sum_{i=1}^{n} \frac{1}{\omega_i}\frac{\partial \mu_i}{\partial \boldsymbol{\beta}}\frac{\partial \mu_i}{\partial \boldsymbol{\beta}'}\bigg|_{\boldsymbol{\beta}_0}. \tag{2.44}$$

A consistent estimate for the variance matrix is obtained by evaluating $\mathbf{A}$ at $\hat{\mu}_i = \mu(\mathbf{x}_i, \hat{\boldsymbol{\beta}}_{\mathsf{LEFN}})$ and $\hat{\omega}_i = \omega(\hat{\mu}_i, \widetilde{\alpha})$. One can, of course, guard against possible misspecification of ω_i in the same way that possible misspecification of v_i was handled in the previous subsection.

The negative binomial model with mean μ and variance $\mu + \alpha\mu^2$ is an example of an LEFN model. The QGPMLE of this model is considered in Section 3.3.

2.4.4 Generalized Linear Models

Generalized linear models (GLMs), introduced by Nelder and Wedderburn (1972), are closely related to the LEFN model. Differences include a notational one due to the use of an alternative parameterization of the exponential family, as well as several simplifications, including the use of more restrictive parameterizations of the conditional mean and variance than (2.29) and (2.39). A very useful simple summary of GLMs is presented in McCullagh and Nelder (1989, chapter 2).

A density $f_{\mathsf{GLM}}(y|\theta,\phi)$ is a member of a linear exponential family with *canonical* (or *natural*) *parameter* θ and *nuisance parameter* ϕ if

$$f_{\mathsf{GLM}}(y|\theta,\phi) = \exp\left\{\frac{\theta y - b(\theta)}{a(\phi)} + c(y,\phi)\right\}. \tag{2.45}$$

The function $b(\cdot)$ is such that

$$\mathsf{E}[y] = b'(\theta), \tag{2.46}$$

where $b'(\theta) = \partial b(\theta)/\partial\theta$. The function $a(\phi)$ is such that

$$\mathsf{V}[y] = a(\phi)b''(\theta), \tag{2.47}$$

where $b''(\theta) = \partial^2 b(\theta)/\partial\theta^2$. Usually $a(\phi) = \phi$. The function $c(\cdot)$ is a normalizing constant. Different functional forms for $a(\cdot)$ and $b(\cdot)$ lead to different GLMs. Note that the functions $a(\cdot)$, $b(\cdot)$, and $c(\cdot)$ for the GLM are different from the functions $a(\cdot)$, $b(\cdot)$, and $c(\cdot)$ for the LEF and LEFN.

As an example, the Poisson is the case $b(\theta) = \exp(\theta)$, $a(\phi) = 1$, and $c(y,\phi) = \ln y!$. Then (2.46) and (2.47) yield $\mathsf{E}[y] = \mathsf{V}[y] = \exp(\theta)$.

Regressors are introduced in the following way. Define the *linear predictor*

$$\eta = \mathbf{x}'\boldsymbol{\beta}. \tag{2.48}$$

The *link function* $\eta = \eta(\mu)$ relates the linear predictor to the mean μ. For example, the Poisson model with mean $\mu = \exp(\mathbf{x}'\boldsymbol{\beta})$ corresponds to the log link function $\eta = \ln\mu$. A special case of the link function of particular interest is the *canonical link function*, when

$$\eta = \theta. \tag{2.49}$$

For the Poisson the log link function is the canonical link function, because $b(\theta) = \exp(\theta)$ implies $\mu = b'(\theta) = \exp(\theta)$, so $\eta = \ln\mu = \ln(\exp(\theta)) = \theta$.

The concept of link function can cause confusion. It is more natural to consider the inverse of the link function, which is the conditional mean function. Thus, for example, the log link function is best thought of as being an exponential conditional mean function. The canonical link function is most easily thought of as leading to the density (2.45) being evaluated at $\theta = \mathbf{x}'\boldsymbol{\beta}$.

The MLE based on a GLM, $\hat{\boldsymbol{\beta}}_{\mathsf{GLM}}$, maximizes

$$\mathcal{L}_{\mathsf{GLM}} = \sum_{i=1}^{n}\left\{\frac{\theta(\mathbf{x}_i'\boldsymbol{\beta})y_i - b(\theta(\mathbf{x}_i'\boldsymbol{\beta}))}{a(\phi)} + c(y_i,\phi)\right\}. \tag{2.50}$$

The first-order conditions can be reexpressed as

$$\sum_{i=1}^{n}\frac{1}{\omega_i}(y_i - \mu_i)\frac{\partial\mu_i}{\partial\boldsymbol{\beta}} = \mathbf{0}, \tag{2.51}$$

where the variance function

$$\omega_i = a(\phi)\,\upsilon(\mu(\mathbf{x}_i'\boldsymbol{\beta})), \tag{2.52}$$

and

$$v(\mu_i) = b''(\theta_i). \tag{2.53}$$

The first-order conditions (2.51) for the GLM are of similar form to the first-order conditions (2.42) for the LEFN. This is because these two models are essentially the same. To link the two models, invert $\mu = b'(\theta)$ to obtain $\theta = d(\mu)$. Then (2.45) can be rewritten as

$$f_{\mathsf{GLM}}(y|\mu,\phi) = \exp\left\{\frac{-b(d(\mu))}{a(\phi)} + c(y,\phi) + \frac{d(\mu)}{a(\phi)}y\right\}, \tag{2.54}$$

which is clearly of the form (2.40), with the restriction that in (2.40) $a(\mu,\phi) = a_1(\mu)/a_2(\phi)$ and $c(\mu,\phi) = c_1(\mu)/c_2(\phi)$. This simplification implies that the GLM variance function $\omega(\mu(\mathbf{x}_i'\boldsymbol{\beta}),\phi)$ is multiplicative in ϕ, so that the first-order conditions can be solved for $\boldsymbol{\beta}$ without knowledge of ϕ. This is not necessarily a trivial simplification. For example, for Poisson with a nuisance parameter the GLM model specifies the variance function to be of multiplicative form $a(\phi)\mu$. The LEFN model, however, allows variance functions such as the quadratic form $\mu + \phi\mu^2$.

The same asymptotic theory as in section 2.4.3, therefore, holds. Assume $(y_i, \mathbf{x}_i)$ is independent over i, and the conditional mean and variance of y_i are correctly specified, so $\mathsf{E}[y_i|\mathbf{x}_i] = \mu(\mathbf{x}_i,\boldsymbol{\beta}_0)$ and $\mathsf{V}[y_i|\mathbf{x}_i] = a(\phi_0)\,v(\mu(\mathbf{x}_i,\boldsymbol{\beta}_0))$. Then $\hat{\boldsymbol{\beta}}_{\mathsf{GLM}} \xrightarrow{p} \boldsymbol{\beta}_0$, so the MLE is consistent for $\boldsymbol{\beta}_0$. Also

$$\sqrt{n}(\hat{\boldsymbol{\beta}}_{\mathsf{GLM}} - \boldsymbol{\beta}_0) \xrightarrow{d} \mathsf{N}[\mathbf{0}, \mathbf{A}^{-1}] \tag{2.55}$$

where

$$\mathbf{A} = \lim_{n\to\infty}\frac{1}{n}\sum_{i=1}^{n}\frac{1}{\omega_i}\frac{\partial\mu_i}{\partial\boldsymbol{\beta}}\frac{\partial\mu_i}{\partial\boldsymbol{\beta}'}\bigg|_{\boldsymbol{\beta}_0}. \tag{2.56}$$

A consistent estimate of the variance matrix is obtained by evaluating $\mathbf{A}$ at $\hat{\mu}_i = \mu(\mathbf{x}_i, \hat{\boldsymbol{\beta}}_{\mathsf{GLM}})$ and $\hat{\omega}_i = a(\hat{\phi})\,v(\hat{\mu}_i)$. The standard estimate of ϕ is obtained from

$$a(\hat{\phi}) = \frac{1}{n-k}\sum_{i=1}^{n}\frac{(y_i - \hat{\mu}_i)^2}{v(\hat{\mu}_i)}. \tag{2.57}$$

Usually $a(\phi) = \phi$.

In summary, the basic GLM model is based on the same density as the LEF and LEFN models presented in sections 2.4.2 through 2.4.3. However, it uses a different parameterization of the LEF, canonical and not mean, that is less natural if interest lies in modeling the conditional mean. The only real difference in the models is that the basic GLM model of Nelder and Wedderburn (1972) imposes some simplifying restrictions on the LEFN model that Gourieroux et al. (1984a) consider.

First, the conditional mean function $\mu(\mathbf{x}_i, \boldsymbol{\beta})$ is restricted to be a function of a linear combination of the regressors and so is of the simpler form $\mu(\mathbf{x}_i'\boldsymbol{\beta})$. This specialization to a single-index model simplifies interpretation of coefficients (see section 3.5) and permits computation of $\hat{\boldsymbol{\beta}}_{\mathsf{GLM}}$ using an iterative weighted least squares procedure that is detailed in McCullagh and Nelder (1989, chapter 2) and is presented for the Poisson model in section 3.8. Thus GLMs can be implemented even if one only has access to an OLS procedure. Given the computational facilities available at the time the GLM model was introduced, this was a considerable advantage.

Second, a particular parameterization of the conditional mean function, one that corresponds to the canonical link, is preferred. It can be shown that then $\partial\mu_i/\partial\boldsymbol{\beta} = \upsilon(\mu_i)\mathbf{x}_i$, so the first-order conditions (2.51) simplify to

$$\sum_{i=1}^{n}(y_i - \mu_i)\mathbf{x}_i = \mathbf{0}, \tag{2.58}$$

which makes computation especially easy. The QGPMLE for the LEFN defined in (2.42) does not take advantage of this simplification and instead solves

$$\sum_{i=1}^{n}\frac{1}{\upsilon(\widetilde{\mu}_i)}(y_i - \mu_i)\upsilon(\mu_i)\mathbf{x}_i = \mathbf{0}.$$

It is, however, asymptotically equivalent.

Third, the GLM variance function is of the simpler form $\omega_i = a(\phi)\,\upsilon(\mu_i)$, which is multiplicative in the nuisance parameter. Then one can estimate $\hat{\boldsymbol{\beta}}_{\mathsf{GLM}}$ without first estimating ϕ. A consequence is that with this simplification the QGPMLE of $\boldsymbol{\beta}$ equals the QMLE. Both can be obtained by using a maximum likelihood routine, with correct standard errors (or t-statistics) obtained by multiplying (or dividing) the standard maximum likelihood output by the square root of the scalar $a(\hat{\phi})$, which is easily estimated using (2.57).

2.4.5 Extensions

The LEFN and GLM densities permit more flexible models of the variance than the basic LEF density. Extensions to the LEF density that permit even greater flexibility in modeling the variance, particularly regression models for the variance, are *extended quasilikelihood* (Nelder and Pregibon, 1987), *double exponential families* (Efron, 1986), exponential dispersion models (Jorgensen, 1987), and *varying dispersion models* (Smyth, 1989). A survey is provided by Jorgensen (1997).

The presentation of GLMs has been likelihood based, in that the estimator of $\boldsymbol{\beta}$ maximizes a log-likelihood function, albeit one possibly misspecified. An alternative way to present the results is to take as starting point the first-order conditions (2.51)

$$\sum_{i=1}^{n}\frac{1}{\omega_i}(y_i - \mu_i)\frac{\partial\mu_i}{\partial\boldsymbol{\beta}} = \mathbf{0}. \tag{2.59}$$

One can define an estimator of $\boldsymbol{\beta}$ to be the solution to these equations without defining an underlying objective function whose derivative with respect to $\boldsymbol{\beta}$ is (2.59). These estimating equations have many properties similar to those of a log-likelihood derivative, and accordingly the left-hand side of (2.59) is called a *quasiscore function*. For completeness one can attempt to integrate this to obtain a *quasilikelihood function*. Accordingly, the solution to (2.59) is called the *quasilikelihood* (QL) *estimator*. For further details, see McCullagh and Nelder (1989, chapter 9).

It follows that the estimator of $\boldsymbol{\beta}$ in the GLM model can be interpreted either (1) as a quasi-MLE or pseudo-MLE, meaning that it is an MLE based on a possibly misspecified density, or (2) as a QL estimator, meaning that it is the solution to estimating equations that look like those from maximization of an unspecified log-likelihood function. It should be clear that in general the terms QML and QL have different meanings.

The recognition that it is sufficient to simply define the QL estimating equations (2.59) has led to generalizations of (2.59) and additional estimating equations to permit, for example, more flexible models of the variance functions that do not require specification of the density.

2.4.6 Generalized Estimating Equations

Suppose we generalize the GLM model to one with $\mathsf{E}[y_{ij}|\mathbf{x}_{ij}] = \mu_{ij} = \mu(\mathbf{x}_{ij}, \boldsymbol{\beta})$, where data are independent over i but correlated over j. Leading examples are longitudinal data on individual i in time period j, and grouped data on, for example, individual j in village i. Then we can consistently estimate $\boldsymbol{\beta}$, using the preceding methods that ignore the correlation over j, and then we can obtain correct standard errors that control for the correlation over j using the robust inference methods given in section 2.7.

More efficient estimation is possible if one models the correlation over j and estimates by a method analogous to a nonlinear version of feasible generalized least squares. Let $\mathbf{y}_i = [y_{i1}, \ldots, y_{iJ}]'$, $\boldsymbol{\mu}_i(\boldsymbol{\beta}) = [\mu_{i1}, \ldots, \mu_{iJ}]'$, and $\mathbf{X}_i = [\mathbf{x}_{i1}, \ldots, \mathbf{x}_{iJ}]'$, where for simplicity we considered balanced grouping though the extension to J_i observations in the ith group is immediate. Let $\boldsymbol{\Sigma}_i(\boldsymbol{\alpha}, \boldsymbol{\beta})$ denote the working variance matrix that is a model for $\mathsf{V}[\mathbf{y}_i|\mathbf{X}_i]$. For example, if data are equicorrelated then $\mathsf{Cov}\big[y_{ij}, y_{ik}|\mathbf{X}_i\big] = \rho\sigma_{ij}\sigma_{ik}$, where σ_{ij}^2 is a model for $\mathsf{V}\big[y_{ij}|\mathbf{x}_{ij}\big]$ such as $\phi\mu_{ij}$ for overdispersed Poisson (in this example, $\boldsymbol{\alpha} = [\rho\ \phi]'$). Then Liang and Zeger (1986) proposed the *generalized estimating equations estimator* (GEE) estimator $\hat{\boldsymbol{\beta}}_{\mathsf{GEE}}$, which solves

$$\sum_{i=1}^{n} \frac{\partial \boldsymbol{\mu}_i(\boldsymbol{\beta})}{\partial \boldsymbol{\beta}'} \hat{\boldsymbol{\Sigma}}_i^{-1} (\mathbf{y}_i - \boldsymbol{\mu}_i(\boldsymbol{\beta})) = \mathbf{0}, \tag{2.60}$$

where $\hat{\boldsymbol{\Sigma}}_i = \boldsymbol{\Sigma}_i(\hat{\boldsymbol{\alpha}}, \hat{\boldsymbol{\beta}})$ and $\hat{\boldsymbol{\alpha}}$ and $\hat{\boldsymbol{\beta}}$ are first-stage estimates of $\boldsymbol{\alpha}$ and $\boldsymbol{\beta}$. These first-order conditions are a natural extension of the weighted linear least squares first-order conditions (2.22), and if $\mathbf{\Sigma}_i = \mathsf{V}[\mathbf{y}_i|\mathbf{X}_i]$ are those for the optimal GMM estimator derived below in section 2.5.3.

Given correct specification of $\mu(\mathbf{x}_{ij}, \boldsymbol{\beta})$, $\hat{\boldsymbol{\beta}}_{\mathsf{GEE}} \xrightarrow{p} \boldsymbol{\beta}_0$ and $\sqrt{n}(\hat{\boldsymbol{\beta}}_{\mathsf{GEE}} - \boldsymbol{\beta}_0) \xrightarrow{d} \mathsf{N}[\mathbf{0}, \mathbf{A}^{-1}\mathbf{B}\mathbf{A}'^{-1}]$, where

$$\mathbf{A} = \lim_{n\to\infty} \frac{1}{n} \sum_{i=1}^{n} \frac{\partial \boldsymbol{\mu}_i}{\partial \boldsymbol{\beta}} \mathbf{\Sigma}_i^{-1} \frac{\partial \boldsymbol{\mu}_i}{\partial \boldsymbol{\beta}'}\bigg|_{\boldsymbol{\theta}_0}, \tag{2.61}$$

$\boldsymbol{\theta}_0 = [\boldsymbol{\alpha}\ \boldsymbol{\beta}]'$, and

$$\mathbf{B} = \lim_{n\to\infty} \frac{1}{n} \sum_{i=1}^{n} \frac{\partial \boldsymbol{\mu}_i}{\partial \boldsymbol{\beta}} \mathbf{\Sigma}_i^{-1} \mathsf{V}[\mathbf{y}_i|\mathbf{X}_i] \mathbf{\Sigma}_i^{-1} \frac{\partial \boldsymbol{\mu}_i}{\partial \boldsymbol{\beta}'}\bigg|_{\boldsymbol{\theta}_0}. \tag{2.62}$$

A consistent estimate for $\mathbf{A}$ is obtained by evaluation of (2.61) at $\hat{\boldsymbol{\theta}}$, whereas for $\mathbf{B}$ we use

$$\hat{\mathbf{B}} = \frac{1}{n} \sum_{i=1}^{n} \frac{\partial \boldsymbol{\mu}_i}{\partial \boldsymbol{\beta}}\bigg|_{\hat{\boldsymbol{\beta}}} \hat{\mathbf{\Sigma}}_i^{-1} (\mathbf{y}_i - \hat{\boldsymbol{\mu}}_i)(\mathbf{y}_i - \hat{\boldsymbol{\mu}}_i)' \frac{\partial \boldsymbol{\mu}_i}{\partial \boldsymbol{\beta}}\bigg|_{\hat{\boldsymbol{\beta}}}. \tag{2.63}$$

In the special case that $\mathsf{V}[\mathbf{y}_i|\mathbf{X}_i] = \mathbf{\Sigma}_i$, results simplify to $\sqrt{n}(\hat{\boldsymbol{\beta}}_{\mathsf{GEE}} - \boldsymbol{\beta}_0) \xrightarrow{d} \mathsf{N}[\mathbf{0}, \mathbf{A}^{-1}]$.

2.5 MOMENT-BASED MODELS

The first-order conditions (2.6) for the Poisson MLE can be motivated by noting that the specification of the conditional mean, $\mathsf{E}[y_i|\mathbf{x}_i] = \exp(\mathbf{x}_i'\boldsymbol{\beta})$, implies the unconditional population moment condition

$$\mathsf{E}[(y_i - \exp(\mathbf{x}_i'\boldsymbol{\beta}))\mathbf{x}_i] = \mathbf{0}.$$

A *method of moments estimator* for $\boldsymbol{\beta}$ is the solution to the corresponding sample moment condition

$$\sum_{i=1}^{n} (y_i - \exp(\mathbf{x}_i'\boldsymbol{\beta}))\mathbf{x}_i = \mathbf{0}.$$

In this example, the number of moment conditions equals the number of parameters, so a numerical solution for $\hat{\boldsymbol{\beta}}$ is possible. This is a special case of the estimating equations approach, presented in section 2.5.1. More generally, there may be more moment conditions than parameters. Then we use the generalized method of moments estimator, which minimizes a quadratic function of the moment conditions and is presented in section 2.5.2.

2.5.1 Estimating Equations

We consider the q population moment conditions

$$\mathsf{E}[\mathbf{g}_i(y_i, \mathbf{x}_i, \boldsymbol{\theta})] = \mathbf{0}, \qquad i = 1, \ldots, n, \tag{2.64}$$

where $\mathbf{g}_i$ is a $q \times 1$ vector with the same dimension as $\boldsymbol{\theta}$. The estimator $\hat{\boldsymbol{\theta}}_{\mathsf{EE}}$ solves the corresponding *estimating equations*

$$\sum_{i=1}^{n} \mathbf{g}_i(y_i, \mathbf{x}_i, \boldsymbol{\theta}) = \mathbf{0}, \tag{2.65}$$

a system of q equations in q unknowns. The term "estimating equations" here should not be confused with the term GEE introduced in the preceding section.

If (2.64) holds at $\boldsymbol{\theta}_0$ and regularity conditions are satisfied,

$$\sqrt{n}(\hat{\boldsymbol{\theta}}_{\mathsf{EE}} - \boldsymbol{\theta}_0) \xrightarrow{d} \mathsf{N}[\mathbf{0}, \mathbf{A}^{-1}\mathbf{B}\mathbf{A}'^{-1}], \tag{2.66}$$

where

$$\mathbf{A} = \lim_{n\to\infty} \frac{1}{n} \mathsf{E}\left[\sum_{i=1}^{n} \left. \frac{\partial \mathbf{g}_i(y_i, \mathbf{x}_i, \boldsymbol{\theta})}{\partial \boldsymbol{\theta}'} \right|_{\boldsymbol{\theta}_0} \right] \tag{2.67}$$

$$\mathbf{B} = \lim_{n\to\infty} \frac{1}{n} \mathsf{E}\left[\sum_{i=1}^{n} \mathbf{g}_i(y_i, \mathbf{x}_i, \boldsymbol{\theta}) \mathbf{g}_i(y_i, \mathbf{x}_i, \boldsymbol{\theta})' \Big|_{\boldsymbol{\theta}_0} \right], \tag{2.68}$$

assuming independent observations.

The variance matrix in (2.66) is consistently estimated by $\hat{\mathbf{A}}^{-1}\hat{\mathbf{B}}\hat{\mathbf{A}}'^{-1}$, where $\hat{\mathbf{A}}$ and $\hat{\mathbf{B}}$ are any consistent estimates of $\mathbf{A}$ and $\mathbf{B}$. Such estimators are called sandwich estimators, because $\hat{\mathbf{B}}$ is sandwiched between $\hat{\mathbf{A}}^{-1}$ and $\hat{\mathbf{A}}'^{-1}$. Throughout the book we use the term *robust sandwich* (RS) estimator for the special case when the consistent estimators of $\mathbf{A}$ and $\mathbf{B}$ are

$$\hat{\mathbf{A}} = \frac{1}{n} \sum_{i=1}^{n} \left. \frac{\partial \mathbf{g}_i(y_i, \mathbf{x}_i, \boldsymbol{\theta})}{\partial \boldsymbol{\theta}'} \right|_{\hat{\boldsymbol{\theta}}_{\mathsf{EE}}}, \tag{2.69}$$

and, assuming independence of the data over i,

$$\hat{\mathbf{B}} = \frac{1}{n} \sum_{i=1}^{n} \mathbf{g}_i(y_i, \mathbf{x}_i, \hat{\boldsymbol{\theta}}_{\mathsf{EE}}) \mathbf{g}_i(y_i, \mathbf{x}_i, \hat{\boldsymbol{\theta}}_{\mathsf{EE}})'. \tag{2.70}$$

This has the special property that it is *robust* to misspecification of the dgp, in the sense that the expectations in (2.67) and (2.68) have been dropped. Section 2.7.1 presents adaptations of $\hat{\mathbf{B}}$ for correlated observations.

For example, the OLS estimator sets $\mathbf{g}_i(y_i, \mathbf{x}_i, \boldsymbol{\beta}) = (y_i - \mathbf{x}_i'\boldsymbol{\beta})\mathbf{x}_i$; see (2.22) with $v_i = 1$. Then

$$\mathbf{B} = \frac{1}{n} \sum_i \mathsf{E}[(y_i - \mathbf{x}_i'\boldsymbol{\beta})^2]\mathbf{x}_i\mathbf{x}_i'.$$

A consistent estimator of $\mathbf{B}$ that makes no assumptions on $\mathsf{E}[(y_i - \mathbf{x}_i'\boldsymbol{\beta})^2]$ is

$$\hat{\mathbf{B}} = \frac{1}{n} \sum_i (y_i - \mathbf{x}_i'\hat{\boldsymbol{\beta}})^2 \mathbf{x}_i\mathbf{x}_i'.$$

White (1980), building on work by Eicker (1967), proposed this estimator to guard against heteroskedasticity in models assuming homoskedasticity. Huber (1967) and White (1982) proposed the sandwich estimator, see (2.19), to guard against misspecification of the density in the maximum likelihood framework. The robust sandwich estimator is often called the Huber estimator or Eicker-White estimator.

The estimating equation approach is general enough to include maximum likelihood and GLM as special cases. GEE presented in section 2.4.6 is a mild extension. Optimal estimating equations based on the first few moments of the dependent variable are given in Chapter 11. Such extensions have tended to be piecemeal and assume that the number of moment conditions equals the number of parameters. A very general framework, widely used in econometrics but rarely used in other areas of statistics, is generalized methods of moments. This is now presented.

2.5.2 Generalized Methods of Moments

We consider the r population moment (or orthogonality) conditions

$$\mathsf{E}[\mathbf{h}_i(y_i, \mathbf{x}_i, \boldsymbol{\theta})] = \mathbf{0}, \qquad i = 1, \ldots, n, \tag{2.71}$$

where $\mathbf{h}_i$ is an $r \times 1$ vector and $r \geq q$, so that the number of moment conditions potentially exceeds the number of parameters. Overidentified instrumental variables estimation, see section 2.5.4, is a leading example.

Hansen (1982) proposed the *generalized methods of moments* (GMM) estimator $\hat{\boldsymbol{\theta}}_{\mathsf{GMM}}$, which makes the sample moment corresponding to (2.71) as small as possible in the quadratic norm

$$\left[\sum_{i=1}^{n} \mathbf{h}_i(y_i, \mathbf{x}_i, \boldsymbol{\theta})\right]' \mathbf{W}_n \left[\sum_{i=1}^{n} \mathbf{h}_i(y_i, \mathbf{x}_i, \boldsymbol{\theta})\right], \tag{2.72}$$

where $\mathbf{W}_n$ is a possibly stochastic symmetric positive definite $r \times r$ *weighting matrix*, which converges in probability to a nonstochastic matrix $\mathbf{W}$. The GMM estimator is calculated as the solution to the first-order conditions

$$\left[\sum_{i=1}^{n} \frac{\partial \mathbf{h}_i'}{\partial \boldsymbol{\theta}}\right] \mathbf{W}_n \left[\sum_{i=1}^{n} \mathbf{h}_i\right] = \mathbf{0}, \tag{2.73}$$

where $\mathbf{h}_i = \mathbf{h}_i(y_i, \mathbf{x}_i, \boldsymbol{\theta})$. The solution will generally require an iterative technique. The parameter $\boldsymbol{\theta}$ is identified if (2.73) has a unique solution.

Under suitable assumptions $\hat{\boldsymbol{\theta}} \xrightarrow{p} \boldsymbol{\theta}_0$, where $\boldsymbol{\theta}_0$ is the value of $\boldsymbol{\theta}$ that minimizes the probability limit of n^{-2} times the objective function (2.72). Also

$$\sqrt{n}(\hat{\boldsymbol{\theta}}_{\mathsf{GMM}} - \boldsymbol{\theta}_0) \xrightarrow{d} \mathsf{N}[\mathbf{0}, \mathbf{A}^{-1}\mathbf{B}\mathbf{A}'^{-1}], \tag{2.74}$$

where the formulas for $\mathbf{A}$ and $\mathbf{B}$ are

$$\mathbf{A} = \mathbf{H}'\mathbf{W}\mathbf{H} \tag{2.75}$$

$$\mathbf{B} = \mathbf{H}'\mathbf{W}\mathbf{S}\mathbf{W}\mathbf{H} \tag{2.76}$$

where

$$\mathbf{H} = \lim_{n\to\infty} \frac{1}{n}\mathsf{E}\left[\sum_{i=1}^{n} \left.\frac{\partial \mathbf{h}_i}{\partial \boldsymbol{\theta}'}\right|_{\boldsymbol{\theta}_0}\right] \tag{2.77}$$

$$\mathbf{S} = \lim_{n\to\infty} \frac{1}{n}\mathsf{E}\left[\sum_{i=1}^{n} \mathbf{h}_i\mathbf{h}_i'|_{\boldsymbol{\theta}_0}\right]. \tag{2.78}$$

The expression for $\mathbf{S}$ assumes independent observations. Note that substitution of (2.75) and (2.76) into (2.74) yields an expression for the variance matrix of the GMM estimator of the same form as (2.24) for the linear WLS estimator.

For a given choice of population moment condition $\mathbf{h}_i(y_i, \mathbf{x}_i, \boldsymbol{\theta})$ in (2.71), the optimal choice of weighting matrix $\mathbf{W}_n$ in (2.72) is the inverse of a consistent estimator $\hat{\mathbf{S}}$ of $\mathbf{S}$. The *two-step* GMM *estimator* $\hat{\boldsymbol{\theta}}_{\mathsf{GMM}}^{2step}$ minimizes

$$\left[\sum_{i=1}^{n} \mathbf{h}_i(y_i, \mathbf{x}_i, \boldsymbol{\theta})\right]' \hat{\mathbf{S}}^{-1} \left[\sum_{i=1}^{n} \mathbf{h}_i(y_i, \mathbf{x}_i, \boldsymbol{\theta})\right]. \tag{2.79}$$

Then

$$\sqrt{n}(\hat{\boldsymbol{\theta}}_{\mathsf{GMM}}^{2step} - \boldsymbol{\theta}_0) \overset{d}{\to} \mathsf{N}[\mathbf{0}, \mathbf{A}^{-1}], \tag{2.80}$$

where

$$\mathbf{A} = \mathbf{H}'\mathbf{S}^{-1}\mathbf{H}. \tag{2.81}$$

A standard procedure is, first, to estimate the model by GMM with weighting matrix $\mathbf{W}_n = \mathbf{I}_r$ to obtain initial consistent estimates of $\boldsymbol{\theta}$ and hence $\hat{\mathbf{S}}$, and, second, to estimate again by GMM using $\mathbf{W}_n = \hat{\mathbf{S}}^{-1}$.

It is important to note that this optimality is limited, because it is for a given moment condition $\mathbf{h}_i(y_i, \mathbf{x}_i, \boldsymbol{\theta})$. Some choices of $\mathbf{h}_i(y_i, \mathbf{x}_i, \boldsymbol{\theta})$ are better than others; see section 2.5.3. If the distribution is completely specified, the MLE is optimal and

$$\mathbf{h}_i(y_i, \mathbf{x}_i, \boldsymbol{\theta}) = \frac{\partial \ln f(y_i|\mathbf{x}_i, \boldsymbol{\theta})}{\partial \boldsymbol{\theta}}.$$

This is pursued further in section 2.5.3.

To operationalize, these results require consistent estimates of $\mathbf{H}$ and $\mathbf{S}$. For $\mathbf{H}$ use

$$\hat{\mathbf{H}} = \frac{1}{n}\sum_{i=1}^{n} \left.\frac{\partial \mathbf{h}_i}{\partial \boldsymbol{\theta}'}\right|_{\hat{\boldsymbol{\theta}}_{\mathsf{GMM}}}. \tag{2.82}$$

When observations are independent over i, one uses

$$\hat{\mathbf{S}} = \frac{1}{n}\sum_{i=1}^{n} \mathbf{h}_i \mathbf{h}_i' |_{\hat{\theta}_{\mathsf{GMM}}}. \tag{2.83}$$

See section 2.7 for extensions of $\hat{\mathbf{S}}$ to correlated observations.

The GMM results simplify if $r = q$, in which case we have the estimating equations presented in the previous subsection. Then $\mathbf{h}_i(y_i, \mathbf{x}_i, \boldsymbol{\theta}) = \mathbf{g}_i(y_i, \mathbf{x}_i, \boldsymbol{\theta})$, $\mathbf{H} = \mathbf{A}$, $\mathbf{B} = \mathbf{S}$, where $\mathbf{B}$ assumes independence over i, and the results are invariant to the choice of weighting matrix $\mathbf{W}_n$. The estimating equations estimator defined by (2.65) is the GMM estimator that minimizes

$$\left[\sum_{i=1}^{n} \mathbf{g}_i(y_i, \mathbf{x}_i, \boldsymbol{\theta})\right]' \left[\sum_{i=1}^{n} \mathbf{g}_i(y_i, \mathbf{x}_i, \boldsymbol{\theta})\right].$$

Unlike the general case $r > q$, this quadratic objective function takes value 0 at the optimal value of $\boldsymbol{\theta}$ when $r = q$.

In the case that $r > q$ the GMM objective function is nonzero, leading to the *overidentifying restrictions* (OIR) test statistic

$$\mathsf{T}_{\mathsf{OIR}} = \frac{1}{n}\left[\sum_{i=1}^{n} \mathbf{h}_i(y_i, \mathbf{x}_i, \hat{\boldsymbol{\theta}}_{\mathsf{GMM}}^{2\text{step}})\right]' \hat{\mathbf{S}}^{-1} \left[\sum_{i=1}^{n} \mathbf{h}_i(y_i, \mathbf{x}_i, \hat{\boldsymbol{\theta}}_{\mathsf{GMM}}^{2\text{step}})\right], \tag{2.84}$$

which is n^{-1} times the objective function (2.72). Hansen (1982) showed that $\mathsf{T}_{\mathsf{OIR}}$ is asymptotically $\chi^2(r-q)$ under $H_0 : \mathsf{E}[\mathbf{h}_i(y_i, \mathbf{x}_i, \boldsymbol{\theta}_0)] = \mathbf{0}$. Too large a value of $\mathsf{T}_{\mathsf{OIR}}$ is interpreted as rejection of the moment conditions (2.71). It is important to note that this test can be used only if estimation is by two-step GMM.

2.5.3 Optimal GMM

We have already considered a limited form of optimal GMM. Given a choice of $\mathbf{h}(y_i, \mathbf{x}_i, \boldsymbol{\theta})$ in (2.71), the optimal GMM estimator is $\hat{\boldsymbol{\theta}}_{\mathsf{GMM}}^{2step}$ defined in (2.79), which uses as a weighting matrix a consistent estimate of $\mathbf{S}$ defined in (2.78).

Now we consider the more difficult question of the optimal specification of $\mathbf{h}(y_i, \mathbf{x}_i, \boldsymbol{\theta})$ in the cross-sectional case or panel case where $y_i, \mathbf{x}_i$ are iid. This is analyzed by Chamberlain (1987) and Newey (1990a), with an excellent summary given in Newey (1993). Suppose interest lies in estimation based on the conditional moment restriction

$$\mathsf{E}[\boldsymbol{\rho}(y_i, \mathbf{x}_i, \boldsymbol{\theta})|\mathbf{x}_i] = \mathbf{0}, \qquad i = 1, \ldots, n, \tag{2.85}$$

where $\boldsymbol{\rho}(\cdot)$ is a residual-type $s \times 1$ vector function.

For example, let $s = 2$ with the components of $\boldsymbol{\rho}(\cdot)$ being $\rho_1(y_i, \mathbf{x}_i, \boldsymbol{\theta}) = y_i - \mu_i$ and $\rho_2(y_i, \mathbf{x}_i, \boldsymbol{\theta}) = (y_i - \mu_i)^2 - \sigma_i^2$, where $\mu_i = \mu(\mathbf{x}_i, \boldsymbol{\theta})$ and $\sigma_i^2 = \omega(\mathbf{x}_i, \boldsymbol{\theta})$ are specified conditional mean and variance functions.

Typically s is less than the number of parameters, so GMM estimation based on (2.85) is not possible. Instead we introduce an $r \times s$ matrix of functions $\mathbf{D}(\mathbf{x}_i)$, where $r \geq q$, and note that by the law of iterated expectations,

$$\mathsf{E}[\mathbf{D}(\mathbf{x}_i)\boldsymbol{\rho}(y_i, \mathbf{x}_i, \boldsymbol{\theta})] = \mathbf{0}, \qquad i = 1, \ldots, n. \tag{2.86}$$

$\boldsymbol{\theta}$ can be estimated by GMM based on (2.86), because there are now $r \geq q$ moment conditions.

The variance of the GMM estimator can be shown to be minimized, given (2.85), by choosing $\mathbf{D}(\mathbf{x}_i)$ equal to the $q \times s$ matrix

$$\mathbf{D}^*(\mathbf{x}_i) = \mathsf{E}\left[\left.\frac{\partial \boldsymbol{\rho}(y_i, \mathbf{x}_i, \boldsymbol{\theta})'}{\partial \boldsymbol{\theta}}\right| \mathbf{x}_i\right] \left\{\mathsf{E}\left[\boldsymbol{\rho}(y_i, \mathbf{x}_i, \boldsymbol{\theta})\boldsymbol{\rho}(y_i, \mathbf{x}_i, \boldsymbol{\theta})'|\mathbf{x}_i\right]\right\}^{-1}.$$

Premultiplication of $\mathbf{D}^*(\mathbf{x}_i)$ by an $s \times s$ matrix of constants (not depending on $\mathbf{x}_i$) yields an equivalent optimal estimator. It follows that the optimal choice of $\mathbf{h}(y_i, \mathbf{x}_i, \boldsymbol{\theta})$ for GMM estimation, given (2.85), is

$$\mathbf{h}_i^*(y_i, \mathbf{x}_i, \boldsymbol{\theta}) = \mathsf{E}\left[\left.\frac{\partial \boldsymbol{\rho}(y_i, \mathbf{x}_i, \boldsymbol{\theta})'}{\partial \boldsymbol{\theta}}\right| \mathbf{x}_i\right] \times \left\{\mathsf{E}\left[[\boldsymbol{\rho}(y_i, \mathbf{x}_i, \boldsymbol{\theta})\boldsymbol{\rho}(y_i, \mathbf{x}_i, \boldsymbol{\theta})'|\mathbf{x}_i]\right]\right\}^{-1} \boldsymbol{\rho}(y_i, \mathbf{x}_i, \boldsymbol{\theta}). \tag{2.87}$$

Note that here $r = k$, so $\mathbf{h}_i^*(y_i, \mathbf{x}_i, \boldsymbol{\theta})$ is $q \times q$ and the estimating equation results of section 2.5.1 are applicable. The optimal GMM estimator is the solution to

$$\sum_{i=1}^{n} \mathbf{h}_i^*(y_i, \mathbf{x}_i, \boldsymbol{\theta}) = \mathbf{0}.$$

The limit distribution is given in (2.66)–(2.68), with $\mathbf{g}_i(\cdot) = \mathbf{h}_i^*(\cdot)$.

This optimal GMM estimator is applied, for example, to models with specified conditional mean and variance functions in Chapter 11.7.2.

2.5.4 Instrumental Variables Estimators

Hansen's GMM estimator is based on earlier work on instrumental variables by Sargan (1958) in the linear case and Amemiya (1974) in the nonlinear case.

Consider the linear regression model $y_i = \mathbf{x}_i'\boldsymbol{\beta} + u_i$. The OLS estimator is inconsistent if $\mathsf{E}[u_i|\mathbf{x}_i] = 0$, because then regressors are correlated with the error. To obtain a consistent estimator, we assume the existence of *instruments* $\mathbf{z}_i$ satisfying $\mathsf{E}[u_i|\mathbf{z}_i] = 0$, which in turn implies $\mathsf{E}[\mathbf{z}_i u_i] = \mathsf{E}[\mathbf{z}_i(y_i - \mathbf{x}_i\boldsymbol{\beta})] = \mathbf{0}$.

In the case that $\dim[\mathbf{z}] = \dim[\mathbf{x}]$, the model is said to be *just identified*, and the corresponding sample moment conditions $\sum_{i=1}^{n} \mathbf{z}_i(y_i - \mathbf{x}_i\boldsymbol{\beta}) = \mathbf{0}$ can be solved to yield the linear *instrumental variables* (IV) estimator

$$\hat{\boldsymbol{\beta}}_{\text{IV}} = \left[\sum_{i=1}^{n} \mathbf{z}_i\mathbf{x}_i'\right]^{-1} \sum_{i=1}^{n} \mathbf{z}_i y_i.$$

If dim[$\mathbf{z}$] > dim[$\mathbf{x}$], the model is said to be *overidentified* and estimation is by GMM, with $\mathbf{h}_i(\cdot) = \sum_{i=1}^n \mathbf{z}_i(y_i - \mathbf{x}_i\boldsymbol{\beta})$. The *two-stage least squares estimator* (2SLS) is obtained using the weighting matrix $\mathbf{W}_n = \left(\frac{1}{n}\sum_{i=1}^n \mathbf{z}_i\mathbf{z}_i'\right)^{-1}$ in (2.72). If errors are heteroskedastic the optimal GMM estimator uses $\mathbf{W}_n = \left(\frac{1}{n}\sum_{i=1}^n \hat{u}_i^2\mathbf{z}_i\mathbf{z}_i'\right)^{-1}$, where $\hat{u}_i = y_i - \mathbf{x}_i'\hat{\boldsymbol{\beta}}$ and $\hat{\boldsymbol{\beta}}$ is a consistent estimator such as 2SLS.

In many applications only one component of $\mathbf{x}_i$ is correlated with u_i. Then IV estimation requires finding at least one instrument for that component, while the remaining components of $\mathbf{x}_i$ can be used as instruments for themselves. If there are more instruments than necessary, then too large a value of the OIR test in (2.84) is interpreted as rejection of the validity of the excess instruments.

For the nonlinear regression model $y_i = \mu(\mathbf{x}_i'\boldsymbol{\beta}) + u_i$, Amemiya (1974) proposed the *nonlinear two-stage least squares* or *nonlinear instrumental variables* (NLIV) estimator, which minimizes

$$\left[\sum_{i=1}^n (y_i - \mu(\mathbf{x}_i'\boldsymbol{\beta}))\mathbf{z}_i'\right]\left[\sum_{i=1}^n \mathbf{z}_i\mathbf{z}_i'\right]^{-1}\left[\sum_{i=1}^n (y_i - \mu(\mathbf{x}_i'\boldsymbol{\beta}))\mathbf{z}_i\right], \qquad (2.88)$$

where $\mathbf{z}_i$ is an $r \times 1$ set of instruments such that $\mathsf{E}[y_i - \mu(\mathbf{x}_i'\boldsymbol{\beta})|\mathbf{z}_i] = 0$. This is a GMM estimator where $\mathbf{h}_i(\cdot) = (y_i - \mu(\mathbf{x}_i'\boldsymbol{\beta}))\mathbf{z}_i$ in (2.72). The weighting matrix in (2.88) is optimal if $\mathsf{V}[y_i|\mathbf{z}_i] = \sigma^2$, because then $\mathbf{S} = \sigma^2 \lim \frac{1}{n}\sum_{i=1}^n \mathbf{z}_i\mathbf{z}_i'$, and the variance of the estimator from (2.81) is $\mathbf{H}'\mathbf{S}^{-1}\mathbf{H}$ where $\mathbf{H} = \lim \frac{1}{n}\sum_{i=1}^n \mathbf{z}_i\partial\mu_i/\partial\boldsymbol{\beta}'$. For further details, see Burguette, Gallant, and Souza (1982), Hansen (1982), and Newey (1990a). This estimator is used, for example, in Chapter 10.4, which also considers extension to heteroskedastic errors.

2.5.5 Sequential 2-Step Estimators

The GMM framework is quite general. One example of its application is to sequential two-step estimators. Consider the case in which a model depends on vector parameters $\boldsymbol{\theta}_1$ and $\boldsymbol{\theta}_2$, and the model is estimated sequentially: (1) Obtain a root-n consistent estimate $\widetilde{\boldsymbol{\theta}}_1$ of $\boldsymbol{\theta}_1$ that solves $\sum_{i=1}^n \mathbf{h}_{1i}(y_i, \mathbf{x}_i, \widetilde{\boldsymbol{\theta}}_1) = \mathbf{0}$, and (2) obtain a root-$n$ consistent estimate $\hat{\boldsymbol{\theta}}_2$ of $\boldsymbol{\theta}_2$ given $\widetilde{\boldsymbol{\theta}}_1$ that solves $\sum_{i=1}^n \mathbf{h}_{2i}(y_i, \mathbf{x}_i, \widetilde{\boldsymbol{\theta}}_1, \hat{\boldsymbol{\theta}}_2) = \mathbf{0}$.

In general, the distribution of $\hat{\boldsymbol{\theta}}_2$ given estimation of $\widetilde{\boldsymbol{\theta}}_1$ differs from, and is more complicated than, the distribution of $\hat{\boldsymbol{\theta}}_2$ if $\boldsymbol{\theta}_1$ is known. Statistical inference is invalid if it fails to take into account this complication. Newey (1984) proposed obtaining the distribution of $\hat{\boldsymbol{\theta}}_2$ by noting that $(\boldsymbol{\theta}_1, \boldsymbol{\theta}_2)$ jointly solve the equations

$$\sum_{i=1}^n \mathbf{h}_{1i}(y_i, \mathbf{x}_i, \boldsymbol{\theta}_1) = \mathbf{0}$$

$$\sum_{i=1}^n \mathbf{h}_{2i}(y_i, \mathbf{x}_i, \boldsymbol{\theta}_1, \boldsymbol{\theta}_2) = \mathbf{0}.$$

This is simply a special case of

$$\sum_{i=1}^{n} \mathbf{h}_i(y_i, \mathbf{x}_i, \boldsymbol{\theta}) = \mathbf{0},$$

defining $\boldsymbol{\theta} = (\boldsymbol{\theta}_1' \quad \boldsymbol{\theta}_2')'$ and $\mathbf{h}_i = (\mathbf{h}_{1i}' \quad \mathbf{h}_{2i}')'$. This is a GMM estimator with $\mathbf{W}_n = \mathbf{W} = \mathbf{I}$. Applying (2.74) with $\mathbf{A}$ and $\mathbf{B}$ partitioned similarly to $\boldsymbol{\theta}$ and $\mathbf{h}_i$ yields a variance matrix for $\hat{\boldsymbol{\theta}}_2$ that is quite complicated, even though simplification occurs because $\partial \mathbf{h}_{1i}(\boldsymbol{\theta})/\partial \boldsymbol{\theta}_2' = \mathbf{0}$. The expression is given in Newey (1984), Murphy and Topel (1985), Pagan (1986), and several texts. See also Pierce (1982). It is often easier to bootstrap; see section 2.6.4.

A well-known exception to the need to take account of the randomness due to estimation of $\widetilde{\boldsymbol{\theta}}_1$ is feasible GLS, where $\boldsymbol{\theta}_1$ corresponds to the first-round estimates used to consistently estimate the variance matrix, and $\boldsymbol{\theta}_2$ corresponds to the second-round feasible GLS estimates of the regression parameters for the conditional mean.

Such simplification occurs whenever $\mathsf{E}[\partial \mathbf{h}_{2i}(\boldsymbol{\theta})/\partial \boldsymbol{\theta}_1] = \mathbf{0}$. This simplification holds for the GLM and LEFN models. To see this for LEFN, from (2.42) with $\hat{\boldsymbol{\theta}}_2 = \hat{\boldsymbol{\beta}}_{\mathsf{LEFN}}$ and $\widetilde{\boldsymbol{\theta}}_1 = \widetilde{\alpha}$

$$\mathbf{h}_{2i}(\boldsymbol{\theta}) = \frac{1}{\widetilde{\omega}_i(\widetilde{\boldsymbol{\theta}}_1)}(y_i - \mu_i(\boldsymbol{\theta}_2))\frac{\partial \mu_i(\boldsymbol{\theta}_2)}{\partial \boldsymbol{\theta}_2},$$

it follows that $\mathsf{E}[\partial \mathbf{h}_{2i}(\boldsymbol{\theta})/\partial \boldsymbol{\theta}_1'] = \mathbf{0}$.

This simplification also arises in the ML framework for jointly estimated $\boldsymbol{\theta}_1$ and $\boldsymbol{\theta}_2$ if the information matrix is block-diagonal. Then the variance of $\hat{\boldsymbol{\theta}}_{1,\mathsf{ML}}$ is simply the inverse of $-\mathsf{E}[\partial^2 \mathcal{L}(\boldsymbol{\theta})/\partial \boldsymbol{\theta}_1 \partial \boldsymbol{\theta}_1'] = \mathsf{E}[\partial(\partial \mathcal{L}(\boldsymbol{\theta})/\partial \boldsymbol{\theta}_1)/\partial \boldsymbol{\theta}_1']$. An example is the negative binomial distribution with quadratic variance function; see Chapter 3.3.1.

2.5.6 Empirical Likelihood

The moment-based approaches do not require complete specification of the conditional density. The empirical likelihood approach, due to Owen (1988), is an alternative estimation procedure based on the same moment conditions as those in (2.64) or (2.71).

For scalar iid random variable y with density $f(y)$, a completely nonparametric approach seeks to estimate the density $f(y)$ evaluated at each of the sample values of y. Let $\pi_i = f(y_i)$ denote the probability that the ith observation on y takes the realized value y_i. Then the goal is to maximize the so-called empirical likelihood function $\prod_i \pi_i$, or equivalently to maximize the empirical log-likelihood function $n^{-1} \sum_i \ln \pi_i$, subject to the constraint that $\sum_i \pi_i = 1$. Some algebra yields the estimated density function $\hat{f}(y)$ with mass $1/n$ at each of the realized values y_i, $i = 1, \ldots, n$, and estimate $\hat{F}(y)$ that is the usual empirical distribution function.

For regression, a moment-based estimator adds the additional structure that $\mathsf{E}[\mathbf{g}_i(y_i, x_i, \boldsymbol{\theta})] = \mathbf{0}$. For simplicity we consider the just-identified case where

$\mathbf{g}(\cdot)$ and $\boldsymbol{\theta}$ are of the same dimension, although the method extends to GMM. The maximum empirical likelihood (EL) estimator $\hat{\boldsymbol{\theta}}_{\text{MEL}}$ maximizes the empirical likelihood function $n^{-1}\sum_i \ln \pi_i$ subject to the constraint $\sum_i \pi_i = 1$ and the additional sample constraint that $\sum_i \pi_i \mathbf{g}_i(y_i, x_i, \theta) = 0$. Thus we maximize with respect to $\pi = (\pi_1, \ldots, \pi_n)'$, η, $\boldsymbol{\lambda}$, and $\boldsymbol{\theta}$, using the Lagrangian

$$\mathcal{L}_{\text{EL}}(\boldsymbol{\pi}, \eta, \boldsymbol{\lambda}, \boldsymbol{\theta}) = \frac{1}{n}\sum_{i=1}^{n} \ln \pi_i - \eta\left(\sum_{i=1}^{n} \pi_i - 1\right) - \boldsymbol{\lambda}'\sum_{i=1}^{n} \pi_i \mathbf{g}_i(y_i, \mathbf{x}_i, \boldsymbol{\theta}), \tag{2.89}$$

where the Lagrangian multipliers are a scalar η and column vector $\boldsymbol{\lambda}$ of the same dimension as $g(\cdot)$.

Qin and Lawless (1994) show that $\hat{\boldsymbol{\theta}}_{\text{MEL}}$ has the same limit distribution as $\hat{\boldsymbol{\theta}}_{\text{EE}}$ given in (2.66). The difference is that inference is based on sample estimates

$$\hat{\mathbf{A}} = \sum\nolimits_{i=1}^{n} \hat{\pi}_i \left.\frac{\partial \mathbf{h}_i'}{\partial \boldsymbol{\theta}}\right|_{\hat{\boldsymbol{\theta}}} \tag{2.90}$$

$$\hat{\mathbf{B}} = \sum\nolimits_{i=1}^{n} \hat{\pi}_i \mathbf{h}_i(\hat{\boldsymbol{\theta}})\mathbf{h}_i(\hat{\boldsymbol{\theta}})',$$

that weight by the estimated probabilities $\hat{\pi}_i$ rather than the proportions $1/n$ (see (2.69) and (2.70)). This leads to different finite sample properties, and in some examples EL outperforms the estimating equations and GMM estimators.

EL is computationally more burdensome, however, so is not implemented very often. The most commonly used estimator is the *continuous updating estimator* (CUE) of Hansen, Heaton, and Yarron (1996) that minimizes

$$\left(\sum_{i=1}^{n} \mathbf{h}_i(y_i, \mathbf{x}_i, \boldsymbol{\theta})\right)' \mathbf{S}(\boldsymbol{\theta})^{-1} \left(\sum_{i=1}^{n} \mathbf{h}_i(y_i, \mathbf{x}_i, \boldsymbol{\theta})\right),$$

where $\mathbf{S}(\boldsymbol{\theta})$ is a model for $\text{Var}[\sum_i \mathbf{h}_i(\boldsymbol{\theta})]$. From (2.72) this is GMM with a weighting matrix $\mathbf{W}_N = \mathbf{S}(\boldsymbol{\theta})^{-1}$ that also depends on $\boldsymbol{\theta}$, and the CUE can be shown to be a generalized EL estimator. For further details, see the survey by Imbens (2002) and Cameron and Trivedi (2005, pp. 202–206).

2.6 TESTING

We consider the three classical tests for likelihood-based models, Wald tests for general models, and conditional moment tests. We additionally present the bootstrap, without and with asymptotic refinement.

2.6.1 Likelihood-Based Models

There is a well-developed theory for testing hypotheses in models in which the likelihood function is specified. Then there are three "classical" statistical techniques for testing hypotheses – the likelihood ratio, Wald, and Lagrange multiplier (or score) tests.

Let the null hypothesis be

$$H_0 : \mathbf{r}(\boldsymbol{\theta}_0) = \mathbf{0},$$

where $\mathbf{r}$ is an $h \times 1$ vector of possibly nonlinear restrictions on $\boldsymbol{\theta}$, $h \leq q$. Let the alternative hypothesis be

$$H_a : \mathbf{r}(\boldsymbol{\theta}_0) \neq \mathbf{0}.$$

For example, $r_l(\boldsymbol{\theta}) = \theta_3\theta_4 - 1$ if the lth restriction is $\theta_3\theta_4 = 1$. We assume the restrictions are such that the $h \times q$ matrix $\partial\mathbf{r}(\boldsymbol{\theta})/\partial\boldsymbol{\theta}'$, with the lj^{th} element $\partial r_l(\boldsymbol{\theta})/\partial\theta_j$, has full rank h. This is the analog of the assumption of linearly independent restrictions in the case of linear restrictions.

Let $\mathsf{L}(\boldsymbol{\theta})$ denote the likelihood function, $\hat{\boldsymbol{\theta}}_u$ denote the unrestricted MLE that maximizes $\mathcal{L}(\boldsymbol{\theta}) = \ln \mathsf{L}(\boldsymbol{\theta})$, and $\tilde{\boldsymbol{\theta}}_r$ denote the *restricted* MLE under H_0 that maximizes $\mathcal{L}(\boldsymbol{\theta}) - \boldsymbol{\lambda}'\mathbf{r}(\boldsymbol{\theta})$, where $\boldsymbol{\lambda}$ is an $h \times 1$ vector of Lagrangian multipliers.

We now present the three standard test statistics. Under the regularity conditions they are all asymptotically $\chi^2(h)$ under H_0, and H_0 is rejected at significance level α if the computed test statistic exceeds $\chi^2(h;\alpha)$.

The *likelihood ratio* (LR) *test statistic* is

$$\mathsf{T}_{\mathsf{LR}} = -2[\mathcal{L}(\tilde{\boldsymbol{\theta}}_r) - \mathcal{L}(\hat{\boldsymbol{\theta}}_u)]. \tag{2.91}$$

The motivation for T_{LR} is that, if H_0 is true, the unconstrained and constrained maxima of the likelihood function should be the same and $\mathsf{T}_{\mathsf{LR}} \simeq 0$. The test is called the likelihood ratio test because T_{LR} equals -2 times the logarithm of the likelihood ratio $\mathsf{L}(\tilde{\boldsymbol{\theta}}_r)/\mathsf{L}(\hat{\boldsymbol{\theta}}_u)$.

The *Wald test statistic* is

$$\mathsf{T}_{\mathsf{W}} = \mathbf{r}(\hat{\boldsymbol{\theta}}_u)' \left\{ \left.\frac{\partial\mathbf{r}(\boldsymbol{\theta})'}{\partial\boldsymbol{\theta}}\right|_{\hat{\boldsymbol{\theta}}_u} \left[\frac{1}{n}\hat{\mathbf{A}}(\hat{\boldsymbol{\theta}}_u)^{-1}\right] \left.\frac{\partial\mathbf{r}(\boldsymbol{\theta})}{\partial\boldsymbol{\theta}'}\right|_{\hat{\boldsymbol{\theta}}_u} \right\}^{-1} \mathbf{r}(\hat{\boldsymbol{\theta}}_u), \tag{2.92}$$

where $\hat{\mathbf{A}}(\hat{\boldsymbol{\theta}}_u)$ is a consistent estimator of the variance matrix defined in (2.13) evaluated at the unrestricted MLE. This tests how close $\mathbf{r}(\hat{\boldsymbol{\theta}}_u)$ is to the hypothesized value of $\mathbf{0}$ under H_0. By a first-order Taylor series expansion of $\mathbf{r}(\hat{\boldsymbol{\theta}}_u)$ about $\boldsymbol{\theta}_0$, it can be shown that under H_0, $\mathbf{r}(\hat{\boldsymbol{\theta}}_u) \overset{a}{\sim} \mathsf{N}[\mathbf{0}, \mathsf{V}_r]$ where V_r is the matrix in braces in (2.92). This leads to the chi-square statistic (2.92).

The *Lagrange multiplier* (LM) *test statistic* is

$$\mathsf{T}_{\mathsf{LM}} = \sum_{i=1}^{n} \left.\frac{\partial \ln f_i}{\partial\boldsymbol{\theta}'}\right|_{\tilde{\boldsymbol{\theta}}_r} \left[\frac{1}{n}\tilde{\mathbf{A}}(\tilde{\boldsymbol{\theta}}_r)\right]^{-1} \sum_{i=1}^{n} \left.\frac{\partial \ln f_i}{\partial\boldsymbol{\theta}}\right|_{\tilde{\boldsymbol{\theta}}_r}, \tag{2.93}$$

where $\tilde{\mathbf{A}}(\tilde{\boldsymbol{\theta}}_r)$ is a consistent estimator of the variance matrix defined in (2.13) evaluated at the restricted MLE. Motivation of T_{LM} is given next.

To motivate T_{LM}, first define the score vector

$$\mathbf{s}(\boldsymbol{\theta}) = \frac{\partial \mathcal{L}}{\partial \boldsymbol{\theta}} = \sum_{i=1}^{n} \frac{\partial \ln f_i}{\partial \boldsymbol{\theta}}. \tag{2.94}$$

For the unrestricted MLE the score vector $\mathbf{s}(\hat{\boldsymbol{\theta}}_{\mathbf{u}}) = \mathbf{0}$. These are just the first-order conditions (2.11) that define the estimator. If H_0 is true, then this maximum should also occur at the restricted MLE, because imposing the constraint will then have little impact on the estimated value of $\boldsymbol{\theta}$. That is, $\mathbf{s}(\widetilde{\boldsymbol{\theta}}_r) = \mathbf{0}$. T_{LM} measures the closeness of this derivative to zero. The distribution of T_{LM} follows from $\mathbf{s}(\widetilde{\boldsymbol{\theta}}_r) \overset{a}{\sim} N[\mathbf{0}, \frac{1}{n}\mathbf{A}]$ under H_0. Using this motivation, T_{LM} is called the *score test* because $\mathbf{s}(\boldsymbol{\theta})$ is the score vector.

An alternative motivation for T_{LM} is to measure the closeness to zero of the expected value of the Lagrange multipliers of the constrained optimization problem for the restricted MLE. Maximizing $\mathcal{L}(\mathbf{y}, \boldsymbol{\theta}) - \boldsymbol{\lambda}'\mathbf{r}(\boldsymbol{\theta})$, the first-order conditions with respect to $\boldsymbol{\theta}$ imply $\mathbf{s}(\widetilde{\boldsymbol{\theta}}_r) = \mathbf{R}(\widetilde{\boldsymbol{\theta}}_r)\widetilde{\boldsymbol{\lambda}}$, where $\mathbf{R}(\boldsymbol{\theta}) = \left[\partial \mathbf{r}(\boldsymbol{\theta})'/\partial\boldsymbol{\theta}\right]$. Tests based on $\mathbf{s}(\widetilde{\boldsymbol{\theta}}_r)$ are equivalent to tests based on the estimated Lagrange multipliers $\widetilde{\boldsymbol{\lambda}}$ because $\mathbf{R}(\widetilde{\boldsymbol{\theta}}_r)$ is of full rank. So T_{LM} is also called the *Lagrange multiplier test*. Throughout this book, we refer to T_{LM} as the LM test. It is exactly the same as the score test, an alternative label widely used in the statistics literature.

In addition to being asymptotically $\chi^2(h)$ under H_0, all three test statistics are noncentral $\chi^2(h)$ with the same noncentrality parameter under local alternatives $H_a : \mathbf{r}(\boldsymbol{\theta}) = n^{-1/2}\boldsymbol{\delta}$, where $\boldsymbol{\delta}$ is a vector of constants. So they all have the same local power. The choice of which test statistic to use is, therefore, mainly one of convenience in computation or small sample performance.

T_{LR} requires estimation of $\boldsymbol{\theta}$ under both H_0 and H_a. If this is easily done, then the test is very simple to implement because one need only read off the log-likelihood statistics routinely printed out, subtract, and multiply by 2. T_{W} requires estimation only under H_a and is best to use if the unrestricted model is easy to estimate. T_{LM} requires estimation only under H_0 and is attractive if the restricted model is easy to estimate.

An additional attraction of the LM test is easy computation. Let $\mathbf{s}_i(\widetilde{\boldsymbol{\theta}}_{\mathbf{r}})$ be the ith component of the summation forming the score vector (2.94) for the unrestricted density evaluated at the restricted MLE. An asymptotically equivalent version of T_{LM} can be computed as the uncentered explained sum of squares, or n times the uncentered R^2, from the *auxiliary* OLS *regression*

$$1 = \mathbf{s}_i(\widetilde{\boldsymbol{\theta}}_{\mathbf{r}})'\boldsymbol{\gamma} + u_i. \tag{2.95}$$

The uncentered explained sum of squares from regression of $\mathbf{y}$ on $\mathbf{X}$ is $\mathbf{y}'\mathbf{X}(\mathbf{X}'\mathbf{X})^{-1}\mathbf{X}'\mathbf{y}$, and the uncentered R^2 is $\mathbf{y}'\mathbf{X}(\mathbf{X}'\mathbf{X})^{-1}\mathbf{X}'\mathbf{y}/\mathbf{y}'\mathbf{y}$.

2.6.2 Wald Tests

The preceding results are restricted to hypothesis tests based on MLEs. The Wald test can be extended to any consistent estimator $\hat{\boldsymbol{\theta}}$ that does not impose the restrictions being tested. The only change in (2.92) is that $\hat{\boldsymbol{\theta}}$ replaces $\hat{\boldsymbol{\theta}}_u$ and $\mathsf{V}[\hat{\boldsymbol{\theta}}]$ replaces $\frac{1}{n}\hat{\mathbf{A}}(\hat{\boldsymbol{\theta}}_u)^{-1}$. We test $H_0 : \mathbf{r}(\boldsymbol{\theta}_0) = \mathbf{0}$ using

$$\mathsf{T_W} = \mathbf{r}(\hat{\boldsymbol{\theta}})' \left\{ \left.\frac{\partial \mathbf{r}(\boldsymbol{\theta})'}{\partial \boldsymbol{\theta}}\right|_{\hat{\boldsymbol{\theta}}} \mathsf{V}[\hat{\boldsymbol{\theta}}] \left.\frac{\partial \mathbf{r}(\boldsymbol{\theta})}{\partial \boldsymbol{\theta}'}\right|_{\hat{\boldsymbol{\theta}}} \right\}^{-1} \mathbf{r}(\hat{\boldsymbol{\theta}}), \tag{2.96}$$

which is $\chi^2(h)$ under H_0. Reject $H_0 : \mathbf{r}(\boldsymbol{\theta}_0) = \mathbf{0}$ against $H_a : \mathbf{r}(\boldsymbol{\theta}_0) \neq \mathbf{0}$ if $\mathsf{T_W} > \chi^2_\alpha(h)$. Although such generality is appealing, a weakness of the Wald test is that in small samples it is not invariant to the parameterization of the model, whereas LR and LM tests are invariant.

For multiple exclusion restrictions, such as testing whether a set of indicator variables for occupation or educational level is jointly statistically significant, $H_0 : \mathbf{R}\boldsymbol{\theta} = \mathbf{0}$, where $\mathbf{R}$ is an $h \times q$ matrix whose rows each have entries of zero except for one entry of unity corresponding to one of the components of $\boldsymbol{\beta}$ that is being set to zero. Then $\mathbf{r}(\boldsymbol{\theta}) = \mathbf{R}\boldsymbol{\theta}$, and one uses (2.96) with $\mathbf{r}(\hat{\boldsymbol{\theta}}) = \mathbf{R}\hat{\boldsymbol{\theta}}$ and $\partial \mathbf{r}(\boldsymbol{\theta})'/\partial \boldsymbol{\theta} = \mathbf{R}$. This is the analog of the F-test in the linear model under normality.

The usual t-test for significance of the jth regressor is the square root of the Wald chi-square test. To see this, note that for $H_0 : \theta_j = 0$, $\mathbf{r}(\boldsymbol{\theta}) = \theta_j$, $\partial \mathbf{r}(\boldsymbol{\theta})/\partial \boldsymbol{\theta}$ is a $q \times 1$ vector with unity in the jth row and zeroes elsewhere, and (2.96) yields $\mathsf{T_W} = \hat{\theta}_j\, [\hat{\mathsf{V}}_{jj}]^{-1}\, \hat{\theta}_j$, where $\hat{\mathsf{V}}_{jj}$ is the jth diagonal entry in $\mathsf{V}[\hat{\boldsymbol{\theta}}]$ or the estimated variance of $\hat{\theta}_j$. The square root of $\mathsf{T_W}$,

$$\mathsf{T_Z} = \frac{\hat{\theta}_j}{\sqrt{\hat{\mathsf{V}}_{jj}}}, \tag{2.97}$$

is standard normal (the square root of $\chi^2(1)$). We reject H_0 against $H_a : \theta_j \neq 0$ at significance level α if $|\mathsf{T_Z}| > z_{\alpha/2}$.

This test is usually called a t-test, following the terminology for the corresponding test in the linear model with normal homoskedastic errors, though its distribution is only obtained using asymptotic theory that yields the standard normal. For nonlinear models such as count models, some packages use standard normal critical values, whereas others use critical values from the $t(n-q)$ distribution. The latter leads to more conservative inference because the $t(n-q)$ has fatter tails. For similar reasons some packages report an F-test statistic version of the test in (2.96), with $F = \mathsf{T_W}/h$, and use critical values from the $F(h, n-q)$ distribution.

The test statistic $\mathsf{T_Z}$ can be used in one-sided tests. Reject $H_0 : \theta_j = 0$ against $H_a : \theta_j > 0$ at significance level α if $\mathsf{T_Z} > z_\alpha$, and reject $H_0 : \theta_j = 0$ against $H_a : \theta_j < 0$ at significance level α if $\mathsf{T_Z} < -z_\alpha$.

The Wald approach can be adapted to obtain the distribution of nonlinear functions of parameter estimates, such as individual predictions of the conditional mean. Suppose interest lies in the function $\boldsymbol{\lambda} = \mathbf{r}(\boldsymbol{\theta})$, and we have available the estimator $\hat{\boldsymbol{\theta}} \overset{a}{\sim} \mathsf{N}[\boldsymbol{\theta}, \mathsf{V}[\hat{\boldsymbol{\theta}}]]$. By the *delta method*

$$\hat{\boldsymbol{\lambda}} = \mathbf{r}(\hat{\boldsymbol{\theta}}) \overset{a}{\sim} \mathsf{N}[\boldsymbol{\lambda}, \mathsf{V}[\hat{\boldsymbol{\lambda}}]], \tag{2.98}$$

where

$$\mathsf{V}[\hat{\boldsymbol{\lambda}}] = \left.\frac{\partial r(\boldsymbol{\theta})'}{\partial \boldsymbol{\theta}}\right|_{\hat{\boldsymbol{\theta}}} \mathsf{V}[\hat{\boldsymbol{\theta}}] \left.\frac{\partial r(\boldsymbol{\theta})}{\partial \boldsymbol{\theta}'}\right|_{\hat{\boldsymbol{\theta}}}. \tag{2.99}$$

This can be used in the obvious way to get standard errors and construct confidence intervals for $\boldsymbol{\lambda}$. For example, if λ is scalar, then a 95% confidence interval for λ is $\hat{\lambda} \pm 1.96\, se(\hat{\lambda})$, where $se(\hat{\lambda})$ equals the square root of the scalar in the right-hand side of (2.99).

The LM and LR hypothesis tests have been extended to GMM estimators by Newey and West (1987b). See this reference or Cameron and Trivedi (2005) or Wooldridge (2010) for further details.

2.6.3 Conditional Moment Tests

The results so far have been restricted to tests of hypotheses on the parameters. The moment-based framework can be used to instead perform tests of model specification. A model may impose a number of moment conditions, not all of which are used in estimation. For example, the Poisson regression model imposes the constraint that the conditional variance equals the conditional mean, which implies

$$\mathsf{E}[(y_i - \exp(\mathbf{x}_i'\boldsymbol{\theta}))^2 - y_i] = 0.$$

Because this constraint is not imposed by the MLE, the Poisson model could be tested by testing the closeness to zero of the sample moment

$$\sum_{i=1}^{n} \{(y_i - \exp(\mathbf{x}_i'\hat{\boldsymbol{\theta}}))^2 - y_i)\}.$$

Such tests, called *conditional moment tests*, provide a general framework for model specification tests. These tests were introduced by Newey (1985) and Tauchen (1985) and are given a good presentation in Pagan and Vella (1989). They nest hypothesis tests such as Wald, LM, and LR, and specification tests such as information matrix tests. This unifying element is emphasized in White (1994).

Suppose a model implies the population moment conditions

$$\mathsf{E}[\mathbf{m}_i(y_i, \mathbf{x}_i, \boldsymbol{\theta}_0)] = \mathbf{0}, \qquad i = 1, \ldots, n, \tag{2.100}$$

where $\mathbf{m}_i(\cdot)$ is an $r \times 1$ vector function. Let $\hat{\boldsymbol{\theta}}$ be a root-n consistent estimator that converges to $\boldsymbol{\theta}_0$, obtained by a method that does not impose the moment condition (2.100). The notation $\mathbf{m}_i(\cdot)$ denotes moments used for the tests, whereas $\mathbf{g}_i(\cdot)$ and/or $\mathbf{h}_i(\cdot)$ denote moments used in estimation.

The correct specification of the model can be tested by testing the closeness to zero of the corresponding sample moment

$$\mathbf{m}(\hat{\boldsymbol{\theta}}) = \sum_{i=1}^{n} \mathbf{m}_i(y_i, \mathbf{x}_i, \hat{\boldsymbol{\theta}}). \tag{2.101}$$

Suppose $\hat{\boldsymbol{\theta}}$ is the solution to the first-order conditions

$$\sum_{i=1}^{n} \mathbf{g}_i(y_i, \mathbf{x}_i, \hat{\boldsymbol{\theta}}) = \mathbf{0}.$$

If $\mathsf{E}[\mathbf{g}_i(y_i, \mathbf{x}_i, \boldsymbol{\theta}_0)] = \mathbf{0}$ and (2.100) holds, then

$$n^{-1/2}\mathbf{m}(\hat{\boldsymbol{\theta}}) \xrightarrow{d} \mathsf{N}[\mathbf{0},\ \mathbf{V}_m] \tag{2.102}$$

where

$$\mathbf{V}_m = \mathbf{HJH}', \tag{2.103}$$

$$\mathbf{J} = \lim_{n\to\infty} \frac{1}{n}\mathsf{E}\left[\begin{array}{cc} \sum_{i=1}^{n} \mathbf{m}_i\mathbf{m}_i' & \sum_{i=1}^{n} \mathbf{m}_i\mathbf{g}_i' \\ \sum_{i=1}^{n} \mathbf{g}_i\mathbf{m}_i' & \sum_{i=1}^{n} \mathbf{g}_i\mathbf{g}_i' \end{array}\Bigg|_{\boldsymbol{\theta}_0}\right], \tag{2.104}$$

the vectors $\mathbf{m}_i = \mathbf{m}_i(y_i, \mathbf{x}_i, \boldsymbol{\theta})$ and $\mathbf{g}_i = \mathbf{g}_i(y_i, \mathbf{x}_i, \boldsymbol{\theta})$, and

$$\mathbf{H} = \left[\mathbf{I}_r \quad -\mathbf{CA}^{-1}\right], \tag{2.105}$$

$$\mathbf{C} = \lim_{n\to\infty} \frac{1}{n}\mathsf{E}\left[\sum_{i=1}^{n} \frac{\partial \mathbf{m}_i}{\partial \boldsymbol{\theta}'}\Bigg|_{\boldsymbol{\theta}_0}\right], \tag{2.106}$$

$$\mathbf{A} = \lim_{n\to\infty} \frac{1}{n}\mathsf{E}\left[\sum_{i=1}^{n} \frac{\partial \mathbf{g}_i}{\partial \boldsymbol{\theta}'}\Bigg|_{\boldsymbol{\theta}_0}\right]. \tag{2.107}$$

The formula for $\mathbf{V}_m$ is quite cumbersome because there are two sources of stochastic variation in $\mathbf{m}(\hat{\boldsymbol{\theta}})$ – the dependent variable y_i and the estimator $\hat{\boldsymbol{\theta}}$. See Section 2.8.5 for details.

The *conditional moment* (CM) *test* statistic

$$\mathsf{T}_{\mathsf{CM}} = n\,\mathbf{m}(\hat{\boldsymbol{\theta}})'\hat{\mathbf{V}}_m^{-1}\mathbf{m}(\hat{\boldsymbol{\theta}}), \tag{2.108}$$

where $\hat{\mathbf{V}}_m$ is consistent for $\mathbf{V}_m$, is asymptotically $\chi^2(r)$. Moment condition (2.100) is rejected at significance level α if the computed test statistic exceeds $\chi^2(r;\alpha)$. Rejection is interpreted as indicating model misspecification, although it is not always immediately apparent in what direction the model is misspecified.

Although the CM test is in general difficult to implement due to the need to obtain the variance $\mathbf{V}_m$, it is simple to compute in two leading cases. First, if the moment $\mathbf{m}_i(\cdot)$ satisfies

$$\mathsf{E}\left[\frac{\partial \mathbf{m}_i}{\partial \boldsymbol{\theta}'}\right] = \mathbf{0}, \tag{2.109}$$

then from section 2.8.5 $\mathbf{V}_m = \lim \frac{1}{n}\mathsf{E}[\sum_i \mathbf{m}_i\mathbf{m}_i']$, which can be consistently estimated by $\hat{\mathbf{V}}_m = \frac{1}{n}\sum_i \hat{\mathbf{m}}_i\hat{\mathbf{m}}_i'$. For cross-section data, this means

$$\mathsf{T}_{\mathsf{CM}} = \sum_{i=1}^{n} \hat{\mathbf{m}}_i' \left[\sum_{i=1}^{n} \hat{\mathbf{m}}_i\hat{\mathbf{m}}_i'\right]^{-1} \sum_{i=1}^{n} \hat{\mathbf{m}}_i. \tag{2.110}$$

This can be computed as the uncentered explained sum of squares, or as n times the uncentered R^2, from the auxiliary regression

$$1 = \mathbf{m}_i(y_i, \mathbf{x}_i, \hat{\boldsymbol{\theta}})'\boldsymbol{\gamma} + u_i. \tag{2.111}$$

If $\mathsf{E}[\mathbf{m}_i\mathbf{m}_i']$ is known, the statistic

$$\sum_{i=1}^{n} \hat{\mathbf{m}}_i' \left[\sum_{i=1}^{n} \mathsf{E}[\mathbf{m}_i\mathbf{m}_i']|_{\hat{\boldsymbol{\theta}}}\right]^{-1} \sum_{i=1}^{n} \hat{\mathbf{m}}_i$$

is an alternative to (2.110).

A second case in which the CM test is easily implemented is if $\hat{\boldsymbol{\theta}}$ is the MLE. Then it can be shown that an asymptotically equivalent version of the CM test can be calculated as the uncentered explained sum of squares or, equivalently, as n times the uncentered R^2, from the auxiliary regression

$$1 = \mathbf{m}_i(y_i, \mathbf{x}_i, \hat{\boldsymbol{\theta}})'\gamma_1 + \mathbf{s}_i(y_i, \mathbf{x}_i, \hat{\boldsymbol{\theta}})'\gamma_2 + u_i, \tag{2.112}$$

where $\mathbf{s}_i$ is the ith component of the score vector (2.94) and uncentered R^2 is defined after (2.95). This auxiliary regression is a computational device with no physical interpretation. It generalizes the regression (2.95) for the LM test. Derivation uses the *generalized information matrix equality* that

$$\mathsf{E}\left[\frac{\partial \mathbf{m}_i(\boldsymbol{\theta})}{\partial \boldsymbol{\theta}'}\right] = -\mathsf{E}\left[\mathbf{m}_i(\boldsymbol{\theta})\frac{\partial \ln f_i(\boldsymbol{\theta})}{\partial \boldsymbol{\theta}'}\right], \tag{2.113}$$

provided $\mathsf{E}[\mathbf{m}_i(\boldsymbol{\theta})] = \mathbf{0}$. The resulting test is called the *outer product of the gradient* (OPG) form of the test because it sums $\mathbf{m}_i(\boldsymbol{\theta}) \times \partial \ln f_i(\boldsymbol{\theta})/\partial \boldsymbol{\theta}'$ evaluated at $\hat{\boldsymbol{\theta}}$.

A leading example of the CM test is the *information matrix* (IM) *test* of White (1982). This tests whether the information matrix equality holds, or equivalently whether the moment condition

$$\mathsf{E}\left[\text{vech}\left(\frac{\partial^2 \ln f_i}{\partial \boldsymbol{\theta}\partial \boldsymbol{\theta}'} + \frac{\partial \ln f_i}{\partial \boldsymbol{\theta}}\frac{\partial \ln f_i}{\partial \boldsymbol{\theta}'}\right)\right] = \mathbf{0}$$

is satisfied, where $f_i(y, \boldsymbol{\theta})$ is the specified density. The *vector-half operator* vech($\cdot$) stacks the components of the symmetric $q \times q$ matrix into a $q(q+1)/2 \times 1$ column vector. The OPG form of the IM test is especially advantageous in this example because otherwise one needs to obtain $\partial \mathbf{m}_i(\boldsymbol{\theta})/\partial \boldsymbol{\theta}'$, which entails third derivatives of the log-density.

Despite their generality, CM tests other than the three classical tests (Wald, LM, and LR) are rarely exploited in applied work, for three reasons. First, the tests are unconventional in that there is no explicit alternative hypothesis. Rejection of the moment condition may not indicate how one should proceed to improve the model. Second, implementation of the CM test is in general difficult, except for the MLE case in which a simple auxiliary regression can be run. But this OPG form of the test has been shown to have poor small sample properties in some leading cases (although a bootstrap with asymptotic refinement, see section 2.6.5, may correct for this). Third, with real data and a large sample, testing at a fixed significance level that does not vary with sample size will always lead to rejection of sample moment conditions implied by a model and to a conclusion that the model is inadequate. A similar situation also exists in more classical testing situations. With a large enough sample, regression coefficients will always be significantly different from zero. But in this latter case this may be precisely the news that researchers want to hear.

Two model specification tests that are used more often are the Hausman test, detailed in Chapter 5.6.6, and the overidentifying restrictions test presented in section 2.5.2.

2.6.4 Bootstrap

The bootstrap, introduced by Efron (1979), is a method to obtain the distribution of a statistic by resampling from the original data set. An introductory treatment is given by Efron and Tibsharani (1993). Horowitz (2001) gives a comprehensive treatment with an emphasis on common regression applications. Here we focus on methods applicable to cross-section count data regression models, using the bootstrap pairs procedure under the assumption that $(y_i, \mathbf{x}_i)$ is iid.

We begin with the use of the bootstrap to estimate standard errors. Let $\hat{\theta}_j$ denote the estimator of the jth component of the parameter vector $\boldsymbol{\theta}$. The *bootstrap pairs* procedure is as follows:

1. Form a new pseudosample of size n, $(y_l^*, \mathbf{x}_l^*)$, $l = 1, \ldots, n$, by sampling with replacement from the original sample $(y_i, \mathbf{x}_i)$, $i = 1, \ldots, n$.
2. Obtain the estimator, say $\hat{\boldsymbol{\theta}}_1$ with j^{th} component $\hat{\theta}_{j,1}$, using the pseudosample data.
3. Repeat steps 1 and 2 B times giving B estimates $\hat{\theta}_{j,1}, \ldots, \hat{\theta}_{j,B}$.

4. Estimate the standard deviation of $\hat{\theta}_j$ using the usual formula for the sample standard deviation of $\hat{\theta}_{j,1}, \ldots, \hat{\theta}_{j,B}$, or

$$\mathsf{se}_{\mathsf{Boot}}[\hat{\theta}_j] = \sqrt{\frac{1}{B-1}\sum_{b=1}^{B}(\hat{\theta}_{j,b} - \overline{\theta}_j)^2} \qquad (2.114)$$

where $\overline{\theta}_j$ is the usual sample mean $\overline{\theta}_j = (1/B)\sum_{b=1}^{B}\hat{\theta}_{j,b}$.

The estimated standard error is the square root of $\hat{\mathsf{V}}_{\mathsf{Boot}}[\hat{\theta}_j]$, where $\hat{\mathsf{V}}_{\mathsf{Boot}}[\hat{\boldsymbol{\theta}}] = \frac{1}{B-1}\sum_{b=1}^{B}(\hat{\boldsymbol{\theta}}_b - \overline{\boldsymbol{\theta}})(\hat{\boldsymbol{\theta}}_b - \overline{\boldsymbol{\theta}})'$.

This bootstrap is very easy to implement, given a resampling algorithm and a way to save parameter estimates from the B simulations. The bootstrap (with $B \to \infty$) yields standard errors asymptotically equivalent to robust sandwich standard errors. For standard error estimation, $B = 400$ is usually more than adequate; see, for example, Cameron and Trivedi (2005, p. 361). Given the bootstrap standard error we can perform Wald hypothesis tests and calculate confidence intervals in the usual way.

The bootstrap is useful when it is difficult to otherwise compute the standard error of an estimator. For example, for the sequential two-step estimation problem of section 2.5.5 the distribution of $\hat{\boldsymbol{\theta}}_2$ is complicated by first-step estimation of $\widetilde{\boldsymbol{\theta}}_1$. Then in each of the B resamples we perform both steps of the two-step estimation, yielding B second-step estimates $\hat{\boldsymbol{\theta}}_{2,1}, \ldots, \hat{\boldsymbol{\theta}}_{2,B}$ and hence the bootstrap standard error of $\hat{\boldsymbol{\theta}}_2$.

The bootstrap is easily adapted to statistics other than an estimator – replace $\hat{\boldsymbol{\theta}}$ by the statistic of interest. For example, a bootstrap can be used to compute the standard error of $\hat{\theta}_2 \times \hat{\theta}_3$, or of the predicted exponential conditional mean $\hat{\mathsf{E}}[y|\mathbf{x} = \mathbf{x}^*] = \exp(\mathbf{x}^{*\prime}\hat{\boldsymbol{\beta}})$, without resort to the delta method of section 2.6.2. As another example, the bootstrap can estimate the standard deviation of a standard error. Let $\mathsf{se}[\hat{\theta}_j]$ be a standard error computed using, for example, the robust sandwich method. Then bootstrap B times to obtain B different values of $\mathsf{se}[\hat{\theta}_j]$ and use the standard deviation of these values. And the bootstrap can be used to estimate other features of the distribution of a statistic than its standard deviation.

For most purposes bootstrap pairs (or nonparametric) resampling is adequate, but there are other ways to resample. A *parametric bootstrap* holds the regressors $\mathbf{x}_i$ fixed and generates resamples y_i by making draws from a fitted parametric model $f(y_i|\mathbf{x}_i, \hat{\boldsymbol{\theta}})$. A *residual bootstrap* that holds the regressors fixed and resamples residuals is relevant to linear regression but not for count models that are nonlinear. A leave-one-out *jackknife* creates n resamples by dropping in turn each of the n observations in the sample.

A key requirement for validity of the bootstrap is that resampling be done on a quantity that is iid. The bootstrap pairs procedure for independent cross-section data ensures this, resampling jointly the pairs $(y_i, \mathbf{x}_i)$ that are assumed to be iid. For correlated observations the resampling methods are more complex, except for clustered observations (see section 2.7.1) for which one resamples over entire clusters rather than over observations. And it is assumed that the estimator is one where the usual asymptotic theory applies so that $\sqrt{N}(\hat{\boldsymbol{\theta}} - \boldsymbol{\theta}_0)$ has a limit normal distribution and the statistic of interest is a smooth function of $\hat{\boldsymbol{\theta}}$.

2.6.5 Bootstrap with Asymptotic Refinement

An appropriately constructed bootstrap can additionally provide improved estimation of the distribution of a statistic in finite samples, in the sense that as $n \to \infty$ the bootstrap estimator converges faster than the usual first-order asymptotic theory. These gains occur because in some cases it is possible to construct the bootstrap as a numerical method to implement an Edgeworth expansion, which is a more refined asymptotic theory than the usual first-order theory. A key requirement for improved finite-sample performance of the bootstrap is that the statistic being considered is *asymptotically pivotal*, which means that the asymptotic distribution of the statistic does not depend on unknown parameters.

We present a version of the bootstrap for hypothesis tests that yields improved small-sample performance. Consider testing the hypothesis $H_0 : \theta_j = \theta_{j0}$ against $H_0 : \theta_j \neq \theta_{j0}$, where estimation is by a standard method such as Poisson QMLE. The t statistic used is $t_j = (\hat{\theta}_j - \theta_{j0})/s_j$, where s_j is the robust sandwich standard error estimate for $\hat{\theta}_j$ which assumes that $(y_i, \mathbf{x}_i)$ is iid. On the basis of first-order asymptotic theory, we would reject H_0 at level α if $|t_j| > z_{\alpha/2}$.

The *percentile-t method* tests H_0 as follows:

1. Form a new pseudosample of size n, $(y_l^*, \mathbf{x}_l^*)$, $l = 1, \ldots, n$, by sampling with replacement from the original sample $(y_i, \mathbf{x}_i)$, $i = 1, \ldots, n$.
2. Obtain the estimator $\hat{\theta}_{j,1}$, the standard error $s_{j,1}$, and the t statistic $t_{j,1} = (\hat{\theta}_{j,1} - \hat{\theta}_j)/s_{j,1}$ for the pseudosample data.
3. Repeat steps 1 and 2 B times, yielding $t_{j,1}, \ldots, t_{j,B}$.
4. Order these B t statistics and calculate $t_{j,[\alpha/2]}$ and $t_{j,[1-\alpha/2]}$, the lower and upper $\alpha/2$ percentiles of $t_{j,1}, \ldots, t_{j,B}$.
5. Reject H_0 at level α if t_j, the t statistic from the original sample, falls outside the interval $(t_{j,[\alpha/2]}, t_{j,[1-\alpha/2]})$.

Note that in step 2 the t statistic is centered around the original sample estimate $\hat{\theta}_j$ and not the hypothesized value θ_{j0}. The reason is that the bootstrap treats the sample as the population and resamples from this population. The corresponding $100(1-\alpha)\%$ bootstrap confidence interval is $(\hat{\theta}_j + t^*_{j,[\alpha/2]} s_j, \hat{\theta}_j + t^*_{j,[1-\alpha/2]} s_j)$.

This bootstrap procedure leads to an improved finite-sample performance in the following sense. Let α be the nominal size for a test procedure. Usual asymptotic theory produces t tests with actual size $\alpha + O(n^{-1/2})$, whereas this bootstrap produces t tests with actual size $\alpha + O(n^{-1})$. This refinement is possible because it is the t statistic, whose asymptotic distribution does not depend on unknown parameters, that is bootstrapped. For both hypothesis tests and confidence intervals, the number of iterations should be larger than for standard error estimation and such that $\alpha(B+1)$ is an integer, say $B = 999$ for a 5% test.

The percentile-t method can potentially overcome the poor finite-sample performance of the auxiliary regressions used to implement LM tests and conditional moment tests, and the lack of invariance of the Wald test to transformations of $\boldsymbol{\theta}$. In practice, however, the bootstrap is mainly used without asymptotic refinement to compute standard errors as in section 2.6.4.

An alternative bootstrap method for tests and confidence intervals is the *percentile method*. This calculates $\theta_{j,[\alpha/2]}$ and $\theta_{j,[1-\alpha/2]}$, the lower and upper $\alpha/2$ percentiles of $\hat{\theta}_{j,1}, \ldots, \hat{\theta}_{j,B}$. Then reject $H_0 : \theta_j = \theta_{j0}$ against $H_a : \theta_j = \theta_{j0}$ if θ_{j0} does not lie in $(\theta_{j,[\alpha/2]}, \theta_{j,[1-\alpha/2]})$, and use $(\theta_{j,[\alpha/2]}, \theta_{j,[1-\alpha/2]})$ as the $100(1-\alpha)\%$ confidence interval. This alternative procedure is asymptotically valid, but is no better than using the usual asymptotic theory because it is based on the distribution of $\hat{\theta}_j$, which unlike t_j depends on unknown parameters. Similarly, using the usual hypothesis tests and confidence intervals, with the one change that s_j is replaced by a bootstrap estimate, is asymptotically valid, but is no better than the usual first-order asymptotic methods.

A bootstrap with asymptotic refinement can also be used for correction of bias in an asymptotically consistent estimator. This is rarely done in practice, however, because the resulting *bias-corrected* estimate is considerably noisier than the original biased estimator; see Efron and Tibsharani (1993, p. 138).

2.7 ROBUST INFERENCE

By robust inference we mean computation of standard errors and of subsequent Wald test statistics and confidence intervals that are robust to departures from some of the assumptions made to motivate the estimator $\hat{\boldsymbol{\theta}}$. In particular, we relax the assumption of independence across observations to consider within-cluster correlation, time-series correlation, and spatial correlation. Doing so leads to alternative consistent estimators of the middle matrix $\mathbf{B}$ in the sandwich variance matrix $\mathbf{A}^{-1}\mathbf{B}\mathbf{A}'^{-1}$ or, for GMM, the matrix $\mathbf{S}$ defined in (2.78).

Correlation of observations can also lead to specification and estimation of different models. In the time-series case, for example, a dynamic model introduces lagged dependent variables as regressors to model $\mathrm{E}[y_t|y_{t-1}, \mathbf{x}_t]$, rather than $\mathrm{E}[y_t|\mathbf{x}_t]$. Such additional complications are considered elsewhere (e.g., in Chapter 7 for time-series correlation). Here we suppose the model and estimator are unchanged and focus on consistent estimation of the variance matrix of the estimator.

2.7.1 Estimating Equations

We assume that the general theory results for the estimating equations estimator $\hat{\boldsymbol{\theta}}$ of section 2.5.1, which solves $\sum_{i=1}^{n} \mathbf{g}_i(y_i, \mathbf{x}_i, \boldsymbol{\theta}) = \mathbf{0}$, hold. So $\hat{\boldsymbol{\theta}}$ is consistent for $\boldsymbol{\theta}_0$, $\sqrt{n}(\hat{\boldsymbol{\theta}} - \boldsymbol{\theta}_0) \xrightarrow{d} \mathsf{N}[\mathbf{0}, \mathbf{A}^{-1}\mathbf{B}\mathbf{A}'^{-1}]$, and $\mathbf{A}$ and $\hat{\mathbf{A}}$ are as defined in (2.67) and (2.69). The one point of departure is to permit some correlation across observations, which leads to changes in $\mathbf{B}$ and $\hat{\mathbf{B}}$.

Given completely unstructured correlation across observations, (2.68) becomes

$$\mathbf{B} = \lim_{n\to\infty} \frac{1}{n} \mathsf{E}\left[\sum_{i=1}^{n} \sum_{j=1}^{n} \mathbf{g}_i(y_i, \mathbf{x}_i, \boldsymbol{\theta}_0)\mathbf{g}_j(y_j, \mathbf{x}_j, \boldsymbol{\theta}_0)' \right]. \tag{2.115}$$

Some structure is needed to simplify (2.115). Essentially we need

$$\mathbf{B} = \lim_{n\to\infty} \frac{1}{n} \sum_{i=1}^{n} \sum_{j=1}^{n} w_{ij} \mathsf{E}\left[\mathbf{g}_i(y_i, \mathbf{x}_i, \boldsymbol{\theta}_0)\mathbf{g}_j(y_j, \mathbf{x}_j, \boldsymbol{\theta}_0)'\right], \tag{2.116}$$

where the weights w_{ij} are zero, or asymptote to zero, for a substantially large fraction of the observations. Different types of correlation lead to different weights w_{ij}.

In section 2.5.1, we assumed independence of observations, in which case $w_{ij} = 1$ if $i = j$ and $w_{ij} = 0$ otherwise. From (2.70) the resulting estimate of $\mathbf{B}$ is $\hat{\mathbf{B}}_{\text{Het}} = \frac{1}{n}\sum_{i=1}^{n} \mathbf{g}_i(\hat{\boldsymbol{\theta}})\mathbf{g}_i(\hat{\boldsymbol{\theta}})'$; a common finite-sample adjustment is to divide by $n - q$ rather than n. This is called a *heteroskedastic-robust* estimator because it does not require assumptions on the exact form of $\mathsf{E}\left[\mathbf{g}_i(\boldsymbol{\theta}_0)\mathbf{g}_i(\boldsymbol{\theta}_0)'\right]$.

One alternative to independent observations is that observations fall into clusters, where observations in different clusters are independent, but observations within the same cluster are no longer independent. For example, individuals may be grouped into villages, with correlation within villages but not across villages.

The first-order conditions (2.65) can be summed within each cluster and reexpressed as

$$\sum\nolimits_{c=1}^{C} \mathbf{g}_c(\hat{\boldsymbol{\theta}}) = \mathbf{0}, \tag{2.117}$$

where c denotes the cth cluster, there are C clusters, and $\mathbf{g}_c(\boldsymbol{\theta}) = \sum_{i:i\in c} \mathbf{g}_i(y_i, \mathbf{x}_i, \boldsymbol{\theta})$. The key assumption is that $\mathsf{E}[\mathbf{g}_i(\boldsymbol{\theta})\mathbf{g}_j(\boldsymbol{\theta})'] = \mathbf{0}$ if i and j are in different clusters. Then $\mathbf{B} = \lim\limits_{n\to\infty} \frac{1}{n}\sum_{c=1}^{C} \mathsf{E}\left[\mathbf{g}_c(\boldsymbol{\theta}_0)\mathbf{g}_c(\boldsymbol{\theta}_0)'\right]$, and we use the *cluster-robust* estimator

$$\hat{\mathbf{B}}_{\text{Clus}} = \frac{C}{C-1}\frac{1}{n} \sum\nolimits_{c} \mathbf{g}_c(\hat{\boldsymbol{\theta}})\mathbf{g}_c(\hat{\boldsymbol{\theta}})'. \tag{2.118}$$

This estimator was proposed by Liang and Zeger (1986), and the scaling $C/(C-1)$ is a more recent ad hoc degrees of freedom correction. The estimator assumes that the number of clusters $C \to \infty$. Cameron and Miller (2011) and Cameron, Gelbach, and Miller (2011) present an easily computed extension to clustering in two or more nonnested dimensions.

The estimator (2.118) is also applicable to panel data where observations are independent across the n individuals but are correlated over time for a given individual. Then the term *panel-robust* estimator is sometimes used, and we require a long panel with $n \to \infty$. Arellano (1987) showed that (2.118) can also be used to obtain a panel-robust variance estimator for the fixed effects estimator in the linear panel model.

For time series data it is simplest to assume that only observations up to m periods apart are correlated, as is the case for a vector moving average process of order m. Then $w_{ij} = 0$ if $|i-j| > m$, and (2.115) simplifies to $\mathbf{B} = \mathbf{\Omega}_0 + \sum_{j=1}^{m}(\mathbf{\Omega}_j + \mathbf{\Omega}_j')$, where $\mathbf{\Omega}_j = \lim_{n\to\infty} \frac{1}{n}\mathsf{E}\left[\sum_{i=j+1}^{n} \mathbf{g}_i(\boldsymbol{\theta})\mathbf{g}_{i-j}(\boldsymbol{\theta})'\right]$. Newey and West (1987a) proposed the *heteroskedastic and autocorrelation-consistent* (HAC) estimator

$$\hat{\mathbf{B}} = \hat{\mathbf{\Omega}}_0 + \sum_{j=1}^{m}(1 - \frac{j}{m+1})(\hat{\mathbf{\Omega}}_j + \hat{\mathbf{\Omega}}_j'), \tag{2.119}$$

where

$$\hat{\mathbf{\Omega}}_j = \frac{1}{n}\sum_{i=j+1}^{n} \mathbf{g}_i(\hat{\boldsymbol{\theta}})\mathbf{g}_{i-j}(\hat{\boldsymbol{\theta}})'. \tag{2.120}$$

This estimator of $\mathbf{B}$ is the obvious estimator of this quantity, aside from multiplication by $(1 - j/(m+1))$ that ensures that $\hat{\mathbf{B}}$ is positive definite.

For cross-section data that are spatially correlated, dependence decays with economic or physical "distance" between observations, rather than time. Thus Conley (1999) proposes the *spatial-consistent estimator*

$$\hat{\mathbf{B}} = \sum_{i=1}^{n}\sum_{j=1}^{n} K_n(s_i, s_j)\mathbf{g}_i(\hat{\boldsymbol{\theta}})\mathbf{g}_{i-j}(\hat{\boldsymbol{\theta}})',$$

where s_i and s_j are the locations of individuals i and j, and $K_n(s_i, s_j)$ is a kernel function that weights pairs of observations so that nearby observations receive weight close to 1 and observations far apart receive weight of 0. A simple example is a cutoff function so $K_n(s_i, s_j) = 1$ if $|s_i - s_j| < m$ and $K_n(s_i, s_j) = 0$ otherwise. Conley (1999) also proposes alternatives that ensure that $\hat{\mathbf{B}}$ is positive definite.

For panel data that are spatially correlated, Driscoll and Kraay (1998) present a robust variance matrix estimator that treats each time period as a cluster, additionally allows observations in different time periods to be correlated for a finite time difference, and assumes a long panel with $T \to \infty$.

2.7.2 Generalized Method of Moments

The robust estimators of the variance matrix for the estimating equations estimator extend easily to GMM. Indeed the already-cited estimators of Newey and West (1987a), Driscoll and Kraay (1998), and Conley (1999) were proposed for the GMM estimator.

From section 2.5.2, the GMM estimator minimizes $\left[\sum_{i=1}^{n} \mathbf{h}_i(y_i, \mathbf{x}_i, \boldsymbol{\theta})\right]'$ $\mathbf{W}_n \left[\sum_{i=1}^{n} \mathbf{h}_i(y_i, \mathbf{x}_i, \boldsymbol{\theta})\right]$. The limit variance matrix is $\mathbf{A}^{-1}\mathbf{B}\mathbf{A}'^{-1}$ as in (2.74), where $\mathbf{A} = \mathbf{H}'\mathbf{W}\mathbf{H}$ and $\mathbf{B} = \mathbf{H}'\mathbf{W}\mathbf{S}\mathbf{W}\mathbf{H}$. We continue to estimate $\mathbf{H}$ by $\hat{\mathbf{H}}$ given in (2.82), and estimate $\mathbf{W}_n$ by $\mathbf{W}$. With correlated observations

$$\mathbf{S} = \lim_{n\to\infty} \frac{1}{n} \mathsf{E}\left[\sum_{i=1}^{n}\sum_{j=1}^{n} \mathbf{h}_i(y_i, \mathbf{x}_i, \boldsymbol{\theta})\mathbf{h}_j(y_j, \mathbf{x}_j, \boldsymbol{\theta})'\right].$$

Then for the various forms of correlation we use estimator $\hat{\mathbf{S}}$ that is the same as $\hat{\mathbf{B}}$ given in section 2.7.1, with $\mathbf{g}_i(\hat{\boldsymbol{\theta}})$ replaced by $\mathbf{h}_i(\hat{\boldsymbol{\theta}})$.

2.8 DERIVATION OF RESULTS

Formal proofs of convergence in probability of an estimator $\hat{\boldsymbol{\theta}}$ to a fixed value $\boldsymbol{\theta}_*$ are generally difficult and not reproduced here. A clear treatment is given in Amemiya (1985, chapter 4), references to more advanced treatment are given in Cameron and Trivedi (2005) and Wooldridge (2010), and a comprehensive treatment is given in Newey and McFadden (1994). If $\hat{\boldsymbol{\theta}}$ maximizes or minimizes an objective function, then $\boldsymbol{\theta}_*$ is the value of $\boldsymbol{\theta}$ that maximizes the probability limit of the objective function, where the objective function is appropriately scaled to ensure that the probability limit exists. For example, for maximum likelihood the objective function is the sum of n terms and is therefore divided by n. Then $\hat{\boldsymbol{\theta}}$ converges to $\boldsymbol{\theta}_*$, which maximizes plim $\frac{1}{n}\sum_{i=1}^{n} \ln f_i$, where the probability limit is taken with respect to the dgp which is not necessarily f_i.

It is less difficult and more insightful to obtain the asymptotic distribution of $\hat{\boldsymbol{\theta}}$. This is first done in a general framework, with specialization to likelihood-based models, generalized linear models, and moment-based models in remaining subsections.

2.8.1 General Framework

A framework that covers the preceding estimators, except GMM, is that the estimator $\hat{\boldsymbol{\theta}}$ of the $q \times 1$ parameter vector $\boldsymbol{\theta}$ is the solution to the equations

$$\sum_{i=1}^{n} \mathbf{g}_i(\boldsymbol{\theta}) = \mathbf{0}, \tag{2.121}$$

where $\mathbf{g}_i(\boldsymbol{\theta}) = \mathbf{g}_i(y_i, \mathbf{x}_i, \boldsymbol{\theta})$ is a $q \times 1$ vector, and we suppress dependence on the dependent variable and regressors. In typical applications (2.121) are the first-order conditions from maximization or minimization of a scalar objective function, and $\mathbf{g}_i$ is the vector of first derivatives of the ith component of the objective function with respect to $\boldsymbol{\theta}$. The first-order conditions (2.6) for the Poisson MLE are an example of (2.121).

By an exact first-order Taylor series expansion of the left-hand side of (2.121) about $\boldsymbol{\theta}_*$, the probability limit of $\hat{\boldsymbol{\theta}}$, we have

$$\sum_{i=1}^{n} \mathbf{g}_i(\boldsymbol{\theta}_*) + \sum_{i=1}^{n} \left.\frac{\partial \mathbf{g}_i(\boldsymbol{\theta})}{\partial \boldsymbol{\theta}'}\right|_{\boldsymbol{\theta}_{**}} (\hat{\boldsymbol{\theta}} - \boldsymbol{\theta}_*) = \mathbf{0}, \tag{2.122}$$

for some $\boldsymbol{\theta}_{**}$ between $\hat{\boldsymbol{\theta}}$ and $\boldsymbol{\theta}_*$. Solving for $\hat{\boldsymbol{\theta}}$ and rescaling by $\sqrt{n}$ yields

$$\sqrt{n}(\hat{\boldsymbol{\theta}} - \boldsymbol{\theta}_*) = -\left(\frac{1}{n}\sum_{i=1}^{n} \left.\frac{\partial \mathbf{g}_i(\boldsymbol{\theta})}{\partial \boldsymbol{\theta}'}\right|_{\boldsymbol{\theta}_{**}}\right)^{-1} \frac{1}{\sqrt{n}}\sum_{i=1}^{n} \mathbf{g}_i(\boldsymbol{\theta}_*) \tag{2.123}$$

where it is assumed that the inverse exists.

It is helpful at this stage to recall the proof of the asymptotic normality of the OLS estimator in the linear regression model. In that case

$$\sqrt{n}(\hat{\boldsymbol{\theta}} - \boldsymbol{\theta}_*) = \left(\frac{1}{n}\sum_{i=1}^{n} \mathbf{x}_i\mathbf{x}_i'\right)^{-1} \frac{1}{\sqrt{n}}\sum_{i=1}^{n} \mathbf{x}_i(y_i - \mathbf{x}_i'\boldsymbol{\theta})$$

which is of the same form as (2.123). We therefore proceed in the same way as in the OLS case, where the first term in the right-hand side converges in probability to a fixed matrix and the second term in the right-hand side converges in distribution to the normal distribution.

Specifically, assume the existence of the $q \times q$ matrix

$$\mathbf{A} = -\operatorname{plim} \frac{1}{n}\sum_{i=1}^{n} \left.\frac{\partial \mathbf{g}_i(\boldsymbol{\theta})}{\partial \boldsymbol{\theta}'}\right|_{\boldsymbol{\theta}_*}, \tag{2.124}$$

where $\mathbf{A}$ is positive definite for a minimization problem and negative definite for a maximization problem. Also assume

$$\frac{1}{\sqrt{n}}\sum_{i=1}^{n} \mathbf{g}_i(\boldsymbol{\theta}_*) \xrightarrow{d} \mathsf{N}[\mathbf{0}, \mathbf{B}], \tag{2.125}$$

where

$$\mathbf{B} = \operatorname{plim} \frac{1}{n}\sum_{i=1}^{n}\sum_{j=1}^{n} \mathbf{g}_i(\boldsymbol{\theta})\mathbf{g}_j(\boldsymbol{\theta})'\big|_{\boldsymbol{\theta}_*} \tag{2.126}$$

is a positive definite $q \times q$ matrix.

From (2.124) through (2.126), $\sqrt{n}(\hat{\boldsymbol{\theta}} - \boldsymbol{\theta}_*)$ in (2.123) is an $\mathsf{N}[\mathbf{0}, \mathbf{B}]$ distributed random variable premultiplied by minus the inverse of a random matrix that converges in probability to a matrix $\mathbf{A}$. Under appropriate conditions

$$\sqrt{n}(\hat{\boldsymbol{\theta}} - \boldsymbol{\theta}_*) \xrightarrow{d} \mathsf{N}[\mathbf{0}, \mathbf{A}^{-1}\mathbf{B}\mathbf{A}'^{-1}], \tag{2.127}$$

or

$$\hat{\boldsymbol{\theta}} \overset{a}{\sim} \mathsf{N}[\boldsymbol{\theta}_*, \frac{1}{n}\mathbf{A}^{-1}\mathbf{B}\mathbf{A}'^{-1}]. \tag{2.128}$$

The assumption (2.124) is verified by a law of large numbers because the right-hand side of (2.124) is an average. The assumption (2.125) is verified by a multivariate central limit theorem because the left-hand side of (2.125) is a rescaling of an average. This average is centered around zero (see below) and hence

$$\mathsf{V}\left[\frac{1}{\sqrt{n}}\sum_{i=1}^{n}\mathbf{g}_i\right]=\mathsf{E}\left[\frac{1}{n}\sum_i\sum_j\mathbf{g}_i\mathbf{g}_j{}'\right],$$

which is finite if there is not too much correlation between $\mathbf{g}_i$ and $\mathbf{g}_j$, $i\neq j$. Note that the definition of $\mathbf{B}$ in (2.126) permits correlation across observations, and the result (2.128) can potentially be applied to time series data.

Finally, note that by (2.123) convergence of $\hat{\boldsymbol{\theta}}$ to $\boldsymbol{\theta}_*$ requires centering around zero of $\frac{1}{\sqrt{n}}\sum_i \mathbf{g}_i(\boldsymbol{\theta}_*)$. An informal proof of convergence for estimators defined by (2.121) is therefore to verify that $\mathsf{E}_*\left[\sum_i \mathbf{g}_i(\boldsymbol{\theta}_*)\right]=\mathbf{0}$, where the expectation is taken with respect to the dgp.

2.8.2 Likelihood-Based Models

For the MLE given in section 2.3, (2.123) becomes

$$\sqrt{n}(\hat{\boldsymbol{\theta}}_{\mathsf{ML}}-\boldsymbol{\theta}_*)=-\left(\frac{1}{n}\sum_{i=1}^{n}\left.\frac{\partial^2\ln f_i}{\partial\boldsymbol{\theta}\partial\boldsymbol{\theta}'}\right|_{\boldsymbol{\theta}_{**}}\right)^{-1}\frac{1}{\sqrt{n}}\sum_{i=1}^{n}\left.\frac{\partial\ln f_i}{\partial\boldsymbol{\theta}}\right|_{\boldsymbol{\theta}_*}. \tag{2.129}$$

where $f_i=f(y_i|\mathbf{x}_i,\boldsymbol{\theta})$. An informal proof of consistency of $\hat{\boldsymbol{\theta}}$ to $\boldsymbol{\theta}_0$, that is $\boldsymbol{\theta}_*=\boldsymbol{\theta}_0$, requires $\mathsf{E}[\,\partial\ln f_i/\partial\boldsymbol{\theta}|_{\boldsymbol{\theta}_0}]=\mathbf{0}$. This is satisfied if the density is correctly specified, so the expectation is taken with respect to $f(y_i|\mathbf{x}_i,\boldsymbol{\theta}_0)$, and the density satisfies the fourth regularity condition.

To see this, note that any density $f(y|\boldsymbol{\theta})$ satisfies $\int f(y|\boldsymbol{\theta})dy=1$. Differentiating with respect to $\boldsymbol{\theta}$, $\frac{\partial}{\partial\boldsymbol{\theta}}\int f(y|\boldsymbol{\theta})dy=\mathbf{0}$. If the range of y does not depend on $\boldsymbol{\theta}$, the derivative can be taken inside the integral and $\int(\partial f(y|\boldsymbol{\theta})/\partial\boldsymbol{\theta})dy=\mathbf{0}$, which can be reexpressed as $\int(\partial\ln f(y|\boldsymbol{\theta})/\partial\boldsymbol{\theta})\,f(y|\boldsymbol{\theta})dy=\mathbf{0}$, since $\partial\ln f(y|\boldsymbol{\theta})/\partial\boldsymbol{\theta}=(\partial f(y|\boldsymbol{\theta})/\partial\boldsymbol{\theta})(1/f(y|\boldsymbol{\theta}))$. Then $\mathsf{E}[\partial\ln f(y|\boldsymbol{\theta})/\partial\boldsymbol{\theta}]=\mathbf{0}$, where E is taken with respect to $f(y|\boldsymbol{\theta})$.

The variance matrix of $\hat{\boldsymbol{\theta}}_{\mathsf{ML}}$ is $\frac{1}{n}\mathbf{A}^{-1}\mathbf{B}\mathbf{A}^{-1}$ where $\mathbf{A}$ and $\mathbf{B}$ are defined in (2.124) and (2.126) with $\mathbf{g}(y_i|\mathbf{x}_i,\boldsymbol{\theta})=\partial\ln f(y_i|\mathbf{x}_i,\boldsymbol{\theta})/\partial\boldsymbol{\theta}$. Simplification occurs if the density is correctly specified and the range of y does not depend on $\boldsymbol{\theta}$. Then the information matrix equality $\mathbf{A}=\mathbf{B}$ holds.

To see this, differentiating $\int(\partial\ln f(y|\boldsymbol{\theta})/\partial\boldsymbol{\theta})\,f(y|\boldsymbol{\theta})dy=\mathbf{0}$ with respect to $\boldsymbol{\theta}$ yields

$$\mathsf{E}\left[\partial^2\ln f(y|\boldsymbol{\theta})/\partial\boldsymbol{\theta}\partial\boldsymbol{\theta}'\right]=-\mathsf{E}\left[\partial\ln f(y|\boldsymbol{\theta})/\partial\boldsymbol{\theta}\;\partial\ln f(y|\boldsymbol{\theta})/\partial\boldsymbol{\theta}'\right]$$

after some algebra, where E is taken with respect to $f(y|\boldsymbol{\theta})$.

If the density is misspecified, it is no longer the case that such simplifications occur, and the results of section 2.8.1 for $g(\boldsymbol{\theta}) = \ln f(y_i|\mathbf{x}_i, \boldsymbol{\theta})$ yield the result given in section 2.3.4.

2.8.3 Generalized Linear Models

For the QMLE for the LEF given in section 2.4.2, (2.123) becomes

$$\begin{aligned}
&\sqrt{n}(\hat{\boldsymbol{\beta}}_{\mathsf{LEF}} - \boldsymbol{\beta}_0) && (2.130)\\
&= -\left[\frac{1}{n}\sum_{i=1}^{n}\frac{1}{v_i}\left\{-\frac{\partial\mu_i}{\partial\boldsymbol{\beta}}\frac{\partial\mu_i}{\partial\boldsymbol{\beta}'} + (y_i-\mu_i)\frac{\partial^2\mu_i}{\partial\boldsymbol{\beta}\partial\boldsymbol{\beta}'} - \frac{y_i-\mu_i}{v_i}\frac{\partial\mu_i}{\partial\boldsymbol{\beta}}\frac{\partial v_i}{\partial\boldsymbol{\beta}'}\right\}\Bigg|_{\boldsymbol{\beta}_0}\right]^{-1}\\
&\quad\times\frac{1}{\sqrt{n}}\sum_{i=1}^{n}\frac{1}{v_i}\{y_i-\mu_i\}\frac{\partial\mu_i}{\partial\boldsymbol{\beta}}\Bigg|_{\boldsymbol{\beta}_0}.
\end{aligned}$$

An informal proof of convergence of $\hat{\boldsymbol{\beta}}_{\mathsf{LEF}}$ to $\boldsymbol{\beta}_0$ is that the second term in the right-hand side is centered around $\mathbf{0}$ if $\mathsf{E}\left[y_i - \mu(\mathbf{x}_i, \boldsymbol{\beta}_0)\right] = 0$, or that the conditional mean is correctly specified. The first term on the right-hand side converges in probability to

$$\mathbf{A} = \lim \frac{1}{n}\sum_{i=1}^{n}\frac{1}{v_i}\frac{\partial\mu_i}{\partial\boldsymbol{\beta}}\frac{\partial\mu_i}{\partial\boldsymbol{\beta}'}\Bigg|_{\boldsymbol{\beta}_0}$$

because $\mathsf{E}\left[y_i - \mu(\mathbf{x}_i, \boldsymbol{\beta}_0)\right] = 0$, and the second term converges in distribution to the normal distribution with variance matrix

$$\mathbf{B} = \lim \frac{1}{n}\sum_{i=1}^{n}\frac{\omega_i}{v_i^2}\frac{\partial\mu_i}{\partial\boldsymbol{\beta}}\frac{\partial\mu_i}{\partial\boldsymbol{\beta}'}\Bigg|_{\boldsymbol{\beta}_0},$$

where $\omega_i = \mathsf{E}\left[(y_i - \mu(\mathbf{x}_i, \boldsymbol{\beta}_0))^2\right]$. Then $\mathsf{V}[\hat{\boldsymbol{\beta}}_{\mathsf{LEF}}] = \frac{1}{n}\mathbf{A}^{-1}\mathbf{B}\mathbf{A}^{-1}$.

For the QGPMLE for the LEFN density in section 2.4.3, we have

$$\begin{aligned}
&\sqrt{n}(\hat{\boldsymbol{\beta}}_{\mathsf{LEFN}} - \boldsymbol{\beta}_0)\\
&= -\left[\frac{1}{n}\sum_{i=1}^{n}\frac{1}{\widetilde{\omega}_i}\left\{-\frac{\partial\mu_i}{\partial\boldsymbol{\beta}}\frac{\partial\mu_i}{\partial\boldsymbol{\beta}'} + (y_i-\mu_i)\frac{\partial^2\mu_i}{\partial\boldsymbol{\beta}\partial\boldsymbol{\beta}'}\right\}\Bigg|_{\boldsymbol{\beta}_0}\right]^{-1}\\
&\quad\times\frac{1}{\sqrt{n}}\sum_{i=1}^{n}\frac{1}{\widetilde{\omega}_i}(y_i-\mu_i)\frac{\partial\mu_i}{\partial\boldsymbol{\beta}}\Bigg|_{\boldsymbol{\beta}_0}, && (2.131)
\end{aligned}$$

where $\widetilde{\omega}_i = \omega(\mu(\mathbf{x}_i, \widetilde{\boldsymbol{\beta}}), \widetilde{\alpha})$. Then v_i in $\mathbf{A}$ and $\mathbf{B}$ given earlier is replaced by ω_i, which implies $\mathbf{A} = \mathbf{B}$.

Derivation for the estimator in the GLM of section 2.4.4 is similar.

2.8.4 Moment-Based Models

Results for estimating equations given in section 2.5.1 follow directly from section 2.8.1.

The GMM estimator given in section 2.5.2 solves the equations

$$\left[\frac{1}{n}\sum_{i=1}^{n}\frac{\partial \mathbf{h}_i(y_i,\mathbf{x}_i,\boldsymbol{\theta})'}{\partial \boldsymbol{\theta}}\right]\mathbf{W}_n\left[\frac{1}{\sqrt{n}}\sum_{i=1}^{n}\mathbf{h}_i(y_i,\mathbf{x}_i,\boldsymbol{\theta})\right]=\mathbf{0}, \tag{2.132}$$

on multiplying by an extra scaling parameter $n^{-3/2}$. Taking a Taylor series expansion of the third term similar to (2.122) yields

$$\left[\frac{1}{n}\sum_{i=1}^{n}\frac{\partial \mathbf{h}_i'}{\partial \boldsymbol{\theta}}\right]\mathbf{W}_n\left[\frac{1}{\sqrt{n}}\sum_{i=1}^{n}\mathbf{h}_i|_{\boldsymbol{\theta}_*}+\frac{1}{n}\sum_{i=1}^{n}\left.\frac{\partial \mathbf{h}_i}{\partial \boldsymbol{\theta}'}\right|_{\boldsymbol{\theta}_{**}}\sqrt{n}(\hat{\boldsymbol{\theta}}_{\mathsf{GMM}}-\boldsymbol{\theta}_*)\right]=\mathbf{0},$$

where $\mathbf{h}_i=\mathbf{h}_i(y_i,\mathbf{x}_i,\boldsymbol{\theta})$. Solving yields

$$\begin{aligned}\sqrt{n}(\hat{\boldsymbol{\theta}}_{\mathsf{GMM}}-\boldsymbol{\theta}_*)&=\left(\left[\frac{1}{n}\sum_{i=1}^{n}\frac{\partial \mathbf{h}_i'}{\partial \boldsymbol{\theta}}\right]\mathbf{W}_n\left[\frac{1}{n}\sum_{i=1}^{n}\left.\frac{\partial \mathbf{h}_i}{\partial \boldsymbol{\theta}'}\right|_{\boldsymbol{\theta}_*}\right]\right)^{-1}\\&\quad\times\left[\frac{1}{n}\sum_{i=1}^{n}\frac{\partial \mathbf{h}_i'}{\partial \boldsymbol{\theta}}\right]\mathbf{W}_n\frac{1}{\sqrt{n}}\sum_{i=1}^{n}\mathbf{h}_i|_{\boldsymbol{\theta}_{**}}.\end{aligned} \tag{2.133}$$

Equation (2.133) is the key result for obtaining the variance of the GMM estimator. It is sufficient to obtain the probability limit of the first five terms and the limit distribution of the last term in the right-hand side of (2.133). Both $\frac{1}{n}\sum_i \partial\mathbf{h}_i/\partial\boldsymbol{\theta}'$ and $\frac{1}{n}\sum_i \partial\mathbf{h}_i/\partial\boldsymbol{\theta}'|_{\boldsymbol{\theta}_*}$ converge in probability to the matrix $\mathbf{H}$ defined in (2.77), and by assumption plim $\mathbf{W}_n=\mathbf{W}$. By a central limit theorem $\frac{1}{\sqrt{n}}\sum_i \mathbf{h}_i|_{\boldsymbol{\theta}_{**}}$ converges in distribution to $\mathsf{N}[\mathbf{0},\mathbf{S}]$ where

$$\mathbf{S}=\lim_{n\to\infty}\frac{1}{n}\mathsf{E}\left[\sum_i\sum_j \mathbf{h}_i\mathbf{h}_j'|_{\boldsymbol{\theta}_*}\right].$$

Thus from (2.133) $\sqrt{n}(\hat{\boldsymbol{\theta}}_{\mathsf{GMM}}-\boldsymbol{\theta}_*)$ has the same limit distribution as $(\mathbf{H}'\mathbf{W}\mathbf{H})^{-1}\mathbf{H}'\mathbf{W}$ times a random variable that is $\mathsf{N}[\mathbf{0},\mathbf{S}]$. Equivalently, $\sqrt{n}(\hat{\boldsymbol{\theta}}-\boldsymbol{\theta}_*)\xrightarrow{d}\mathsf{N}[\mathbf{0},\mathbf{A}^{-1}\mathbf{B}\mathbf{A}^{-1}]$, where $\mathbf{A}=\mathbf{H}'\mathbf{W}\mathbf{H}$ and $\mathbf{B}=\mathbf{H}'\mathbf{W}\mathbf{S}\mathbf{W}\mathbf{H}$.

The optimal GMM estimator can be motivated by noting that the variance is exactly the same matrix form as that of the linear WLS estimator given in (2.24), with $\mathbf{X}=\mathbf{H}$, $\mathbf{V}^{-1}=\mathbf{W}$, and $\boldsymbol{\Omega}=\mathbf{S}$. For given $\mathbf{X}$ and $\boldsymbol{\Omega}$ the linear WLS variance is minimized by choosing $\mathbf{V}=\boldsymbol{\Omega}$. By the same matrix algebra, for given $\mathbf{H}$ and $\mathbf{S}$ the GMM variance is minimized by choosing $\mathbf{W}=\mathbf{S}^{-1}$. Analogously to feasible GLS, one can equivalently use $\mathbf{W}_n=\hat{\mathbf{S}}^{-1}$, where $\hat{\mathbf{S}}$ is consistent for $\mathbf{S}$.

2.8.5 Conditional Moment Tests

For the distribution of the conditional moment test statistic (2.101), we take a first-order Taylor series expansion about $\boldsymbol{\theta}_0$

$$\frac{1}{\sqrt{n}}\mathbf{m}(\hat{\boldsymbol{\theta}}) = \frac{1}{\sqrt{n}}\sum_{i=1}^{n}\mathbf{m}_i(\boldsymbol{\theta}_0) + \frac{1}{n}\sum_{i=1}^{n}\frac{\partial \mathbf{m}_i(\boldsymbol{\theta}_0)}{\partial \boldsymbol{\theta}'}\sqrt{n}(\hat{\boldsymbol{\theta}} - \boldsymbol{\theta}_0), \quad (2.134)$$

where $\mathbf{m}_i(\boldsymbol{\theta}_0) = \mathbf{m}_i(y_i, \mathbf{x}_i, \boldsymbol{\theta}_0)$ and $\partial \mathbf{m}_i(\boldsymbol{\theta}_0)/\partial \boldsymbol{\theta}' = \partial \mathbf{m}_i(y_i, \mathbf{x}_i, \boldsymbol{\theta}_0)/\partial \boldsymbol{\theta}'\big|_{\boldsymbol{\theta}_0}$. We suppose that $\hat{\boldsymbol{\theta}}$ is the solution to the first-order conditions

$$\sum_{i=1}^{n}\mathbf{g}_i(\hat{\boldsymbol{\theta}}) = \mathbf{0},$$

where $\mathbf{g}_i(\boldsymbol{\theta}) = \mathbf{g}_i(y_i, \mathbf{x}_i, \boldsymbol{\theta})$. Replacing $\sqrt{n}(\hat{\boldsymbol{\theta}} - \boldsymbol{\theta}_0)$ in (2.134) by the right-hand side of (2.123) yields

$$\frac{1}{\sqrt{n}}\mathbf{m}(\hat{\boldsymbol{\theta}}) = \frac{1}{\sqrt{n}}\sum_{i=1}^{n}\mathbf{m}_i(\boldsymbol{\theta}_0) \quad (2.135)$$

$$\times - \frac{1}{n}\sum_{i=1}^{n}\frac{\partial \mathbf{m}_i(\boldsymbol{\theta}_0)}{\partial \boldsymbol{\theta}'}\left(\frac{1}{n}\sum_{i=1}^{n}\frac{\partial \mathbf{g}_i(\boldsymbol{\theta}_0)}{\partial \boldsymbol{\theta}'}\right)^{-1}\frac{1}{\sqrt{n}}\sum_{i=1}^{n}\mathbf{g}_i(\boldsymbol{\theta}_0).$$

It follows from some algebra that

$$\frac{1}{\sqrt{n}}\mathbf{m}(\hat{\boldsymbol{\theta}}) \overset{LD}{=} \left[\mathbf{I}_r \quad -\mathbf{C}\mathbf{A}^{-1}\right]\begin{bmatrix}\frac{1}{\sqrt{n}}\sum_{i=1}^{n}\mathbf{m}_i(\boldsymbol{\theta}_0)\\ \frac{1}{\sqrt{n}}\sum_{i=1}^{n}\mathbf{g}_i(\boldsymbol{\theta}_0)\end{bmatrix}, \quad (2.136)$$

where $\overset{LD}{=}$ means has the same limit distribution as the right-hand expression, and

$$\mathbf{C} = \lim_{n\to\infty}\frac{1}{n}\mathsf{E}\left[\sum_{i=1}^{n}\frac{\partial \mathbf{m}_i}{\partial \boldsymbol{\theta}'}\bigg|_{\boldsymbol{\theta}_0}\right],$$

and

$$\mathbf{A} = \lim_{n\to\infty}\frac{1}{n}\mathsf{E}\left[\sum_{i=1}^{n}\frac{\partial \mathbf{g}_i}{\partial \boldsymbol{\theta}'}\bigg|_{\boldsymbol{\theta}_0}\right].$$

Equation (2.136) is the key to obtaining the distribution of the CM test statistic. By a central limit theorem the second term in the right-hand side of (2.136) converges to $\mathsf{N}[\mathbf{0}, \mathbf{J}]$, where

$$\mathbf{J} = \lim_{n\to\infty}\frac{1}{n}\mathsf{E}\left[\begin{array}{cc}\sum_{i=1}^{n}\mathbf{m}_i\mathbf{m}_i' & \sum_{i=1}^{n}\mathbf{m}_i\mathbf{g}_i'\\ \sum_{i=1}^{n}\mathbf{g}_i\mathbf{m}_i' & \sum_{i=1}^{n}\mathbf{g}_i\mathbf{g}_i'\end{array}\right]\Bigg|_{\boldsymbol{\theta}_0}.$$

It follows that $n^{-1/2}\mathbf{m}(\hat{\boldsymbol{\theta}}) \stackrel{d}{\rightarrow} \mathsf{N}[\mathbf{0}, \mathbf{V}_m]$, where

$$\mathbf{V}_m = \mathbf{HJH}',$$

$\mathbf{J}$ is defined already, and

$$\mathbf{H} = \begin{bmatrix} \mathbf{I}_r & -\mathbf{CA}^{-1} \end{bmatrix}.$$

The CM test can be operationalized by dropping the expectation and evaluating the expressions above at $\hat{\boldsymbol{\theta}}$.

In the special case in which (2.109) holds, that is, $\mathsf{E}[\partial \mathbf{m}_i/\partial \boldsymbol{\theta}'] = \mathbf{0}$, $\mathbf{C} = \mathbf{0}$ so $\mathbf{V}_m = \mathbf{HJH}' = \lim \frac{1}{n}\mathsf{E}[\sum_{i=1}^n \mathbf{m}_i \mathbf{m}_i']$ leading to the simplification (2.110). For the OPG auxiliary regression (2.112) if $\hat{\boldsymbol{\theta}}$ is the MLE, see, for example, Pagan and Vella (1989).

2.9 BIBLIOGRAPHIC NOTES

Standard references for estimation theory for cross-sectional data are T. Amemiya (1985), Cameron and Trivedi (2005), Wooldridge (2010), and Greene (2011). A comprehensive treatment is given in Newey and McFadden (1994). For maximum likelihood estimation, see also Hendry (1995). The two-volume work by Gourieroux and Monfort (1995) presents estimation and testing theory in considerable detail, with considerable emphasis on the QML framework.

Reference to GLM is generally restricted to the statistics literature, even though it nests many common nonlinear regression models, including the linear, logit, probit, and Poisson regression models. Key papers are Nelder and Wedderburn (1972), Wedderburn (1974), and McCullagh (1983); the standard reference is McCullagh and Nelder (1989). The book by Fahrmeier and Tutz (1994) presents the GLM framework in a form amenable to econometricians.

Ziegler, Kastner, and Blettner (1998) provide many references to the GLM and GEE literatures. The estimating equation approach is summarized by Carroll et al. (2006). For GMM, emphasized in the econometrics literature, key papers are Hansen (1982) and Newey and West (1987a). Detailed textbook treatments of GMM include Davidson and MacKinnon (1993, chapter 17), Ogaki (1993), Hamilton (1994, chapter 14), and Hayashi (2000). Ziegler (2011) presents both GLM/GEE and GMM approaches.

2.10 EXERCISES

2.1 Let the dgp for $\mathbf{y}$ be $\mathbf{y} = \mathbf{X}\boldsymbol{\beta}_0 + \mathbf{u}$, where $\mathbf{X}$ is nonstochastic and $\mathbf{u} \sim (\mathbf{0}, \boldsymbol{\Omega})$. Show by substituting out $\mathbf{y}$ that the WLS estimator defined in (2.23) can be expressed as $\hat{\boldsymbol{\beta}}_{\mathsf{WLS}} = \boldsymbol{\beta}_0 + (\mathbf{X}'\mathbf{V}^{-1}\mathbf{X})^{-1}\mathbf{X}'\mathbf{V}^{-1}\mathbf{u}$. Hence, obtain $\mathsf{V}[\hat{\boldsymbol{\beta}}_{\mathsf{WLS}}]$ given in (2.24).

2.2 Let y have the LEF density $f(y|\mu)$ given in (2.26), where the range of the y does not depend on $\mu \equiv \mathsf{E}[y]$. Show by differentiating with respect to μ the identity $\int f(y|\mu)dy = 1$ that $\mathsf{E}[a'(\mu) + c'(\mu)y] = 0$. Hence obtain $\mathsf{E}[y]$ given in (2.27). Show by differentiating with respect to μ the identity $\int yf(y|\mu)dy = \mu$ that $\mathsf{E}[a'(\mu)y + c'(\mu)y^2] = 1$. Hence obtain $\mathsf{V}[y]$ given in (2.28).

2.3 For the LEF log-likelihood defined in (2.29) and (2.30), obtain the first-order conditions for the MLE $\hat{\boldsymbol{\beta}}_{\mathsf{ML}}$. Show that these can be reexpressed as (2.31) using (2.27) and (2.28). From (2.31) obtain the first-order conditions for the MLE of the Poisson regression model with exponential mean function.

2.4 Consider the geometric density $f(y|\mu) = \mu^y(1+\mu)^{-y-1}$, where $y = 0, 1, 2, \ldots$ and $\mu = \mathsf{E}[y]$. Write this density in the LEF form (2.26). Hence obtain the formula for $\mathsf{V}[y]$ using (2.28). In the regression case in which $\mu_i = \exp(\mathbf{x}_i'\boldsymbol{\beta})$, obtain the first-order conditions for the MLE for $\boldsymbol{\beta}$. Give the distribution for this estimator, assuming correct specification of the variance.

2.5 Consider the geometric density $f(y|\mu) = \mu^y(1+\mu)^{-y-1}$, where $y = 0, 1, 2, \ldots$ and $\mu = \mathsf{E}[y]$. Write this density in the canonical form of the LEF (2.45). Hence obtain the formula for $\mathsf{V}[y]$ using (2.47). Obtain the canonical link function for the geometric, verifying that it is not the log link function. In the regression case with the canonical link function obtain the first-order conditions for the MLE for $\boldsymbol{\beta}$. Give the distribution for this estimator, assuming correct specification of the variance.

2.6 Models with exponential mean function $\exp(\mathbf{x}_i'\boldsymbol{\beta})$, where $\boldsymbol{\beta}$ and $\mathbf{x}_i$ are $k \times 1$ vectors, satisfy $\mathsf{E}[(y_i - \exp(\mathbf{x}_i'\boldsymbol{\beta}))\mathbf{x}_i] = \mathbf{0}$. Obtain the first-order conditions for the GMM estimator that minimizes (2.72), where $h(y_i, \mathbf{x}_i, \boldsymbol{\beta}) = (y_i - \exp(\mathbf{x}_i'\boldsymbol{\beta}))\mathbf{x}_i$ and $\mathbf{W}$ is $k \times k$ of rank k. Show that these first-order conditions are a full-rank $k \times k$ matrix transformation of the first-order conditions (2.6) for the Poisson MLE. What do you conclude?

2.7 For the Poisson regression model with exponential mean function $\exp(\mathbf{x}_i'\boldsymbol{\beta})$, consider tests for exclusion of the subcomponent $\mathbf{x}_{2i}$ of $\mathbf{x}_i = [\mathbf{x}_{1i}', \mathbf{x}_{2i}']$, which are tests of $\boldsymbol{\beta}_2 = \mathbf{0}$. Obtain the test statistic T_{LM} given (2.93). State how to compute an asymptotically equivalent version of the LM test using an auxiliary regression. State how to alternatively implement Wald and LR tests of $\boldsymbol{\beta}_2 = \mathbf{0}$.

2.8 Show that variance-mean equality in the Poisson regression model with exponential mean implies that $\mathsf{E}[\{y_i - \exp(\mathbf{x}_i'\boldsymbol{\beta})^2 - y_i\}^2] = 0$. Using this moment condition, obtain the conditional moment test statistic T_{CM} given (2.110), first showing that the simplifying condition (2.109) holds if $y_i \sim \mathsf{P}[\exp(\mathbf{x}_i'\boldsymbol{\beta})]$. State how to compute T_{CM} by an auxiliary regression. Does $\hat{\boldsymbol{\beta}}$ need to be the MLE here, or will any $\sqrt{n}$–consistent estimator do?

CHAPTER 3

Basic Count Regression

3.1 INTRODUCTION

This chapter is intended to provide a self-contained treatment of basic cross-section count data regression analysis. It is analogous to a chapter in a standard statistics text that covers both homoskedastic and heteroskedastic linear regression models.

The most commonly used count models are Poisson and negative binomial. For readers interested only in these models, it is sufficient to read sections 3.1 to 3.5, along with preparatory material in sections 1.2 and 2.2.

As indicated in Chapter 2, the properties of an estimator vary with the assumptions made on the dgp. By correct specification of the conditional mean or variance or density, we mean that the functional form and explanatory variables in the specified conditional mean or variance or density are those of the dgp.

The simplest regression model for count data is the Poisson regression model. For the Poisson MLE, the following can be shown:

1. Consistency requires correct specification of the conditional mean. It does not require that the dependent variable y be Poisson distributed.
2. Valid statistical inference using default computed maximum likelihood standard errors and t statistics requires correct specification of both the conditional mean and variance. This requires equidispersion, that is, equality of conditional variance and mean, but not Poisson distribution for y.
3. Valid statistical inference using appropriately computed standard errors is still possible if data are not equidispersed, provided the conditional mean is correctly specified.
4. More efficient estimators than Poisson MLE can be obtained if data are not equidispersed.

Properties 1 through 4 are similar to those of the OLS estimator in the classical linear regression model, which is the MLE if errors are iid normal. The Poisson restriction of equidispersion is directly analogous to homoskedasticity in the linear model. If errors are heteroskedastic in the linear model, one would use alternative standard errors to those from the usual OLS output (property 3) and preferably estimate by WLS (property 4).

In practice count data are often overdispersed, with conditional variance exceeding the conditional mean. There are then several ways to proceed.

1. Poisson quasi-MLE (QMLE) with corrected standard errors.
2. Poisson quasi-generalized pseudo-MLE (QGPMLE) with corrected standard errors. Estimation is based on the same conditional mean as for the Poisson quasi-MLE, but on a different conditional variance function.
3. MLE of a better parametric model for counts, such as the negative binomial.

Approaches 1 and 2 are valid provided the conditional mean is correctly specified, usually as $\exp(\mathbf{x}'\boldsymbol{\beta})$. Analysis focuses on how this conditional mean changes as regressors change. Approach 2 has the advantage over approach 1 of yielding potentially more efficient estimation; however, with cross-sectional data this gain can often be relatively small, so approach 1 is used more often.

Approach 3 generally requires that the parametric model is correctly specified, not just the conditional mean. This stronger assumption can lead to more efficient estimation and allows one to analyze how conditional probabilities such as $\Pr[y \geq 4|\mathbf{x}]$, rather than just the conditional mean, change as regressors change.

It is important that such modifications to the Poisson MLE be made. Count data are often very overdispersed, which causes default computed Poisson ML t statistics to be considerably overinflated. This can lead to very erroneous and overly optimistic conclusions of statistical significance of regressors.

The various Poisson regression estimators are presented in section 3.2, and negative binomial regression estimators are given in section 3.3. Tests for overdispersion are presented in section 3.4. Practical issues of interpretation of coefficients with an exponential, rather than linear, specification of the conditional mean, as well as use of estimates for prediction, are presented in section 3.5.

An alternative approach to count data is to assume an underlying continuous latent process, with higher counts arising as the continuous variable passes successively higher thresholds. Ordered probit and related discrete choice models are presented in section 3.6. Least squares methods are the focus of section 3.7. These are nonlinear least squares with exponential conditional mean function, and OLS with the dependent variable a transformation of the count data y to reduce heteroskedasticity and asymmetry. Given the simplicity of standard count models, there is generally no reason to use these least squares methods that ignore the count nature of the data.

For completeness, many different models, regression parameter estimators, and standard error estimators are presented in this chapter. The models emphasized are the Poisson and two variants of the negative binomial – NB1 and

NB2. The estimators considered include MLE, QMLE, and QGPMLE. An acronym such as NB2 MLE is shorthand for the NB2 model maximum likelihood estimator. Throughout this chapter, the methods presented are applied to a regression model for the number of doctor visits, introduced in section 3.2.6.

For many analyses the Poisson QMLE with corrected standard errors, the negative binomial MLE (NB2), and the ordered probit MLE are sufficient. The most common departures from these standard count models, such as hurdle models, zero-inflated models, the Poisson-Normal mixture model, and finite mixtures models, are presented in the next chapter.

3.2 POISSON MLE, QMLE, AND GLM

Many of the algebraic results presented in this chapter need to be modified if the conditional mean function is not exponential.

3.2.1 Poisson MLE

From section 1.2.3, the Poisson regression model specifies that y_i given $\mathbf{x}_i$ is Poisson distributed with density

$$f(y_i|\mathbf{x}_i) = \frac{e^{-\mu_i}\mu_i^{y_i}}{y_i!}, \qquad y_i = 0, 1, 2, \ldots \tag{3.1}$$

and mean parameter

$$\mathsf{E}[y_i|\mathbf{x}_i] = \mu_i = \exp(\mathbf{x}_i'\boldsymbol{\beta}). \tag{3.2}$$

The specification (3.2) is called the *exponential mean function*. The model comprising (3.1) and (3.2) is usually referred to as the Poisson regression model, a terminology we also use, although more precisely it is the Poisson regression model with exponential mean function. In the statistics literature the model is also called a *log-linear model*, because the logarithm of the conditional mean is linear in the parameters: $\ln \mathsf{E}[y_i|\mathbf{x}_i] = \mathbf{x}_i'\boldsymbol{\beta}$.

Given independent observations, the log-likelihood is

$$\ln \mathsf{L}(\boldsymbol{\beta}) = \sum_{i=1}^{n}\{y_i\mathbf{x}_i'\boldsymbol{\beta} - \exp(\mathbf{x}_i'\boldsymbol{\beta}) - \ln y_i!\}. \tag{3.3}$$

The Poisson MLE $\hat{\boldsymbol{\beta}}_{\mathsf{P}}$ is the solution to the first-order conditions

$$\sum_{i=1}^{n}(y_i - \exp(\mathbf{x}_i'\boldsymbol{\beta}))\mathbf{x}_i = \mathbf{0}. \tag{3.4}$$

Note that if the regressors include a constant term then the residuals $y_i - \exp(\mathbf{x}_i'\hat{\boldsymbol{\beta}}_{\mathsf{P}})$ sum to zero by (3.4).

The standard method for computation of $\hat{\boldsymbol{\beta}}_{\mathsf{P}}$ is the Newton-Raphson iterative method. Convergence is guaranteed, because the log-likelihood function is globally concave. In practice often fewer than 10 iterations are needed. The

Newton-Raphson method can be implemented by iterative use of OLS as presented in section 3.8.

If the dgp for y_i is indeed Poisson with mean (3.2), we can apply the usual maximum likelihood theory as in Chapter 2.3.2. This yields

$$\hat{\boldsymbol{\beta}}_{\mathsf{P}} \overset{a}{\sim} \mathsf{N}\left[\boldsymbol{\beta}, \mathsf{V}_{\mathsf{ML}}[\hat{\boldsymbol{\beta}}_{\mathsf{P}}]\right] \tag{3.5}$$

where

$$\mathsf{V}_{\mathsf{ML}}[\hat{\boldsymbol{\beta}}_{\mathsf{P}}] = \left(\sum_{i=1}^{n} \mu_i \mathbf{x}_i \mathbf{x}_i'\right)^{-1}, \tag{3.6}$$

using $\mathsf{E}[\partial^2 \ln\mathsf{L}/\partial\boldsymbol{\beta}\partial\boldsymbol{\beta}'] = -\sum_{i=1}^{n} \mu_i \mathbf{x}_i \mathbf{x}_i'$. Strictly speaking, we should assume that the dgp evaluates $\boldsymbol{\beta}$ at the specific value $\boldsymbol{\beta}_0$ and replace $\boldsymbol{\beta}$ by $\boldsymbol{\beta}_0$ in (3.5). This more formal presentation is used in Chapter 2. In the rest of the book we use a less formal presentation, provided the estimator is indeed consistent.

Most statistical programs use Hessian maximum likelihood standard errors (MLH) using (3.6) evaluated at $\hat{\mu}_i = \exp(\mathbf{x}_i'\hat{\boldsymbol{\beta}}_{\mathsf{P}})$. By the information matrix equality, one can instead use the summed outer product of the first derivatives (see Chapter 2.3.2), leading to the maximum likelihood outer product (MLOP) estimator

$$\mathsf{V}_{\mathsf{MLOP}}[\hat{\boldsymbol{\beta}}_{\mathsf{P}}] = \left(\sum_{i=1}^{n} (y_i - \mu_i)^2 \mathbf{x}_i \mathbf{x}_i'\right)^{-1}, \tag{3.7}$$

evaluated at $\hat{\mu}_i$. A general optimization routine may provide standard errors based on (3.7), which asymptotically equals (3.6) if data are equidispersed.

In practice, data are not equidispersed, in which case neither (3.6) nor (3.7) should be used.

3.2.2 Poisson Quasi-MLE

The assumption of a Poisson distribution is stronger than necessary for statistical inference based on $\hat{\boldsymbol{\beta}}_{\mathsf{P}}$ defined by (3.4). As discussed in Chapter 2.4.2, whose results are used extensively here, consistency holds for the MLE of any specified LEF density such as the Poisson, provided the conditional mean function (3.2) is correctly specified. An intuitive explanation is that consistency requires the left-hand side of the first-order conditions (3.4) to have expected value zero. This is the case if $\mathsf{E}[y_i|\mathbf{x}_i] = \exp(\mathbf{x}_i'\boldsymbol{\beta})$, because then $\mathsf{E}[(y_i - \exp(\mathbf{x}_i'\boldsymbol{\beta}))\mathbf{x}_i] = \mathbf{0}$. Furthermore, according to this first-order condition y need not even be a count. It is sufficient that $y \geq 0$.

Given this robustness to distributional assumptions, we can continue to use $\hat{\boldsymbol{\beta}}_{\mathsf{P}}$ even if the dgp for y_i is not the Poisson. If an alternative dgp is entertained, the estimator defined by the Poisson maximum likelihood first-order conditions (3.4) is called the *Poisson quasi-MLE* (QMLE) or the *Poisson pseudo-*MLE. This terminology means that the estimator is like the Poisson MLE in that the Poisson model is used to motivate the first-order condition defining the estimator, but it is unlike the Poisson MLE in that the dgp used to obtain the distribution of the estimator need not be the Poisson. Here we assume the Poisson mean, but relax the Poisson restriction of equidispersion.

The Poisson QMLE $\hat{\boldsymbol{\beta}}_{\mathsf{P}}$ is defined to be the solution to (3.4). If (3.2) holds then

$$\hat{\boldsymbol{\beta}}_{\mathsf{P}} \overset{a}{\sim} \mathsf{N}\left[\boldsymbol{\beta}, \mathsf{V}_{\mathsf{QML}}[\hat{\boldsymbol{\beta}}_{\mathsf{P}}]\right], \tag{3.8}$$

where

$$\mathsf{V}_{\mathsf{QML}}[\hat{\boldsymbol{\beta}}_{\mathsf{P}}] = \left(\sum_{i=1}^{n} \mu_i \mathbf{x}_i \mathbf{x}_i'\right)^{-1} \left(\sum_{i=1}^{n} \omega_i \mathbf{x}_i \mathbf{x}_i'\right) \left(\sum_{i=1}^{n} \mu_i \mathbf{x}_i \mathbf{x}_i'\right)^{-1} \tag{3.9}$$

and

$$\omega_i = \mathsf{V}[y_i | \mathbf{x}_i] \tag{3.10}$$

is the conditional variance of y_i. Implementation of (3.9) depends on what functional form, if any, is assumed for ω_i.

Poisson QMLE with Robust Standard Errors

The variance matrix (3.9) can be consistently estimated without specification of a functional form for ω_i. We need to estimate for unknown ω_i the middle term in $\mathsf{V}_{\mathsf{QML}}[\hat{\boldsymbol{\beta}}_{\mathsf{P}}]$ defined in (3.9). Formally, a consistent estimate of $\lim \frac{1}{n} \sum_{i=1}^{n} \mathsf{E}[(y_i - \mu_i)^2 | \mathbf{x}_i] \mathbf{x}_i \mathbf{x}_i'$ is needed. It can be shown that if $(y_i, \mathbf{x}_i)$ are iid this $k \times k$ matrix is consistently estimated by $\frac{1}{n} \sum_{i=1}^{n} (y_i - \hat{\mu}_i)^2 \mathbf{x}_i \mathbf{x}_i'$, even though it is impossible to consistently estimate each of the n scalars ω_i by $(y_i - \hat{\mu}_i)^2$. This yields the variance matrix estimate

$$\mathsf{V}_{\mathsf{RS}}[\hat{\boldsymbol{\beta}}_{\mathsf{P}}] = \left(\sum_{i=1}^{n} \mu_i \mathbf{x}_i \mathbf{x}_i'\right)^{-1} \left(\sum_{i=1}^{n} (y_i - \mu_i)^2 \mathbf{x}_i \mathbf{x}_i'\right) \left(\sum_{i=1}^{n} \mu_i \mathbf{x}_i \mathbf{x}_i'\right)^{-1}, \tag{3.11}$$

which is evaluated at $\hat{\mu}_i$. Usually the variance matrix is multiplied by the finite-sample adjustment $n/(n-k)$.

The estimator (3.11) is the RS estimator discussed in sections 2.5.1 and 2.7. It builds on the work by Eicker (1967), who obtained a similar result in the nonregression case, and White (1980), who obtained this result in the OLS regression case and popularized its use in econometrics. See Robinson (1987a) for a history of this approach and for further references. This method is used extensively throughout this book, in settings much more general than the Poisson QMLE with cross-section data. As shorthand we refer to standard errors as *robust standard errors* whenever a similar approach is used to obtain standard errors without specifying functional forms for the second moments of the dependent variable.

An alternative way to proceed when the variance function ω_i is not specified is to *bootstrap*. This approach estimates properties of the distribution of $\hat{\boldsymbol{\beta}}_{\mathsf{P}}$ and performs statistical inference on $\boldsymbol{\beta}$ by resampling from the original data set. Bootstrapping the pairs $(y_i, \mathbf{x}_i)$, see Chapter 2.6.4, leads to standard errors asymptotically equivalent to those based on (3.11).

For equidispersed data, with $\mathrm{E}[(y_i - \mu_i)^2 | \mathbf{x}_i] = \mu_i$, the RS variance estimator is similar to the MLH and MLOP estimators because then $\sum_i (y_i - \mu_i)^2 \mathbf{x}_i \mathbf{x}_i' \simeq \sum_i \mu_i \mathbf{x}_i \mathbf{x}_i'$. But in the common case that data are overdispersed, with $\mathrm{E}[(y_i - \mu_i)^2 | \mathbf{x}_i] > \mu_i$, the correct RS variance estimator will be larger than the incorrect MLH and MLOP estimators. These differences become larger the greater the overdispersion.

3.2.3 NB1 and NB2 Variance Functions

In the Poisson regression model y_i has mean $\mu_i = \exp(\mathbf{x}_i'\boldsymbol{\beta})$ and variance μ_i. We now consider alternative models for the variance, because data almost always reject the restriction that the variance equals the mean, while maintaining the assumption that the mean is $\exp(\mathbf{x}_i'\boldsymbol{\beta})$.

It is natural to continue to model the conditional variance ω_i defined in (3.10) as a function of the mean, with

$$\omega_i = \omega(\mu_i, \alpha) \tag{3.12}$$

for some specified function $\omega(\cdot)$ and where α is a scalar parameter. Most models specialize this to the general variance function

$$\omega_i = \mu_i + \alpha \mu_i^p, \tag{3.13}$$

where the constant p is specified. Analysis is usually restricted to two special cases, in addition to the Poisson case of $\alpha = 0$.

First, the NB1 *variance function* sets $p = 1$. Then the variance

$$\omega_i = (1 + \alpha)\mu_i, \tag{3.14}$$

is a multiple of the mean. In the GLM framework this is usually rewritten as

$$\omega_i = \phi \mu_i, \tag{3.15}$$

where $\phi = 1 + \alpha$.

Second, the NB2 *variance function* sets $p = 2$. Then the variance is quadratic in the mean

$$\omega_i = \mu_i + \alpha \mu_i^2. \tag{3.16}$$

In both cases the *dispersion parameter* α is a parameter to be estimated.

Cameron and Trivedi (1986), in the context of negative binomial models, used the terminology NB1 model to describe the case $p = 1$ and NB2 model to describe the case $p = 2$. Here we have extended this terminology to the variance function itself.

These variance functions can be used in two ways. First, they can lead to estimators that are more efficient than the Poisson QMLE; see subsequent sections. Second, they can be used to obtain alternative standard errors for the Poisson QMLE, as outlined in the remainder of this section. Common practice is to nonetheless use the robust sandwich form (3.11) because it does not require assuming a model for ω_i.

Poisson QMLE with Poisson Variance Function

If the conditional variance of y_i is that for the Poisson, so $\omega_i = \mu_i$, then the variance matrix (3.9) simplifies to (3.6). Thus the usual Poisson ML inference is valid provided the first two moments are those for the Poisson.

Poisson QMLE with NB1 Variance Function

The simplest generalization of $\omega_i = \mu_i$ is the NB1 variance function (3.15). Since $\omega_i = \phi\mu_i$ the variance matrix in (3.9) simplifies to

$$\mathsf{V}_{\mathsf{NB1}}[\hat{\boldsymbol{\beta}}_{\mathsf{P}}] = \phi\left(\sum_{i=1}^{n}\mu_i\mathbf{x}_i\mathbf{x}_i'\right)^{-1} = \phi\mathsf{V}_{\mathsf{ML}}[\hat{\boldsymbol{\beta}}_{\mathsf{P}}], \tag{3.17}$$

where $\mathsf{V}_{\mathsf{ML}}[\hat{\boldsymbol{\beta}}_{\mathsf{P}}]$ is the maximum likelihood variance matrix given in (3.6). Thus, the simplest way to handle overdispersed or underdispersed data is to begin with the computed Poisson maximum likelihood output. Then, multiply maximum likelihood output by ϕ to obtain the correct variance matrix, multiply by $\sqrt{\phi}$ to obtain the correct standard errors, and divide by $\sqrt{\phi}$ to get the correct t statistics.

The standard estimator of ϕ is

$$\hat{\phi}_{\mathsf{NB1}} = \frac{1}{n-k}\sum_{i=1}^{n}\frac{(y_i - \hat{\mu}_i)^2}{\hat{\mu}_i}. \tag{3.18}$$

The motivation for this estimator is that variance function (3.15) implies $\mathsf{E}[(y_i - \mu_i)^2] = \phi\mu_i$ and hence $\phi = \mathsf{E}[(y_i - \mu_i)^2/\mu_i]$. The corresponding sample moment is (3.18), where division by $(n-k)$ rather than n is a degrees-of-freedom correction. This approach to estimation is the GLM approach presented in Chapter 2.4.3; see also section 3.2.4. Poisson regression packages using the GLM framework often use (3.17) for standard errors; most others instead use (3.6) as the default.

Poisson QMLE with NB2 Variance Function

A common alternative specification for the variance of y_i is the NB2 variance function (3.16). Then because $\omega_i = \mu_i + \alpha\mu_i^2$, the variance matrix (3.9) becomes

$$\mathsf{V}_{\mathsf{NB2}}[\hat{\boldsymbol{\beta}}_{\mathsf{P}}] = \left(\sum_{i=1}^{n}\mu_i\mathbf{x}_i\mathbf{x}_i'\right)^{-1}\left(\sum_{i=1}^{n}(\mu_i + \alpha\mu_i^2)\mathbf{x}_i\mathbf{x}_i'\right)\left(\sum_{i=1}^{n}\mu_i\mathbf{x}_i\mathbf{x}_i'\right)^{-1}. \tag{3.19}$$

This does not simplify and computation requires matrix routines. One of several possible estimators of α is

$$\hat{\alpha}_{\mathsf{NB2}} = \frac{1}{n-k}\sum_{i=1}^{n}\frac{\{(y_i - \hat{\mu}_i)^2 - \hat{\mu}_i\}}{\hat{\mu}_i^2}. \tag{3.20}$$

The motivation for this estimator of α is that (3.16) implies $\mathsf{E}[(y_i - \mu_i)^2 - \mu_i] = \alpha\mu_i^2$ and hence $\alpha = \mathsf{E}[\{(y_i - \mu_i)^2 - \mu_i\}/\mu_i^2]$. The corresponding sample moment with degrees-of-freedom correction is (3.20). This estimator was proposed by Gourieroux et al. (1984a, 1984b).

Alternative estimators of ϕ and α for NB1 and NB2 variance functions are given in Cameron and Trivedi (1986). In practice, studies do not present estimated standard errors for $\hat{\phi}_{\mathsf{NB1}}$ and $\hat{\alpha}_{\mathsf{NB2}}$, although they can be obtained using the delta method given in Chapter 2.6.2. A series of papers by Dean (1993, 1994) and Dean, Eaves and Martinez (1995) considers different estimators for the dispersion parameter and consequences for variance matrix estimation. Some different estimators of α_{NB2} are discussed after Table 3.4.

3.2.4 Poisson GLM

Generalized linear models are defined in section 2.4.4. For the Poisson with mean function (3.2), which is the canonical link function for this model, the Poisson GLM density is

$$f(y_i|\mathbf{x}_i) = \exp\left\{\frac{\mathbf{x}_i'\boldsymbol{\beta}\ y_i - \exp(\mathbf{x}_i'\boldsymbol{\beta})}{\phi} + c(y_i, \phi)\right\}, \tag{3.21}$$

where $c(y_i, \phi)$ is a normalizing constant. Then $\mathsf{V}[y_i] = \phi\mu_i$, which is the NB1 variance function.

The Poisson GLM estimator $\hat{\boldsymbol{\beta}}_{\mathsf{PGLM}}$ maximizes with respect to $\boldsymbol{\beta}$ the corresponding log-likelihood, with first-order conditions

$$\sum_{i=1}^{n}\frac{1}{\phi}(y_i - \exp(\mathbf{x}_i'\boldsymbol{\beta}))\mathbf{x}_i = \mathbf{0}. \tag{3.22}$$

These first-order conditions coincide with (3.4) for the Poisson QML, except for scaling by the constant ϕ. Consequently $\hat{\boldsymbol{\beta}}_{\mathsf{PGLM}} = \hat{\boldsymbol{\beta}}_{\mathsf{P}}$, and the variance matrix is the same as (3.17) for the Poisson QML with NB1 variance function

$$\mathsf{V}[\hat{\boldsymbol{\beta}}_{\mathsf{PGLM}}] = \phi\left(\sum_{i=1}^{n}\mu_i\mathbf{x}_i\mathbf{x}_i'\right)^{-1}. \tag{3.23}$$

To implement this last result for statistical inference on $\boldsymbol{\beta}$, GLM practitioners use the consistent estimate $\hat{\phi}_{\mathsf{NB1}}$ defined in (3.18).

A more obvious approach to estimating the nuisance parameter ϕ is to maximize the log-likelihood based on (3.21) with respect to both $\boldsymbol{\beta}$ and ϕ. Differentiation with respect to ϕ requires an expression for the normalizing constant $c(y_i, \phi)$, however, and the restriction that probabilities sum to unity

$$\sum_{y_i=0}^{\infty} \exp\left\{\frac{1}{\phi}(\mathbf{x}_i'\boldsymbol{\beta}\, y_i - \exp(\mathbf{x}_i'\boldsymbol{\beta})) + c(y_i, \phi)\right\} = 1$$

has no simple solution for $c(y_i, \phi)$. One therefore uses the estimator (3.18), which is based on assumptions about the first two moments rather than the density. More generally the density (3.21) is best thought of as merely giving a justification for the first-order conditions (3.22), rather than as a density that would be used, for example, to predict the probabilities of particular values of y.

3.2.5 Poisson EE

A quite general estimation procedure is to use the estimating equation presented in Chapter 2.4.5,

$$\sum_{i=1}^{n} \frac{1}{\omega_i}(y_i - \mu_i)\frac{\partial \mu_i}{\partial \boldsymbol{\beta}} = \mathbf{0},$$

which generalizes linear WLS. Consider a specific variance function of the form $\omega_i = \omega(\mu_i, \alpha)$, and let $\widetilde{\alpha}$ be a consistent estimator of α. For the exponential mean function $\partial \mu_i / \partial \boldsymbol{\beta} = \mu_i \mathbf{x}_i$, so $\hat{\boldsymbol{\beta}}_{\mathsf{EE}}$ solves the first-order conditions

$$\sum_{i=1}^{n} \frac{1}{\omega(\mu_i, \widetilde{\alpha})}(y_i - \mu_i)\mu_i \mathbf{x}_i = \mathbf{0}. \tag{3.24}$$

If the variance function is correctly specified, then it follows that

$$\mathsf{V}_{\mathsf{EE}}[\hat{\boldsymbol{\beta}}_{\mathsf{EE}}] = \left(\sum_{i=1}^{n} \frac{1}{\omega(\mu_i, \alpha)} \mu_i^2 \mathbf{x}_i \mathbf{x}_i'\right)^{-1}. \tag{3.25}$$

Because this estimator is motivated by specification of the first two moments, it can also be viewed as a method of moments estimator, a special case of GMM whose more general framework is unnecessary here because the number of equations (3.24) equals the number of unknowns. The first-order conditions nest as special cases those for the Poisson MLE and GLM, which replace $\omega(\mu_i, \widetilde{\alpha})$ by μ_i or $\phi\mu_i$.

3.2.6 Example: Doctor Visits

Consider the following example of the number of doctor visits in the past two weeks for a single-adult sample of size 5,190 from the Australian Health Survey 1977–78. This and several other measures of health service utilization such as days in hospital and number of medicines taken were analyzed in Cameron, Trivedi, Milne, and Piggott (1988) in the light of an economic model of joint determination of health service utilization and health insurance choice.

Table 3.1. *Doctor visits: Actual frequency distribution*

Count	0	1	2	3	4	5	6	7	8	9
Frequency	4,141	782	174	30	24	9	12	12	5	1
Relative frequency	0.798	0.151	0.033	0.006	0.005	0.002	0.002	0.002	0.001	0.000

The particular data presented here were also studied by Cameron and Trivedi (1986). The analysis of this example in this chapter (see also sections 3.3.5, 3.4.2, 3.5.2, and 3.7.5) is more detailed and covers additional methods.

Table 3.1 summarizes the dependent variable *DVISITS*. There are few large counts, with 98% of the sample taking values of 0, 1, or 2. The mean number of doctor visits is 0.302 with variance 0.637. The raw data are therefore overdispersed, although inclusion of regressors may eliminate the overdispersion.

The variables are defined and summary statistics given in Table 3.2. Regressors can be grouped into four categories: (1) socioeconomic: *SEX, AGE, AGESQ, INCOME*; (2) health insurance status indicators: *LEVYPLUS, FREEPOOR*, and *FREEREPA*, with *LEVY* (government Medibank health insurance) the omitted category; (3) recent health status measures: *ILLNESS, ACTDAYS*; and (4) long-term health status measures: *HSCORE, CHCOND1, CHCOND2*.

The Poisson maximum likelihood estimates defined by (3.4) are given in the first column of Table 3.3. These estimates are by definition identical to the Poisson QML estimates. Various estimates of the standard errors are given

Table 3.2. *Doctor visits: Variable definitions and summary statistics*

Variable	Definition	Mean	Standard deviation
DVISITS	Number of doctor visits in past two weeks	0.302	0.798
SEX	Equals 1 if female	0.521	0.500
AGE	Age in years divided by 100	0.406	0.205
AGESQ	*AGE* squared	0.207	0.186
INCOME	Annual income in tens of thousands of dollars	0.583	0.369
LEVYPLUS	Equals 1 if private health insurance	0.443	0.497
FREEPOOR	Equals 1 if free government health insurance due to low income	0.043	0.202
FREEREPA	Equals 1 if free government health insurance due to old age, disability, or veteran status	0.210	0.408
ILLNESS	Number of illnesses in past two weeks	1.432	1.384
ACTDAYS	Number of days of reduced activity in past two weeks due to illness or injury	0.862	2.888
HSCORE	General health questionnaire score using Goldberg's method	1.218	2.124
CHCOND1	Equals 1 if chronic condition not limiting activity	0.403	0.491
CHCOND2	Equals1 if chronic condition limiting activity	0.117	0.321

Table 3.3. *Doctor visits: Poisson* QMLE *with different standard error estimates*

	Coefficient	Standard errors						*t* statistic
Variable	Poisson PMLE	RS	MLH	MLOP	NB1	NB2	Boot	RS
ONE	−2.224	0.254	0.190	0.144	0.219	0.207	0.271	−8.74
SEX	0.157	0.079	0.056	0.041	0.065	0.062	0.076	1.98
AGE	1.056	1.364	1.001	0.750	1.153	1.112	1.391	0.77
AGESQ	−0.849	1.460	1.078	0.809	1.242	1.210	1.477	−0.58
INCOME	−0.205	0.129	0.088	0.062	0.102	0.096	0.129	−1.59
LEVYPLUS	0.123	0.095	0.072	0.056	0.083	0.077	0.100	1.29
FREEPOOR	−0.440	0.290	0.180	0.116	0.207	0.188	0.293	−1.52
FREEREPA	0.080	0.126	0.092	0.070	0.106	0.102	0.132	0.63
ILLNESS	0.187	0.024	0.018	0.014	0.021	0.021	0.024	7.81
ACTDAYS	0.127	0.008	0.005	0.004	0.006	0.006	0.008	16.33
HSCORE	0.030	0.014	0.010	0.007	0.012	0.012	0.014	2.11
CHCOND1	0.114	0.091	0.067	0.051	0.077	0.071	0.087	1.26
CHCOND2	0.141	0.123	0.083	0.059	0.096	0.092	0.120	1.15
−ln L	3355.5							

Note: Different standard error estimates are due to different specifications of ω, the conditional variance of y. RS, unspecified ω robust sandwich estimate; MLH, $\omega = \mu$ Hessian estimate; MLOP, $\omega = \mu$ summed outer product of first derivatives estimate; NB1, $\omega = \phi\mu = (1+\alpha)\mu$ where $\alpha = 0.382$; NB2, $\omega = \mu + \alpha\mu^2$ where $\alpha = 0.286$; and Boot, unspecified ω bootstrap estimate.

in the remainder of the table, under different assumptions about the variance of y, where throughout it is assumed that the conditional mean is correctly specified as in (3.2). Standard errors are presented rather than t statistics to allow comparison with the precision of alternative estimators given in later tables.

The column labeled RS lists the preferred robust sandwich estimates given in (3.11). The column labeled Boot lists bootstrap standard errors, based on bootstrap pairs resampling with 400 repetitions, that are generally within 5% of the RS standard errors, as expected.

The column labeled MLH gives standard errors, which are often the default standard errors following Poisson regression, that are based on the inverse of the Hessian (3.6). They are on average 70% of the RS standard errors, a consequence of the overdispersion in the data as discussed at the end of section 3.2.2.

If instead one uses the summed outer product of the first derivatives, the resulting MLOP standard errors using (3.7) are in this example on average an additional 25% lower than MLH standard errors. Comparison of (3.6) and (3.7) shows that this difference in standard errors is consistent with $\mathsf{E}[(y_i - \mu_i)^2|\mathbf{x}_i] = \phi\mu_i$, where $1/\sqrt{\phi} \simeq 0.75$ or $\alpha = (\phi - 1) \simeq 0.78$. More generally for overdispersed data the MLOP standard errors will be biased downward even more than the usual MLH standard errors (3.6).

The columns labeled MLH, NB1, and NB2 specify that the variance of y equals, respectively, the mean, a multiple of the mean, and a quadratic function of the mean. The standard errors NB1 are 1.152 times MLH standard errors, because $\hat{\phi}_{\text{NB1}} = 1.328$ using (3.18), which has square root 1.152. The standard errors NB2 are obtained using (3.9), where (3.20) yields $\hat{\alpha}_{\text{NB2}} = 0.286$. These estimated values of α are not reported in the table, because they are not used in forming an estimate of $\boldsymbol{\beta}$. Those values are used only to obtain standard errors.

Although the NB1 and NB2 standard errors are preferred to MLH, they are still substantially downward biased compared to the preferred RS standard errors. In principle specifying a model for ω_i does have the advantage of yielding more precise standard error estimates (i.e., a smaller variance of $\hat{\mathsf{V}}[\hat{\boldsymbol{\beta}}_{\text{P}}]$). In results not reported in Table 3.3, we used the bootstrap method to estimate the standard deviation of the standard error estimates as discussed in section 2.6.4. The standard deviation of NB1 standard errors was on average 3% of the average NB1 standard error, whereas the standard deviation of RS standard errors was on average 6% of the average RS standard error. This lower variability for NBI standard errors is not enough to compensate for the downwards bias.

Other count applications yield similar results. In the usual case in which data are overdispersed, the MLH (and MLOP) standard errors are smaller than RS standard errors and should not be used. The differences can be much greater than in this example if data are greatly overdispersed. One should never use MLH or MLOP here. The NB1 (or NB2) standard errors are an improvement, and historically NB1 standard errors were very appealing due to the computational advantage of being a simple rescaling of MLH standard errors often reported by ML routines. But it is safest and common practice to use RS standard errors.

The final column of Table 3.3 gives t statistics based on the RS standard errors. By far the most statistically significant determinants of doctor visits in the past two weeks are recent health status measures – the number of illnesses and days of reduced activity in the past two weeks – with positive coefficients, confirming that sicker people are more likely to visit a doctor. The long-term health status measure *HSCORE* and the socioeconomic variable *SEX* are also statistically significant at significance level 5%. Discussion of the impact of these variables on the number of doctor visits is deferred to section 3.5.

3.3 NEGATIVE BINOMIAL MLE AND QGPMLE

The Poisson QML estimator handles overdispersion or underdispersion by moving away from complete distributional specification. Instead estimation is based on specification of the first moment and inference that adjusts for departures from equidispersion.

Here we consider two methods that can lead to more efficient estimation. The first is to specify a more appropriate parametric model to account for overdispersion. In this chapter we consider the standard model that accommodates overdispersion, the negative binomial, which is motivated and derived in Chapter 4.

The second method is a less parametric method that bases estimation on specification of the first two moments, where the variance is modeled using the NB2 variance.

3.3.1 NB2 Model and MLE

The most common implementation of the *negative binomial* is the NB2 model, with mean μ and NB2 variance function $\mu + \alpha\mu^2$ defined in (3.16). It has density

$$f(y|\mu,\alpha) = \frac{\Gamma(y+\alpha^{-1})}{\Gamma(y+1)\Gamma(\alpha^{-1})}\left(\frac{\alpha^{-1}}{\alpha^{-1}+\mu}\right)^{\alpha^{-1}}\left(\frac{\mu}{\alpha^{-1}+\mu}\right)^{y},$$
$$\alpha \geq 0,\ y = 0,1,2,\ldots. \quad (3.26)$$

This reduces to the Poisson if $\alpha = 0$ (see section 3.3.4).

The function $\Gamma(\cdot)$ is the gamma function, defined in Appendix B, where it is shown that $\Gamma(y+a)/\Gamma(a) = \prod_{j=0}^{y-1}(j+a)$, if y is an integer. Thus,

$$\ln\left(\frac{\Gamma(y+\alpha^{-1})}{\Gamma(\alpha^{-1})}\right) = \sum_{j=0}^{y-1}\ln(j+\alpha^{-1}). \quad (3.27)$$

Substituting (3.27) into (3.26), the log-likelihood function for exponential mean $\mu_i = \exp(\mathbf{x}_i'\boldsymbol{\beta})$ is therefore

$$\ln \mathsf{L}(\alpha,\beta) = \sum_{i=1}^{n}\left\{\left(\sum\nolimits_{j=0}^{y_i-1}\ln(j+\alpha^{-1})\right) - \ln y_i! \right. \quad (3.28)$$
$$\left. -(y_i+\alpha^{-1})\ln(1+\alpha\exp(\mathbf{x}_i'\boldsymbol{\beta})) + y_i\ln\alpha + y_i\mathbf{x}_i'\boldsymbol{\beta}\right\}.$$

The NB2 MLE $(\hat{\boldsymbol{\beta}}_{\mathsf{NB2}}, \hat{\alpha}_{\mathsf{NB2}})$ is the solution to the first-order conditions

$$\sum_{i=1}^{n}\frac{y_i-\mu_i}{1+\alpha\mu_i}\mathbf{x}_i = \mathbf{0}$$
$$\sum_{i=1}^{n}\left\{\frac{1}{\alpha^2}\left(\ln(1+\alpha\mu_i) - \sum\nolimits_{j=0}^{y_i-1}\frac{1}{(j+\alpha^{-1})}\right) + \frac{y_i-\mu_i}{\alpha(1+\alpha\mu_i)}\right\} = 0. \quad (3.29)$$

Given correct specification of the distribution

$$\begin{bmatrix}\hat{\boldsymbol{\beta}}_{\mathsf{NB2}}\\ \hat{\alpha}_{\mathsf{NB2}}\end{bmatrix} \overset{a}{\sim} \mathsf{N}\left(\begin{bmatrix}\boldsymbol{\beta}\\ \alpha\end{bmatrix}, \begin{bmatrix}\mathsf{V}_{\mathsf{ML}}[\hat{\boldsymbol{\beta}}_{\mathsf{NB2}}] & \mathsf{Cov}_{\mathsf{ML}}[\hat{\boldsymbol{\beta}}_{\mathsf{NB2}}, \hat{\alpha}_{\mathsf{NB2}}]\\ \mathsf{Cov}_{\mathsf{ML}}[\hat{\boldsymbol{\beta}}_{\mathsf{NB2}}, \hat{\alpha}_{\mathsf{NB2}}] & \mathsf{V}_{\mathsf{ML}}[\hat{\alpha}_{\mathsf{NB2}}]\end{bmatrix}\right) \quad (3.30)$$

where

$$\mathsf{V}_{\mathsf{ML}}[\hat{\boldsymbol{\beta}}_{\mathsf{NB2}}] = \left(\sum_{i=1}^{n} \frac{\mu_i}{1+\alpha\mu_i}\mathbf{x}_i\mathbf{x}_i'\right)^{-1}, \tag{3.31}$$

$$\mathsf{V}_{\mathsf{ML}}[\hat{\alpha}_{\mathsf{NB2}}] = \left(\sum_{i=1}^{n} \frac{1}{\alpha^4}\left(\ln(1+\alpha\mu_i) - \sum_{j=0}^{y_i-1}\frac{1}{(j+\alpha^{-1})}\right)^2 + \frac{\mu_i}{\alpha^2(1+\alpha\mu_i)}\right)^{-1}, \tag{3.32}$$

and

$$\mathsf{Cov}_{\mathsf{ML}}[\hat{\boldsymbol{\beta}}_{\mathsf{NB2}}, \hat{\alpha}_{\mathsf{NB2}}] = \mathbf{0}. \tag{3.33}$$

This result is obtained by noting that the information matrix is block-diagonal, because differentiating the first term in (3.29) with respect to α yields

$$\mathsf{E}\left[\frac{\partial^2 \ln \mathsf{L}}{\partial \boldsymbol{\beta}\partial\alpha}\right] = \mathsf{E}\left[-\sum_{i=1}^{n}\frac{y_i-\mu_i}{(1+\alpha\mu_i)^2}\mu_i\mathbf{x}_i\mathbf{x}_i'\right] = \mathbf{0}, \tag{3.34}$$

as $\mathsf{E}[y_i|\mathbf{x}_i] = \mu_i$. This simplifies analysis because then the general result in Chapter 2.3.2 for the maximum likelihood variance matrix specializes to

$$\begin{bmatrix} \mathsf{E}\left[\frac{\partial^2 \ln \mathsf{L}}{\partial\boldsymbol{\beta}\partial\boldsymbol{\beta}'}\right] & \mathbf{0} \\ \mathbf{0} & \mathsf{E}\left[\frac{\partial^2 \ln \mathsf{L}}{\partial\alpha^2}\right]\end{bmatrix}^{-1} = \begin{bmatrix} \left[\mathsf{E}\left[\frac{\partial^2 \ln \mathsf{L}}{\partial\boldsymbol{\beta}\partial\boldsymbol{\beta}'}\right]\right]^{-1} & \mathbf{0} \\ \mathbf{0} & \left[\mathsf{E}\left[\frac{\partial^2 \ln \mathsf{L}}{\partial\alpha^2}\right]\right]^{-1}\end{bmatrix}.$$

Many packages offer this negative binomial model as a standard option. Alternatively, one can use a maximum likelihood routine with user-provided log-likelihood function and possibly derivatives. In this case, potential computational problems can be avoided by using the form of the log-likelihood function given in (3.28) or by using the log-gamma function rather than first calculating the gamma function and then taking the natural logarithm. If instead one tries to directly compute the gamma functions, numerical calculation of $\Gamma(z)$ with large values of z may cause an overflow, for example, if $z > 169$ in the matrix program GAUSS. The restriction to α positive can be ensured by instead estimating $\alpha^* = \ln\alpha$ and then obtaining $\alpha = \exp(\alpha^*)$.

3.3.2 NB2 Model and QGPMLE

The NB2 density can be reexpressed as

$$f(y|\mu,\alpha) = \exp\left\{-\alpha^{-1}\ln(1+\alpha\mu) + \ln\left(\frac{\Gamma(y+\alpha^{-1})}{\Gamma(y+1)\Gamma(\alpha^{-1})}\right) + y\ln\left(\frac{\alpha\mu}{1+\alpha\mu}\right)\right\}. \tag{3.35}$$

If α is known this is an LEF density defined in Chapter 2.4.2 with $a(\mu) = -\alpha^{-1}\ln(1+\alpha\mu)$ and $c(\mu) = \ln(\alpha\mu/(1+\alpha\mu))$. Because $a'(\mu) = -1/(1+\alpha\mu)$ and $c'(\mu) = 1/\mu(1+\alpha\mu)$ it follows that $\mathsf{E}[y] = \mu$ and $\mathsf{V}[y] = \mu + \alpha\mu^2$, which is the NB2 variance function.

If α is unknown, this is an LEFN density defined in Chapter 2.4.3. Given a consistent estimator $\widetilde{\alpha}$ of α, such as (3.20), the QGPMLE $\hat{\boldsymbol{\beta}}_{\mathsf{QGPML}}$ maximizes

$$\ln \mathsf{L}_{\mathsf{LEFN}} = \sum_{i=1}^{n} \left\{ -\widetilde{\alpha}^{-1} \ln(1+\widetilde{\alpha}\mu_i) + y_i \ln\left(\frac{\widetilde{\alpha}\mu_i}{1+\widetilde{\alpha}\mu_i} \right) + b(y_i, \widetilde{\alpha}) \right\}. \tag{3.36}$$

For exponential conditional mean (3.4), the first-order conditions are

$$\sum_{i=1}^{n} \frac{y_i - \mu_i}{1+\widetilde{\alpha}\mu_i} \mathbf{x}_i = \mathbf{0}. \tag{3.37}$$

Using results from Chapter 2.4.3, or noting that (3.37) is a special case of the estimating equation (3.24), $\hat{\boldsymbol{\beta}}_{\mathsf{QGPML}}$ is asymptotically normal with mean $\mathbf{0}$ and variance

$$\mathsf{V}[\hat{\boldsymbol{\beta}}_{\mathsf{QGPML}}] = \left(\sum_{i=1}^{n} \frac{\mu_i}{1+\alpha\mu_i} \mathbf{x}_i \mathbf{x}_i' \right)^{-1}. \tag{3.38}$$

This equals $\mathsf{V}_{\mathsf{ML}}[\hat{\boldsymbol{\beta}}_{\mathsf{NB2}}]$ defined in (3.31), so that $\hat{\boldsymbol{\beta}}_{\mathsf{QGPML}}$ is fully efficient for $\boldsymbol{\beta}$ if the density is NB2, although $\widetilde{\alpha}$ is not necessarily fully efficient for α.

The NB2 density with α known is a member of the GLM class. So we could alternatively call $\hat{\boldsymbol{\beta}}_{\mathsf{QGPML}}$ the NB2 GLM estimator.

The GLM literature sometimes uses a specification of the conditional mean other than exponential. Given (3.31) with α known, the canonical link function (see Chapter 2.4.4) is $\eta = \ln(\frac{\alpha\mu}{1+\alpha\mu})$, leading to conditional mean function

$$\mathsf{E}[y_i|\mathbf{x}_i] = \mu_{\mathsf{GLM},i} = \frac{\exp(\mathbf{x}_i'\boldsymbol{\beta})}{\widetilde{\alpha}(1-\exp(\mathbf{x}_i'\boldsymbol{\beta}))}, \tag{3.39}$$

This functional form has the advantage that maximization of (3.36) when $\mu_i = \mu_{\mathsf{GLM},i}$ leads to first-order conditions $\sum_{i=1}^{n}(y_i - \mu_{\mathsf{GLM},i})\mathbf{x}_i = \mathbf{0}$ that can be interpreted as the residual is orthogonal to the regressors. But the parameters $\boldsymbol{\beta}$ in specification (3.39) are not easily interpreted, with $\partial\mu_{\mathsf{GLM},i}/\partial\boldsymbol{\beta} = (\mu_{\mathsf{GLM},i} + \widetilde{\alpha}\mu_{\mathsf{GLM},i}^2)\mathbf{x}_i$. And the computational method needs to ensure that $\mathbf{x}_i'\boldsymbol{\beta} < \mathbf{0}$ so $\mu_{\mathsf{GLM},i} > 0$. This limits the usefulness of this parameterization.

3.3.3 NB1 Model and MLE

Cameron and Trivedi (1986) considered a more general class of negative binomial models with mean μ_i and variance function $\mu_i + \alpha\mu_i^p$. The NB2 model, with $p = 2$, is the standard formulation of the negative binomial model. Models with other values of p have the same density as (3.26), except that α^{-1} is replaced everywhere by $\alpha^{-1}\mu^{2-p}$.

The NB1 model, which sets $p = 1$, is also of interest because it has the same variance function, $(1+\alpha)\mu_i = \phi\mu_i$, as that used in the GLM approach. The NB1 log-likelihood function is

$$\ln \mathsf{L}(\alpha, \beta) = \sum_{i=1}^{n} \left\{ \left(\sum_{j=0}^{y_i-1} \ln(j + \alpha^{-1} \exp(\mathbf{x}_i'\boldsymbol{\beta}) \right) - \ln y_i! \right.$$
$$\left. - (y_i + \alpha^{-1} \exp(\mathbf{x}_i'\boldsymbol{\beta})) \ln(1+\alpha) + y_i \ln \alpha \right\}.$$

The NB1 MLE solves the associated first-order conditions

$$\sum_{i=1}^{n} \left\{ \left(\sum_{j=0}^{y_i-1} \frac{\alpha^{-1}\mu_i}{(j+\alpha^{-1}\mu_i)} \right) \mathbf{x}_i + \alpha^{-1}\mu_i \mathbf{x}_i \right\} = \mathbf{0}$$

$$\sum_{i=1}^{n} \frac{1}{\alpha^2} \left\{ - \left(\sum_{j=0}^{y_i-1} \frac{\mu_i}{(j+\alpha^{-1})} \right) - \alpha^{-2}\mu_i \ln(1+\alpha) - \frac{\alpha}{1+\alpha} + y_i\alpha \right\} = 0.$$

Estimation based on the first two moments of the NB1 density yields the Poisson GLM estimator, which we also call the NB1 GLM estimator.

3.3.4 Discussion

One can clearly consider negative binomial models other than NB1 and NB2. The generalized event count model, presented in Chapter 4.11.1, includes the negative binomial with mean μ_i and variance function $\mu_i + \alpha\mu_i^p$, where p is estimated rather than set to the value 1 or 2. And we could generalize α to $\alpha_i = \exp(\mathbf{z}_i'\boldsymbol{\alpha})$.

The NB2 model has a number of special features not shared by other models in this class, including block diagonality of the information matrix, being a member of the LEF if α is known, robustness to distributional misspecification, and nesting as a special case of the geometric distribution when $\alpha = 1$.

The NB2 MLE is robust to distributional misspecification, due to membership in the LEF for specified α. Thus, provided the conditional mean is correctly specified, the NB2 MLE is consistent for $\boldsymbol{\beta}$. This can be seen by directly inspecting the first-order conditions for $\boldsymbol{\beta}$ given in (3.29), whose left-hand side has an expected value of zero if the mean is correctly specified. This follows because $\mathsf{E}[y_i - \mu_i|\mathbf{x}_i] = 0$.

The associated maximum likelihood standard errors of the NB2 MLE are, however, generally inconsistent if there is any distributional misspecification. First, they are inconsistent if (3.16) does not hold, so the variance function is incorrectly specified. Second, even if the variance function is correctly specified, in which case it can be shown that $\mathsf{V}[\hat{\boldsymbol{\beta}}_{\mathsf{NB2}}]$ is again that given in (3.31), failure of the negative binomial assumption leads to evaluation of (3.31) at an inconsistent estimate of α. From (3.29) consistency of $\hat{\alpha}_{\text{NB2}}$ requires both $\mathsf{E}[y_i - \mu_i|\mathbf{x}_i] = 0$ and, in expectation, $\sum_{i=1}^{n}\{\ln(1+\alpha\mu_i) - \sum_{j=0}^{y_i-1} 1/(j+\alpha^{-1})\} = 0$. This last condition holds only if in fact y is negative binomial.

For this reason it is best to use robust standard errors for the NB2 MLE. Essentially we treat the NB2 MLE as a quasi-MLE and use the variance matrix estimate given in section 3.2.4.

Negative binomial models other than NB2 are not at all robust to distributional misspecification. Then, consistency of the MLE for $\boldsymbol{\beta}$ requires that the data are negative binomial. Correct specification of the mean and variance is not enough.

Another variation in negative binomial models is to allow the overdispersion parameter to depend on regressors. For example, for the NB2 model α can be replaced by $\alpha_i = \exp(\mathbf{z}_i'\boldsymbol{\gamma})$, which is ensured to be positive. Then the first-order conditions with respect to $\boldsymbol{\gamma}$ are similar to (3.29), with the term in the sum in the second line of (3.29) multiplied by $\partial\alpha_i/\partial\boldsymbol{\gamma}_i$. In this book we present a range of models with the dispersion parameter entering in different ways, but for simplicity we generally restrict the dispersion parameter to be constant. Jorgensen (1997) gives a general treatment of models with the dispersion parameter that depends on regressors.

It is not immediately clear that the Poisson is a special case of the negative binomial. To see this for NB2, we use the gamma recursion and let $a = \alpha^{-1}$. The NB2 density (3.26) is

$$\begin{aligned} f(y) &= \left(\prod_{j=0}^{y-1}(j+a)\right)\frac{1}{y!}\left(\frac{a}{a+\mu}\right)^a\left(\frac{1}{a+\mu}\right)^y\mu^y \\ &= \left(\prod_{j=0}^{y-1}\frac{j+a}{a+\mu}\right)\left(\frac{a}{a+\mu}\right)^a\mu^y\frac{1}{y!} \\ &= \left(\prod_{j=0}^{y-1}\frac{\frac{j}{a}+1}{1+\frac{\mu}{a}}\right)\left(\frac{1}{1+\frac{\mu}{a}}\right)^a\mu^y\frac{1}{y!} \\ &\to 1\; e^{-\mu}\;\mu^y\frac{1}{y!} \qquad \text{as } a\to\infty. \end{aligned}$$

The second equality uses $(1/(a+\mu))^y = \prod_{j=0}^{y-1} 1/(a+\mu)$. The third equality involves some rearrangement. The last equality uses $\lim_{a\to\infty}(1+x/a)^a = e^x$. The final expression is the Poisson density. So Poisson is the special case of NB2 where $\alpha = 0$.

If a negative binomial model or other parametric model correctly specifies the distribution, it has the advantage of enabling predictions about individual probabilities. This allows analysis of tail behavior, for example, rather than only the conditional mean.

Table 3.4. *Doctor visits: NB2 and NB1 model estimators and standard errors*

	Estimators				Standard errors			
	NB2		NB1		NB2		NB1	
Variable	MLE	QGP	MLE	GLM	MLE	QGP	MLE	GLM
ONE	−2.190	−2.203	−2.202	−2.224	0.249	0.252	0.228	0.254
SEX	0.217	0.188	0.164	0.157	0.074	0.076	0.071	0.079
AGE	−0.216	0.511	0.279	1.056	1.367	1.362	1.208	1.364
AGESQ	0.609	−0.227	0.021	−0.849	1.473	1.459	1.315	1.460
INCOME	−0.142	−0.174	−0.135	−0.205	0.122	0.126	0.110	0.129
LEVYPLUS	0.118	0.113	0.212	0.123	0.091	0.094	0.084	0.095
FREEPOOR	−0.497	−0.461	−0.538	−0.440	0.254	0.276	0.254	0.290
FREEREPA	0.145	0.100	0.208	0.080	0.121	0.124	0.113	0.126
ILLNESS	0.214	0.198	0.196	0.187	0.024	0.024	0.022	0.024
ACTDAYS	0.144	0.132	0.112	0.127	0.009	0.008	0.007	0.008
HSCORE	0.038	0.034	0.036	0.030	0.014	0.014	0.013	0.013
CHCOND1	0.099	0.104	0.132	0.114	0.083	0.087	0.080	0.091
CHCOND2	0.190	0.159	0.174	0.141	0.117	0.120	0.107	0.123
α	1.077	0.286	0.455	0.328	0.117		0.057	
−ln L	3198.7	3226.8						

Note: NB1 variance function is $\omega = \phi\mu = (1+\alpha)\mu$; NB2 is $\omega = \mu + \alpha\mu^2$.

3.3.5 Example: Doctor Visits (Continued)

Table 3.4 presents estimates and standard errors using the most common generalizations of Poisson regression for count data. In all cases we present robust sandwich standard errors. If these standard errors are needed for the NB1 MLE, however, then there is a more fundamental problem of estimator inconsistency, whereas the other estimators retain consistency provided the conditional mean is correctly specified.

The various estimators of the regression coefficients $\boldsymbol{\beta}$ tell a consistent story, although they differ for some variables by more than 20% across the different estimation methods. The signs of *AGE* and *AGESQ* vary across estimators, although all can be shown to imply that the number of doctor visits increases with age for ages in the range of 20–60 years.

The robust sandwich standard errors reveal some efficiency gain to using estimators other than Poisson (equivalent to NB1 GLM in the final column). The efficiency gain is greatest for NB1 MLE with standard errors that are 10%–15% lower than Poisson. From results not given in Table 3.4, the preferred robust standard errors are larger than the default nonrobust standard errors, which assume correct specification of the variance function. The difference is 5% for the ML estimators and a substantial 25% for the QGPML estimators.

The first column of Table 3.4 gives the NB2 MLE for overdispersed data, whereas the second column gives the NB2 QGPMLE. The difference between the two is due to different estimates of α: $\hat{\alpha} = 1.077$ for the MLE and $\hat{\alpha} = 0.286$

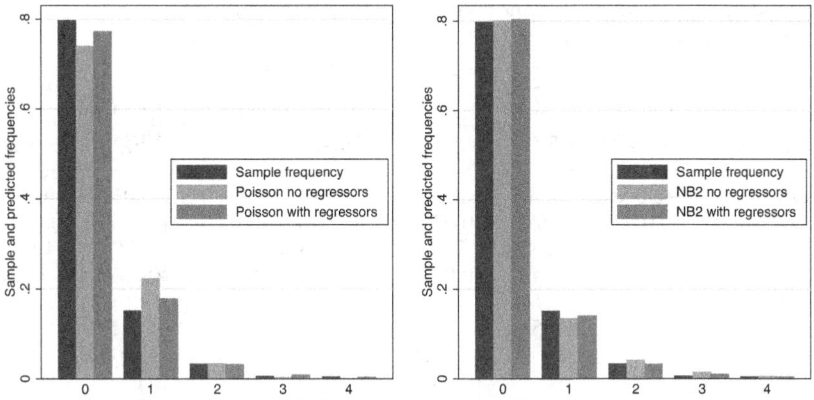

Figure 3.1. Doctor visits: Poisson and NB2 predicted probabilities.

for the QGPMLE estimate from first estimating $\boldsymbol{\beta}$ by Poisson regression and then using (3.20). For NB2 QGPMLE or, equivalently, NB2 GLM, there are many different ways to estimate α. The GLM literature suggests choosing α so that the Pearson statistic or deviance statistic, see Chapter 5.3, equals $n - k$. Using the Pearson statistic this leads to $\hat{\alpha} = 1.080$, which is considerably different from the value 0.286 used in Table 3.4; using the deviance statistic yields $\hat{\alpha} = 0$ because the deviance is less than $n - k$ for all $\hat{\alpha} \geq 0$.

The third column gives the NB1 MLE for overdispersed data, whereas the fourth column gives the NB1 QGPMLE. For the NB1 variance function, the maximum likelihood and moment-based estimates $\hat{\alpha} = 1 - \hat{\phi}$ of, respectively, 0.455 and 0.328 are quite similar, unlike for the NB2 variance function. The moment-based estimates for NB1 use (3.18). Another moment-based method, that given in Cameron and Trivedi (1986, p. 46), yields $\hat{\alpha} = 0.218$. Such differences in estimates of α have received little attention in the literature, in part because interest lies in estimation of $\boldsymbol{\beta}$, with α a nuisance parameter. But these estimates can lead to different results.

Within the maximum likelihood framework, the NB2 model is preferred here to NB1 because it has considerably higher log-likelihood, $-3198.7 > -3226.6$, with the same number of parameters. In practice, most studies use either NB2 MLE or Poisson QMLE (equivalent to NB1 GLM).

Fully parametric models permit computation of predicted probabilities rather than just the predicted conditional mean. Let $p_{ij} = \Pr[y_i = j|\mathbf{x}_i, \boldsymbol{\theta}]$ and $\hat{p}_{ij} = \Pr[y_i = j|\mathbf{x}_i, \hat{\boldsymbol{\theta}}]$. For example, for the Poisson model $p_{ij} = e^{-\exp(\mathbf{x}_i'\boldsymbol{\beta})} \exp(j\mathbf{x}_i'\boldsymbol{\beta})/j!$ A useful parametric model check is to compare the average fitted probabilities $\hat{p}_j = \frac{1}{n}\sum_{i=1}^{n} \hat{p}_{ij}$ for $j = 0, 1, 2, \ldots$ with the corresponding sample frequencies.

Figure 3.1 compares sample frequencies to predicted probabilities from the Poisson model (using (3.1)) and negative binomial (NB2) model (using (3.26)), without and with regressors. From the first panel, the intercept-only

Table 3.5. *Simulation: Estimators under alternative dgps*

	Estimators and standard errors						
	Data generated from NB2			Data generated from NB1			
Variable	Poiss	NB1	NB2	Poiss	NB1	NB2	NB2def
ONE	−0.008	0.497	−0.009	0.019	0.014	0.017	0.017
	(0.037)	(0.028)	(0.034)	(0.026)	(0.024)	(0.027)	(0.022)
x	2.021	1.218	2.022	1.978	1.985	1.981	1.981
	(0.062)	(0.043)	(0.057)	(0.037)	(0.034)	(0.039)	(0.035)
α for NB1		1.941			.	0.141	
		(0.051)			(1.62)	(1.15)	
α for NB2			7.224			0.601	0.601
			(0.181)			(0.015)	(0.016)
−ln L	−37,654	−21,449	−21,381	−24,982	−21,399	−21,666	−21,666

Note: Simulation with intercept 0, slope 2, $\alpha = 2$, and $n = 10{,}000$ observations. Poiss is Poisson MLE, NB1 is NB1 MLE, and NB2 is NB2 MLE. Reported standard errors are all robust sandwich, except for the last column, which uses the MLE default standard errors.

Poisson clearly underpredicts zero counts and overpredicts ones. The inclusion of regressors reduces but not does not eliminate this difference. This is the "excess zeros" phenomenon that often arises when the Poisson is applied to overdispersed data. From the second panel, the negative binomial does not have this weakness, at least for these data, and the inclusion of regressors leads to better prediction of counts greater than zero. Such comparisons also form the basis of a formal goodness-of-fit test; see Chapter 5.3.4.

3.3.6 Simulation: Estimator properties given known dgp

We consider estimation given a known data generating process with 10,000 observations. The purpose of the exercise is to demonstrate the theoretical result that Poisson QMLE and NB2 MLE estimators are consistent provided only that the conditional mean is correctly specified, whereas the NB1 MLE additionally requires correct specification of the conditional variance (and even then it would additionally require correct specification of the distribution).

In the first three columns of Table 3.5, the data are generated from an NB2 model with mean $\mu_i = \exp(0 + 2x_i)$ and variance $\mu_i + 2\mu_i^2$, where $x_i \sim$ U[0, 1]. The counts have mean 3.2 and variance 35.2, and they range from 0 to 102. From the second column, the NB1 MLE is inconsistent, with slope coefficient 1.218 rather than 2.0. Comparing the first and second columns, the NB2 MLE is somewhat more efficient than Poisson, with a standard error for the slope coefficient of 0.057 compared to 0.062.

In the final four columns of Table 3.5, the data are generated from an NB1 model with mean $\mu_i = \exp(0 + 2x_i)$ and variance $\mu_i + 2\mu_i$, where $x_i \sim$ U[0, 1]. The counts have mean 3.2 and variance 12.4, and they range from

0 to 27. All three estimators, including the NB2 MLE, are consistent for the regression parameters. The NB1 MLE is most efficient as expected, with a standard error for the slope coefficient of 0.034 compared to 0.037 for Poisson and 0.039 for NB2 MLE. The final column of Table 3.5 shows that the default standard errors for the NB2 MLE are smaller than the robust standard errors, being 0.035 versus 0.039 for the slope coefficient. This demonstrates that, although the NB2 MLE remains consistent if the variance function is misspecified, robust rather than default standard errors should then be used. This mirrors the result for the Poisson MLE.

3.4 OVERDISPERSION TESTS

Failure of the Poisson assumption of equidispersion has similar qualitative consequences to failure of the assumption of homoskedasticity in the linear regression model, but the magnitude of the effect on reported standard errors and t statistics can be much larger. To see this, suppose $\omega_i = 4\mu_i$. Then by equation (3.17) the variance matrix of the Poisson QML estimator is four times the reported maximum likelihood variance matrix using (3.6). As a result the default MH Poisson maximum likelihood t statistics need to be deflated by a factor of two. Overdispersion as large as $\omega_i = 4\mu_i$ arises, for example, in the recreational trips data (Chapter 1.3) and in health services data on length of hospitalization.

The easiest way to deal with overdispersion is to view the Poisson MLE as a quasi-MLE and to base inference on the robust standard errors given in section 3.2.2. The tests for overdispersion presented in this section should be employed if instead one wants to take a full parametric approach and view the estimator as an MLE and base inference on default ML standard errors.

3.4.1 Overdispersion Tests

Data are overdispersed if the conditional variance exceeds the conditional mean. An indication of the magnitude of overdispersion or underdispersion can be obtained simply by comparing the sample mean and variance of the dependent count variable. Subsequent Poisson regression somewhat decreases the conditional variance of the dependent variable. The average of the conditional mean will be unchanged, however, because the average of the fitted means equals the sample mean. This follows because Poisson residuals sum to zero if a constant term is included. If the sample variance is less than the sample mean, the data necessarily are even more underdispersed once regressors are included. If the sample variance is more than twice the sample mean, then data are likely to remain overdispersed after the inclusion of regressors. This is particularly so for cross-section data, for which regressors usually explain less than half the variation.

The standard models for overdispersion have already been presented. These are the NB1 variance function $\omega_i = \mu_i + \alpha\mu_i$ as in (3.14) or NB2 variance function $\omega_i = \mu_i + \alpha\mu_i^2$ as in (3.16). If one takes the partially parametric mean-variance approach (GLM), it is easier to use the NB1 variance function, which leads to the Poisson QMLE. If one takes the fully parametric negative binomial approach, it is customary to use NB2.

A sound practice is to estimate both Poisson and negative binomial models if software is readily available. The Poisson is the special case of the negative binomial with $\alpha = 0$. The null hypothesis $H_0 : \alpha = 0$ can be tested against the alternative $\alpha > 0$ using the hypothesis test methods presented in Chapter 2.6.1. An LR test uses -2 times the difference in the fitted log-likelihoods of the two models. Alternatively a Wald test can be performed, using the reported t statistic for the estimated α in the negative binomial model.

The distribution of these statistics is nonstandard, due to the restriction that α cannot be less than zero. This complication is usually not commented on, a notable exception being Lawless (1987). One way to see problems that arise is to consider constructing a Monte Carlo experiment to obtain the distribution of the test statistic. We would draw samples from the Poisson, because this is the model under the null hypothesis of no overdispersion. Roughly half of the time the data will be underdispersed. Then the negative binomial MLE for α is zero, the negative binomial parameter estimates equal the Poisson estimates, and the LR test statistic takes a value of 0. Clearly this test statistic is not χ^2 distributed, because half its mass is at zero. Similar problems arise for the Wald test statistic. A general treatment of hypothesis testing at boundary values is given by Moran (1971). The asymptotic distribution of the LR test statistic has probability mass of one-half at zero and a half-$\chi^2(1)$ distribution above zero. This means that if testing at level δ, where $\delta > 0.5$, one rejects H_0 if the test statistic exceeds $\chi^2_{1-2\delta}(1)$ rather than $\chi^2_{1-\delta}(1)$. The Wald test is usually implemented as a t-test statistic, which here has mass of one-half at zero and a normal distribution for values above zero. In this case one continues to use the usual one-sided test critical value of $z_{1-\delta}$. Essentially the only adjustment that needs to be made is an obvious one to the χ^2 critical values, which arises due to performing a one-sided rather than two-sided test.

If a package program for negative binomial regression is unavailable, one can still test for overdispersion by estimating the Poisson model, constructing fitted values $\hat{\mu}_i = \exp(\mathbf{x}_i'\hat{\boldsymbol{\beta}})$, and performing the auxiliary OLS regression (without constant)

$$\frac{(y_i - \hat{\mu}_i)^2 - y_i}{\hat{\mu}_i} = \alpha\hat{\mu}_i + u_i, \tag{3.40}$$

where u_i is an error term. The reported t statistic for α is asymptotically normal under the null hypothesis of no overdispersion against the alternative of overdispersion of the NB2 form. To test overdispersion of the NB1 form, replace (3.40) with

$$\frac{(y_i - \hat{\mu}_i)^2 - y_i}{\hat{\mu}_i} = \alpha + u_i. \tag{3.41}$$

These auxiliary regression tests coincide with the score or LM test for Poisson against negative binomial. They additionally test for underdispersion, can be given a more general motivation based on using only the specified mean and variance (see Chapter 5), and can be made robust to non-Poisson third and fourth moments of y under the null by using robust sandwich standard errors for the test of $\alpha = 0$.

Beyond rejection or nonrejection of the null hypothesis of equidispersion, interest may lie in interpreting the magnitude of departures from equidispersion. Estimates of α for the NB1 variance function $(1 + \alpha)\mu_i$ are easily interpreted, with underdispersion when $\alpha < 0$, modest overdispersion when, say, $0 < \alpha < 1$, and considerable overdispersion if, say, $\alpha > 1$. For the NB2 variance function $\mu_i + \alpha\mu_i^2$ underdispersion also occurs if $\alpha < 0$. The NB2 variance function can be inappropriate for underdispersed data, because the estimated variance is negative for observations with $\alpha < -1/\mu_i$. For interpretation of the magnitude of overdispersion, it is helpful to rewrite the NB2 variance as $(1 + \alpha\mu_i)\ \mu_i$. Then values of considerable overdispersion arise when, say, $\alpha\mu_i > 1$, because then the multiplier $1 + \alpha\mu_i > 2$. Thus a value of α equal to 0.5 would indicate modest overdispersion if the dependent variable took mostly values of 0, 1, and 2, but great overdispersion if counts of 10 or more were often observed.

Most often count data are overdispersed rather than underdispersed, and tests for departures from equidispersion are usually called overdispersion tests. Note that the negative binomial model can only accommodate overdispersion.

3.4.2 Example: Doctor Visits (Continued)

From Table 3.1 the data before inclusion of regressors are overdispersed, with a variance-mean ratio of $0.637/0.302 = 2.11$. The only question is whether this overdispersion disappears on inclusion of regressors.

We first consider tests of Poisson against NB2 at a significance level of 1%. Given the results in Tables 3.3 and 3.4, the LR test statistic is $2(3355.5 - 3198.7) = 313.6$, which exceeds the 1% critical value of $\chi^2_{.98}(1) = 5.41$. The Wald test statistic from Table 3.4 is $1.077/0.117 = 9.21$, which exceeds the 1% critical value of $z_{.99} = 2.33$. Finally, the LM test statistic computed using the auxiliary regression (3.40) is 7.50 and exceeds the 1% critical value of $z_{.99} = 2.33$. Therefore all three tests strongly reject the null hypothesis of Poisson, indicating the presence of overdispersion. Note that these tests are asymptotically equivalent, yet there is quite a difference in their realized values of $\sqrt{313.6} = 17.69$, 9.21, and 7.50.

Similar test statistics for Poisson against NB1 are $\mathsf{T}_{\mathsf{LR}} = 2 \times (3355.5 - 3226.8) = 257.4$, $\mathsf{T}_{\mathsf{W}} = 0.455/0.057 = 7.98$, and $\mathsf{T}_{\mathsf{LM}} = 6.54$ on running the auxiliary regression (3.41). These results again strongly reject equidispersion.

Clearly some control is necessary for overdispersion. Possibilities include the Poisson QMLE with RS standard errors, see Table 3.3, which corrects the standard errors for overdispersion, and the various estimators presented in Table 3.4, most notably NB2 MLE.

3.5 USE OF REGRESSION RESULTS

The techniques presented to date allow estimation of count data models and performance of tests of statistical significance of regression coefficients. We have focused on tests on individual coefficients. Tests of joint hypotheses such as overall significance can be performed using the Wald, LM, and LR tests presented in Chapter 2.6. Confidence intervals for functions of parameters can be formed using the delta method in Chapter 2.6.2.

We now turn to interpretation of regression coefficients and prediction of the dependent variable.

3.5.1 Marginal Effects and Interpretation of Coefficients

An important issue is interpretation of regression coefficients. For example, what does $\hat{\beta}_j = 0.2$ mean? This is straightforward in the linear regression model $\mathsf{E}[y|\mathbf{x}] = \mathbf{x}'\boldsymbol{\beta}$. Then $\partial\mathsf{E}[y|\mathbf{x}]/\partial x_j = \beta_j$, so $\hat{\beta}_j = 0.2$ means that a one-unit change in the jth regressor increases the conditional mean by 0.2 units.

We consider the exponential conditional mean

$$\mathsf{E}[y|\mathbf{x}] = \exp(\mathbf{x}'\boldsymbol{\beta}), \tag{3.42}$$

where for exposition the subscript i is dropped. We consider the effect of changes in x_j, the jth regressor.

Marginal Effects

Differentiating with respect to a continuous variate x_j yields the *marginal effect* (ME)

$$\mathsf{ME}_j = \frac{\partial\mathsf{E}[y|\mathbf{x}]}{\partial x_j} = \beta_j \exp(\mathbf{x}'\boldsymbol{\beta}). \tag{3.43}$$

For example, if $\hat{\beta}_j = 0.2$ and $\exp(\mathbf{x}_i'\hat{\boldsymbol{\beta}}) = 2.5$, then a one-unit change in the jth regressor increases the expectation of y by 0.5 units. This is interpreted as the marginal effect of a unit change in x_j. Calculated values differ across individuals, however, due to different values of $\mathbf{x}$. This variation makes interpretation more difficult.

One procedure is to estimate the typical value by first aggregating over all individuals and then calculating the average response. The *average marginal effect* (AME) is defined as

$$\mathsf{AME}_j = \frac{1}{n}\sum_{i=1}^{n} \frac{\partial\mathsf{E}[y_i|\mathbf{x}_i]}{\partial x_{ij}} = \frac{1}{n}\sum_{i=1}^{n} \beta_j \exp(\mathbf{x}_i'\boldsymbol{\beta}). \tag{3.44}$$

In the special case that one uses the Poisson MLE or QMLE, and the regression includes an intercept term, this expression simplifies to

$$\mathsf{AME}_j = \beta_j \bar{y} \tag{3.45}$$

because the first-order conditions (3.4) imply $\sum_i \exp(\mathbf{x}_i'\boldsymbol{\beta}) = \sum_i y_i$. The preceding gives the sample-average ME. For a population-average ME one uses population weights in computing (3.44).

A second procedure is to calculate the response for the individual with average characteristics. The *marginal effect at the mean* (MEM) is defined as

$$\mathsf{MEM}_j = \left.\frac{\partial \mathsf{E}[y|\mathbf{x}]}{\partial x_j}\right|_{\bar{\mathbf{x}}} = \beta_j \exp(\bar{\mathbf{x}}'\boldsymbol{\beta}). \tag{3.46}$$

Because $\exp(\cdot)$ is a convex function, by Jensen's inequality the average of $\exp(\cdot)$ evaluated at several points exceeds $\exp(\cdot)$ evaluated at the average of the same points. So (3.46) gives responses smaller than (3.44). Due to the need for less calculation, it is common in nonlinear regression to report responses at the sample mean of regressors. It is conceptually better, however, to report the average response (3.44) over all individuals. And it is actually easier to do so in the special case of Poisson with intercept included, because (3.45) can be used.

A third procedure is to calculate (3.43) for select values of $\mathbf{x}$ of particular interest. The *marginal effect at a representative value* (MER) is defined as

$$\mathsf{MER}_j = \left.\frac{\partial \mathsf{E}[y|\mathbf{x}]}{\partial x_j}\right|_{\mathbf{x}^*} = \beta_j \exp(\mathbf{x}^{*\prime}\boldsymbol{\beta}). \tag{3.47}$$

This is perhaps the best method. The values of $\mathbf{x}^*$ of particular interest vary from application to application.

These marginal effects are interpreted as the effect of a one-unit change in x_j, which is not free of the units used to measure x_j. One method is to scale by the sample mean of x_j, so use $\hat{\beta}_j\bar{x}_j$, which given (3.43) is a measure of the elasticity of $\mathsf{E}[y|\mathbf{x}]$ with respect to x_j. An alternative method is to consider *semi-standardized coefficients* that give the effect of a one-standard-deviation change in x_j, so use $\hat{\beta}_j s_j$, where s_j is the standard deviation of x_j. Such adjustments, of course, need to be made even for the linear regression model.

Treatment Effects

An additional complication arises from the presence of binary regressors. Often these regressors are a *treatment variable*, a regressor equal to 1 if the individual is treated and 0 if not. An example is having health insurance versus not having it – the *LEVYPLUS* regressor in the doctor visits example. Of interest is the treatment effect, the marginal effect on the conditional mean as the indicator variable, say d, changes from 0 to 1.

Let $\mathbf{x} = [d\ \mathbf{z}]$, where $\mathbf{z}$ denotes all regressors other than the binary regressor d. Then the *treatment effect* is defined by

$$\begin{aligned}\mathsf{TE} &= \mathsf{E}[y|d=1,\mathbf{z}] - \mathsf{E}[y|d=0,\mathbf{z}] \\ &= \exp(\beta_d + \mathbf{z}'\boldsymbol{\beta}_z) - \exp(\mathbf{z}'\boldsymbol{\beta}_z).\end{aligned} \tag{3.48}$$

This method of computation is sometimes called the finite-difference method and for a nonlinear conditional mean leads to somewhat different estimate than the earlier calculus method.

The treatment effect varies with z. The *average treatment effect* (ATE) is obtained by averaging TE across the individuals in the sample. The *average treatment effect on the treated* (ATET) is obtained by instead averaging TE over those individuals who actually received the treatment (i.e., for the subsample for which $d = 1$).

Direct Interpretation of Coefficients

It is useful to note that direct interpretation of the coefficients is possible without such additional computations.

- For exponential conditional mean models, the coefficient β_j equals the proportionate change in the conditional mean if the jth regressor changes by one unit. This follows from rewriting (3.43) as $\partial \mathsf{E}[y|\mathbf{x}]/\partial x_j = \beta_j \mathsf{E}[y|\mathbf{x}]$, using (3.42), and hence

$$\beta_j = \frac{\partial \mathsf{E}[y|\mathbf{x}]}{\partial x_j} \frac{1}{\mathsf{E}[y|\mathbf{x}]}. \tag{3.49}$$

  This is a *semi-elasticity*, which can alternatively be obtained by rewriting (3.42) as $\ln \mathsf{E}[y|\mathbf{x}] = \mathbf{x}'\boldsymbol{\beta}$ and differentiating with respect to x_j.
- The sign of the response $\partial \mathsf{E}[y|\mathbf{x}]/\partial x_j$ is given by the sign of β_j, because the response is β_j times the scalar $\exp(\mathbf{x}'\boldsymbol{\beta})$, which is always positive.
- If one regression coefficient is twice as large as another, then the effect of a one-unit change of the associated regressor is double that of the other. This result follows from

$$\frac{\partial \mathsf{E}[y|\mathbf{x}]/\partial x_j}{\partial \mathsf{E}[y|\mathbf{x}]/\partial x_k} = \frac{\beta_j \exp(\mathbf{x}'\boldsymbol{\beta})}{\beta_k \exp(\mathbf{x}'\boldsymbol{\beta})} = \frac{\beta_j}{\beta_k}. \tag{3.50}$$

A *single-index model* is one for which $\mathsf{E}[y|\mathbf{x}] = g(\mathbf{x}'\boldsymbol{\beta})$, for monotonic function $g(\cdot)$. The last two properties hold more generally for any single-index model.

For a binary regressor the treatment effect given in (3.48) can instead be interpreted in relative terms as

$$\frac{\mathsf{E}[y|d=1,\mathbf{z}]}{\mathsf{E}[y|d=0,\mathbf{z}]} = \frac{\exp(\beta_d + \mathbf{z}'\boldsymbol{\beta}_z)}{\exp(\mathbf{z}'\boldsymbol{\beta}_z)} = \exp(\beta_d).$$

So the conditional mean is $\exp(\beta_d)$ times larger if the indicator variable is unity rather than zero. If we instead used calculus methods, then (3.49) predicts a proportionate change of β_d or, equivalently, that the conditional mean is $(1 + \beta_d)$ times larger. This is a good approximation for small β_d, say $\beta_d < 0.1$, because then $\exp(\beta_d) \simeq 1 + \beta_d$. More generally for a discrete one-unit change in any regressor x_j the exponential mean is $\exp(\beta_j)$ times larger, so some programs present exponentiated coefficients.

Logarithmic and Interactive Regressors

Sometimes regressors enter logarithmically in (3.42). For example, we may have

$$\begin{aligned}\mathsf{E}[y|\mathbf{x}] &= \exp(\beta_1 \ln(x_1) + \mathbf{x}_2'\boldsymbol{\beta}_2) \\ &= x_1^{\beta_1} \exp(\mathbf{x}_2'\boldsymbol{\beta}_2).\end{aligned}$$

Then β_1 is an elasticity, giving the percentage change in $\mathsf{E}[y|\mathbf{x}]$ for a 1% change in x_1. This formulation is particularly appropriate if x_1 is a measure of exposure, such as number of miles driven if modeling the number of automobile accidents or population size if modeling the incidence of a disease. Then we expect β_1 to be close to unity.

The conditional mean function may include interaction terms. For example, suppose

$$\mathsf{E}[y|\mathbf{x}] = \exp(\beta_1 + \beta_2 x_2 + \beta_2 x_3 + \beta_4 x_2 x_3).$$

Then the proportionate change in the conditional mean due to a one-unit change in x_3 equals $(\beta_2 + \beta_4 x_2)$, because

$$\frac{\partial \mathsf{E}[y|\mathbf{x}]}{\partial x_3} \frac{1}{\mathsf{E}[y|\mathbf{x}]} = (\beta_2 + \beta_4 x_2).$$

Thus even the semi-elasticity measuring the effect of changes in x_3 varies according to the value of regressors – here x_2.

3.5.2 Example: Doctor Visits (Continued)

Various measures of the magnitude of the response of the number of doctor visits to changes in regressors are given in Table 3.6. These measures are based on the Poisson QML estimates given earlier in Table 3.3.

The coefficient estimates given in the column labeled QMLE using (3.49) can be interpreted as giving the proportionate change in the number of doctor visits due to a one-unit change in the regressors (though it is more precise to use $e^{\beta_j} - 1$ rather than β_j). If we consider *ACTDAYS*, the most highly statistically significant regressor, an increase of one day of reduced activity in the preceding two weeks leads to a .127 proportionate change or 12.7% change in the expected number of doctor visits. More complicated is the effect of age, which appears through both *AGE* and *AGESQ*. For a 40-year-old person, *AGE* = 0.4 and *AGESQ* = 0.16 and an increase of one year or 0.01 units leads to a $0.01 \times (1.056 - 2 \times 0.849 \times 0.40) = 0.0038$ proportionate change or .38% change in the expected number of doctor visits.

The columns AME and MEM give two different measures of the change in the number of doctor visits due to a one-unit change in regressors. First,

Table 3.6. *Doctor visits: Poisson QMLE mean effects and scaled coefficients*

Variable	Coefficient QMLE	Mean effect AME	Mean effect MEM	Mean effect OLS	Scaled Coeffs Elast	Scaled Coeffs SSC	Summary Statistics Mean	Summary Statistics Standard deviation
ONE	−2.224							
SEX	1.157	0.047	0.035	0.034	0.082	0.078	0.521	0.500
AGE	1.056	0.319	0.241	0.203	0.430	0.216	0.406	0.205
AGESQ	−0.849	−0.256	−0.193	−0.062	−0.176	−0.157	0.207	0.186
INCOME	−0.205	−0.062	−0.047	−0.057	−0.120	−0.076	0.583	0.369
LEVYPLUS	0.123	0.037	0.028	0.035	0.055	0.061	0.443	0.497
FREEPOOR	−0.440	−0.133	−0.100	−0.103	−0.019	−0.089	0.043	0.202
FREEREPA	0.080	0.024	0.018	0.033	0.017	0.033	0.210	0.408
ILLNESS	0.187	0.056	0.043	0.060	0.268	0.259	1.432	1.384
ACTDAYS	0.127	0.038	0.029	0.103	0.109	0.366	0.862	2.888
HSCORE	0.030	0.009	0.007	0.017	0.037	0.064	1.218	2.124
CHCOND1	0.114	0.034	0.026	0.004	0.046	0.056	0.403	0.491
CHCOND2	0.141	0.043	0.032	0.042	0.016	0.045	0.117	0.321

Note: AME, average over sample of effect of y of a one-unit change in x; MEM, effect on y of a one-unit change in x evaluated at average regressors; OLS, OLS coefficients; Elast, coefficients scaled by sample mean of x; SSC, coefficients scaled by standard deviation of x.

the column AME gives the average of the individual responses, using (3.45). Second, the column MEM gives the response for the individual with regressors equal to the sample mean values, computed using (3.46). The preferred AME estimates are about 30% larger than those of the "representative" individual, a consequence of the convexity of the exponential mean function. An increase of one day of reduced activity in the preceding two weeks, for example, leads on average to an increase of 0.038 doctor visits. The t statistics for the estimates of AME and MEM, not presented, are essentially the same as those for $\hat{\beta}_j$, which is to be expected because they are just multiples of β_j. When the marginal effect for age is correctly computed allowing for the quadratic in age, the AME is .076 and the MEM is 0.086. And the AME and MEM are slightly larger for binary regressors when we calculate using the treatment effect in (3.48) rather than the ME in (3.43).

The column OLS gives coefficient estimates from OLS of y on $\mathbf{x}$. Like the preceding two columns, it gives estimates of the effects of a one-unit change of x_j on $\mathsf{E}[y|\mathbf{x}]$, the difference being in different functional forms for the conditional mean. The three columns are generally similar, with OLS closer to the AME as is usually the case.

All these measures consider the effect of a one-unit change in x_j, but it is not always clear whether such a change is a large or small change. The Elast column gives $\hat{\beta}_j\overline{x}_j$, where $\overline{x}_j$ is given in the second-to-last column of the table. Given the exponential mean function, this measures the elasticity of $\mathsf{E}[y|\mathbf{x}]$ with respect to changes in regressors. The SSC column gives $\hat{\beta}_j s_j$, which instead scales by the standard deviation of the regressors, given in the last column of the table. Both measures highlight the importance of the health status measures *ILLNESS*, *ACTDAYS*, and *HSCORE* much more clearly than do the raw coefficient estimates and the estimates of $\partial \mathsf{E}[y|\mathbf{x}]/\partial x_j$. For example, the estimates imply that a 1% increase in illness days leads to a .268% increase in expected doctor visits, and a 1-standard-deviation increase in activity days leads to a 0.259% increase in expected doctor visits.

3.5.3 Prediction

We begin by considering, for an individual observation with $\mathbf{x} = \mathbf{x}_p$, prediction of the conditional mean $\mu_p = \mathsf{E}[y|\mathbf{x} = \mathbf{x}_p]$.

For the exponential conditional mean function, the mean prediction is

$$\hat{\mu}_p = \exp(\mathbf{x}_p'\hat{\boldsymbol{\beta}}). \tag{3.51}$$

A 95% confidence interval, which allows for the imprecision in the estimate $\hat{\boldsymbol{\beta}}$, can be obtained using the delta method given in Chapter 2.6.2. Consider estimation procedures that additionally estimate a scalar nuisance parameter, α, to accommodate overdispersion. Because $\partial \mu_p/\partial \boldsymbol{\beta} = \mu_p \mathbf{x}_p$ and $\partial \mu_p/\partial \alpha = \mathbf{0}$, we obtain

$$\mu_p \in \hat{\mu}_p \pm z_{.025}\sqrt{\hat{\mu}_p^2 \mathbf{x}_p' \mathsf{V}[\hat{\boldsymbol{\beta}}]\mathbf{x}_p}, \tag{3.52}$$

for estimator $\hat{\boldsymbol{\beta}} \overset{a}{\sim} \mathsf{N}[\boldsymbol{\beta}, \mathsf{V}[\hat{\boldsymbol{\beta}}]]$. As expected, greater precision in the estimation of $\boldsymbol{\beta}$ leads to a narrower confidence interval.

One may also wish to predict the actual value of y, rather than its predicted mean. This is considerably more difficult, because the randomness of y needs to be accounted for, in addition to the randomness in the estimator $\hat{\boldsymbol{\beta}}$. For low counts in particular, individual values are poorly predicted due to the intrinsic randomness of y.

For an individual observation with $\mathbf{x} = \mathbf{x}_p$, and using the exponential conditional mean function, the individual prediction is

$$\hat{y}_p = \exp(\mathbf{x}_p'\hat{\boldsymbol{\beta}}). \tag{3.53}$$

Note that while $\hat{y}_p$ equals $\hat{\mu}_p$, it is being used as an estimate of y_p rather than μ_p. If we consider variance functions of the form (3.12), the estimated variance of y_p is $\omega(\hat{\mu}_p, \hat{\alpha})$. Adding this to the earlier variance due to imprecision in the estimate $\hat{\boldsymbol{\beta}}$, the variance of $\hat{y}_p$ is consistently estimated by $\omega(\hat{\mu}_p, \hat{\alpha}) + \hat{\mu}_p^2 \mathbf{x}_p' \mathsf{V}[\hat{\boldsymbol{\beta}}]\mathbf{x}_p$. A two-standard error interval is

$$y_p \in \hat{y}_p \pm 2\sqrt{\omega(\hat{\mu}_p, \hat{\alpha}) + \hat{\mu}_p^2 \mathbf{x}_p' \mathsf{V}[\hat{\boldsymbol{\beta}}]\mathbf{x}_p}. \tag{3.54}$$

This can be used as a guide, but is not formally a 95% confidence interval because y_p is not normally distributed even in large samples.

The width of this interval is increasing in $\hat{\mu}_p$, because $\omega(\hat{\mu}_p, \hat{\alpha})$ is increasing in $\hat{\mu}_p$. The interval is quite wide, even for low counts. For example, consider the Poisson distribution, so $\omega(\mu, \alpha) = \mu$, and assume a large sample size, so that $\boldsymbol{\beta}$ is very precisely estimated and $\mathsf{V}[\hat{\boldsymbol{\beta}}] \simeq \mathbf{0}$. Then (3.54) becomes $y_p \in \hat{y}_p \pm 2\sqrt{\hat{y}_p}$. Even for $\hat{y}_p = 4$ this yields (0, 8). If y is Poisson and $\boldsymbol{\beta}$ is known it is better to directly use this knowledge, rather than the approximation (3.54). Then because $\Pr\left[1 \leq y_p \leq 8\right] = 0.0397$ when y_p is Poisson distributed with $\mu_p = 4$, a 96.03% confidence interval for y_p is $[1, 8]$.

Interest can also lie in predicting the probabilities of particular values of y occurring, for an individual observation with $\mathbf{x} = \mathbf{x}_p$. Let p_k denote the probability that $y_p = k$ if $\mathbf{x} = \mathbf{x}_p$. For the Poisson, for example, this is estimated by $\hat{p}_k = \exp(-\hat{\mu}_p)\hat{\mu}_p^k/k!$, where $\hat{\mu}_p$ is given in (3.51). The delta method can be used to obtain a confidence interval for p_k. Most parametric models include an overdispersion parameter α. Because $\partial p_k/\partial\alpha \neq 0$ the delta method does not yield an interval as simple as (3.52).

A fourth quantity that might be predicted is the change in the conditional mean if the jth regressor changes by one unit. From (3.43) the predicted change is $\hat{\beta}_j\hat{\mu}_p$. Again the delta method can be used to form a confidence interval.

As the sample size gets very large, the variance of $\hat{\boldsymbol{\beta}}$ goes to zero, and the confidence intervals for predictions of the conditional mean, individual probabilities, and changes in predicted probabilities collapse to a point. The confidence intervals for predictions of individual values of y, however, can remain wide as demonstrated earlier. Thus, within-sample individual predictions of y differ considerably from the actual values of y, especially for small counts. If interest lies in assessing the usefulness of model predictions, it can be better to consider predictions at a more aggregated level. This is considered in Chapter 5.

The preceding methods require the use of a computer program that saves the variance matrix of the regression parameters and permits matrix multiplication. It is generally no more difficult to use the bootstrap instead.

More problematic in practice is deciding what values of $\mathbf{x}_p$ to use in prediction. If, for example, there are just two regressors, which are both binary regressors, one would simply predict for each of the four distinct values of $\mathbf{x}$. But in practice there are many different distinct values of $\mathbf{x}$, and it may not be obvious which few values to focus on. Such considerations also arise in the linear regression model.

3.6 ORDERED AND OTHER DISCRETE-OUTCOME MODELS

Count data can alternatively be modeled using discrete outcome methods surveyed in Cameron and Trivedi (2005). This is particularly natural if most observed counts take values zero and one, with few counts in excess of one. Then one might simply model whether the count is zero or nonzero, using a binary outcome model such as logit or probit.

Extending this to just a few low observed counts, say 0, 1, and 2 or more, the natural discrete outcome model is an ordered multinomial model, usually ordered probit or ordered logit, that accounts for the natural ordering of the discrete outcomes. These models treat the data as generated by a continuous unobserved latent variable, which on crossing a threshold leads to an increase of one in the observed outcome. This is a representation of the dgp quite different from the Poisson process that leads to the Poisson density for counts. In theory, one should use an ordered model for data for which the underlying dgp is felt to be a continuous latent variable.

Such alternatives to the count model may provide a better fit to the data in cases where the support of the distribution is limited to a few values (e.g., the fertility example in Chapter 1). For time series of counts it can be difficult to formulate reasonable dynamic models, and one possibility is a dynamic ordered multinomial model; see Jung, Kukuk, and Liesenfeld (2006). In addition, a dynamic ordered multinomial model can be applied to an integer-valued dependent variable that may take negative values; see the end of section 3.6.2.

3.6.1 Binary Outcome Models

Let y_i be the count variable of interest. Define the indicator variable

$$\begin{aligned} d_i &= 1 \qquad \text{if } y_i > 0 \\ &= 0 \qquad \text{if } y_i = 0. \end{aligned} \tag{3.55}$$

This equals the count variable except that y_i values of 2 or more are recoded to 1. Other partitions are possible, such as $y_i > k$ and $y_i \leq k$.

The general form of such a *binary outcome* model is

$$\begin{aligned} &\Pr[d_i = 1] = F(\mathbf{x}_i'\boldsymbol{\beta}) \\ &\Pr[d_i = 0] = 1 - F(\mathbf{x}_i'\boldsymbol{\beta}), \end{aligned} \tag{3.56}$$

where $0 < F(\cdot) < 1$. It is customary to let the probability be a transformation of a linear combination of the regressors rather than to use the more general functional form $F(\mathbf{x}_i, \boldsymbol{\beta})$. By construction the probabilities sum to one. A parsimonious way to write the density given by (3.56) is $F(\mathbf{x}_i'\boldsymbol{\beta})^{d_i}\ (1 - F(\mathbf{x}_i'\boldsymbol{\beta}))^{1-d_i}$, which leads to the log-likelihood function

$$\ln \mathsf{L} = \sum_{i=1}^{n} d_i \ln F(\mathbf{x}_i'\boldsymbol{\beta}) + (1 - d_i)\ln(1 - F(\mathbf{x}_i'\boldsymbol{\beta})). \tag{3.57}$$

Different binary outcome models correspond to different choices of the function $F(\cdot)$. Standard choices include the *logit model*, which corresponds to $F(z) = \exp(z)/(1 + \exp(z))$, and the *probit model*, which corresponds to $F(z) = \Phi(z)$, where $\Phi(\cdot)$ is the standard normal cdf.

There is a loss of efficiency due to combining all counts in excess of zero into a single category. Suppose y_i are Poisson distributed with mean μ_i. Then $\Pr[d_i = 1] = \Pr[y_i > 0] = 1 - \exp(-\mu_i)$, so for $\mu_i = \exp(\mathbf{x}_i'\boldsymbol{\beta})$

$$F(\mathbf{x}_i'\boldsymbol{\beta}) = 1 - \exp(-\exp(\mathbf{x}_i'\boldsymbol{\beta})). \tag{3.58}$$

The *binary Poisson* MLE $\hat{\boldsymbol{\beta}}_{\mathsf{BP}}$ maximizes (3.57) with this specification of $F(\cdot)$. The model is identical to the *complementary log-log model*, used as an alternative to the logit and probit models when the outcome is mostly 0 or mostly 1. It can be shown that

$$\mathsf{V}[\hat{\boldsymbol{\beta}}_{\mathsf{BP}}] = \left(\sum_{i=1}^{n} c_i \mu_i \mathbf{x}_i \mathbf{x}_i' \right)^{-1},$$

where $c_i = \mu_i \exp(-\mu_i)/(1 - \exp(-\mu_i))$. $\mathsf{V}[\hat{\boldsymbol{\beta}}_{\mathsf{BP}}]$ exceeds $\left(\sum_{i=1}^{n} \mu_i \mathbf{x}_i \mathbf{x}_i'\right)^{-1}$, the variance matrix of the Poisson MLE from (3.6), because $c_i < 1$ for $\mu_i > 0$. As expected, the relevant efficiency loss is increasing in the Poisson mean. For example, for $\mu_i = 0.5$ and 1, respectively, $c_i = 0.77$ and 0.58. In the Poisson iid case with $\mu = 0.5$, the standard error of $\hat{\boldsymbol{\beta}}_{\mathsf{BP}}$ is $\sqrt{1/0.77} = 1.14$ times that of the Poisson MLE, even though less than 10% of the counts will exceed unity.

3.6.2 Ordered Multinomial Models

The Poisson and negative binomial models treat discrete data as being the result of an underlying point process. One could instead model the number of doctor visits, for example, as being due to a continuous process that on crossing a threshold leads to a visit to a doctor. Crossing further thresholds leads to additional doctor visits. Before specializing to threshold models such as ordered probit, we first present general results for multinomial models.

Suppose the count variable y_i takes values $0, 1, 2, \ldots, m$. Define the $m+1$ indicator variables

$$\begin{aligned} d_{ij} &= 1 \quad y_i = j \\ &= 0 \quad y_i \neq j. \end{aligned} \tag{3.59}$$

Also define the corresponding probabilities

$$\Pr\left[d_{ij} = 1\right] = p_{ij}, \qquad j = 0, \ldots, m, \tag{3.60}$$

where p_{ij} may depend on regressors and parameters. Then the density function for the ith observation can be written as

$$f(y_i) = f(d_{i0}, d_{i1}, \ldots, d_{im}) = \prod_{j=0}^{m} p_{ij}^{d_{ij}}, \tag{3.61}$$

and the log-likelihood function is

$$\ln \mathsf{L} = \sum_{i=1}^{n} \sum_{j=0}^{m} d_{ij} \ln p_{ij}. \tag{3.62}$$

Different multinomial models arise from different specifications of the probabilities p_{ij}. The most common is the multinomial logit model, which specifies $p_{ij} = \exp(\mathbf{x}_i'\boldsymbol{\beta}_j)/(\sum_{k=0}^{m} \mathbf{x}_i'\boldsymbol{\beta}_k)$ in (3.62). This model is inappropriate for count data for which the outcome – the number of occurrences of an event – is naturally ordered. One way to see this is to note that a property of multinomial logit is that the relative probabilities of any two outcomes are independent of the probabilities of other outcomes. For example, the probability of one doctor visit, conditional on the probability of zero or one visit, would not depend on the probability of two visits. But one expects that this conditional probability will be higher the higher the probability of two visits. It is better to use a multinomial model that explicitly incorporates the ordering of the data.

The *ordered probit* model, presented for example in Maddala (1983), introduces a latent (unobserved) random variable

$$y_i^* = \mathbf{x}_i'\boldsymbol{\beta} + \varepsilon_i, \tag{3.63}$$

where ε_i is $\mathsf{N}[0, 1]$. The observed discrete data variable y_i is generated from the unobserved y_i^* in the following way:

$$y_i = j \quad \text{if} \quad \alpha_j < y_i^* \leq \alpha_{j+1}, \qquad j = 0, \ldots, m, \tag{3.64}$$

where $\alpha_0 = -\infty$ and $\alpha_{m+1} = \infty$. It follows that

$$\begin{aligned} p_{ij} &= \Pr[\alpha_j < y_i^* \leq \alpha_{j+1}] \\ &= \Pr[\alpha_j - \mathbf{x}_i'\boldsymbol{\beta} < \varepsilon_i \leq \alpha_{j+1} - \mathbf{x}_i'\boldsymbol{\beta}] \\ &= \Phi(\alpha_{j+1} - \mathbf{x}_i'\boldsymbol{\beta}) - \Phi(\alpha_j - \mathbf{x}_i'\boldsymbol{\beta}), \end{aligned} \tag{3.65}$$

where $\Phi(\cdot)$ is the standard normal cdf, $j = 0, 1, 2, \ldots, m$, and $\alpha_{m+1} = \infty$. The log-likelihood function (3.62) with probabilities (3.65) is

$$\ln \mathsf{L} = \sum_{i=1}^{n} \sum_{j=0}^{m} d_{ij} \ln \left[\Phi(\alpha_{j+1} - \mathbf{x}_i'\boldsymbol{\beta}) - \Phi(\alpha_j - \mathbf{x}_i'\boldsymbol{\beta}) \right]. \tag{3.66}$$

Estimation of $\boldsymbol{\beta}$ and $\alpha_1, \ldots, \alpha_m$ by maximum likelihood is straightforward. Identification requires a normalization, such as 0, for one of $\alpha_1, \ldots, \alpha_m$ or for the intercept term in $\boldsymbol{\beta}$.

If there are many counts and/or few observations for a given count, then some aggregation of count data may be necessary. For example, if there are few observations larger than 3, one might have categories of 0, 1, 2, and 3 or more. As an alternative to the ordered probit, one can use the *ordered logit* model, in which case $\Phi(\cdot)$ is replaced by the logistic cdf $L(z) = e^z/(1 + e^z)$. More generally if ε_i in (3.63) has a nonnormal distribution the log-likelihood function is (3.66) with $\Phi(\cdot)$ replaced by the cdf of ε_i.

The ordered discrete outcome model has the additional advantage of being applicable to count data that are negative. Such data may arise if instead of directly modeling a count variable, one differences and models the change in the count. For example, some U.S. stock prices are a count, because they are measured in units of a tick, or one-eighth of a dollar. Hausman, Lo, and MacKinlay (1992) model price changes in consecutive time-stamped trades of a given stock using the ordered probit model, generalized to allow ε_i to be $\mathsf{N}[0, \sigma_i^2]$, where the variance σ_i^2 is itself modeled by a regression equation.

3.6.3 Binomial Model

In some cases the count has an upper value. For example, we might model the number of children in a family who attend college. The natural model for these data is the binomial.

We suppose that y_i is the number of successes in N_i trials, each with probability of success p_i. Then y_i is binomial distributed with density

$$f(y_i|p_i) = \frac{N_i!}{p_i!(1-p_i)!} p_i^{y_i}(1-p_i)^{N_i-y_i}, \tag{3.67}$$

where $p_i = F(\mathbf{x}_i'\boldsymbol{\beta})$ is most often modeled as a logit or probit model.

In this model the conditional mean

$$\mathsf{E}[y_i|\mathbf{x}_i] = N_i F(\mathbf{x}_i'\boldsymbol{\beta}). \tag{3.68}$$

The binomial with known number of trials is a member of the linear exponential family, so consistent estimation of $\boldsymbol{\beta}$ requires only correct specification of the conditional mean, and inference is best based on a robust estimate of the variance matrix. The model is included in statistical software for generalized linear models.

3.7 OTHER MODELS

In this section we consider whether least squares methods might be usefully applied to count data y. Three variations of least squares are considered. The first is linear regression of y on $\mathbf{x}$, making no allowance for the count nature of the data aside from using heteroskedasticity robust standard errors. The second is linear regression of a nonlinear transformation of y on $\mathbf{x}$, for which the transformation leads to a dependent variable that is close to homoskedastic and symmetric. Third, we consider nonlinear least squares regression with the conditional mean of y specified to be $\exp(\mathbf{x}'\boldsymbol{\beta})$. The section finishes with a discussion of estimation using duration data, rather than count data, if the data are generated by a Poisson process.

3.7.1 OLS without Transformation

The OLS estimator is clearly inappropriate because it specifies a conditional mean function $\mathbf{x}'\boldsymbol{\gamma}$ that may take negative values and a variance function that is homoskedastic. If the conditional mean function is, in fact, $\exp(\mathbf{x}'\boldsymbol{\beta})$, the OLS estimator is inconsistent for $\boldsymbol{\beta}$, and the computed OLS output gives the wrong asymptotic variance matrix.

Nonetheless, OLS estimates in practice give results qualitatively similar to those for Poisson and other estimators using the exponential mean. The ratio of OLS slope coefficients is often similar to the ratio of Poisson slope coefficients, with the OLS slope coefficients approximately $\bar{y}$ times the Poisson slope coefficients, and the most highly statistically significant regressors from OLS regression, using usual OLS output t statistics, are in practice the most highly significant using Poisson regression. This is similar to comparing different models for binary data such as logit, probit, and OLS. In all cases the conditional mean is restricted to be of form $g(\mathbf{x}'\boldsymbol{\beta})$, which is a monotonic transformation of a linear combination of the regressors. The only difference across models is the choice of function g, which leads to a different scaling of the parameters $\boldsymbol{\beta}$.

A first-order Taylor series expansion of the exponential mean $\exp(\mathbf{x}'\boldsymbol{\beta})$ around the sample mean $\overline{y}$ – that is, around $\mathbf{x}'\boldsymbol{\beta} = \ln \overline{y}$ – yields $\exp(\mathbf{x}'\boldsymbol{\beta}) = \overline{y} + \overline{y}(\mathbf{x}'\boldsymbol{\beta} - \ln \overline{y})$. For models with intercept, this can be rewritten as $\exp(\beta_1 + \mathbf{x}_2'\boldsymbol{\beta}_2) = \gamma_1 + \mathbf{x}_2'\boldsymbol{\gamma}_2$, where $\gamma_1 = \overline{y} + \beta_1\overline{y} - \ln \overline{y}$ and $\boldsymbol{\gamma}_2 = \boldsymbol{\beta}_2\overline{y}$. So linear mean slope coefficients are approximately $\overline{y}$ times the exponential slope coefficients. This approximation will be more reasonable the less dispersed the predicted values $\exp(\mathbf{x}_i'\hat{\boldsymbol{\beta}})$ are about $\overline{y}$.

The OLS estimator can be quite useful for preliminary data analysis, such as determining key variables, in simple count models. Dealing with more complicated count models for which no off-the-shelf software is readily available would be easier if one could first ignore the count aspect of the data and do the corresponding adjustment to OLS. For example, if the complication is endogeneity, then do linear two-stage least squares as a potential guide to the impact of endogeneity. But experience is sufficiently limited that one cannot advocate this approach.

3.7.2 OLS with Log Transformation

For skewed continuous data such as that on individual income or on housing prices, a standard transformation is the *log transformation*. For example, if y is log-normal distributed then $\ln y$ is by definition exactly normally distributed, so the log transformation induces constant variance and eliminates skewness.

The *log transformation* may also be used for count data that are often skewed. However, there is usually no benefit to doing so. First, the assumption that $\mathsf{E}[\ln y|\mathbf{x}] = \mathbf{x}'\boldsymbol{\beta}$ does not imply that $\mathsf{E}[y|\mathbf{x}] = \exp(\mathbf{x}'\boldsymbol{\beta})$, a problem called retransformation bias, and interest usually lies in marginal effects and predictions for $\mathsf{E}[y|\mathbf{x}]$ rather than $\mathsf{E}[\ln y|\mathbf{x}]$. Second, if some counts are 0, then $\ln 0$ is not defined. A standard solution is to add a constant term, such as 0.5, and to model $\ln(y + 0.5)$ by OLS. This model has been criticized by King (1989b) as performing poorly numerically, and marginal effects with respect to $\mathsf{E}[\ln(y + 0.5)|\mathbf{x}]$ are not of interest.

A common application of the log transformation is to a dependent variable that is a rate (y/z), such as state-level data on murders (y) per 100,000 people (z). One approach is OLS regression of $\ln(y/z)$ on $\mathbf{x}$. If data are available for both y and z, rather than just on the rate y/z, then it is better to explicitly model y as a count. Then $\mathsf{E}[y|\mathbf{x}, z] = z \times \exp(\mathbf{x}'\boldsymbol{\beta}) = \exp(\ln z + \mathbf{x}'\boldsymbol{\beta})$, and we perform count regression of y on $\ln z$ and $\mathbf{x}$ with the coefficient of $\ln z$ restricted to unity, a restriction that count regression commands often permit. Furthermore, this restriction can be easily tested. The regressor z is a measure of exposure, discussed in section 3.5.1.

Going the other way, for a right-skewed continuous nonnegative dependent variable, there are reasons for using the Poisson QMLE rather than the more customary OLS in a log-linear model. A leading example is expenditure data with some zero observations, in which case there is a problem for the log-linear model because $\ln 0$ is not defined. Modeling $\ln(y + 0.5)$ has already been criticized. A two-part model that separately models whether $y = 0$ and then models $\ln y$ when $y \geq 0$ is another possible approach, but marginal effects are difficult to obtain. Poisson regression for the model $\ln \mathsf{E}[y|\mathbf{x}] = \mathbf{x}'\boldsymbol{\beta}$ has no problem with $y = 0$, does not require that y be a count (see section 3.2.2), and easily yields marginal effects (see section 3.5.1). How well this model works will vary with the application. Santos Silva and Tenreyro (2006) advocate the use of Poisson regression rather than log-linear modeling with application to the gravity model for international trade flows.

3.7.3 OLS with Other Transformations

An alternative transformation is the *square root transformation.* Following McCullagh and Nelder (1989, p. 236), let $y = \mu(1 + \varepsilon)$. Then a fourth-order Taylor series expansion around $\varepsilon = 0$ yields

$$y^{1/2} \simeq \mu^{1/2}\left(1 + \frac{1}{2}\varepsilon - \frac{1}{8}\varepsilon^2 + \frac{1}{16}\varepsilon^3 - \frac{5}{128}\varepsilon^4\right).$$

For the Poisson, $\varepsilon = (y - \mu)/\mu$ has first four moments 0, $1/\mu$, $1/\mu^2$, and $(3/\mu^2 + 1/\mu^3)$. It follows that $\mathsf{E}[\sqrt{y}] \simeq \sqrt{\mu}(1 - 1/8\mu + O(1/\mu^2))$, $\mathsf{V}[\sqrt{y}] \simeq (1/4)(1 + 3/8\mu + O(1/\mu^2))$, and $\mathsf{E}[(\sqrt{y} - \mathsf{E}[\sqrt{y}])^3] \simeq -(1/16\sqrt{\mu})(1 + O(1/\mu))$. Thus if y is Poisson then $\sqrt{y}$ is close to homoskedastic and is close to symmetric. The skewness index is the third central moment divided by variance raised to the power 1.5. Here it is approximately $-(1/16\sqrt{\mu})/(1/4)^{1.5} = -(1/2\sqrt{\mu})$. By comparison for the Poisson y is heteroskedastic with variance μ and asymmetric with skewness index $1/\sqrt{\mu}$. The square-root transformation works quite well for large μ.

One therefore models $\sqrt{y}$ by OLS, regressing $\sqrt{y_i}$ on $\mathbf{x}_i$. The usual OLS t statistics can be used for statistical inference. More problematic is the interpretation of coefficients. These give the impact of a one-unit change in x_j on $\mathsf{E}[\sqrt{y}]$ rather than $\mathsf{E}[y]$, and by Jensen's inequality $\mathsf{E}[y] \neq (\mathsf{E}[\sqrt{y}])^2$. A similarly problem arises in prediction, although the method of Duan (1983) can be used to predict $\mathsf{E}[y_i]$, given the estimated model for $\sqrt{y_i}$.

3.7.4 Nonlinear Least Squares

The nonlinear least squares (NLS) estimator with exponential mean minimizes the sum of squared residuals $\sum_i (y_i - \exp(\mathbf{x}_i'\boldsymbol{\beta}))^2$. The estimator $\hat{\boldsymbol{\beta}}_{\mathsf{NLS}}$ is the solution to the first-order conditions

$$\sum_{i=1}^{n} \mathbf{x}_i(y_i - \exp(\mathbf{x}_i'\boldsymbol{\beta}))\exp(\mathbf{x}_i'\boldsymbol{\beta}) = \mathbf{0}. \tag{3.69}$$

This estimator is consistent if the conditional mean of y_i is $\exp(\mathbf{x}_i'\boldsymbol{\beta})$. It is inefficient, however, because the errors are certainly not homoskedastic, and the usual reported NLS standard errors are inconsistent. $\hat{\boldsymbol{\beta}}_{\text{NLS}}$ is asymptotically normal with variance

$$\mathsf{V}[\hat{\boldsymbol{\beta}}_{\text{NLS}}] = \left(\sum_{i=1}^{n} \mu_i^2 \mathbf{x}_i\mathbf{x}_i'\right)^{-1} \left(\sum_{i=1}^{n} \omega_i \mu_i^2 \mathbf{x}_i\mathbf{x}_i'\right) \left(\sum_{i=1}^{n} \mu_i^2 \mathbf{x}_i\mathbf{x}_i'\right)^{-1}, \tag{3.70}$$

where $\omega_i = \mathsf{V}[y_i|\mathbf{x}_i]$. The robust sandwich estimate of $\mathsf{V}[\hat{\boldsymbol{\beta}}_{\text{NLS}}]$ is (3.70), with μ_i and ω_i replaced by $\hat{\mu}_i$ and $(y_i - \hat{\mu}_i)^2$.

The NLS estimator can, therefore, be used, but more efficient estimates can be obtained using the estimators given in sections 3.2 and 3.3.

3.7.5 Example: Doctor Visits (Continued)

Coefficient estimates of binary Poisson, ordered probit, OLS, OLS of transformations of y (both $\ln(y + 0.1)$ and $\sqrt{y}$), Poisson QMLE, and NLS with exponential mean are presented in Table 3.7. The associated t statistics reported are based on RS standard errors, except for binary Poisson and ordered probit. The skewness and kurtosis measures given are for model residuals $z_i - \hat{z}_i$, where z_i is the dependent variable, for example, $z_i = \sqrt{y_i}$, and are estimates of, respectively, the third central moment divided by s^3 and the fourth central moment divided by s^4, where s^2 is the estimated variance. For the standard normal distribution the kurtosis measure is 3.

We begin with estimation of a binary outcome model for the recoded variable $d = 0$ if $y = 0$ and $d = 1$ if $y \geq 1$. To allow direct comparison with Poisson estimates, we estimate the nonstandard binary Poisson model introduced in section 3.6.1. Compared with Poisson estimates in the Poiss column, the BP results for health status measures are similar, although for the statistically insignificant socioeconomic variables *AGE*, *AGESQ*, and *INCOME* there are sign changes. Similar sign changes for *AGE* and *AGESQ* occur in Table 3.4 and are discussed there. The log-likelihood for BP exceeds that for Poisson, but this comparison is meaningless due to the different dependent variable. Logit and probit, not reported, lead to similar log-likelihood and qualitatively similar estimates to those from binary Poisson, so differences between binary Poisson and Poisson can be attributed to aggregating all positive counts into one value.

The ordered probit model normalizes the error variance to 1. To enable comparison with OLS estimates, we multiply these by $s = 0.714$, the estimated standard deviation of the residual from OLS regression. Also, because only one observation took the value 9, it was combined into a category of 8 or more. The intercept was normalized to 0, and the rescaled threshold parameter estimates

Table 3.7. *Doctor visits: Alternative estimates and* t *ratios*

	Estimators and t statistics						
	Discrete outcome		OLS of transformations			Exponential mean	
Variable	BP	OrdProb	y	$\ln y$	$\sqrt{y}$	Poiss	NLS
ONE	−2.312	–	0.028	−2.115	0.070	−2.224	−2.234
	(9.76)		(0.38)	(21.43)	(1.54)	(8.74)	(6.13)
SEX	0.206	0.094	0.034	0.081	0.034	0.157	−0.057
	(2.87)	(2.97)	(1.47)	(2.73)	(2.47)	(1.98)	(0.42)
AGE	−1.001	−0.381	0.203	−0.566	−0.161	1.056	3.626
	(0.78)	(0.65)	(0.45)	(0.97)	(0.60)	(0.77)	(1.82)
AGESQ	1.474	0.612	−0.062	0.877	0.292	−0.849	−3.676
	(1.05)	(0.95)	(0.12)	(1.31)	(0.94)	(0.58)	(1.70)
INCOME	0.002	−0.044	−0.057	−0.019	−0.017	−0.205	−0.394
	(0.02)	(0.88)	(1.64)	(0.43)	(0.80)	(1.59)	(2.02)
LEVYPLUS	0.222	0.098	0.035	0.080	0.039	0.123	0.214
	(2.55)	(2.62)	(1.61)	(2.58)	(2.40)	(1.29)	(1.47)
FREEPOOR	−0.572	−0.247	−0.103	−0.182	−0.081	−0.440	−0.232
	(2.46)	(2.55)	(2.17)	(3.17)	(2.99)	(1.52)	(0.54)
FREEREPA	0.346	0.127	0.033	0.139	0.054	0.080	−0.003
	(2.82)	(2.39)	(0.77)	(2.45)	(2.06)	(0.63)	(0.02)
ILLNESS	0.207	0.107	0.060	0.110	0.048	0.187	0.140
	(8.90)	(9.75)	(6.03)	(8.52)	(8.11)	(7.81)	(3.63)
ACTDAYS	0.099	0.072	0.103	0.106	0.054	0.127	0.121
	(12.04)	(15.45)	(10.59)	(13.56)	(13.05)	(16.33)	(14.19)
HSCORE	0.044	0.023	0.017	0.029	0.013	0.030	0.023
	(2.87)	(3.36)	(2.37)	(3.30)	(3.17)	(2.11)	(1.03)
CHCOND1	0.120	0.044	0.004	0.022	0.009	0.114	0.079
	(1.50)	(1.26)	(0.20)	(0.70)	(0.61)	(1.25)	(0.55)
CHCOND2	0.227	0.097	0.042	0.102	0.043	0.141	−0.055
	(2.08)	(1.92)	(0.90)	(1.80)	(1.62)	(1.15)	(0.31)
−ln L	2303.8	3138.1				3355.5	
Skewness			3.6	1.2	1.4	3.1	
Kurtosis			26.4	4.0	5.5	25.6	

Note: BP, MLE for binary Poisson; OrdProb, MLE for rescaled Ordered Probit; y, OLS for y; $\ln y$, OLS for $\ln(y+0.1)$; $\sqrt{y}$, OLS for $\sqrt{y}$; Poiss, Poisson QMLE; NLS, NLS with exponential mean. The t statistics are robust sandwich for all but BP and OrdProb. Skewness and kurtosis are for model residuals.

are .98, 1.58, 2.07, 2.21, 2.38, 2.47, 2.67, and 2.98, with t statistics all in excess of 9. Despite the rescaling there is still considerable difference from the OLS estimates. It is meaningful to compare the ordered probit log-likelihood with that of other count data models; the change of one observation from 9 to 8 or more in the ordered probit should have little effect. The log-likelihood is higher for this model than for NB2, because $-3138.1 > -3198.7$, although six more parameters are estimated.

The log transformation $\ln(y+0.1)$ was chosen on grounds of smaller skewness and kurtosis than $\ln(y+0.2)$ or $\ln(y+0.4)$. The skewness and kurtosis are somewhat smaller for ln y than $\sqrt{y}$. Both transformations appear quite successful in moving toward normality, especially compared with residuals from OLS or Poisson regression with y as dependent variables. Much of this gain appears even before the inclusion of regressors, because the inclusion of regressors reduces skewness and kurtosis by about 20% in this example. All models give similar results regarding the statistical significance of regressors, although interpretation of the magnitude of the effect of regressors is more difficult if the dependent variable is $\ln(y + 0.1)$ or $\sqrt{y}$.

The NLS estimates for exponential mean lead to similar conclusions as Poisson for the health-status variables, but quite different conclusions for socioeconomic variables, with considerably larger coefficients and t statistics for *AGE*, *AGESQ*, and *INCOME* and a sign change for *SEX*.

3.7.6 Exponential Duration Model

For a Poisson point process the number of events in a given interval of time is Poisson distributed. The duration of a spell, the time from one occurrence to the next, is exponentially distributed. Here we consider modeling durations rather than counts.

Suppose that for each individual in a sample of n individuals we observe the duration of one complete spell, generated by a Poisson point process with rate parameter γ_i. Then t_i has exponential density $f(t_i) = \gamma_i \exp(-\gamma_i t_i)$ with mean $\mathsf{E}[t_i] = 1/\gamma_i$. For regression analysis it is customary to specify $\gamma_i = \exp(\mathbf{x}_i'\boldsymbol{\beta})$. The exponential MLE, $\hat{\boldsymbol{\beta}}_{\mathsf{E}}$, maximizes the log-likelihood function

$$\ln \mathsf{L} = \sum_{i=1}^{n} \mathbf{x}_i'\boldsymbol{\beta} - \exp(\mathbf{x}_i'\boldsymbol{\beta})t_i. \tag{3.71}$$

The first-order conditions can be expressed as

$$\sum_{i=1}^{n} (1 - \exp(\mathbf{x}_i'\boldsymbol{\beta})t_i)\mathbf{x}_i = \mathbf{0}, \tag{3.72}$$

and application of the usual maximum likelihood theory yields

$$\mathsf{V}_{\mathsf{ML}}[\hat{\boldsymbol{\beta}}_{\mathsf{E}}] = \left(\sum_{i=1}^{n} \mathbf{x}_i\mathbf{x}_i' \right)^{-1}. \tag{3.73}$$

If instead we modeled the number of events from a Poisson point process with rate parameter $\gamma_i = \exp(\mathbf{x}_i'\boldsymbol{\beta})$ we obtain

$$\mathsf{V}_{\mathsf{ML}}[\hat{\boldsymbol{\beta}}_{\mathsf{P}}] = \left(\sum_{i=1}^{n} \gamma_i\mathbf{x}_i\mathbf{x}_i' \right)^{-1}.$$

The two variance matrices coincide if $\gamma_i = 1$. Thus if we choose intervals for each individual so that individuals on average experience one event such as a doctor visit, the count data have the same information content, in terms of precision of estimation of $\boldsymbol{\beta}$, as observing for each individual one completed spell such as the time between successive visits to the doctor. More simply, one count conveys the same information as the length of one complete spell.

3.8 ITERATIVELY REWEIGHTED LEAST SQUARES

Most of the models and estimators in this book require special statistical packages for nonlinear models. An exception is the Poisson QMLE, which can be computed in the following way.

In general at the sth iteration, the Newton-Raphson method updates the current estimate $\hat{\boldsymbol{\beta}}_s$ by the formula $\hat{\boldsymbol{\beta}}_{s+1} = \hat{\boldsymbol{\beta}}_s - \hat{\mathbf{H}}_s^{-1}\hat{\mathbf{g}}_s$, where $\mathbf{g} = \partial \ln \mathrm{L}/\partial \boldsymbol{\beta}$ and $\mathbf{H} = \partial^2 \ln \mathrm{L}/\partial \boldsymbol{\beta}\partial \boldsymbol{\beta}'$ are evaluated at $\hat{\boldsymbol{\beta}}_s$. Here this becomes

$$\hat{\boldsymbol{\beta}}_{s+1} = \hat{\boldsymbol{\beta}}_s + \left[\sum_{i=1}^{n} \hat{\mu}_{is}\mathbf{x}_i\mathbf{x}_i'\right]^{-1} \sum_{i=1}^{n} \mathbf{x}_i(y_i - \hat{\mu}_{is}),$$

where we consider the exponential mean function so $\hat{\mu}_{is} = \exp(\mathbf{x}_i'\hat{\boldsymbol{\beta}}_s)$. This can be rewritten as

$$\hat{\boldsymbol{\beta}}_{s+1} = \left[\sum_{i=1}^{n} (\sqrt{\hat{\mu}_{is}}\mathbf{x}_i)(\sqrt{\hat{\mu}_{is}}\mathbf{x}_i)'\right]^{-1}$$
$$\times \sum_{i=1}^{n} (\sqrt{\hat{\mu}_{is}}\mathbf{x}_i)\left\{\sqrt{\hat{\mu}_{is}}\frac{(y_i - \hat{\mu}_{is})}{\hat{\mu}_{is}} + \sqrt{\hat{\mu}_{is}}\mathbf{x}_i'\hat{\boldsymbol{\beta}}_s\right\},$$

which is the formula for the OLS estimator from the regression

$$\sqrt{\hat{\mu}_{is}}\left\{\frac{(y_i - \hat{\mu}_{is})}{\hat{\mu}_{is}} + \mathbf{x}_i'\hat{\boldsymbol{\beta}}_s\right\} = (\sqrt{\hat{\mu}_{is}}\mathbf{x}_i)'\boldsymbol{\beta}_{s+1} + u_i, \tag{3.74}$$

where u_i is an error term. Thus the Poisson QMLE can be calculated by this iterative OLS regression. Equivalently, it can be estimated by WLS regression of $\{((y_i - \hat{\mu}_{is})/\hat{\mu}_{is}) + \mathbf{x}_i'\hat{\boldsymbol{\beta}}_s\}$ on $\mathbf{x}_i$, where the weights $\sqrt{\hat{\mu}_{is}}$ change at each iteration.

For the general conditional mean function $\hat{\mu}_i = \mu(\mathbf{x}_i, \boldsymbol{\beta})$, the method of scoring, which replaces $\mathbf{H}$ by $\mathrm{E}[\mathbf{H}]$, yields a similar regression, with the dependent variable $\hat{\mu}_{is}^{-1/2}\{(y_i - \hat{\mu}_{is}) + \partial\mu_i/\partial\boldsymbol{\beta}' |_{\hat{\boldsymbol{\beta}}_s} \hat{\boldsymbol{\beta}}_s\}$ and regressor $\hat{\mu}_{is}^{-1/2}\partial\mu_i/\partial\boldsymbol{\beta}' |_{\hat{\boldsymbol{\beta}}_s}$.

3.9 BIBLIOGRAPHIC NOTES

This chapter provides considerable detail on the early literature on basic count regression. Over time the literature has settled on two basic estimators for count data: the Poisson QMLE (with robust sandwich standard errors) and the NB2 MLE. The latter estimator is only applicable to count data that are overdispersed, even after inclusion of regressors, but this is often the case.

An early application of the Poisson regression model is by Jorgenson (1961). In the statistical literature much of the work on the Poisson uses the GLM approach. The key reference is McCullagh and Nelder (1989), with Poisson regression detailed in Chapter 6. In biostatistics a brief survey of clinical trials and epidemiological studies is provided by Kianifard and Gallo (1995). An early review of count models in marketing is Morrison and Schmittlein (1988). In econometrics early influential papers were by Gourieroux et al. (1984a, 1984b). The first paper presented the general theory for LEFN models; the second paper specialized analysis to count data. Cameron and Trivedi (1986) presented both LEFN and negative binomial maximum likelihood approaches, with a detailed application to the doctor visits data used in this chapter. The paper by Hausman, Hall, and Griliches (1984) is also often cited. It considers the more difficult topic of panel count data and is discussed in Chapter 9.

The books by Maddala (1983), Gourieroux and Monfort (1995), Cameron and Trivedi (2005, 2011), Wooldridge (2010), and Greene (2011) provide brief to chapter-length treatments of Poisson regression. Surveys by Winkelmann and Zimmermann (1995), Cameron and Trivedi (1996), and Winkelmann (1995) cover the material in Chapter 3 and also some of the material in Chapter 4. The survey by Gurmu and Trivedi (1994) provides a condensed treatment of many aspects of count data regression. Wooldridge (1997b) emphasizes robust QML estimation. The survey by Trivedi and Munkin (2010) focuses on recent developments. The book by Winkelmann (2008) provides detailed analysis of count models for cross-section data and a briefer treatment of models for time-series, multivariate, and longitudinal data.

3.10 EXERCISES

3.1 The first-order conditions for the Poisson QMLE are $\sum_{i=1}^{n} \mathbf{g}_i(\boldsymbol{\beta}) = \mathbf{0}$, where $\mathbf{g}_i(\boldsymbol{\beta}) = (y_i - \exp(\mathbf{x}_i'\boldsymbol{\beta}))\mathbf{x}_i$. Find $\mathsf{E}\left[\sum_{i=1}^{n} \partial \mathbf{g}_i(\boldsymbol{\beta})/\partial \boldsymbol{\beta}'\right]$ and $\mathsf{E}[\sum_{i=1}^{n} \mathbf{g}_i(\boldsymbol{\beta})\mathbf{g}_i(\boldsymbol{\beta})']$ if y_i has mean μ_i and variance ω_i. Hence verify that the asymptotic variance is (3.9), using the general results in Chapter 2.8.1.

3.2 Obtain the expression for the asymptotic variance of $\hat{\phi}_{\text{NB1}}$ defined in (3.18), using the delta method given in Chapter 2.6.2.

3.3 The geometric model is the special case of NB2 if $\alpha = 1$. Give the density of the geometric model if $\mu_i = \exp(\mathbf{x}_i'\boldsymbol{\beta})/[1 - \exp(\mathbf{x}_i'\boldsymbol{\beta})]$, and obtain the first-order conditions for the MLE of $\boldsymbol{\beta}$. This functional form for the conditional mean corresponds to the canonical link function.

3.4 Using a similar approach to that of Exercise 3.1, obtain the asymptotic variance for the QGPMLE of the NB2 model defined as the solution to (3.37) if in fact y_i has variance ω_i rather than $(\mu_i + \alpha\mu_i^2)$. Hence, give the RS estimator for the variance matrix.

3.5 For regression models with exponential conditional mean function, use the delta method in Chapter 2.6.2 to obtain the formula for a 95% confidence interval for the change in the conditional mean if the jth regressor changes by one unit.

3.6 For the ordered probit model, give the log-likelihood function if $\varepsilon_i \sim \mathsf{N}[0, \sigma_i^2]$ rather than $\varepsilon_i \sim \mathsf{N}[0, 1]$.

3.7 Consider the NLS estimator that minimizes $\sum_i (y_i - \exp(\mathbf{x}_i'\boldsymbol{\beta}))^2$. Show that the first-order conditions for $\boldsymbol{\beta}$ are those shown in (3.69). Using a similar approach to that of Exercise 3.1, show that the asymptotic variance of the NLS estimator is (3.70).

CHAPTER 4

Generalized Count Regression

4.1 INTRODUCTION

The most commonly used models for count regression, Poisson and negative binomial, were presented in Chapter 3. In this chapter we introduce richer models for count regression using cross-section data. For some of these models the conditional mean retains the exponential functional form. Then the Poisson QMLE and NB2 ML estimators remain consistent, although they may be inefficient and may not be suitable for predicting probabilities, rather than the conditional mean. For many of these models, however, the Poisson and NB2 estimators are inconsistent. Then alternative methods are used, ones that generally rely heavily on parametric assumptions.

One reason for the failure of the Poisson regression is that the Poisson process has unobserved heterogeneity that contributes additional randomness. This leads to mixture models, the negative binomial being only one example. A second reason is the failure of the Poisson process assumption and its replacement by a more general stochastic process.

Some common departures from the standard Poisson regression are as follows.

1. *Failure of the mean-equals-variance restriction*: Frequently the conditional variance of data exceeds the conditional mean, which is usually referred to as *extra-Poisson variation* or *overdispersion* relative to the Poisson model. Overdispersion may result from neglected or unobserved heterogeneity that is inadequately captured by the covariates in the conditional mean function. It is common to allow for random variation in the Poisson conditional mean by introducing a multiplicative error term. This leads to families of mixed Poisson models.
2. *Truncation and censoring*: The observed counts may be left truncated (zero truncation is quite common) leading to small counts being excluded, or right censored, by aggregating counts exceeding some value.
3. *The "excess zeros" or "zero inflation" problem*: The observed data may show a higher relative frequency of zeros or of some other integer

than is consistent with the Poisson model. The higher relative frequency of zeros is a feature of all Poisson mixtures obtained by convolution.

4. *Multimodality*: Observed univariate count distributions are sometimes bimodal or multimodal. If this is also a feature of the conditional distribution of counts, perhaps because observations may be drawn from different populations, then extensions of the Poisson, notably finite mixture models, are desirable.
5. *Correlated observations*: Even in a cross-section setting, observations may be correlated due to clustering or spatial correlation. In a time series, correlation is even more likely.
6. *Waiting times between event occurrences not exponential*: Then the process is not a Poisson process, and stochastic process theory leads to count models other than the Poisson.
7. *Trends*: The mean rate of event occurrence, the intensity function, may have a trend or some other deterministic form of time dependence that violates the simple Poisson process assumption.
8. *Simultaneity and sample selection*: Some covariates may be jointly determined with the dependent variable or the included observations may be subject to a sample selection rule.

The first five considerations generally maintain the assumption of an underlying Poisson process, while making concessions to the characteristics of observed data. These considerations are dealt with in detail in this chapter, except for the correlation in the case of time series. This chapter also deals with the sixth departure, which is a failure of the Poisson process assumption. The remaining two complications are presented in, respectively, Chapters 7 and 10.

Section 4.2 presents unobserved heterogeneity and mixture models in detail, a Poisson-gamma mixture leading to the NB model being a leading example. Sections 4.3 and 4.4 consider the case in which the range of observed counts is restricted by either truncation or censoring. Sections 4.5 and 4.6 present two classes of modified count models, hurdle models and zero-inflated models, that give special treatment to zero counts. These models combine elements of both mixing and truncation. Hierarchical models that allow for quite rich modeling of unobserved heterogeneity are presented in section 4.7, whereas finite mixture or latent class models are presented in section 4.8. Models with cross-section dependence are presented in section 4.9. Count models based on an underlying duration model of the time between occurrences of an event are presented in section 4.10. In section 4.11, we present additional models for counts that generally do not rely on mixing. A brief application is provided in section 4.8, with more extensive applications provided in Chapter 6.

4.2 MIXTURE MODELS

Regression models partially control for heterogeneity, since specifying $\mu = \exp(\mathbf{x}'\boldsymbol{\beta})$ allows different individuals to have different Poisson means. In practice there is still unexplained heterogeneity, and many parametric models for counts are obtained by additionally introducing unobserved heterogeneity into the Poisson model.

Unobserved heterogeneity is usually introduced as a multiple of the Poisson mean. Then $y|\mu, \nu \sim \mathsf{P}[\mu\nu]$, and the random heterogeneity term $\nu > 0$ is integrated out to obtain the distribution of y given μ. Such models are called doubly stochastic Poisson by Cox (1955) and a Cox process by Kingman (1993), and they are the basis for the random effects Poisson for panel data presented in Chapter 9.

Several leading models set $\mathsf{E}[\nu] = 1$, the multiplicative analog to the linear model case of additive heterogeneity $y = \mu + u$ with $\mathsf{E}[u] = 0$. When $\mathsf{E}[\nu] = 1$, the mean remains $\mathsf{E}[y] = \mu$ while $\mathsf{V}[y] > \mu$ so that the count data become overdispersed. Different distributions for ν lead to different generalizations of the Poisson, and we present several leading models including the negative binomial, analyzed here in more detail than in Chapter 3. We also discuss estimation when the integral has no closed-form solution.

The multiplicative heterogeneity model is a special case of a mixture model. General results from the mixture literature, including conditions for identification, are also presented in this section. Finally, additive heterogeneity, less appropriate for count data that are necessarily positive, is briefly discussed.

4.2.1 Unobserved Heterogeneity and Overdispersion

We consider mixing based on multiplicative heterogeneity that on average (formally, in expectation) leaves the Poisson mean unchanged. Then there is increased variability in the Poisson mean that in turn can be expected to increase the variability of the count. Intuitively this will lead to overdispersion and to increased probabilities of the occurrence of low counts and of high counts. In this subsection, we confirm that this is indeed the case.

Multiplicative Heterogeneity

The starting point is the Poisson with multiplicative unobserved heterogeneity ν_i,

$$y_i|\mu_i, \nu_i \sim \mathsf{P}[\mu_i\nu_i], \tag{4.1}$$

so $\mathsf{E}[y_i|\mu_i, \nu_i] = \mu_i\nu_i$ and $\mathsf{V}[y_i|\mu_i, \nu_i] = \mu_i\nu_i$ using the equidispersion of the Poisson.

For count regression μ_i depends on observables. The standard Poisson model specifies an exponential conditional mean and includes an intercept term. Then $\mu_i = \exp(\beta_0 + \mathbf{x}_{1i}'\boldsymbol{\beta}_1)$ and

$$\begin{aligned}\mu_i\nu_i &= \exp(\beta_0 + \mathbf{x}_{1i}'\boldsymbol{\beta}_1)\nu_i \\ &= \exp(\beta_0 + \mathbf{x}_{1i}'\boldsymbol{\beta}_1 + \ln \nu_i) \\ &= \exp((\beta_0 + u_i) + \mathbf{x}_{1i}'\boldsymbol{\beta}),\end{aligned} \tag{4.2}$$

where $u_i = \ln \nu_i$. For an exponential conditional mean model, multiplicative heterogeneity is equivalent to a random intercept model.

The random heterogeneity term ν_i is assumed to be iid with the following properties:

$$\begin{aligned} \mathsf{E}[\nu_i|\mu_i] &= \mathsf{E}[\nu_i] \\ \mathsf{E}[\nu_i] &= 1 \\ \mathsf{V}[\nu_i|\mu_i] &= \sigma^2_{\nu_i}. \end{aligned} \tag{4.3}$$

The first assumption in (4.3) is that ν_i is mean independent of μ_i. Then from (4.2) ν_i enters the conditional mean function only through the random intercept term; we refer to this as the separability property of unobserved heterogeneity because the random effect is separable from that due to variation in x_i. If instead ν_i enters through a random slope coefficient, the conditional mean contains terms that are products of x_i and ν_i so that x_i and ν_i are no longer separable. The second assumption in (4.3) is just a normalization that is needed since the model depends on μ_i and ν_i only through the product $\mu_i\nu_i$, and $\mu_i\nu_i$ is unchanged if we double ν_i and halve μ_i. Alternatively, from (4.2) it is clear that with an exponential conditional mean we can only identify the sum $\beta_0+ \mathsf{E}[u_i]$. The third assumption in (4.3) potentially allows $\mathsf{V}[\nu_i|\mu_i]$ to vary with μ_i.

Overdispersion

Under assumptions (4.3) the first two moments of y_i conditional on μ_i, but unconditional on ν_i, are

$$\mathsf{E}[y_i|\mu_i] = \mu_i, \tag{4.4}$$

using $\mathsf{E}_{y|\mu}[y|\mu] = \mathsf{E}_{\nu|\mu}[\mathsf{E}_{y|\mu,\nu}[y|\mu,\nu]] = \mathsf{E}_{\nu|\mu}[\mu\nu] = \mu\mathsf{E}_{\nu|\mu}[\nu] = \mu \times 1 = \mu$, and

$$\mathsf{V}\left[y_i|\mu_i\right] = \mu_i + \sigma^2_{\nu_i}\mu_i^2, \tag{4.5}$$

using $\mathsf{V}_{y|\mu}[y|\mu] = \mathsf{E}_{\nu|\mu}[\mathsf{V}_{y|\mu,\nu}[y|\mu,\nu]] + \mathsf{V}_{\nu|\mu}[\mathsf{E}_{y|\mu,\nu}[y|\mu,\nu]] = \mathsf{E}_{\nu|\mu}[\nu\mu] + \mathsf{V}_{\nu|\mu}[\nu\mu] = \mu\mathsf{E}_{\nu|\mu}[\nu] + \mu^2\mathsf{V}_{\nu|\mu}[\nu]$.

By the results of Chapter 2.4, consistency of the Poisson MLE requires only the correct specification of $\mu_i = \mathsf{E}[y_i|x_i]$. Result (4.4) implies that the Poisson MLE with a correctly specified conditional mean, usually $\mu_i = \exp(\mathbf{x}_i'\boldsymbol{\beta})$, remains a consistent estimator in the presence of unobserved heterogeneity satisfying (4.3). This result will not necessarily hold under alternative specifications of unobserved heterogeneity, such as nonseparability considered later.

The unobserved heterogeneity has induced overdispersion, however, because by (4.5), $\mathsf{V}\left[y_i|\mu_i\right] \geq \mu_i$. Inference based on the Poisson QMLE needs to be based on standard errors that control for this overdispersion.

The overdispersion appears to be of the NB2 form defined in Chapter 3.2.3, with variance a quadratic function of the mean. This is indeed the case if $\sigma^2_{\nu_i} = \alpha$, which occurs often. But more generally we can let $\sigma^2_{\nu_i} = \alpha g(\mu_i)$ for some specified function $g(\cdot)$. For example if $\sigma^2_{\nu_i} = \alpha\mu_i^{k-2}$ then $\mathsf{V}[y_i|\mu_i] = \mu_i + \alpha\mu_i^k$. Examples given later include overdispersion of NB1 form and of cubic form $\mathsf{V}[y_i|\mu_i] = \mu_i + \alpha\mu_i^3$.

The presence of overdispersion opens up the possibility of more efficient estimation. The moment conditions (4.4) and (4.5) provide the basis for sequential quasi-likelihood estimation (McCullagh, 1983; Gourieroux et al., 1984b; Cameron and Trivedi, 1986) and moment estimation (Moore, 1986); see Chapter 2.5.

Poisson Mixtures

More often, however, a fully parametric approach is taken, with a distribution for the unobserved heterogeneity specified. Let $f(y_i|\mu_i, \nu_i)$ denote the Poisson probability mass function parameterized by $\mu_i\nu_i$ (rather than μ_i), so

$$f(y|\mu, \nu) = \frac{e^{-\mu\nu}(\mu\nu)^y}{y!}, \tag{4.6}$$

and let $g(\nu_i|\mu_i)$ denote the probability density function of $\nu_i|\mu_i$. The mixed marginal density of $y_i|\mu_i$ is derived by integrating with respect to ν_i, so

$$h(y|\mu) = \int f(y|\mu, \nu)g(\nu|\mu)d\nu. \tag{4.7}$$

The precise form of this mixed Poisson distribution depends on the specific choice of $g(\nu|\mu)$, although in all cases mixing leads to overdispersion. If $g(\cdot)$ and $f(\cdot)$ are conjugate families, the resulting compound model is expressible in a closed form.

The class of mixture models generated through different assumptions about ν are also known as mixed and/or multilevel models. There is a large literature devoted to this class of models; see, for example, Skrondal and Rabe-Hesketh (2004). There are many special cases such as latent class models, regime switching models, and hidden Markov models that have been found useful in empirical analysis.

Table 4.1 summarizes several leading examples of mixture models for counts that are presented in this chapter. The first five models are mixtures of the Poisson with unobserved heterogeneity that is separable so that $\mathsf{E}[y] = \mu$, where usually $\mu = \exp(\mathbf{x}'\boldsymbol{\beta})$. These models are presented in this section. The remaining three models have a more general form of mixing that changes the conditional mean. These latter models are presented in later sections of this chapter.

Table 4.1. *Selected mixture models for count data*

Distribution	$f(y) = \Pr[Y = y]$		Mean; variance
1. Poisson	$e^{-\mu}\mu^y/y!$		μ; μ
2. NB1	As in NB2 below with α^{-1} replaced by $\alpha^{-1}\mu$		μ; $(1+\alpha)\mu$
3. NB2	$\frac{\Gamma(\alpha^{-1}+y)}{\Gamma(\alpha^{-1})\Gamma(y+1)}\left(\frac{\alpha^{-1}}{\alpha^{-1}+\mu}\right)^{\alpha^{-1}}\left(\frac{\mu}{\mu+\alpha^{-1}}\right)^{y}$		μ; $(1+\alpha\mu)\mu$
4. P-IG	See section 4.2.5		μ; $\mu+\mu^k/\tau$
5. Poisson-normal	$\int_{-\infty}^{\infty}\exp\left(-e^{\mathbf{x}'\boldsymbol{\beta}+\sigma v}\right)\left(e^{\mathbf{x}'\boldsymbol{\beta}+\sigma v}\right)^{y}\frac{1}{y!}\frac{1}{\sqrt{2\pi}}e^{-\frac{v^2}{2}}dv$		μ; $\mu+(e^{\sigma^2}-1)e^{\sigma^2}\mu^2$
6. Hurdle	$f_1(0)$	if $y=0$,	$\frac{1-f_1(0)}{1-f_2(0)}\mu_2$
	$\frac{1-f_1(0)}{1-f_2(0)}f_2(y)$	if $y\geq 1$.	
7. Zero-inflated	$\pi+(1-\pi)f_2(0)$	if $y=0$,	$(1-\pi)\mu_2$;
	$(1-\pi)f_2(y)$	if $y\geq 1$.	$(1-\pi)(\sigma_2^2+\pi\mu_2^2)$
8. Finite mixture	$\sum_{j=1}^{m}\pi_j f_j(y\vert\boldsymbol{\theta}_j)$		$\sum_j \pi_j\mu_j$;
			$\sum_j \pi_j[\mu_j+\sigma_j^2]$
			$-(\sum_j \pi_j\mu_j)^2$

Excess Zeros

In the remainder of this subsection we assume v_i is independent of μ_i, so $g(v|\mu) = g(v)$, and that mixing leads to no change in the conditional mean, so $\mathsf{E}[y|\mu] = \mu$.

Then for any nondegenerate heterogeneity distribution $g(\cdot)$, mixing defined in (4.7) increases the proportion of zero counts

$$h(0|\mu_i) > f(0|\mu_i, v_i). \tag{4.8}$$

Feller (1943) and Mullahy (1997a) provide proofs of this result. Mullahy shows that this property follows from strict convexity of $f(0|\mu, v)$ in μv, and the result generalizes to many count models other than the Poisson, such as the negative binomial. In most instances the frequency of $y = 1$ is less in the mixture distribution than in the parent distribution, so $h(1|\mu_i) < f(1|\mu_i, v_i)$. Finally, the mixture exhibits thicker right tail than the parent distribution. These properties of the Poisson mixtures may be used for constructing specification tests of departures from the Poisson (Mullahy, 1997a).

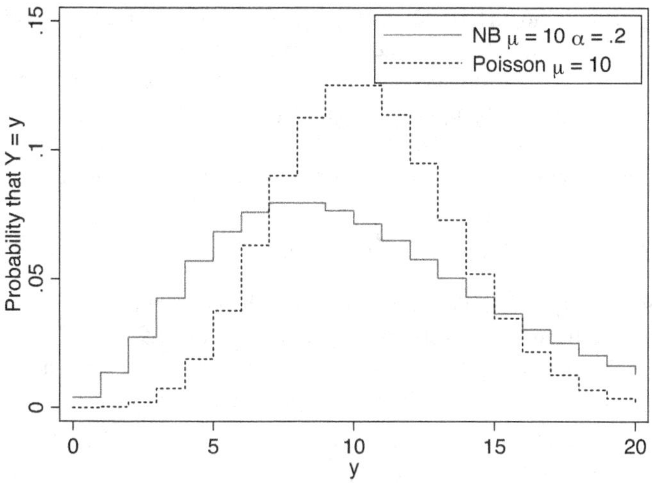

Figure 4.1. Two Crossing Theorem: Negative binomial compared with the Poisson.

The result of mixing leading to fatter left and right tails is a special case of a general result on exponential family mixtures referred to as the Two Crossings Theorem by Shaked (1980).

Two Crossings Theorem: For the random variable y, continuous or discrete, let $f(y|\mu, \nu)$ denote an exponential family conditional (on ν) model density and let $\mathsf{E}[\nu] = 1$. Then the mixed (marginal with respect to ν) distribution $h(y|\mu) = \mathsf{E}_\nu[f(y|\mu, \nu)]$ will have heavier tails than $f(y|\mu, \nu)$ in the sense that the sign pattern of marginal minus the conditional, $h(y|\mu) - f(y|\mu, \nu)$, is $\{+, -, +\}$ because y increases on its support.

That is, for the same mean, any marginal distribution must "cross" the original conditional distribution twice, first from above and then from below. The first crossing accounts for a relative excess of zeros, and the second accounts for greater thickness of the right tail. A sketch of the proof of this theorem is given in section 4.12.

Figure 4.1 illustrates these results, comparing $h(\cdot)$ for a Poisson-gamma mixture or negative binomial with $f(\cdot)$ for a Poisson, where both have the same mean 10. As expected, the mixture has a higher probability of zeroes, fatter right tail, and greater variation.

4.2.2 Negative Binomial as Poisson-Gamma Mixture

The interpretation and derivation of the negative binomial as a *Poisson-gamma mixture* are an old result (Greenwood and Yule, 1920), but the precise parameterization of the gamma may lead to different variance functions (Cameron and Trivedi, 1986). Here we work with the most popular parameterization, which leads to a quadratic-in-mean variance function.

Begin by specifying the heterogeneity term ν to have a two-parameter gamma distribution $g(\nu|\delta,\phi) = (\phi^\delta/\Gamma(\delta))\nu^{\delta-1}e^{-\nu\phi}$, $\delta > 0$, $\phi > 0$. Then $\mathsf{E}[\nu] = \delta/\phi$, and $\mathsf{V}[\nu] = \delta/\phi^2$. To satisfy the normalization that $\mathsf{E}[\nu] = 1$ in (4.3), set $\delta = \phi$ so that ν has the one-parameter gamma distribution

$$g(\nu|\delta) = \frac{\delta^\delta}{\Gamma(\delta)}\nu^{\delta-1}e^{-\nu\delta},\ \delta > 0, \tag{4.9}$$

with $\mathsf{V}[\nu] = 1/\delta$. The Poisson model with heterogeneity is $y|\mu,\nu \sim \mathsf{P}[\mu\nu]$, and it is convenient to work with $\theta \equiv \mu\nu$. So in (4.9) transform from ν to θ using $\theta = \mu\nu$ and the Jacobian transformation term $1/\mu$. The pdf for the heterogeneous term $\theta = \mu\nu$ is then the gamma density

$$\begin{aligned} g(\theta|\delta,\mu) &= \frac{1}{\mu}\frac{\delta^\delta}{\Gamma(\delta)}\left(\frac{\theta}{\mu}\right)^{\delta-1} e^{-\frac{\theta}{\mu}\delta} \\ &= \frac{\left(\frac{\delta}{\mu}\right)^\delta}{\Gamma(\delta)}\theta^{\delta-1}e^{-\frac{\delta\theta}{\mu}}. \end{aligned} \tag{4.10}$$

The Poisson-gamma mixture is

$$\begin{aligned} h(y|\mu,\delta) &= \int f(y|\theta)g(\theta|\delta,\mu)d\theta \\ &= \int \frac{e^{-\theta}\theta^y}{y!} \times \frac{\left(\frac{\delta}{\mu}\right)^\delta}{\Gamma(\delta)}\theta^{\delta-1}e^{-\frac{\delta\theta}{\mu}}d\theta \\ &= \frac{\left(\frac{\delta}{\mu}\right)^\delta}{\Gamma(\delta)y!}\int \theta^{y+\delta-1}\exp\left(-\left(1+\frac{\delta}{\mu}\right)\theta\right)d\theta \\ &= \frac{\left(\frac{\delta}{\mu}\right)^\delta}{\Gamma(\delta)\Gamma(y+1)}\left(1+\frac{\delta}{\mu}\right)^{-(\delta+y)}\Gamma(\delta+y). \end{aligned} \tag{4.11}$$

The last line uses the following properties of the gamma function:

$$\begin{aligned} \Gamma(a) &= \int_0^\infty t^{a-1}e^{-t}dt, && \text{for any } a > 0 \\ \Gamma(a-1) &= a!, && \text{for any } a > 0 \\ b^{-a}\Gamma(a) &= \int_0^\infty t^{a-1}e^{-bt}dt, && \text{for any } b > 0. \end{aligned}$$

Defining $\alpha = 1/\delta$ and rearranging, (4.11) becomes

$$h(y|\mu,\alpha) = \frac{\Gamma(\alpha^{-1}+y)}{\Gamma(\alpha^{-1})\Gamma(y+1)}\left(\frac{\alpha^{-1}}{\alpha^{-1}+\mu}\right)^{\alpha^{-1}}\left(\frac{\mu}{\alpha^{-1}+\mu}\right)^y. \tag{4.12}$$

This is the negative binomial distribution originally defined in Chapter 3.3.1. As noted there,

$$\mathsf{E}[y|\mu,\alpha] = \mu$$
$$\mathsf{V}[y|\mu,\alpha] = \mu(1+\alpha\mu) > \mu, \qquad \text{since } \alpha > 0.$$

Maximum likelihood estimation of this model is discussed in Chapter 3.3. Note that the Poisson QMLE is also consistent here, since $\mathsf{E}[y|\mu] = \mu$ due to the separability assumption on unobserved heterogeneity.

4.2.3 Other Characterizations of the Negative Binomial

The characterization of the NB distribution as a Poisson-gamma mixture is only one of a number of chance mechanisms that can generate that distribution. Boswell and Patil (1970) list 13 distinct stochastic mechanisms for generating the NB. These include the NB as a waiting-time distribution, as a Poisson sum of a logarithmic series of random variables, as a linear birth and death process, as the equilibrium of a Markov chain, and as a group-size distribution.

An Alternative Parameterization of NB

A parametrization of the NB used in many references, as well as in algorithms to generate draws from the negative binomial, is the following:

$$\Pr[Y=y] = \binom{r+y-1}{r-1}(1-p)^r p^y, \quad y = 0,1,2,\ldots, \tag{4.13}$$

where $0 < p < 1$ and $r > 0$. Then $\mathsf{E}[Y] = pr/(1-p)$ and $\mathsf{V}[Y] = pr/(1-p)^2$. Letting $r = 1/\alpha$ and $p = \alpha\mu/(1+\alpha\mu)$ yields the parameterization of the NB given in (4.12).

NB as a Poisson-Stopped Log-Series Distribution

Many counted events can occur in spells (e.g., doctor visits arise from each of several spells of illness). Then there is a distribution of the random variable S (spells) and a distribution of events (R_j) within the jth spell, conditional on a spell occurring.

Suppose the number of spells is generated according to the Poisson distribution with parameter λ, while the number of events (which may be dependent) within a given spell is generated by a log-series probability law defined as

$$\Pr[R=r] = \frac{a\theta^r}{r}, \quad r = 1,2,\ldots,$$

where $0 < \theta < 1$ and $a = -1/(\ln(1-\theta))$. The first two moments are $\mathsf{E}[R] = a\theta/(1-\theta)$ and $\mathsf{V}[R] = a\theta(1-a\theta)/(1-\theta)^2$. A feature of the log-series distribution is that the probability mass function has a maximum at $R = 1$, and $\Pr[R = r+1]$ always exceeds $\Pr[R = r]$.

Combining the Poisson and log-series assumptions gives the marginal distribution

$$\Pr[Y = y] = \binom{a\lambda + y - 1}{a\lambda - 1}(1-\theta)^{a\lambda}\theta^{y}, \quad y = 0, 1, 2, \ldots, \tag{4.14}$$

where $Y = \sum_{j=1}^{S} R_j = S + \sum_{j=1}^{S}(R_j - 1)$, and R_j denotes the number of events in spell j. This is a negative binomial density, equaling (4.13) with $r = a\lambda$ and $p = \theta$. The first two moments are $\mathsf{E}[Y] = a\lambda\theta/(1-\theta)$ and $\mathsf{V}[Y] = a\lambda\theta/(1-\theta)^2 = \mathsf{E}[Y]/(1-\theta)$. Because the variance is linear in the mean, this is the NB1 negative binomial. The standard NB1 regression model can be obtained by assuming $\lambda = \exp(\mathbf{x}'\boldsymbol{\beta})$ and treating both a and θ as constants.

This mixture distribution is also known as the Poisson-stopped log-series distribution or stopped-sum distribution; see Johnson, Kemp, and Kotz (2005). The zeros of this model imply that a spell did not occur; conversely, if a spell does occur then the event of interest must be observed at least once because the log-series distribution has no mass at zero. This condition is potentially restrictive. For example, in practice a spell of illness may occur, but no doctor visit may result.

Santos Silva and Windmeijer (2001) implement this model for data on the total number of doctor visits, so only Y is observed and not its subcomponents $R_j, j = 1, \ldots, S$. They compare results with those from a hurdle model, defined in section 4.5, which can be viewed as a single-spell model. They point out that additional flexibility can result from further parameterizing θ as a logistic function of the form $\exp(\mathbf{z}'\boldsymbol{\gamma})/(1 + \exp(\mathbf{z}'\boldsymbol{\gamma}))$, where $\mathbf{z}$ are regressors that affect doctor visits once a spell occurs, whereas $\mathbf{x}$ are factors influencing the number of spells.

NB as a limiting Polya Distribution

A special mention should be made of Eggenberger and Polya's (1923) derivation of the NB as a limit of an urn scheme. The idea here is that an urn scheme can be used to model true contagion in which the occurrence of an event affects the probability of later events.

Consider an urn containing N balls of which a fraction p are red and fraction $q = 1 - p$ are black. A random sample of size n is drawn. After each draw the ball drawn is replaced and $k = \theta N$ balls of the *same* color are added to the urn. Let Y be the number of red balls in n trials. Then the distribution of Y is the Polya distribution defined for $y = 0, 1, \ldots, n$ as

$$\Pr[Y = y] = \binom{n}{y}\frac{p(p+\theta)\cdots[p+(y-1)\theta]\,q(q+\theta)\cdots[q+(n-y)\theta]}{1(1+\theta)\cdots[1+(n-1)\theta]},$$

Let $n \to \infty$, $p \to 0$, $\theta \to 0$, with $np \to \eta$ and $\theta n \to b\eta$, for some constant b. Then the limit of the Polya distribution can be shown to be the NB with parameters $k \equiv 1/b$ and $1/(1+b\eta)$; that is,

$$\Pr[Y=y] = \binom{k+y-1}{k-1}\left(\frac{\eta-k}{\eta}\right)^k\left(\frac{k}{\eta}\right)^y, \tag{4.15}$$

where for convenience we take $k = \theta N$ to be an integer (see Feller, 1968; pp. 118–145; Boswell and Patil, 1970). This is (4.13) with $r = k$ and $p = k/\eta$.

The Polya urn scheme can be interpreted in terms of the occurrence of social or economic events. Suppose an event of interest, such as an accident, corresponds to drawing a red ball from an urn. Suppose that subsequent to each such occurrence, social or economic behavior increases the probability of the next occurrence. This is analogous to a scheme for replacing the red ball drawn from the urn in such a way that the proportion of red balls to black balls increases. If, however, after a ball is drawn, more balls of the opposite color are added to the urn, then the drawing of a ball reduces the probability of a repeat occurrence at the next draw.

Of course, there can be many possible replacement schemes in the urn problem. Which scheme one uses determines the nature of the dependence between one event and the subsequent ones. This shows that the NB distribution can arise due to occurrence dependence, called true contagion. Alternatively, as seen earlier the NB distribution can arise when different individuals experience an event with constant but different probabilities. This is analogous to individuals having separate urns with red and black balls in different proportions. For each person the probability of drawing a red ball is constant, but there is a distribution across individuals. In the aggregate one observes apparent dependence, called spurious contagion, due to heterogeneity. This topic is pursued further in section 4.10.1.

4.2.4 Poisson-Lognormal Mixture

The Poisson-gamma mixture is popular because it leads to a closed-form solution, the NB, for the mixture distribution. But heterogeneity distributions other than the gamma may be more natural choices and may provide a Poisson mixture model that better fits the data.

An obvious choice is the lognormal distribution for ν_i, with $\ln \nu \sim \mathsf{N}[0, \sigma^2]$. In the case of an exponential model for μ, similar to (4.2) we have

$$\begin{aligned} \mu\nu &= \exp(\mathbf{x}'\boldsymbol{\beta} + \ln \nu) \\ &= \exp(\mathbf{x}'\boldsymbol{\beta} + \sigma\varepsilon), \end{aligned}$$

where $\varepsilon \sim \mathsf{N}[0, 1]$. The marginal distribution obtained by integrating out ε is

$$\begin{aligned} h(y|\mathbf{x}, \boldsymbol{\beta}, \sigma) &= \textstyle\int_{-\infty}^{\infty} f(y|\mathbf{x}, \boldsymbol{\beta}, \sigma, \varepsilon)g(\varepsilon)d\varepsilon \\ &= \textstyle\int_{-\infty}^{\infty} f(y|\mathbf{x}, \boldsymbol{\beta}, \sigma, \varepsilon)\frac{1}{\sqrt{2\pi}}e^{-\varepsilon^2/2}d\varepsilon \qquad (4.16) \\ &= \textstyle\int_{-\infty}^{\infty} \exp\left(-e^{\mathbf{x}'\boldsymbol{\beta}+\sigma\varepsilon}\right)\left(-e^{\mathbf{x}'\boldsymbol{\beta}+\sigma\varepsilon}\right)^y \frac{1}{y!}\frac{1}{\sqrt{2\pi}}e^{-\varepsilon^2/2}d\varepsilon \end{aligned}$$

since $f(\cdot)$ is the $\mathsf{P}[\mu\nu]$ and $g(\varepsilon)$ is the standard normal density. The model is variously called the *Poisson-lognormal model*, because ν is lognormal, or the *Poisson-normal model*, because ε is normal distributed.

There is no closed-form solution for $h(y|\mathbf{x}, \boldsymbol{\beta}, \sigma)$, unlike for the Poisson-gamma, but numerical methods work well. There are several ways to proceed.

Simulated Maximum Likelihood

Because the marginal distribution is a mathematical expectation with respect to density $g(\varepsilon)$, the expression in (4.16) can be approximated by making draws of ε from $g(\varepsilon)$, here the $\mathsf{N}[0, 1]$ distribution, evaluating $f(y|\mathbf{x}, \boldsymbol{\beta}, \sigma, \varepsilon)$ at these draws, and averaging. Then for the ith observation with S draws $\varepsilon_i^{(1)}, \ldots, \varepsilon_i^{(S)}$, we use

$$\hat{h}(y_i|\mathbf{x}_i, \boldsymbol{\beta}, \sigma) = \frac{1}{S}\sum_{s=1}^{S} f(y_i|\mathbf{x}_i, \boldsymbol{\beta}, \sigma, \varepsilon_i^{(s)}). \tag{4.17}$$

This method is also called Monte Carlo integration.

The simulated likelihood function can be built up using n terms like that above. The *maximum simulated likelihood estimator* maximizes the simulated log-likelihood function

$$\begin{aligned}\hat{\mathcal{L}}(\boldsymbol{\beta}, \sigma) &= \sum_{i=1}^{n} \ln \hat{h}(y_i|\mathbf{x}_i, \boldsymbol{\beta}, \sigma) \qquad (4.18)\\ &= \sum_{i=1}^{n} \ln\left(\frac{1}{S}\sum_{s=1}^{S} f(y_i|\mathbf{x}_i, \boldsymbol{\beta}, \sigma, \varepsilon_i^{(s)})\right).\end{aligned}$$

It is known that if S is held fixed while $n \to \infty$, the resulting estimator of $(\boldsymbol{\beta}, \sigma)$ is biased (McFadden and Ruud, 1994; Gourieroux and Monfort, 1997). However, if S and n tend to infinity in such a way that $\sqrt{S}/n \to 0$, the estimator is both consistent and asymptotically efficient.

An active area of research is concerned with generating $\varepsilon_i^{(s)}$ such that the effective size of S becomes large while the nominal number of draws remains manageable, because this avoids the computationally burdensome task of letting S go to infinity. For example, quasi-random draws of $\varepsilon_i^{(s)}$ are substituted for pseudorandom draws. Another modification is to incorporate a bias correction term in the approximation to the density using finite S (Munkin and Trivedi, 1999). For simple problems such as this example where draws are with respect to the univariate standard normal, however, there is no practical problem in simply using a very high value of S.

The simulation approach is flexible and potentially very useful in the context of truncated or censored models with unobserved heterogeneity and also for structural models. An example is given in Chapter 10.3.1.

An alternative simulation procedure to maximum simulated likelihood is to use Bayesian methods. They are treated separately in Chapter 12, and Bayesian analysis of the Poisson-lognormal model is presented in Chapter 12.5.2.

Gaussian Quadrature

The integral in (4.16) may be evaluated using *numerical integration*, rather than simulation. An alternative estimation procedure may be based on numerical integration. For univariate integrals, a method that works well is Gaussian quadrature, specifically Gauss-Hermite quadrature for (4.16) since integration is over $(-\infty, \infty)$. In general m-point *Gauss-Hermite quadrature* approximates $\int_{-\infty}^{\infty} e^{-x^2} r(x)$ by the weighted sum $\sum_{j=1}^{m} w_j r(x_j)$, where the weights w_j and evaluation points x_j are given in tables in Stroud and Secrest (1966), for example, or can be generated using computer code given in Press, Teukolsky, Vetterling, and Flannery (2008).

To apply this method here we transform from ε to $x = \varepsilon/\sqrt{2}$ with Jacobian $\sqrt{2}$, and rewrite the second line in (4.16) as

$$h(y|\mathbf{x}, \boldsymbol{\beta}, \sigma) = \int_{-\infty}^{\infty} e^{-x^2} \frac{1}{\sqrt{\pi}} f(y|\mathbf{x}, \boldsymbol{\beta}, \sigma, \sqrt{2}x)dx.$$

The Gauss-Hermite approximation for the ith observation is

$$\hat{h}(y_i|\mathbf{x}_i, \boldsymbol{\beta}, \sigma) = \sum_{j=1}^{m} w_j \frac{1}{\sqrt{\pi}} f(y|\mathbf{x}_i, \boldsymbol{\beta}, \sigma, \sqrt{2}x_j). \tag{4.19}$$

Note that scalar evaluation points x_j here should not be confused with the regressors $\mathbf{x}_i$. The MLE then maximizes $\sum_{i=1}^{n} \ln \hat{h}(y_i|\mathbf{x}_i, \boldsymbol{\beta}, \sigma)$. An example is given in Chapter 10.3.1.

This approach was used by Hinde (1982) and Winkelmann (2004) for the Poisson-lognormal model. Many software packages offer this model as an option. Given the relative simplicity of one-dimensional numerical integration, any distribution for unobserved heterogeneity that a user finds appealing can be handled by numerical integration so there is no compelling justification for sticking to closed-form mixtures.

4.2.5 Poisson-Inverse Gaussian Mixture

The statistical literature has documented that the NB model has limitations in modeling long-tailed data. For example, based on research on word frequencies, Yule (1944) conjectures that the compound Poisson model provides a way to tackle the problem, but he rejects the NB model because of its poor performance. Sichel (1971) suggests a distribution suitable for modeling long-tailed frequency data that has been mainly used in biology (Ord and Whitmore, 1986) and linguistics (Sichel, 1974, 1975). However, these studies involve only the univariate version, whereas we are interested in the regression model.

A special case of the Sichel distribution is the *Poisson-inverse Gaussian* (P-IG) distribution. Willmot (1987) discusses the properties and interpretations of the P-IG model and its attractiveness relative to the NB model in some univariate actuarial applications. Although on a priori grounds P-IG appears to be a useful model in a variety of contexts, it has not been used much in econometrics.

The random variable ν is positive-valued and follows the inverse Gaussian (IG) distribution, with pdf

$$f(\nu|\tau,\lambda) = \sqrt{\frac{\tau}{2\pi\nu^3}}\exp\left[-\frac{\tau(\nu-\lambda)^2}{2\lambda^2\nu}\right], \tag{4.20}$$

where $\nu > 0$, $\tau > 0$, and $\lambda > 0$. The IG is a two-parameter exponential family with shape similar to a chi-squared distribution. As the shape parameter $\tau \to \infty$, the IG becomes more normal in shape. It can be shown that the IG distribution has mean λ and variance λ^3/τ. Alternative equivalent parameterizations and other properties of the IG are given in Tweedie (1957).

We present a P-IG regression model due to Guo and Trivedi (2002). As usual $y|\mu,\nu \sim \mathsf{P}[\mu\nu]$. To normalize $\mathsf{E}[\nu] = 1$, see (4.3), set $\lambda = 1$ in (4.20); then $\mathsf{V}[\nu] = 1/\tau$. The P-IG model is obtained as

$$\Pr[Y = y] = \int_0^\infty \frac{\exp(-\nu\mu)(\nu\mu)^y}{y!} f(\nu|\tau,\lambda)d\nu. \tag{4.21}$$

The resulting mean and variance are $\mathsf{E}[Y] = \mu$ and $\mathsf{V}[Y] = \mu + \mu^3/\tau$, and the P-IG approaches the standard Poisson as $\tau \to \infty$. In contrast to the quadratic variance of the standard NB model, this version of the P-IG has a cubic variance function that permits a high rate of increment that may better model highly overdispersed data.

From Chapter 3.3.3 a range of NB models can be obtained according to how the overdispersion parameter is parameterized. A range of P-IG models can similarly be obtained. Guo and Trivedi (2002) reparameterize the overdispersion parameter τ in (4.20) as μ^2/η. Then the P-IG distributed random variable has mean μ, variance $\mu + \mu\eta$, and probability mass function

$$\begin{aligned}\Pr[Y=0] &= \exp\left[\tfrac{\mu}{\eta}\left(1-\sqrt{1+2\eta}\right)\right]\\ \Pr[Y=y] &= \Pr[Y=0]\tfrac{\mu^y}{\Gamma(y+1)}(1+2\eta)^{-y/2}\times\\ &\quad\times\textstyle\sum_{j=0}^{y-1}\tfrac{\Gamma(y+j)}{\Gamma(y-j)\Gamma(j+1)}\left(\tfrac{\eta}{2\mu}\right)^j(1+2\eta)^{-j/2},\end{aligned} \tag{4.22}$$

for $y = 1, 2, \ldots$. The details of the derivation, which involves considerable algebra, are omitted; see Willmot (1987). Because the score equations for the P-IG model are very nonlinear, an iterative solution algorithm is needed to implement ML estimation. Application is to two data sets, one on the number of patent applications and one on the number of patents awarded.

Dean, Lawless, and Willmot (1989) use an alternative version of the P-IG model for which the variance of Y is $\mu + \eta\mu^2$. Then in (4.22) replace η by $\eta\mu$.

They argue that this model provides an attractive framework for handling longer tailed distributions. They applied the model to the number of motor insurance claims made in each of 315 risk groups.

4.2.6 General Mixture Results

The statistics literature contains many examples of generalized count models generated by mixtures. An historical account can be found in Johnson, Kemp, and Kotz (2005). In addition to NB, which is the most popular mixture, and P-IG and Poisson-lognormal, already covered, other mixtures include discrete lognormal (Shaban, 1988), generalized Poisson (Consul and Jain, 1973; Consul, 1989), and Gauss-Poisson (Johnson, Kemp, and Kotz, 2005). Additional flexibility due to the presence of parameters of the mixing distribution generally improves the fit of the resulting distribution to observed data.

It is useful to distinguish between *continuous mixtures* (*convolutions*) and *finite mixtures.*

Definition. *Suppose $F(y|\boldsymbol{\theta})$ is a parametric distribution depending on $\boldsymbol{\theta}$, and let $\pi(\boldsymbol{\theta}|\boldsymbol{\alpha})$ define a continuous mixing distribution. Then a convolution (continuous mixture) is defined by $F(y|\boldsymbol{\alpha}) = \int_{-\infty}^{\infty} \pi(\boldsymbol{\theta}|\boldsymbol{\alpha})F(y|\boldsymbol{\theta})d\boldsymbol{\theta}$.*

Definition. *If $F_j(y|\boldsymbol{\theta}_j)$, $j = 1, 2, \ldots, m$, is a distribution function then $F(y|\pi_j) = \sum_{j=1}^{m} \pi_j F_j(y|\boldsymbol{\theta}_j)$, $0 < \pi_j < 1$, $\sum_{j=1}^{m} \pi_j = 1$, defines an m-component finite mixture.*

Note that although these definitions are stated in terms of mixtures of the cdf rather than the pdf, definitions in terms of the latter are feasible (and have been used so far in section 4.2). The second definition is a special case of the first, if $\pi(\boldsymbol{\theta}|\boldsymbol{\alpha})$ is discrete and assigns positive probability to a finite number of parameter values $\boldsymbol{\theta}_1, \ldots, \boldsymbol{\theta}_m$. In this case π_j, the mixing proportion, is the probability that an observation comes from the jth population. By contrast, in a continuous mixture the parameter $\boldsymbol{\theta}$ of the conditional density is subject to chance variation described by a density with an infinite number of support points. Estimation and inference for convolutions involve the parameters $(\boldsymbol{\theta}, \boldsymbol{\alpha})$ and for finite mixtures the parameters $(\pi_j, \boldsymbol{\theta}_j; j = 1, \ldots, m)$.

The Poisson with multiplicative heterogeneity is the special case where $\boldsymbol{\theta} = (\theta_1, \theta_2) = (\mu, \nu)$ and a mixing distribution is only specified for θ_2. The more general formulation here leads naturally to hurdle, zero-inflated, and finite mixture models that are presented in subsequent sections.

4.2.7 Identification

Identifiability, or unique characterization, of mixtures should be established before estimation and inference. A mixture is *identifiable* if there is a unique correspondence between the mixture and the mixing distribution, usually in the presence of some a priori constraints (Teicher, 1961).

The probability generating function (pgf) of a mixed Poisson model, denoted $P(z)$, can be expressed as the convolution integral

$$P(z) = \int_0^\infty \exp(\mu(z-1))\pi(\mu)d\mu, \tag{4.23}$$

where $\exp(\mu(z-1))$ is the pgf of the Poisson distribution and $\pi(\mu)$ is the assumed distribution for μ.

Mixture models, being akin to "reduced-form" models, are subject to an identification problem. The same distribution can be obtained from more than one mixture. For example, the negative binomial mixture can be generated as a Poisson-gamma mixture by allowing the Poisson parameter μ to have a gamma distribution (see section 4.2.2). It can also be generated by taking a random sum of independent random variables in which the number of terms in the sum has a Poisson distribution; if each term is discrete and has a logarithmic distribution and if the number of terms has a Poisson distribution, then the mixture is negative binomial (Daley and Vere-Jones, 1988). In this case, identification may be secured by restricting the conditional event distribution to be Poisson. This follows from the uniqueness property of exponential mixtures (Jewel, 1982).

A practical consideration is that in applied work, especially that based on small samples, it may be difficult to distinguish between alternative mixing distributions, and the choice may be largely based on the ease of computation. Most of the issues are analogous to those that have been discussed extensively in the duration literature; see Lancaster (1990, chapter 7) for the proportional hazard model. In the examples in the duration literature, finiteness of the mean of the mixing distribution is required for identifiability of the mixture. For counts, it is possible to obtain Poisson-gamma (NB) and Poisson-inverse Gaussian (P-IG) mixtures with the same mean and variance, see sections 3.3.1 and 4.2.5, so the two mixing distributions can only be distinguished using information about higher moments. This may be empirically challenging given typical sample sizes.

Although more flexible count distributions are usually derived by mixing, it may sometimes be appropriate to directly specify flexible functional forms for counts, without the intermediate step of introducing a distribution of unobserved heterogeneity. For example, this may be convenient in aggregate time-series applications. In microeconometric applications, however, mixing seems a natural way of handling heterogeneity.

4.2.8 Computational Considerations

Under certain conditions, parametric mixture distributions based on the Poisson are straightforward to estimate using standard optimizers. The relative simplicity of optimization in this case is a consequence of unimodality of the likelihood function. This property of unimodality is realized under the conditions of the following theorem due to Holgate (1970).

Definition. *Let $g(\mu)$ be the probability density function of an absolutely continuous random variable. Then the nonnegative integer-valued random variable with probability mass function* $\Pr[Y = y] = \int \frac{e^{-\mu}\mu^y}{y!} g(\mu) d\mu,\ \ y \geq 0$, *is a unimodal lattice variable.*

By this theorem continuous mixtures based on the Poisson, such as the NB and P-IG distributions, have unimodal likelihood functions. Hence an application of a standard gradient-based algorithm will globally maximize the likelihood function. This is an attractive feature of the parametric mixtures considered earlier, even examples such as Poisson-lognormal that do not have a closed-form solution for the mixture density.

4.2.9 Consequences of Misspecified Heterogeneity

In the duration literature, in which the shape of the hazard function is of central interest, there has been an extensive discussion of how misspecified unobserved heterogeneity can lead to inconsistent estimates of the hazard function; see Heckman and Singer (1984) and Lancaster (1990, pp. 294–305) for a summary.

Under the assumption of correct conditional mean specification, the consequences of misspecified separable heterogeneity were dealt with in Chapters 2 and 3 in the context of an uncensored distribution. The Poisson and NB2 MLEs remain consistent, since they are in the LEFN, although a robust variance matrix estimator needs to be used. The MLE in other NB models, P-IG and Poisson-lognormal models, will be inconsistent.

When the data distribution is truncated or censored, the robustness of Poisson and NB2 to misspecification disappears. Suppose the dgp is NB2, a Poisson-gamma mixture, but the data are truncated at zero. Then the appropriate model is the zero-truncated NB2, and the MLE for the zero-truncated Poisson is inconsistent; see section 4.3.2. The neglect of unobserved heterogeneity is no longer relatively innocuous and is not remedied by the use of a robust variance matrix estimator. A similar conclusion will apply to finite mixture-type models in which unobserved heterogeneity does not enter separably. A fuller treatment of these issues is provided in Sections 4.3–4.8.

4.2.10 Additive Heterogeneity

The preceding discussion of Poisson mixtures focused on multiplicative continuous unobserved heterogeneity. The additive case has also been considered.

A Poisson model with a conditional mean function that incorporates *additive* independent nonnegative unobserved heterogeneity $\nu > 0$ is

$$\begin{aligned} y|\mu, \nu &\sim \mathsf{P}[\mu + \nu] \\ &\sim \mathsf{P}[\mu] + \mathsf{P}[\nu]. \end{aligned}$$

The second line uses the result that the sum of independent Poissons is Poisson with mean the sum of the means. Taking expectation with respect to ν, so that conditioning is just on μ, additive heterogeneity increases the mean of y from μ to $\mu + \mathsf{E}[\nu]$, and the variance from μ to $\mu + \mathsf{E}[\nu] + \mathsf{V}[\nu]$. Unlike the multiplicative case with $\mathsf{E}[\nu] = 1$, the mean necessarily increases. The variance also increases and by more than the mean so there is overdispersion.

As an example, suppose $\nu \sim \mathsf{Gamma}[\alpha, \beta]$, a two-parameter gamma that is not restricted to have $\mathsf{E}[\nu] = 1$. Then

$$y|\mu \sim \mathsf{P}[\mu] + \mathsf{P}[\mathsf{Gamma}[\alpha, \beta]],$$

a marginal distribution that goes by the name Delaporte distribution. A special case is when $\alpha = 1$, so $\mathsf{E}[\nu] = 1$, when we have a sum of Poisson and negative binomial. See Vose (2008) for additional detail and examples and Fé (2012) for theory and application to production frontier analysis.

4.3 TRUNCATED COUNTS

Models for incompletely observed count data due to truncation and censoring are presented, respectively, in this section and the subsequent section. Failure to control for truncation or censoring leads to inconsistent parameter estimation, because the observed data will have a conditional distribution and mean that differ from those of the underlying dgp for the count. Linking the two relies on strong parametric assumptions. Analysis usually needs to be based on richer models than the Poisson, such as NB, though for illustrative purposes we often use the Poisson.

Truncated and censored count models are discrete counterparts of truncated and censored models for continuous variables, particularly the Tobit model for normally distributed data, that have been used extensively in the economics literature (Maddala, 1983; Amemiya, 1984). Sample selection counterparts of these models are discussed in Chapter 10. Truncated models are also a component of the hurdle model presented in section 4.5.

4.3.1 Standard Truncated Models

Truncation arises if observations (y_i, x_i) in some range of y_i are totally lost. In particular, for some studies involving count data, inclusion in the sample requires that sampled individuals have been engaged in the activity of interest or, as Johnson and Kotz (1969, p. 104) put it, "the observational apparatus become[s] active only when a specified number of events (usually one) occurs." Examples of left truncation include the number of bus trips made per week in surveys taken on buses, the number of shopping trips made by individuals sampled at a mall, and the number of unemployment spells among a pool of unemployed. Right truncation occurs if high counts are not observed. For simplicity we focus on left truncation. Analogous results for right truncation can be derived by adapting those for left truncation.

The following general framework for truncated count models is used. Let

$$\begin{aligned} h(y|\boldsymbol{\theta}) &= \Pr[Y = y] \\ H(y|\boldsymbol{\theta}) &= \Pr[Y \leq y] \end{aligned} \tag{4.24}$$

denote the pdf and cdf of the nonnegative integer-valued random variable Y, where $\boldsymbol{\theta}$ is a parameter vector.

Left Truncation

If realizations of y less than a positive integer, say r, are omitted, then the ensuing distribution is called *left truncated* or *truncated from below*. The left-truncated (at r) count distribution is given by

$$f(y|\boldsymbol{\theta}, y \geq r) = \frac{h(y|\boldsymbol{\theta})}{1 - H(r-1|\boldsymbol{\theta})}, \qquad y = r, r+1, \ldots. \tag{4.25}$$

A special case is the *left-truncated* NB2 *model*. Then $h(y|\boldsymbol{\theta})$ is the NB2 density defined in (4.12) and $\boldsymbol{\theta} \equiv (\mu, \alpha)$. The resulting truncated NB2 density has truncated mean γ and truncated variance σ^2 defined by

$$\begin{aligned} \gamma &= \mu + \delta \\ \sigma^2 &= \mu + \alpha\mu^2 - \delta(\gamma - r) \\ \delta &= \mu\,[1 + \alpha(r-1)]\,\lambda(r-1|\mu, \alpha) \\ \lambda(r-1|\mu, \alpha) &= h(r-1|\mu, \alpha)/[1 - H(r-1|\mu, \alpha)], \end{aligned} \tag{4.26}$$

see Gurmu and Trivedi (1992).

For the NB2 model, the relation between the truncated mean and the mean of the parent distribution can be expressed as

$$\mathsf{E}[y|y \geq r] = \mathsf{E}[y] + \delta,$$

where $\delta > 0$. The truncated mean exceeds the untruncated mean. Furthermore, the difference between the truncated and untruncated means depends on model parameters directly and via $\lambda(r-1|\mu, \alpha)$. The adjustment factor δ plays a useful role, analogous to the Mill's ratio in continuous models, in the estimation and testing of count models. For the variance

$$\mathsf{V}[y|y \geq r] = \mathsf{V}[y] - \delta(\gamma - r),$$

so truncation decreases the variance, since the truncated mean γ necessarily exceeds the truncation point r.

The Poisson is the limiting case of NB2 as $\alpha \to 0$. The *left-truncated Poisson* density is

$$f(y|\mu, y \geq r) = \frac{e^{-\mu}\mu^y}{\left[1 - \sum_{j=0}^{r-1} \frac{e^{-\mu}\mu^j}{j!}\right] y!}, \qquad y = r, r+1, \ldots, \tag{4.27}$$

with truncated mean γ and truncated variance σ^2 given by

$$\begin{aligned} \gamma &= \mu + \delta \\ \sigma^2 &= \mu - \delta(\gamma - r) \\ \delta &= \mu\lambda(r-1|\mu) \\ \lambda(r-1|\mu) &= h(r-1|\mu)/1 - H(r-1|\mu). \end{aligned} \tag{4.28}$$

As for the NB2 density, truncation leads to a higher mean and to a lower variance. Note that the truncated variance does not equal the truncated mean.

Comparing the Poisson to NB2, the truncated means differ, whereas they were equal in the untruncated case. This shows that when truncation occurs, the method of moments estimators based on the truncated mean will be highly dependent on correct specification of the untruncated distribution $H(y|\boldsymbol{\theta})$. The left-truncated NB2 incorporates overdispersion in the sense that the variance of truncated NB2, given in (4.26), exceeds that for truncated Poisson.

Left Truncation at Zero

The most common form of truncation in count models is left truncation at zero, in which case $r = 1$ (Gurmu, 1991). That is, the observation apparatus is activated only by the occurrence of an event.

In general the zero-truncated moments are related to the untruncated moments $\mathsf{E}[y]$ and $\mathsf{V}[y]$ by

$$\mathsf{E}[y|y>0] = \frac{\mathsf{E}[y]}{\Pr[y>0]} \tag{4.29}$$

$$\mathsf{V}[y|y>0] = \frac{\mathsf{V}[y]}{\Pr[y>0]} - \frac{\Pr[y=0](\mathsf{E}[y])^2}{(\Pr[y>0])^2},$$

where the second expression is obtained after some algebra given in section 4.12. Left truncation at zero decreases the mean, and the first two truncated moments depend crucially on $\Pr[y > 0]$ which varies with the parametric model for y.

For *Poisson-without-zeros* the truncated mean given in (4.28) simplifies to

$$\mathsf{E}[y|y>0] = \frac{\mu}{1-e^{-\mu}}, \tag{4.30}$$

using $r = 1$ so $\lambda(r-1) = \lambda(0) = h(0)/(1-h(0)) = e^{-\mu}/(1-e^{-\mu})$ as $h(0) = e^{-\mu}$. Alternatively we could use (4.29) and $\Pr[y > 0] = 1 - e^{-\mu}$. Also

$$\mathsf{V}[y|y>0] = \frac{\mu}{1-e^{-\mu}}\left[1 - \frac{\mu e^{-\mu}}{1-e^{-\mu}}\right], \tag{4.31}$$

since from (4.28) the truncated variance $\sigma^2 = \mu - \delta(\gamma - 1) = \mu - \delta(\mu + \delta - 1) = \mu(1-\delta) + \delta(1-\delta) = \gamma(1-\delta)$. As noted already $\mathsf{E}[y|y > 0]$ depends on the parametric model, which here sets $\Pr[y = 0] = e^{-\mu}$. Misspecification of this probability leads to a misspecified truncated mean, leading to inconsistent estimation of the untruncated mean μ.

For NB2-*without-zeroes*, $h(0) = (1+\alpha\mu)^{-\alpha^{-1}}$, so $\lambda(0) = (1+\alpha\mu)^{-\alpha^{-1}}/(1-(1+\alpha\mu)^{-\alpha^{-1}})$. The truncated mean given in (4.26) simplifies to

$$\mathsf{E}[y|y>0] = \frac{\mu}{1-(1+\alpha\mu)^{-\alpha^{-1}}} \tag{4.32}$$

and, the truncated variance $\sigma^2 = \mu + \alpha\mu^2 - \delta(\gamma - 1) = \gamma(1-\delta) + \alpha\mu^2$ is

$$\mathsf{V}[y|y>0] = \frac{\mu}{1-(1+\alpha\mu)^{-\alpha^{-1}}}\left[1+\alpha\mu - \frac{\mu(1+\alpha\mu)^{-\alpha^{-1}}}{1-(1+\alpha\mu)^{-\alpha^{-1}}}\right]. \tag{4.33}$$

In practice the process generating zero counts may not be the same as that for the positive counts. In the truncated case there are no data at all for zeros, so there is no alternative but to assume the processes are the same. Then analysis is given a restricted interpretation. For example, if hunting trips are observed only for those who hunt at least once, then we interpret the truncated zeros as being due to hunters who this year chose not to hunt, rather than being for the entire population, which will include many people who never hunt. The untruncated mean μ is interpreted as the mean number of trips by hunters.

Right Truncation

With *right truncation*, or *truncation from above*, we lose data when y is greater than c. For the general count model with cdf $H(y|\boldsymbol{\theta})$, the right truncated density is

$$f(y|\boldsymbol{\theta}, y \leq c) = \frac{h(y|\boldsymbol{\theta})}{H(c|\boldsymbol{\theta})}, \qquad y = 0, 1, \ldots, c. \tag{4.34}$$

For the Poisson the right-truncated mean γ_1 is

$$\gamma_1 = \mu - \mu h(c|\boldsymbol{\theta})/H(c|\boldsymbol{\theta}). \tag{4.35}$$

Other moments of right-truncated distributions can be obtained from the corresponding moments of the left-truncated distribution, already given, by simply replacing $r-1$, δ, and λ in (4.26) or (4.28) by, respectively, c, δ_1, and λ_1. Right truncation results in a smaller mean and variance relative to the parent distribution. Detailed analysis of the moments of left- and right-truncated negative binomial models is given in Gurmu and Trivedi (1992).

It is actually more common to observe right censoring, often due to aggregation of counts above a specified value, than right truncation. Censoring is studied in section 4.4.

4.3.2 Maximum Likelihood Estimation

Standard count estimators such as those for Poisson and NB2 models are inconsistent in the presence of truncation. For example, for the iid Poisson the mean of the truncated-at-zero sample will converge to its expected value of $\mu/(1-e^{-\mu})$, see (4.30), and therefore is inconsistent for the desired untruncated mean μ. Since the expression for the untruncated mean depends on parametric assumptions, the standard approach is to be fully parametric and estimate by maximum likelihood.

We consider ML estimation of left-truncated Poisson models with mean μ_i, usually $\exp(\mathbf{x}_i'\boldsymbol{\beta})$. Using (4.27), the log-likelihood based on n independent observations is

$$\mathcal{L}(\boldsymbol{\beta}) = \sum_{i=1}^{n} \left[y_i \ln \mu_i - \mu_i - \ln \left(1 - e^{-\mu_i} \sum_{j=0}^{r-1} \mu_i^j / j! \right) - \ln y_i! \right]. \tag{4.36}$$

The MLE of $\boldsymbol{\beta}$ is the solution of the following equation:

$$\sum_{i=1}^{n} [y_i - \mu_i - \delta_i] \mu_i^{-1} \frac{\partial \mu_i}{\partial \boldsymbol{\beta}} = \mathbf{0}, \tag{4.37}$$

where, from (4.28), $\delta_i = \mu_i h(r, \mu_i)/[1 - H(r-1, \mu_i)]$. The information matrix is

$$\mathcal{I}(\boldsymbol{\beta}) = -\mathsf{E}\left[\frac{\partial^2 \mathcal{L}(\boldsymbol{\beta})}{\partial \boldsymbol{\beta} \partial \boldsymbol{\beta}'} \right] = \sum_{i=1}^{n} [\mu_i - \delta_i (\mu_i + \delta_i - r)] \mu_i^{-2} \frac{\partial \mu_i}{\partial \boldsymbol{\beta}} \frac{\partial \mu_i}{\partial \boldsymbol{\beta}'}. \tag{4.38}$$

The MLE $\hat{\boldsymbol{\beta}}$ is asymptotically normal with mean $\boldsymbol{\beta}$ and covariance matrix $\mathcal{I}(\boldsymbol{\beta})^{-1}$.

Rewriting (4.37) as

$$\sum_{i=1}^{n} \left[\frac{y_i - \mu_i - \delta_i}{\sqrt{\mu_i}} \right] \left(\frac{1}{\sqrt{\mu_i}} \frac{\partial \mu_i}{\partial \boldsymbol{\beta}} \right) = \mathbf{0}, \tag{4.39}$$

and recalling from (4.28) that the truncated mean is $\mu_i + \delta_i$, this score equation is an orthogonality condition between the standardized truncated residual and standardized gradient vector of the conditional mean. This interpretation parallels that for the normal truncated regression. For the exponential specification $\mu_i = \exp(\mathbf{x}_i \boldsymbol{\beta})$, $\partial \mu_i / \partial \boldsymbol{\beta} = \mu_i \mathbf{x}_i$, and (4.37) reduces to an orthogonality condition between the regressors $\mathbf{x}_i$ and the residual, here $y_i - \mu_i - \delta_i$ given truncation. This qualitative similarity with the untruncated Poisson arises because the *truncated* Poisson is also an LEF density.

The coefficients $\boldsymbol{\beta}$ can be directly interpreted in terms of the untruncated mean. Thus if $\mu_i = \exp(\mathbf{x}_i'\boldsymbol{\beta})$, the untruncated conditional mean $\mathsf{E}[y_i|\mathbf{x}_i]$ changes by a proportion $\hat{\beta}_j$ in response to a one-unit change in x_{ij}. Rearranging (4.29), for left-truncation at zero the untruncated mean

$$\mathsf{E}[y|\mathbf{x}] = \mathsf{E}[y|y > 0, \mathbf{x}] \Pr[y > 0|\mathbf{x}].$$

So the change in the untruncated mean may be decomposed into a part that affects the mean of the currently untruncated part of the distribution and a part that affects the probability of truncation

$$\frac{\partial \mathsf{E}[y|\mathbf{x}]}{\partial x_j} = \frac{\partial \mathsf{E}\,[y|y>0,\mathbf{x}]}{\partial x_j} \Pr[y>0|\mathbf{x}] + \mathsf{E}\,[y|y>0,\mathbf{x}] \frac{\partial \Pr[y>0|\mathbf{x}]}{\partial x_j}.$$

In ML estimation of truncated models, a misspecification of the underlying distribution leads to inconsistency due to the presence of the adjustment factor δ_i. Suppose that the counts in the parent distribution are conditionally NB2 distributed and $\alpha > 0$. If the misspecified distribution is the truncated Poisson, rather than the truncated NB2, then the conditional mean is misspecified (compare (4.28) with (4.26)) and the MLE will be inconsistent. To reiterate, ignoring overdispersion in the truncated count model leads to inconsistency, not just inefficiency. Thus, the result that neglected overdispersion does not affect the consistency property of the correctly specified *untruncated* Poisson conditional mean function does not carry over to the truncated Poisson.

The negative binomial is a better starting point for analyzing data that would be overdispersed in the absence of truncation. For the NB2 model, $\Pr[y=0] = (1+\alpha\mu)^{-1/\alpha}$. So for the most common case of left truncation at zero ($r = 1$), the *truncated* NB2 *model* has log-likelihood

$$\begin{aligned}\mathcal{L}(\boldsymbol{\beta},\alpha) = \sum_{i=1}^{n} \Big\{ \Big(\sum\nolimits_{j=0}^{y_i-1} \ln(j+\alpha^{-1}) \Big) - \ln y_i! + y_i \ln\alpha + y_i \mathbf{x}_i'\boldsymbol{\beta} \qquad (4.40)\\ -(y_i+\alpha^{-1})\ln(1+\alpha\exp(\mathbf{x}_i'\boldsymbol{\beta})) - \ln\big(1-(1+\alpha\exp(\mathbf{x}_i'\boldsymbol{\beta}))^{-1/\alpha}\big)\Big\}.\end{aligned}$$

4.4 CENSORED COUNTS

Censoring arises if observations $(y_i, \mathbf{x}_i)$ are available for a restricted range of y_i, but those for $\mathbf{x}_i$ are always observed. This is in contrast to truncation, where all data are lost for some range of values of y_i. Hence, censoring involves the loss of information that is less serious than with truncation.

As with truncation, however, failure to control for censoring leads to inconsistent parameter estimation, and adjustments to yield consistent estimates rely on fully parametric models.

4.4.1 Standard Censored Models

Censoring of count observations may arise from aggregation or may be imposed by survey design; see, for example, Terza (1985). Alternatively, censored samples may result if high counts are not observed, the case we consider here.

The counts y^* are assumed to be generated by density $h(y^*|\boldsymbol{\theta}) = h(y^*|\mathbf{x},\boldsymbol{\theta})$, such as Poisson or NB2, with corresponding cdf $H(c|\boldsymbol{\theta}) = \Pr[y^* \le c|\boldsymbol{\theta}]$. But these latent counts are not completely observed. Instead, for *right censoring* at c we observe the dependent variable y_i, where

$$y_i = \begin{cases} y_i^* & \text{if } y_i^* < c \\ c & \text{if } y_i^* \ge c, \end{cases} \qquad (4.41)$$

where c is a known positive integer.

Define a binary censoring indicator as follows:

$$d_i = \begin{cases} 1 & \text{if } y_i^* < c, \\ 0 & \text{if } y_i^* \geq c. \end{cases} \tag{4.42}$$

The probability of censoring the ith observation is then

$$\Pr[y_i^* \geq c\,] = \Pr[d_i = 0] = 1 - \Pr[d_i = 1] = 1 - \mathsf{E}[d_i]. \tag{4.43}$$

It is assumed that the regressors $\mathbf{x}_i$ are observed for all $i = 1, \ldots, n$.

For $y^* < c$ we observe y and the density of y is simply the usual $h(y|\boldsymbol{\theta})$. For $y^* \geq c$ we observe only this fact, which occurs with probability $\Pr[y^* \geq c|\boldsymbol{\theta}] = 1 - \Pr[y^* \leq c - 1|\boldsymbol{\theta}] = 1 - H(c - 1|\boldsymbol{\theta})$. A parsimonious way to combine these two terms is the following:

$$f(y|\boldsymbol{\theta}) = h(y|\boldsymbol{\theta})^d [1 - H(c - 1|\boldsymbol{\theta})]^{1-d}, \tag{4.44}$$

where d is the indicator variable defined in (4.42).

The censored conditional mean is

$$\begin{aligned}\mathsf{E}[y] &= \Pr[y^* < c] \times \mathsf{E}[y^*|y^* < c] + \Pr[y^* \geq c] \times c \\ &= H(c - 1|\boldsymbol{\theta}) \times \mathsf{E}[y^*|y^* < c] + (1 - H(c - 1|\boldsymbol{\theta})) \times c,\end{aligned}$$

where $\mathsf{E}[y^*|y^* < c]$ is the right-truncated mean discussed in section 4.3.

4.4.2 Maximum Likelihood Estimation

Given (4.44), the log-likelihood function for n independent observations is

$$\begin{aligned}\mathcal{L}(\boldsymbol{\theta}) &= \sum_{i=1}^{n} \ln \left\{ h(y_i|\mathbf{x}_i, \boldsymbol{\theta})^{d_i} \left[1 - H(c - 1|\mathbf{x}_i, \boldsymbol{\theta})\right]^{1-d_i} \right\} \\ &= \sum_{i=1}^{n} \{ d_i \ln h(y_i|\mathbf{x}_i, \boldsymbol{\theta}) + (1 - d_i) \ln[1 - H(c - 1|\mathbf{x}_i, \boldsymbol{\theta})] \}.\end{aligned} \tag{4.45}$$

Maximization of the log-likelihood is straightforward using gradient-based methods such as Newton-Raphson.

The coefficients $\boldsymbol{\theta}$ are directly interpreted in terms of the uncensored model. Thus if $\mu_i = \exp(\mathbf{x}_i'\boldsymbol{\beta})$, the uncensored conditional mean $\mathsf{E}[y_i|\mathbf{x}_i]$ changes by a proportion $\hat{\beta}_j$ in response to a one-unit change in x_{ij}.

For the *right-censored Poisson model* the ML first-order conditions can be shown to be

$$\sum_{i=1}^{n} \left[d_i(y_i - \mu_i) + (1 - d_i)\delta_i \right] \mu_i^{-1} \frac{\partial \mu_i}{\partial \boldsymbol{\beta}} = \mathbf{0}, \tag{4.46}$$

where

$$\delta_i = \mu_i h(c - 1|\mu_i) / \left[1 - H(c - 1|\mu_i)\right]$$

is the adjustment factor associated with the left-truncated Poisson model. Because $(y_i - \mu_i)$ is the error for the uncensored Poisson model and $\delta_i = \mathsf{E}[y_i - \mu_i | y_i \geq c]$, see (4.28), the term $[d_i(y_i - \mu_i) + (1 - d_i)\delta_i]$ in (4.46) is interpreted as the generalized error (Gourieroux et al., 1987a) for the right-censored Poisson model. The score equations imply that the vector of generalized residuals is orthogonal to the vector of exogenous variables.

The ML procedure is easily adapted for *interval-recorded counts*. For example, if the number of sexual partners is recorded as 0, 1, $2-5$, and more than 5, then the corresponding contributions to the likelihood function are, respectively, $\Pr[y = 0]$, $\Pr[y = 1]$, $\Pr[2 \leq y \leq 5]$, and $\Pr[y \geq 5]$.

ML estimation of censored count models raises issues similar to those in Tobit models (Terza, 1985; Gurmu, 1993). Applications of censored count models include provision of hospital beds for emergency admissions (Newell, 1965) and the number of shopping trips (Terza, 1985; Okoruwa, Terza, and Nourse, 1988).

4.4.3 Expectation-Maximization Algorithm

An alternative way to estimate $\boldsymbol{\theta}$ in (4.45) is to use the *expectation-maximization* (EM) *algorithm* based on *expected likelihood.*

Expected likelihood, $\mathcal{EL}(\boldsymbol{\theta})$, treats d_i as unobserved random variables and takes the expected value of $\mathcal{L}(\boldsymbol{\theta})$ with respect to d_i, conditional on data $y_i, \mathbf{x}_i$, and parameters $\boldsymbol{\theta}$. Since (4.45) is linear in d_i, this is equivalent here to replacing d_i in (4.45) by $\mathsf{E}[d_i | y_i, \mathbf{x}_i, \boldsymbol{\theta}]$. Then

$$\begin{aligned} \mathcal{EL}(\boldsymbol{\theta}) = \sum_{i=1}^{n} \{ & \mathsf{E}[d_i | \mathbf{x}_i, \boldsymbol{\theta}] \ln h(y_i | \mathbf{x}_i, \boldsymbol{\theta}) \\ & + (1 - \mathsf{E}[d_i | \mathbf{x}_i, \boldsymbol{\theta}]) \ln (1 - H(c - 1 | \mathbf{x}_i, \boldsymbol{\theta})\}. \end{aligned} \tag{4.47}$$

The expected likelihood is a weighted sum of pdf and cdf with $\mathsf{E}[d_i | y_i, \mathbf{x}_i, \boldsymbol{\theta}]$, the probability of data being censored, as the weight.

In the EM algorithm, the estimates are obtained iteratively by replacing $\mathsf{E}[d_i | \mathbf{x}_i, \boldsymbol{\theta}]$ in (4.47) by

$$\mathsf{E}[d_i | \mathbf{x}_i, \boldsymbol{\theta}] = \mathsf{E}[y_i^* < c | \mathbf{x}_i, \boldsymbol{\theta}] = H(c - 1 | \mathbf{x}_i, \boldsymbol{\theta}),$$

evaluated at the current estimate of $\boldsymbol{\theta}$, and then maximizing the expected likelihood (4.47) with respect to $\boldsymbol{\theta}$. Given the new estimate of $\boldsymbol{\theta}$, the expected value of d_i, can be recomputed, the expected likelihood remaximized, and so on.

The EM method is often used to estimate more complicated censored models and finite mixtures models.

4.5 HURDLE MODELS

The Poisson mixture models of section 4.2 can accommodate overdispersion. But it is still common for there to be *excess zeros*, meaning more zero counts in the data than predicted by a fitted parametric model such as Poisson or NB2.

Modified count data models, proposed by Mullahy (1986), are *two-part models* that specify a process for the zero counts that differs from that for the positive counts. We present hurdle models in this section and zero-inflated count models in the next section. Both models can be viewed as discrete mixtures, rather than the continuous mixtures emphasized in section 4.2.

4.5.1 Two-Part Structure of Hurdle Models

Suppose individuals participate in an activity, with positive counts, once a threshold is crossed. The threshold is not crossed with probability $f_1(0)$, in which case we observe a count of 0. If the threshold is crossed, we observe positive counts, with probabilities coming from the count density $f_2(y)$ with the associated truncated density $f_2(y)/(1 - f_2(0))$ that needs to be multiplied by $(1 - f_1(0))$ to ensure probabilities sum to one.

This leads to the *hurdle model*

$$\Pr[y = j] = \begin{cases} f_1(0) & \text{if } j = 0 \\ \dfrac{1 - f_1(0)}{1 - f_2(0)} f_2(j) & \text{if } j > 0. \end{cases} \tag{4.48}$$

The model collapses to the standard model only if $f_1(0) = f_2(0)$. The model clearly allows for excess zeros if $f_1(0) > f_2(0)$, though in principle it can also model too few zeros if $f_1(0) < f_2(0)$. The density $f_2(\cdot)$ is a count density such as Poisson or NB2, whereas $f_1(\cdot)$ could also be a count data density, or more simply, the probabilities $f_1(0)$ and $1 - f_1(0)$ may be estimated from logit or probit models.

An alternative motivation for this model is that zero counts arise from two sources. For instance, in response to the question, "How many times did you go fishing in the last two months?" zero responses would be recorded from never fishers and from fishers who happened not to fish in the past two months. Then $f_1(0)$ incorporates zeros from both of these sources, whereas $f_2(0)$ corresponds only to fishers who happened not to fish in the past two months. This interpretation is also related to the with-zeros model, covered in the next section.

The model (4.48) can be viewed as a finite mixture model, see section 4.2.6, with two components. The mixture weights for the two components are $f_1(0)$ and $1 - f_1(0)$. One component is a degenerate probability mass function $f_3(y)$ with $f_3(j) = 1$ if $j = 0$ and $f_3(j) = 0$ if $j > 0$. The other component is the zero-truncated probability mass function $f_2(y)/(1 - f_2(0))$. By contrast, for the finite mixture models presented in section 4.8, none of the component densities are degenerate.

The hurdle model is a modified count model in which the two processes generating the zeros and the positives are not constrained to be the same. In the context of a censored normal density (the Tobit model) the hurdle model was proposed by Cragg (1971). The basic idea is that a binomial probability governs the binary outcome of whether a count variate has a zero or a positive realization. If the realization is positive, the hurdle is crossed, and the conditional distribution of the positives is governed by a truncated-at-zero count data model. Mullahy (1986) provided the general form of hurdle count regression models, together with application to daily consumption of various beverages. The hurdle model is the dual of the split-population survival time model (Schmidt and Witte, 1989) where the probability of an eventual 'death' and the timing of 'death' depend separately on individual characteristics.

The moments of the hurdle model are determined by the probability of crossing the threshold and by the moments of the zero-truncated density. The conditional mean is

$$\mathsf{E}[y|\mathbf{x}] = \Pr[y > 0|\mathbf{x}] \times \mathsf{E}_{y>0}[y|y > 0, \mathbf{x}], \tag{4.49}$$

where the second expectation is taken relative to the zero-truncated density. Given (4.48), it follows that

$$\mathsf{E}[y|\mathbf{x}] = \frac{1 - f_1(0|\mathbf{x})}{1 - f_2(0|\mathbf{x})}\mu_2(\mathbf{x}), \tag{4.50}$$

where $\mu_2(\mathbf{x})$ is the untruncated mean in density $f_2(y|\mathbf{x})$. Following estimation, marginal effects can be obtained by differentiating (4.50) with respect to $\mathbf{x}$.

In section 4.12, it is shown that the hurdle model variance is

$$\mathsf{V}[y] = \frac{1 - f_1(0|\mathbf{x})}{1 - f_2(0|\mathbf{x})}\sigma_2^2(\mathbf{x}) + \frac{(1 - f_1(0|\mathbf{x}))(f_1(0|\mathbf{x}) - f_2(0|\mathbf{x}))}{(1 - f_2(0|\mathbf{x}))^2}(\mu_2(\mathbf{x}))^2, \tag{4.51}$$

where $\sigma_2^2(\mathbf{x})$ is the untruncated variance in density $f_2(y|\mathbf{x})$.

4.5.2 Maximum Likelihood Estimation

Define a binary censoring indicator as follows:

$$d_i = \begin{cases} 1 & \text{if } y_i > 0, \\ 0 & \text{if } y_i = 0. \end{cases} \tag{4.52}$$

Then the density for the typical observation can be written as

$$\begin{aligned} f(y) &= f_1(0)^{1-d} \times \left[\frac{1 - f_1(0)}{1 - f_2(0)} f_2(y)\right]^d \\ &= \left[f_1(0)^{1-d}(1 - f_1(0))^d\right] \times \left[\frac{f_2(y)}{1 - f_2(0)}\right]^d . \end{aligned} \tag{4.53}$$

Introducing regressors, let the probability of zero be $f_1(0|\mathbf{x}, \boldsymbol{\theta}_1)$ and the positives come from density $f_2(y|\mathbf{x}, \boldsymbol{\theta}_2)$. Then using (4.53) the log-likelihood function for the observations splits into two components, thus,

$$\begin{aligned}\mathcal{L}(\boldsymbol{\theta}_1, \boldsymbol{\theta}_2) &= \mathcal{L}_1(\boldsymbol{\theta}_1) + \mathcal{L}_2(\boldsymbol{\theta}_2) \\ &= \sum_{i=1}^{n}[(1-d_i)\ln f_1(0|\mathbf{x}_i, \boldsymbol{\theta}_1) + d_i \ln(1 - f_1(0|\mathbf{x}_i, \boldsymbol{\theta}_1))] \\ &\quad + \sum_{i=1}^{n} d_i[\ln f_2(y_i|\mathbf{x}_i, \boldsymbol{\theta}_2) - \ln(1 - f_2(0|\mathbf{x}_i, \boldsymbol{\theta}_2))].\end{aligned} \tag{4.54}$$

Here $\mathcal{L}_1(\boldsymbol{\theta}_1)$ is the log-likelihood for the binary process that splits the observations into zeros and positives, and $\mathcal{L}_2(\boldsymbol{\theta}_2)$ is the log-likelihood function for the truncated count model for the positives. Because the two mechanisms are specified to be independent, conditional on regressors, the joint likelihood can be maximized by separately maximizing each component. Practically this means that the hurdle model can be estimated using software that may not explicitly include the hurdles option.

Although marginal effects for $\mathsf{E}[y|\mathbf{x}]$ are based on (4.50), interest may also lie in marginal effects for each of the two components: the probability $1 - f(0)$ of crossing the hurdle and the mean $\mathsf{E}[y|\mathbf{x}]$ of the count density $f_2(y|\mathbf{x})$; see section 4.5.3 for an example.

In his original paper, Mullahy (1986) specified $f_1(\cdot)$ and $f_2(\cdot)$ to be from the same family and with the same regressors, so that a test of $\boldsymbol{\theta}_1 = \boldsymbol{\theta}_2$ is a test of whether there is a need for a hurdle model, since the two processes are the same if $\boldsymbol{\theta}_1 = \boldsymbol{\theta}_2$. Mullahy also proposed a Hausman test and for implementation considered both Poisson and geometric specifications. For the geometric $\Pr[y = 0|\mathbf{x}] = (1+\mu_1)^{-1}$, so that if $\mu_1 = \exp(\mathbf{x}'\boldsymbol{\beta}_1)$, the binary hurdle process is a logit model.

For a hurdle model with zero counts determined by NB2 with mean $\mu_{1i} = \exp(\mathbf{x}_i'\boldsymbol{\beta}_1)$ and dispersion parameter α_1, and positive counts determined by NB2 with mean $\mu_{2i} = \exp(\mathbf{x}_i'\boldsymbol{\beta}_2)$ and dispersion parameter α_2, we have

$$\begin{aligned} f_1(0|\mathbf{x}_i, \boldsymbol{\beta}_1, \alpha_1) &= \left(1+\alpha_1\mu_{1i}\right)^{-1/\alpha_1} \\ f_2(0|\mathbf{x}_i, \boldsymbol{\beta}_1, \alpha_2) &= \left(1+\alpha_2\mu_{2i}\right)^{-1/\alpha_2} \\ f_2(y_i|\mathbf{x}_i, \boldsymbol{\beta}_2, \alpha_2) &= \frac{\Gamma(y_i+\alpha_2^{-1})}{\Gamma(\alpha_2^{-1})\Gamma(y_i+1)}\left(\frac{\alpha_2^{-1}}{\alpha_2^{-1}+\mu_{2i}}\right)^{\alpha_2^{-1}}\left(\frac{\mu_{2i}}{\alpha_2^{-1}+\mu_{2i}}\right)^{y_i}\end{aligned} \tag{4.55}$$

Common practice now is to presume that there is a need for a hurdle model and to specify different processes for $f_1(\cdot)$ and $f_2(\cdot)$. The binary process for $f_1(0)$ is specified to be a standard logit or probit model, whereas $f_2(\cdot)$ is an NB density, or Poisson if a Poisson hurdle is sufficient to handle any overdispersion present in a nonhurdle model. Then one estimates a logit model, yielding $\hat{\boldsymbol{\beta}}_1$ for whether or not the count is positive, and separately estimates a left-truncated-at-zero NB model, yielding estimates $\hat{\boldsymbol{\beta}}_2$ and $\hat{\alpha}_2$. The log-likelihood for the truncated NB2 is given in (4.40).

Winkelmann (2004) proposed an extension of the hurdle model that permits correlation across the two components. Specifically the hurdle is crossed if the latent variable $\mathbf{x}'_{1i}\boldsymbol{\beta}_1 + \varepsilon_{1i} > 0$, the base count density is Poisson with mean $\exp(\mathbf{x}'_{2i}\boldsymbol{\beta}_2 + \varepsilon_{2i})$, and $(\varepsilon_{1i}, \varepsilon_{2i})$ are joint normally distributed. The model simplifies to a probit first part and Poisson lognormal second part if ε_{1i} and ε_{2i} are uncorrelated.

4.5.3 Example: Hurdles and Two-Part Decision Making

Pohlmeier and Ulrich (1995) developed a count model of the two-part decision-making process in the demand for health care in West Germany. The model postulates that "while at the first stage it is the patient who determines whether to visit the physician (contact analysis), it is essentially up to the physician to determine the intensity of the treatment (frequency analysis)" (Pohlmeier and Ulrich, 1995, p. 340). Thus the analysis is in the principal-agent framework where the physician (the agent) determines utilization on behalf of the patient (the principal). This contrasts with the approach in which the demand for health care is determined primarily by the patient.

Pohlmeier and Ulrich estimate an NB1 hurdle model. They use cross-section data from the West German Socioeconomic Panel. The binary outcome model of the contact decision separates the full sample (5,096) into those who did or did not have zero demand for physician (or specialist) consultations during the period under study. The second stage estimates the left-truncated-at-zero NB1 model for those who had at least one physician (or specialist) consultation (2,125 or 1,640). The authors point out that under the then-prevalent system in West Germany, the insured individual was required to initiate the demand for covered services by first obtaining a sickness voucher from the sickness fund each quarter. The demand for specialist services was based on a referral from a general practitioner to the specialist. The authors argue that such an institutional set-up supports a hurdle model that allows initial contact and subsequent frequency to be determined independently.

The authors test the Poisson hurdle and NB1 hurdle models against the Poisson and reject the latter. Then they test the first two models against a less-restrictive NB1 hurdle model, which is preferred to all restrictive alternatives on the basis of Wald and Hausman specification tests. The authors report "important differences between the two-part decision making stages." For example, the physician-density variable, which reflects accessibility to service, has no significant impact on the contact decision. But it does have a significant positive impact on the frequency decision in the general practitioner and specialist equations.

4.6 ZERO-INFLATED COUNT MODELS

Zero-inflated count models provide an alternative way to model count data with excess zeros.

4.6.1 Zero-Inflated Models

Suppose the base count density is $f_2(y)$, using the same notation as for the hurdle model, but this predicts too few zeros. An obvious remedy is to add a separate component that inflates the probability of a zero by, say, π. Then the *zero-inflated model* or *with-zeroes model* specifies

$$\Pr[y = j] = \begin{cases} \pi + (1-\pi) f_2(0), & \text{if } j = 0 \\ (1-\pi) f_2(j) & \text{if } j > 0. \end{cases} \tag{4.56}$$

In (4.56) the proportion of zeros, π, is added to the baseline distribution, and the probabilities from the base model $f_2(y)$ are decreased by the proportion $(1-\pi)$ to ensure that probabilities sum to one. The probability π may be set as a constant or may depend on regressors via a binary outcome model such as logit or probit. One possible justification for this model is that observations are misrecorded, with the misrecording concentrated exclusively in the zero class.

The model (4.56) can be viewed as a finite mixture model with two components. The mixture weights for the two components are π and $1-\pi$. One component is a degenerate probability mass function $f_1(y)$ with $f_1(j) = 1$ if $j = 0$ and $f_1(j) = 0$ if $j > 0$. The other component is the untruncated probability mass function $f_2(y)$.

For this model the uncentered moments $\mathsf{E}[y^k]$ are simply $(1-\pi)$ times the corresponding moments in the base density $f_2(y)$; see section 4.12. Thus the mean

$$\mathsf{E}[y] = (1-\pi)\mu_2, \tag{4.57}$$

where μ_2 is the mean in the base density $f_2(y)$. In regression application both π and μ_2 are functions of $\mathbf{x}$, and following estimation marginal effects can be obtained by differentiating (4.50) with respect to $\mathbf{x}$.

The variance

$$\begin{aligned} \mathsf{V}[y] &= \mathsf{E}[y^2] + (\mathsf{E}[y])^2 \\ &= (1-\pi)(\sigma_2^2 - \mu_2^2) + [(1-\pi)\mu_2]^2 \\ &= (1-\pi)(\sigma_2^2 + \pi\mu_2^2), \end{aligned} \tag{4.58}$$

where σ_2^2 is the variance in the base density $f_2(y)$; see section 4.12.2.

4.6.2 Maximum Likelihood Estimation

Define a binary censoring indicator in the same way as for the hurdle model:

$$d_i = \begin{cases} 1 & \text{if } y_i > 0, \\ 0 & \text{if } y_i = 0. \end{cases} \tag{4.59}$$

Note that $d_i = 0$ with probability π_i and $d_i = 1$ with probability $1 - \pi_i$. Then the density for the typical observation can be written as

$$f(y) = [\pi + (1-\pi) f_2(0)]^{1-d} \times [(1-\pi) f_2(y)]^d. \tag{4.60}$$

Introducing regressors, let $\pi = \pi(\mathbf{x}, \boldsymbol{\theta}_1)$ and the base density be $f_2(y|\mathbf{x}, \boldsymbol{\theta}_2)$. Then using (4.60) the log-likelihood function is

$$\mathcal{L}(\boldsymbol{\theta}_1, \boldsymbol{\theta}_2) = \sum_{i=1}^{n} (1 - d_i) \ln[\pi(\mathbf{x}_i, \boldsymbol{\theta}_1) + (1 - \pi(\mathbf{x}_i, \boldsymbol{\theta}_1)) f_2(0|\mathbf{x}_i, \boldsymbol{\theta}_2)]$$
$$+ \sum_{i=1}^{n} d_i \ln[(1 - \pi(\mathbf{x}_i, \boldsymbol{\theta}_1)) f_2(y_i|\mathbf{x}_i, \boldsymbol{\theta}_2)]. \qquad (4.61)$$

In his original paper, Mullahy (1986) specified $f_2(\cdot)$ to be Poisson or geometric, and $\pi = f_1(0)$, where $f_1(\cdot)$ is also Poisson or geometric. Mullahy also noted that, regardless of the models used, in the intercept-only case (so no regressors appear in either part of the model) the with-zeros model is equivalent to the hurdle model, with estimation yielding the same log-likelihood and fitted probabilities.

Lambert (1992) considered an application to soldering components onto printed wiring boards, with y the number of leads improperly soldered. Lambert proposed a with-zeroes model where the probability π is logit and the base count density is Poisson. Then

$$\pi_i(\mathbf{x}_i, \boldsymbol{\beta}_1) = \exp(\mathbf{x}_i'\boldsymbol{\beta}_1)/[1 + \exp(\mathbf{x}_i'\boldsymbol{\beta}_1)]$$
$$f_2(y_i|\mathbf{x}_i, \boldsymbol{\beta}_2) = \exp(-\exp(\mathbf{x}_i'\boldsymbol{\beta}_2))(\exp(\mathbf{x}_i'\boldsymbol{\beta}_2))^{y_i}/y_i!$$

Lambert suggested using the EM algorithm to maximize the likelihood. As in the censored example of section 4.4.3, the indicator variables d_i in (4.61) are replaced by current-round estimates of their expected values, the resulting expected likelihood function is then maximized with respect to $\boldsymbol{\theta}$, yielding new estimates of $\mathsf{E}[d_i]$, and so on. Alternatively, the usual Newton-Raphson method may be used.

Although the logistic functional form is convenient, generalizations of the logistic such as Prentice's F distribution (Stukel, 1988) may also be used. For economics data it is often the case that there is still overdispersion, and it is better to use the NB model as the base count density. The increased generality comes at the cost of a more heavily parameterized model, some of whose parameters can be subject to difficulties of identification. Consequently, convergence in ML estimation may be slow.

Both the hurdle and zero-inflated models are motivated by a desire to explain excess zeros in the data. As already noted in section 4.5.1, the hurdle model can also explain too few zeros. The with-zeros model can similarly explain too few zeros, if $\pi < 0$. This is possible because the model remains valid, with probabilities between 0 and 1 that sum to 1 provided $-f_2(0)/(1 - f_2(0)) \leq \pi \leq 1$. If π is parameterized as logit or probit, however, then necessarily $0 \leq \pi \leq 1$, and only excess zeros can be explained.

4.7 HIERARCHICAL MODELS

The Poisson-mixture models of section 4.2 can be viewed as Poisson models with an intercept that varies across observations according to a specified continuous distribution. Additionally the slope parameters may similarly vary. A natural setting for such richer models is when data come from a sample survey collected using a multilevel design. An example is state-level data that are further broken down by counties, or province-level data clustered by communes; see Chang and Trivedi (2003).

When multilevel covariate information is available, hierarchical modeling becomes feasible. Features of these models include relaxation of the assumption that observations are statistically independent and heterogeneity in slope coefficients, not just intercepts. Such models have been widely applied to the generalized linear mixed model (GLMM) class of which Poisson regression is a member.

4.7.1 Two-Level Poisson-Lognormal Model

Wang, Yau, and Lee (2002) consider a hierarchical Poisson mixture regression to account for the inherent correlation of outcomes of patients clustered within hospitals. In their set-up the data are in m clusters, each cluster has n_j observations, $j = 1, \ldots, m$, and there are $n = \sum_j n_j$ observations in all.

The standard Poisson-lognormal mixture of section 4.2.4 can be interpreted as a *one-level hierarchical model*, with

$$\begin{aligned} y_{ij} &\sim \mathsf{P}[\mu_{ij}] \\ \ln \mu_{ij} &= \mathbf{x}'_{ij}\boldsymbol{\beta} + \varepsilon_{ij}, \quad \varepsilon_{ij} \sim \mathsf{N}[0, \sigma_\varepsilon^2], \end{aligned} \tag{4.62}$$

for observations $i = 1, \ldots, n_j$, in clusters $j = 1, \ldots, m$.

A *two-level hierarchical model*, also known as a *hierarchical Poisson mixture*, that incorporates covariate information at both levels is as follows:

$$\begin{aligned} y_{ij} &\sim \mathsf{P}[\mu_{ij}] \\ \ln \mu_{ij} &= \mathbf{x}'_{ij}\boldsymbol{\beta}_j + \varepsilon_{ij}, \quad \varepsilon_{ij} \sim \mathsf{N}[0, \sigma_\varepsilon^2], \\ \beta_{lj} &= \mathbf{w}'_{lj}\boldsymbol{\gamma} + v_{lj}, \quad v_{lj} \sim \mathsf{N}[0, \sigma_v^2], \end{aligned} \tag{4.63}$$

for observations $i = 1, \ldots, n_j$, in clusters $j = 1, \ldots, m$, and for the coefficients of regressors $l = 1, \ldots, L$.

In this two-level model the regression coefficients $\boldsymbol{\beta}_j$ now vary by clusters. The cluster-specific variables $\mathbf{w}_{kj}$ enter at the second level to determine these first-level parameters $\boldsymbol{\beta}_j$ with l^{th} component β_{lj}. The errors v_{lj} induce correlation of observations within the same cluster. The parameter vector $\boldsymbol{\gamma}$, also called the *hyperparameter*, is the target of statistical inference. Both classical (Wang et al., 2002) and Bayesian analyses can be applied.

4.7.2 Three-Level Poisson-Lognormal Model

We consider a specific three-level example. Students are grouped (nested) in classes which make up schools. The lowest level labeled i is made of students. These students are grouped in classes, the second level labeled j. These classes are in turn grouped to form schools, the highest level labeled k.

A three-level hierarchical model with regressors at the first and third levels is

$$\begin{aligned} y_{ijk} &\sim \mathsf{P}[\mu_{ijk}] \\ \ln \mu_{ijk} &= \mathbf{x}'_{ijk}\boldsymbol{\beta}_k + \varepsilon_{ijk}, \quad \varepsilon_{ijk} \sim \mathsf{N}[0, \sigma^2_{\varepsilon}], \\ \beta_{lk} &= \mathbf{w}'_{lk}\boldsymbol{\gamma} + \ v_{lk}, \quad v_{ljk} \sim \mathsf{N}[0, \sigma^2_{v}], \end{aligned} \tag{4.64}$$

for regressors $l = 1, \ldots, L$. Stacking $\boldsymbol{\beta}'_k = [\beta_{1k} \cdots \beta_{Lk}]'$, $\mathbf{W}_k = [\mathbf{w}_{1k} \cdots \mathbf{w}_{Lk}]'$, and $\mathbf{v}'_k = [v_{1k} \cdots v_{Lk}]'$, and then substituting $\boldsymbol{\beta}_k$ in (4.64) yields

$$\ln \mu_{ijk} = \mathbf{x}'_{ijk}\mathbf{W}_k\boldsymbol{\gamma} + \mathbf{x}'_{ijk}\mathbf{v}_k + \varepsilon_{ijk}. \tag{4.65}$$

As a specific example, let $k = 1, 2;\ j = 1, 2;$ and $i = 1, 2, 3$. This means the data are obtained from two schools, each of them has two classes, and each class contains three students. Let y_{ijk} be the response variable for each student. For each level, the data can be stacked recursively as

$$\mathbf{y}^{(2)}_{jk} = \begin{bmatrix} y_{1jk} \\ y_{2jk} \\ y_{3jk} \end{bmatrix} \ j = 1, 2; \quad \mathbf{y}^{(3)}_{k} = \begin{bmatrix} y^{(2)}_{1k} \\ y^{(2)}_{2k} \end{bmatrix} \ k = 1, 2; \quad \mathbf{y} = \begin{bmatrix} y^{(3)}_{1} \\ y^{(3)}_{2} \end{bmatrix}.$$

The conditional density for class j in school k is denoted $f(\mathbf{y}^{(2)}_{jk}|\mathbf{v}_k, \varepsilon_{ijk})$, the density for each school is denoted $f(\mathbf{y}^{(3)}_{k}|\varepsilon_{ijk})$, and the marginal density for the whole sample is $f(\mathbf{y})$. Under appropriate independence assumptions at each level, we can write the three densities as follows. For class j in school k,

$$f(\mathbf{y}^{(2)}_{jk}|\mathbf{v}_k, \varepsilon_{ijk}) = \prod_{i=1}^{3} f(y_{ijk}|\mathbf{v}_k, \varepsilon_{ijk}). \tag{4.66}$$

For school k

$$\begin{aligned} f(\mathbf{y}^{(3)}_{jk}|\varepsilon_{ijk}) &= \int \prod_{j=1}^{2} f(\mathbf{y}_{jk(2)}|\mathbf{v}_k, \varepsilon_{ijk})g(\mathbf{v}_k|\varepsilon_{ijk})d\mathbf{v}_k \\ &= \int \prod_{j=1}^{2}\prod_{i=1}^{3} f(\mathbf{y}_{ijk}|\mathbf{v}_k, \varepsilon_{ijk})g(\mathbf{v}_k|\varepsilon_{ijk})d\mathbf{v}_k. \end{aligned} \tag{4.67}$$

And aggregating across all,

$$\begin{aligned} f(\mathbf{y}_{jk}) &= \int \prod_{k=1}^{2} f(\mathbf{y}^{(3)}_{k}|\varepsilon_{ijk})h(\varepsilon_{ijk})d\varepsilon_{ijk} \\ &= \left[\prod_{k=1}^{2} \int \prod_{j=1}^{2}\left(\prod_{i=1}^{3} \int f(\mathbf{y}_{ijk}|\mathbf{v}_k, \varepsilon_{ijk})g(\mathbf{v}_k|\varepsilon_{ijk})d\mathbf{v}_k\right)\right] h(\varepsilon_{ijk})d\varepsilon_{ijk}. \end{aligned} \tag{4.68}$$

The log-likelihood function for a single observation is $\ln f(\mathbf{y})$. ML estimation is computationally demanding because of the integrals. Standard methods of handling these integrals numerically include Monte Carlo simulation (similar to that in section 4.2.4 for the Poisson-Normal mixture), quadrature (section 4.2.4) and Laplace approximation. See Skrondal and Rabe-Hesketh (2004, ch.6.3.2) for additional details.

Note that this model leads to a change in the conditional mean. From (4.65) we have

$$\begin{aligned}\mathsf{E}[y_{ijk}|\mathbf{x}_{ijk}, \mathbf{W}_k] &= \mathsf{E}[\exp(\mathbf{x}'_{ijk}\mathbf{W}_k\boldsymbol{\gamma} + \mathbf{x}'_{ijk}\mathbf{v}_k + \varepsilon_{ijk})] \\ &= \mathsf{E}[\exp(\mathbf{x}'_{ijk}\mathbf{W}_k\boldsymbol{\gamma})]\mathsf{E}[\exp(\mathbf{x}'_{ijk}\mathbf{v}_k)]\mathsf{E}[\exp(\varepsilon_{ijk})] \\ &= \exp(\mathbf{x}'_{ijk}\mathbf{W}_k\boldsymbol{\gamma})\exp(\sigma_v^2\mathbf{x}'_{ijk}\mathbf{x}_{ijk}/2)\exp(\sigma_\varepsilon^2/2) \\ &= \exp(\mathbf{x}'_{ijk}\mathbf{W}_k\boldsymbol{\gamma} + (\sigma_v^2\mathbf{x}'_{ijk}\mathbf{x}_{ijk} + \sigma_\varepsilon^2)/2),\end{aligned}$$

using, for example, $\mathbf{x}'_{ijk}\mathbf{v}_k \sim \mathsf{N}[0, \sigma_v^2\mathbf{x}'_{ijk}\mathbf{x}_{ijk}]$ and $\mathsf{E}[\exp(\mathbf{z})] = \exp(\mu + \sigma^2/2)$ for $\mathbf{z} \sim \mathsf{N}[\mu, \sigma^2]$.

An alternative approach developed by Lee and Nelder (1996) in the context of the class of generalized linear models, which includes the Poisson-lognormal and Poisson-gamma as special cases, uses the concept of *hierarchical likelihood* or *h-likelihood*, which involves a particular approximation to the marginal likelihood. Here one works with the joint density of $(\mathbf{v}_k, \varepsilon_{ijk})$ after defining a new vector of random variables $\xi_{ijk} = \mathbf{x}'_{ijk}\mathbf{v}_k + \varepsilon_{ijk}$. Again, given appropriate independence assumptions, the marginal density is

$$f(\mathbf{y}) = \prod_{k=1}^{2}\prod_{j=1}^{2}\prod_{i=1}^{3}\int f(y_{ijk}|\xi_{ijk})f(\xi_{ijk})d\xi_{ijk} \tag{4.69}$$

The h-likelihood of Lee and Nelder (1996) corresponding to equation (4.69),

$$\sum_{k=1}^{2}\sum_{j=1}^{2}\sum_{i=1}^{3}\left[\ln\left[\int f(y_{ijk}|\xi_{ijk})f(\xi_{ijk})d\xi_{ijk}\right]\right] \tag{4.70}$$

is approximated by

$$\sum_{k=1}^{2}\sum_{j=1}^{2}\sum_{i=1}^{3}\{\ln f(y_{ijk}|\mathbf{v}_k, \varepsilon_{ijk}) + \ln g(\mathbf{v}_k|\varepsilon_{ijk}) + \ln h(\varepsilon_{ijk})\}. \tag{4.71}$$

In this set-up the random effects $\mathbf{v}_k$ and ε_{ijk} considered as if they were additional parameters, and the numerical integration is thereby avoided.

4.8 FINITE MIXTURES AND LATENT CLASS ANALYSIS

The formal definition of a finite mixture is given in section 4.2.6. The hurdle and zero-inflated models can be viewed as two-component finite mixture models, but the mixing is of a limited form that focuses on a separate treatment of the zeros. We now consider general finite mixture models where each component density may give positive probability to all possible values of the counts and there may be more than two component densities.

4.8.1 Finite Mixtures

In a finite mixture model a random variable is postulated as a draw from a superpopulation that is an additive mixture of C distinct populations with component (subpopulation) densities $f_1(y|\boldsymbol{\theta}_1), \ldots, f_C(y|\boldsymbol{\theta}_1)$, in proportions $\pi_1, \ldots, \pi_C$, where $\pi_j \geqslant 0$, $j = 1, \ldots, C$, and $\sum_{j=1}^{C} \pi_j = 1$. The mixture density for the ith observation is given by

$$f(y_i|\boldsymbol{\Theta}) = \sum_{j=1}^{C} \pi_j f_j(y_i|\boldsymbol{\theta}_j). \tag{4.72}$$

In general the π_j are unknown and are estimated along with all other parameters, denoted $\boldsymbol{\Theta} = (\boldsymbol{\theta}_1, \ldots, \boldsymbol{\theta}_C, \pi_1, \ldots, \pi_{C-1})$. For identifiability, we use the labeling restriction that $\pi_1 \geq \pi_2 \geq \ldots . \geq \pi_C$. This can always be satisfied by postestimation rearrangement, so this restriction is generally not imposed in estimation.

The uncentered (or raw) moments of a finite mixture distribution are simply the weighted sum of the corresponding moments of the component densities:

$$\mathsf{E}[y^r] = \sum_{j=1}^{C} \pi_j \mathsf{E}[y^r|f_j(\cdot)], \tag{4.73}$$

where $\mathsf{E}[y^r|f_j(\cdot)]$ denotes the rth uncentered moment given density $f_j(y)$ (Johnson, Kemp, and Kotz, 2005). The central moments of the mixture can than be derived using standard relations between the raw and central moments. Note that this result means that mixing will change the conditional mean, unless $\mathsf{E}[y]$ is the same in all component densities, whereas the continuous heterogeneous mixing of section 4.2.1 did not change the mean.

Finite mixtures of some standard univariate count models are discussed in Titterington, Smith, and Makov (1985). An attraction of a finite mixture model is that it can model data that are multimodal. As an example, Figure 4.2 shows the probability mass function for the univariate mixture of $0.50 \times \mathsf{P}[0.2] + 0.50 \times \mathsf{P}[6]$, where $\mathsf{P}[\mu]$ denotes the Poisson with mean μ. The mixture is multimodal – a common feature of mixtures in which the components are well separated.

The mixture model has a long history in statistics; see McLachlan and Basford (1988), Titterington et al. (1985), and Everitt and Hand (1981). For fixed C the model is essentially a flexible parametric formulation. The constraint

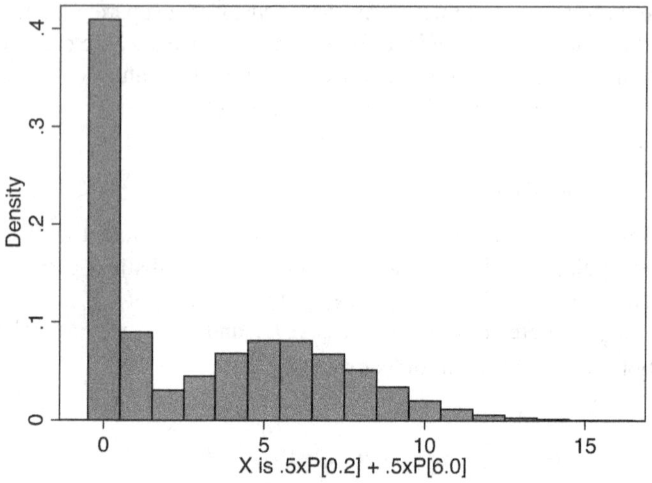

Figure 4.2. Two-component mixture of Poissons.

$0 < \pi_j < 1$ can be imposed by using a multinomial model, with

$$\pi_j = \frac{\exp(\lambda_j)}{1 + \sum_{k=1}^{C-1} \exp(\lambda_k)}, \qquad j = 1, \ldots, C - 1, \tag{4.74}$$

and $\pi_C = \sum_{k=1}^{C-1} \pi_k$.

Although there have been many univariate applications, there are also many applications with regression components; see McLachlan and Peel (2000). In regression applications the component densities $f_j(y_i|\boldsymbol{\theta}_j) = f_j(y_i|\mathbf{x}_i, \boldsymbol{\theta}_j)$ depend on regressors. Although in principle the component distributions may come from different parametric families, in practice it is usual to restrict them to come from the same families. In the analysis that follows, we consider only proportions π_j that do not vary with regressors. Even then the finite mixture class offers a flexible way of specifying mixtures. Extension to the case where the proportions π_j may also depend on regressors is deferred to section 4.8.7.

An interesting special case is the *random intercept model*. For the jth component density let $\boldsymbol{\theta}_j' = [\theta_{1j}' \ \boldsymbol{\theta}_2']$, where θ_{1j} is the intercept, which can vary with j, whereas slope parameters $\boldsymbol{\theta}_2$ are restricted to be the same across components. That is, subpopulations are assumed to differ randomly only with respect to their location parameter. This is sometimes referred to as a semiparametric representation of unobserved heterogeneity, because unlike in section 4.2.1, a parametric assumption about the distribution of unobserved heterogeneity is avoided. Discrete specifications of heterogeneity is used in Laird (1978), Lindsay (1983), and Heckman and Singer (1984).

The more general model allows for full heterogeneity by permitting all parameters in the C components to differ. This case is sometimes referred to as a mixture with random effects in the intercept and the slope parameters (Wedel, DeSarbo, Bult, and Ramaswamy, 1993).

A key advantage of a finite mixture model, rather than a continuous mixing distribution, is that it provides a natural and intuitively attractive representation of heterogeneity in a finite, usually small, number of latent classes, each of which may be regarded as a "type" or a "group"; see section 4.8.2. The choice of the number of components in the mixture determines the number of "types," whereas the choice of the functional form for the density can accommodate heterogeneity within each component.

The latent class interpretation is not essential, however, and the finite mixture can always be viewed simply as a way of flexibly and parsimoniously modeling the data, with each mixture component providing a local approximation to some part of the true distribution. As such the approach is an alternative to nonparametric estimation.

Other advantages of using a discrete rather than a continuous mixing distribution include the following. A finite mixture is semiparametric because it does not require any distributional assumptions for the mixing variable. From the results of Laird (1978) and Heckman and Singer (1984), finite mixtures may provide good numerical approximations even if the underlying mixing distribution is continuous. And the commonly used continuous mixing densities for parametric count models can be restrictive and computationally intractable except when the conditional (kernel) and mixing densities are from conjugate families that yield a marginal density with closed functional form. By contrast, there are several practical methods available for estimating finite mixture models (Böhning, 1995).

4.8.2 Latent Class Analysis

The finite mixture model is related to *latent class analysis* (Aitkin and Rubin, 1985; Wedel et al., 1993). In practice, the number of latent components has to be estimated, but initially we assume that the number of components in the mixture, C, is given.

Let $\mathbf{d}_i = (d_{i1}, \ldots, d_{iC})$ define indicator (dummy) variables such that

$$d_{ij} = \begin{cases} 1 & \text{if } y_i \text{ is drawn from latent class } j \text{ with density } f(y_i|\boldsymbol{\theta}_j) \\ 0 & \text{if } y_i \text{ is not drawn from latent class } j, \end{cases} \tag{4.75}$$

$j = 1, \ldots, C$, where $\sum_{j=1}^{C} d_{ij} = 1$. That is, each observation may be regarded as a draw from one of the C latent classes or "types," each with its own distribution. The classes are called latent classes because they are hidden and not directly observed.

It follows from (4.75) that $y_i|\mathbf{d}_i, \boldsymbol{\theta}$ has density

$$f(y_i|\mathbf{d}_i, \boldsymbol{\theta}) = \prod_{j=1}^{C} f(y_i|\boldsymbol{\theta}_j)^{d_{ij}}, \tag{4.76}$$

where $\boldsymbol{\theta} = (\boldsymbol{\theta}_1, \ldots, \boldsymbol{\theta}_C)$. The unobserved variables $\mathbf{d}_i$ are assumed to be iid (over i) with multinomial distribution

$$g(\mathbf{d}_i|\boldsymbol{\pi}) = \prod_{j=1}^{C} \pi_j^{d_{ij}}, \quad 0 < \pi_j < 1, \quad \sum_{j=1}^{C} \pi_j = 1, \tag{4.77}$$

where $\boldsymbol{\pi}' = (\pi_1, \ldots, \pi_C)$. Using $f(y_i, \mathbf{d}_i) = f(y_i|\mathbf{d}_i)g(\mathbf{d}_i)$, (4.76) and (4.77) imply that

$$f(y_i, \mathbf{d}_i|\boldsymbol{\theta}, \boldsymbol{\pi}) = \prod_{j=1}^{C} \pi_j^{d_{ij}} f_j(y|\boldsymbol{\theta}_j)^{d_{ij}}, \quad 0 < \pi_j < 1, \quad \sum_{j=1}^{C} \pi_j = 1, \tag{4.78}$$

Hence the complete data likelihood function is

$$\mathsf{L}(\boldsymbol{\theta}, \boldsymbol{\pi}|\mathbf{y}, \mathbf{d}) = \prod_{i=1}^{n}\prod_{j=1}^{C} \pi_j^{d_{ij}} [f_j(\mathbf{y}|\boldsymbol{\theta}_j)]^{d_{ij}}, \quad 0 < \pi_j < 1, \quad \sum_{j=1}^{C} \pi_j = 1. \tag{4.79}$$

The likelihood expression in (4.79) is referred to as the full-data or complete-data likelihood because it depends on both the observed data and the latent class indicators. By contrast, the likelihood based only on observed data is referred to as the marginal likelihood. Note that for simplicity we have suppressed dependence on regressors $\mathbf{x}_i$ in the preceding expressions.

Since d_{ij} is unobserved, it is relevant to also consider the marginal density of y_i. In general $f(y_i) = \sum_{\mathbf{d}_i} f(y_i, \mathbf{d}_i)$, and summing (4.78) over all possible values of the mutually exclusive indicators $\mathbf{d}_i$ yields

$$f(y_i|\mathbf{x}_i, \boldsymbol{\theta}, \boldsymbol{\pi}) = \sum_{j=1}^{C} \pi_j f_j(y_i|\mathbf{x}_i, \boldsymbol{\theta}_j), \tag{4.80}$$

where we now introduce regressors $\mathbf{x}_i$. This is the C-component mixture density given in (4.72).

The probability mass function $f(d_{ij}|y_i)$ of d_{ij} conditional on y_i, is also of interest. This gives the so-called *posterior probability* that the ith observation is in class j, denoted z_{ij}, with

$$\begin{aligned} z_{ij} &= \Pr[\text{observation } i \text{ belongs to class } j|y_i] \\ &= \Pr[d_{ij} = 1|y_i] \\ &= \Pr[d_{ij} = 1|y_i]/f(y_i) \\ &= \pi_j f_j(y_i|\mathbf{x}_i, \boldsymbol{\theta}_j)/f(y_i|\mathbf{x}_i, \boldsymbol{\theta}, \boldsymbol{\pi}) \\ &= \frac{\pi_j f_j(y_i|\mathbf{x}_i, \boldsymbol{\theta}_j)}{\sum_{j=1}^{C} \pi_j f_j(y_i|\mathbf{x}_i, \boldsymbol{\theta}_j)}, \end{aligned} \tag{4.81}$$

where the last equality uses (4.80). The average value of z_{ij} over i is the marginal probability that a randomly chosen individual belongs to subpopulation j. Given (4.81) this is simply π_j, so

$$\mathsf{E}\left[z_{ij}\right] = \pi_j. \tag{4.82}$$

4.8.3 Nonparametric Maximum Likelihood

An important distinction is made between whether or not the number of components C is known. In the case where C is unknown, C is also a parameter that needs to be estimated. Let $\mathsf{L}(\boldsymbol{\pi}, \boldsymbol{\theta}, C\,|\mathbf{y})$ denote the likelihood based on (4.72) with C distinct parametrically specified components. The probability distribution $f(y_i|\hat{\boldsymbol{\theta}}; \hat{C})$ that maximizes $\mathcal{L}(\boldsymbol{\pi}, \boldsymbol{\theta}, C|\mathbf{y})$ is called the semiparametric maximum likelihood estimator. Lindsay (1995) has discussed its properties.

In some cases C may be given. Then the problem is to maximize $\mathcal{L}(\boldsymbol{\pi}, \boldsymbol{\Theta}|C, \mathbf{y})$. It is easier to handle estimation by maximizing log-likelihood for a selection of values of C and then using model-selection criteria to choose among estimated models.

4.8.4 Estimation

A widely recommended procedure for estimating the finite mixture model is the EM procedure. This seeks to obtain the MLE based on the likelihood $\mathsf{L}(\boldsymbol{\theta}, \boldsymbol{\pi}|\mathbf{y}, \mathbf{d})$, with the unobserved variables d_{ij} treated as missing data as follows.

If the d_{ij} were observable, from (4.79) the log-likelihood of the model would be

$$\mathcal{L}(\boldsymbol{\pi}, \boldsymbol{\theta}|\mathbf{y}, \mathbf{d}) = \sum_{i=1}^{n}\sum_{j=1}^{C} d_{ij}\left[\ln f_j(y_i|\boldsymbol{\theta}_j) + \ln \pi_j\right], \tag{4.83}$$

where we suppress dependence of $f_j(\cdot)$ on regressors $\mathbf{x}_i$. Let $\hat{\boldsymbol{\pi}}$ and $\hat{\boldsymbol{\theta}}$ be the current estimates of $\boldsymbol{\pi}$ and $\boldsymbol{\theta}$. The E-step of the EM algorithm replaces $\mathcal{L}(\boldsymbol{\pi}, \boldsymbol{\theta}|\mathbf{y})$ by its expected value conditional on observed data and $\hat{\boldsymbol{\pi}}$ and $\hat{\boldsymbol{\theta}}$. Since (4.83) is linear in d_{ij}, this entails replacing the unobserved d_{ij} with its expected value given y_i, $\hat{\boldsymbol{\pi}}$, and $\hat{\boldsymbol{\theta}}$. Now $\mathsf{E}\left[d_{ij}|y_i\right] = \Pr[d_{ij} = 1|y_i] = z_{ij}$. So the expected log-likelihood replaces d_{ij} in (4.83) by $\hat{z}_{ij}$ equal to z_{ij} defined in (4.81) evaluated at $\hat{\boldsymbol{\pi}}$ and $\hat{\boldsymbol{\theta}}$. The resulting expected log-likelihood is

$$\mathcal{EL}(\boldsymbol{\theta}, \boldsymbol{\pi}|\mathbf{y}, \hat{\boldsymbol{\pi}}, \hat{\boldsymbol{\theta}}) = \sum_{i=1}^{n}\sum_{j=1}^{C} \hat{z}_{ij}\left[\ln f_j(y_i|\boldsymbol{\theta}_j) + \ln \pi_j\right]. \tag{4.84}$$

The M-step of the EM procedure then maximizes (4.84) with respect to $\boldsymbol{\pi}$ and $\boldsymbol{\theta}$. This leads to

$$\hat{\pi}_j = \frac{1}{n}\sum_{i=1}^{n} \hat{z}_{ij}, \quad j = 1, \ldots, C, \tag{4.85}$$

while $\hat{\boldsymbol{\theta}}_1, \ldots, \hat{\boldsymbol{\theta}}_C$ solve the first-order conditions

$$\sum_{j=1}^{C} \sum_{i=1}^{n} \hat{z}_{ij} \frac{\partial \ln f_j(y_i|\boldsymbol{\theta}_j)}{\partial \boldsymbol{\theta}_j} = \mathbf{0}. \tag{4.86}$$

The algorithm returns to the E-step, recomputing $\hat{z}_{ij}$ given the new parameter estimates $\hat{\boldsymbol{\pi}}$ and $\hat{\boldsymbol{\theta}}$, and then to the M-step, re-maximizing (4.84) given the new $\hat{z}_{ij}$, and so on.

The EM algorithm may be slow to converge, especially if the starting values are poorly chosen. Other approaches such as the Newton-Raphson or Broyden-Fletcher-Goldfarb-Shanno may also be worth considering. These work directly with the original finite mixture density representation (4.72) and maximize with respect to $\boldsymbol{\pi}$ and $\boldsymbol{\theta}$ the log-likelihood function

$$\mathcal{L}(\boldsymbol{\pi}, \boldsymbol{\theta}|\mathbf{y}) = \sum_{i=1}^{n} \ln \left(\sum_{j=1}^{C} \pi_j f_j(y_i|\boldsymbol{\theta}_j) \right). \tag{4.87}$$

To ensure $0 < \pi_j < 1$ and $\sum_{j=1}^{C} \pi_j = 1$, the reparameterization of π_j given in (4.74) can be used. A discussion of reliable algorithms is in Böhning (1995). The applications in this section and in Chapter 6 use the Newton-Raphson algorithm to maximize (4.87).

For C given, the covariance matrix of estimated parameters is obtained by specializing the results in Chapter 2.3.2. Alternative estimators for the variance, including the sandwich and robust sandwich estimators, are also available. However, they should be used with caution, especially if C is likely to be misspecified.

Finally we note that in the special case of a $C-$component finite mixture model with exponential conditional means for each component and only the intercept varying across the components, the overall conditional mean

$$\begin{aligned}
\mathsf{E}[y_i|\mathbf{x}_i] &= \sum_{j=1}^{C} \pi_j \exp(\delta_j + \mathbf{w}_i'\boldsymbol{\gamma}) \\
&= \exp\left(\ln\left(\textstyle\sum_{j=1}^{C} \pi_j \exp(\delta_j)\right) + \mathbf{w}_i'\boldsymbol{\gamma}\right).
\end{aligned}$$

The intercept of the constrained finite mixture involves a weighted combination of the intercepts, with weights determined by the population proportions. Hence in the case of Poisson mixtures, the usual Poisson QMLE yields consistent estimates of the slope parameters $\boldsymbol{\gamma}$, though only an estimate of a weighted component of the intercepts and not the individual components that we need to estimate subpopulation means.

4.8.5 Unknown C

If the number of components is unknown, then effectively this is the case of an unknown mixing distribution. The justification for using a discrete mixing distribution then follows from Lindsay (1995), who has shown that, provided the data set has sufficient variation and the likelihoods $f(y|\boldsymbol{\theta})$ are bounded in $\boldsymbol{\theta}$, a discrete mixing distribution with the number of components (support points) at most equal to the number of distinct data points in the sample is observationally equivalent to the true unknown mixing distribution. This result implies that a potentially difficult problem of nonparametric estimation of a mixing density can be reduced to an estimation problem in finite dimension. This still leaves open the question of whether a specified value of C in a particular application is sufficiently large. We revisit this issue in the context of an empirical example in Chapter 6.

4.8.6 Inference on C

The preceding account has not dealt with the choice of the parameter C. Two approaches have been widely considered in the literature.

The first is to use a likelihood ratio test to test whether $C = C^*$ versus $C = C^* + 1$. This is equivalent to the test of $H_0 : \pi_{C^*+1} = 0$ versus $H_1 : \pi_{C^*+1} \neq 0$. Unfortunately, the likelihood ratio test statistic in this case does not have the standard chi-square distribution because the regularity conditions for likelihood-based inference are violated.

For example, let $C = 2$, and suppose we wish to test

$$H_0 : f(y_i|\boldsymbol{\Theta}) = \mathsf{P}[\theta_1],$$

where $\mathsf{P}[\theta]$ denotes Poisson with mean θ, against

$$H_1 : f(y_i|\boldsymbol{\Theta}) = (1-\pi)\mathsf{P}[\theta_1] + \pi\mathsf{P}[\theta_2],$$

where $\boldsymbol{\Theta} = (\pi, \theta_1, \theta_2) \in [0, 1] \times (0, \infty) \times (0, \infty)$, with $\theta_1 \neq \theta_2$. The null holds not only if $\pi = 0$ but also if $\theta_1 = \theta_2$. That is, the parameter space under H_0 is

$$\boldsymbol{\Theta}_0 = (\pi, \theta_1, \theta_2) \in \{[0] \times (0, \infty) \times (0, \infty)\} \cup \{[0, 1] \times (0, \infty)\}.$$

This is the entire (θ_1, θ_2) space when $\pi = 0$, and the line segment $\pi \in [0, 1]$ when $\theta_1 = \theta_2$. Under the null hypothesis the parameter π is not identifiable because it is on the boundary of the parameter space. The standard assumption for likelihood-based testing assumes regularity conditions stated in Chapter 2.3. This includes the condition that $\boldsymbol{\Theta}$ is in the interior of parameter space. Hence the standard asymptotic distribution theory does not apply. As a result, when testing a model against a more general mixture model with h additional parameters, the critical values and p-values for the LR test statistic cannot be obtained from the usual $\chi^2(h)$ distribution.

One solution is to use a parametric bootstrap to obtain the critical values of the LR test statistic. This bootstrap, where the bootstrap resamples are obtained by draws from the fitted parametric model under H_0, is detailed in Chapter 6.6.2. Feng and McCulloch (1996) prove that the MLE in the more general H_1 model converges to the subset of the parameter set with distributions indistinguishable from those under H_0 (i.e., same density values). This should ensure that this bootstrap will yield an empirical estimate, without asymptotic refinement, of the distribution of the LR test statistic, though to date there appears to be no formal proof of this conjecture.

A second solution is to apply the results of Vuong (1989), see Chapter 5.4.2, that yield a sum of weighted $\chi^2(1)$ distributions for the LR test statistic. Lo, Mendell, and Rubin (2001) prove that this is applicable in the special case of mixture of normals with the same variance, though Jeffries (2003) argues that Lo et al. (2001) make assumptions that are too strong to cover cases of practical interest.

A third approach is to use information criteria, see Chapter 5.4.1, that penalize the log-likelihood in models with more parameters. In that case it is common to first fix the largest value of C one is prepared to accept, often a small number like two, three, or four. The model is estimated for all values of $C \leq C^*$. The model with the smallest BIC or related measure is chosen. Chapter 6 contains examples and further discussion.

Much of the evidence to support these methods is based on simulation evidence, primarily for mixture models based on the normal distribution. A recent example is Nylund, Asparouhov, and Muthén (2007). The consensus appears to be that bootstrapped critical values do well, and among information criteria BIC does reasonably well. BIC is often used because it has the advantage of simplicity and computational speed compared to bootstrapping or using Vuong's method.

In an unconstrained model, adding components can easily lead to overparameterization in two senses. The total number of parameters may be large. Then the number of components may also be too large relative to the information in the data. Because the problem is akin to classification into "types," doing so reliably requires that interindividual differences are large in the relevant sense. In the case of a constrained mixture, such as the random intercept model, the resulting increase in the number of parameters is small, but in the unconstrained model in which all parameters are allowed to vary, the total number of parameters can be quite large (see Chapter 6 for an example).

If the model is overparameterized, one can expect difficulties in estimation due to a flat log-likelihood. This problem is further compounded by the possible multimodality of the mixture likelihood. Thus it is possible for an estimation algorithm to converge to a local rather than a global maximum. In practical applications of finite mixtures, therefore, it is important to test for the presence of mixture components. If the evidence for mixture is weak, identification is a problem. It is also important to test whether a global maximum of the likelihood is attained. These important issues are reconsidered in Chapter 6.

4.8.7 Extensions

The MLE of finite mixture models is overly influenced by outlying observations. For example, for count data in the nonregression case, one unusually large count may lead to a model with an extra component simply to handle this outlier. In this subsection we present an alternative, more robust estimator than the MLE. We also consider extension to models where the mixing proportions π_j may vary with regressors.

Minimum Distance Estimators

This chapter has emphasized ML estimation. The recent biometric literature has proposed alternative, more robust minimum distance estimators for finite mixture models.

The Hellinger distance criterion has been found attractive because it is less influenced by outlying observations. For mixture models, it is defined as

$$D(\boldsymbol{\pi}, \boldsymbol{\theta}) = \sum\nolimits_{k=0}^{m} \left[\left(\overline{p}_k\right)^{1/2} - \left(\frac{1}{n} \sum\nolimits_{i=1}^{n} f(y_i = k|\mathbf{x}_i, \boldsymbol{\theta}, \boldsymbol{\pi}) \right)^{1/2} \right]^2 \tag{4.88}$$

where $m = \max\{y_i,\ i = 1, \ldots, n\}$, $\overline{p}_k$ is the fraction of observations with $y_i = k$, and $f(y_i|\mathbf{x}_i, \boldsymbol{\theta}, \boldsymbol{\pi})$ is the mixture density defined in (4.72). In general the squared Hellinger distance between two continuous densities $g(y)$ and $f(y)$ is $\frac{1}{2}\int \left(\sqrt{g(y)} - \sqrt{f(y)}\right)^2 dy$. In (4.88), $g(y)$ is the sample density, $f(y) = \frac{1}{n}\sum_{i=1}^{n} f(y|x_i)$, where $f(y|\mathbf{x}_i)$ is the specified mixture density, and we have adapted to the discrete case. The squared Hellinger distance can be reexpressed as $1 - \int \sqrt{g(y) f(y)}$, so minimizing $D(\boldsymbol{\pi}, \boldsymbol{\theta})$ is equivalent to maximizing

$$D^*(\boldsymbol{\pi}, \boldsymbol{\theta}) = \sum\nolimits_{k=0}^{m} \left(\overline{p}_k \frac{1}{n} \sum\nolimits_{i=1}^{n} f(y_i = k|\mathbf{x}_i, \boldsymbol{\theta}, \boldsymbol{\pi}) \right)^{1/2}.$$

This latter representation is used to obtain the estimating equations for $\boldsymbol{\theta}$ and $\boldsymbol{\pi}$.

Lu, Hui, and Lee (2003), following Karlis and Xekalaki (1998), who consider the nonregression case, present minimum Hellinger distance estimation (MHDE) for finite mixtures of Poisson regressions and also a computational algorithm that has some similarities with the EM algorithm. Xiang, Yau, Van Hui, and Lee (2008) use MHDE for estimating a k-component Poisson regression with random effects. The attraction of MHDE relative to MLE is that it is expected to be more robust to the presence of outliers. It is also claimed to do better when mixture components are not well separated and/or when the model fit is poor.

Smoothly Varying Mixing Proportions

The mixing proportions may depend on observable variables, denoted $\mathbf{z}$. To ensure $0 \leq \pi_j \leq 1$, and $\sum_{j=1}^{C} \pi_j = 1$, let

$$\pi_j(\mathbf{z}_i, \boldsymbol{\alpha}) = \frac{\exp(\mathbf{z}_i'\boldsymbol{\alpha}_j)}{1 + \exp(\mathbf{z}_i'\boldsymbol{\alpha}_1) + \exp(\mathbf{z}_i'\boldsymbol{\alpha}_2) + \cdots + \exp(\mathbf{z}_i'\boldsymbol{\alpha}_{C-1})}.$$

Then the finite mixture density is given by

$$f(y_i|\mathbf{x}_i, \mathbf{z}_i, \boldsymbol{\theta}, \boldsymbol{\alpha}) = \sum_{j=1}^{C} \pi_j(\mathbf{z}_i, \boldsymbol{\alpha}) f_j(y_i|\mathbf{x}_i, \boldsymbol{\theta}_j).$$

The full data likelihood can be written using an expression similar to that in (4.79). Although this generalization provides obvious additional model flexibility, it can make it more difficult to use a latent class interpretation of the finite mixture model.

In a multiperiod pooled or population-averaged sample, it may be realistic to allow for transitions between latent classes. An observation assigned to group j may transit to another group k. In the macroeconomic literature on business cycles, where latent classes are often referred to as regimes, this transition is called regime switching; see Hamilton (1989) and Kim and Nelson (1999). In time-series analysis of count data (see Chapter 7) observations may be treated as mixtures of realizations from different regimes, each regime being characterized by its own probability law. When the transitions between regimes follow a first-order Markov chain the structure is called a hidden Markov model (HMM); see MacDonald and Zucchini (1997). In the analysis given earlier, this imposes a Markovian structure on the d_{ij}, the indicator variable for observation i belonging to class j. The HMM structure can be extended also to panel data; see Chapter 9.10.1.

4.8.8 Example: Patents

Hausman et al. (1984) consider the relationship between the number of patents awarded to a firm and current and past research and development (R&D) expenditures, using a short panel. Here we use the data illustratively. Because the emphasis of this chapter is on cross-section data, we use only the 1979 data (the last year of the panel) on 346 firms as in Hall, Griliches, and Hausman (1986).

Several features of the data indicate that these are difficult data to model. First, from Table 4.2 even though there is a significant proportion of zeros, the right tail of the distribution is very long. The overall mean is about 32 and the maximum is more than 500! Second, there are some very large values that contribute substantially to overdispersion. These two features in combination make it difficult to specify a model with a conditional mean and variance that capture the main features of the data. In that sense the example is pathological.

Table 4.2. *Patents: Frequencies for 1979 data*

Count	0	1	2–5	6–20	21–50	51–100	101–300	300–515
Frequency	76	41	74	51	43	30	27	4
Relative frequency	0.220	0.119	0.214	0.147	0.124	0.087	0.078	0.012

The conditional mean is specified as $\exp(\mathbf{x}_i'\boldsymbol{\beta})$. The regressor of interest is the logarithm of total research over the current and preceding five years, $\ln R\&D$. Given the logarithmic transformation and the exponential conditional mean, the coefficient of $\ln R\&D$ is an elasticity, so that the coefficient should equal unity if there are constant returns to scale. To control for firm-specific effects, the estimated model includes two time-invariant regressors: $\ln SIZE$, the logarithm of firm book value in 1972, which is a measure of firm size, and $DSCI$, which is a sector indicator variable equal to one if the firm is in the science sector. Therefore, the starting point is the Poisson and NB2 models presented in section 4.2. The third model is a two-component finite mixture of NB2, henceforth referred to as FMNB2(2).

Table 4.3 presents ML estimates for three models. It is clear that the basic Poisson regression will completely fail to provide a reasonable fit to the data, due to the overdispersion in the data. This is indeed the case, and a negative binomial model (NB2) leads to the log-likelihood increasing from −3366 to −1128. For both models the coefficient of $\ln R\&D$ is statistically significant but is also statistically significantly less than one. The coefficient varies across

Table 4.3. *Patents: Poisson, NB2, and FMNB2(2) models with 1979 data*

			FMNB2(2)	
Variable	Poisson	NB2	Component 1	Component 2
Intercept	−0.340	−0.811	−1.274	1.516
	(0.190)	(0.180)	(0.161)	(0.191)
ln R&D	0.475	0.816	0.918	0.352
	(0.064)	(0.083)	(0.063)	(0.090)
ln *SIZE*	0.271	0.115	0.065	0.277
	(0.055)	(0.069)	(0.057)	(0.062)
DSCI	0.467	−0.090	0.103	−0.573
	(0.155)	(0.138)	(0.142)	(0.346)
α		0.885	0.651	0.062
		(0.100)	(0.080)	(0.047)
π			0.929	
			(0.022)	
Number of parameters	4	5	11	
log-likelihood	−3366	−1128	−1106	
AIC	6740	2265	2234	
BIC	6755	2284	2277	

Note: Robust standard errors are given in parentheses.

the two models: 0.82 for the NB2 model compared to 0.84 for Poisson (and compared to 0.49 for the NB1 model which is not reported in the table).

Clearly the magnitude is sensitive to the functional form used. As a basis for comparison we note that the fixed effect models for panel count data, as estimated by Hausman et al. (1984) and others, see Chapter 9, yield relatively low estimates of around 0.6. Also from Chapter 9, other authors using GMM methods obtain even lower estimates such as 0.3 that are even more strongly indicative of decreasing returns to scale. These panel applications expand on the model estimated here by including as six separate regressors the natural logarithm of R&D expenditures in the current and each of the past five years, and then calculating the overall elasticity as the sum of the six coefficients. When we estimate this richer specification here, the coefficients for each lagged year are noisily estimated, but the sum of the coefficients is close to that for the ln *R&D* variable in Table 4.3 and there is little gain in the log-likelihood. It is difficult to parse out the individual contribution of each year.

The final columns of Table 4.3 report estimates from a two-component finite mixture of NB2. The first component has slope coefficients similar to the NB2 model and occurs with probability 0.93. The second component has slope coefficients similar to the Poisson model and an overdispersion parameter close to zero. The model fit is improved, with log-likelihood increasing by 22 compared to the NB2 model. Using information criteria presented in Chapter 5.4.1 that penalize for model size, the FMNB2(2) model is preferred to the more parsimonious NB2 model on the grounds of lower AIC and BIC.

4.9 COUNT MODELS WITH CROSS-SECTIONAL DEPENDENCE

The assumption of cross-sectionally independent observations was routine in the early econometric count data literature. Recent theoretical and empirical work pays greater attention to the possibility of dependence even with cross-section data.

Such dependence is likely to arise whenever data come from a complex survey, a common source of data for social science research, because complex surveys are designed to interview individuals in clusters to lower the cost of sampling. For example, rather than randomly choose individuals from the population, a set of villages (or census blocks) is randomly chosen; then within a chosen village (or census block) a set of multiple households is randomly chosen, and possibly all individuals in a chosen household are interviewed. This sampling scheme is likely to generate correlation within clusters due to variation induced by common unobserved cluster-specific factors.

Cross-sectional dependence also arises when the count outcomes have a spatial dimension, as when the data are drawn from geographical regions. Then outcomes of units that are spatially contiguous may not be independent of each other.

It is essential to control for cross-sectional dependence. Estimator standard errors based on the assumption of independence are likely to be underestimates, and estimator precision can be greatly overestimated. The intuition in the cluster case is that observations are usually positively correlated within clusters, so an additional observation from the same cluster no longer provides a completely independent piece of new information. Furthermore, it is possible that the mean outcome for an individual depends on that of other individuals in the sample. Then the usual conditional mean is no longer correctly specified and the usual estimators are inconsistent.

The issues for clustered data are qualitatively similar to those detailed in Chapter 9 for panel data. In that case an observation is an individual-time pair, and observations may be correlated over time for a given individual. The treatment here is brief, and Chapter 9 provides considerable additional detail. The issues for spatial correlation are qualitatively similar to those for time-series correlation.

4.9.1 Cluster Error Correlation

By *cluster error correlation* we mean that the conditional mean of the count is unaffected by other observations, but counts from the same cluster are conditionally correlated. Thus the usual estimators retain their consistency, but standard errors should be adjusted and more efficient estimation is possible.

Consider a sample falling in C distinct clusters with the cth cluster having n_c observations, $c = 1, \ldots, C$, and the sample size $n = \sum_{c=1}^{C} n_c$. Let $(y_{ic}, \mathbf{x}_{ic})$ denote data for individual i in cluster c. We assume the conditional mean is unchanged, so

$$\mathsf{E}[y_{ic}|y_{jc'}, \mathbf{x}_{ic'}, \mathbf{x}_{jc'}] = \mathsf{E}[y_{ic}|\mathbf{x}_{ic}] \text{ all } c, c' \text{ and } i \neq j. \tag{4.89}$$

We assume that observations from the same cluster are correlated, whereas those in different clusters are uncorrelated. Then

$$\begin{aligned} \mathsf{Cov}[y_{ic}, y_{jc}|\mathbf{x}_{ic}, \mathbf{x}_{jc}] &\neq 0 \\ \mathsf{Cov}[y_{ic}, y_{jc'}|\mathbf{x}_{ic'}, \mathbf{x}_{jc'}] &= 0 \quad \text{if } c \neq c'. \end{aligned} \tag{4.90}$$

The simplest correction, and that most common in econometrics applications, is to use a pooled regression that ignores any clustering, but then obtain cluster-robust standard errors presented in Chapter 2.7. Suppose the estimating equations conditions for $\hat{\boldsymbol{\theta}}$ are $\sum_{c=1}^{C} \sum_{i=1}^{n_c} \mathbf{g}(y_{ic}, \mathbf{x}_{ic}, \hat{\boldsymbol{\theta}}) = \mathbf{0}$. For example, for the usual Poisson QMLE, $\mathbf{g}_{ic}(\cdot) = (y_{ic} - \exp(\mathbf{x}_{ic}'\boldsymbol{\beta}))\mathbf{x}_{ic}$. Then

$$\mathsf{V}[\hat{\boldsymbol{\theta}}] = \left[\sum_{c=1}^{C} \sum_{i=1}^{n_c} \frac{\partial \mathbf{g}_{ic}}{\partial \boldsymbol{\theta}'} \right]^{-1} \left[\sum_{c=1}^{C} \sum_{i=1}^{n_c} \sum_{j=1}^{n_c} \mathbf{g}_{ic} \mathbf{g}_{jc}' \right] \left[\sum_{c=1}^{C} \sum_{i=1}^{n_c} \frac{\partial \mathbf{g}_{ic}'}{\partial \boldsymbol{\theta}} \right]^{-1}, \tag{4.91}$$

and standard errors are obtained by evaluating $\mathsf{V}[\hat{\boldsymbol{\theta}}]$ at $\boldsymbol{\theta} = \hat{\boldsymbol{\theta}}$. This variance matrix estimator requires $C \to \infty$, so is appropriate when there are a large number of clusters.

More efficient estimates can be obtained by specifying a model for the within-cluster correlation and then estimating the regression using methods that make use of this model. For specificity, we consider Poisson regression with exponential conditional mean $\exp(\mathbf{x}'_{ic}\boldsymbol{\beta})$.

The generalized estimating equations (GEE) method specifies a model for within-cluster correlation, such as equicorrelation with $\mathsf{Cor}[y_{ic}, y_{jc}|\mathbf{x}_{ic}, \mathbf{x}_{jc}] = \rho\sigma_{ic}\sigma_{jc}$, where $\sigma^2_{ic} = \mathsf{Var}[y_{ic}|\mathbf{x}_{ic}]$. The GEE estimator uses pooled Poisson regression to obtain consistent but inefficient estimate $\hat{\boldsymbol{\beta}}$, uses this to obtain estimate $\hat{\rho}$ (and estimates of any additional parameters determining σ^2_{ic}), and then efficiently estimates $\boldsymbol{\beta}$ by feasible nonlinear GLS. Standard errors can then be obtained that are robust to misspecification of the model for $\mathsf{Cor}[y_{ic}, y_{jc}]$. Chapters 2.4.6 and 9.3.2 detail GEE estimation.

A random intercept model assumes

$$\begin{aligned}\mathsf{E}[y_{ic}|\mathbf{x}_{ic}, \alpha_c] &= \exp(\alpha_c + \alpha + \mathbf{x}'_{ic}\boldsymbol{\beta}) \\ &= \delta_c \exp(\alpha + \mathbf{x}'_{ic}\boldsymbol{\beta}),\end{aligned} \tag{4.92}$$

where α_c is iid $[0, \sigma^2_\alpha]$ and hence $\delta_c = \exp(\alpha_c)$ is iid $[1, \sigma^2_\delta]$. A distribution for δ_c is specified, and estimation is by MLE. If δ_c is one-parameter gamma distributed, then there is a closed-form expression for the likelihood function (unconditional on δ_c) and ML estimation is straightforward. If instead α_c is normally distributed, so δ_c is lognormal, then ML estimation requires computation of a one-dimensional integral. See sections 9.5.1 and 9.5.2 for further details.

A random parameters model additionally allows for the slope parameters to vary across clusters. Then $\boldsymbol{\beta}$ in (4.92) is replaced by $\boldsymbol{\beta}_c \sim \mathsf{N}[\boldsymbol{\beta}, \Sigma_{\boldsymbol{\beta}}]$, and estimation is by MLE. This is a special case of a hierarchical model presented in section 4.7. Note that in this model the conditional mean is no longer $\mathsf{E}[y_{ic}|\mathbf{x}_{ic}] = \exp(\alpha + \mathbf{x}'_{ic}\boldsymbol{\beta})$; see Chapter 9.5.3.

Demidenko (2007) compares these various approaches for the Poisson regression model.

4.9.2 Spatial Error Correlation

By *spatial error correlation* we mean that the conditional mean of the count is unaffected by other observations, but different observations are conditionally spatially correlated. This is similar to the cluster case, except here the correlation is determined by the spatial distance between observations, rather than whether or not they belong to the same cluster.

We still maintain correct specification of the conditional mean, so $\mathsf{E}[y_i|y_j, \mathbf{x}_i, \mathbf{x}_j] = \mathsf{E}[y_i|\mathbf{x}_i]$. Spatial correlation means that

$$\mathsf{Cov}[y_i, y_j|\mathbf{x}_i, \mathbf{x}_j] = w_{ij}\sigma_i\sigma_j, \quad 0 \le w_{ij} \le 1, \tag{4.93}$$

where $\sigma_i^2 = \mathsf{V}[y_i|\mathbf{x}_i]$ and w_{ij} is a function of the spatial distance between observations i and j. For example, if data are from different regions then the spatial distance is the distance between the regions. We specify $w_{ij} = 1$ when $i = j$ and require that $w_{ij} \rightarrow 0$ as the spatial distance between i and j becomes large.

Then for pooled estimator $\hat{\boldsymbol{\theta}}$ solving $\sum_{i=1}^{n} \mathbf{g}(y_i, \mathbf{x}_i, \hat{\boldsymbol{\theta}}) = \mathbf{0}$, we have

$$\mathsf{V}[\hat{\boldsymbol{\theta}}] = \left[\sum_{i=1}^{n} \frac{\partial \mathbf{g}_i}{\partial \boldsymbol{\theta}'}\right]^{-1} \left[\sum_{i=1}^{n}\sum_{j=1}^{n} w_{ij}\mathbf{g}_i\mathbf{g}_j'\right] \left[\sum_{i=1}^{n} \frac{\partial \mathbf{g}_i'}{\partial \boldsymbol{\theta}}\right]^{-1}, \tag{4.94}$$

and standard errors are obtained by evaluating $\mathsf{V}[\hat{\boldsymbol{\theta}}]$ at $\boldsymbol{\theta} = \hat{\boldsymbol{\theta}}$.

Mechanically the variance matrix (4.91) with clustering is the special case of (4.94) where w_{ij} equals one if observations i and j are in the same cluster, and w_{ij} equals zero otherwise. But the asymptotic theory is quite different. The cluster case requires independence across clusters and $C \rightarrow \infty$. The spatial case requires dampening in w_{ij} analogous to dampening time series correlation as $n \rightarrow \infty$; see Conley (1999).

More efficient estimation is possible by specifying a functional form for w_{ij} and estimating by nonlinear feasible GLS. For example, negative exponential distance decay specifies $w_{ij} = \exp(-\gamma d_{ij})$ for $i \neq j$, where d_{ij} is the distance between observations i and j.

4.9.3 Cluster-Specific Fixed Effects

We now relax the assumption that the conditional mean is unaffected by clustering.

The *cluster-specific fixed effects model* allows the conditional mean to depend in part on the other observations in the cluster. The model specifies

$$\mathsf{E}[y_{ic}|\mathbf{x}_{1c}, \ldots, \mathbf{x}_{n_c,c}, \alpha_c] = \alpha_c \exp(\mathbf{x}_{ic}'\boldsymbol{\beta}). \tag{4.95}$$

The cluster-specific effect α_c may be correlated with $\mathbf{x}_{ic}$, unlike the random intercept model (4.92). As a result, pooled Poisson regression of y_{ic} on $\mathbf{x}_{ic}$ will lead to inconsistent parameter estimates. An alternative estimator is needed.

In Chapter 9.4.1 it is shown that (4.95) implies

$$\mathsf{E}[(y_{ic} - (\lambda_{ic}/\bar{\lambda}_c)\bar{y}_c)|\mathbf{x}_{1c}, \ldots, \mathbf{x}_{n_c,c}, \alpha_c] = 0, \tag{4.96}$$

where $\bar{y}_c = \frac{1}{n_c}\sum_i \lambda_{ic}$, $\lambda_{ic} = \exp(\mathbf{x}_{ic}'\boldsymbol{\beta})$, and $\bar{\lambda}_c = \frac{1}{n_c}\sum_i \lambda_{ic}$. The expression (4.96) has eliminated α_i. The Poisson *cluster-specific fixed effects estimator* of $\boldsymbol{\beta}$ solves the corresponding sample moment conditions

$$\sum_{c=1}^{C}\sum_{i=1}^{n_c} \mathbf{x}_{ic}\left(y_{ic} - \frac{\bar{y}_c}{\bar{\lambda}_c}\lambda_{ic}\right) = \mathbf{0}. \tag{4.97}$$

It is shown in Chapter 9.4.2 that equivalently one can perform regular pooled Poisson regression of y_{ic} on $\mathbf{x}_{ic}$ and a set of mutually exclusive cluster-specific dummy variables.

Expression (4.96) implies

$$\mathsf{E}[y_{ic}|\mathbf{x}_{1c}, \ldots, \mathbf{x}_{n_c,c}] = \left(\frac{\bar{y}_c}{\bar{\lambda}_c}\right) \exp(\mathbf{x}_{ic}'\boldsymbol{\beta}),$$

so we can view the other observations in the cluster as changing the conditional mean by the multiple $\bar{y}_c/\bar{\lambda}_c$.

4.9.4 Spatial Lag Dependence

Spatial lag models permit the conditional mean for a given observation to depend on the outcomes of other observations. The term "spatial lag" is used by analogy to time-series lag dependence, where y_t depends on y_{t-1}.

Griffith and Haining (2006) survey the early literature on spatial Poisson regression and a number of its modern extensions. For simplicity we consider the pure spatial case where there are no exogenous regressors.

Besag (1974) defined a general class of "auto-models" suitable for different types of spatially correlated data. Let N_i denote the set of observations in the neighborhood of i (not including i itself) that determine, in part, y_i. The "auto-Poisson" model specifies y_i as being Poisson distributed with mean μ_i that depends on observations y_j in the neighborhood of i. Thus

$$\begin{aligned} f(y_i|y_j, j \in N_i) &= \frac{e^{-\mu_i}\mu_i^{y_i}}{y_i!} \\ \mu_i &= \exp\left(\alpha_i + \textstyle\sum_{j\in N_i} \beta_{ij} y_j\right), \end{aligned} \tag{4.98}$$

where the parameter α_i is an area-specific effect, and $\beta_{ij} = \beta_{ji}$. The standard set-up specifies $\beta_{ij} = \gamma w_{ij}$, where γ is a spatial autoregressive parameter, w_{ij} ($= 0$ or 1) represents the neighborhood structure, and $w_{ii} = 0$.

The difficulty with this auto-Poisson model is that the restriction $\sum_i f(y_i|y_j, j \in N_i) = 1$ implies $\gamma \leq 0$, which permits only negative spatial correlation. The spatial count literature has evolved in different directions to overcome this difficulty; see Kaiser and Cressie (1997) and Griffith and Haining (2006). One line of development uses spatial weighting dependent on assumptions about the spatial dependence structure.

An alternative approach is based on a Poisson model with a mixture specification. The conditional mean function in (4.98) is defined as

$$\mu_i = \exp(\alpha_i + S_i), \tag{4.99}$$

where S_i is a random effect defined by a conditional spatial autoregressive model. For example,

$$S_i \sim \mathsf{N}[0, \sigma^2(I_n - \tau\mathbf{W})^{-1}], \tag{4.100}$$

where the spatial weighting matrix $\mathbf{W}$ has ij^{th} entry equal to the spatial weight w_{ij}. This model can capture negative or positive dependence between outcomes through the spatial autoregressive parameter τ. The resulting model is essentially a Poisson mixture model with a particular dependence structure. ML estimation of this mixture specification will typically require simulation because the likelihood cannot be expressed in a closed form. Griffith and Haining (2006) suggest Bayesian Markov chain Monte Carlo methods; they compare the performance of several alternative estimators for the Poisson regression applied to georeferenced data.

Note that spatial random effect models such as (4.99) and (4.100) actually change the conditional mean so that it depends on values of other y's in the neighborhood of observation i. It is easiest to see this by analogy to the linear times-series model with an AR(1) error. Then $y_t = \mathbf{x}_t'\boldsymbol{\beta} + u_t$ with $u_t = \rho u_{t-1} + \varepsilon_t$ and ε_t iid with mean zero implies

$$\mathsf{E}[y_t|y_{t-1}, \mathbf{x}_t, \mathbf{x}_{t-1}] = \rho y_{t-1} + \mathbf{x}_t'\boldsymbol{\beta} - \rho\mathbf{x}_{t-1}'\boldsymbol{\beta},$$

so the conditional mean is no longer simply $\mathbf{x}_t'\boldsymbol{\beta}$.

4.10 MODELS BASED ON WAITING TIME DISTRIBUTIONS

Chapter 1.1.3 studied the duality between waiting time distributions and event count distributions in the simple case of a Poisson process for which the waiting times between each event occurrence are independent. Then durations are exponentially distributed and counts are Poisson distributed. We pursue this issue in greater detail here, when richer models for the waiting time distributions are specified, leading to alternative models for the counts.

4.10.1 True versus Apparent Contagion

The main thrust of this chapter has been to presume an underlying count model arising from a Poisson process, with departures from the Poisson arising due to unobserved heterogeneity that leads to overdispersion. An alternative reason for departures from the Poisson, raised in section 4.2.3 for the negative binomial, is that there is dependence between each occurrence of the event.

Such discussion of heterogeneity and overdispersion is related to a longstanding discussion in the biostatistics literature on the important distinction between true contagion and apparent contagion. *True contagion* refers to *dependence* between the occurrence of successive events. The occurrence of an event, such as an accident or illness, may change the probability of the subsequent occurrence of similar events. True positive contagion implies that the occurrence of an event shortens the expected waiting time to the next occurrence of the event, whereas true negative contagion implies that the expected waiting time to the next occurrence of the event is longer. *Apparent contagion* or *spurious contagion* arises when there is no dependence between the occurrence of successive events, but different individuals have different propensities to experience the event, so that when we aggregate over individuals it appears that experiencing an event once makes it more likely that the event will happen again. For example, it may appear in analyzing aggregated data that experiencing an accident once makes it more likely that an accident is more likely to happen again, when in fact the true explanation is one of unobserved heterogeneity – some people are prone to accidents, whereas others are not. Rather than true contagion the contagion is spurious.

Another mode of dynamic dependence is present in the notion that events occur in "spells" that themselves occur independently according to some probability law. Events within a given spell follow a different probability law and may be dependent.

The discussion of accident proneness of individuals in the early statistical literature emphasized the difficulty of distinguishing between true accident proneness and effects of interindividual heterogeneity. In reference to Neyman (1939), Feller (1943) pointed out that the same negative binomial model had been derived by Greenwood and Yule (1920), using the assumption of population heterogeneity and by Eggenberger and Polya (1923), who assumed true contagion. He observed, "Therefore, the possibility of its interpretation in two ways, diametrically opposite in their nature as well as their implications is of greatest statistical significance" (Feller, 1943, p. 389). Neyman (1965, p. 6) emphasized that the distinction between true and apparent contagion would become possible "if one has at one's disposal data on accidents incurred by each individual separately for two periods of six months each"; clearly this refers to longitudinal data. These issues are further pursued in Chapter 9.

4.10.2 Renewal Process

Renewal theory deals with functions of iid nonnegative random variables that represent time intervals between successive events (renewals). The topic was introduced in Chapter 1.1.3.

A *renewal process* is the sequence W_k, where W_k denotes the length of time between the occurrences of events $(k-1)$ and k. Denote the *sum of waiting times* for r events

$$S_r = \sum_{k=1}^{r} W_k, \quad r \geq 1. \tag{4.101}$$

The cdf of S_r, denoted $F_r(t) = \Pr[S_r \leq t]$, gives the probability that the cumulative waiting time for r events to occur or, equivalently, the arrival time of the rth event is less than t.

The associated *counting process*, $\{N(t), t \geq 0\}$, measures the successive occurrences of an event in the time interval $(0, t]$. *Renewal theory* derives properties of random variables associated with the number of events, $N(t)$, or associated with the waiting times, S_r, for a given specification of the distribution of the waiting times.

Given these definitions, $N(t) < r$ if and only if $S_r > t$. It follows that the cdf of the number of events to time t is

$$\begin{aligned} \Pr[N(t) < r] &= \Pr[S_r > t] \\ &= 1 - F_r(t), \end{aligned} \tag{4.102}$$

and the probability mass function of $N(t)$ is

$$\begin{aligned} \Pr[N(t) = r] &= \Pr[N(t) < r + 1] - \Pr[N(t) < r] \\ &= (1 - F_{r+1}(t)) - (1 - F_r(t)) \\ &= F_r(t) - F_{r+1}(t), \end{aligned} \tag{4.103}$$

which is the fundamental relation between the distribution of waiting times and that of event counts. This relation may form the basis of a count model corresponding to an arbitrary waiting time model.

From (4.103) it follows that the mean number of occurrences in the interval $(0, t]$, called the *renewal function*, is

$$\begin{aligned} m(t) &= \mathsf{E}[N(t)] \\ &= \sum_{r=1}^{\infty} r\,[F_r(t) - F_{r+1}(t)] \\ &= \sum_{r=1}^{\infty} F_r(t). \end{aligned} \tag{4.104}$$

The variance of the number of occurrences in the interval $(0, t]$ is

$$\mathsf{V}[N(t)] = \sum_{r=1}^{\infty} r^2\,[F_r(t) - F_{r+1}(t)] - \left(\sum_{r=1}^{\infty} F_r(t)\right)^2. \tag{4.105}$$

4.10.3 Waiting-Time Models for Counts

Let y_i denote the number of event occurrences for individual i in the interval $(0, t]$. Conditioning on exogenous variables $\mathbf{x}_i$, (4.103) implies that the likelihood for n independent observations is defined by

$$\mathsf{L} = \prod_{i=1}^{n} \left[F_{y_i}(t|\mathbf{x}_i) - F_{y_i+1}(t|\mathbf{x}_i)\right]. \tag{4.106}$$

Recall that $F_r(t)$ is the cdf of $S_r = \sum_{k=1}^{r} W_k$. The practical utility of (4.106) depends in part on whether the cdf of S_r can be easily evaluated. In the special case that the waiting times W_k are iid exponential distributed with mean $1/\mu$, S_r has gamma distribution with density given in Chapter 1.1.3 and corresponding cdf $1 - \sum_{k=1}^{r-1} e^{-\mu t} \mu^{tj}/j!$ It follows that $F_r(t) - F_{r+1}(t) = e^{-\mu t} \mu^{tr}/r!$ so setting $t = 1$ the term in brackets in (4.106) is simply the Poisson probability of $r = y_i$ events with Poisson with mean equal to $\mu(\mathbf{x}_i)$.

A broad class of parametric models for the waiting times W_k for individual i, denoted t_i, can be generated by the regression

$$\ln t_i = \mathbf{x}_i'\boldsymbol{\beta} + \alpha\varepsilon_i \tag{4.107}$$

with the specified distribution of ε_i. If ε_i is iid extreme value distributed with density $f(\varepsilon) = e^{\varepsilon}\exp(-e^{\varepsilon})$ and $\alpha = 1$, then by a change of variable it can be shown that t is exponentially distributed with mean $\exp(\mathbf{x}'\boldsymbol{\beta})$. As already noted after (4.106), exponentially distributed waiting times lead to a Poisson model for y_i counts, with the mean in this case of $\exp(-\mathbf{x}'\boldsymbol{\beta})$.

If more generally $\alpha > 0$ is unspecified in (4.107), while ε_i is again extreme value distributed, then the waiting times are Weibull distributed. In this case, there is no closed-form expression for the distribution of counts. However, McShane, Adrian, Bradlow, and Fader (2008) have proposed a generalized model for count data based on an assumed Weibull interarrival process. The Poisson and negative binomial models are special cases of this generalized model. To overcome the computational intractability, the authors derive the Weibull count model, using a polynomial expansion, with integration being carried out on a term-by-term basis. This leads to a computationally tractable regression model for count data that accommodates either overdispersion or underdispersion.

Lee (1997) proposed a model with k^{th} weighting time W_k for individual i generated by

$$\ln t_{ki} = \mathbf{x}_i'\boldsymbol{\beta} + \varepsilon_{ki} - u_i, \tag{4.108}$$

where the ε_{ki} are iid extreme value distributed, while u_i is a common shock across all spells for individual i that is a log-gamma random variable. Specifically, $v = u - \ln\alpha$ has density $f(v) = \exp\left[v/\alpha - \exp(v)\right]/\Gamma(1/\alpha)$ (Kalbfleisch and Prentice, 1980). This choice leads to an NB2 count model with mean $\exp(-\mathbf{x}'\boldsymbol{\beta})$ and overdispersion parameter α (Lee, 1997). Essentially conditional on u_i, the waiting times t_{ki} are exponential, leading to Poisson for the cumulative waiting time y_i, and then integrating out log-gamma u_i leads to NB2 for y_i.

Even when no closed-form expression for the cdf $F_r(t)$ is available, computer-intensive methods could be used as suggested by Lee (1997). Furthermore, the approach can be extended to generalized renewal processes. For example, we may allow the waiting time distribution of the first event to differ from that of subsequent events (Ross, 1996) – a case analogous to the hurdle count model.

4.10.4 Gamma Waiting Times

As an example of generating a count model from a richer waiting time model than the exponential, we present a model due to Winkelmann (1995).

Winkelmann (1995) considered *gamma-distributed* waiting times. Then the density of W is given by

$$f(W|\phi,\alpha) = \frac{\alpha^{\phi}}{\Gamma(\phi)} W^{\phi-1} e^{-\alpha W}, \qquad W > 0. \tag{4.109}$$

This gamma distribution with shape parameter ϕ and scale parameter $1/\alpha$ has mean ϕ/α and variance ϕ/α^2. It is a generalization of the exponential, the special case $\phi = 1$.

The *hazard function* gives the conditional instantaneous probability of the event occurring at time t, given that it has not yet occurred by time t. A benchmark is a constant hazard, the case for the exponential. For the density in (4.109), the hazard function is

$$\begin{aligned} h(W) &= \frac{f(W)}{1 - F(W)} \\ &= \frac{(\alpha^{\phi}/\Gamma(\phi)) W^{\phi-1} e^{-\alpha W}}{\int_W^{\infty} (\alpha^{\phi}/\Gamma(\phi)) v^{\phi-1} e^{-\alpha v} dv} \\ &= \left[\int_W^{\infty} e^{-\alpha(v-W)} (v/W)^{\phi-1} du\right]^{-1} \\ &= \left[\int_0^{\infty} e^{-\alpha u} (1 + u/W)^{\phi-1} du\right]^{-1}, \end{aligned}$$

by change of variables $u = v - W$. The hazard function does not have a closed-form expression. But it can be shown to be monotonically increasing in W for $\phi > 1$, constant (and equal to α) when $\phi = 1$, and monotonically decreasing for $\phi < 1$. A decreasing hazard, for example, implies negative duration dependence, with the spell less likely to end the longer it lasts. An attraction of the gamma specification is that we can determine positive or negative dependence according to whether $\phi > 1$ or $\phi < 1$.

The distribution of S_r is the sum of r iid waiting times, each with density $f(W|\phi, \alpha)$ given in (4.109). In section 4.12, it is shown that S_r is also gamma distributed, with parameters $r\phi$ and α. So the density $f_r(t)$ of S_r is

$$f_r(t|\phi,\alpha) = \frac{\alpha^{r\phi}}{\Gamma(r\phi)} t^{r\phi-1} e^{-\alpha t}, \tag{4.110}$$

and the corresponding cdf $F_r(t) = \Pr[S_r \leq t]$ is

$$\begin{aligned} F_r(t|\phi,\alpha) &= \int_0^t \frac{\alpha^{r\phi}}{\Gamma(r\phi)} s^{r\phi-1} e^{-\alpha s} ds \\ &= \frac{1}{\Gamma(r\phi)} \int_0^{\alpha t} u^{(r\phi)-1} e^{-u} du \\ &\equiv G(r\phi, \alpha t), \end{aligned} \tag{4.111}$$

where the second equality uses the change of variable to $u = \alpha s$. The integral is an incomplete gamma integral that can be numerically evaluated, and $G(r\phi, \alpha t)$ is the incomplete gamma ratio. Using (4.103) and normalizing $t = 1$, it follows that the gamma waiting time distribution model implies count distribution

$$\Pr[N = r] = G(r\phi, \alpha) - G(r\phi + \phi, \alpha) \quad \text{for } r = 0, 1, 2, \ldots, \tag{4.112}$$

where $G(0, \alpha) = 1$. The MLE is obtained by substituting (4.112) into the likelihood function (4.106), where each term in the likelihood involves two incomplete gamma integrals that can be evaluated numerically.

Result (4.112) simplifies to the Poisson with mean α if $\phi = 1$, as expected since then the waiting times are exponential distributed. But the model allows for positive duration dependence if $\phi > 1$ and negative duration dependence if $\phi < 1$ (Winkelmann, 1995). For any $\alpha > 0$, it can be shown that $\phi > 1$ leads to underdispersion and $\phi < 1$ to overdispersion.

4.10.5 Dependence and Dispersion

This relationship between duration dependence and count dispersion is not specific to gamma waiting times. It applies whenever the waiting time hazard function is monotonic in time.

Let W denote the waiting time distribution for one spell and $N(t)$ denote the number of events in the interval $(0, t]$. Cox (1962b, p. 40) shows that

$$\begin{aligned} & W \sim [\mu_w, \sigma_w^2] \\ \Rightarrow\ & N(t) \overset{a}{\sim} \mathsf{N}[t/\mu_w, t\sigma_w^2/\mu_w^3], \end{aligned} \tag{4.113}$$

where the approximation is good if $\mathsf{E}[W]$ is small relative to t.

The coefficient of variation (CV) for the waiting time distribution is

$$\mathsf{CV} = \frac{\sigma_w}{\mu_w}. \tag{4.114}$$

For waiting time distributions with monotonic hazards, it can be shown (Barlow and Proschan, 1965, p. 33) that

$\mathsf{CV} < 1 \Rightarrow$ increasing duration dependence
$\mathsf{CV} > 1 \Rightarrow$ decreasing duration dependence.

In the borderline case that $\mathsf{CV}= 1$, $\sigma_w^2/\mu_w^2 = 1$ so (4.113) simplifies to $N(t) \overset{a}{\sim} \mathsf{N}[t/\mu_w, t/\mu_w]$ and the count is equidispersed. If $\mathsf{CV} < 1$ then $\mathsf{V}[N(t)] < t/\mu_w$ and the count is underdispersed. Thus

$\mathsf{CV} < 1 \Rightarrow$ underdispersion
$\mathsf{CV} > 1 \Rightarrow$ overdispersion.

Thus overdispersion (or underdispersion or equidispersion) in count models is consistent with negative- (or positive- or zero-) duration dependence of waiting times. The intuition is that with negative dependence a spell is less likely to end the longer the spell lasts, so the number of event occurrences will not be very dispersed.

Although this result provides a valuable connection between models of counts and durations, the usefulness of the result for interpreting estimated count models depends on whether the underlying assumption of monotone hazards and the absence of other types of model misspecification is realistic.

4.10.6 Clustering and Dispersion

In this section we have assumed that event occurrences are independent, with any dependence introduced via the waiting time distribution for each spell. An alternative approach is to directly model the probabilities of each occurrence of the event as being correlated.

Consider the binomial-stopped-by Poisson characterization, introduced in Chapter 1.1.4, but relax the independence assumption to allow the sequence of binary outcome variables to be correlated. The count is $Y = \sum_{i=1}^{n} B_i$, where B_i is a binary 0/1 random variable that takes the value 1 with probability π if the event occurs. The *correlated binomial* model, denoted $\mathsf{CB}[n, \pi, \rho]$, assumes that the pairs (B_i, B_j), $i \neq j$, have covariance $\rho\pi(1-\pi)$, $0 \leq \rho < 1$, so all pairs have constant correlation ρ (Dean, 1992; Luceño, 1995).

Assuming a random number of events $n \to \infty$ in a given period and $\pi \to 0$ while $n\pi = \mu$, the $\mathsf{CB}[n, \pi, \rho]$ model generates the *correlated Poisson* model, denoted $\mathsf{CP}[\mu(1-\rho), (1-\rho)(\mu + \rho\mu^2)]$, where the arguments are, respectively, the mean and the variance of the CP model. Note that the variance-mean ratio in this case has the same form as the NB2 model, with ρ replacing the parameter α. Clearly, clumping or correlation of events can generate overdispersion. This phenomenon is of potential interest in time-series count models.

Luceño (1995) noted some limitations of this correlated Poisson model. He proposed a generalized partially correlated Poisson model that assumes events occur in clusters. These clusters are independent of each other, though within each cluster there is dependence due to a correlated binomial model induced by clusters, and the number of clusters $N \to \infty$ while $N\pi$ is constant.

4.11 KATZ, DOUBLE POISSON, AND GENERALIZED POISSON

Multiplicative mixtures lead to overdispersion. However, sometimes it is not evident that overdispersion is present in the data. It is then of interest to consider models that have variance functions flexible enough to cover both over- and underdispersion. In this section we consider several nonmixture models that have this property. Additional cross-sectional models with a similar underlying motivation are discussed in Chapter 11.

4.11.1 The Katz System

Some extensions of the Poisson model that can permit both over- and underdispersion can be obtained by introducing a variance function with additional parameters.

For the overdispersed case Cameron and Trivedi (1986) suggested the variance function

$$\mathsf{V}[y|\mathbf{x}] = \mathsf{E}[y|\mathbf{x}] + \alpha(\mathsf{E}[y|\mathbf{x}])^{2-p}, \quad \alpha > 0. \tag{4.115}$$

This specializes to that for the Poisson, NB1, and NB2 for $p = 2$, 1, and 0, respectively.

To cover overdispersion as well as underdispersion, Winkelmann and Zimmermann (1991), following King (1989b), reparameterized this variance function as

$$\mathsf{V}[y|\mathbf{x}] = \mathsf{E}[y|\mathbf{x}] + \left(\sigma^2 - 1\right)(\mathsf{E}[y|\mathbf{x}])^{k+1}, \quad \sigma^2 > 1, \tag{4.116}$$

where $-p = k - 1$ and $\alpha = \sigma^2 - 1$. The restriction $\sigma^2 - 1 = 0$ yields the Poisson case, and $\sigma^2 > 0$ implies overdispersion. Underdispersion arises if $0 < \sigma^2 < 1$ and additionally $(\mathsf{E}[y|\mathbf{x}])^k \leq 1/(1 - \sigma^2)$.

The variance specification (4.116) arises if the distribution of y is in the *Katz family* of distributions. Katz (1963) studied the system of distributions for nonnegative integer-valued random variables defined by the probability recursion

$$\Pr[y + 1] = \Pr[y]\frac{\omega + \gamma y}{1 + y}, \quad \omega + \gamma y \geq 0,\ \mu > 0,\ \gamma < 1, \tag{4.117}$$

which has mean $\mu = \omega/(1 - \gamma)$ and variance $\omega/(1 - \gamma)^2$. Special cases include the Poisson ($\gamma = 0$) and the negative binomial ($0 < \gamma < 1$). Setting $\omega/(1 - \gamma) = \mu$ and the Katz family variance $\omega/(1 - \gamma)^2$ equal to the right-hand side of (4.116) and solving for (ω, γ) as a function of (σ^2, μ) yields

$$\gamma = \frac{\left(\sigma^2 - 1\right)\mu^k}{\left(\sigma^2 - 1\right)\mu^k + 1}; \quad \omega = \frac{\mu}{\left(\sigma^2 - 1\right)\mu^k + 1}.$$

Substituting these back into (4.116) and solving for the pdf of y yields the so-called generalized event count (GEC[k]) density (Winkelmann and Zimmermann, 1995).

King (1989b) originally proposed this model in the special case that $k = 1$, in which case dispersion is of the NB1 form. Winkelmann and Zimmermann (1991) considered the more general case and proposed treating k as an unknown parameter. The likelihood is based on the probabilities given by the recursion (4.117). The MLE maximizes the log-likelihood with respect to $\boldsymbol{\beta}$, σ^2, and k, where as usual $\mathsf{E}[y|\mathbf{x}] = \exp(\mathbf{x}'\boldsymbol{\beta})$. In the underdispersed case the Katz distribution restricts y to be less than the integer immediately greater than $\mu^{1-k}/(1 - \sigma^2)$, or less than $\mu^{1-k}/(1 - \sigma^2)$ if $\mu^{1-k}/(1 - \sigma^2)$ is an integer. This is a departure from the ML regularity conditions in that the range of y does not depend on parameters.

4.11.2 Double Poisson Model

Overdispersion is far more common than underdispersion. However, there are examples of underdispersion in empirical work, and there is a role for models specifically for underdispersed counts or those that are flexible enough to capture this phenomenon; see Ridout and Besbeas (2004). Efron's (1986) double Poisson model is an example of such a model.

The double Poisson distribution was proposed by Efron (1986) within the context of the double exponential family. This distribution is obtained as an exponential combination of two Poisson distributions, $y \sim \mathsf{P}[\mu]$ and $y \sim \mathsf{P}[y]$. Then

$$f(y, \mu, \phi) = K(\mu, \phi)\phi^{1/2}\left[\frac{e^{-\mu}\mu^{y}}{y!}\right]^{\phi}\left[\frac{e^{-y}y^{y}}{y!}\right]^{1-\phi}, \tag{4.118}$$

where $K(\mu, \phi)$ is a normalizing constant that ensures $f(\cdot)$ sums to 1, with

$$\frac{1}{K(\mu, \phi)} \simeq 1 + \frac{1-\phi}{12\phi\mu}\left(1 + \frac{1}{\phi\mu}\right).$$

On some rearrangement of (4.118), the *double Poisson density* is

$$f(y, \mu, \phi) = K(\mu, \phi)\phi^{1/2}\exp(-\phi\mu)\exp\left(\frac{e^{-y}y^{y}}{y!}\right)\left(\frac{e\mu}{y}\right)^{\phi y}. \tag{4.119}$$

This distribution has mean value approximately μ and variance approximately μ/ϕ (Efron 1986, p. 715). The parameter μ is similar to the Poisson mean parameter, whereas ϕ is a dispersion parameter. The double Poisson model allows for overdispersion ($\phi < 1$), as well as underdispersion ($\phi > 1$), and reduces to the Poisson if $\phi = 1$. Another advantage of the double Poisson regression model is that both the mean and the dispersion parameters may depend on observed explanatory variables. Thus, it is possible to model the mean and dispersion structure separately, as is sometimes done for a heteroskedastic normal linear regression model and for exponential dispersion models considered by Jorgensen (1987).

Efron (1986) shows that the constant $K(\mu, \phi)$ in (4.119) nearly equals 1. Because it is a source of significant nonlinearity, the approximate density obtained by suppressing it is used in approximate maximum likelihood estimation. Ignoring the term $K(\mu, \phi)$, $\ln f(y, \mu, \phi)$ in (4.119) equals $-\phi\mu + \phi y \ln \mu$ plus terms not involving μ. Letting $\mu_i = \exp(\mathbf{x}_i'\boldsymbol{\beta})$, the ML first-order conditions for $\boldsymbol{\beta}$ are

$$\sum_{i=1}^{n} \frac{(y_i - \mu_i)}{(\mu_i/\phi_i)} \frac{\partial \mu_i}{\partial \boldsymbol{\beta}'} = \mathbf{0}. \tag{4.120}$$

If $\phi_i = \phi$ is a constant then (4.120) yields the usual Poisson QMLE. Furthermore, in that case the ML estimate of ϕ is simply the average value of the deviance measure,

$$\hat{\phi} = n^{-1} \sum \{y_i \ln y_i/\hat{\mu}_i - (y_i - \hat{\mu}_i)\}. \tag{4.121}$$

Because the estimating equations for $\boldsymbol{\beta}$ are the same as in the Poisson case, the Poisson QMLE is consistent even if the dgp is double Poisson. A simple way of adjusting its default variance is to scale the estimated variance matrix by multiplying by $1/\hat{\phi}$. However, to calculate the event probabilities the expression in (4.119) should be used.

4.11.3 Neyman's Contagious Distributions

Neyman (1939) developed his type A distribution, as well as generalizations to type B and type C distributions, to handle a form of clustering that is common in the biological sciences.

These distributions can be thought of as compound Poisson distributions that involve two processes. For example, suppose that the number of events (e.g., doctor visits) within a spell (of illness), denoted y, follows the Poisson distribution, and the random number of spells within a specified period, denoted z, follows some other discrete distribution. Then the marginal distribution of events is compound Poisson.

One variant of the Neyman type A with two parameters is obtained when $y \sim \mathsf{P}[\phi z]$ and independently $z \sim \mathsf{P}[\lambda]$. Then

$$\begin{aligned}\Pr[y = k] &= \textstyle\sum_{j=1}^{\infty} \Pr[y = k|z = j] \Pr[z = j] \\ &= \textstyle\sum_{j=1}^{\infty} \frac{e^{-\phi z}(\phi z)^k}{k!} \times \frac{e^{-\lambda}\lambda^j}{j!}.\end{aligned}$$

Johnson, Kemp, and Kotz (2005) give an excellent account of the properties of Neyman type A and further references. In univariate cases Neyman type A and negative binomial are close competitors. There are few regression applications of Neyman type A. A notable exception is Dobbie and Welsh (2001), who apply the model to counts of a particular animal at 151 different sites.

4.11.4 Consul's Generalized Poisson

Consul (1989) has proposed a distribution that can accommodate both over and underdisperion. The distribution is

$$f(y_i|x_i) = \begin{cases} \dfrac{\exp(-\mu - \gamma y) \times (\mu + \gamma y)^{y-1}}{y!}, & y = 0, 1, \ldots, \\ 0 \quad \text{for } y > N, \text{ when } \gamma < 0, & \end{cases} \tag{4.122}$$

where $\max(-1, -\mu/N) < \gamma \leq 1$ and N is the largest positive integer that satisfies the inequality $\mu + N\gamma > 0$ if μ is large. For this distribution $\mathsf{E}[y] = \mu(1 - \gamma)^{-1}$ and $\mathsf{V}[y] = \mu(1 - \gamma)^{-3}$. The distribution is underdispersed if $\max(-1, -\mu/N) < \gamma \leq 0$, equidispersed if $\gamma = 0$, and overdispersed if $0 < \gamma < 1$.

This model does not appear to have been used much in regression contexts. Because for this distribution the range of the random variable depends on the unknown parameter γ, it violates one of the standard conditions for consistency and asymptotic normality of the maximum likelihood estimation.

4.12 DERIVATIONS

4.12.1 Two Crossings Theorem

We sketch a proof of the Two Crossings Theorem, following Shaked (1980). We consider just a one-parameter exponential family, but it nonetheless covers many distributions including the Poisson.

For the random variable y, $y \in R_y$, let $f(y|\theta) = \exp\{a(\theta) + c(\theta)y\}$ denote the exponential family density. We suppose θ is random with nondegenerate density $\pi(\theta)$ and mean $\tau = \int \theta\pi(\theta)d\theta$. This leads to mixture density,

$$h(y) = \int \exp\{a(\theta) + c(\theta)y\}\pi(\theta)d\theta.$$

Ignoring overdispersion, we instead use the original density evaluated at $\tau = \mathsf{E}[\theta]$

$$f(y) \equiv f(y|\tau) = \exp\{a(\tau) + c(\tau)y\}.$$

The Two Crossings Theorem studies the sign pattern of $h(y) - f(y)$, assuming that the parent and the mixture distributions have the same mean; that is,

$$\int yf(y|\tau)dy = \int yh(y)dy. \tag{4.123}$$

Let $S^-(h - f)$ denote the number of sign changes in $h - y$ over the set $\mathcal{R}_y$. The number of sign changes of $h - f$ is the same as the number of sign changes of the function

$$\begin{aligned} r(y) &= \frac{h(y)}{f(y)} - 1 \\ &= \int \exp\{(a(\theta) - a(\tau)) + (c(\theta) - c(\tau))\, y\,\}\pi(\theta)d\theta - 1. \end{aligned}$$

Because $r(y)$ is convex on the set $\mathcal{R}_y$ it must be that $S^-(h - f) \leq 2$. Obviously $S^-(h - f) \neq 0$.

To show that $S^-(h - f) \neq 1$, assume, on the contrary that $S^-(h - f) = 1$. If the sign sequence is $\{+, -\}$ then the cdfs $H(y) = \int_{-\infty}^{y} h(x)dx$ and $F(y) = \int_{-\infty}^{y} f(x)dx$ satisfy $H(y) \geq F(y)$ for all y. This result, together with $h \neq f$, implies

$$\int yh(y)dy < \int yf(y)dy,$$

which contradicts assumption (4.123) of equal means. Similarly if the sequence is $\{-,+\}$, (4.123) is also contradicted. Thus $S^-(h-f) \neq 1$ and hence $S^-(h-f) = 2$. The sign sequence $\{+,-,+\}$ follows from the convexity of $r(y)$.

This proof considers heterogeneity in the canonical parameters θ. This is equivalent to heterogeneity in μ, because there is a one-to-one correspondence between μ and θ, which in turn is equivalent to heterogeneity in μv, where μ is fixed and v is random independent of μ.

4.12.2 Mean and Variance for Modified Count Models

We derive general expressions for the mean and variance for left truncation at zero, hurdle models, and zero-inflated models.

Consider the count random variable with density $f(y)$. The density given left truncation at zero is $f(y)/[1-f(0)]$, using $\Pr[y>0] = 1-f(0)$, with k^{th} uncentered moment

$$\begin{aligned}
\mathsf{E}[y^k|y>0] &= \textstyle\sum_{y=1}^{\infty} y^k \frac{f(y)}{1-f(0)} \\
&= \frac{1}{1-f(0)} \textstyle\sum_{y=0}^{\infty} y^k f(y) \\
&= \frac{1}{1-f(0)} \mathsf{E}[y^k],
\end{aligned}$$

using $\sum_{y=1}^{\infty} y^k f(y) = \sum_{y=0}^{\infty} y^k f(y)$ since $0^k \times f(0) = 0$. The result (4.29) follows immediately for the truncated mean, whereas for the truncated variance

$$\begin{aligned}
\mathsf{V}[y|y>0] &= \mathsf{E}[y^2|y>0] - (\mathsf{E}[y|y>0])^2 \\
&= \frac{1}{1-f(0)} \mathsf{E}[y^2] - \left(\frac{1}{1-f(0)} \mathsf{E}[y]\right)^2 \\
&= \frac{1}{1-f(0)} (\mathsf{V}[y] + (\mathsf{E}[y])^2) - \left(\frac{1}{1-f(0)}\right)^2 (\mathsf{E}[y])^2 \\
&= \frac{1}{1-f(0)} \mathsf{V}[y] - \frac{f(0)}{(1-f(0))^2} (\mathsf{E}[y])^2.
\end{aligned}$$

For the hurdle model with density given in (4.48), the kth uncentered moment is

$$\begin{aligned}
\mathsf{E}[y^k] &= 0^k \times f(0) + \textstyle\sum_{y=1}^{\infty} y^k \frac{1-f_1(0)}{1-f_2(0)} f_2(y) \\
&= \frac{1-f_1(0)}{1-f_2(0)} \textstyle\sum_{y=0}^{\infty} y^k f_2(y) \\
&= \frac{1-f_1(0)}{1-f_2(0)} \mathsf{E}_2[y^k],
\end{aligned}$$

where $\mathsf{E}_2[y^k]$ denotes expectation with respect to $f_2(y)$. The mean given in (4.50) follows immediately. For the variance given in (4.51),

$$\begin{aligned}
\mathsf{V}[y] &= \mathsf{E}[y^2] - (\mathsf{E}[y])^2 \\
&= \frac{1-f_1(0)}{1-f_2(0)} \mathsf{E}_2[y^2] - \left(\frac{1-f_1(0)}{1-f_2(0)} \mathsf{E}_2[y]\right)^2 \\
&= \frac{1-f_1(0)}{1-f_2(0)} (\mathsf{V}_2[y] + (\mathsf{E}_2[y])^2) - \left(\frac{1-f_1(0)}{1-f_2(0)}\right)^2 (\mathsf{E}_2[y])^2 \\
&= \frac{1-f_1(0)}{1-f_2(0)} \mathsf{V}_2[y] + \frac{(1-f_1(0))(f_1(0)-f_2(0))}{(1-f_2(0))^2} (\mathsf{E}_2[y])^2.
\end{aligned}$$

For the zero-inflated model with density given in (4.56), the kth uncentered moment is

$$\begin{aligned}\mathsf{E}[y^k] &= 0 \times [\pi + (1-\pi)f_2(0)] + \textstyle\sum_{y=1}^{\infty} y^k(1-\pi)f_2(y)\\ &= (1-\pi)\textstyle\sum_{y=0}^{\infty} y_2^f k(y)\\ &= (1-\pi)\mathsf{E}_2[y^k].\end{aligned}$$

The result (4.57) follows immediately for the zero-inflated mean, whereas for the corresponding variance given in (4.58)

$$\begin{aligned}\mathsf{V}[y] &= \mathsf{E}[y^2|y>0] - (\mathsf{E}[y|y>0])^2\\ &= (1-\pi)\mathsf{E}[y^2] - ((1-\pi)\mathsf{E}[y])^2\\ &= (1-\pi)(\mathsf{V}[y] + (\mathsf{E}[y])^2) - (1-\pi)^2(\mathsf{E}[y])^2\\ &= (1-\pi)\mathsf{V}[y] + \pi(1-\pi)(\mathsf{E}[y])^2.\end{aligned}$$

4.12.3 Distribution of Sum of Waiting Times

We use the Laplace transform, the analog for nonnegative random variables of the moment generating function. The density $f(W)$ defined in (4.109) has Laplace transform

$$\begin{aligned}\mathsf{L}_W(z) &= \mathrm{E}[e^{zW}]\\ &= \textstyle\int_0^{\infty} e^{zw} f(w)dw\\ &= \textstyle\int_0^{\infty} e^{zw}\frac{\alpha^{\phi}}{\Gamma(\phi)} w^{\phi-1}e^{-\alpha w}dw\\ &= \textstyle\int_0^{\infty} \frac{\alpha^{\phi}}{\Gamma(\phi)} w^{\phi-1}e^{-(\alpha-z)w}dw.\end{aligned}$$

Now change variables to $y = (\alpha - z)w$, $z < \alpha$, so $w = y/(\alpha - z)$ and the Jacobian is $(\alpha - z)^{-1}$. Then

$$\begin{aligned}\mathsf{L}_W(z) &= \int_0^{\infty} \tfrac{\alpha^{\phi}}{\Gamma(\phi)}\left(\tfrac{1}{\alpha-z}\right)^{\phi-1} y^{\phi-1}e^{-y}(\alpha-z)^{-1}dy\\ &= \tfrac{\alpha^{\phi}}{\Gamma(\phi)}\left(\tfrac{1}{\alpha-z}\right)^{\phi}\int_0^{\infty} y^{\phi-1}e^{-y}dy\\ &= \left(\tfrac{\alpha}{\alpha-z}\right)^{\phi}\\ &= \left(\tfrac{1}{1-z/\alpha}\right)^{\phi},\end{aligned}$$

where the second to last equality uses $\Gamma(\phi) = \displaystyle\int_0^{\infty} y^{\phi-1}e^{-y}dy$ and the Laplace transform is defined for $z < \alpha$. The Laplace transform of $S_r = \sum_{j=1}^{r} W_j$, where

W_j are iid with density (4.109), is therefore

$$\begin{aligned} \mathsf{L}_{S_r}(z) &= \mathsf{E}[\exp(zS_r)] \\ &= \mathsf{E}\left[\exp z\left(\sum_{j=1}^{r} W_j\right)\right] \\ &= \prod_{i=1}^{r} \mathrm{E}\left[\exp zW_j\right] \\ &= \prod_{i=1}^{r} \left(\tfrac{1}{1-z/\alpha}\right)^{\phi} \\ &= \left(\tfrac{1}{1-z/\alpha}\right)^{r\phi}. \end{aligned}$$

This is the Laplace transform of the gamma density (4.109) with parameters $r\phi$ and α, so S_r has density

$$f(t|\phi,\alpha) = \frac{\alpha^{r\phi}}{\Gamma(r\phi)} t^{r\phi-1} e^{-\alpha t}.$$

Note that the exponential is the special case of gamma with $\phi = 1$, in which case $f(S_r|\phi,\alpha) = \dfrac{\alpha^r}{(r-1)!} t^{r-1} e^{-\alpha t}$, as given in Chapter 1.1.3.

4.13 BIBLIOGRAPHIC NOTES

Feller (1971) is a classic reference for several topics discussed in this chapter. Cox processes are concisely discussed in Kingman (1993, pp. 65–72). Mullahy (1997a) provides an accessible discussion of the Two Crossings Theorem of Shaked (1980). Gelfand and Dalal (1990, p. 57) extend Shaked's two crossings analysis from a one-parameter exponential family to a two-parameter exponential family by exploiting convexity as in Chapter 8.9. Applications of MSL estimation to Poisson models with lognormal heterogeneity include Hinde (1982), Crepon and Duguet (1997b), and Winkelmann (2004). A quite general treatment of heterogeneity based on simulation is given in Gourieroux and Monfort (1991).

Detailed analysis of the moment properties of truncated count regression models include Gurmu (1991), who considers the zero-truncated Poisson, and Gurmu and Trivedi (1992), who deal with left or right truncation in general and also with tests of overdispersion in truncated Poisson regression. Applications of truncated count regression include Grogger and Carson (1991), Creel and Loomis (1990), and Brännäs (1992). Censored Poisson regression was analyzed in some detail by Gurmu (1993), who obtains estimates using the EM algorithm. T. Amemiya (1985) provides a good exposition of the EM algorithm. Mullahy (1997a) presents both hurdle and zero-inflated models, with an application. Surveys by Winkelmann and Zimmermann (1995), Cameron and Trivedi (1996), Winkelmann (2008), and Greene (2011) and the book by Winkelmann (2008) cover most of the material in sections 4.2 to 4.6.

For hierarchical generalized linear models, see Bryk and Raudenbush (2002) and Skrondal and Rabe-Hesketh (2004). For a recent survey on inference with clustered data, see Cameron and Miller (2011).

For a succinct statement of the deeper issues of identification and inference in finite mixture models, see Lindsay and Stewart (2008), who also provide a discussion of the links between mixture models and some nonlinear time-series models. The monograph by McLachlan and Peel (2000) gives a good treatment of finite mixtures, and Wedel et al. (1993) and Deb and Trivedi (1997) provide econometric applications. Brännäs and Rosenqvist (1994) and Böhning (1995) have outlined the computational properties of alternative algorithms. Finite mixtures can be handled in a Bayesian framework using Markov chain Monte Carlo methods as in Robert (1996) and Frühwirth-Schnatter (2001).

Chapters 4 and 5 in Lancaster (1990) contain material complementary to that here on models based on waiting time distributions. Lee (1997) provides a further development and important extensions of this approach, using simulated maximum likelihood to estimate the model. Gourieroux and Visser (1997) also use the duration model to define a count model, and they consider the heterogeneity distribution arising from the presence of both individual and spell-specific components. Luceño (1995) examines several models in which clustering leads to overdispersion.

4.14 EXERCISES

4.1 The Katz family of distributions is defined by the probability recursion

$$\frac{\Pr[y+1]}{\Pr[y]} = \frac{\mu + \gamma y}{1+y} \quad \text{for } y = 0, 1, \ldots, \text{ and } \mu + \gamma y \geq 0.$$

Show that this yields overdispersed distributions for $0 < \gamma < 1$ and underdispersed distributions for $\gamma < 0$.

4.2 Using the NB2 density show that the density collapses to that of the Poisson as the variance of the mixing distribution approaches zero.

4.3 The Poisson-lognormal mixture is obtained by considering the following model in which μ is normally distributed with mean $\mathbf{x}'\boldsymbol{\beta}$ and variance 1. That is, given $y|\mu_\nu \sim \mathsf{P}[\mu_\nu]$, $\ln \mu_\nu = \mathbf{x}'\boldsymbol{\beta} + \sigma\nu$, and $\nu \sim \mathsf{N}[0, 1]$, show that although the Poisson-lognormal mixture cannot be written in a closed form, the *mean* of the mixture distribution is shifted by a constant. Show that the first two moments of the marginal distribution are

$$\mathsf{E}[y|\mathbf{x}] = \exp\left(\mathbf{x}'\boldsymbol{\beta} + \frac{1}{2}\sigma^2\right)$$

$$\mathsf{V}[y|\,\mathbf{x}\,] = \exp(2\mathbf{x}'\boldsymbol{\beta})\left[\exp(2\sigma^2) - \exp(\sigma^2)\right] + \exp\left(\mathbf{x}'\boldsymbol{\beta} + \frac{1}{2}\sigma^2\right).$$

4.4 Compare the variance function obtained in 4.3 with the quadratic variance function in the NB2 case.

4.5 **(a)** Suppose y takes values 0, 1, 2, ..., with density $f(y)$ and mean μ. Find $\mathsf{E}[y|y > 0]$.
(b) Suppose y takes values 0, 1, 2, ..., with hurdle density given by

$$\Pr[y = 0] = f_1(0) \text{ and } \Pr[y = k] = (1 - f_1(0)) / (1 - f_2(0))\, f_2(0),$$
$$k = 1, 2, \ldots$$

where the density $f_2(y)$ has untruncated mean μ_2, i.e. $\sum_{k=0}^{\infty} kf(k) = \mu_2$. Find $\mathsf{E}[y]$.
(c) Introducing regressors, suppose the zeros are given by a logit model and positives by a Poisson model; that is,

$$f_1(0) = 1/\left[1 + \exp(\mathbf{x}'\boldsymbol{\beta}_1)\right]$$
$$f(k) = \exp[-\exp(\mathbf{x}'\boldsymbol{\beta}_2)][\exp(\mathbf{x}'\boldsymbol{\beta}_2)^k / y!],\, k = 1, 2, \ldots;$$

give an expression for $\mathsf{E}[y|\mathbf{x}]$.
(d) Hence obtain an expression for $\partial \mathsf{E}[y|\mathbf{x}]/\partial \mathbf{x}$ for the hurdle model.

4.6 Derive the information matrix for μ and ϕ in the double Poisson case. Show how its block-diagonal structure may be exploited in devising a computer algorithm for estimating these parameters.

4.7 Let y denote the zero-truncated Poisson-distributed random variable with density

$$f(y|\mu) = \mu^y e^{-\mu} / \left[y!(1 - e^{-\mu})\right], \quad \mu > 0.$$

Let μ be a random variable with distribution

$$g(\mu) = c(1 - e^{-\mu})e^{-\theta\mu}\mu^{\eta-1}/y!$$

where the normalizing constant $c = \Gamma(\eta)\theta^{-\eta}\left[1 + [\theta/(1+\theta)]^{\eta}\right]$. Show that the marginal distribution of y is zero-truncated NB distribution. (This example is due to Boswell and Patil (1970), who emphasized that in this case the mixing distribution is not a gamma distribution.)

CHAPTER 5

Model Evaluation and Testing

5.1 INTRODUCTION

It is desirable to analyze count data using a cycle of model specification, estimation, testing, and evaluation. This cycle can go from specific to general models – for example, it can begin with Poisson and then test for the negative binomial – or one can use a general-to-specific approach, beginning with the negative binomial and then testing the restrictions imposed by Poisson. In terms of inclusion of regressors in a given count model, either approach might be taken; for the choice of the count data model itself, other than simple choices such as Poisson or negative binomial, the former approach is most often useful. For example, if the negative binomial model is inadequate, there is a very wide range of models that might be considered, rendering a general-to-specific approach difficult to implement.

The preceding two chapters have presented the specification and estimation components of this cycle for cross-section count data. In this chapter we focus on the testing and evaluation aspects of this cycle. This includes residual analysis, goodness-of-fit measures, and model specification tests, in addition to classical statistical inference.

Residual analysis, based on a range of definitions of the residual for heteroskedastic data such as counts, is presented in section 5.2. A range of measures of goodness of fit, including pseudo R-squareds and a chi-square goodness-of-fit statistic, is presented in section 5.3. Discrimination among nonnested models is the subject of section 5.4. Many of the methods are illustrated using a regression model for the number of takeover bids, which is introduced in section 5.2.5. The remainder of the chapter considers in detail previously introduced tests of model specification. Tests for overdispersion are analyzed in section 5.5, and conditional moment tests are presented in section 5.6.

The presentation here is in places very detailed. For the practitioner, the use of simple residuals such as Pearson residuals is well established and can be quite informative. For overall model fit in fully parametric models, chi-square goodness-of-fit measures are straightforward to implement. The standard methods for discriminating between nonnested models have been adapted to count data. In testing for overdispersion in the Poisson model, the overdispersion tests presented in Chapter 3.4 often are adequate. The current chapter gives a more theoretical treatment. Conditional moment tests are easily implemented, but their interpretation if they are applied to count data is quite subtle due to the inherent heteroskedasticity.

The treatment of many of these topics, as with estimation, varies according to whether we use a fully parametric maximum likelihood framework or a moment approach based on specification of the first one or two conditional moments of the dependent variable. Even within this classification results may be specialized, notably maximum likelihood methods to LEF and moment methods to GLMs. Also, most detailed analysis is restricted to cross-section data. Many of the techniques presented here have been developed only for such special cases, and their generality is not always clear. There is considerable scope for generalization and application to a broader range of count data models.

5.2 RESIDUAL ANALYSIS

Residuals measure the departure of fitted values from actual values of the dependent variable. They can be used to detect model misspecification; outliers, or observations with poor fit; and influential observations, or observations with a big impact on the fitted model.

Residual analysis, particularly visual analysis, can potentially indicate the nature of misspecification and ways that it may be corrected, as well as provide a feel for the magnitude of the effect of the misspecification. By contrast, formal statistical tests of model misspecification can be black boxes, producing only a single number that is then compared to a critical value. Moreover, if one tests at the same significance level (usually 5%) without regard to sample size, any model using real data will be rejected with a sufficiently large sample even if it does fit the data well.

For linear models a residual is easily defined as the difference between actual and fitted values. For nonlinear models the very definition of a residual is not unique. Several residuals have been proposed for the Poisson and other GLMs. These residuals do not always generalize in the presence of common complications, such as censored or hurdle models, for which it may be more fruitful to appeal to residuals proposed in the duration literature. We present many candidate definitions of residuals because there is no one single residual that can be used in all contexts.

We also give a brief treatment of detection of outliers and influential observations. This topic is less important in applications in which data sets are large and the relative importance of individual observations is small. And if data sets are small and an influential or outlying observation is detected, it is not always clear how one should proceed. Dropping the observation or adapting the model simply to better fit, one observation creates concerns of data mining and overfitting.

5.2.1 Pearson, Deviance, and Anscombe Residuals

The natural residual is the *raw residual*

$$r_i = (y_i - \hat{\mu}_i), \tag{5.1}$$

where the fitted mean $\hat{\mu}_i$ is the conditional mean $\mu_i = \mu(\mathbf{x}_i'\boldsymbol{\beta})$ evaluated at $\boldsymbol{\beta} = \hat{\boldsymbol{\beta}}$. Asymptotically this residual behaves as $(y_i - \mu_i)$, because $\hat{\boldsymbol{\beta}} \xrightarrow{p} \boldsymbol{\beta}$ implies $\hat{\mu}_i \xrightarrow{p} \mu_i$. For the classical linear regression model with normally distributed homoskedastic error $(y - \mu) \sim \mathsf{N}[0, \sigma^2]$, so that in large samples the raw residual has the desirable properties of being symmetrically distributed around zero with constant variance. For count data, however, $(y - \mu)$ is heteroskedastic and asymmetric. For example, if $y \sim \mathsf{P}[\mu]$ then $(y - \mu)$ has variance μ and third moment μ. So the raw residual even in large samples is heteroskedastic and asymmetric.

For count data there is no one residual that has zero mean, constant variance, and symmetric distribution. This leads to the use of several different residuals according to which of these properties is felt to be most desirable.

The obvious correction for heteroskedasticity is the *Pearson residual*,

$$p_i = \frac{(y_i - \hat{\mu}_i)}{\sqrt{\hat{\omega}_i}}, \tag{5.2}$$

where $\hat{\omega}_i$ is an estimate of the variance ω_i of y_i. The sum of the squares of these residuals is the Pearson statistic, defined in (5.16). For the Poisson, GLM, and NB2 models, respectively, one uses $\omega = \mu$, $\omega = \alpha\mu$, and $\omega = \mu + \alpha\mu^2$. In large samples this residual has zero mean and is homoskedastic (with variance unity), but it is asymmetrically distributed. For example, if y is Poisson then $\mathsf{E}[((y - \mu)/\sqrt{\mu})^3] = 1/\sqrt{\mu}$.

If y is generated by an LEF density one can use the *deviance residual*, which is

$$d_i = sign(y_i - \hat{\mu}_i)\sqrt{2\{l(y_i) - l(\hat{\mu}_i)\}}, \tag{5.3}$$

where $l(\hat{\mu})$ is the log-density of y evaluated at $\mu = \hat{\mu}$ and $l(y)$ is the log-density evaluated at $\mu = y$. A motivation for the deviance residual is that the sum of the squares of these residuals is the deviance statistic, defined in (5.18), which is the generalization for LEF models of the sum of raw residuals in the linear model. Thus, for the normal distribution with σ^2 known, $d_i = (y_i - \mu_i)/\sigma$, the usual standardized residual. For the Poisson this residual equals

$$d_i = sign(y_i - \hat{\mu}_i)\sqrt{2\{y_i \ln(y_i/\hat{\mu}_i) - (y_i - \hat{\mu}_i)\}}, \tag{5.4}$$

where $y \ln y = 0$ if $y = 0$. Most other count data models are not GLMs, so this residual cannot be used. A notable exception is the NB2 model with α known. Then

$$d_i = sign(y_i - \hat{\mu}_i)\sqrt{2\{y_i \ln(y_i/\hat{\mu}_i) - (y_i + \alpha^{-1})\ln\left((y_i + \alpha^{-1})/(\hat{\mu}_i + \alpha^{-1})\right)\}}. \tag{5.5}$$

The *Anscombe residual* is defined to be the transformation of y that is closest to normality, which is then standardized to mean zero and variance 1. This transformation has been obtained for LEF densities. If y is Poisson distributed, the function $y^{2/3}$ is closest to normality, and the Anscombe residual is

$$a_i = \frac{1.5(y_i^{2/3} - \mu_i^{2/3})}{\mu_i^{1/6}}. \tag{5.6}$$

The Pearson, deviance, and Anscombe residuals for the Poisson can all be reexpressed as $\sqrt{\mu}$ times a function of y/μ. Specifically, $p = \sqrt{\mu}(c - 1)$, $d = \sqrt{\mu}\sqrt{2(c \ln c - (c-1))}$, and $a = \sqrt{\mu}1.5(c^{2/3} - 1)$. McCullagh and Nelder (1989, p. 39) tabulate these residuals for selected values of $c = y/\mu$. All three residuals are zero when $y = \mu$ (i.e., $c = 1$) and are increasing in y/μ. There is very little difference between the deviance and Anscombe residuals. The Pearson residuals are highly correlated with the other two residuals, but are scaled quite differently, being larger for $c > 0$ and smaller in absolute value for $c < 0$. For example, $p = 5\sqrt{\mu}$, $d = 3.39\sqrt{\mu}$, and $a = 3.45\sqrt{\mu}$ when $y = 5\mu$, whereas $p = -0.80\sqrt{\mu}$, $d = -0.98\sqrt{\mu}$, and $a = -0.99\sqrt{\mu}$ for $y = 0.2\mu$. Pierce and Schafer (1986) consider these residuals in some detail.

5.2.2 Generalized Residuals

Cox and Snell (1968) define a *generalized residual* to be any function

$$R_i = R_i(\mathbf{x}_i, \hat{\boldsymbol{\theta}}, y_i), \tag{5.7}$$

subject to some weak restrictions. This quite broad definition includes Pearson, deviance, and Anscombe residuals as special cases.

Many other possible residuals satisfy (5.7). For example, consider a count data model with conditional mean function $\mu(\mathbf{x}_i, \boldsymbol{\theta})$ and multiplicative error, that is, $y_i = \mu(\mathbf{x}_i, \boldsymbol{\theta})\varepsilon_i$, where $\mathsf{E}[\varepsilon_i|\mathbf{x}_i] = 1$. Solving for $\varepsilon_i = y_i/\mu(\mathbf{x}_i, \boldsymbol{\theta})$ suggests the residual $R_i = y_i/\mu(\mathbf{x}_i, \hat{\boldsymbol{\theta}})$. An additive error leads one instead to the raw residual $y_i - \mu(\mathbf{x}_i, \hat{\boldsymbol{\theta}})$ presented in the previous section.

Another way to motivate a generalized residual is to make comparison to least squares first-order conditions. For single-index models with log-density $l_i = l(y_i, \eta(\mathbf{x}_i, \boldsymbol{\theta}))$ the first-order conditions are

$$\sum_{i=1}^{n} \frac{\partial \eta_i}{\partial \boldsymbol{\theta}} \frac{\partial l_i}{\partial \eta_i} = \mathbf{0}. \tag{5.8}$$

Comparison with $\sum_{i=1}^{n} \mathbf{x}_i(y_i - \mathbf{x}_i'\boldsymbol{\beta}) = \mathbf{0}$ for the linear model suggests interpreting $\partial\eta_i/\partial\boldsymbol{\theta}$ as the regressors and using

$$R_i = \partial l_i/\partial \eta_i \tag{5.9}$$

as a generalized residual. For the Poisson model with $\eta_i = \mu_i = \mu(\mathbf{x}_i, \boldsymbol{\theta})$, this leads to the residual $(y_i - \mu_i)/\mu_i$. The Pearson residual $(y_i - \mu_i)/\sqrt{\mu_i}$ arises if R_i is standardized to have unit variance. This last result, that the Pearson residual equals $\partial l_i/\partial\eta_i/\sqrt{\mathsf{V}[\partial l_i/\partial\eta_i]}$, holds for all LEF models. More problematic is how to proceed in models more general than single-index models. For the NB2 and ordered probit models the log-density is of the form $l_i = l(y_i, \eta(\mathbf{x}_i, \boldsymbol{\beta}), \boldsymbol{\alpha})$, and one might again use $R_i = \partial l_i/\partial\eta_i$ by considering the first-order conditions with respect to $\boldsymbol{\beta}$ only.

For regression models based on a normal latent variable, several authors have proposed residuals. Chesher and Irish (1987) propose using $R_i = \mathsf{E}[\varepsilon_i^*|y_i]$ as the residual where $\varepsilon_i^* = y_i^* - \mu_i$, y_i^* is an unobserved variable that is distributed as $\mathsf{N}[\mu_i, \sigma^2]$, and the observed variable $y_i = g(y_i^*)$. Different functions $g(\cdot)$ lead to probit, censored tobit, and grouped normal models. This approach can be applied to ordered probit (or logit) models for count data. Gourieroux, Monfort, Renault, and Trognon (1987a) generalize this approach to LEF densities. Thus, let the log of the LEF density of the latent variable be $l_i^* = l^*(y_i^*, \eta(\mathbf{x}_i, \boldsymbol{\theta}))$. If y_i^* was observed one could use $R_i^* = \partial l_i^*/\partial\eta_i$ as a generalized residual. Instead one uses $R_i = \mathsf{E}[R_i^*|y_i]$. An interesting result for LEF densities is that $R_i = \partial l_i/\partial\eta_i$, where l_i is the log-density of the observed variable. Thus the same residual is obtained by applying (5.9) to the latent variable model and then conditioning on observed data as is obtained by directly applying (5.9) to the observed variable.

A count application is the left-truncated Poisson model, studied by Gurmu and Trivedi (1992), whose results were summarized in Chapter 4.3. Then $y_i^* \sim \mathsf{P}[\mu_i]$ and we observe $y_i = y_i^*$ if $y_i^* \geq r$. For the latent variable model $R_i^* = (y_i^* - \mu_i)/\mu_i$. Since $\mathsf{E}[y_i^*|y_i^* \geq r] = \mu_i + \delta_i$, where the correction factor δ_i is given in section 4.3.1, the residual for the observed variable is $R_i = \mathsf{E}[R_i^*|y_i^* \geq r] = (y_i - \mu_i - \delta_i)/\mu_i$. Alternatively, inspection of the ML first-order conditions given in Chapter 4.3.2 also leads to this residual.

5.2.3 Using Residuals

Perhaps the most fruitful way to use residuals is by plotting residuals against other variables of interest. Such plots include residuals plotted against predicted values of the dependent variable, for example to see whether the fit is poor for small or large values of the dependent variable; against omitted regressors, to see whether there is any relationship in which case the residuals should be included; and against included regressors, to see whether regressors should enter through a different functional form than that specified.

For the first plot it is tempting to plot residuals against the actual value of the dependent variable, but such a plot is not informative for count data. To see this, consider this plot using the raw residual. Because $\mathsf{Cov}[y - \mu, y] = \mathsf{V}[y]$, which equals μ for Poisson data, there is a positive relationship between $y - \mu$ and y. Such plots are more useful for the linear regression model under classical assumptions, in which case $\mathsf{V}[y]$ is a constant and any pattern in the relationship between $y - \mu$ and y is interpreted as indicating heteroscedasticity. For counts we instead plot residuals against predicted means and note that $\mathsf{Cov}[y - \mu, \mu] = 0$. A variation is to plot the actual value of y against the predicted value. This plot is difficult to interpret, however, if the dependent variable takes only a few values.

If there is little variation in predicted means, the residuals may also be lumpy due to lumpiness in y, making it difficult to interpret plots of the residuals against the fitted mean. A similar problem arises in the logit and other discrete choice models. Landwehr, Pregibon, and Shoemaker (1984) propose graphical smoothing methods (see also Chesher and Irish, 1987). For the probit model based on a normal latent variable, Gourieroux, Monfort, Renault, and Trognon (1987b) propose the use of simulated residuals as a way to overcome lumpiness, but this adds considerable noise. The approach could be applied to ordered probit and logit models for count data for which the underlying latent variable is discrete. An ad hoc solution is to "jitter" the residuals by adding random uniform noise.

Even if the variables being plotted are not lumpy, it can still be difficult to detect a relationship, and it is preferable to perform a nonparametric regression of R on x, where R denotes the residual being analyzed and x is the variable it is being plotted against. One can then plot the predictions $\hat{R}$ against x, where $\hat{R}$ is the estimate of the potentially nonlinear mean $\mathsf{E}[R|x]$.

There are a number of methods for such nonparametric regression. Let y_i be a dependent variable (in the preceding discussion a model residual), and x_i be a regressor (in the preceding discussion the dependent variable, fitted mean, or model regressor). We wish to estimate $\mathsf{E}[y_l|x_l]$, where the evaluation points x_l may or may not be actual sample values of x. The nonparametric estimator of the regression function is

$$\hat{y}_l = \hat{\mathsf{E}}[y_l|x_l] = \left(\sum_{i=1}^{n} w_{il} y_i\right) \Big/ \left(\sum_{i=1}^{n} w_{il}\right), \tag{5.10}$$

where the weights w_{il} are decreasing functions of $|x_i - x_l|$. Different methods lead to different weighting functions, with kernel, local linear, lowess, and nearest-neighbors methods particularly popular.

An overall test of adequacy of a model may be to see how close the residuals are to normality. This can be done by a normal scores plot, which orders the residuals r_i from smallest to largest, and plots them against the values predicted if the residuals were exactly normally distributed; that is, plot the ordered r_i against

$$rnorm_i = \bar{r} + s_r\,\Phi^{-1}((i - .5)/n), \tag{5.11}$$

$i = 1, \ldots, n$, where s_r is the sample standard deviation of r and Φ^{-1} is the inverse of the standard normal cdf. If the residuals are exactly normal this produces a straight line. Davison and Gigli (1989) advocate using such normal scores plots with deviance residuals to check distributional assumptions.

5.2.4 Small Sample Corrections and Influential Observations

The preceding motivations for the various residuals have implicitly treated $\hat{\mu}_i$ as μ_i, ignoring estimation error in $\hat{\mu}_i$. Estimation error can lead to quite different small-sample behavior between, for example, the raw residual $(y_i - \hat{\mu}_i)$ and $(y_i - \mu_i)$, just as it does in the linear regression model.

In the linear model, $\mathbf{y} = \mathbf{X}'\boldsymbol{\beta} + \mathbf{u}$, it is a standard result that the OLS residual vector $(\mathbf{y} - \hat{\boldsymbol{\mu}}) = (\mathbf{I} - \mathbf{H})\mathbf{u}$, where $\mathbf{H} = \mathbf{X}(\mathbf{X}'\mathbf{X})^{-1}\mathbf{X}'$. Under classic assumptions that $\mathsf{E}[\mathbf{u}\mathbf{u}'] = \sigma^2\mathbf{I}$ it follows that $\mathsf{E}[(\mathbf{y} - \hat{\boldsymbol{\mu}})(\mathbf{y} - \hat{\boldsymbol{\mu}})'] = \sigma^2(\mathbf{I} - \mathbf{H})$. Therefore, $(y_i - \hat{\mu}_i)$ has variance $(1 - h_{ii})\sigma^2$, where h_{ii} is the ith diagonal entry of $\mathbf{H}$. For very large n, $h_{ii} \to 0$ and the OLS residual has variance σ^2 as expected. But for small n the variance may be quite different, and it is best to use the studentized residual $(y_i - \hat{\mu}_i)/\sqrt{(1 - h_{ii})s^2}$, where $s^2 = \frac{1}{(n-k)}\sum_i \hat{u}_i^2$ and studentization means dividing by the standard deviation of the residual. Some authors additionally replace s^2 by s_i^2 obtained from regression with the ith observation deleted.

The matrix $\mathbf{H}$ also appears in detecting influential observations. Because in the linear model estimated by OLS $\hat{\boldsymbol{\mu}} = \hat{\mathbf{y}} = \mathbf{H}\mathbf{y}$, $\mathbf{H}$ is called the hat matrix. If h_{ii}, the ith diagonal entry in $\mathbf{H}$, is large, then the design matrix $\mathbf{X}$, which determines $\mathbf{H}$, is such that y_i has a big influence on its own prediction.

Pregibon (1981) generalized this analysis to the logit model. The logit results in turn have been extended to GLMs (see, e.g., Williams, 1987, and McCullagh and Nelder, 1989, for a summary). For GLMs the hat matrix is

$$\mathbf{H} = \mathbf{W}^{1/2}\mathbf{X}(\mathbf{X}'\mathbf{W}\mathbf{X})^{-1}\mathbf{X}'\mathbf{W}^{1/2}, \tag{5.12}$$

where $\mathbf{W} = \mathsf{Diag}[w_i]$, a diagonal matrix with i^{th} entry w_i, and $w_i = (\partial\mu_i/\partial\mathbf{x}_i'\boldsymbol{\beta})^2/\mathsf{V}[y_i]$. For the Poisson with exponential mean function $w_i = \mu_i$, so $\mathbf{H}$ is easily calculated. As in the linear model the $n \times n$ matrix $\mathbf{H}$ is idempotent with trace equal to its rank k, the number of regressors. So the average value of h_{ii} is k/n, and values of h_{ii} in excess of $2k/n$ are viewed as having high leverage. The *studentized Pearson residual* is

$$p_i^* = p_i/\sqrt{1 - h_{ii}} \tag{5.13}$$

and the *studentized deviance residual* is

$$d_i^* = d_i/\sqrt{1 - h_{ii}}. \tag{5.14}$$

Note that some authors call p_i^* and d_i^* standardized residuals. Other small-sample corrections for generalized residuals are given by Cox and Snell (1968). See also Davison and Snell (1991), who consider GLM and more general residuals.

A practical problem in implementing these methods is that $\mathbf{H}$ is of dimension $n \times n$, so that if one uses the obvious matrix commands to compute $\mathbf{H}$ the data set cannot be too large, due to the need to compute a matrix with n^2 elements. Some ingenuity may be needed to calculate the diagonal entries in $\mathbf{H}$.

Table 5.1. *Takeover bids: Actual frequency distribution*

Count	0	1	2	3	4	5	6	7	8	9	10
Frequency	9	63	31	12	6	1	2	1	0	0	1
Relative frequency	0.071	0.500	0.246	0.095	0.048	0.008	0.016	0.008	0.001	0.000	0.008

These asymptotic approximations are for small n. Some authors also consider so-called small-m asymptotics, which correct for not having multiple observations on y for each value of the regressors. Such corrections lead to an adjusted deviance residual that is closer to the normal distribution than the deviance residual. For the Poisson the adjusted deviance residual is

$$dadj_i = d_i + 1/(6\sqrt{\mu_i}). \tag{5.15}$$

Pierce and Schafer (1986) find that the adjusted deviance residual is closest to normality, after taking account of the discreteness by making a continuity correction that adds or subtracts 0.5 to or from y, toward the center of the distribution.

5.2.5 Example: Takeover Bids

Jaggia and Thosar (1993) model the number of bids received by 126 U.S. firms that were targets of tender offers during the period from 1978 through 1985 and were actually taken over within 52 weeks of the initial offer. The dependent count variable is the number of bids after the initial bid (*NUMBIDS*) received by the target firm.

Data on the number of bids are given in Table 5.1. Less than 10% of the firms received zero bids, one-half of the firms received exactly one bid (after the initial bid), a further one-quarter received exactly 2 bids, and the remainder of the sample received between 3 and 10 bids. The mean number of bids is 1.738 and the sample variance is 2.050, see Table 5.2. There is only a small amount of overdispersion ($2.050/1.738 = 1.18$), which can be expected to disappear as regressors are added.

The variables are defined and summary statistics given in Table 5.2. Regressors can be grouped into three categories: (1) defensive actions taken by management of the target firm: *LEGLREST*, *REALREST*, *FINREST*, *WHITEKNT*; (2) firm-specific characteristics: *BIDPREM*, *INSTHOLD*, *SIZE*, *SIZESQ*); and (3) intervention by federal regulators: *REGULATN*. The defensive action variables are expected to decrease the number of bids, except for *WHITEKNT*, which may increase bids because it is itself a bid. With greater institutional holdings it is expected that outside offers are more likely to be favorably received, which encourages more bids. As the size of the firm increases there are expected to be more bids, up to a point where the firm gets so large that few others are capable of making a credible bid. This is captured by the quadratic in firm size. Regulator intervention is likely to discourage bids.

Table 5.2. *Takeover bids: Variable definitions and summary statistics*

Variable	Definition	Mean	Standard deviation
NUMBIDS	Number of takeover bids	1.738	1.432
LEGLREST	Equals 1 if legal defense by lawsuit	0.429	0.497
REALREST	Equals 1 if proposed changes in asset structure	0.183	0.388
FINREST	Equals 1 if proposed changes in ownership structure	0.103	0.305
WHITEKNT	Equals 1 if management invitation for friendly third-party bid	0.595	0.493
BIDPREM	Bid price divided by price 14 working days before bid	1.347	0.189
INSTHOLD	Percentage of stock held by institutions	0.252	0.186
SIZE	Total book value of assets in billion of dollars	1.219	3.097
SIZESQ	*SIZE* squared	10.999	59.915
REGULATN	Equals 1 if intervention by federal regulator	0.270	0.446

The Poisson QML estimates are given in Table 5.3, along with robust sandwich standard errors and t-statistics. The estimated value of the overdispersion parameter α is 0.746, which is considerably less than unity. To test for underdispersion, we can use the LM tests of Chapter 3.4. These yield t statistics of -1.19 for underdispersion of the NB2 form and -3.03 for underdispersion of the NB1 form, so at level 0.05 with critical value 1.645 we reject equidispersion against underdispersion of the NB1 form, though not the NB2 form. We continue to use Poisson QML estimates, with robust standard errors to correct for underdispersion.

The defensive action variables are generally statistically insignificant at 5%, except for *LEGLREST*, which actually has an unexpected positive effect.

Table 5.3. *Takeover bids: Poisson QMLE with robust sandwich standard errors and* t *ratios*

	Poisson QMLE		
Variable	Coefficient	Standard errors	t statistic
ONE	0.986	0.414	2.38
LEGLREST	0.260	0.125	2.08
REALREST	−0.196	0.182	−1.08
FINREST	0.074	0.264	0.28
WHITEKNT	0.481	0.106	4.52
BIDPREM	−0.678	0.298	−2.28
INSTHOLD	−0.362	0.323	−1.12
SIZE	0.179	0.062	2.86
SIZESQ	−0.008	0.003	−2.72
REGULATN	−0.029	0.142	−0.21
$-\ln \mathsf{L}$	185.0		

Table 5.4. *Takeover bids: Descriptive statistics for various residuals*

Residual	Mean	Standard deviation	Skewness	Kurtosis	Minimum.	10%	90%	Maximum
r	0.000	1.23	1.38	7.48	−3.23	−1.29	1.33	5.57
p	0.002	0.83	1.12	4.99	−1.61	−0.96	1.01	3.03
p*	−0.003	0.89	1.12	5.19	−1.87	−0.99	1.05	3.11
d	−0.090	0.84	0.29	3.88	−2.27	−1.11	0.93	2.40
d*	−0.099	0.89	0.30	4.03	−2.37	−1.27	0.94	2.67
dadj	0.044	0.84	0.25	3.82	−2.17	−1.03	1.03	2.52
a	−0.097	0.85	0.21	3.93	−2.41	−1.11	0.93	2.42

Note: r, raw; p, Pearson; p*, studentized Pearson; d, deviance; d*, studentized deviance; dadj, adjusted deviance; a, Anscombe residual.

Although the coefficient of *WHITEKNT* is statistically different from zero at 5%, its coefficient implies that the number of bids increases by $0.481 \times 1.738 \simeq 0.84$ of a bid. This effect is not statistically significantly different from unity. (If a white-knight bid has no effect on bids by other potential bidders we expect it to increase the number of bids by one.) The firm-specific characteristics with the exception of *INSTHOLD* are statistically significant with the expected signs. *BIDPREM* has a relatively modest effect, with an increase in the bid premium of 0.2, which is approximately one standard deviation of *BIDPREM*, or 20%, leading to a decrease of $0.2 \times 0.678 \times 1.738 \simeq 0.24$ in the number of bids. Bids first increase and then decrease as firm size increases. Government regulator intervention has very little effect on the number of bids.

Summary statistics for different definitions of residuals from the same Poisson QML estimates are given in Table 5.4. These residuals are the raw, Pearson, deviance, and Anscombe residuals defined in, respectively, (5.1), (5.2), (5.4), and (5.6); small-sample corrected or studentized Pearson and deviance residuals (5.13) and (5.14) obtained by division by $\sqrt{1-h_{ii}}$; and the adjusted deviance residual (5.15).

Table 5.5. *Takeover bids: Correlations of various residuals*

Residual	r	p	p*	d	d*	dadj	a
r	1.000						
p	0.976	1.000					
p*	0.983	0.998	1.000				
d	0.956	0.984	0.981	1.000			
d*	0.964	0.982	0.984	0.998	1.000		
dadj	0.955	0.983	0.980	0.999	0.997	1.000	
a	0.951	0.980	0.977	0.999	0.997	0.999	1.000

Note: r, raw; p, Pearson; p*, studentized Pearson; d, deviance; d*, studentized deviance; dadj, adjusted deviance; a, Anscombe residual.

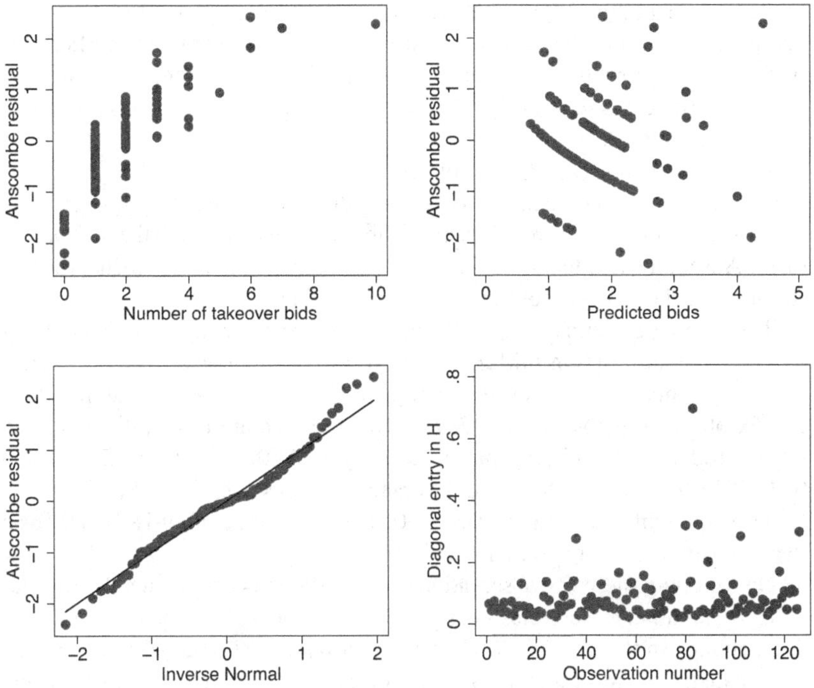

Figure 5.1. Takeover bids: Residual plots.

The various residuals are intended to be closer to normality – that is, with no skewness and kurtosis equal to 3 – than the raw residual if the data are $\mathsf{P}[\mu_i]$. For these real data, which are not exactly $\mathsf{P}[\mu_i]$, this is the case for all except the raw residual, which has quite high kurtosis. The deviance and Anscombe residuals are clearly preferred to the Pearson residual on these criteria. Studentizing makes little difference and is expected to make little difference for most observations, because the average $h_{ii} = 10/126 = 0.079$ leading to a small correction. For this sample even the second largest value of $h_{ii} = 0.321$ only leads to division of Pearson and deviance residuals by 0.82, not greatly different from unity.

The similarity between the residuals is also apparent from Table 5.5, which gives correlations among the various residuals, which are especially high between deviance and Anscombe residuals. The correlations between the residuals are all in excess of 0.9, and small-sample corrected residuals have a correlation of 0.998 or more with the corresponding uncorrected residual.

We conclude that for this sample the various residuals should all tell a similar story. We focus on the Anscombe residual a_i, because it is the closest to normality. Results using deviance residuals will be very similar. Various residual plots are presented in Figure 5.1.

Panel 1 of Figure 5.1 plots the Anscombe residual against the dependent variable. This shows the expected positive relationship, explained earlier. It is better to plot against the predicted mean, which is done in panel 2. It is difficult to visually detect a relationship, and a fitted nonparametric regression curve shows no relationship.

A normal scores or Q-Q plot of ordered Anscombe residuals against predicted quantiles if the residual is normally distributed, see (5.11), is given in panel 3 of Figure 5.1. The relationship is close to linear, although the high values of the Anscombe residuals are above the line, suggesting higher-than-expected residuals for large values of the dependent variable.

The hat matrix defined in (5.12) can be used for detecting influential observations. For the Poisson QMLE, a plot of the ith diagonal entry h_{ii} against observation number is given in panel 4 of Figure 5.1. For this sample there are six observations with $h_{ii} > 3k/n = 0.24$. These are observations 36, 80, 83, 85, 102, and 126 with h_{ii} of, respectively, 0.28, 0.32, 0.70, 0.32, 0.28, and 0.30. If instead we use the OLS leverage measures (i.e., $\mathbf{H} = \mathbf{X}(\mathbf{X}'\mathbf{X})^{-1}\mathbf{X}'$), the corresponding diagonal entries are 0.18, 0.27, 0.45, 0.58, 0.18, and 0.16 so that one would come to similar conclusions.

On dropping these six observations the coefficients of the most highly statistically significant variables change by around 30%. The major differences are a change in the sign of *SIZESQ* and that both *SIZE* and *SIZESQ* become very statistically insignificant. Further investigation of the data reveals that these six observations are for the six largest firms and that the size distribution has a very fat tail with the kurtosis of *SIZE* equal to 31, explaining the high leverage of these observations. The leverage measures very strongly alert one to the problem, but the solution is not so clear. Dropping the observations with large *SIZE* is not desirable if one wants to test the hypothesis that, other things being equal, very large firms attract fewer bids than medium-sized firms. Different functional forms for *SIZE* might be considered, such as log(*SIZE*) and its square, or an indicator variable for large firms might be used.

For this example there is little difference in the usefulness of the various standardized residuals. The sample size with 126 observations regressors is relatively small for statistical inference based on asymptotic theory, especially with 10 regressors, yet is sufficiently large that small-sample corrections made virtually no difference to the residuals. Using the hat matrix to detect influential observations was useful in suggesting possible changes to the functional form of the model.

5.3 GOODNESS OF FIT

In the preceding section the focus was on evaluating the performance of the model in fitting individual observations. Now we consider the overall performance of the model. Common goodness-of-fit measures for GLMs are the Pearson and deviance statistics, which are weighted sums of residuals. These can be used to form pseudo R-squared measures, with those based on deviance statistics preferred. A final measure is comparison of average predicted probabilities of counts with empirical relative frequencies, using a chi-square goodness-of-fit test that controls for estimation error in the regression coefficients.

5.3.1 Pearson Statistic

A standard measure of goodness of fit for any model of y_i with mean μ_i and variance ω_i is the *Pearson statistic*

$$P = \sum_{i=1}^{n} \frac{(y_i - \hat{\mu}_i)^2}{\hat{\omega}_i}, \tag{5.16}$$

where $\hat{\mu}_i$ and $\hat{\omega}_i$ are estimates of μ_i and ω_i. If the mean and variance are correctly specified, then $\mathsf{E}[\sum_{i=1}^{n}(y_i - \mu_i)^2/\omega_i] = n$, because $\mathsf{E}[(y_i - \mu_i)^2/\omega_i] = 1$. In practice P is compared with $(n - k)$, reflecting a degrees-of-freedom correction due to estimation of μ_i.

The simplest count application is to the Poisson regression model. This sets $\omega_i = \mu_i$, so that

$$P_{\mathsf{P}} = \sum_{i=1}^{n} \frac{(y_i - \hat{\mu}_i)^2}{\hat{\mu}_i}. \tag{5.17}$$

In the GLM literature it is standard to interpret $P_{\mathsf{P}} > n - k$ as evidence of overdispersion, that is, the true variance exceeds the mean, which implies $\mathsf{E}[(y_i - \mu_i)^2/\mu_i] > 1$; $P_{\mathsf{P}} < n - k$ indicates underdispersion. Note that this interpretation presumes the correct specification of μ_i. In fact, $P_{\mathsf{P}} \neq n - k$ may instead indicate misspecification of the conditional mean.

In practice even the simplest count data models make some correction for overdispersion. In the GLM literature the variance is often modeled as a multiple of the mean as in Chapter 3.2.4. Then $\hat{\omega}_i = \hat{\phi}\hat{\mu}_i$, where $\hat{\phi} = \hat{\phi}_{\mathsf{NB1}} = (n-k)^{-1}\sum_{i=1}^{n}\{(y_i - \hat{\mu}_i)^2/\hat{\mu}_i\}$. But this implies P always equals $(n - k)$, so P is no longer a useful diagnostic. If instead one uses the NB2 model, $\hat{\omega}_i = \hat{\mu}_i + \hat{\alpha}\hat{\mu}_i^2$, with $\hat{\alpha}$ the ML estimate of α from Chapter 3.3.1, then P is still a useful diagnostic for detecting overdispersion. Again, departures of P from $(n - k)$ may actually reflect misspecification of the conditional mean.

Some references to the Pearson statistic suggest that it is asymptotically chi-square distributed, but this is only true in the special case of grouped data with multiple observations for each μ_i. McCullagh (1986) gives the distribution in the more common case of ungrouped data, in which case one needs to account for the dependence of $\hat{\mu}_i$ on $\hat{\boldsymbol{\beta}}$. This distribution can be obtained by appealing to the results of CM tests given in Chapter 2.6.3. Thus $\mathsf{T}_{\mathsf{P}} = P'\hat{\mathsf{V}}_P^{-1}P \overset{a}{\sim} \chi^2(1)$, where the formula for the variance V_P is quite cumbersome. For the NB2 model estimated by maximum likelihood, one can use the simpler OPG form of the CM test. Then an asymptotically equivalent version of T_{P} is n times the uncentered R^2 from auxiliary regression of 1 on the $k + 2$ regressors

$(y_i - \hat{\mu}_i)^2/(\hat{\mu}_i + \hat{\alpha}\hat{\mu}_i{}^2)$, $(y_i - \hat{\mu}_i)/(1 + \hat{\alpha}\hat{\mu}_i)\mathbf{x}_i$ and $\{\hat{\alpha}^{-2}[\ln(1 + \hat{\alpha}\hat{\mu}_i) - \sum_{j=0}^{y_i-1}(j + \hat{\alpha}^{-1})] + (y_i - \hat{\mu}_i)/\alpha(1 + \hat{\alpha}\hat{\mu}_i)\}$. Studies very seldom implement such a formal test statistic.

5.3.2 Deviance Statistic

A second measure of goodness of fit, restricted to GLMs, is the deviance. Let $\mathcal{L}(\boldsymbol{\mu}) \equiv \ln \mathrm{L}(\boldsymbol{\mu})$ denote the log-likelihood function for an LEF density, defined in Chapter 2.4, where $\boldsymbol{\mu}$ is the $n \times 1$ vector with i^{th} entry μ_i. Then the fitted log-likelihood is $\mathcal{L}(\hat{\boldsymbol{\mu}})$, and the maximum log-likelihood achievable, that in a full model with n parameters, can be shown to be $\mathcal{L}(\mathbf{y})$, where $\hat{\boldsymbol{\mu}}$ and $\mathbf{y}$ are the $n \times 1$ vectors with i^{th} entries $\hat{\mu}_i$ and y_i. The *deviance* is defined to be

$$D(\mathbf{y}, \hat{\boldsymbol{\mu}}) = 2\{\mathcal{L}(\mathbf{y}) - \mathcal{L}(\hat{\boldsymbol{\mu}})\}, \tag{5.18}$$

which is twice the difference between the maximum log-likelihood achievable and the log-likelihood of the fitted model.

GLMs additionally introduce a dispersion parameter ϕ, with variance scaled by $a(\phi)$. Then the log-likelihood is $\mathcal{L}(\boldsymbol{\mu}, \phi)$, and the *scaled deviance* is defined to be

$$SD(\mathbf{y}, \hat{\boldsymbol{\mu}}, \phi) = 2\{\mathcal{L}(\mathbf{y}, \phi) - \mathcal{L}(\hat{\boldsymbol{\mu}}, \phi)\}. \tag{5.19}$$

For GLM densities $SD(\mathbf{y}, \hat{\boldsymbol{\mu}}, \phi)$ equals a function of $\mathbf{y}$ and $\hat{\boldsymbol{\mu}}$ divided by $a(\phi)$. It is convenient to multiply $SD(\mathbf{y}, \hat{\boldsymbol{\mu}}, \phi)$ by the dispersion factor $a(\phi)$, and the *deviance* is defined as

$$D(\mathbf{y}, \hat{\boldsymbol{\mu}}) = 2a(\phi)\{\mathcal{L}(\mathbf{y}, \phi) - \mathcal{L}(\hat{\boldsymbol{\mu}}, \phi)\}. \tag{5.20}$$

The left-hand side of (5.20) is not a function of ϕ because the terms in ϕ in the right-hand side cancel. McCullagh (1986) gives the distribution of the deviance.

For the linear regression model under normality, the deviance equals the residual sum of squares $\sum_{i=1}^{n}(y_i - \hat{\mu}_i)^2$. This has led to the deviance being used in the GLM framework as a generalization of the sum of squares. This provides the motivation for the deviance residual defined in Chapter 5.2.1. To compare sequences of nested GLMs, the analysis of deviance generalizes the analysis of variance.

For the Poisson model

$$D_{\mathrm{P}} = \sum_{i=1}^{n} \left\{ y_i \ln\left(\frac{y_i}{\hat{\mu}_i}\right) - (y_i - \hat{\mu}_i) \right\}, \tag{5.21}$$

where $y \ln y = 0$ if $y = 0$. This statistic is also called the *G-squared statistic*; see Bishop, Feinberg, and Holland (1975). Because Poisson residuals sum to zero if an intercept is included and the exponential mean function is used, D_{P}

can more easily be calculated as $\sum_i y_i \ln(y_i/\hat{\mu}_i)$. For the NB2 model with α known,

$$D_{\text{NB2}} = \sum_{i=1}^{n} \left\{ y_i \ln\left(\frac{y_i}{\hat{\mu}_i}\right) - (y_i + \alpha^{-1}) \ln\left(\frac{y_i + \alpha^{-1}}{\hat{\mu}_i + \alpha^{-1}}\right) \right\}. \tag{5.22}$$

5.3.3 Pseudo R-Squared Measures

There is no universal definition of R-squared in nonlinear models. A number of measures can be proposed. This indeterminedness is reflected in the use of "pseudo" as a qualifier. Pseudo R-squareds usually have the property that, on specialization to the linear model, they coincide with an interpretation of the linear model R-squared.

The attractive features of the linear model R-squared measure disappear in nonlinear regression models. In the linear regression model, the starting point for obtaining R^2 is decomposition of the total sum of squares. In general,

$$\begin{aligned} \sum_{i=1}^{n}(y_i - \bar{y})^2 = \sum_{i=1}^{n}(y_i - \hat{\mu}_i)^2 + \sum_{i=1}^{n}(\hat{\mu}_i - \bar{y})^2 \\ + 2\sum_{i=1}^{n}(y_i - \hat{\mu}_i)(\hat{\mu}_i - \bar{y}). \end{aligned} \tag{5.23}$$

The first three summations are, respectively, the total sum of squares (TSS), residual sum of squares (RSS), and explained sum of squares (ESS). The final summation is zero for OLS estimates of the linear regression model if an intercept is included. It is nonzero, however, for virtually all other estimators and models, including the Poisson and NLS with exponential conditional mean. This leads to different measures according to whether one uses $R^2 = 1-\text{RSS/TSS}$ or $R^2 = \text{ESS/TSS}$. Furthermore, because estimators such as the Poisson MLE do not minimize RSS, these R-squareds need not necessarily increase as regressors are added, and even if an intercept is included, the first one may be negative and the second may exceed unity.

The deviance is the GLM generalization of the sum of squares, as noted in the previous subsection. Cameron and Windmeijer (1996, 1997) propose a pseudo-R^2 based on decomposition of the deviance. Then

$$D(\mathbf{y}, \bar{\mathbf{y}}) = D(\mathbf{y}, \hat{\boldsymbol{\mu}}) + D(\hat{\boldsymbol{\mu}}, \bar{\mathbf{y}}), \tag{5.24}$$

where $D(\mathbf{y}, \bar{\mathbf{y}})$ is the deviance in the intercept-only model, $D(\mathbf{y}, \hat{\boldsymbol{\mu}})$ is the deviance in the fitted model, and $D(\hat{\boldsymbol{\mu}}, \bar{\mathbf{y}})$ is the explained deviance. One uses

$$R^2_{DEV} = 1 - \frac{D(\mathbf{y}, \hat{\boldsymbol{\mu}})}{D(\mathbf{y}, \bar{\mathbf{y}})}, \tag{5.25}$$

which measures the reduction in the deviance due to the inclusion of regressors. This equals $D(\hat{\boldsymbol{\mu}}, \bar{\mathbf{y}})/D(\mathbf{y}, \bar{\mathbf{y}})$, with the R^2 based instead on the explained deviance. It lies between 0 and 1, increases as regressors are added and can be given an information-theoretic interpretation as the proportionate reduction in Kullback-Liebler divergence due to the inclusion of regressors. Only the last one properties actually requires correct specification of the distribution of y.

This method can be applied to models for which the deviance is defined. For the Poisson linear regression model, the deviance is given in (5.21) leading to

$$R^2_{\text{DEV,P}} = \frac{\sum_{i=1}^n y_i \ln\left(\frac{\hat{\mu}_i}{\bar{y}}\right) - (y_i - \hat{\mu}_i)}{\sum_{i=1}^n y_i \ln\left(\frac{y_i}{\bar{y}}\right)}, \tag{5.26}$$

where $y \ln y = 0$ if $y = 0$. The same measure is obtained for the Poisson GLM with NB1 variance function. For maximum likelihood estimation of the negative binomial with NB2 variance function the *deviance pseudo-*R^2 is

$$R^2_{\text{DEV,NB2}} = 1 - \frac{\sum_{i=1}^n y_i \ln\left(\frac{\hat{\mu}_i}{\bar{y}}\right) - (y_i + \hat{a}) \ln\left(\frac{y_i + \hat{a}}{\hat{\mu}_i + \hat{a}}\right)}{\sum_{i=1}^n y_i \ln\left(\frac{y_i}{\bar{y}}\right) - (y_i + \hat{a}) \ln\left(\frac{y_i + \hat{a}}{\bar{y} + \hat{a}}\right)}, \tag{5.27}$$

where $\hat{a} = 1/\hat{\alpha}$ and $\hat{\alpha}$ is the estimate of α in the fitted model. Note that $R^2_{\text{DEV,P}}$ and $R^2_{\text{DEV,NB2}}$ have different denominators and are not directly comparable. In particular it is possible, and indeed likely for data that are considerably overdispersed, that $R^2_{\text{DEV,P}} > R^2_{\text{DEV,NB2}}$. One can instead modify the deviance pseudo-R^2 measures to have the common denominator $\sum_{i=1}^n y_i \ln(y_i/\bar{y})$, in which case the intercept-only Poisson is being used as the benchmark model. The NB1 model is not a GLM model, although Cameron and Windmeijer (1996) nonetheless propose a deviance-type R^2 measure in this case.

Cameron and Windmeijer (1996) motivate the deviance R^2 measures as measures based on residuals; one uses deviance residuals rather than raw residuals. One can alternatively view these measures in terms of the fraction of the potential log-likelihood gain that is achieved with the inclusion of regressors. Formally,

$$R^2 = \frac{\mathcal{L}_{\text{fit}} - \mathcal{L}_0}{\mathcal{L}_{\max} - \mathcal{L}_0} = 1 - \frac{\mathcal{L}_{\max} - \mathcal{L}_{\text{fit}}}{\mathcal{L}_{\max} - \mathcal{L}_0}, \tag{5.28}$$

where $\mathcal{L}_{\text{fit}}$ and $\mathcal{L}_0$ denote the log-likelihood in the fitted and intercept-only models and $\mathcal{L}_{\max}$ denotes the maximum log-likelihood achievable. In (5.28), $(\mathcal{L}_{\max} - \mathcal{L}_0)$ is the potential log-likelihood gain and $(\mathcal{L}_{\max} - \mathcal{L}_{\text{fit}})$ is the log-likelihood gain achieved. This approach was taken by Merkle and Zimmermann (1992) for the Poisson model. The difficult part in the implementation for more general models is defining $\mathcal{L}_{\max}$. For some models $\mathcal{L}_{\max}$ is unbounded, in which case any model has an R^2 of zero. For GLM models $\mathcal{L}_{\max} = \mathcal{L}(\mathbf{y})$ is finite and the approach is useful.

For nonlinear models some studies have proposed instead using the pseudo-R^2 measure, $R^2 = 1 - (\mathcal{L}_{\text{fit}}/\mathcal{L}_0)$, sometimes called the *likelihood ratio index*.

This is the same as (5.28) in the special case $\mathcal{L}_{\max} = 0$, which is the case for binary and multinomial outcome models. More generally, however, the likelihood ratio index can be considerably less than unity for discrete densities, regardless of how good the fit is, because $\mathcal{L}_{\max} \leq 0$. Also, for continuous densities, problems may arise because $\mathcal{L}_{\text{fit}} > 0$ and negative R^2 values are possible.

Poisson packages usually report $\mathcal{L}_{\text{fit}}$ and $\mathcal{L}_0$. Using (5.28), $R^2_{\text{DEV,P}}$ can be computed if one additionally computes $\mathcal{L}_{\max} = \sum_i \{y_i \log y_i - y_i - \log y_i!\}$. The measure can also be clearly applied to truncated and censored variants of these models.

Cameron and Windmeijer (1996) also consider a similar R^2 measure based on the Pearson residual. For models with variance function $\omega(\mu, \alpha)$,

$$R^2_{\text{PEARSON}} = 1 - \sum_{i=1}^{n} \frac{(y_i - \hat{\mu}_i)^2}{\omega(\hat{\mu}_i, \hat{\alpha})} \Big/ \sum_{i=1}^{n} \frac{(y_i - \hat{\mu}_0)^2}{\omega(\hat{\mu}_0, \hat{\alpha})}, \tag{5.29}$$

where $\hat{\alpha}$ is the estimate of α in the fitted model and $\hat{\mu}_0 = \hat{\mu}_0(\hat{\alpha})$ denotes the predicted mean in the intercept-only model estimated under the constraint that $\alpha = \hat{\alpha}$. For Poisson and NB2 models $\hat{\mu}_0 = \bar{y}$. This measure has the attraction of requiring only mean-variance assumptions and being applicable to a wide range of models. This measure can be negative, however, and can decrease as regressors are added. These weaknesses are not just theoretical, because they are found to arise often in simulations and in applications. Despite its relative simplicity and generality, use of R^2_{PEARSON} is not recommended.

Finally, a simple familiar measure is

$$R^2_{\text{COR}} = \left(\widehat{\text{Cor}}[y_i, \hat{\mu}_i]\right)^2,$$

the squared correlation coefficient between actual and fitted values. This has the disadvantage for count model estimation that it can decrease as regressors are added. It has the advantage of comparability across quite disparate models, with one model definitely better at explaining the data than another if R^2_{COR} is substantially higher.

5.3.4 Chi-Square Goodness of Fit Test

For fully parametric models such as Poisson and negative binomial maximum likelihood, a crude diagnostic is to compare fitted probabilities with actual frequencies, where the fitted frequency distribution is computed as the average over observations of the predicted probabilities fitted for each count.

Suppose the count y_i takes values $0, 1, \ldots, m$, where $m = \max_i(y_i)$. Let the observed frequencies (i.e., the fraction of the sample with $y = j$), be denoted by $\bar{p}_j$ and the corresponding fitted frequencies be denoted by $\hat{p}_j$, $j = 0, \ldots, m$. For the Poisson, for example, $\hat{p}_j = n^{-1} \sum_{i=1}^{n} \exp(-\hat{\mu}_i)\hat{\mu}_i^j / j!$ Comparison of $\hat{p}_j$ with $\bar{p}_j$ can be useful in displaying poor performance of a model, in highlighting ranges of the counts for which the model has a tendency to underpredict or overpredict, and for allowing a simple comparison of the predictive performance of competing models. Figure 3.1 provides an example. Without doing a formal test, however, it is not clear when $\hat{p}_j$ is "close" enough to $\overline{p}_j$ for one to conclude that the model is a good one.

Formal comparison of $\hat{p}_j$ and $\overline{p}_j$ can be done using a CM test. We consider a slightly more general framework, one where the range of y is broken into J mutually exclusive cells, where each cell may include more than one value of y and the J cells span all possible values of y. For example, in data where only low values are observed, the cells may be $\{0\}$, $\{1\}$, $\{2, 3\}$, and $\{4, 5, \ldots\}$. Let $d_{ij}(y_i)$ be an indicator variable with $d_{ij} = 1$ if y_i falls in the jth set and $d_{ij} = 0$ otherwise. Let $p_{ij}(\mathbf{x}_i, \boldsymbol{\theta})$ denote the predicted probability that observation i falls in the jth set, where to begin with we assume the parameter vector $\boldsymbol{\theta}$ is known. Consider testing whether $d_{ij}(y_i)$ is centered around $p_{ij}(\mathbf{x}_i, \boldsymbol{\theta})$,

$$\mathsf{E}[d_{ij}(y_i) - p_{ij}(\mathbf{x}_i, \boldsymbol{\theta})] = 0, \qquad j = 1, \ldots, J, \tag{5.30}$$

or stacking all J moments in obvious vector notation

$$\mathsf{E}[\mathbf{d}_i(y_i) - \mathbf{p}_i(\mathbf{x}_i, \boldsymbol{\theta})] = \mathbf{0}. \tag{5.31}$$

This hypothesis can be tested by testing the closeness to zero of the corresponding sample moment

$$\mathbf{m}(\hat{\boldsymbol{\theta}}) = \sum_{i=1}^{n} (\mathbf{d}_i(y_i) - \mathbf{p}_i(\mathbf{x}_i, \hat{\boldsymbol{\theta}})). \tag{5.32}$$

This is clearly a CM test, presented in Chapter 2.6.3. The CM test statistic is

$$\mathsf{T}_{\chi^2} = \mathbf{m}(\hat{\boldsymbol{\theta}})'\hat{\mathbf{V}}_m^{-}\mathbf{m}(\hat{\boldsymbol{\theta}}), \tag{5.33}$$

where $\hat{\mathbf{V}}_m$ is a consistent estimate of $\mathbf{V}_m$, the asymptotic variance matrix of $\mathbf{m}(\hat{\boldsymbol{\theta}})$, and $\hat{\mathbf{V}}_m^{-}$ is the Moore-Penrose generalized inverse of $\hat{\mathbf{V}}_m$. The generalized inverse is used because the $J \times J$ matrix $\mathbf{V}_m$ may not be of full rank. Under the null hypothesis that the density is correctly specified, that is, that $p_{ij}(\mathbf{x}_i, \boldsymbol{\theta})$ gives the correct probabilities, the test statistic is chi-square distributed with $\mathsf{rank}[\hat{\mathbf{V}}_m]$ degrees of freedom.

The results in Chapter 2.6.3 can be used to obtain $\mathbf{V}_m$, which here usually has $\mathsf{rank}[\mathbf{V}_m] = J - 1$ rather than J as a consequence of the probabilities over all J cells summing to one. This entails considerable algebra, and it is easiest to instead use the asymptotically equivalent OPG form of the test, which is appropriate because fully parametric models are being considered here so that $\hat{\boldsymbol{\theta}}$ will be the MLE. The test is implemented calculating n times the uncentered R^2 from the artificial regression of 1 on the scores $\mathbf{s}_i(y_i, \mathbf{x}_i, \hat{\boldsymbol{\theta}})$ and $d_{ij}(y_i) - p_{ij}(\mathbf{x}_i, \hat{\boldsymbol{\theta}})$, $j = 1, \ldots, J - 1$, where one cell has been dropped due to $\mathsf{rank}[\mathbf{V}_m] = J - 1$. In some cases $\mathsf{rank}[\mathbf{V}_m] < J - 1$. This occurs if the estimator $\hat{\boldsymbol{\theta}}$ is the solution to first-order conditions that set a linear transformation of $\mathbf{m}(\hat{\boldsymbol{\theta}})$ equal to zero or is asymptotically equivalent to such an estimator. An example is the multinomial model, with an extreme case being the binary logit model whose first-order conditions imply $\sum_{i=1}^{n}(d_{ij}(y_i) - p_{ij}(\mathbf{x}_i, \hat{\boldsymbol{\theta}})) = 0$, $j = 0, 1$.

The test statistic (5.33) is called the *chi-square goodness-of-fit test,* because it is a generalization of *Pearson's chi-square test,*

$$\sum_{j=1}^{J} \frac{(n\overline{p}_j - n\hat{p}_j)^2}{n\hat{p}_j}. \tag{5.34}$$

In an exercise it is shown that (5.34) can be rewritten as (5.33) in the special case in which $\mathbf{V}_m$ is a diagonal matrix with i^{th} entry $\sum_{i=1}^{n} p_{ij}(\mathbf{x}_i, \boldsymbol{\theta})$. Although this is the case in the application originally considered by Pearson – y_i is iid and takes only J discrete values and a multinomial MLE is used – in most regression applications the more general form (5.33) must be used. The generalization of Pearson's original chi-square test by Heckman (1984), Tauchen (1985), Andrews (1988a, 1988b), and others is reviewed in Andrews (1988b, pp. 140-141). For simplicity we have considered partition of the range of y into J cells. More generally the partition may be over the range of $(y, \mathbf{x})$.

Example: Takeover Bids (Continued)

We consider goodness-of-fit measures for the Poisson model estimates given in Table 5.3. The Pearson statistic (5.18) is 86.58, much less than its theoretical value of $n - k = 116$, indicating underdispersion. The deviance statistic (5.21) is 88.61. The Poisson deviance in the intercept-only model is 121.86, so R^2 given in (5.26) equals $1 - 88.61/121.86 = 0.27$. The Pearson R^2 given in (5.29) equals $1 - 86.58/147.49 = 0.41$. Note that these two R^2 measures are still valid if the conditional variance equals $\alpha\mu_i$ rather than μ_i and can be easily computed using knowledge of the deviance and Pearson statistics plus the frequency distribution given in Table 5.1. Although experience with these R^2 measures is limited, it seems reasonable to conclude that the fit is quite good for cross-section data. A third R^2 measure, the squared correlation between actual and fitted values, equals 0.26. If one instead runs an OLS regression, the R^2 equals 0.24.

Before performing a formal chi-square goodness-of-fit test, it is insightful to compare predicted relative frequencies $\hat{p}_j$ with actual relative frequencies $\overline{p}_j$. These are given in Table 5.6, where counts of five or more are grouped into the one cell to prevent cell sizes from getting too small. Clearly the Poisson overpredicts greatly the number of zeros and underpredicts the number of ones.

The last column of Table 5.6 gives $n(\overline{p}_j - \hat{p}_j)^2/\hat{p}_j$, which is the contribution of count j to Pearson's chi-square test statistic (5.34). Although this test statistic, whose value is 31.58, is inappropriate due to failure to control for estimation error in $\hat{p}_j$, it does suggest that the major contributors to the formal test will be the predictions for zeroes and ones. The formal chi-square test statistic (5.33) yields a value 48.66 compared to a $\chi^2(5)$ critical value of 9.24 at 5%. The Poisson model is strongly rejected.

Table 5.6. *Takeover bids: Poisson MLE predicted and actual probabilities*

Counts	Actual	Predicted	|Diff|	Pearson
0	0.071	0.213	0.142	11.88
1	0.500	0.298	0.202	17.32
2	0.246	0.233	0.013	0.10
3	0.095	0.137	0.041	1.58
4	0.048	0.068	0.020	0.77
≥5	0.040	0.052	0.012	0.00

Note: Actual, actual relative frequency; predicted, predicted relative frequency; |Diff|, absolute difference between predicted and actual probabilities; Pearson, contribution to Pearson chi-square test.

We conclude that the Poisson is an inadequate fully parametric model, due to its inability to model the relatively few zeroes in the sample. Analysis of the data by Cameron and Johansson (1997) using alternative parametric models – Katz, hurdle, double Poisson, and a flexible parametric model – is briefly discussed in Chapter 11.2.3. Interestingly, none of the earlier diagnostics, such as residual analysis, detected this weakness in the Poisson estimates.

5.4 DISCRIMINATING AMONG NONNESTED MODELS

Two models are *nonnested models* if neither model can be represented as a special case of the other. A further distinction can be made between models that are *overlapping*, in which case some specializations of the two models are equal, and models that are *strictly nonnested*, in which case there is no overlap at all. Models based on the same distribution that have some regressors in common and some regressors not in common are overlapping models. Models with different nonnested distributions and models with different nonnested functional forms for the conditional mean are strictly nonnested. Formal definitions are given in Pesaran (1987) and Vuong (1989).

The usual method of discriminating among models by hypothesis test of the parameter restrictions that specialize one model to the other – for example whether the dispersion parameter is zero in moving from the negative binomial to Poisson – is no longer available.

Instead, for likelihood-based models two approaches are taken. First, beginning with Akaike (1973) likelihood-based models are compared on the basis of the fitted log-likelihood with penalty given for a lack of parsimony. This is similar to comparing linear models estimated by OLS on the basis of adjusted R^2. Second, beginning with Cox (1961, 1962a), hypothesis testing is performed, but in a nonstandard framework where the usual LR test no longer applies. We focus on these approaches because most richer count models that we wish to discriminate between, such as hurdle, zero-inflated, and finite mixtures, are likelihood-based.

The Cox approach has spawned a variety of related procedures, including extensions beyond the likelihood framework. These extensions, often limited to linear models, are not presented here. A brief review of these extensions is given in Davidson and MacKinnon (1993, chapter 11). Artificial nesting, proposed by Davidson and MacKinnon (1984), leads to J-tests and related tests. The encompassing principle, proposed by Mizon and Richard (1986), leads to a quite general framework for testing one model against a competing nonnested model. White (1994) and Lu and Mizon (1996) link this approach with CM tests. Wooldridge (1990b) derived encompassing tests for the conditional mean in nonlinear regression models with heteroskedasticity, including GLMs such as the Poisson estimated by QML.

5.4.1 Information Criteria

For comparison of nonnested models based on maximum likelihood, several authors beginning with Akaike (1973) have proposed model selection criteria based on the fitted log-likelihood function. Because we expect the log-likelihood to increase as parameters are added to a model, these criteria penalize models with larger k, the number of parameters in the model. This penalty function may also be a function of n, the number of observations.

Akaike (1973) proposed the *Akaike information criterion*,

$$\mathsf{AIC} = -2 \ln \mathsf{L} + 2k, \tag{5.35}$$

with the model with lowest AIC preferred. The term *information criterion* is used because the log-likelihood is closely related to the Kullback-Liebler information criterion. Modifications to AIC include the *Bayesian information criterion*,

$$\mathsf{BIC} = -2 \ln \mathsf{L} + (\ln n)\, k, \tag{5.36}$$

proposed by Schwarz (1978), and the *consistent Akaike information criterion*,

$$\mathsf{CAIC} = -2 \ln \mathsf{L} + (1 + \ln n)\, k. \tag{5.37}$$

These three criteria give increasingly large penalties in k and n. As an example, suppose we wish to compare two models where one model has one more parameter than the other, so $\Delta k = 1$, and the sample size is $n = 1{,}000$, so $\ln n = 6.9$. For the larger model to be preferred it needs to increase $2 \ln \mathsf{L}$ by 1.0 if one uses AIC, 6.9 if BIC is used, and 7.9 if CAIC is used. By comparison if the two models were nested and a likelihood ratio test was formed, the larger model is preferred at significance level 5% if $2 \ln \mathsf{L}$ increases by 3.84. The AIC, BIC, and CAIC in this example correspond to p values of, respectively, 0.317, 0.009, and 0.005.

5.4.2 Tests of Nonnested Models

There is a substantial literature on discrimination among nonnested models on the basis of hypothesis tests, albeit they are nonstandard tests. Consider choosing between two nonnested models – model F_θ with density $f(y_i|\mathbf{x}_i, \boldsymbol{\theta})$ and model G_γ with density $g(y_i|\mathbf{x}_i, \boldsymbol{\gamma})$.

The LR statistic for the model F_θ against G_γ is

$$\mathsf{LR}(\hat{\boldsymbol{\theta}}, \hat{\boldsymbol{\gamma}}) \equiv \mathcal{L}_f(\hat{\boldsymbol{\theta}}) - \mathcal{L}_g(\hat{\boldsymbol{\gamma}}) = \sum_{i=1}^{n} \ln \frac{f(y_i|\mathbf{x}_i, \hat{\boldsymbol{\theta}})}{g(y_i|\mathbf{x}_i, \hat{\boldsymbol{\gamma}})}. \tag{5.38}$$

In the special case where the models are nested, $F_\theta \subset G_\gamma$, we get the usual result that 2 times $\mathsf{LR}(\hat{\boldsymbol{\theta}}, \hat{\boldsymbol{\gamma}})$ is chi-square distributed under the null hypothesis that $G_\gamma = F_\theta$. Here we consider the case of nonnested models in which $F_\theta \not\subseteq G_\gamma$ and $G_\gamma \not\subseteq F_\theta$. Then the chi-square distribution is no longer appropriate.

Cox (1961, 1962a) proposed solving this problem by applying a central limit theorem under the assumption that F_θ is the true model. This approach is difficult to implement because it requires analytically obtaining $\mathsf{E}_f[\ln(f(y_i|\mathbf{x}_i, \boldsymbol{\theta})/g(y_i|\mathbf{x}_i, \boldsymbol{\gamma}))]$, where E_f denotes expectation with respect to the density $f_i(y_i|\mathbf{x}_i, \theta)$. Furthermore, if a similar test statistic is obtained with the roles of F_θ and G_γ reversed, it is possible to find both that model F_θ is rejected in favor of G_γ and that model G_γ is rejected in favor of F_θ.

Vuong (1989) instead discriminated between models on the basis of their distance from the true data-generating process, which has unknown density $h_0(y_i|\mathbf{x}_i)$, where distance is measured using the Kullback-Liebler information criterion. He proposed the use of the statistic

$$\begin{aligned} \mathsf{T}_{\mathsf{LR,NN}} &= \frac{1}{\sqrt{n}} \sum_{i=1}^{n} \ln \frac{f(y_i|\mathbf{x}_i, \hat{\boldsymbol{\theta}})}{g(y_i|\mathbf{x}_i, \hat{\boldsymbol{\gamma}})} \bigg/ \hat{\omega} \\ &= \frac{1}{\sqrt{n}} \mathsf{LR}(\hat{\boldsymbol{\theta}}, \hat{\boldsymbol{\gamma}}) \Big/ \hat{\omega}, \end{aligned} \tag{5.39}$$

where

$$\hat{\omega}^2 = \frac{1}{n} \sum_{i=1}^{n} \left(\ln \frac{f(y_i|\mathbf{x}_i, \hat{\boldsymbol{\theta}})}{g(y_i|\mathbf{x}_i, \hat{\boldsymbol{\gamma}})} \right)^2 - \left(\frac{1}{n} \sum_{i=1}^{n} \ln \frac{f(y_i|\mathbf{x}_i, \hat{\boldsymbol{\theta}})}{g(y_i|\mathbf{x}_i, \hat{\boldsymbol{\gamma}})} \right)^2 \tag{5.40}$$

is an estimate of the variance of $\frac{1}{\sqrt{n}}\mathsf{LR}(\hat{\boldsymbol{\theta}}, \hat{\boldsymbol{\gamma}})$. An alternative asymptotically equivalent statistic instead uses $\tilde{\omega}$ in (5.39), where $\tilde{\omega}^2 = \frac{1}{n}\sum_{i=1}^{n} \left(\ln(f(y_i|\mathbf{x}_i, \hat{\boldsymbol{\theta}})/g(y_i|\mathbf{x}_i, \hat{\boldsymbol{\gamma}})) \right)^2$.

For strictly nonnested models,

$$\mathsf{T}_{\mathsf{LR,NN}} \xrightarrow{d} \mathsf{N}[0, 1] \tag{5.41}$$

under

$$H_0 : \mathsf{E}_h \left[\ln \frac{f(y_i|\mathbf{x}_i, \boldsymbol{\theta})}{g(y_i|\mathbf{x}_i, \boldsymbol{\gamma})} \right] = 0, \tag{5.42}$$

where E_h denotes expectation with respect to the (unknown) dgp $h(y_i|\mathbf{x}_i)$. One therefore rejects at significance level .05 the null hypothesis of equivalence of the models in favor of F_θ being better (or worse) than G_γ if $\mathsf{T}_{\mathsf{LR,NN}} > z_{.05}$ (or if $\mathsf{T}_{\mathsf{LR,NN}} < -z_{.05}$). The null hypothesis is not rejected if $|\mathsf{T}_{\mathsf{LR,NN}}| \leq z_{.025}$.

Tests of overlapping models are more difficult to implement than tests of strictly nonnested models because there is a possibility that $f(y_i|\mathbf{x}_i, \boldsymbol{\theta}_*) = g(y_i|\mathbf{x}_i, \boldsymbol{\gamma}_*)$, where $\boldsymbol{\theta}_*$ and $\boldsymbol{\gamma}_*$ are the pseudotrue values of $\boldsymbol{\theta}$ and $\boldsymbol{\gamma}$. To eliminate the possibility of equality, Vuong (1989) shows that

$$\Pr\left[n\hat{\omega}^2 \leq x\right] - M_{p+q}(x; \hat{\lambda}^2) \overset{as}{\to} 0, \tag{5.43}$$

for any $x > 0$, under

$$H_0^\omega : \mathsf{V}_0 \left[\ln \frac{f(y_i|\mathbf{x}_i, \boldsymbol{\theta})}{g(y_i|\mathbf{x}_i, \boldsymbol{\gamma})} \right] = 0, \tag{5.44}$$

where E_h denotes expectation with respect to the (unknown) dgp $h(y_i|\mathbf{x}_i)$ and $\boldsymbol{\theta}$ and $\boldsymbol{\gamma}$ have dimensions p and q. $M_{p+q}(x; \hat{\lambda}^2)$ denotes the cdf of the weighted sum of chi-squared variables $\sum_{j=1}^{p+q} \hat{\lambda}_j^2 Z_j^2$, where Z_j^2 are iid $\chi^2(1)$ and $\hat{\lambda}_j^2$ are the squares of the eigenvalues of the sample analog of the matrix $\mathbf{W}$ defined in Vuong (1989, p. 313). One therefore rejects H_0^ω if $n\hat{\omega}^2$ exceeds the critical value obtained using (5.43). If H_0^ω is not rejected, it is concluded that the data cannot discriminate between F_θ and G_γ. If H_0^ω is rejected then proceed to discriminate between F_θ and G_γ on the basis of the same test of H_0 as used in the case of strictly nested models.

Vuong (1989, p. 322) also considers the case in which one of the overlapping models is assumed to be correctly specified, an approach qualitatively similar to Cox (1961, 1962a), in which case

$$\Pr\left[n\hat{\omega}^2 \leq x\right] - M_{p+q}(x; \hat{\lambda}) \overset{as}{\to} 0, \tag{5.45}$$

can be used as the basis for a two-sided test.

Example: Takeover Bids (Continued)

The Poisson model is clearly not the correct parametric model for the takeover bids data because from section 5.2.5 the data are underdispersed and from Table 5.6 the model greatly underpredicts zeroes and overpredicts ones. The standard parametric generalization of the Poisson, the negative binomial, is not estimable because the data are underdispersed.

Instead we consider a Poisson hurdle model and zero-inflated Poisson as alternatives to the Poisson, where the original list of regressors is used for both parts of the hurdle and zero-inflated Poisson models. For additional alternative models for these data, see Cameron and Johansson (1997).

Table 5.7. *Takeover bids: Information criteria for alternative parametric models*

	Poisson	Hurdle Poisson	Zero-inflated Poisson
n	126	126	126
k	10	20	20
lnL	−184.9	−159.5	−180.0
AIC	−389.9	−359.0	−400.0
BIC	−418.3	−415.7	−456.7

Notes: n is number of observations; k is number of regressors; AIC is Akaike information criterion; BIC is Bayesian information criterion.

Results are given in Table 5.7. Larger (less negative) values of $\ln L$, AIC, and BIC are preferred. Both alternatives to the Poisson increase the log-likelihood as expected, but involve more parameters. These alternative models are still preferred using AIC, which has a small penalty for additional parameters. Using BIC, which has a large penalty, only the hurdle Poisson is preferred to Poisson, but not by much.

As an example of the test of Vuong we consider comparison of the Poisson and ZIP models. The test statistic $T_{\mathsf{LR,NN}}$ defined in (5.39), with ZIP density f and Poisson density g, equals 2.05. This favors the ZIP model with p value $\Pr[Z > 2.05] = 0.020$. So the ZIP model is clearly favored at significance level 5%.

5.5 TESTS FOR OVERDISPERSION

Hypothesis tests on regression coefficients and dispersion parameters in count data models involve straightforward application of the theory in Chapter 2.6. A general approach is the Wald test. If the maximum likelihood framework, for example a negative binomial model, is used, one can additionally use LR and LM tests. This theory has already been applied in Chapter 3 to basic count data models and is not presented here.

In the remainder of this chapter we instead tests for adequate model specification. In this section we consider in detail the tests of overdispersion introduced in Chapter 3.5. In section 5.6 we consider in further detail conditional moment tests introduced in Chapter 2.6.3.

5.5.1 LM Test for Overdispersion against Katz System

Tests for overdispersion are tests of the variance-mean equality imposed by the Poisson against the alternative that the variance exceeds the mean. They are implemented by tests of the Poisson with mean μ_i and variance μ_i, against the negative binomial with mean

$$\mathsf{E}[y_i|\mathbf{x}_i] = \mu_i = \mu_i(\mathbf{x}_i, \boldsymbol{\beta}) \tag{5.46}$$

and variance

$$\mathsf{V}[y_i|\mathbf{x}_i] = \mu_i + \alpha g(\mu_i), \tag{5.47}$$

for specified function $g(\mu_i)$. Usually $g(\mu_i) = \mu_i$ or $g(\mu_i) = \mu_i^2$. The null hypothesis is

$$H_0 : \alpha = 0. \tag{5.48}$$

Such tests are easily implemented as LR, Wald, or LM tests of H_0 against $H_a : \alpha > 0$. These are presented in Chapter 3. As noted there, the usual critical values of the LR and Wald cannot be used, and adjustment needs to be made because the null hypothesis $\alpha = 0$ lies on the boundary of the parameter space for the negative binomial, which does not permit underdispersion.

Tests for underdispersion, or variance less than the mean, can be constructed in a similar manner. One needs a distribution that permits underdispersion and nests the Poisson. The Katz system, defined in Chapter 4, has this property. For overdispersed data it equals the commonly used negative binomial. For underdispersed data, however, the Katz system model is not offered as a standard model in count data packages. The LM test, which requires estimation only under the null hypothesis of Poisson, is particularly attractive for the underdispersed case.

Derivation of the LM test of Poisson against the Katz system, due to Lee (1986), who considered $g(\mu_i) = \mu_i$ and $g(\mu_i) = \mu_i^2$, is not straightforward. In section 5.7 it is shown that for the Katz system density with mean μ_i and variance $\mu_i + \alpha g(\mu_i)$

$$\left.\frac{\partial \mathcal{L}}{\partial \boldsymbol{\beta}}\right|_{\alpha=0} = \sum_{i=1}^{n} \mu_i^{-1}(y - \mu_i)\frac{\partial \mu_i}{\partial \boldsymbol{\beta}} \tag{5.49}$$

$$\left.\frac{\partial \mathcal{L}}{\partial \alpha}\right|_{\alpha=0} = \sum_{i=1}^{n} \frac{1}{2}\mu_i^{-2} g(\mu_i)\{(y_i - \mu_i)^2 - y_i\}.$$

The LM test is based on these derivatives evaluated at the restricted MLE, which is $\hat{\boldsymbol{\theta}} = (\hat{\boldsymbol{\beta}}' \ \hat{\alpha})' = (\hat{\boldsymbol{\beta}}' \ 0)'$, where $\hat{\boldsymbol{\beta}}$ is the Poisson MLE. But the first summation in (5.49) equals the derivative with respect to $\boldsymbol{\beta}$ of the Poisson log-likelihood with general conditional mean $\mu_i(\boldsymbol{\beta})$, so $\partial \mathcal{L}/\partial \boldsymbol{\beta}|_{\boldsymbol{\beta}=\hat{\boldsymbol{\beta}},\alpha=0} = \mathbf{0}$ and hence

$$\left.\frac{\partial \mathcal{L}}{\partial \boldsymbol{\theta}}\right|_{\boldsymbol{\beta}=\hat{\boldsymbol{\beta}},\alpha=0} = \begin{bmatrix} \mathbf{0} \\ \sum_{i=1}^{n} \hat{\mu}_i^{-2} g(\hat{\mu}_i)\frac{1}{2}\{(y_i - \hat{\mu}_i)^2 - y_i\} \end{bmatrix}. \tag{5.50}$$

To construct the LM test defined earlier in Chapter 2.6.1, we additionally need a consistent estimate of the variance matrix, which by the information matrix equality is the limit of

$$T^{-1}\mathsf{E}[(\partial \mathcal{L}/\partial \boldsymbol{\theta}|_{\alpha=0})(\partial \mathcal{L}/\partial \boldsymbol{\theta}'|_{\alpha=0})].$$

Now under the null hypothesis that $y \sim \mathsf{P}[\mu]$,

$$\begin{aligned} \mathsf{E}[(y-\mu)^2] &= \mu \\ \mathsf{E}[\{(y-\mu)^2 - y\}(y-\mu)] &= 0 \\ \mathsf{E}[\{(y-\mu)^2 - y\}^2] &= 2\mu^2. \end{aligned} \tag{5.51}$$

This implies

$$\mathsf{E}\left[\left.\frac{\partial^2 \mathcal{L}}{\partial \boldsymbol{\theta} \partial \boldsymbol{\theta}'}\right|_{\alpha=0}\right] = \begin{bmatrix} \sum_{i=1}^n \mu_i^{-1} \frac{\partial \mu_i}{\partial \boldsymbol{\beta}} \frac{\partial \mu_i}{\partial \boldsymbol{\beta}'} & \mathbf{0} \\ \mathbf{0} & \sum_{i=1}^n \frac{1}{2} \mu_i^{-2} g^2(\mu_i) \end{bmatrix}. \tag{5.52}$$

Given (5.50) and (5.52) the LM test statistic in Chapter 2.6.1 is constructed. This will be a $(k+1) \times (k+1)$ matrix with zeroes everywhere except for the last diagonal entry. Taking the square root of this scalar yields

$$\mathsf{T}_{\mathsf{LM}} = \left[\sum_{i=1}^n \frac{1}{2} \hat{\mu}_i^{-2} g^2(\hat{\mu}_i)\right]^{-1/2} \sum_{i=1}^n \frac{1}{2} \hat{\mu}_i^{-2} g(\hat{\mu}_i)\{(y_i - \hat{\mu}_i)^2 - y_i\}. \tag{5.53}$$

Because the negative binomial is the special case $\alpha > 0$ of the Katz system, this statistic is the LM test for Poisson against both negative binomial overdispersion and Katz system underdispersion. At significance level .05, for example, the null hypothesis of equidispersion is rejected against the alternative hypothesis of overdispersion if $\mathsf{T}_{\mathsf{LM}} > z_{.05}$, of underdispersion if $\mathsf{T}_{\mathsf{LM}} < -z_{.05}$, and of over- or underdispersion if $|\mathsf{T}_{\mathsf{LM}}| > z_{.025}$.

Clearly one can obtain different LM test statistics by nesting the Poisson in other distributions. In particular, Gurmu and Trivedi (1993) nest the Poisson in the double Poisson, a special case of nesting the LEF in the extended LEF that they more generally consider, and obtain a test statistic for overdispersion that is a function of the deviance statistic.

The LM test for Poisson against the negative binomial has been extended to positive Poisson models by Gurmu (1991) and to left- and right-truncated Poisson models by Gurmu and Trivedi (1992). These extensions involve a number of complications, including non-block-diagonality of the information matrix so that the off-diagonal elements in the generalization of (5.52) are nonzero.

5.5.2 Auxiliary Regressions for LM Test

As is usual for test statistics, there are many asymptotically equivalent versions under H_0 of the overdispersion test statistic T_{LM} given in (5.53). Several of these versions can be easily calculated from many different auxiliary OLS regressions. Like T_{LM} they are distributed as $\mathsf{N}[0,1]$, or $\chi^2(1)$ on squaring, under H_0.

The auxiliary OPG regression for the LM test given in Chapter 2.6.1 uses the uncentered explained sums of squares from OLS regression of 1 on $\frac{1}{2}\hat{\mu}_i^{-2}g(\hat{\mu}_i)\{(y_i-\hat{\mu}_i)^2-y_i\}$ and $\hat{\mu}_i^{-1}(y_i-\hat{\mu}_i)\,\partial\mu_i/\partial\boldsymbol{\beta}\big|_{\hat{\boldsymbol{\beta}}}$. The square root of this is asymptotically equivalent to T_{LM}.

An asymptotically equivalent variant of this auxiliary regression is to use the uncentered explained sums of squares from OLS regression of 1 on $\frac{1}{2}\hat{\mu}_i^{-2}g(\hat{\mu}_i)\{(y_i-\hat{\mu}_i)^2-y_i\}$ alone. This simplification is possible since $\frac{1}{2}\hat{\mu}_i^{-2}g(\hat{\mu}_i)\{(y_i-\hat{\mu}_i)^2-y_i\}$ and $\hat{\mu}_i^{-1}(y_i-\hat{\mu}_i)\,\partial\mu_i/\partial\boldsymbol{\beta}\big|_{\hat{\boldsymbol{\beta}}}$ are asymptotically uncorrelated because for the Poisson $\mathsf{E}[\{(y_i-\mu_i)^2-y_i\}(y_i-\mu_i)]=0$ by (5.51) and $\sum_{i=1}^{n}\hat{\mu}_i^{-1}(y_i-\hat{\mu}_i)\,\partial\mu_i/\partial\boldsymbol{\beta}\big|_{\hat{\boldsymbol{\beta}}}=\mathbf{0}$ by the first-order conditions for the H_0 Poisson MLE. The square root of the explained sum of squares is

$$\mathsf{T}^{*}_{\mathsf{LM}}=\left[\sum_{i=1}^{n}\left(\frac{1}{2}\right)^2\hat{\mu}_i^{-4}g^2(\hat{\mu}_i)\{(y_i-\hat{\mu}_i)^2-y_i\}^2\right]^{-1/2} \times\sum_{i=1}^{n}\frac{1}{2}\hat{\mu}_i^{-2}g(\hat{\mu}_i)\{(y_i-\hat{\mu}_i)^2-y_i\}, \tag{5.54}$$

using the result that the square root of the uncentered explained sum of squares from regression of y_i^* on the scalar x_i is $(\sum_i x_i^2)^{-1/2}\sum_i x_i y_i^*$. This test is asymptotically equivalent to T_{LM}, as for the Poisson

$$\mathsf{E}\left[\frac{1}{2}\mu_i^{-2}\{(y_i-\mu_i)^2-y_i\}^2\right]=1, \tag{5.55}$$

by the last equation in (5.51).

In the special case $g(\mu_i)=\mu_i^l$, Cameron and Trivedi (1986) proposed using an alternative variant of T_{LM}. For general $g(\mu_i)$, this variant becomes

$$\mathsf{T}^{**}_{\mathsf{LM}}=\left[\frac{1}{n}\sum_{i=1}^{n}\frac{1}{2}\hat{\mu}_i^{-2}\{(y-\hat{\mu}_i)^2-y_i\}^2\right]^{-1/2}\left[\sum_{i=1}^{n}\frac{1}{2}\hat{\mu}_i^{-2}g^2(\hat{\mu}_i)\right]^{-1/2} \times\sum_{i=1}^{n}\frac{1}{2}\hat{\mu}_i^{-2}g(\hat{\mu}_i)\{(y_i-\hat{\mu}_i)^2-y_i\}. \tag{5.56}$$

This is asymptotically equivalent to T_{LM} because the first term in the parentheses has plim unity by (5.55). This can be computed as the square root of n times the uncentered explained sum of squares from the OLS regression of $\frac{1}{2}\hat{\mu}_i^{-1}\{(y_i-\hat{\mu}_i)^2-y_i\}$ against $\frac{1}{2}\hat{\mu}_i^{-1}g(\hat{\mu}_i)$.

In general the t-test from the regression $y_i^*=\alpha x_i+u_i$, where x_i is a scalar, can be shown to equal $(1/s)(\sum_i x_i^2)^{-1/2}\sum_i x_i y_i^*$, where s is the standard error of this regression. For the regression,

$$(\sqrt{2}\hat{\mu}_i)^{-1}\{(y_i-\hat{\mu}_i)^2-y_i\}=(\sqrt{2}\hat{\mu}_i)^{-1}g(\hat{\mu}_i)\alpha+v_i, \tag{5.57}$$

it follows that the t-statistic for $\alpha = 0$ is

$$\mathsf{T}^{***}_{\mathsf{LM}} = \left[s^2 \sum_{i=1}^{n} \frac{1}{2} \hat{\mu}_i^{-2} g^2(\hat{\mu}_i) \right]^{-1/2} \sum_{i=1}^{n} \frac{1}{2} \hat{\mu}_i^{-2} g(\hat{\mu}_i) \{ (y_i - \hat{\mu}_i)^2 - y_i \}, \tag{5.58}$$

where

$$s^2 = \frac{1}{n-1} \sum_{i=1}^{n} (\sqrt{2}\hat{\mu}_i)^{-2} \{ (y_i - \hat{\mu}_i)^2 - y_i - g(\hat{\mu}_i)\hat{\alpha} \}^2. \tag{5.59}$$

This test is asymptotically equivalent to T_{LM}, because $\text{plim}\, s^2 = 1$ under H_0 on setting $\alpha = 0$ and using the moment condition (5.55). This is the regression given in Chapter 3.4, on elimination of $\frac{1}{2}$ from both sides of the regression and letting $g(\mu_i) = \mu_i^2$ for tests against NB2 and $g(\mu_i) = \mu_i$ for tests against NB1.

In principle the LM test statistic can be computed using any of the many auxiliary regressions, because they are all asymptotically equivalent under H_0. In practice, however, the computed values can differ significantly. This is made clear by noting that asymptotic equivalence is established using assumptions, such as $\mathsf{E}[\{(y_i - \mu_i)^2 - y_i\}^2] = 2\mu_i$, which hold only under H_0.

The regression (5.57) has a physical interpretation, in addition to being a computational device. It can be viewed as a WLS regression based on testing whether $\alpha = 0$ in the population moment condition

$$\mathsf{E}[(y_i - \mu_i)^2 - y_i] = \alpha g(\mu_i). \tag{5.60}$$

This moment condition is implied by the alternative hypothesis given by (5.46) and (5.47). Tests based on (5.60) of overdispersion or underdispersion, under much weaker stochastic assumptions than Poisson against the Katz system, were proposed by Cameron and Trivedi (1985, 1990a). Their testing approach is presented in section 5.6.

5.5.3 LM test against Local Alternatives

Cox (1983) proposed a quite general method to construct the LM test statistic without completely specifying the density under the alternative hypothesis. The general result is presented before considering specialization to overdispersion tests.

Let y have density $f(y|\lambda)$, where the scalar parameter λ is itself a random variable, distributed with density $p(\lambda|\mu, \tau)$ where μ and τ denote the mean and variance of λ. This mixture distribution approach has already been presented in Chapter 4. For example, if y is Poisson distributed with parameter λ where λ is gamma distributed, then y is negative binomial distributed, conditional on the gamma distribution parameters.

Interest lies in the distribution of y given μ and τ:

$$h(y|\mu, \tau) = \int f(y|\lambda)p(\lambda|\mu, \tau)d\lambda. \tag{5.61}$$

A second-order Taylor series expansion of $f(y|\lambda)$ about $\lambda = \mu$ yields

$$\begin{aligned} h(y|\mu, \tau) = \int \{ & f(y|\mu) + f'(y|\mu)(\lambda - \mu) \\ & + \frac{1}{2} f''(y|\mu)(\lambda - \mu)^2) + R\} p(\lambda|\mu, \tau)d\lambda, \end{aligned} \tag{5.62}$$

where $f'(\cdot)$ and $f''(\cdot)$ denote the first and second derivatives, respectively, and R is a remainder term. Cox (1983) considered only small departures of λ from its mean of μ, specifically $\mathsf{V}[\lambda] = \tau = \delta/\sqrt{n}$, where δ is finite nonzero. After considerable algebra, given in section 5.7, this can be reexpressed as

$$h(y|\mu, \tau) = f(y|\mu) \exp\left[\frac{1}{2}\tau \left\{\frac{\partial^2 \ln f(y|\mu)}{\partial \mu^2} + \left(\frac{\partial \ln f(y|\mu)}{\partial \mu}\right)^2\right\}\right] + O(n^{-1}). \tag{5.63}$$

Cox (1983) considered LM (or score) tests against this approximation to the alternative hypothesis density, which from (5.63) reduces to the null hypothesis density $f(y|\mu)$ if $\tau = 0$.

For application to the Poisson we suppose $\mathsf{V}[\lambda] = \tau = \delta g(\mu)/\sqrt{n}$, which implies $\mathsf{V}[y] = \mu + \delta g(\mu)/\sqrt{n}$. Then in (5.47) we are considering local alternatives $\alpha = \delta/\sqrt{n}$. The log-likelihood under local alternatives is $\mathcal{L} = \sum_{i=1}^{n} \ln h(y_i|\mu_i, \tau)$ and

$$\left.\frac{\partial \mathcal{L}}{\partial \alpha}\right|_{\alpha=0} = \sum_{i=1}^{n} \frac{1}{2} g(\mu_i) \left\{\frac{\partial^2 \ln f(y_i|\mu_i)}{\partial \mu_i^2} + \left(\frac{\partial \ln f(y_i|\mu_i)}{\partial \mu_i}\right)^2\right\}. \tag{5.64}$$

If $f(y_i|\mu_i)$ is the Poisson density, this yields

$$\left.\frac{\partial \mathcal{L}}{\partial \alpha}\right|_{\alpha=0} = \sum_{i=1}^{n} \frac{1}{2} g(\mu_i)\mu_i^{-2} \left\{(y_i - \mu_i)^2 - y_i\right\}, \tag{5.65}$$

which is exactly the same as the second term in (5.49). The first term, $\partial \mathcal{L}/\partial \boldsymbol{\beta}|_{\alpha=0}$ is also the same as in (5.49), and the LM test statistic is T_{LM}, given in (5.53).

The approach of Cox (1983) demonstrates that T_{LM} in (5.53) is valid for testing Poisson against all local alternatives satisfying (5.46) and (5.47), not just the Katz system. The general form (5.63) for the density under local alternatives is clearly related to the information matrix equality. In section 5.6.5 we make the connection between the Cox test and the information matrix test.

5.5.4 Finite-Sample Corrections

A brief discussion of finite sample performance of hypothesis tests in the Poisson and negative binomial models is given by Lawless (1987). He concluded that the LR test was preferable for tests on regression coefficients, although none of the methods worked badly in small samples. There was, however, considerable small-sample bias in testing dispersion parameters.

The standard general procedure to handle small-sample bias in statistical inference is the bootstrap with asymptotic refinement, detailed in Chapter 2.6.5. Here we consider LM tests for overdispersion, using an approach due to Dean and Lawless (1989a) that differs from the Edgeworth expansion and bootstrap. This method can be applied to any GLM, not just the Poisson.

Consider the LM test statistic for Poisson against NB2 given in (5.53). The starting point is the result in McCullagh and Nelder (1983, appendix C) that, for GLM density with mean μ_i and variance $\mathsf{V}[y_i]$, the residual $(y_i - \hat{\mu}_i)$ has approximate variance $(1 - h_{ii})\mathsf{V}[y_i]$, where h_{ii} is the ith diagonal entry of the hat matrix $\mathbf{H}$ defined in (5.12). Applying this result to the Poisson, it follows that

$$\mathsf{E}[(y_i - \hat{\mu}_i)^2 - y_i] \simeq (1 - \hat{h}_{ii})\hat{\mu}_i - \hat{\mu}_i \simeq -h_{ii}\hat{\mu}_i. \tag{5.66}$$

This leads to small-sample bias under $H_0 : \mathsf{E}[(y_i - \mu_i)^2 - y_i]$, which can be corrected by adding $\hat{h}_{ii}\hat{\mu}_i$ to components of the sum in the numerator of (5.53), yielding the adjusted LM test statistic,

$$\mathsf{T}^{\mathsf{a}}_{\mathsf{LM}} = \left[\sum_{i=1}^{n} \frac{1}{2}\hat{\mu}_i^{-2} g^2(\hat{\mu}_i)\right]^{-1/2} \sum_{i=1}^{n} \frac{1}{2}\hat{\mu}_i^{-2} g(\hat{\mu}_i)\{(y_i - \hat{\mu}_i)^2 - y_i + \hat{h}_{ii}\hat{\mu}_i\}. \tag{5.67}$$

For the Poisson with exponential mean function, $\hat{h}_{ii}$ is the ith diagonal entry in

$$\mathbf{W}^{1/2}\mathbf{X}(\mathbf{X}'\mathbf{W}\mathbf{X})^{-1}\mathbf{X}'\mathbf{W}^{1/2}$$

where $\mathbf{W} = \mathsf{Diag}[\hat{\mu}_i]$ and $\mathbf{X}$ is the matrix of regressors.

Dean and Lawless (1989a) considered tests of Poisson against NB2 overdispersion, $g(\mu_i) = \mu_i^2$. The method has also been applied to other GLM models. Application to overdispersion in the binomial model is relatively straightforward and is presented in Dean (1992). Application to a truncated Poisson model, also a GLM, is considerably more complex and is given by Gurmu and Trivedi (1992). For data left truncated at r, meaning only $y_i \geq r$ is observed, the adjusted LM test for Poisson against negative binomial is

$$\mathsf{T}^{\mathsf{a}}_{\mathsf{LM}} = [\hat{\mathcal{I}}^{\alpha\alpha}]^{-1/2} \sum_{i=1}^{n} \frac{1}{2}\hat{\mu}_i^{-2} g(\hat{\mu}_i)\{(y_i - \hat{\mu}_i)^2 - y_i$$
$$+ (2y_i - \hat{\mu}_i - r + 1)\lambda(r - 1, \hat{\mu}_i)\hat{\mu}_i\},$$

where $\hat{\mathcal{I}}^{\alpha\alpha}$ is the scalar subcomponent for α of the inverse of the information matrix $-\mathsf{E}[\partial^2\mathcal{L}/\partial\boldsymbol{\theta}\partial\boldsymbol{\theta}']$ evaluated at $\hat{\boldsymbol{\theta}} = (\hat{\boldsymbol{\beta}}'\ 0)'$, see Gurmu and Trivedi (1992,

p. 350), and $\lambda(r-1,\mu) = f(y,\mu)/1 - F(y,\mu)$, where $f(\cdot)$ and $F(\cdot)$ are the untruncated Poisson density and cdf.

The procedure has a certain asymmetry in that a finite-sample correction is made only to the term in the numerator of the score test statistic. Conniffe (1990) additionally considered correction to the denominator term.

This method for small-sample correction of heteroskedasticity tests is much simpler than using the Edgeworth expansion, which from Honda (1988) is surprisingly complex even for the linear regression model under normality. The method cannot be adapted to tests of the regression coefficients themselves, however, because the score test in this case involves a weighted sum of $(y_i - \hat{\mu}_i)$ and the method of Dean and Lawless (1989a) yields a zero asymptotic bias for $(y_i - \hat{\mu}_i)$. Finite-sample adjustments are most easily done using the bootstrap, which is actually an empirical implementation of an Edgeworth expansion as already noted.

5.6 CONDITIONAL MOMENT SPECIFICATION TESTS

The preceding section presented hypothesis tests for overdispersion that are likelihood based. In this section we instead take a moment-based approach to hypothesis testing, using the CM test framework. The general approach is outlined in section 5.6.1. See also Chapter 2.6.3 for motivation and general theory. The focus is on CM tests of correct specification of the mean and variance. Key results and links to the LM tests presented earlier are given in section 5.6.2. Generalization of these results to general cross-section models is given in section 5.6.3. In section 5.6.4 we present CM tests based on orthogonal polynomials in $(y-\mu)$, an alternative way to use the low-order moments of y. Two commonly used CM tests, the Hausman test and the information test, are presented in, respectively, section 5.6.5 and section 5.6.6.

One conclusion from this section is that many insights gained from the linear regression model with homoskedastic errors require substantial modification before being applied to even the simple Poisson model. This is because the Poisson has complications of both the nonlinear conditional mean function and of heteroskedasticity that is a function of that mean. A better guide is provided by a binary choice model, such as logit or probit. But even this model is too limited since the variance function cannot be misspecified in binary choice models, because it is always the mean times one minus the mean, whereas with count data the variance function is not restricted to that being imposed by the Poisson regression model.

5.6.1 Introduction

Suppose a model implies the population moment condition,

$$\mathsf{E}[\mathbf{m}_i(y_i, \mathbf{x}_i, \boldsymbol{\theta})] = \mathbf{0}, \qquad i = 1, \ldots, n, \tag{5.68}$$

where $\mathbf{m}_i(\cdot)$ is an $r \times 1$ vector function. A CM test of this moment condition is based on the closeness to zero of the corresponding sample moment condition, that is

$$\mathbf{m}(\hat{\boldsymbol{\theta}}) = \sum_{i=1}^{n} \hat{\mathbf{m}}_i,$$

where $\hat{\mathbf{m}}_i = \mathbf{m}_i(y_i, \mathbf{x}_i, \hat{\boldsymbol{\theta}})$. The CM test statistic in general is

$$\sum_{i=1}^{n} \hat{\mathbf{m}}_i' \left\{ \mathsf{V}\left[\sum_{i=1}^{n} \hat{\mathbf{m}}_i \right] \right\}^{-1} \sum_{i=1}^{n} \hat{\mathbf{m}}_i$$

and is asymptotically chi-square distributed.

Two issues arise in applying CM tests. The first issue is the choice of the function $\mathbf{m}_i(\cdot)$. Here we focus on tests based on the first two moments of count data regression models. The second issue is choosing how to implement the test. Several asymptotically equivalent versions are available, some of which can be computed using an auxiliary regression. Here we focus on applications in which the moment condition is chosen so that $\mathbf{m}_i(\cdot)$ satisfies

$$\mathsf{E}\left[\frac{\partial \mathbf{m}_i(y_i, \mathbf{x}_i, \boldsymbol{\theta})}{\partial \boldsymbol{\theta}'} \right] = \mathbf{0}. \tag{5.69}$$

Then from Chapter 2.6.3 the CM test statistic simplifies to

$$\sum_{i=1}^{n} \hat{\mathbf{m}}_i' \left[\sum_{i=1}^{n} \mathsf{E}[\mathbf{m}_i \mathbf{m}_i']\big|_{\hat{\boldsymbol{\theta}}} \right]^{-1} \sum_{i=1}^{n} \hat{\mathbf{m}}_i. \tag{5.70}$$

If $m_i(\cdot)$ is a scalar, taking the square root of (5.70) yields the test statistic

$$\mathsf{T}_{\mathsf{CM}} = \left[\sum_{i=1}^{n} \mathsf{E}[m_i^2]\big|_{\hat{\boldsymbol{\theta}}} \right]^{-1/2} \sum_{i=1}^{n} \hat{m}_i, \tag{5.71}$$

which is asymptotically $\mathsf{N}[0, 1]$ if (5.68) and (5.69) hold. An asymptotically equivalent version is

$$\mathsf{T}_{\mathsf{CM}}^{*} = \left[\sum_{i=1}^{n} \hat{m}_i^2 \right]^{-1/2} \sum_{i=1}^{n} \hat{m}_i. \tag{5.72}$$

Even if (5.69) does not hold, implementation is still simple, provided $\hat{\boldsymbol{\theta}}$ is the MLE. Then a chi-square test statistic is the uncentered explained sum of squares from regression of 1 on $\hat{\mathbf{m}}_i$ and $\hat{\mathbf{s}}_i$, where $\hat{\mathbf{s}}_i = \partial \ln f(y_i|\mathbf{x}_i, \boldsymbol{\theta})/\partial \boldsymbol{\theta}|_{\hat{\boldsymbol{\theta}}}$.

If possible CM tests are compared to the LM test, which is a special case of a CM test and is, of course, the most powerful test if a fully parametric approach is taken. Particular interest lies in tests of correct specification of the conditional mean and variance.

For the Poisson regression, the LM test for exclusion of the subcomponent $\mathbf{x}_{2i}$ of $\mathbf{x}_i = [\mathbf{x}_{1i}', \mathbf{x}_{2i}']'$ model is a CM test of

$$\mathsf{E}[\mathbf{m}_i(y_i, \mathbf{x}_i, \boldsymbol{\beta}_1)] = \mathsf{E}[(y_i - \mu_{1i})\mathbf{x}_{2i}] = \mathbf{0}, \tag{5.73}$$

where $\mu_{1i} = \exp(\mathbf{x}_{1i}'\boldsymbol{\beta}_1)$.

For overdispersion, the test of Poisson with variance $\mu_i = \mu(\mathbf{x}_i'\boldsymbol{\beta})$ against the Katz system with variance function $\mu_i + \alpha g(\mu_i)$ is from (5.50) a CM test of

$$\mathsf{E}[m_i(y_i, \mathbf{x}_i, \boldsymbol{\beta})] = \mathsf{E}[\{(y_i - \mu_i)^2 - y_i\}\mu_i^{-2} g(\mu_i)] = 0. \tag{5.74}$$

Note that the simplifying condition (5.69) holds for $m_i(y_i, \mathbf{x}_i, \boldsymbol{\beta})$ defined in (5.74), provided (5.74) holds and $\mathsf{E}[y_i - \mu_i] = 0$. It can be shown that for the moment condition (5.74) the test statistic (5.71) yields T_{LM} given in (5.53), and the test statistic (5.72) yields $\mathsf{T}^*_{\mathsf{LM}}$ given in (5.54).

CM tests can be obtained under relatively weak stochastic assumptions. Several examples are given here, beginning with one in which a regression provides the motivation or basis for the test, rather than just providing a way to calculate a test statistic.

5.6.2 Regression-Based Tests for Overdispersion

Consider cross-section data $(y_i, \mathbf{x}_i)$ where under the null hypothesis the first two moments are those of the Poisson regression model,

$$H_0 : \mathsf{E}[y_i|\mathbf{x}_i] = \mu_i = \mu(\mathbf{x}_i, \boldsymbol{\beta}), \quad \mathsf{V}[y_i|\mathbf{x}_i] = \mu_i, \tag{5.75}$$

while under the alternative hypothesis

$$H_a : \mathsf{E}[y_i|\mathbf{x}_i] = \mu_i = \mu(\mathbf{x}_i, \boldsymbol{\beta}), \quad \mathsf{V}[y_i|\mathbf{x}_i] = \mu_i + \alpha g(\mu_i), \tag{5.76}$$

where $g(\mu_i)$ is a specified function such as μ_i or μ_i^2. The moments (5.76) are those of, for example, the negative binomial models presented in Chapter 3.3.

The moment condition (5.76) implies

$$H_a : \mathsf{E}[\{(y_i - \mu_i)^2 - y_i\}|\mathbf{x}_i] = \alpha g(\mu_i), \tag{5.77}$$

while H_0 imposes the constraint that $\alpha = 0$. If μ_i is known one could perform a t-test of $\alpha = 0$ based on regression of $(y_i - \mu_i)^2 - y_i$ on $g(\mu_i)$. Two complications are that μ_i is unknown and that the error term in this regression is in general heteroskedastic because the conditional variance of $(y_i - \mu_i)^2 - y_i$ is a function of μ_i, say

$$\omega_i = \omega(\mu_i) = \mathsf{V}[(y_i - \mu_i)^2 - y_i|\mathbf{x}_i]. \tag{5.78}$$

The null hypothesis

$$H_0 : \alpha = 0 \tag{5.79}$$

can be tested by the t-test of $\alpha = 0$ in the LS regression

$$\sqrt{\hat{\omega}_i}\{(y_i - \hat{\mu}_i)^2 - y_i\} = \alpha\sqrt{\hat{\omega}_i} g(\hat{\mu}_i) + u_i, \tag{5.80}$$

where $\hat{\mu}_i = \mu(\mathbf{x}_i'\hat{\boldsymbol{\beta}})$, $\hat{\omega}_i = \omega(\hat{\mu}_i)$, and $\hat{\boldsymbol{\beta}}$ is a consistent estimator of $\boldsymbol{\beta}$ under H_0. The WLS regression is used because it yields the most efficient least squares regression estimator of α and hence the most powerful or optimal test. In principle replacing μ_i by $\hat{\mu}_i$ leads to a more complicated distribution for $\hat{\alpha}$. This is not a problem in this particular application, however, essentially because

$\partial\{(y_i - \mu_i)^2 - y_i\}/\partial\boldsymbol{\beta} = -2(y_i - \mu_i)\,\partial\mu_i/\partial\boldsymbol{\beta}$ has expected value $\mathbf{0}$ so (5.69) holds.

Standard results for OLS yield the t-test statistic $\hat{\alpha}/\sqrt{\hat{\mathsf{V}}[\hat{\alpha}]}$ or

$$\mathsf{T}_{\mathsf{CM}}^{\mathsf{OLS}} = \left[s^2 \sum_{i=1}^{n} \hat{\omega}_i^{-1} g^2(\hat{\mu}_i)\right]^{-1/2} \sum_{i=1}^{n} \hat{\omega}_i^{-1} g(\hat{\mu}_i)\{(y_i - \hat{\mu}_i)^2 - y_i\}, \quad (5.81)$$

where

$$s^2 = \frac{1}{n-1} \sum_{i=1}^{n} \hat{\omega}_i^{-1} \{(y_i - \hat{\mu}_i)^2 - y_i - g(\hat{\mu}_i)\hat{\alpha}\}^2.$$

Under H_0, $\operatorname{plim} s^2 = 1$ as $\alpha = 0$, and one can equivalently use

$$\mathsf{T}_{\mathsf{CM}} = \left[\sum_{i=1}^{n} \hat{\omega}_i^{-1} g^2(\hat{\mu}_i)\right]^{-1/2} \sum_{i=1}^{n} \hat{\omega}_i^{-1} g(\hat{\mu}_i)\{(y - \hat{\mu}_i)^2 - y_i\}. \quad (5.82)$$

Advantages of this approach to testing, in addition to simplicity of use, include the following:

1. If the first four moments of y_i under the null hypothesis are those of the Poisson, T_{CM} equals the optimal LM test statistic for testing Poisson against the Katz system.
2. The test is easily adapted to situations in which assumptions on only the first two moments are made.
3. The test can be given a simple interpretation as a CM test based on the first two moments of y_i.
4. The test is computed from an OLS regression that has interpretation as a model, rather than being merely a computational device to compute the statistic.
5. The approach can be generalized to other testing situations.

These points are made clear in the following discussion.

First, in the fully parametric situation in which $y_i \sim \mathsf{P}[\mu_i]$, using formulas for the first four moments of the Poisson, one obtains $\omega_i = 2\mu_i^2$. But then $\mathsf{T}_{\mathsf{CM}}^{\mathsf{OLS}}$ in (5.81) equals $\mathsf{T}_{\mathsf{LM}}^{***}$, the LM test for Poisson against negative binomial presented in (5.58), whereas T_{CM} in (5.82) equals T_{LM} given in (5.53).

Second, consider adaptation of the test statistic (5.81) in the case in which only the first two moments of y_i are assumed. Then ω_i in (5.78) is unknown. The least squares regression (5.80) is again run, but with weighting function $\hat{\omega}_i$ replaced by $\hat{\upsilon}_i$. One might choose $\hat{\upsilon}_i = 2\hat{\mu}_i^2$, although one should note that it is no longer assumed that $\omega_i = 2\mu_i^2$. Now the error u_i has heteroskedasticity of an unknown functional form, so the t-test of $\alpha = 0$ uses robust sandwich standard errors. The test statistic is

$$\mathsf{T}_{\mathsf{CM}}^{\mathsf{Robust}} = \left[\sum_{i=1}^{n} \hat{u}_i^2 \hat{\upsilon}_i^{-1} g^2(\hat{\mu}_i)\right]^{-1/2} \sum_{i=1}^{n} \hat{\upsilon}_i^{-1} g(\hat{\mu}_i)\{(y_i - \hat{\mu}_i)^2 - y_i\}, \quad (5.83)$$

where $\hat{u}_i^2 = \{(y_i - \hat{\mu}_i)^2 - y_i - \hat{\alpha} g(\hat{\mu}_i)\}^2$ and $\hat{\alpha}$ is the least squares estimator in (5.80) with $\hat{\omega}_i$ replaced by $\hat{v}_i$. Under H_0, $\mathsf{T}_{\mathsf{CM}}^{\mathsf{Robust}}$ is asymptotically $\mathsf{N}[0, 1]$.

In the special case that $\hat{v}_i = 2\hat{\mu}_i^2$ the test statistic (5.83) provides a variant of the LM test statistic for overdispersion given in section 5.5 that is robust to misspecification of the third and fourth moments of y_i. This is directly analogous to the modification of the Breusch-Pagan LM test for heteroskedasticity in the regression model under normality proposed by Koenker (1982) to allow for nonconstant fourth moments of the dependent variable.

Third, consider a CM test for variance-mean equality. The null hypothesis (5.75) implies that

$$\mathsf{E}[\{(y_i - \mu_i)^2 - y_i\}|\mathbf{x}_i] = 0, \tag{5.84}$$

which in turn implies

$$\mathsf{E}[h(\mu_i)\{(y_i - \mu_i)^2 - y_i\}] = 0, \tag{5.85}$$

where we let $h(\mu_i)$ be a specified scalar function. The test based on closeness to zero of $\sum_{i=1}^{n} h(\hat{\mu}_i)\{(y_i - \hat{\mu}_i)^2 - y_i\}$ is a special case of the CM chi-square test statistic given in Chapter 2.6.3. Because $\partial\{(y_i - \mu_i)^2 - y_i\}/\partial\boldsymbol{\beta} = -2(y_i - \mu_i)\,\partial\mu_i/\partial\boldsymbol{\beta}$ has expected value $\mathbf{0}$ so (5.69) holds, one can use T_{CM} given in (5.71)

$$\mathsf{T}_{\mathsf{CM}} = \left[\sum_{i=1}^{n} h^2(\hat{\mu}_i)\hat{\omega}_i\right]^{-1/2} \sum_{i=1}^{n} h(\hat{\mu}_i)\{(y_i - \hat{\mu}_i)^2 - y_i\}, \tag{5.86}$$

where $\omega_i = \mathsf{E}[\{(y_i - \mu_i)^2 - y_i\}^2|\mathbf{x}_i]$. T_{CM} is asymptotically $\mathsf{N}[0, 1]$ under (5.84). This test specifies only a moment condition under H_0. Different choices of function $h(\mu_i)$ will test in different directions away from H_0, with the optimal choice of $h(\mu_i)$ depending on the particular alternative to (5.84). The regression-based approach makes it clear that if the alternative is (5.77) then the optimal choice is $h(\mu_i) = \omega_i^{-1} g(\mu_i)$.

Fourth, consider calculation of overdispersion test statistics by an auxiliary regression. The usual approach is to obtain a CM or LM test statistic and then give ways to calculate the statistic or an asymptotically equivalent variant of the statistic by an auxiliary regression. This regression has no physical interpretation, being merely a computational device. Such tests are best called *regression-implemented* tests, though they are often called *regression-based* because their computation is based on a regression. By comparison the overdispersion test statistic given in this subsection is regression based in the stronger sense that there is a regression motivation for the test statistic.

The final point, that this *regression-based* or *regression-motivated* test approach generalizes to other testing situations, is outlined in the next subsection.

Regression-based tests for overdispersion were proposed by Cameron and Trivedi (1985, 1990a). In addition to the regression (5.80) they also considered the t-test of $\alpha = 0$ in the least squares regression,

$$\sqrt{\hat{\omega}_i}\{(y_i - \hat{\mu}_i)^2 - \hat{\mu}_i\} = \alpha\sqrt{\hat{\omega}_i}g(\hat{\mu}_i) + u_i. \tag{5.87}$$

This replaces y_i by $\hat{\mu}_i$ in the left-hand side, the rationale being that the moment condition (5.76) implies not only (5.77) but also

$$H_a : \mathsf{E}[\{(y_i - \mu_i)^2 - \mu_i\}|\mathbf{x}_i] = \alpha g(\mu_i). \tag{5.88}$$

The test based on the regression (5.87) is more difficult to implement because replacing μ_i by $\hat{\mu}_i$ makes a difference in this regression, since $\partial\{(y_i - \mu_i)^2 - \mu_i\}/\partial\boldsymbol{\beta} = -\{2(y_i - \mu_i) + 1\}\,\partial\mu_i/\partial\boldsymbol{\beta}$ has nonzero expected value. Cameron and Trivedi (1985) show that the analog of (5.82) is

$$\mathsf{T}_{\mathsf{CM},2} = \left[\sum_{i=1}^{n}\sum_{j=1}^{n}\hat{w}_{ij}g(\hat{\mu}_i)g(\hat{\mu}_j)\right]^{-1/2}\sum_{i=1}^{n}\sum_{j=1}^{n}\hat{w}_{ij}g(\hat{\mu}_i)\{(y_j - \hat{\mu}_j)^2 - \hat{\mu}_j\}, \tag{5.89}$$

where w_{ij} is the ij^{th} entry in $\mathbf{W} = [\mathbf{D} - \Delta(\Delta'\mathbf{D}_{\mu}^{-1}\Delta)^{-1}\Delta']^{-1}$, $\mathbf{D}$ and $\mathbf{D}_{\mu}$ are $n \times n$ diagonal matrices with i^{th} entries $(2\mu_i^2 + \mu_i)$ and μ_i, respectively, and Δ is an $n \times k$ matrix with i^{th} row $\partial\mu_i/\partial\boldsymbol{\beta}'$.

The test $\mathsf{T}_{\mathsf{CM},2}$ is a different test from T_{CM} in (5.82), with different power properties. In particular, $\mathsf{T}_{\mathsf{CM},2}$ is the LM test, and hence the most powerful test, for testing $\mathsf{N}[\mu_i, \mu_i]$ against $\mathsf{N}[\mu_i, \mu_i + \alpha g(\mu_i)]$. It has already been shown that T_{CM} is the LM test for $\mathsf{P}[\mu_i]$ against NB2. In addition to greater power in the standard set-up for overdispersion tests, T_{CM} has the advantage of being easier to implement.

5.6.3 Regression-Based CM Tests

Suppose that a specified model imposes the conditional moment restriction

$$\mathsf{E}[r(y_i, \mathbf{x}_i, \boldsymbol{\theta})|\mathbf{x}_i] = 0, \tag{5.90}$$

where for simplicity $r(\cdot)$ is a scalar function. Suppose we wish to test this restriction against the specific alternative conditional expected value for $r(y_i, \mathbf{x}_i, \boldsymbol{\theta})$,

$$\mathsf{E}[r(y_i, \mathbf{x}_i, \boldsymbol{\theta})|\mathbf{x}_i] = \mathbf{g}(\mathbf{x}_i, \boldsymbol{\theta})'\boldsymbol{\alpha}, \tag{5.91}$$

where $\mathbf{g}(\cdot)$ and $\boldsymbol{\alpha}$ are $p \times 1$ vectors.

The moment condition (5.90) can be tested against (5.91) by test of $\boldsymbol{\alpha} = \mathbf{0}$ in the regression

$$r(y_i, \mathbf{x}_i, \boldsymbol{\theta}) = \mathbf{g}(\mathbf{x}_i, \boldsymbol{\theta})'\boldsymbol{\alpha} + \varepsilon_i. \tag{5.92}$$

The most powerful test of $\boldsymbol{\alpha} = \mathbf{0}$ is based on the efficient GLS estimator, which for data independent over i is the WLS estimator,

$$\hat{\boldsymbol{\alpha}} = \left[\sum_{i=1}^{n} \frac{1}{\sigma^2(\mathbf{x}_i, \boldsymbol{\theta})} \mathbf{g}(\mathbf{x}_i, \boldsymbol{\theta})\mathbf{g}(\mathbf{x}_i, \boldsymbol{\theta})'\right]^{-1} \sum_{i=1}^{n} \frac{1}{\sigma^2(\mathbf{x}_i, \boldsymbol{\theta})} \mathbf{g}(\mathbf{x}_i, \boldsymbol{\theta}) r(y_i, \mathbf{x}_i, \boldsymbol{\theta}), \tag{5.93}$$

where

$$\sigma^2(\mathbf{x}_i, \boldsymbol{\theta}) = \mathsf{E}_0[r(y_i, \mathbf{x}_i, \boldsymbol{\theta})^2 | \mathbf{x}_i] \tag{5.94}$$

is the conditional variance of $r(y_i, \mathbf{x}_i, \boldsymbol{\theta})$ under the null hypothesis model.

Tests based on $\hat{\boldsymbol{\alpha}}$ are equivalent to tests based on any full rank transformation of $\hat{\boldsymbol{\alpha}}$. Most simply, multiply by $\sum_{i=1}^{n} \sigma^{-2}(\mathbf{x}_i, \boldsymbol{\theta})\mathbf{g}(\mathbf{x}_i, \boldsymbol{\theta}) r(y_i, \mathbf{x}_i, \boldsymbol{\theta})$. This is equivalent to a CM test of the unconditional moment condition,

$$\mathsf{E}[\mathbf{m}(y_i, \mathbf{x}_i, \boldsymbol{\theta})] = \mathsf{E}\left[\frac{1}{\sigma^2(\mathbf{x}_i, \boldsymbol{\theta})} \mathbf{g}(\mathbf{x}_i, \boldsymbol{\theta}) r(y_i, \mathbf{x}_i, \boldsymbol{\theta})\right] = \mathbf{0}. \tag{5.95}$$

Note that the unconditional moment (5.95) for the CM test is obtained as a test of the conditional moment condition (5.90) against the alternative (5.91).

An example already considered is testing variance-mean equality, in which case $r(y_i, \mathbf{x}_i, \boldsymbol{\beta}) = (y_i - \mu_i)^2 - y_i$ and $g(\mathbf{x}_i, \boldsymbol{\beta})'\alpha = \alpha\mu_i^2$ in the case of overdispersion of the NB2 form. If the Poisson assumption is maintained under the null, then $\sigma^2(\mathbf{x}_i, \boldsymbol{\beta}) = 2\mu_i^2$ and the CM test based on the unconditional moment (5.95) simplifies to a test of

$$\mathsf{E}[(y_i - \mu_i)^2 - y_i] = 0. \tag{5.96}$$

Now consider a CM test based on an unconditional moment condition that can be partitioned into a product of the form,

$$\mathsf{E}[m^*(y_i, \mathbf{x}_i, \boldsymbol{\theta})] = \mathsf{E}[\mathbf{g}^*(\mathbf{x}_i, \boldsymbol{\theta}) r^*(y_i, \mathbf{x}_i, \boldsymbol{\theta})], \tag{5.97}$$

where y_i only appears through the scalar function $r^*(\cdot)$. A simple interpretation of this test is that it is testing failure of the conditional moment condition,

$$\mathsf{E}[r^*(y_i, \mathbf{x}_i, \boldsymbol{\theta}) | \mathbf{x}_i] = 0, \tag{5.98}$$

in the direction $\mathbf{g}^*(\mathbf{x}_i, \boldsymbol{\theta})$. A much more specific interpretation of the direction of the test, using (5.95), is that it is testing (5.98) against the conditional moment condition,

$$\mathsf{E}[r^*(y_i, \mathbf{x}_i, \boldsymbol{\theta}) | \mathbf{x}_i] = \sigma^{*2}(\mathbf{x}_i, \boldsymbol{\theta}) \mathbf{g}^*(\mathbf{x}_i, \boldsymbol{\theta})' \boldsymbol{\alpha}, \tag{5.99}$$

where

$$\sigma^{*2}(\mathbf{x}_i, \boldsymbol{\theta}) = \mathsf{V}[r^*(y_i, \mathbf{x}_i, \boldsymbol{\theta}) | \mathbf{x}_i].$$

Considering again the test of overdispersion of the NB2 form, it is not immediately apparent what form of overdispersion is being tested by the CM test of (5.96). Using (5.99), however, the test can be viewed as a test against the alternative $\mathsf{E}[(y_i - \mu_i)^2 - y_i] = \alpha{\mu_i}^2$, where this interpretation uses the result that the null hypothesis Poisson model implies $\mathsf{V}[(y_i - \mu_i)^2 - y_i] = 2\mu_i^2$.

To summarize, in the usual case in which interest lies in the expected value of a scalar function $r(y_i, \mathbf{x}_i, \boldsymbol{\theta})$ of the dependent variable, CM tests of an explicit null against an explicit alternative conditional expected value of $r(y_i, \mathbf{x}_i, \boldsymbol{\theta})$ are easily developed. Going the other way, a CM test for zero unconditional expected value of the product of $r(y_i, \mathbf{x}_i, \boldsymbol{\theta})$ and a specified function of $\mathbf{x}_i$ and $\boldsymbol{\theta}$ can be interpreted as a test of an explicit null against an explicit alternative conditional expected value of $r(y_i, \mathbf{x}_i, \boldsymbol{\theta})$.

This approach, called *regression-based CM tests* by Cameron and Trivedi (1990c), therefore provides a link between the standard formulation of CM tests as model misspecification tests in no particular direction and formal hypothesis tests that test against an explicit alternative hypothesis. Several applications and extension to the case in which $\mathbf{r}(y_i, \mathbf{x}_i, \boldsymbol{\theta})$ is a vector are given in Cameron and Trivedi (1990c).

The preceding discussion looks only at the formation of the moment for the CM test. For actual implementation one can either run regression (5.92) – that is, replace $\boldsymbol{\theta}$ by $\hat{\boldsymbol{\theta}}$ and use weights $\sigma^{-1}(\mathbf{x}_i, \hat{\boldsymbol{\theta}})$ – or form a test based on the sample analog of (5.95), $\sum_{i=1}^{n} \sigma^{-2}(\mathbf{x}_i, \hat{\boldsymbol{\theta}})\mathbf{g}(\mathbf{x}_i, \hat{\boldsymbol{\theta}})r(y_i, \mathbf{x}_i, \hat{\boldsymbol{\theta}})$. In either case the distribution is most easily obtained if condition (5.69) holds, which is the case if in addition to (5.90)

$$\mathsf{E}\left[\frac{\partial r(y_i, \mathbf{x}_i, \boldsymbol{\theta})}{\partial \boldsymbol{\theta}'}\right] = \mathbf{0}. \tag{5.100}$$

For example, in testing variance-mean equality, the choice $r_i(y_i, \mathbf{x}_i, \boldsymbol{\theta}) = \{(y_i - \mu_i)^2 - \mu_i\}$ satisfies (5.100).

In other testing situations it is also possible to construct tests in which (5.100) holds. In particular, for testing that the correct model for heteroskedasticity is a specified function $v(\mu_i)$ against the alternative that $\mathsf{V}[y_i|\mathbf{x}_i] = v(\mu_i) + \alpha g(\mu_i)$, where $\mathsf{E}[y_i|\mathbf{x}_i] = \mu_i = \mu(\mathbf{x}_i, \boldsymbol{\beta})$, the choice

$$r(y_i, \mathbf{x}_i, \boldsymbol{\beta}) = (y_i - \mu_i)^2 - \frac{\partial v(\mu_i)}{\partial \mu_i}(y_i - \mu_i) - v(\mu_i), \tag{5.101}$$

satisfies (5.100). The regression-based CM test using (5.101) not only is easy to implement but also coincides with the LM test of Poisson against negative binomial, the LM test of binomial against the beta-binomial, and the Breusch and Pagan (1979) test of normal homoskedastic error against normal heteroskedastic error (in which case $v(\mu_i) = v(\mu_i, \sigma^2) = \sigma^2$). For details, see Cameron (1991).

Further examples of regression-based CM tests are given here.

5.6.4 Orthogonal Polynomial Tests

In the CM test framework it is natural to focus on tests of correct specification of the first few conditional moments of the dependent variable. One possibility is to construct a sequence of tests based on whether the expected values of y_i^k equal those imposed by the model, for $k = 1, 2, 3, \ldots$. Another is to use central moments, in which case the sequence is of the expected values of $(y_i - \mu_i)^k$ for $k = 1, 2, 3, \ldots$.

An alternative approach, proposed by Cameron and Trivedi (1990b), is to consider a sequence of *orthogonal polynomial* functions in y_i, in which case terms in the sequence are uncorrelated. One can additionally consider *orthonormal polynomials*, for which terms are orthogonal and normalized to have unit variance.

For the Poisson density, the orthonormal polynomials are called the Gram-Charlier polynomial series, with these first three terms:

$$\begin{aligned} Q_1(y) &= (y-\mu)/\sqrt{\mu} \\ Q_2(y) &= \{(y-\mu)^2 - y\}/\sqrt{2}\mu \\ Q_3(y) &= \{(y-\mu)^3 - 3(y-\mu)^2 - (3\mu-2)(y-\mu) + 2\mu\}/\sqrt{6\mu^3}. \end{aligned} \tag{5.102}$$

Further terms are obtained using $Q_j(y) = \partial\{(1+z)^y \exp(-\mu z)\}/\partial z|_{z=0}$. Note that $\mathsf{E}[Q_j(y)] = 0$. These polynomials have the property of orthogonality – that is, $\mathsf{E}[Q_j(y)Q_k(y)] = 0$ for $j \neq k$ – and are normalized so that $\mathsf{E}[Q_j^2(y)] = 1$. These properties hold if y is $\mathsf{P}[\mu]$, and also hold for the first j terms under the weaker assumption that y has the same first $j+k$ moments as $\mathsf{P}[\mu]$ for orthogonality, and the same first $2j$ moments as $\mathsf{P}[\mu]$ for orthonormality.

The orthonormal polynomials can be used directly for CM tests. Thus, to test correct specification of the jth moment of y_i, assuming correct specification of the first $(j-1)$ moments, use

$$\mathsf{E}[\mathbf{m}_j(y_i, \mathbf{x}_i, \boldsymbol{\beta})] = \mathsf{E}[Q_j(y_i, \mathbf{x}_i, \boldsymbol{\beta})\,\mathbf{g}_j(\mathbf{x}_i, \boldsymbol{\beta})] = \mathbf{0}, \tag{5.103}$$

where for regression applications $Q_j(y_i, \mathbf{x}_i, \boldsymbol{\beta})$ is $Q_j(y)$ in (5.102) evaluated at $\mu_i = \mu(\mathbf{x}_i, \boldsymbol{\beta})$. Using the regression-based interpretation of the CM test in (5.98) and (5.99), this is a test of $\mathsf{E}[Q_j(y_i, \mathbf{x}_i, \boldsymbol{\beta})|\mathbf{x}_i] = 0$ against $\mathsf{E}[Q_j(y_i, \mathbf{x}_i, \boldsymbol{\beta})|\mathbf{x}_i] = \mathbf{g}(\mathbf{x}_i, \boldsymbol{\beta})'\boldsymbol{\alpha}_j$, where simplification occurs because $\mathsf{V}[Q_j(y_i)] = 1$ due to orthonormality. Define the jth central moment of the Poisson to be $\mu'_j = \mathsf{E}[(y-\mu)^j]$. Then equivalently, given correct assumption of the first $(j-1)$ moments, the CM test (5.103) is a test of $\mathsf{E}[(y_i - \mu_i)^j] = \mu'_j$ against $\mathsf{E}[(y_i - \mu_i)^j] = \mu'_j + \mathbf{g}(\mathbf{x}_i, \boldsymbol{\beta})'\boldsymbol{\alpha}_j$. These CM tests are easy to compute, because for the Poisson the orthonormal polynomials satisfy $\mathsf{E}[\partial Q_j(y)/\partial\mu] = 0$, so (5.69) holds.

The orthonormal polynomials can also be used to construct LM tests for the Poisson against series expansions around a baseline Poisson density. A property of orthogonal polynomials is that a general density $g(y)$ can be represented as the following series expansion around the baseline density $f(y)$:

$$g(y) = f(y)\left[1 + \sum_{j=1}^{\infty} a_j Q_j(y)\right],$$

where $a_j = \int Q_j(y)g(y)dy$ and $g(y)$ is assumed to satisfy the boundedness condition $\int \{g(y)/f(y)\}^2 f(y)dy < \infty$. Then, because $\ln g(y) = \ln f(y) + \ln\left[1 + \sum_{j=1}^{\infty} a_j Q_j(y)\right]$,

$$\left.\frac{\partial \ln g(y)}{\partial a_j}\right|_{a_1=0,a_2=0,\ldots} = Q_j(y), \quad j = 1, 2, \ldots.$$

This suggests a sequence of score tests for the null hypothesis that the Poisson density is correctly specified:

$$\mathsf{E}[Q_j(y_i, \mathbf{x}_i, \boldsymbol{\beta})] = 0, \quad j = 1, 2, \ldots. \tag{5.104}$$

For the Poisson, the tests based on orthonormal polynomials equal the standard LM tests for correct specification of the first two moments. Thus tests based on $Q_1(y_i) = (y_i - \mu_i)$ coincide with the LM test (5.73) for excluded variables, whereas tests based on $Q_2(y_i) = (y_i - \mu_i)^2 - y_i$ correspond to the LM test (5.74) for Poisson against the Katz system. Non-Poisson skewness can be tested using $Q_3(y_i)$ given in (5.102). This is essentially the same test as that of Lee (1986) for the Poisson against truncated Gram-Charlier series expansion.

The CM tests proposed here differ in general from those obtained by considering only the jth central moment. For example, consider a CM test of the second moment of the Poisson, conditional on correct specification of the first. Then, because $\mathsf{E}[(y - \mu)^2] = \mu$ for $y \sim \mathsf{P}[\mu]$, the CM test based on the second central moment is

$$\mathsf{E}[\{(y_i - \mu(\mathbf{x}_i, \boldsymbol{\beta}))^2 - \mu(\mathbf{x}_i, \boldsymbol{\beta})\}\, \mathbf{g}_2(\mathbf{x}_i, \boldsymbol{\beta})] = \mathbf{0}.$$

The test based on the second orthogonal polynomial from (5.102) is

$$\mathsf{E}[\{(y_i - \mu(\mathbf{x}_i, \boldsymbol{\beta}))^2 - y_i\}\, \mathbf{g}_2(\mathbf{x}_i, \boldsymbol{\beta})] = \mathbf{0}.$$

As pointed out at the end of section 5.6.2, for overdispersion tests with $\mathbf{g}_2(\mathbf{x}_i, \boldsymbol{\beta}) = \mu_i^{-2} g(\mu_i)$, these different moment conditions lead to quite different tests.

Key properties of orthogonal and orthonormal polynomials are summarized in Cameron and Trivedi (1993) and also discussed in Chapter 8. The discussion here has focused on the Poisson. The properties that the resulting CM tests coincide with standard LM tests for mean and variance and that the tests are easy to implement as (5.69) holds carry over to the LEF with quadratic variance function (QVF). For the LEF-QVF the variance is a quadratic function of the mean, $\mathsf{V}[y] = v_0 + v_1\mu + v_2\mu^2$, where various possible choices of the coefficients v_0, v_1, and v_2 lead to six exponential families, five of which – the normal, Poisson, gamma, binomial, and negative binomial – constitute the Meixner class. This is discussed in detail in Cameron and Trivedi (1990b).

Finally, although results are especially straightforward for LEF-QVF densities, one can construct orthogonal polynomial sequences for any assumed model. In particular, letting $\mu = \mathsf{E}[y]$ and $\mu_j = \mathsf{E}[(y-\mu)^j]$ for $j = 2, 3$, the first two orthogonal polynomials are

$$\begin{aligned} P_1(y) &= y - \mu \\ P_2(y) &= (y-\mu)^2 - \frac{\mu_3}{\mu_2}(y-\mu) - \mu_2. \end{aligned} \tag{5.105}$$

These can be used for tests of the specified conditional mean and variance in general settings.

5.6.5 Information Matrix Tests

The IM test, introduced in Chapter 2.6.3, is a CM test for fully parametric models of whether the information matrix equality holds, that is, whether

$$\mathsf{E}\left[\text{vech}\left(\frac{\partial^2 \ln f_i}{\partial\boldsymbol{\theta}\partial\boldsymbol{\theta}'} + \frac{\partial \ln f_i}{\partial\boldsymbol{\theta}}\frac{\partial \ln f_i}{\partial\boldsymbol{\theta}'}\right)\right] = \mathbf{0}, \tag{5.106}$$

where $f_i = f_i(y_i, \mathbf{x}_i, \boldsymbol{\theta})$ is the density. This can be applied in a straightforward manner to any specified density for count data, such as negative binomial and hurdle, and is easily computed using the OPG regression given in Chapter 2.6.3.

An interesting question is what fundamental features of the model are being tested by an IM test. Chesher (1984) showed that, quite generally, the IM test is a test for random parameter heterogeneity; Hall (1987) showed that for the linear model under normality, subcomponents of the IM test were tests of heteroskedasticity, symmetry, and kurtosis. Here we focus on Poisson regression.

For the Poisson regression model with exponential mean function, substitution into (5.106) of the first and second derivatives of the log-density with respect to $\boldsymbol{\theta}$ yields

$$\mathsf{E}[\{(y_i - \mu_i)^2 - \mu_i\}\,\text{vech}(\mathbf{x}_i\mathbf{x}_i')] = \mathbf{0}. \tag{5.107}$$

The IM test is a CM test of (5.107). If only the component of the IM test on the intercept term is considered, the IM test coincides with the LM test of overdispersion of form $\mathsf{V}[y_i] = \mu_i + \alpha\mu_i^2$, because the test is based on $\sum_{i=1}^n (y_i - \hat{\mu}_i)^2 - \hat{\mu}_i$ which equals $\sum_{i=1}^n (y_i - \hat{\mu}_i)^2 - y_i$ since the Poisson first-order conditions imply $\sum_{i=1}^n (y_i - \hat{\mu}_i) = 0$.

Considering all components of the IM test, (5.107) is of the form (5.97). Using $\mathsf{E}[\{(y_i - \mu_i)^2 - \mu_i\}^2] = 2\mu_i^2$, this leads to the interpretation of the IM test as a test of $\mathsf{E}[\{(y_i - \mu_i)^2 - \mu_i\}|\mathbf{x}_i] = 0$ against $\mathsf{E}[\{(y_i - \mu_i)^2 - \mu_i\}|\mathbf{x}_i] = \mu_i^2\,\text{vech}(\mathbf{x}_i\mathbf{x}_i')'\boldsymbol{\alpha}$. So the IM test is a test against overdispersion of form $\mathsf{V}[y_i] = \mu_i + \mu_i^2\text{vech}(\mathbf{x}_i\mathbf{x}_i')'\boldsymbol{\alpha}$. This result is analogous to the result of Hall (1987) for the regression parameters subcomponent for the linear regression model under normality. Note, however, that $\text{vech}(\mathbf{x}_i\mathbf{x}_i')'\boldsymbol{\alpha}$ is weighted by μ_i^2, whereas in the linear model the test is simply against $\mathsf{V}[y_i] = \sigma^2 + \text{vech}(\mathbf{x}_i\mathbf{x}_i')'\boldsymbol{\alpha}$.

The result (5.107) holds only for an exponential conditional mean. If the conditional mean is of general form $\mu_i = \mu(\mathbf{x}_i'\boldsymbol{\beta})$ then the IM test can be shown to be a CM test of

$$\mathsf{E}\left[\frac{y_i - \mu_i}{\mu_i}\text{vech}\left(\frac{\partial^2\mu_i}{\partial\boldsymbol{\beta}\partial\boldsymbol{\beta}'}\right)\right] + \mathsf{E}\left[\left\{\frac{(y_i - \mu_i)^2 - y_i}{\mu_i^2}\right\}\text{vech}(\mathbf{x}_i\mathbf{x}_i')\right] = \mathbf{0}. \tag{5.108}$$

This CM test is a test of both the specification of the variance, through the second term, and the more fundamental condition that the conditional mean is misspecified. Similar results for LEF models in general are given in Cameron and Trivedi (1990d).

The Poisson regression model is a special case of a model in which the underlying distribution of y_i depends only on the scalar μ_i, which is then parameterized to depend on a function of k regression parameters, meaning that the density is of the special form $f(y_i|\mu(\mathbf{x}_i, \boldsymbol{\beta}))$. It follows that $\partial \ln f_i/\partial\boldsymbol{\beta} = (\partial \ln f_i/\partial\mu_i)(\partial\mu_i/\partial\boldsymbol{\beta})$ and the IM test (5.106) can be expressed as a test of

$$\mathsf{E}\left[\frac{\partial \ln f_i}{\partial\mu_i}\text{vech}\left(\frac{\partial^2\mu_i}{\partial\boldsymbol{\beta}\partial\boldsymbol{\beta}'}\right)\right] + \mathsf{E}\left[\left\{\frac{\partial^2 \ln f_i}{\partial\mu_i^2} + \left(\frac{\partial \ln f_i}{\partial\mu_i}\right)^2\right\}\text{vech}\left(\frac{\partial\mu_i}{\partial\boldsymbol{\beta}}\frac{\partial\mu_i}{\partial\boldsymbol{\beta}'}\right)\right]$$
$$= \mathbf{0}. \tag{5.109}$$

There are clearly two components to the IM test. The first component is a test of whether $\mathsf{E}[\partial \ln f_i/\partial\mu_i] = 0$, required for consistency of the MLE that solves $\sum_{i=1}^n \partial \ln f_i/\partial\mu_i\,(\partial\mu_i/\partial\boldsymbol{\beta}) = \mathbf{0}$. The second component is a test of the IM equality in terms of the underlying parameter μ_i rather than the regression parameter $\boldsymbol{\beta}$.

Setting $\tau = \text{vech}((\partial\mu_i/\partial\boldsymbol{\beta})(\partial\mu_i/\partial\boldsymbol{\beta}'))'\boldsymbol{\alpha}/\sqrt{n}$ in (5.63) yields an LM test of $\boldsymbol{\alpha} = \mathbf{0}$, which equals the second component of the IM test. It follows that the second component of the IM test is an LM test that the density is $f(y_i|\mu_i)$, where $\mu_i = \mu(\mathbf{x}_i, \beta)$, against the alternative that it is $f(y_i|\lambda_i)$ where λ_i is a random variable with mean μ_i and $\text{vech}((\partial\mu_i/\partial\boldsymbol{\beta})(\partial\mu_i/\partial\boldsymbol{\beta}'))'\boldsymbol{\alpha}/\sqrt{n}$. Cameron and Trivedi (1990d) show that the complete IM test, with both components, is a test against the alternative that y_i has density $f(y_i|\lambda_i)$ where λ_i is a random variable with mean $\mu_i + \text{vech}(\partial^2\mu_i/\partial\boldsymbol{\beta}\partial\boldsymbol{\beta}')/\sqrt{n}$ and variance $\text{vech}((\partial\mu_i/\partial\boldsymbol{\beta})(\partial\mu_i/\partial\boldsymbol{\beta}'))'\boldsymbol{\alpha}/\sqrt{n}$. So the IM test additionally tests for misspecification of the mean function.

Interpretations of the IM test have ignored the first component of the IM test because they have focused on the linear model where $\mu_i = \mathbf{x}_i'\boldsymbol{\beta}$, in which case $\partial^2\mu_i/\partial\boldsymbol{\beta}\partial\boldsymbol{\beta}' = \mathbf{0}$. In general the components can be negative or positive, and so they may be offsetting. So even if the first component is large in magnitude, indicating a fundamental misspecification, a model may pass the IM test. This indicates the usefulness, in nonlinear settings, of determining the moment conditions being tested by the IM test.

5.6.6 Hausman Tests

One way to test for simultaneity in a single linear regression equation with iid errors is to compare the OLS and two-stage least squares estimators. If there is simultaneity the two estimators differ in probability limit, because OLS is inconsistent. If there is no simultaneity the two estimators will have the same probability limit, because both are consistent. Tests based on such comparisons between two different estimators are called Hausman tests, after Hausman (1978), or Wu-Hausman tests or even Durbin-Wu-Hausman tests after Wu (1973) and Durbin (1954), who also proposed such tests.

Consider two estimators $\tilde{\boldsymbol{\theta}}$ and $\hat{\boldsymbol{\theta}}$ where

$$\begin{aligned} H_0&: \ \text{plim}(\tilde{\boldsymbol{\theta}} - \hat{\boldsymbol{\theta}}) = \mathbf{0} \\ H_a&: \ \text{plim}(\tilde{\boldsymbol{\theta}} - \hat{\boldsymbol{\theta}}) \neq \mathbf{0}. \end{aligned} \tag{5.110}$$

The *Hausman test statistic* of H_0 is

$$\mathsf{T}_\mathsf{H} = n\,(\tilde{\boldsymbol{\theta}} - \hat{\boldsymbol{\theta}})'\mathbf{V}^{-1}_{\tilde{\theta}-\hat{\theta}}(\tilde{\boldsymbol{\theta}} - \hat{\boldsymbol{\theta}}), \tag{5.111}$$

which is $\chi^2(q)$ under H_0, where it is assumed that $\sqrt{n}(\tilde{\boldsymbol{\theta}} - \hat{\boldsymbol{\theta}}) \xrightarrow{d} \mathsf{N}[\mathbf{0}, \mathbf{V}_{\tilde{\theta}-\hat{\theta}}]$, under H_0. In some applications $\mathbf{V}_{\tilde{\theta}-\hat{\theta}}$ is of less than full rank, in which case $\mathbf{V}^{-1}_{\tilde{\theta}-\hat{\theta}}$ is replaced by the generalized-inverse of $\mathbf{V}_{\tilde{\theta}-\hat{\theta}}$ and the degrees of freedom are $\mathsf{rank}[\mathbf{V}_{\tilde{\theta}-\hat{\theta}}]$. We reject H_0 if T_H exceeds the chi-square critical value.

The Hausman test is easy in principle but difficult in practice due to the need to obtain a consistent estimate of the variance matrix $\mathbf{V}_{\tilde{\theta}-\hat{\theta}}$. Initial implementations focused on the special case in which $\tilde{\boldsymbol{\theta}}$ is the fully efficient estimator under the null, $\mathbf{V}_{\tilde{\theta}-\hat{\theta}} = \mathbf{V}_{\hat{\theta}} - \mathbf{V}_{\tilde{\theta}}$. Therefore it is easily computed as the difference between the variance matrices of the two estimators (see Hausman, 1978, or Amemiya, 1985a, p. 146). An example is linear regression with possible correlation between regressors and the error, with $\tilde{\boldsymbol{\theta}}$ the OLS estimator that is the efficient estimator under the null of no correlation and spherical errors, whereas $\hat{\boldsymbol{\theta}}$ is the two-stage least squares estimator that maintains consistency under the alternative. Hausman (1978) gives examples in which the Hausman test can be computed by a standard test for the significance of regressors in an augmented regression, but these results are confined to linear models.

More recently, applied research has focused on robust inference, in which case $\tilde{\boldsymbol{\theta}}$ is no longer fully efficient. Then one can perform B bootstraps and use the variance of the B values $(\tilde{\boldsymbol{\theta}}_b - \hat{\boldsymbol{\theta}}_b)$, $b = 1, \ldots, B$, to estimate $\mathbf{V}_{\tilde{\theta}-\hat{\theta}}$.

Holly (1982) considered Hausman tests for nonlinear models in the likelihood framework and compared the Hausman test with standard likelihood-based hypothesis tests such as the LM test. Partition the parameter vector as $\boldsymbol{\theta} = (\boldsymbol{\theta}_1', \boldsymbol{\theta}_2')'$, where the null hypothesis $H_0 : \boldsymbol{\theta}_1 = \boldsymbol{\theta}_{10}$ applies to the first component of $\boldsymbol{\theta}$, and the second component $\boldsymbol{\theta}_2$ is a nuisance parameter vector. Restricted and unrestricted maximum likelihood estimation of $\boldsymbol{\theta}$ provides two estimates $\widetilde{\boldsymbol{\theta}}_2$ and $\hat{\boldsymbol{\theta}}_2$ of the nuisance parameter. Suppose the alternative hypothesis is $H_0 : \boldsymbol{\theta}_1 = \boldsymbol{\theta}_{10} + \boldsymbol{\delta}$, in which case classical hypothesis tests are tests of $H_0 : \boldsymbol{\delta} = \mathbf{0}$. Holly (1982) showed that by comparison the Hausman test based on the difference $\widetilde{\boldsymbol{\theta}}_2 - \hat{\boldsymbol{\theta}}_2$ is a test of $H_0 : \mathcal{I}_{22}^{-1}\mathcal{I}_{21}\boldsymbol{\delta} = \mathbf{0}$, where $\mathcal{I}_{ij} = \mathsf{E}\left[\partial^2\mathcal{L}(\boldsymbol{\theta}_1, \boldsymbol{\theta}_2)/\partial\boldsymbol{\theta}_i\partial\boldsymbol{\theta}_j\right]$. Holly (1987) extended the analysis to the nonlinear hypothesis $\mathbf{h}(\boldsymbol{\theta}_1) = \mathbf{0}$, with Hausman tests based on linear combinations $\mathbf{D}(\widetilde{\boldsymbol{\theta}} - \hat{\boldsymbol{\theta}})$ rather than just the subcomponent $(\widetilde{\boldsymbol{\theta}}_2 - \hat{\boldsymbol{\theta}}_2)$. Furthermore, the model under consideration may be potentially misspecified, covering the QMLE based on an exponential family density.

The Hausman test can be used in many count applications, particularly ones analogous to those in the linear setting such as testing for endogeneity. It should be kept in mind, however, that the test is designed for situations in which at least one of the estimators is inconsistent under the alternative. Consider a Hausman test of Poisson against the NB2 model, where in the latter model the conditional mean is correctly specified although there is overdispersion ($\alpha \neq 0$). The Poisson MLE is fully efficient under the null hypothesis. Because the Poisson regression coefficients $\boldsymbol{\beta}$ maintain their consistency under the alternative hypothesis; however, common sense suggests that a Hausman test of the difference between NB2 and Poisson estimates of $\boldsymbol{\beta}$ will have no power. In terms of Holly's framework, $\theta_1 = \alpha$, $\boldsymbol{\theta}_2 = \boldsymbol{\beta}$, and the Hausman test is a test of $H_0 : \mathcal{I}_{22}^{-1}\mathcal{I}_{21}\boldsymbol{\delta} = \mathbf{0}$. But here $\mathcal{I}_{21} = \mathbf{0}$ from Chapter 3.2. In fact, if the Poisson and NB2 estimates of $\boldsymbol{\beta}$ are compared, then large values of the Hausman test statistic reflect more fundamental misspecification, that of the conditional mean.

5.7 DERIVATIONS

5.7.1 Test of Poisson against Katz System

The Katz system density with mean μ and variance $\mu + \alpha g(\mu)$ can be written as

$$f(y) = f(0)\prod_{l=1}^{y}\frac{\mu + \alpha\mu^{-1}g(\mu)(y-l)}{[1+\alpha\mu^{-1}g(\mu)]\,(y-l+1)}, \qquad \text{for } y = 1, 2, \ldots, \tag{5.112}$$

where $f(0)$ is the density for $y = 0$. This density generalizes slightly the Chapter 4 density, which sets $g(\mu) = \mu\,\mu^{k_2}$, and changes the index from j to $l = y - j + 1$. The log-density is

$$\ln f(y) = \sum_{l=1}^{y} \ln(\mu + \alpha\mu^{-1}g(\mu)(y-l)) - \ln(1+\alpha\mu^{-1}g(\mu)) - \ln(y - l + 1) + \ln f(0). \tag{5.113}$$

Then

$$\frac{\partial \ln f(y)}{\partial \alpha} = \left\{ \sum_{l=1}^{y} \frac{\mu^{-1}g(\mu)(y-l)}{\mu + \alpha\mu^{-1}g(\mu)(y-l)} - \frac{\mu^{-1}g(\mu)}{1+\alpha\mu^{-1}g(\mu)} \right\} + \frac{\partial \ln f(0)}{\partial \alpha}.$$

Specializing to $H_0 : \alpha = 0$,

$$\begin{aligned}\left.\frac{\partial \ln f(y)}{\partial \alpha}\right|_{\alpha=0} &= \left\{ \sum_{l=1}^{y} \mu^{-2}g(\mu)(y-l) - \mu^{-1}g(\mu) \right\} + \left.\frac{\partial \ln f(0)}{\partial \alpha}\right|_{\alpha=0} \\ &= \mu^{-2}g(\mu)\{y(y-1)/2 - \mu y\} + \left.\frac{\partial \ln f(0)}{\partial \alpha}\right|_{\alpha=0},\end{aligned}$$

using $\sum_{l=1}^{y}(y-l) = y(y-1)/2$. Because $\mathsf{E}[\partial \ln f(y)/\partial\alpha] = 0$ under the usual maximum likelihood regularity conditions, a derivative of the form $\partial \ln f(y)/\partial\alpha = h(y) + \partial \ln f(0)/\partial\alpha$ implies $\mathsf{E}[\partial \ln f(0)/\partial\alpha] = -\mathsf{E}[h(y)]$. Therefore

$$\begin{aligned}\left.\frac{\partial \ln f(y)}{\partial \alpha}\right|_{\alpha=0} &= \mu^{-2}g(\mu)\{y(y-1)/2 - \mu y - \mathsf{E}[y(y-l)/2 - \mu y]\} \\ &= \mu^{-2}g(\mu)\{y(y-1)/2 - \mu y - (\mu^2/2 - \mu^2)\} \\ &= \mu^{-2}g(\mu)\frac{1}{2}\{(y-\mu)^2 - y\}.\end{aligned}$$

Similar manipulations lead to

$$\left.\frac{\partial \ln f(y)}{\partial \boldsymbol{\beta}}\right|_{\alpha=0} = \mu^{-1}(y-\mu)\frac{\partial\mu}{\partial\boldsymbol{\beta}}.$$

Using $\mathcal{L} = \sum_{i=1}^{n} \ln f(y)$ leads directly to (5.49).

5.7.2 LM test against Local Alternatives

We begin with

$$h(y|\mu,\tau) = \int \{f(y|\mu) + f'(y|\mu)(\lambda-\mu) + \frac{1}{2}f''(y|\mu)(\lambda-\mu)^2) + R\}p(\lambda|\mu,\tau)d\lambda.$$

Because λ has mean μ and variance τ and τ is $O(n^{-1})$, $\mathsf{E}[\lambda-\mu] = 0$ and $\mathsf{E}[(\lambda-\mu)^2] = \tau + O(n^{-1})$. This implies

$$h(y|\mu,\tau) = f(y|\mu) + 0 + \frac{1}{2}f''(y|\mu)\tau + O(n^{-1}).$$

Now

$$\frac{\partial \ln f(y|\mu)}{\partial \mu} = \frac{f'(y|\mu)}{f(y|\mu)},$$

and

$$\frac{\partial^2 \ln f(y|\mu)}{\partial \mu^2} = \frac{f''(y|\mu)}{f(y|\mu)} - \frac{f'(y|\mu)^2}{f(y|\mu)^2}$$
$$= \frac{f''(y|\mu)}{f(y|\mu)} - \left(\frac{\partial \ln f(y|\mu)}{\partial \mu}\right)^2,$$

which implies

$$f''(y|\mu) = f(y|\mu)\left\{\frac{\partial^2 \ln f(y|\mu)}{\partial \mu^2} + \left(\frac{\partial \ln f(y|\mu)}{\partial \mu}\right)^2\right\}.$$

Making this substitution yields

$$h(y|\mu,\tau) = f(y|\mu)\left[1 + \frac{1}{2}\tau\left\{\frac{\partial^2 \ln f(y|\mu)}{\partial \mu^2} + \left(\frac{\partial \ln f(y|\mu)}{\partial \mu}\right)^2\right\} + O(n^{-1})\right],$$

and using $\exp(x) \simeq 1 + x$ for small x,

$$h(y|\mu,\tau) = f(y|\mu)\exp\left[\frac{1}{2}\tau\left\{\frac{\partial^2 \ln f(y|\mu)}{\partial \mu^2} + \left(\frac{\partial \ln f(y|\mu)}{\partial \mu}\right)^2\right\}\right] + O(n^{-1}).$$

5.8 BIBLIOGRAPHIC NOTES

Key references on residuals include Cox and Snell (1968), who consider a very general definition of residuals; Pregibon (1981), who extends many of the techniques for normal model residuals to logit model residuals; McCullagh and Nelder (1989, chapter 12), who summarize extensions and refinements of Pregibon's work to GLMs; and Davison and Snell (1991), who consider both GLMs and more general models. Discussion of the Pearson and deviance statistics is given in any GLM review, such as McCullagh and Nelder (1989) or Firth (1991). In the econometrics literature, generalized residuals and simulated residuals were proposed by Gourieroux, Monfort, Renault, and Trognon (1987a, 1987b) and are summarized in Gourieroux and Monfort (1995). The material on R-squared measures is based on Cameron and Windmeijer (1996, 1997). A comprehensive treatment of the chi-square goodness-of-fit test is given in Andrews (1988a, 1988b), with the latter reference providing the more accessible treatment. The LR test for nonnested likelihood-based models was proposed by Vuong (1989).

There is a long literature on overdispersion tests. Attention has focused on the LM test of Poisson against negative binomial, introduced in the iid case by Collings and Margolin (1985) and in the regression case by Lee (1986) and Cameron and Trivedi (1986). Small-sample corrections were proposed by

Dean and Lawless (1989a). A more modern approach is to use a bootstrap with asymptotic refinement. Efron and Tibshirani (1993) provide an introduction, and Horowitz (1997) covers the regression case in considerable detail. Treatments of overdispersion tests under weaker stochastic assumptions are given by Cox (1983), Cameron and Trivedi (1985, 1990a), Breslow (1990), and Wooldridge (1991a, 1991b). White (1994) considers various specializations that arise for statistical inference with LEF densities and covers the CM test framework in considerable detail.

5.9 EXERCISES

5.1 Show that if $y_i \sim \mathsf{P}[\mu_i]$ the log-density of y_i is maximized with respect to μ_i by $\mu_i = y_i$. Conclude that $\mathbf{y}$ maximizes $\mathcal{L}(\boldsymbol{\mu})$ for the Poisson. Hence show that for the Poisson the deviance statistic defined in (5.18) specializes to (5.21), and the deviance residual is (5.3).

5.2 Show that if y_i has the LEF density defined in Chapter 2.4.2, then the log-density of y_i is maximized with respect to μ_i by $\mu_i = y_i$. Conclude that $\mathbf{y}$ maximizes $\mathcal{L}(\boldsymbol{\mu})$ for LEF densities. Hence, obtain (5.22) for the deviance of the NB2 density with α known, using the result in Chapter 3.3.2 that this density is a particular LEF density.

5.3 For discrete data y_i that take only two values, 0 or 1, the appropriate model is a binary choice model with $\Pr[y_i = 1] = \mu_i$ and $\Pr[y_i = 0] = 1 - \mu_i$. The density $f(y_i) = \mu_i^{y_i}(1-\mu_i)^{y_i}$ is an LEF density. Show that the deviance R^2 in this case simplifies to $R^2 = 1 - (\mathcal{L}_{\text{fit}}/\mathcal{L}_0)$ rather than the more general form (5.28).

5.4 Show that Pearson's chi-square test statistic given in (5.34) can be rewritten using the notation of section 5.3.4 as

$$\left(\sum_{i=1}^{n}(\mathbf{d}_i(y_i) - \mathbf{p}_i(\mathbf{x}_i, \hat{\boldsymbol{\theta}}))\right)' \mathbf{D} \left(\sum_{i=1}^{n}(\mathbf{d}_i(y_i) - \mathbf{p}_i(\mathbf{x}_i, \hat{\boldsymbol{\theta}}))\right).$$

Conclude that the test statistic (5.34) is only chi-square distributed in the special case in which $\mathbf{V}_m^-$ in (5.33) equals $\mathbf{D}$. Hint: $\sum_{i=1}^n d_{ij}(y_i) = n\overline{p}_j$ and $\sum_{i=1}^n p_{ij}(\mathbf{x}_i, \hat{\boldsymbol{\theta}}) = n\hat{p}_j$, where $\mathbf{D}$ is a diagonal matrix with i^{th} entry $\left(\sum_{i=1}^n p_{ij}(\mathbf{x}_i, \hat{\boldsymbol{\theta}})\right)^{-1}$.

5.5 Obtain the general formula for the t statistic for $\alpha = 0$ in the linear regression $y_i^* = \alpha x_i + u_i$, where x_i is a scalar and an intercept is not included. Hence obtain the regression-based overdispersion test statistic given in (5.81).

5.6 Consider testing whether $\boldsymbol{\beta}_2 = \mathbf{0}$ in the regression model $\mathsf{E}[y_i|\mathbf{x}_i] = \exp(\mathbf{x}_{1i}'\boldsymbol{\beta}_1 + \mathbf{x}_{2i}'\boldsymbol{\beta}_2)$. Show that a first-order Taylor series expansion around

$\boldsymbol{\beta}_2 = \mathbf{0}$ yields $\mathsf{E}[y_i] = \mu_{1i} + \mu_{1i}\mathbf{x}_{2i}'\boldsymbol{\beta}_2$, where $\mu_{1i} = \exp(\mathbf{x}_{1i}'\boldsymbol{\beta}_1)$ and for small $\boldsymbol{\beta}_2$ the remainder term is ignored. Hence test $\boldsymbol{\beta}_2 = \mathbf{0}$ using (5.95) for the regression-based CM test of $H_0 : \mathsf{E}[y_i - \mu_{1i}|\mathbf{x}_i] = 0$ against $H_a : \mathsf{E}[y_i - \mu_{1i}|\mathbf{x}_i] = \mu_{1i}\mathbf{x}_{2i}'\boldsymbol{\beta}_2$. Show that when $y_i \sim \mathsf{P}[\mu_{1i}]$ under H_0 (5.95) is the same as (5.73), the moment condition for the LM test of exclusion restrictions.

5.7 For $y_i \sim \mathsf{P}[\mu_i = \mu(\mathbf{x}_i'\boldsymbol{\beta})]$ show that the IM test is a test of moment condition (5.108) and specializes to (5.107) if $\mu_i = \exp(\mathbf{x}_i'\boldsymbol{\beta})$.

CHAPTER 6

Empirical Illustrations

6.1 INTRODUCTION

In this chapter we provide a detailed discussion of empirical models for three examples based on four cross-sectional data sets. The first example analyzes the demand for medical care by the elderly in United States and shares many features of health utilization studies based on cross-section data. The second example is an analysis of recreational trips. The third is an analysis of completed fertility – the total number of children born to a woman with a complete history of births.

Figure 6.1 presents histograms for the four count variables studied; the first two histograms exclude the highest percentile for readability. Physician visits appear roughly negative binomial, with a mild excess of zeros. Recreational trips have a very large excess of zeroes. Completed fertility in both cases is bimodal, with modes at 0 and 2. Different count data models will most likely be needed for these different datasets.

The applications presented in this chapter emphasize fully parametric models for counts, an issue discussed in section 6.2. Sections 6.3 to 6.5 deal, in turn, with each of the three empirical applications. The health care example in section 6.3 is the most extensive example and provides a lengthy treatment of model fitting, selecting, and interpreting, with focus on a finite mixture model. The recreational trips example in section 6.4 pays particular attention to special treatment of zero counts versus positive counts. The completed fertility illustration in section 6.5 is a nonregression example that emphasizes fitting a distribution that is bimodal. Section 6.6 pursues a methodological question concerning the distribution of the LR test under nonstandard conditions, previously raised in Chapter 4.8.5.

The emphasis of this chapter is on practical aspects of modeling. Each application involves several competing models that are compared and evaluated using model diagnostics and goodness-of-fit measures. Although the Poisson regression model is the most common starting point in count data analysis, it is usually abandoned in favor of a more general mixed Poisson model. This usually occurs after diagnostic tests reveal overdispersion. But in many cases

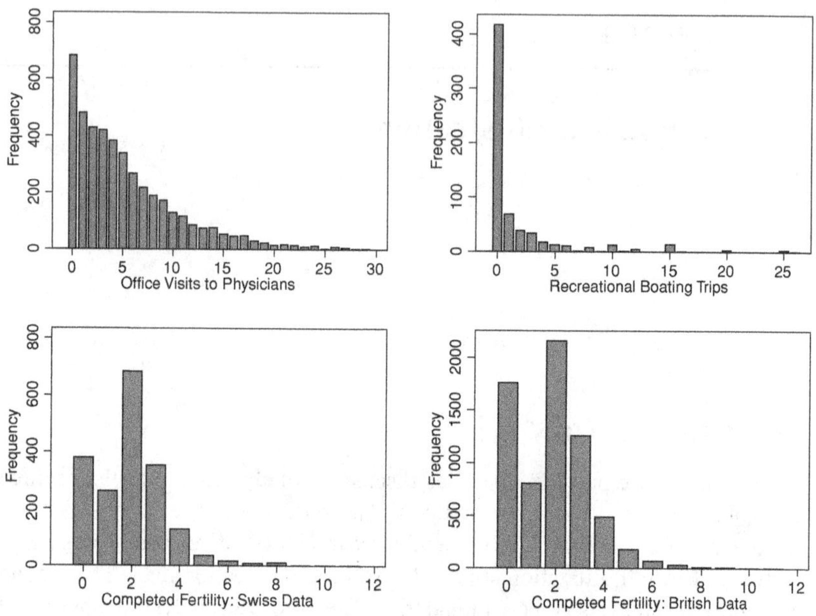

Figure 6.1. Histograms for the four count variables.

this mechanical approach can produce misleading results. Overdispersion may be a consequence of many diverse factors. An empirical model that simply controls for overdispersion does not shed any light on its source. Tests of overdispersion do not unambiguously suggest remedies, since they may have power against many commonplace misspecifications. Rejection of the null against a specific alternative does not imply that the alternative itself is valid. Hence misspecifications and directions for model revision should be explored with care.

6.2 BACKGROUND

Before presenting the empirical studies, we survey two general modeling issues. The first is the distinction between modeling only the conditional mean versus modeling the full distribution of counts. The second concerns behavioral interpretation of count models, an issue of importance to econometricians who emphasize the distinction between reduced form and structural models.

6.2.1 Fully Parametric Estimation

Event count models may have two different uses. In some cases, the main interest lies in modeling the conditional expectation of the count and in making inferences about key parameters (e.g., price elasticity). Different models and estimation methods may yield similar results with respect to the conditional mean, even though they differ in the goodness of fit.

In other cases, the entire frequency distribution of events will be relevant. An interesting example is Dionne and Vanesse (1992), in which the entire distribution of auto accidents is used to derive insurance premium tables as a function of accident history and individual characteristics. Another example is the probability distribution of the number of patient days in the hospital as a function of patient characteristics. These probabilities might be needed to generate the expected costs of hospital stays.

If the objective is to make conditional predictions about the expected number of events, the focus would be on the conditional mean function. But if the focus is on the conditional *probability* of a given number of events, the frequency distribution itself is relevant. In the former case, features such as overdispersion may affect prediction intervals, but not mean prediction. In the latter case overdispersion will affect estimated cell probabilities. Hence parametric methods are attractive in the latter case, whereas robustness of the estimate is more important in the former.

This neat separation of modeling just the conditional mean versus modeling the distribution is not possible, however, if consistent estimation of the conditional expectation also requires fully parametric models. For example, if a zero-inflated model or hurdle model is needed, the conditional mean is no longer simply $\exp(\mathbf{x}'\boldsymbol{\beta})$, and it will vary with the parametric model for the probabilities. Chapter 3 focused on conditional mean modeling in which other aspects of the distribution, notably the variance, are modeled only to improve estimator efficiency. Chapter 4, by contrast, presents many parametric models in which consistent estimation of the conditional mean parameters requires fully parametric methods. To conclude, the attention given to features such as variance function modeling will vary on a case-by-case basis.

To make the foregoing discussion concrete, consider the issue of how to treat the joint presence of excess zero observations and of long right tails relative to the Poisson regression. One interpretation of this condition is that it indicates unobserved heterogeneity. Hence it is a problem of modeling the variance function. The Two Crossings Theorem of Chapter 4.2 supports this interpretation. An alternative interpretation is that the excess zeros reflect behavior. Many individuals do not record (experience) positive counts because they do not "participate" in a relevant activity. That is, optimizing behavior generates corner solutions. On this interpretation, the presence of excess zeros may well be a feature of the *conditional mean function, not the variance function.* When one adds to this the presence of unobserved heterogeneity, it is concluded that *both* the conditional expectation and variance function are involved. So the use of an overdispersed Poisson model, such as the negative binomial, without also allowing for the additional nonlinearity generated by an excess of corner solutions, will yield inconsistent estimates of the parameters.

The empirical illustrations in this chapter emphasize modeling the conditional distribution, not just the conditional mean.

6.2.2 Behavioral Interpretation

The Poisson and negative binomial distributions and extensions that are employed in this chapter can be viewed as reduced-form models. We do not attempt to give a structural interpretation to the results, though note that counts that arise from individual decision-making can be given a behavioral interpretation within the random utility framework used in binary choice models.

Consider a single probabilistic event such as a doctor consultation. Denote by U_1 the utility of seeking care, and by U_0 the utility of not seeking care. Both U_0 and U_1 are latent variables, modeled as

$$U_{1i} = \mathbf{x}_i'\boldsymbol{\beta}_1 + \varepsilon_{1i}$$
$$U_{0i} = \mathbf{x}_i'\boldsymbol{\beta}_0 + \varepsilon_{0i},$$

where $\mathbf{x}_i$ is the vector of individual attributes, and ε_{1i} and ε_{0i} are random errors. Then for individual i who seeks care, we have

$$U_{1i} > U_{0i} \quad \Rightarrow \quad \varepsilon_{0i} - \varepsilon_{1i} < \mathbf{x}_i'(\boldsymbol{\beta}_1 - \boldsymbol{\beta}_0).$$

Individual i seeks care when $U_{1i} > U_{0i}$, and we then observe $y_i = 1$; otherwise, we observe $y_i = 0$. The probability that $y_i = 1$, denoted π_i, is given by $\Pr[\varepsilon_{0i} - \varepsilon_{1i} < \mathbf{x}_i'(\boldsymbol{\beta}_1 - \boldsymbol{\beta}_0)]$. Thus, the probability of the decision to seek care is characterized by a binary outcome model.

Next consider repeated events of the same kind. If the events are independent and there is a fixed number, N, of repetitions, the event distribution is the usual binomial $\mathsf{B}[N, p]$. Suppose instead that N is random and follows the Poisson distribution. By application of the Poisson-stopped binomial result, see Chapter 1.1.4, the distribution of the number of successes will be Poisson distributed. This argument justifies the framework of count data models for the study of repeated events based on event counts.

This argument can be generalized in a straightforward manner to allow for serial correlation of events or unobserved heterogeneity, both of which imply overdispersion. The framework also generalizes to the multinomial case where the observed event is one of k outcomes. For example, the individual may choose to visit any one of k possible recreational sites, and such a choice outcome may be repeated a random number of times.

6.3 ANALYSIS OF DEMAND FOR HEALTH CARE

Count models are used extensively in modeling health care utilization. Discrete measures of health care use such as the number of doctor visits are often more easily available than continuous data on health expenditures. The measures are usually obtained from national health or health expenditure surveys that also provide information on key covariates, such as measures of health insurance, health status, income, education, and many sociodemographic variables.

This lengthy application begins with a discussion of the data, proposes a range of count models, and discusses methods to discriminate between the various models. Then the fitted models are compared on the basis of the various criteria, and the preferred model, a finite mixtures model, is analyzed in greater detail.

6.3.1 NMES Data

This example draws on Deb and Trivedi (1997), who modeled count data on medical care utilization by the elderly aged 66 years and over in the United States, using six mutually exclusive measures of utilization. The article compared the performance of negative binomial (NB), two-part hurdles negative binomial (NBH), and finite mixture negative binomial (FMNB) models, where in all cases both NB1 and NB2 versions of NB are specified, in a study of the demand for medical care.

A sample of 4,406 cases was obtained from the National Medical Expenditure Survey conducted in 1987 and 1988 (NMES). The data provide a comprehensive picture of how Americans use and pay for health services. Here only one measure, that dealing with the *office visits to physicians* (*OFP*), is considered. A feature of these data is that they *do not* include a high proportion of zero counts, but do reflect a high degree of unconditional overdispersion.

The NMES is based on a representative, national probability sample of the civilian, non-institutionalized population and individuals admitted to long-term care facilities during 1987. Under the survey design of the NMES, more than 38,000 individuals in 15,000 households across the United States were interviewed quarterly about their health insurance coverage, the services they used, and the cost and source of payments of those services. In addition to health care data, NMES provides information on health status and various socioeconomic measures.

Access to health insurance is an important determinant of health care use. In the sample analyzed here, all people are old enough to be covered by Medicare. This national social insurance program provides basic insurance coverage for hospital (Part A) and physician, outpatient home health, and preventive services (Part B). Dental, vision, hearing, long-term nursing care, and, at the time of these data, pharmaceutical drugs (except those administered in hospital) are not covered. Two forms of supplemental insurance are available. Medicaid, a government-provided program for those with low income, is not generous but covers some services not covered by Medicare. Private insurance can be obtained in a variety of ways, notably as a stand-alone Medigap policy, through retiree health insurance, or through an employer if the respondent or spouse is still working.

This last form of insurance may be an endogenous regressor, because those with high expected health needs (y) may be more likely to obtain supplementary insurance (x). Deb and Trivedi (1997) note that basic Medicare covers the bulk of health care costs for this group, that some private insurance is also employment-related, and that stand-alone Medigap policies provide strong financial incentives to choose it at the time of joining Medicare at age 65,

Table 6.1. *OFP visits: Actual frequency distribution (n = 4,406)*

Number of visits	0	1	2	3	4	5	6	7	8	9	10	11	12	13+
Frequency	683	481	428	420	383	338	268	217	188	171	128	115	86	500

rather than later as a consequence of a serious health event. These factors reduce the extent of any endogeneity. Private insurance is treated as exogenous here. Chapter 10.5 presents an empirical example where insurance is treated as endogenous.

Table 6.1 presents the frequency distribution of physician office visits. The zero counts account for only about 15% of the visits. There is a long right tail – 10% of the sample have 13 or more visits and 1% have more than 30. The first panel of Figure 6.1 gives a histogram, truncated at 30 visits, that suggests there are some excess zeroes, in addition to the long right tail.

Table 6.2 presents definitions and summary statistics for the explanatory variables. The health measures include self-perceived measures of health (*EXCLHLTH* and *POORHLTH*), the number of chronic diseases and conditions

Table 6.2. *OFP visits: Variable definitions and summary statistics*

Variable	Definition	Mean	Standard deviation
OFP	Number of physician office visits	5.77	6.76
OFNP	Number of nonphysician office visits	1.62	5.32
OPP	Number of physician outpatient visits	0.75	3.65
OPNP	Number of nonphysician outpatient visits	0.54	3.88
EMR	Number of emergency room visits	0.26	0.70
HOSP	Number of hospitalizations	0.30	0.75
EXCLHLTH	Equals 1 if self-perceived health is excellent	0.08	0.27
POORHLTH	Equals 1 if self-perceived health is poor	0.13	0.33
NUMCHRON	Number of chronic conditions	1.54	1.35
ADLDIFF	Equals 1 if the person has a condition that limits activities of daily living	0.20	0.40
NOREAST	Equals 1 if the person lives in northeastern U.S.	0.19	0.39
MIDWEST	Equals 1 if the person lives in the midwestern U.S.	0.26	0.44
WEST	Equals 1 if the person lives in the western U.S.	0.18	0.39
AGE	Age in years divided by 10	7.40	0.63
BLACK	Equals 1 if the person is African American	0.12	0.32
MALE	Equals 1 if the person is male	0.40	0.49
MARRIED	Equals 1 if the person is married	0.55	0.50
SCHOOL	Number of years of education	10.29	3.74
FAMINC	Family income in $10,000s	2.53	2.92
EMPLOYED	Equals 1 if the person is employed	0.10	0.30
PRIVINS	Equals 1 if the person is covered by private health insurance	0.78	0.42
MEDICAID	Equals 1 if the person is covered by Medicaid	0.09	0.29

(*NUMCHRON*), and a measure of disability status (*ADLDIFF*). To control for regional differences, we use regional dummies *NOREAST*, *MIDWEST*, and *WEST*. The demographic variables include *AGE*, race (*BLACK*), gender (*MALE*), marital status (*MARRIED*), and education (*SCHOOL*). The economic variables are family income (*FAMINC*) and employment status (*EMPLOYED*). Finally, the insurance variables are indicator variables for supplementary private insurance (*PRIVINS*) and security-net public insurance (*MEDICAID*). *MEDICAID* is available to low-income individuals only and should not be confused with Medicare, which is available to all those over age 65.

6.3.2 Models for Demand for Medical Care

The data are very highly overdispersed with variance of $6.76^2 = 45.7$, that is eight times the mean of 5.77. So we begin with NB1 and NB2 regression models.

Several empirical studies of health utilization suggest that a better starting point is the two-part hurdle model of Chapter 4.5 (Pohlmeier and Ulrich, 1995; Gurmu, 1997; Geil et al., 1997). It is typically superior to specifications in which the two separate origins of zero observations are not recognized. The hurdle model may be interpreted as a principal-agent type model in which the first part specifies the individual's decision to seek care and the second part models the subsequent number of visits recommend by the doctor.

An alternative model, the finite mixture model, has several additional attractive features. It allows for additional population heterogeneity, but avoids the sharp dichotomy between the population of "users" and "non-users" embodied in the hurdle model. The latent class interpretation of the model splits the population into classes that may not be well captured by proxy variables such as self-perceived health status and chronic health conditions. A two-component finite mixture model can lead to a dichotomy between latent "healthy" and "ill" groups, whose demands for health care are characterized by, respectively, low mean and low variance, and high mean and high variance, for a given level of the exogenous regressors.

6.3.3 NB, Hurdle, and Finite Mixtures Models

Three fully parametric models are compared in this subsection.

The first model is the NB model of Chapter 3.3 with conditional mean $\mathsf{E}[y_i|\mathbf{x}_i] = \mu_i = \exp(\mathbf{x}_i'\beta)$. The variance is $\mathsf{V}[y_i|\mathbf{x}_i] = \mu_i + \alpha\mu_i$ for the NB1 model and is $\mathsf{V}[y_i|\mathbf{x}_i] = \mu_i + \alpha\mu_i^2$ for the NB2 model, where $\alpha > 0$ is an overdispersion parameter. The density for the NB2 model is

$$f(y_i|\mathbf{x}_i, \boldsymbol{\beta}, \alpha) = \frac{\Gamma(y_i+\alpha^{-1})}{\Gamma(\alpha^{-1})\Gamma(y_i+1)} \left(\frac{\alpha^{-1}}{\alpha^{-1}+\mu_i}\right)^{\alpha^{-1}} \left(\frac{\mu_i}{\alpha^{-1}+\mu_i}\right)^{y_i} \tag{6.1}$$
$$\mu_i = \exp(\mathbf{x}_i'\boldsymbol{\beta}),$$

whereas the NB1 model replaces all occurrences of α^{-1} in (6.1) with $\alpha^{-1}\mu_i$.

The second model is a hurdle negative binomial model (NBH); see section 4.5. Then

$$\Pr[y_i = k|\mathbf{x}_i] = \begin{cases} f_1(0|\mathbf{x}_i) & \text{if } k = 0 \\ \dfrac{1 - f_1(0|\mathbf{x}_i)}{1 - f_2(0|\mathbf{x}_i)} f_2(k|\mathbf{x}_i) & \text{if } k > 0. \end{cases} \tag{6.2}$$

Deb and Trivedi (1997) followed Mullahy (1986) and used an NB model for both parts. So the first component is a binary model for $\Pr[y = 0]$, where y is NB distributed, so for NB2

$$f_1(0|\mathbf{x}_i) = \Pr[y_i = 0|\mathbf{x}_i] = (1 + \alpha_1 \exp(\mathbf{x}_i'\boldsymbol{\beta}_1))^{-1/\alpha}, \tag{6.3}$$

and for NB1 replace both occurrences of α_1^{-1} in (6.3) with $\alpha_1^{-1}\exp(\mathbf{x}_i'\boldsymbol{\beta}_1)$. For the second component, the base density for the positive counts $f_2(y|\mathbf{x}_i)$ is an NB density $f(y_i|\mathbf{x}_i, \boldsymbol{\beta}_2, \alpha_2)$ as defined in (6.1) for the NB2, while NB1 replaces all occurrences of α_2^{-1} with $\alpha_2^{-1}\mu_{2i}$. The same regressors $\mathbf{x}_i$ are used in the two parts of this model. This model has twice as many parameters as an NB model.

The third model is the C-component finite mixture density, see Chapter 4.8, specified as

$$f(y_i|\boldsymbol{\theta}_1, \ldots, \boldsymbol{\theta}_j, \pi_1, \ldots, \pi_{C-1}) = \sum_{j=1}^{C} \pi_j f_j(y_i|\boldsymbol{\theta}_j), \tag{6.4}$$

where $\pi_1 \geq \pi_2 \geq \ldots . \geq \pi_C$, $\pi_C = (1 - \sum_{j=1}^{C-1} \pi_j)$, are mixing probabilities estimated along with all other parameters, and $\boldsymbol{\theta}_j$ are the parameters of the component densities $f_j(y|\boldsymbol{\theta}_j)$. The finite mixture negative binomial model (FMNB-C) with C components specifies the components densities $f_j(y_i|\boldsymbol{\theta}_j) = f(y_i|\boldsymbol{\beta}_j, \alpha_j)$, $j = 1, \ldots, C$, to be NB densities. Again the NB2 density is defined in (6.1), while the NB1 replaces all occurrences of α_j^{-1} with $\alpha_j^{-1}\mu_i$. The same regressors $\mathbf{x}_i$ are used in each component of this model.

A variant of the third model constrains slope coefficients to be equal. Thus letting $\mathbf{x}'\boldsymbol{\beta} = \alpha + \mathbf{x}'\boldsymbol{\gamma}$, we have

$$\mu_{ij} = \exp(\alpha_j + \mathbf{x}_i'\boldsymbol{\gamma}_j) = \exp(\alpha_j)\exp(\mathbf{x}_i'\boldsymbol{\gamma}),$$

so only the intercept varies across model. This model is called a constrained FMNB-C model, denoted CFMNB-C. If $\dim[\boldsymbol{\beta}] = k$, this model has C intercepts, $(k - 1)$ slopes, C overdispersion parameters, and $C - 1$ mixing probability parameters for a total of $k + 3C$ parameters, which is not much more than NB with $k + 1$ parameters. By contrast the unconstrained FMNB-C has an additional $(C - 1)(k - 1)$ parameters than CFMNB-C due to an additional $(C - 1)k$ slope parameters. This can be a big difference for the example here, with 17 regressors including the intercept. In addition, the constrained version of this model, as noted in Chapter 4.8.1, can be interpreted as providing a discrete specification of unobserved heterogeneity.

6.3.4 Model Selection and Comparison

Before interpreting model estimates, we first choose between the various possible models on the basis of hypothesis tests, where possible, and information criteria. We also use goodness-of-fit criteria to evaluate whether the preferred model(s) provides a good fit to the data. In the case of finite mixture models, it is presumed that there is a finite mixture; consideration of methods to determine whether there is a finite mixture is deferred to the next subsection.

The analysis here is fully parametric. The obvious method to use to discriminate between models is a likelihood ratio test. This is possible when models are nested and the LR test statistic has the usual $\chi^2(p)$ distribution, where p is the difference in the number of parameters in the model. Here this can be applied for testing NB against NBH, in which case the test statistic is $\chi^2(k+1)$ distributed, and for testing CFMNB-C against FMNB-C, in which case the test statistic is $\chi^2(Ck-k)$ distributed.

Even though a C-component model is nested in a (C–1)-component model, the standard LR test cannot be applied, because the null hypothesis is on the boundary of the parameter space; see Chapter 4.8.5. For example, the two-component mixture density $\pi f(y|\boldsymbol{\theta}_1)+(1-\pi)f(y|\boldsymbol{\theta}_2)$ reduces to a one-component model if either $\pi = 1$ or $\boldsymbol{\theta}_1 = \boldsymbol{\theta}_2$. Thus the standard LR test cannot be applied to test NB against FMNB-C or to test FMNB (CFMNB) with $C-1$ components against C components.

Comparison of any model with NB1 density against any model with NB2 density necessarily involves comparing nonnested models, as does comparing NBH models against FMNB and CFMNB models. Then it is standard to use the information criteria presented in Chapter 5.4.1. We use the Akaike information criterion (AIC) and the Bayesian information criterion (BIC). Note that if the only difference between any two models is whether NB is specified as NB1 or NB2, the two models will have the same number of parameters so that smaller AIC or BIC simply means higher fitted log-likelihood.

Böhning et al. (1994) have used simulation analysis to examine the distribution of the likelihood ratio test for several distributions involving a boundary hypothesis. This showed that the use of the usual χ^2 critical values under-rejects the false null, leading to finite mixture models with a value of C that is too small. Leroux (1992) proves that under regularity conditions the maximum penalized likelihood approach (e.g., the use of AIC or BIC) leads to a consistent estimator of the true finite mixture model. This means that as $n \to \infty$ the number of components will not be underestimated.

A possible strategy for model selection can be summarized as follows:

- Fix maximum $C = C^*$. Use AIC and BIC to compare the sequence of models FMNB-C*, ..., FMNB-1.
- Use the LR test to compare FMNB-C with CFMNB-C.
- Use the LR test to compare NBH and NB.
- Use AIC and BIC to compare FMNB-C (or CFMNB-C) with NBH.

Table 6.3. *OFP visits: Likelihood ratio tests*

Null model	Alternative	NB1	NB2	5% Critical Value
NB	NBH	59.8	183.4	$\chi^2(17) = 28.0$
NB	CFMNB-2	116.7	106.7	$\chi^2(3) = 7.8$
NB	FMNB-2	166.9	136.0	$\chi^2(19) = 30.1$
CFMNB-2	FMNB-2	50.2	29.2	$\chi^2(16) = 26.3$
CFMNB-2	CFMNB-3	0.002	0.003	$\chi^2(3) = 7.8$
FMNB-2	FMNB-3	11.3	0.002	$\chi^2(19) = 30.1$

Note: The LR test statistic for the null against the alternative is shown.

This strategy can be used separately for models based on NB1 and models based on NB2. To compare any NB1 model with any NB2 model, the models are nonnested, but models in the same class have the same degrees of freedom so can be simply chosen on the basis of the fitted log-likelihood.

Finally, to evaluate the goodness of fit of the model selected after this multilevel comparison of models, one may use the chi-square diagnostic test introduced in Chapter 5.3.4. It compares actual and fitted cell frequencies of events where cells are formed for the integer values of y (or a range of values of y). Let $\overline{p}(y)$ denote the actual frequency of the value y in the sample. And let $\hat{p}(y) = \frac{1}{n}\sum_{i=1}^{n} f(y|\mathbf{x}_i, \hat{\boldsymbol{\theta}})$ denote the fitted probability of the value y. Stack $\overline{p}(y)$ into the vector $\overline{\mathbf{p}}$. For example, if the cells are formed for y taking values 0, 1, 2, and 3 or more, then $\overline{\mathbf{p}}$ has entries $\overline{p}(0)$, $\overline{p}(1)$, $\overline{p}(2)$, and $1 - \overline{p}(0) - \overline{p}(1) - \overline{p}(2)$. Similarly stack $\hat{p}(y)$ into the vector $\hat{\mathbf{p}}$. Then the chi-square goodness-of-fit statistic is

$$\mathsf{T}_{\mathsf{GoF}} = (\overline{\mathbf{p}} - \hat{\mathbf{p}})'\hat{\mathbf{V}}^{-}(\overline{\mathbf{p}} - \hat{\mathbf{p}}), \tag{6.5}$$

where $\hat{\mathbf{V}}$ is a consistent estimate of the asymptotic variance matrix of $(\overline{\mathbf{p}} - \hat{\mathbf{p}})$, and $\hat{\mathbf{V}}^{-}$ is the Moore-Penrose generalized-inverse of $\hat{\mathbf{V}}_m$. The generalized-inverse is used because the $J \times J$ matrix $\mathbf{V}_m$ may not be of full rank. Under the null hypothesis that the density $f(y|\mathbf{x}_i, \boldsymbol{\theta})$ is correctly specified, the test statistic in this application is $\chi^2(q-1)$ distributed where q is the number of cells used in the test. This statistic can be computed using the auxiliary regression presented in Chapter 5.3.4.

The proposed model comparison and selection approach is subject to the usual criticism that it fails to control for pretest bias. If data availability permits, this approach should be used on a training sample. Final results should then be based on the preferred model, reestimated using a sample that is independent of the training sample.

Table 6.3 presents likelihood ratio tests of NB versus NBH, the unicomponent model versus the two-component models (CFMNB-2 and FMNB-2), and the two-component models versus the corresponding three-component models. The NB model is rejected in favor of NBH for both NB1 and NB2 models. The NB model is also rejected in favor of two-component or three-component

Table 6.4. *OFP visits: Information criteria (AIC and BIC)*

NB1 or NB2	Model	k	lnL	AIC	BIC	T_{GoF}
NB1	NB	18	−12156	24348	24463	32.3
	NBH	36	−12126	24323	24546	37.4
	CFMNB-2	21	−12096	24238	**24372**[c,d]	6.0
	FMNB-2	37	−12073	**24220**[a,b]	24456	11.2
	CFMNB-3	24	−12098	24244	24397	–
	FMNB-3	56	−12066	24246	24604	74.3
NB2	NB	18	−12202	24440	24555	58.0
	NBH	36	−12108	**24291**[a]	24515	21.8
	CFMNB-2	21	−12149	24340	**24474**[c]	98.0
	FMNB-2	37	−12134	24342	24579	123.8
	CFMNB-3	24	−12149	24346	24499	–
	FMNB-3	56	−12133	24380	24738	137.0

Note:
[a] Model preferred by the AIC within NB1 class or NB2 class.
[b] Model preferred by the AIC overall.
[c] Model preferred by the BIC within NB1 class or NB2 class.
[d] Model preferred by the BIC overall.
[e] T_{GoF} is the χ^2 (5) goodness-of-fit test.

mixture models, even though the misuse of the usual critical values for the LR test in this nonstandard setting means that we have used conservative critical values.

Though there are problems with the LR test for choosing between two- and three-component mixtures, it is interesting to note that we do not get a significant LR statistic for any pairwise comparison between two- and three-component finite mixture, regardless, whether the NB1 or NB2 mixtures are used. Essentially adding a third component leads to an improvement in the log-likelihood only in the case of FMNB for NB1. Finally, within the two-component models, the LR test of CFMNB-2 against FMNB-2 model leads to rejection of the null model at 5% significance level for both NB1 and NB2.

Table 6.4 presents values of the log-likelihood, AIC, and BIC for all the models. For any model the NB1 version has higher log-likelihood than the NB2 version. For both NB1 and NB2, the three-component mixture models lead to very little gain in the log-likelihood compared to the two-component mixtures. For NB1 models the two-component mixture models are preferred to the hurdle or standard model. The AIC criterion favors FMNB-2, whereas the BIC criterion, which has a much bigger penalty for model size, favors the more parsimonious constrained specification CFMNB-2. Finally for NB2 models NBH is preferred using AIC, and CFMNB-2 is preferred using BIC.

Clearly a two-component finite mixture NB1 specification is preferred. Based on AIC and BIC there is some ambiguity about whether to use the constrained version, but since the constrained version was rejected using an LR test the preferred model is the NB1 version of FMNB-2. This evidence in favor of a two-component mixture also allows us to interpret the two populations as being healthy and ill.

6.3.5 Is There a Mixture?

Specification and ML estimation of the NB and NBH models are straightforward. By contrast, there are important issues in evaluating fitted finite mixture models. First, is there evidence for the presence of more than one component? That is, is mixing present? Second, is a global maximum attained in estimation? These issues were not fully addressed in Chapter 4.8.

Lindsay and Roeder (1992) developed diagnostic tools for checking these properties for the case of exponential mixtures. The key idea behind the diagnostic for the presence of a mixture component is the Two Crossings Theorem in Chapter 4.2.4.

Let $f(y_i|\mathbf{x}_i, \boldsymbol{\theta})$ denote the density of a one-component (or nonmixture) model and $f(y_i|\mathbf{x}_i, C) = f(y_i|\mathbf{x}_i, \boldsymbol{\theta}, \boldsymbol{\pi})$ denote the density of a C-component mixture model. Then the fitted cell frequencies for the one-component and C-component models are, respectively,

$$\hat{p}(y|\hat{\boldsymbol{\theta}}) = \tfrac{1}{n}\textstyle\sum_{i=1}^{n} f(y|\mathbf{x}_i, \hat{\boldsymbol{\theta}}),\ y = 0, 1, 2, \ldots$$

$$\hat{p}(y|\hat{C}) = \tfrac{1}{n}\textstyle\sum_{i=1}^{n} f(y|\mathbf{x}_i, \hat{C}),\ y = 0, 1, 2, \ldots$$

By an extension of Shaked's Two Crossings Theorem, the differences $(\hat{p}(y|\hat{\boldsymbol{\theta}}) - \hat{p}(y|\hat{C}))$ or equivalently, $((\hat{p}(y|\hat{\boldsymbol{\theta}})/\hat{p}(y|\hat{C}) - 1)$ will show a $\{+, -, +\}$ sign pattern.

This is the basis of the first of two diagnostic tools developed by Lindsay and Roeder. Their *directional gradient* is

$$d(y, \hat{C}, \hat{\boldsymbol{\theta}}) = \frac{\hat{p}(y|\hat{\boldsymbol{\theta}})}{\hat{p}(y|\hat{C})} - 1, \qquad y = 0, 1, 2, \ldots. \tag{6.6}$$

Convexity of the graph of $d(y, \hat{C}, \hat{\boldsymbol{\theta}})$ against y, ideally with confidence bands, is interpreted as evidence in favor of a mixture. Note, however, that such convexity may be observed for more than one value of C, leaving open the issue of which value to select for C. If instead a plot of $d(y, \hat{C}, \hat{\boldsymbol{\theta}})$ against y is a horizontal line though zero, then a one-component model is favored.

Lindsay and Roeder also suggest the use of the *gradient function*

$$D(\hat{C}, \boldsymbol{\theta}) = \sum_{y \in S} d(y, \hat{C}, \boldsymbol{\theta}) p(y), \tag{6.7}$$

where $d(y, \hat{C}, \boldsymbol{\theta})$ is evaluated at a range of values of $\boldsymbol{\theta}$, S is the set of distinct values taken by y in the sample, and $p(y) = \frac{1}{n}\sum_{i=1}^{n} \mathbf{1}[y_i = y]$ is the proportion of the sample observations taking value y. This is simply the sample average of $d(y_i, \hat{C}, \boldsymbol{\theta})$ and can be interpreted as a quantitative measure of the deviation in fitted cell probabilities induced by the mixture as $\boldsymbol{\theta}$ varies. The mixing distribution is an ML estimator if $D(\hat{C}, \boldsymbol{\theta}) \leq 0$ for all $\boldsymbol{\theta}$ and $D(\hat{C}, \boldsymbol{\theta}) = 0$ for all

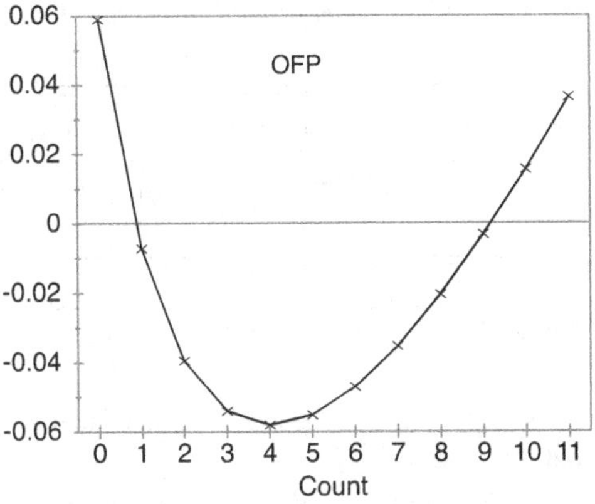

Figure 6.2. OFP visits: Directional gradients.

$\boldsymbol{\theta}$ in the support of $\hat{C}$ (Lindsay and Roeder, 1992, p. 787). In the case of scalar θ one can plot $D(\hat{C}, \theta)$ against θ. A limitation of the gradient function is that it can be overly influenced by tail behavior that is noisily estimated, making the components $d(y, \hat{C}, \boldsymbol{\theta})$ erratic and $D(\hat{C}, \boldsymbol{\theta})$ hard to interpret. Recognizing this feature, Lindsay and Roeder suggest that a truncated gradient statistic is preferred for unbounded densities, but there is little guidance on how the statistic should be truncated.

The diagnostic tools of Lindsay and Roeder are intended for exponential mixtures. Strictly speaking, they do not apply to NB mixtures (though the two crossings result is extended to two-parameter exponential families by Gelfand and Dalal, 1990). In this chapter the directional gradient is nonetheless applied to the FMNB-C model in a heuristic and exploratory fashion.

Figure 6.2 presents the directional gradient function defined in (6.6) for FMNB-2 compared with NB for NB1 models. This appears to satisfy the convexity requirement.

6.3.6 Component Means and Probabilities in the Finite Mixtures Model

Given selection of a preferred model, the NB1 version of the two-component finite mixtures model (FMNB-2), we next evaluate it. In this subsection we evaluate the predicted probabilities and fitted means, both overall and separately, for the two components. In the next subsection we evaluate the role of individual regressors.

Goodness of fit: Table 6.5 presents the sample frequency distribution of the count variable along with the sample averages of the estimated cell frequencies from the selected model. The discrepancy between the actual and fitted cell frequencies is never greater than 1%. Discrepancies between actual and predicted frequencies based on NB and NBH models (not presented) are usually much

Table 6.5. *OFP visits: FMNB-2 NB1 model, actual and fitted distributions, and goodness-of-fit test*

Count	0	1	2	3	4	5	6	7	8	9	10+
Actual	15.5	10.9	9.7	9.5	8.7	7.7	6.1	4.9	4.3	3.9	18.9
Fitted	15.1	11.5	10.5	9.4	8.2	7.0	6.0	5.1	4.2	3.6	19.1
T_{GoF}						14.80					

Note: The fitted frequencies are the sample averages of the cell frequencies estimated from the FMNB-2 NB1 model. T_{GoF} is the $\chi^2(13)$ goodness-of-fit test.

larger. The $\chi^2(13)$ goodness-of-fit statistic, based on cells of $0, 1, \ldots, 12$ and 13 or more, does not reject the model.

Component means: The probability π_1 for the first component is 0.91, which is large relative to its estimated standard error of 0.02 (see Table 6.7). This reinforces the evidence supporting the two-population hypothesis.

The individual means in the two-component NB1 model are $\mu_{ji} = \exp(\mathbf{x}_i'\boldsymbol{\beta}_j)$, and the individual variances are $\sigma_{ji}^2 = \mu_{ji}(1+\alpha_j)$, for $j = 1, 2$, and $i = 1, \ldots, n$. Then for each individual the fitted mean across the two components is

$$\mathsf{E}[y_i|\mathbf{x}_i] = \overline{\mu}_i = \sum_{j=1}^{2} \pi_j \mu_{ji}, \tag{6.8}$$

where here $\pi_1 = 0.91$ and $\pi_2 = 0.09$. The model implies an overall variance of

$$\mathsf{V}[y_i|\mathbf{x}_i] = \sum\nolimits_{j=1}^{2} \pi_j[\mu_{ji}^2 + \sigma_{ji}^2] - \left(\sum\nolimits_j \pi_j \mu_{ji}\right)^2 \tag{6.9}$$

$$= \left(\sum\nolimits_{j=1}^{2} \pi_j \mu_{ji}^2(1 + \alpha_j \mu_{ji}^{p-2})\right) + \overline{\mu}_i - \overline{\mu}_i^{\,2}, \tag{6.10}$$

where $p = 1$ for NB1 and $p = 2$ for NB2.

Table 6.6 summarizes the variability in the two components. The first component (with $\pi_1 = 0.91$) clearly captures healthier low-use individuals who have on average 5.6 visits to a physician and a variance that is five times the

Table 6.6. *OFP visits: FMNB-2 NB1 model fitted means and variances*

	π_1	Mean	Variance
Low-use group	0.912	5.54	24.88
High-use group	0.088	8.18	142.23
Mixture		5.77	41.36

Note: The mixture mean and variance are calculated using (6.8) and (6.9), respectively.

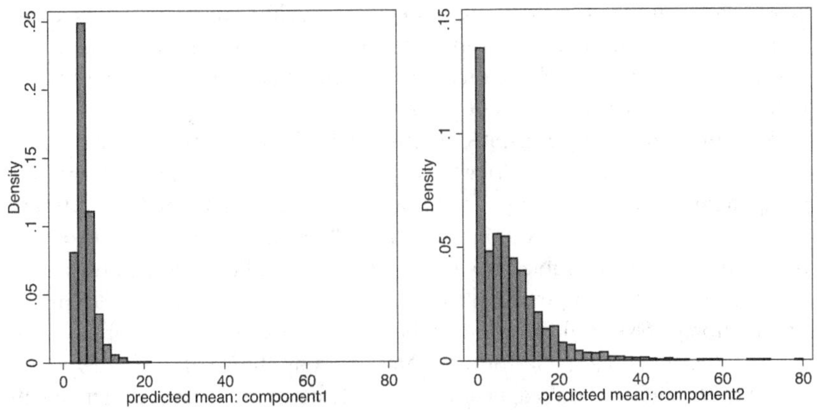

Figure 6.3. Comparison of the fitted means for the two latent classes.

mean. The remaining individuals seek care on average 8.2 times and have a fitted variance that is close to 20 times the mean. The overall mixture mean equals the sample mean of 5.77, and the overall mixture variance of 35.79 is close to the sample variance of 45.05.

Figure 6.3 presents histograms of the fitted means $\hat{\mu}_{1i}$ and $\hat{\mu}_{2i}$. These show clearly that the first component captures individuals who are relatively low users. The second component has both a higher mean and a longer right tail.

Component probabilities: It is also interesting to look at the fitted probabilities from the two-component densities. Figure 6.4 shows a greater proportion of both zeros and high values for the ill population. It appears that, although most healthy individuals see a doctor a few times a year for health maintenance and minor illnesses, a larger fraction of the ill do not seek any (preventive?) care. Those ill individuals who do see a doctor (these individuals have

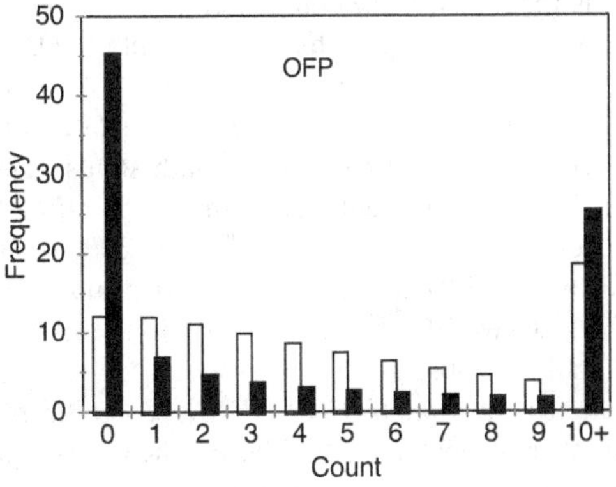

Figure 6.4. Component densities from the FMNB-2 NB1 model.

sickness-events) do so much more often than healthy individuals. Since preventive care usually takes the form of a visit to a doctor in an office setting, one would not expect such a pattern of differences between the healthy and ill to arise for the other measures of utilization.

Sensitivity to outliers: As noted in Chapter 4.8.7, when finite mixture models are applied to counts with long right tails there is a danger that spurious components are added simply to fit these outliers, leading to model overfitting. With this in mind all observations of $OFP > 70$ were replaced by $OFP = 70$ in the results detailed in Tables 6.5–6.7. This change led to just one observation (with $OFP = 89$) being recoded to 70 (slightly more than the second highest observation of 68). With this change the mean for the low-use group was 5.55 visits, rather than 5.21, and for the high-use group the mean was 8.18, rather than 11.79. Thus, it would appear that a single observation could dramatically increase the estimated differences between the components. Several of the regression coefficients in the high-use group also changed substantially. With this change the log-likelihood was −12,077 rather than −12,092, and predicted frequencies of counts of 0, 1, and 2 were closer to sample frequencies. By contrast, for the standard NB1 model the effect was much smaller, with log-likelihood of −12,153 when $OFP = 89$ was recoded to 70 compared to −12,156 without recoding.

6.3.7 The Role of Regressors in the Finite Mixtures Model

Table 6.7 presents estimated coefficients and their standard errors for the preferred model, the NB1 version of the two-component finite mixtures model. The following discussion underscores the advantage of a finite mixture formulation, because it highlights the differential response of different groups to changes in covariates.

Semi-elasticities: Since the conditional mean in each of the two components is of exponential functional form, the regression coefficients in each component can be interpreted as semi-elasticities. In interpreting the results, bear in mind that the coefficients for the high-use component are much less precisely estimated, since the probability of being in this group is only 0.09.

- Consistent with previous studies, both the number of chronic conditions (*NUMCHRON*) and self-perceived health status (*EXCLHLTH* and *POORHLTH*) are important determinants of *OFP*. An additional chronic condition increases office visits by 20% in the low-use (healthy) group and by 14% in the high-use (ill) group. The presence of *EXCLHLTH* reduces *OFP* by around 75% in the high-use group and only 25% in the low-use group.
- Medicaid insurance coverage (*MEDICAID*) is a significant determinant of the number of doctor visits. The coefficient is significantly *positive* in the component density for the low-use group. For the high-use group the coefficient is large and negative, though is statistically

Table 6.7. *OFP visits: FMNB-2 NB1 model estimates and* t*-ratios*

	OFP			OFP	
Variable	Low users	High users	Variable (cont.)	Low users	High users
EXCLHLTH	−0.250	−0.751	*MARRIED*	0.047	−0.513
	(4.31)	(1.02)		(1.34)	(1.55)
POORHLTH	0.234	0.039	*SCHOOL*	0.014	0.152
	(3.59)	(0.06)		(2.73)	(1.84)
NUMCHRON	0.186	0.139	*FAMINC*	−0.000	−0.005
	(14.62)	(1.33)		(0.06)	(0.27)
ADLDIFF	−0.016	0.546	*EMPLOYED*	−0.059	0.372
	(0.37)	(2.04)		(1.04)	(0.54)
NOREAST	0.083	0.178	*PRIVINS*	0.254	3.020
	(1.72)	(0.37)		(4.79)	(1.12)
MIDWEST	0.018	0.042	*MEDICAID*	0.353	−3.147
	(0.46)	(0.12)		(5.69)	(0.97)
WEST	0.089	0.236	*ONE*	0.778	1.862
	(1.82)	(0.49)		(3.48)	(0.71)
AGE	0.028	−0.611	α	3.492	16.390
	(1.07)	(2.47)		(15.44)	(6.86)
BLACK	−0.079	−1.068	π	0.912	0.088
	(1.12)	(0.98)		(32.65)	(3.16)
MALE	−0.136	0.120			
	(3.92)	(0.46)	$-\ln L$	12077	

insignificant. For the healthy group, the price (insurance) effect of the *MEDICAID* dummy outweighs the income effect because *MEDICAID* is a health insurance plan for the poor.

- Persons with supplementary private insurance (*PRIVINS*) seek care from physicians in office more often than individuals without supplementary coverage. The effect is extraordinarily large for the high-use group, with a coefficient of 3.02 that implies a proportional increase of $e^{3.02} - 1$. For the low-use group the coefficient is much lower, but the coefficient of 0.25 still implies that having private insurance increases *OFP* by approximately 25% in the low-use group.
- The effect of income on *OFP* usage is negligible. Furthermore, being employed (*EMPLOYED*), which may also capture income effects, is statistically insignificant. One explanation for the negligible income effect is that the overall generosity of Medicare irrespective of family income, combined with Social Security income, which guarantees a certain level of income, makes utilization insensitive to marginal changes in income.

Marginal effects: The preceding discussion focused on interpreting coefficients as semielasticities. The marginal effect for the jth component with respect to the kth regressor is

$$\mathsf{ME}_{ijk} = \frac{\partial \mathsf{E}_j[y_i|\mathbf{x}_i]}{\partial x_{ik}} = \beta_{jk}\mu_{ji}, \tag{6.11}$$

so the predicted changes in *OFP* in the high-use group will be larger, for given size of β_{jk}, since μ_{ij} is larger for the high-use group.

The total effect of a change in a regressor on *OFP* is obtained as the weighted average of the component marginal effects, where the weights are the component probabilities. Thus the marginal effect of a change in the kth regressor is

$$\mathsf{ME}_{ik} = \frac{1}{n}\sum_{j=1}^{2} \mathsf{ME}_{ijk} = \pi_1\beta_{1k}\mu_{1i} + (1-\pi_1)\beta_{2k}\mu_{2i}, \tag{6.12}$$

and the average marginal effect $\mathsf{AME}_k = \frac{1}{n}\sum_{i=1}^{n} \mathsf{ME}_{ik}$.

Without assigning individuals to their respective subpopulations, there is no obvious way to compare marginal effects in the subpopulations; a convention is required for doing so. One that has been used in the literature calculates the AME parameter using the full sample, evaluated separately for each set of component-specific coefficients. Equivalently, given (6.11), one can simply use $\frac{1}{n}\sum_{i=1}^{n} \hat{\beta}_{jk}\hat{\mu}_{ji}$, where $\hat{\mu}_{ji} = \exp(\mathbf{x}_i'\hat{\boldsymbol{\beta}}_j)$.

Another alternative is to make a similar comparison using the MEM parameter. We apply the first approach to our preferred FMNB model. The results help gauge differences in responses in the two groups. For example, for the insurance variable *PRIVINS* the marginal effect in the low-use group is 1.32 visits, whereas it is 8.09 visits in the high-use group.

Overdispersion: The fitted variance in the low-use group is $\mu_{1i}(1+3.49) = 4.49\mu_{1i}$ and for the high-use group is $17.39\mu_{2i}$. So *OFP* is much more dispersed for the high-use group because of both the higher mean μ_{2i} and the higher multiplier of 17.39.

6.3.8 Comparison with the Hurdle Model

The finite mixture model results may be compared with those for the more commonly used hurdle model.

Table 6.8 presents the results from the NB2 version of the NBH model. In a departure from the analysis of Deb and Trivedi (1997) we follow the now more common approach of estimating a logit model for the hurdle, rather than the NB binary model given in (6.3). This leads to a slightly higher log-likelihood of $-12{,}110$ compared to $-12{,}108$ in Table 6.4.

The key regressors for the FMNB-2 model are also the most important for the NBH model. Both the probability of crossing the hurdle and the number of positive visits are decreasing with excellent health (*EXCLHLTH*), increasing with the number of chronic conditions (*NUMCHRON*), increasing with

Table 6.8. *OFP visits: NB2 hurdle model estimates and* t*-ratios*

	Zeros		Positives	
Variable	Coefficient	$\lvert t \rvert$	Coefficient	$\lvert t \rvert$
EXCLHLTH	−0.329	2.31	−0.378	4.32
POORHLTH	0.071	0.42	0.333	5.86
NUMCHRON	0.557	10.55	0.143	10.59
ALDIFF	−0.188	1.45	0.129	2.50
NOREAST	0.129	1.03	0.104	1.97
MIDWEST	0.101	0.88	−0.016	0.34
WEST	0.202	1.51	0.123	2.45
AGE	0.190	2.35	−0.075	2.33
BLACK	−0.327	2.45	0.002	0.02
MALE	−0.464	4.71	0.004	0.10
MARRIED	0.247	2.38	−0.092	2.10
SCHOOL	0.054	4.11	0.022	3.82
FAMINC	0.007	0.36	−0.002	0.38
EMPLOYED	−0.012	0.07	0.030	0.40
PRIVINS	0.762	6.50	0.227	4.01
MEDICAID	0.554	3.05	0.185	2.78
ONE	−1.475	2.28	1.631	6.06
α			0.744	18.40
−ln L	12110			
T_{GoF}	21.86			

insurance (*MEDICAID* and *PRIVINS*), and invariant with respect to income (*INCOME*) and employment (*EMPLOYED*).

The NBH and FMNB models can be compared on the basis of average marginal effects. The expression for the conditional mean of the hurdle model is given in Chapter 4.5.1. Suppressing the individual subscript, the marginal effect with respect to a change in the kth regressor is

$$\begin{aligned}\frac{\partial \mathsf{E}[y|\mathbf{x}]}{\partial x_k} &= \partial(\Pr[y > 0|\mathbf{x}]\mathsf{E}[y|\mathbf{x}, y > 0])/\partial x_k \\ &= \mathsf{E}[y|\mathbf{x}, y > 0]\frac{\partial \Pr[y > 0|\mathbf{x}]}{\partial x_k} + \Pr[y > 0|\mathbf{x}]\frac{\partial \mathsf{E}[y|\mathbf{x}, y > 0]}{\partial x_k}\end{aligned} \tag{6.13}$$

where $\mathsf{E}[y|\mathbf{x}, y > 0] = \mu_2(\mathbf{x})(1 - f_1(0|\mathbf{x}))/(1 - f_2(0|\mathbf{x}))$, $\mu_2(\mathbf{x}) = \exp(\mathbf{x}'\boldsymbol{\beta}_2)$, and $\Pr[y > 0|\mathbf{x}] = 1 - f_1(0|\mathbf{x})$. The two terms in (6.13) reflect, respectively, the direct effect due to an individual moving from a non-user to user category and the indirect effect on the usage of those already in the user category. Note that the calculation of this expression involves both parts of the hurdle model, since the truncation probability is given by the binary ('zero') part of the model and the truncated mean by the "positive part" of the model. Calculation is more complicated than for the FMNB-2 model. A suitable overall measure of partial response is the sample average marginal effect $n^{-1}\sum_i \partial \mathsf{E}[y_i|\mathbf{x}_i]/\partial x_{ik}$.

6.3.9 Economic Significance

Once a finite mixture model has been estimated, the posterior probability that observation i belongs to category j can be calculated for all (i, j) pairs using (4.81). Each observation may then be assigned to its highest probability class. The crispness of the resulting classification by "types" or "groups" will vary in practice depending on the differences between the respective mean rates of utilization. Large and significant differences will induce a sharp classification of data.

Further insights into the data can be obtained by examining the distribution of *related* variables in different categories. These variables may be explanatory variables already included in the regression or may be other variables that are analyzed separately. For example, we could make useful summary statements about the sociodemographic characteristics of the identified groups. An example is from marketing. One might wish to identify the characteristics of the group of frequent purchasers of some item.

Such an *ex post* analysis was applied to the data used in the present illustration. Deb and Trivedi (1997) augmented the data used here with information derived from similar analyses of five additional count measures of health care utilization from the same NMES data source. These measures were number of nonphysician office visits (*OFNP*), number of physician hospital outpatient visits (*OPP*), number of nonphysician hospital outpatient visits (*OPNP*), number of emergency room visits (*EMR*), and number of hospital stays (*HOSP*). Separate models were specified and estimated for each utilization measure. Finally, a preferred specification of the model was determined using the criteria presented in this chapter.

One interesting issue is whether individuals assigned to the high-use group on one measure (e.g., *OFP*), also get classified similarly on a different measure, such as *OFNP*. According to our analysis of *OFP*, 91% of the sample falls in the low-utilization category and the remaining 9% in the high-utilization category. Using posterior probability calculations, every individual in the sample of size 4,406 was assigned to one of these two categories. Similar assignment was also made using the finite mixture models for the other five count variables. Of the 99 users classified as high users of *OFP*, 66 (67%) were also classified as high users of *OFNP*, and 77 (78%) as high users of *OFNP*. However, fewer than 10 (10%) of these 99 were classified as high users of *EMR* or *HOSP*. These intuitively sensible results suggest that *OFP*, *OPNP*, and *OFNP* are closely related. Hence a joint analysis of these measures using a finite mixtures model may be worthwhile.

A second issue is whether high users are concentrated in particular sociodemographic groups. Here it is found that of the 516 *BLACK* members of the sample, 416 (81%) are classified as high users of *OFNP* and 462 (90%) are classified as high users of *OPNP*. With respect to *OFP*, however, only 14 (14%) of the 99 high users are *BLACK*. Two-way frequency tables were also calculated for the high users of *OPP* and *OPNP* against *MEDICAID*. Interestingly, these showed that of the 402 individuals on *MEDICAID*, 271 (67%) were classified as high users of *OFNP* and 364 (91%) as high users of *OPNP*. Of the 402 on *MEDICAID*, 142 (35%) are *BLACK*, so these results seem internally coherent.

We end this discussion here; additional detailed results and interpretations are given in Deb and Trivedi (1997). This illustration has shown that for these data the latent class framework may be more realistic and fruitful than the hurdle framework, mainly because the hurdle-model dichotomy between one population at risk and the other not at risk may be too sharp. At the same time caution should be exercised to avoid overfitting the data if a finite mixture (or latent class) model is used. For example, if measurement errors raise the proportion of high counts, there will be a tendency to overestimate the number of components. A minimum distance estimator that may downweight these high counts is given in Chapter 4.8.7.

Our next example deals with modeling of recreational trips. In that case the hurdle model seems empirically more appropriate than the latent class model.

6.4 ANALYSIS OF RECREATIONAL TRIPS

In the literature on environmental and resource economics, count models have been widely used to model recreational trips. A readily available measure of an individual's demand for some recreational activity is the frequency with which he or she engages in a particular activity, such as boating, fishing, or hiking. An important issue in these studies is the sensitivity of resource usage to entrance charges and travel costs. The latter are inputs into calculations of the changes in consumer welfare resulting from reduced access or higher costs.

The data used in analyses of trips are often derived from sample surveys of potential users of such resources who are asked to recall their usage in some past period. Note that if instead the data are derived from an *on-site survey* of those who actually used the resource or facility during some time period, the data are *not* a random sample from the population. The resulting complications due to endogenous stratification are discussed in Chapter 10.

In this section we consider a case study based on data that come from a survey that covers actual and potential users of the resource. This illustration draws heavily from the article by Gurmu and Trivedi (1996), which provides further details about several aspects that are dealt with briefly here. This illustration considers a broader range of econometric specifications than was the case for the NMES data.

6.4.1 Recreational Trip Data

The ideas and techniques of earlier chapters are illustrated by estimating a recreation demand function, due to Ozuna and Gomaz (1995), for the number

Table 6.9. *Recreational trips: Actual frequency distribution*

Number of trips	0	1	2	3	4	5	6	7	8	9	10
Frequency	417	68	38	34	17	13	11	2	8	1	13
Number of trips	11	12	15	16	20	25	26	30	40	50	88
Frequency	2	5	14	1	3	3	1	3	3	1	1

of recreational boating trips (*TRIPS*) to Lake Somerville, East Texas, in 1980. The data are a subset of that collected by Sellar, Stoll, and Chavas (1985) through a survey administered to 2,000 registered leisure boat owners in 23 counties in eastern Texas. All subsequent analyses are based on a sample of 659 observations.

Table 6.9 presents the frequency distribution of the data on the number of trips. There is a very high proportion of zeros – more than 65% of the respondents reported taking no trips in the survey period. At the same time there is a relatively long tail, with 50 respondents reporting taking 10 or more trips. There is also some clustering at 10 and 15 trips, creating a rough impression of multimodality. Furthermore, the presence of responses in "rounded" categories like 20, 25, 30, 40, and 50 raises a suspicion that the respondents in these categories may not accurately recall the frequency of their visits. The unusual large excess of zeros and long right tail are clear from the histogram (truncated at 30) given in the second panel of Figure 6.1.

Table 6.10 presents variable definitions and summary statistics. As is clear from Table 6.9, the data on trips are very overdispersed with variance of $6.292^2 = 38.6$, that is, many times the mean of 2.2. The regressors are the quality of the recreational site (*SO*), whether the individual water skied at the site (*SKI*), household income (*I*), an indicator for whether a user fee was paid (*FC3*), and costs of visiting Lake Somerville (*C3*) or two alternative recreational

Table 6.10. *Recreational trips: Variable definitions and summary statistics*

Variable	Definition	Mean	Standard deviation
TRIPS	Number of recreational boating trips in 1980 by a sampled group	2.244	6.292
SO	Facility's subjective quality ranking on a scale of 1 to 5	1.419	1.812
SKI	Equals 1 if engaged in water skiing at the lake	0.367	0.482
I	Household income of the head of the group ($1,000/year)	3.853	1.852
FC3	Equals 1 if user's fee paid at Lake Somerville	0.020	0.139
C1	Dollar expenditure when visiting Lake Conroe	55.42	46.68
C3	Dollar expenditure when visiting Lake Somerville	59.93	46.38
C4	Dollar expenditure when visiting Lake Houston	55.99	46.13

sites (*C1*, *C4*). The three cost variables are highly correlated, with pairwise correlations in excess of 0.96. For a more comprehensive description of the data and the method used for calculating the costs of the visit, the reader is referred to Sellar et al. (1985).

6.4.2 Initial Specifications

The models and methods of Chapter 3 provide a starting point. In modeling this data set, we focus on the choice of the parametric family and estimation method, the representation of unobserved heterogeneity in the sample, and evaluation of the fitted model. Our example is intended to provide alternatives to the commonly followed approach in which one settles on the negative binomial model after pretesting for overdispersion. In this example with a very large excess of zeros it is likely that a hurdle or zero-inflated model is more appropriate.

Following Ozuna and Gomaz (1995), we begin with Poisson and negative binomial models (NB2) regression models with conditional mean function

$$\mu_i \equiv \mathsf{E}[TRIPS_i] = \exp(\mathbf{x}_i'\boldsymbol{\beta}), \tag{6.14}$$

where the vector $\mathbf{x} = (SO,\ SKI,\ I,\ FC3,\ C1,\ C3,\ C4)$.

The first few columns of Table 6.11 present results for Poisson regression, along with t-statistics. The column labeled $|t^{def}|$ gives t-statistics based on default standard errors. These t-statistics are misleadingly small, being three to four times larger than the preferred robust t-statistics given in the column labeled $|t|$ that correct for overdispersion.

The Pearson and deviance statistics divided by their degrees of freedom are, respectively, 6.30 and 3.54, indicating considerable overdispersion because they are well in excess of one. This is confirmed using the overdispersion tests of Chapter 3.4. OLS regression without intercept of $((y_i - \hat{\mu}_i)^2 - y_i)/\hat{\mu}_i$ on $\hat{\mu}_i$ yields a t-statistic of 2.88 for the coefficient of $\hat{\mu}_i$, so the null of no overdispersion is rejected against the alternative of overdispersion of NB2 form. The test against overdispersion of NB2 form also rejects the null of no overdispersion.

From data not given in the table, the deviance statistic falls from 4849.7 in an intercept-only model to 2305.8 when regressors are included, so the deviance R^2 defined in Chapter 5.3.3 equals 0.48. An alternative R^2 based on the squared correlation between *TRIPS* and its predicted mean from Poisson regression equals 0.17.

A simple test of adequacy of the Poisson model as a parametric model for these data is based on a comparison of observed zero outcomes and the proportion of zeroes predicted under the null model. Let $\mathbf{1}[\cdot]$ denote the 0/1 indicator function. Since $\mathrm{E}[\mathbf{1}[y_i = 0|\mathbf{x}_i]] = \Pr[y_i = 0|\mathbf{x}_i] = \exp(-\mu_i)$ for the Poisson, where $\mu_i = \exp(\mathbf{x}_i'\boldsymbol{\beta})$, it follows that $\mathrm{E}[\mathbf{1}[y_i = 0] - \exp(-\mu_i)] = 0$. Mullahy's test is based on the corresponding sample moment

$$\hat{m} = \frac{1}{n}\sum_{i=1}^{n}(\mathbf{1}[y_i = 0] - \exp(-\hat{\mu}_i)) \equiv \frac{1}{n}\sum_{i=1}^{n}\hat{\delta}_i, \tag{6.15}$$

where the last equality defines $\hat{\delta}$. The distribution of $\hat{m}$ can be obtained using the results for conditional moment tests given in Chapter 2.8.5. This yields

$$\mathsf{T_Z} = \frac{\sqrt{n}\hat{m}}{\sqrt{\frac{1}{n}\sum_i \hat{v}_i^2}} \overset{d}{\to} \mathsf{N}[0, 1], \tag{6.16}$$

where

$$\hat{v}_i = \hat{\delta}_i - \left[\frac{1}{n}\sum_{i=1}^{n} \mathbf{1}[y_i = 0]\mathbf{x}_i \hat{u}_i\right]' \left(-\sum_{i=1}^{n} \hat{\mu}_i \mathbf{x}_i \mathbf{x}_i'\right)^{-1} \mathbf{x}_i \hat{u}_i, \tag{6.17}$$

where $\hat{u}_i = y_i - \hat{\mu}_i$ and we use $\mu_i = \exp(\mathbf{x}_i'\boldsymbol{\beta})$. The excess zero test statistic for the trip data is $\mathsf{T_Z} = 6.87$, which is highly statistically significant, suggesting that the null Poisson be rejected. This is not surprising because from Table 6.14 the sample average predicted probability of a zero is 0.42, much lower than the sample frequency of 0.63. An asymptotically equivalent version of the test replaces $\mathbf{1}[y_i = 0]\mathbf{x}_i \hat{u}_i$ in (6.17) with $-\hat{\mu}_i \exp(-\hat{\mu}_i)\mathbf{x}_i$.

A more complete chi-square goodness-of-fit test based on cells for $0, \ldots, 4$ and 5 or more trips based on (6.5) leads to a value of 252.6, much larger than the $\chi^2(5)$ critical value, again indicating a poor fit of the Poisson to the data. Table 6.14 also reveals that the Poisson model overpredicts counts in the mid-range (1 to 14) and underpredicts the high counts.

Table 6.11 also presents negative binomial (NB2) estimates. The results strongly favor the NB2 model compared to Poisson. The NB2 overdispersion parameter is highly statistically significant ($t = 7.41$). The log-likelihood rises from $-1{,}529$ to -825, leading to a $\chi^2(1)$-distributed LR test statistic of 1408 that strongly rejects the Poisson. And the BIC falls from 3,111 to 1,709. The default standard errors for NB2 based on the Hessian are smaller than the robust sandwich standard errors, leading to default t-statistics that are about 30% larger than the preferred robust t-statistics.

The NB2 estimates indicate that the number of trips increases substantially with the perceived quality of the site and for those who water ski and who pay a user's fee at the site, while income has little effect and is statistically insignificant. The NB2 estimates are plausible in that they indicate substitution from other sites ($C1$ and $C4$) toward Lake Somerville as travel costs rise, while trips to Lake Somerville decline as travel costs to Lake Somerville ($C3$) rise. The cost variables have standard deviations of around 45, so the effect of a one-standard-deviation change in costs is substantial.

The presence of overdispersion, although consistent with the NB2 specification, does not in itself imply that the NB2 specification is adequate; rejection of the null against a specified alternative does not necessarily imply that the alternative is the correct one. Predicted probabilities given in Table 6.14 indicate that the NB2 model does reasonably well. For example, the average predicted probability of a zero is 0.642 compared to the sample frequency of 0.632. The statistic $\mathsf{T_{GoF}}$ given in Table 6.11 is 23.5. Although this is a substantial

Table 6.11. *Recreational trips: Poisson and NB2 model estimates and t-ratios*

	Poisson			NB2		
Variable	Coefficient	$\lvert t^{def}\rvert$	$\lvert t\rvert$	Coefficient	$\lvert t^{def}\rvert$	$\lvert t\rvert$
ONE	0.265	2.83	0.61	−1.122	5.08	3.79
SO	0.472	27.60	9.65	0.722	15.93	12.37
SKI	0.419	7.31	2.16	0.612	4.07	3.06
I	−0.111	5.68	2.21	−0.026	0.58	0.48
FC3	0.898	11.37	3.63	0.669	1.85	2.09
C1	−0.003	1.10	0.23	0.048	3.01	1.70
C3	−0.042	25.47	3.62	−0.093	11.21	6.38
C4	0.036	13.34	3.85	0.039	3.32	2.19
α	–			1.37	9.43	7.41
Fitted mean	2.24			8.97		
− lnL	1529			825		
BIC	3111			1709		
Pearson	6.30					
Deviance	3.54					
T_Z	6.9					
T_{GoF}	252.6			23.5		

Note: t-Statistics based on robust standard errors, except t^{def} is based on default standard errors. Fitted mean is the average predicted mean. Pearson and Deviance are, respectively, the Pearson and deviance statistics divided by degrees of freedom. T_z is test for correct proportion of zeros. T_{GoF} is a chi-squared goodness-of-fit test with five degrees of freedom.

improvement on the Poisson, the model is still rejected since the 5% critical value for $\chi^2(5)$ is 11.07.

From Table 6.14 the NB2 model overpredicts high counts, with the average probability of 15 or more trips equal to 0.042 compared to a sample frequency of 0.023. More remarkably, from Table 6.11 the average fitted mean from NB2 estimation is 8.97, which is, much higher than the sample mean of *TRIPS* of 2.24. This substantial overprediction is due to several observations having large predicted means; whereas the median predicted number of trips is 0.24, one observation had a predicted mean of 3,902.38. Such a great difference between the sample mean and average fitted mean from an NB2 model is highly unusual. For example, for the section 6.3 example the average fitted mean from an NB2 model is 5.83 compared with a sample mean of *OFP* of 5.77.

One possible explanation is that the unobserved heterogeneity distribution is misspecified. An NB1 model may do better because the variance is linear in μ rather than quadratic. This is the case here, with the average fitted mean from NB1 estimation of 2.24, which is essentially the same as the sample frequency. Note, however, that the NB1 model has a log-likelihood of −834 that is higher than −825 for NB2.

A second possibility is that a more flexible conditional mean function is needed. Inclusion of quadratic terms and interactions ($C1SQ$, $C3SQ$, $C4SQ$,

Table 6.12. *Recreational trips: Finite mixture model estimates and* t-*ratios*

	Finite Mixtures Poisson two-component				Finite Mixtures NB2 two-component			
	Low users		High users		Low users		High users	
Variable	Coefficient	$\|t\|$	Coefficient	$\|t\|$	Coefficient	$\|t\|$	Coefficient	$\|t\|$
ONE	−1.865	6.00	1.220	2.19	−1.877	9.13	1.006	1.04
SO	0.659	15.47	0.281	2.91	0.890	19.90	−0.095	0.43
SKI	0.557	3.05	0.852	5.97	0.449	2.50	1.369	3.55
I	−0.097	0.43	0.092	1.28	−0.049	1.02	−0.035	0.23
FC3	0.970	3.43	0.158	0.84	1.070	2.97	−0.130	0.43
C1	0.001	0.03	0.054	4.40	−0.000	0.01	0.186	7.50
C3	−0.064	7.85	−0.064	7.91	−0.050	5.24	−0.253	9.52
C4	0.057	4.50	0.002	0.24	0.047	3.11	0.049	3.42
α					0.825	7.58	0.192	1.88
π	0.887		0.113	5.04	0.876	15.52	0.124	2.20
Fitted mean	1.55		10.11		2.77		29,313	
−ln L		916				786		
BIC		1,994				1,695		
T_{GoF}		43.7				20.6		

$C1C3$, $C1C4$, and $C3C4$) for the three cost variables in the conditional mean function for NB2 leads to improvement for the NB2 model in both the log-likelihood, now −802, and BIC, now 1702. The average probability of 15 or more trips is now slightly higher at 0.048. But the resulting average fitted mean is now 2.80, close to the sample frequency of 2.24, and the maximum fitted mean is 70.3.

A third explanation is that a standard NB model cannot simultaneously model both the large excess of zeros and the large overdispersion in the right tail. The most natural model then is a hurdle model or a zero-inflated model.

6.4.3 Modified Count Models

Plausible alternatives to the models considered above are hurdle models, zero-inflated models, and finite mixture models, that lead to changes in the conditional mean *and* the conditional variance specification.

Table 6.12 presents estimates of two-component Poisson and NB2 finite mixture models. The Poisson finite mixture model is a great improvement on the Poisson model, with log-likelihood rising from −1529 to −916, but it still is not as high as for the NB2 model.

The two-component finite mixture NB2 model has a lower BIC than that for the NB2 model. Giving a latent class interpretation, most individuals are low users and for these individuals the results are comparable to those for the NB2 model. For the 12.4% who are high users the statistically significant coefficients have the expected sign and are larger in magnitude, implying a larger multiplicative effect of the regressors. The FMNB2 model is still rejected by the chi-square goodness-of-fit test. We interpret this result to mean that the characterization of the dgp as a mechanism that samples two subpopulations, one of relatively high users and the other of low users of the recreational site, leaves some features of the data unexplained.

Furthermore, the average fitted mean for the high users is unbelievably high, at 29,313, and even with the two largest predicted means dropped, the average fitted mean is 103. So the problem of the standard NB2 model overpredicting, discussed in detail in the preceding subsection, is compounded in a mixture setting. The overprediction is much less in an NB1 two-component mixture model (not reported in a table), with average fitted means of 8.84 for higher users and 1.42 for lower users. However, even then the overall average sample mean is still too high at 6.46 (using the estimated probability of being a high user of 0.68), and the log-likelihood is higher at −794 compared to −786 in the NB2 case. If instead quadratic and interaction terms in the cost variables are included as additional regressors, there is still great overprediction, unlike the earlier case for standard NB2.

Finally, we consider hurdle and zero-inflated models, which a priori were the most natural models for these data. Regression results for the NB2 hurdle model are given in Table 6.13. For these data the binary regressor $FC3$ only equals 1 if $TRIPS > 0$. As a result it is not identified in a binary model for whether or not $TRIPS = 0$, so we drop $FC3$ from the binary part of the NB2 hurdle model. For both the hurdle portion of the model and the positives the coefficients have the same sign, and this sign is the same as that for the NB2 model. So the regressors have qualitatively the same effect as in the NB2 model. Few of the variables for the zeros part of the model are significant, which suggests that most of the explanatory power of the covariates derives from their impact on positive counts.

The model fit is substantially improved. The log-likelihood is −721 compared to −825 for the NB2 model and −786 for the finite mixture model, and the corresponding values of BIC are 1,553, 1,709, and 1,695. The log-likelihood for the NB2 hurdle model is also much higher than that for the Poisson hurdle model (−1200) and the NB1 hurdle model (−782). In terms of fitted frequency distribution (see Table 6.14), the NBH model does extremely well, for both zero counts and high counts. *Indeed* NBH *is the only model that is not rejected by the goodness-of-fit test.* This result is especially interesting since the difference between NBH and FMNB is quantitative rather than qualitative – NBH views the data as a mixture with respect to zeros only, whereas FMNB views the data as a mixture with respect to zeros and positives. The opposite conclusion was reached in regard to the NMES data. Finally, Gurmu and Trivedi (1996) find that for these data the NBH model is also superior to flexible parametric models based on series expansions that are presented in Chapter 12.

In Table 6.13 we also include an estimate of the zero-inflated NB2 model, see section 4, in which the probability of a nonzero count is further modeled as a logit function with the same regressors as in the main model. The motivation is that the sample under analysis may represent a mixture of at least two types, those who never choose boating as recreation and those who do, but some of the latter might simply happen not to have had a positive number of boating trips

Table 6.13. *Recreational trips: With zeros and hurdle model estimates and t-ratios*

	NB2 hurdle				Zero-inflated NB2			
	Zeros (Inflate)		NB2		Zeros		Positives	
Variable	Coefficient	$\|t\|$	Coefficient	$\|t\|$	Coefficient	$\|t\|$	Coefficient	$\|t\|$
ONE	−4.614	1.58	0.842	2.30	20.975	7.36	1.092	3.74
SO	18.134	1.17	0.172	2.39	−38.636	19.29	0.171	3.01
SKI	3.832	0.92	0.622	3.36	−16.066	14.82	0.493	3.37
I	−0.068	0.16	−0.057	2.34	−0.203	0.60	−0.069	1.66
FC3	–	–	0.576	2.29	−11.430	4.72	0.547	2.48
C1	0.263	1.29	0.057	5.88	−0.024	1.55	0.040	2.16
C3	−0.491	1.13	−0.078	0.78	0.078	3.61	−0.066	6.31
C4	0.216	1.58	0.012	2.30	−0.063	2.93	0.021	1.82
α	1.206	3.81	1.700	3.87	–		0.832	8.69
Fitted mean		3.77				2.72		
− lnL		721				719		
BIC		1553				1549		

in the sample period. The results are once again plausible, with coefficients in the main NB2 component of the model having the same sign as for the NB2 model. The model for zeros is a logit model for whether $TRIPS = 0$ rather than $TRIPS > 0$, so we expect the coefficients in the model for zeros to have the reverse sign of those in the main NB2 part of the model; this is indeed the case for all but income, which is statistically insignificant. The ZINB model actually has somewhat higher log-likelihood and lower BIC than the NB2 hurdle model. The average fitted mean of 2.72 is close to the sample mean of 2.24.

In terms of goodness of fit measured by either log-likelihood or the BIC, the hurdles NB2 or zero-inflated NB2 models are best. This result can be interpreted as follows. Although there is considerable unobserved heterogeneity among those who use the recreational facility, there is also a significant proportion in the population for whom the "optimal" solution is a corner solution.

Table 6.14. *Recreational trips: Actual and fitted cumulative frequencies*

	0	1	2	3	4	5	6–8	9–11	12–14	15–17	18–62	63+
Frequency												
Observed	0.632	0.103	0.057	0.052	0.026	0.020	0.032	0.024	0.008	0.023	0.021	0.002
Poisson	0.420	0.221	0.103	0.062	0.045	0.034	0.061	0.026	0.012	0.006	0.010	0.002
NB	0.642	0.122	0.050	0.030	0.021	0.016	0.033	0.020	0.013	0.009	0.034	0.008
Zero-inflated NB	0.656	0.072	0.054	0.040	0.031	0.024	0.047	0.026	0.015	0.010	0.024	0.003
Hurdle NB	0.623	0.107	0.063	0.043	0.030	0.023	0.041	0.022	0.013	0.009	0.023	0.003
Cumulative												
Observed	0.632	0.736	0.794	0.845	0.871	0.891	0.923	0.947	0.955	0.977	0.998	1.000
Poisson	0.420	0.640	0.744	0.805	0.850	0.885	0.945	0.971	0.983	0.988	0.998	1.000
NB	0.642	0.764	0.815	0.845	0.866	0.883	0.915	0.935	0.948	0.958	0.992	1.000
Zero-inflated NB	0.656	0.728	0.782	0.822	0.853	0.877	0.924	0.949	0.964	0.974	0.997	1.000
Hurdle NB	0.623	0.730	0.793	0.836	0.866	0.889	0.930	0.952	0.965	0.974	0.997	1.000

That is, they may consistently have a zero demand for recreational boating. Since no theoretical model was provided for explaining zero demand for recreational boating, the potential importance of excess zeros was not emphasized. Consumer choice theory (and common sense) predicts the occurrence of zero solutions; see Pudney (1989, chapter 4.3). However, a priori reasoning cannot in itself predict whether their relative frequency will be greater or less than that implied by the Poisson model. This is an empirical issue, whose resolution depends also on how the sample data were obtained. Theory may still help in suggesting variables that explain the proportion of such corner solutions.

As in the case of the NMES data, several other finite mixture models were also estimated. The diagnostic tests and the BIC criteria show all finite mixture models to be inferior to the hurdles or zero-inflated NB2 models. Thus, in contrast to the NMES data, the outcome supports the idea that the sample is drawn from two subpopulations of non-users and users, rather than from two subpopulations of low and high users.

6.4.4 Economic Implications

The model discrimination and selection exercise relied heavily on statistical criteria in both applications. Are we simply fine-tuning the model, or are the resulting changes economically meaningful? Only a partial answer can be given here.

In the modeling of recreational trips, it is reasonable to suppose that a random sample will include nonparticipants because of taste differences among individuals. A hurdle model is worthwhile in this case because parameter estimates and welfare analysis should be based only on the participants' responses.

By contrast, in the case of NMES data on the elderly, the notion of nonparticipants is implausible. Given differences in health status of individuals as well as other types of unobserved heterogeneity, however, the distinction between high users and low users is reasonable. This feature can explain the superior performance of the finite mixture NB model in section 6.3.

6.5 ANALYSIS OF FERTILITY DATA

A number of studies that have analyzed the role of socioeconomic factors and social norms on fertility have pointed out that fertility data exhibit a number of features that call for a modeling strategy that goes beyond the standard Poisson and negative binomial regressions.

For example, Melkersson and Rooth (2000) report inflation at frequencies of zero and two in their analysis of Swedish fertility data. As we see later, recent fertility data from Switzerland and Germany also exhibit a similar feature of a double mode, one at zero and another at two. In contrast to such data from economically advanced West European countries, the data from developing countries may not exhibit this feature. For example, Miranda (2010) analyzes Mexican fertility data from a 1997 survey that do not contain an excess of two

Table 6.15. *Number of children: Actual and fitted frequency distribution for Swiss data (n = 1,878)*

Children	0	1	2	3	4	5	6	7	8	9	10	11
Frequency	379	262	684	353	128	35	16	8	10	1	1	1
Percent	20.2	14.0	36.4	18.8	6.8	1.9	0.9	0.4	0.5	0.1	0.1	0.1
NB fitted	15.6	27.8	25.9	16.8	8.5	3.6	1.3	0.4	0.1	0.0	0.0	0.0

outcomes and exhibit overdispersion. He suggests that data from developing countries may show an excess of large counts. Planned fertility studies that ask how many children a family would like to have generate samples with just a few realizations (Miranda, 2008). Furthermore, over time societies may transit from high to low fertility rates as social norms and economic choices favor smaller family units. It is convenient to think of fertility data as characterized by limited support, in the sense that just a few event counts dominate the frequency distribution.

In contrast to the two preceding sections that provided detailed count data regression analysis for two data sets, this section is limited to univariate (unconditional) descriptive analysis of fertility data from two samples. The goal is simply to highlight some features of fertility data that create a modeling challenge. Covariates will provide relatively low explanatory power, as is typical for cross-section data, so that their introduction will most likely not be sufficient to explain the essential features of the count data observed here. Instead, models richer than Poisson or negative binomial are needed to explain the bimodality in the data, even with covariates.

6.5.1 Fertility Data

The first data set on completed fertility comes from the Swiss Household Panel W1 (1999). The mean number of children in this sample ($n = 1{,}878$) of women over age 45 years is 1.94 and the standard deviation is 1.45. There is moderate overdispersion with sample variance of 2.11 not much greater than the mean. Table 6.15 presents the frequency distribution for the data. It shows a half-mode at 0 and a more pronounced mode at 2, similar to the features of data in Melkersson and Rooth (2000).

The second data set is a larger sample ($n = 6{,}782$) from the British Household Panel Survey (BHPS). Here the sample mean number of children is 1.85 and the standard deviation is 1.51. There is a bit more overdispersion than in the first data set, with sample variance of 2.28 so the variance is 1.23 times the mean. Overdispersion is still limited compared with that found in fertility data from developing countries. For example, Miranda's ENADID 2006 data on completed fertility in Mexico has a mean of 4.43 children, a standard deviation of 2.75, and a variance of 7.56 that is 1.71 times the sample mean. Table 6.16

Table 6.16. *Number of children: Actual and fitted frequency distribution for British data (n = 6,782)*

Children	0	1	2	3	4	5	6	7	8	9	10	11
Frequency	1764	805	2160	1262	487	181	69	33	12	5	3	1
Percent	26.0	11.9	31.9	18.6	7.2	2.7	1.0	0.5	0.2	0.1	0.0	0.0
NB fitted	19.3	28.3	23.7	14.9	7.8	3.6	1.5	0.6	0.2	0.1	0.0	0.0

presents the frequency distribution for the BHPs data. It shows a half-mode at 0 and a more pronounced mode at 2, similar to the Swiss data.

Although the Poisson or negative binomial regression model may seem a promising starting point for analyzing these fertility data, these distributions are unimodal, which conflicts with the fact that the observed frequency distribution is bimodal. To illustrate this point, we fit the intercept-only NB2 regression to both sets of data, with similar results obtained if instead Poisson models are fitted.

The final rows of Tables 6.15 and 6.16 present the predicted probability of each count. It is clear that NB provides a poor fit at 0, 1, and 2 counts, greatly overpredicting the probability of one child and underpredicting the probability of zero children or two children.

6.5.2 Univariate Models

We use the Swiss data to fit an intercept-only version of several models. Table 6.17 presents measures of fit of the various models.

The negative binomial model will accommodate overdispersion in the data, but the overdispersion here is not great. The NB model leads to a slight improvement over Poisson, and the $\chi^2(1)$ distributed LR test statistic of 6.53 favors the NB model at the 1% significance level.

A two-component Poisson finite mixture model could, in principle, be able to accommodate the bimodality in the data, coming from either Poisson with mean near 0 or Poisson with mean around 2. We found two optima with these data. The first yields $f(y) = 0.988 \times \mathsf{P}[1.88] + 0.012 \times \mathsf{P}[6.57]$, with

Table 6.17. *Swiss fertility data: Fit of various models*

Model	Parameters	Log-likelihood	AIC	BIC
Poisson	1	−3,238	6,478	6,484
Negative binomial	2	−3,235	6,474	6,485
Finite mixture Poisson − 2	3	−3,203	6,409	6,421
Hurdle Poisson	2	−3,203	6,409	6,421
Zero-inflated Poisson	2	−3,203	6,409	6,421
Ordered probit	6	−3,031	6,076	6,109

$\ln L = -3{,}225.7$. This model uses the mixture to better fit the higher counts. The second yields $f(y) = 0.904 \times \mathsf{P}[2.14] + 0.094 \times \mathsf{P}[0.0]$, with $\ln L = -3{,}202.7$. This fitted model is a boundary solution where $\mu_2 = 0$, and it is equivalent to a zero-inflated Poisson that adds .094 to the probability of a zero count. Both optima are an appreciable improvement on Poisson and NB, and the latter optima is preferred because it leads to much higher log-likelihood. We note in passing that both a two-component NB model and a three-component Poisson model led to the same results as the two-component Poisson.

The hurdle Poisson and zero-inflated Poisson are observationally equivalent models in this example with no regressors. We nonetheless estimate both. The two models yield the same fitted log-likelihood value, $\ln L = -3202.7$, and the same predicted probabilities. The predicted frequencies of 0 to 5 children are, respectively, 20.2, 22.7, 24.4, 17.4, 9.3, and 4.0. This is better than the predicted frequencies for NB given in Table 6.15, but the dip at one child is still not fully explained. Melkersson and Rooth (2000) argue that additional modification through inflation at two may be warranted to account for social norms that favor a two-child family over a one-child family. Fitted negative binomial versions of the hurdle and zero-inflated models collapsed to the Poisson versions.

The ordered probit model, see Chapter 3.6.2, is motivated by the feature that the data are indeed ordered. Thinly populated cells should be conveniently aggregated into categories. In this example we aggregate six or more children into one category. For this model without regressors the fitted ordered probit necessarily gives the predicted frequencies equal to the sample frequencies.

The ordered probit model has many more parameters than the other models but has much higher log-likelihood so is strongly preferred on grounds of BIC. The next best fitting models are the equivalent Hurdle Poisson, ZIP, and, in this case, two-component finite mixtures Poisson.

Yet another potential model is a sequential conditional probability (SCP) for more than $j+1$ events conditional on observing j events, with $\Pr[y = j | y > j-1] = \Lambda(\mathbf{x}'\boldsymbol{\beta}_j)$, $j = 1, \ldots, m$, in the case of sequential logit. Each stage in the sequence is a binary outcome model based on a decreasing number of observations as we move up the chain of event counts. Although the model is likely to be highly parameterized, it could provide useful information for specific links in the chain that may have a special significance. For example, interest may lie in the conditional probability of four or more children given three.

This approach can also be used if we want to test the assumption that data are generated by a homogeneous Poisson process. The Poisson model implies, on simplification, that

$$\Pr[y = j | \, y > j-1] = \frac{\exp(\mathbf{x}'\boldsymbol{\beta})^j / j!}{\sum_{k=j}^{\infty} \exp(\mathbf{x}'\boldsymbol{\beta})^k / k!}.$$

A series of binary outcome models with this specification can be estimated, with parameters $\boldsymbol{\beta}_j$ for $j = 1, \ldots, m$, and the Poisson process assumption can be tested by testing the restriction $\boldsymbol{\beta}_1 = \boldsymbol{\beta}_2 = \cdots = \boldsymbol{\beta}_m$. This approach can be adapted to other models, such as NB.

The results of Melkersson and Rooth (2003) and Santos Silva and Covas (2000) cast doubt on the validity of the standard Poisson process assumption for analyzing fertility data. In addition, social norms and dynamics play a role in generating distributions that are inadequately characterized by standard Poisson-type models.

6.6 MODEL SELECTION CRITERIA: A DIGRESSION

The examples in this chapter have illustrated alternative ways of handling "non-Poisson" features of count data. Plausible models include hurdles and finite mixtures models, and potentially even finite mixtures based on hurdle models. Selecting a preferred parametric model can be difficult because the usual LR test cannot be applied, either because it has a nonstandard distribution, such as when determining the number of components in a finite mixtures model, or because models are not nested in each other.

In this section we report the results of a small simulation experiment designed to throw light on the properties of model evaluation criteria used in this chapter, and we discuss using a bootstrap to obtain critical values for the LR test.

6.6.1 A Simulation Analysis of Model Selection Criteria

Data were generated from three models: Poisson, Poisson hurdle (PH), and a two-component finite mixture of Poissons (FMP-2). The conditional mean(s) for the models were, respectively,

Poisson		$\mu = \exp(-1.445 + 3.0x)$
Poisson Hurdle	Zeros	$\mu = \exp(-1.6 + 3.0x)$
	Positives	$\mu = \exp(-1.35 + 3.0x)$
FM Poisson	Component 1	$\mu_1 = \exp(-1.225 + 3.0x)$
$(\pi = 0.75)$	Component 2	$\mu_2 = \exp(-1.5 + 0.75x)$

and the single regressor $x \sim$ Uniform $[0, 1]$. The sample size was $n = 500$. The dgp was calibrated in each case to obtain zeros for about 40% of the observations.

Table 6.18 reports results from 500 simulations. We first consider the size and power of LR and goodness-of-fit tests, followed by the performance of associated information criteria.

The first two entries of the first column of Table 6.18 list the frequency of rejection at nominal significance level of 0.10 using the LR test when the dgp

Table 6.18. *Rejection frequencies at nominal 10% significance level*

		Data-generating process	
Test/criterion	Poisson	Hurdle (PH)	Finite mixture (FMP)
LR: P versus PH	0.100	0.926	0.990
LR: P versus FMP	0.092	0.880	1.000
GoF: P rejected	0.126	0.740	0.988
GoF: PH rejected	0.244	0.100	0.530
GoF: FMP rejected	0.142	0.346	0.118
AIC: P lowest	0.822	0.050	0.000
AIC: PH lowest	0.146	0.800	0.066
AIC: FMP lowest	0.032	0.148	0.934
BIC: P lowest	0.998	0.544	0.060
BIC: PH lowest	0.002	0.400	0.050
BIC: FMP lowest	0.000	0.050	0.890

is the Poisson. The Poisson hurdle and finite mixture Poisson each have two additional parameters, but as noted in Chapter 4.8.6 the LR test statistic has a nonstandard distribution and is not asymptotically $\chi^2(2)$ distributed. Some Monte Carlo studies by other authors suggest that for (noncount) mixture models with two extra parameters the appropriate distribution is somewhere between $\chi^2(1)$ and $\chi^2(2)$. Here we use $\chi^2(1)$ critical values. Testing both Poisson against Poisson hurdle, and Poisson against finite mixture Poisson, leads in this example to a test size close to .10.

The chi-squared goodness-of-fit tests are based on cells for 0, 1, 2, 3, and 4 or more counts. The rejection rates are close to the nominal size, with 0.126 for Poisson when the dgp is Poisson, 0.100 for PH when the dgp is PH , and 0.118 for FMP when the dgp is FMP.

Next consider the power of the LR and GoF tests. First, the LR test of P against PH is more powerful than the GoF-P test, with rejection rate 0.926 > 0.740, and the LR test of P against FMP is more powerful than the GoF-P test, with rejection rate 1.000 > 0.988. This is expected because in these cases the LR test is the most powerful test, and the two tests were found to have approximately correct size in this example. Second, even when the actual dgp differs from that assumed for the alternative hypothesis model, the LR test remains more powerful than the goodness-of-fit test, which has rejection rates of no more than 0.530. These results are, of course, affected by the choice of parameter values but do indicate a loss of power in using goodness-of-fit tests.

The performance of the information criteria is shown in the lower part of Table 6.18. The interpretation of the "rejection frequency" here is somewhat different because the reported figure shows the proportion of times the model had the smallest value of the criterion. For example, when the true model is Poisson, the AIC selects Poisson as the best model in 82% of the cases, and BIC does so in almost every case. Similarly the information criteria generally strongly favor PH when the dgp is PH, and FMP when the dgp is FMP. The one exception is that when the dgp is PH, BIC favors PH in only 40% of the cases, due to its high penalty for additional parameters and actually picks the Poisson as the best model more frequently.

The goodness-of-fit tests and information criteria are useful in evaluating models. Rejection of the null by the LR, GoF, or the AIC would seem to indicate a deficiency of the model. Thus, despite its theoretical limitations, the standard LR test of the one-component model against the mixture alternative may have useful power.

6.6.2 Bootstrapping the LR Test

As detailed in Chapter 4.8.6, the LR test statistic does not have the usual χ^2 distribution when used to test the number of components in a finite mixture model.

If computational cost is not an important consideration, a parametric bootstrap provides a way to obtain better critical values for the LR test. Feng and McCulloch (1996) suggested and analyzed a bootstrap LR test for the null that the number of components in the mixture is $C - 1$ against the alternative that it is C. Their examples are in a univariate iid setting. For the regression case we proceed as follows:

1. Estimate the $(C - 1)$-component model and the C-component mixture model by MLE. Form the LR statistic, denoted LR.
2. Draw a bootstrap pseudo-sample $(y_i^*, \mathbf{x}_i)$ by generating y_i^* from $f(y_{C-1}|\mathbf{x}_i, \hat{\theta}_{C-1})$, the fitted density for the null hypothesis $(C - 1)$-component model. Reestimate both null and alternative models and construct the LR statistic LR^*.
3. Repeat step 2 B times, giving B values of LR statistic, denoted LR_b^*, $b = 1, \ldots, B$.
4. Using the empirical distribution of $\text{LR}_1^*, \ldots, \text{LR}_B^*$, determine the $(1 - \alpha)$ percent quantile as the critical value, denoted $\text{LR}_{[1-\alpha]}$.
5. Reject $H_0 : C^* = C - 1$ against $H_a : C^* = C$ if $\text{LR} > \text{LR}_{[1-\alpha]}$.

Since the bootstrap is being used for testing rather than standard error estimation B should be quite large, say $B = 999$. The bootstrap may take a while because finite mixture models are computationally intensive to estimate. Furthermore the likelihood can be multimodal, so ideally a range of starting values would be used in estimating in each of the bootstrap replications.

6.7 CONCLUDING REMARKS

Most empirical studies generate substantive and methodological questions that motivate subsequent investigations. We conclude by mentioning three issues and lines of investigation worth pursuing.

First, in the context of the recreational trips example, one might question the assumption of independence of sample observations. This assumption is standard in cross-section analysis. However, our data also have a spatial dimension. Even after conditioning, observations may be spatially correlated. This feature will affect the estimated variances of the parameters. Essentially, the assumption of independent observations implies that our sample is more informative than might actually be the case; see Chapter 4.9.

A related issue concerns the stochastic process for events. Many events may belong to a spell of events, and each spell may constitute several correlated events. The spells themselves may follow some stochastic process and may in fact be observable. One might then consider whether to analyze pooled data or to analyze events grouped by spells. An example is the number of doctor visits within a spell of illness (Newhouse, 1993; Santos Silva and Windmeijer, 2001). In many data situations one is uncertain whether the observed events are a part of the same spell or of different spells; see Chapter 4.2.3.

Another issue is joint modeling of several types of events. The empirical examples considered in this chapter involve conditional models for individual events, not joint models for several events. This may be restrictive. In some studies the event of interest generates multiple observations on several counts. A health event, for instance, may lead to hospitalization, doctor consultations, and usage of prescribed medicines, all three being interrelated. The analysis described in the previous paragraph can be extended to this type of situation by considering a mixture of multinomial and count models, leading to multivariate count models discussed in Chapter 8.

6.8 BIBLIOGRAPHIC NOTES

Applications of single-equation count data models in economics, especially in accident analysis, insurance, health, labor, resource, and environmental economics, are now standard; examples are Johansson and Palme (1996), Pohlmeier and Ulrich (1995), Gurmu and Trivedi (1996), and Grogger and Carson (1991). Rose (1990) uses Poisson models to evaluate the effect of regulation on airline safety records. Dionne and Vanasse (1992) use a sample of about 19,000 Quebec drivers to estimate a NB2 model that is used to derive predicted claims frequencies, and hence insurance premiums, from data on different individuals with different individual characteristics and driving records. Schwartz and Torous (1993) combine the Poisson regression approach with the proportional hazard structure. They separately model monthly grouped data on mortgage prepayments and defaults. Cameron and Trivedi (1996) survey this and a number of other count data applications in financial economics.

Lambert (1992) provides an interesting analysis of the number of defects in a manufacturing process, using the Poisson regression with "excess zeros." Nagin and Land (1993) use the Heckman-Singer type nonparametric approach in their mixed Poisson longitudinal data model of criminal careers. Their model is essentially a hurdles-type Poisson model with nonparametric treatment of heterogeneity. Post-estimation they classify observations into groups according to criminal propensity, in a manner analogous to that used in the health utilization example. Panel data applications are featured in Chapter 9. Two interesting marketing applications of latent class (finite mixture) count models are Wedel et al. (1993) and Ramaswamy, Anderson, and DeSarbo (1994). Haab and McConnell (1996) discuss estimation of consumer surplus measures in the presence of excess zeros.

Deb and Trivedi (2002) reexamine the Rand Health Insurance Experiment (RHIE) pooled cross-section data and show that finite mixture models fit the data better than hurdle models. Of course, this conclusion, though not surprising, is specific to their data set. Lourenco and Ferreira (2005) apply the finite mixture model to model doctor visits to public health centers in Portugal using truncated-at-zero samples. Böhning and Kuhnert (2006) study the relationship between mixtures of truncated count distributions and truncated mixture distributions and give conditions for their equivalence. Recent biometric literature offers promise of more robust estimation of finite mixtures via alternative estimators to maximum likelihood.

6.9 EXERCISES

6.1 Using the estimated mixing proportion π_1 in Table 6.6, and the estimated component means in Table 6.6, check whether the sample mean of *OFP* given in Table 6.2 coincides with the fitted mean. Using the first-order conditions for maximum likelihood estimation of NB2, consider whether a two-component finite mixture of the NB2 model will display an analogous property.

6.2 Suppose the dgp is a two-component CFMNB family, with the slope parameters of the conditional mean functions equal, but intercepts left free. An investigator misspecifies the model and estimates a unicomponent Poisson regression model instead. Show that the Poisson MLE consistently estimates the slope parameters.

6.3 In the context of modeling the zeros/positives binary outcome using the NBH specification, compare the following two alternatives from the viewpoint of identifiability of the parameters $(\boldsymbol{\beta}, \boldsymbol{\alpha}_1)$,

$$\Pr[y_i = 0|\mathbf{x}_i] = \begin{cases} 1/(1+\alpha_1\mu_i)^{1/\alpha_1}, & \text{or} \\ 1/(1+\mu_i), \end{cases}$$

where $\mu_i = \exp(\mathbf{x}_i'\boldsymbol{\beta})$.

6.4 Consider how to specify and estimate a two-component finite mixture of the NBH model. Show that this involves a mixture of binomials as well as a mixture of NB families.

6.5 Consider whether the alternative definitions of residuals in Chapter 5 can be extended to finite mixtures of Poisson components.

6.6 Verify the result given in (6.9). To do so, first derive (4.73) for $r = 2$; then derive the central second moment by subtracting off $\overline{\mu}^2$.

CHAPTER 7

Time Series Data

7.1 INTRODUCTION

The previous chapters have focused on models for cross-section regression on a single count dependent variable. We now turn to models for more general types of data – univariate time series data in this chapter, multivariate cross-section data in Chapter 8, and longitudinal or panel data in Chapter 9.

Count data introduce complications of discreteness and heteroskedasticity. For cross-section data, this leads to moving from the linear model to the Poisson regression model. However, this model is often too restrictive when confronted with real data, which are typically overdispersed. With cross-section data, overdispersion is most frequently handled by leaving the conditional mean unchanged and rescaling the conditional variance. The same adjustment is made regardless of whether the underlying cause of overdispersion is unobserved heterogeneity in a Poisson point process or true contagion leading to dependence in the process.

For time series count data, one can again begin with the Poisson regression model. In this case, however, it is not clear how to proceed if dependence is present. For example, developing even a pure time series count model where the count in period t, y_t, depends only on the count in the previous period, y_{t-1}, is not straightforward, and there are many possible ways to proceed. Even restricting attention to a fully parametric approach, one can specify distributions for y_t either conditional on y_{t-1} or unconditional on y_{t-1}. For count data this leads to quite different models, whereas for continuous data the assumption of joint normality leads to both conditional and marginal distributions that are also normal.

Remarkably many time series models have been developed. These models, although conceptually and in some cases mathematically innovative, are generally restrictive. For example, some models restrict serial correlation to being positive. At this stage it is not clear which, if any, of the current models will become the dominant model for time series count data.

A review of linear time series models is given in section 7.2, along with a brief summary of seven different classes of count time series models. In section 7.3

we consider estimation of static regression models, as well as residual-based tests for serial correlation. In sections 7.4 to 7.10 each of the seven models is presented in detail. Basic static and dynamic regression models, controlling for both autocorrelation and heteroskedasticity present in time-series data, are detailed in sections 7.3 to 7.5. The simplest, though not necessarily fully efficient, estimators for these models are relatively straightforward to implement. For many applied studies this will be sufficient. For a more detailed analysis of data, the models of sections 7.6 to 7.8 are particularly appealing. Estimation (efficient estimation in the case of section 7.6) of these models entails complex methods, and implementation requires reading the original papers. Some of the models are applied in section 7.11 to monthly time-series data on the number of contract strikes in U.S. manufacturing, first introduced in section 7.3.4.

7.2 MODELS FOR TIME SERIES DATA

7.2.1 Linear Models

A key characteristic of a time series model is that it generates autocorrelated data sequences. For a continuous dependent variable, there are two well-established dynamic regression models that generate a serially dependent $\{y_t\}$ process.

The first is the *serially correlated error* model, which is the classic linear regression with serially correlated disturbance. This starts with a static regression function

$$y_t = \mathbf{x}_t'\boldsymbol{\beta} + u_t, \tag{7.1}$$

but then assumes that the error term u_t is serially correlated, following, for example, a linear ARMA process. The simplest case is an autoregressive error of order one (AR(1)),

$$u_t = \rho u_{t-1} + \varepsilon_t, \tag{7.2}$$

where ε_t is iid $(0, \sigma^2)$. Then the model can be rewritten as

$$y_t = \rho y_{t-1} + \mathbf{x}_t'\boldsymbol{\beta} - \mathbf{x}_{t-1}'\boldsymbol{\beta}\rho + \varepsilon_t, \tag{7.3}$$

which is an autoregressive model with nonlinear restrictions imposed on the parameters.

Because the dynamics of the $\{y_t|\mathbf{x}_t\}$ process are driven by the autoregressive parameter ρ through the latent process (7.2), this model is referred to as a *parameter-driven model* in the time-series count data literature (Cox, 1981; Zeger, 1988).

The second approach is to directly model the dynamic behavior of y. Beginning with *pure time series*, where the only explanatory variables are lagged values of the dependent variable, the standard class of linear dynamic models is the *autoregressive moving average model* of orders p and q, or ARMA(p, q) model. In the ARMA(p, q) model, the current value of y is the weighted sum of the past p values of y and the current and past q values of an iid error

$$y_t = \rho_1 y_{t-1} + \cdots + \rho_p y_{t-p} + \varepsilon_t + \gamma_1 \varepsilon_{t-1} + \cdots + \gamma_q \varepsilon_{t-q}, \qquad (7.4)$$

where ε_t is iid $(0, \sigma^2)$. Again following Zeger (1988), this dynamic model is called an *observation-driven model.* Stationarity of the time series $\{y_t\}$ implies restrictions on the parameters.

For linear *time-series regression* an *autoregressive model* or *dynamic model* includes both exogenous regressors and lagged dependent variables in the regression function. An example is

$$y_t = \rho y_{t-1} + \mathbf{x}_t'\boldsymbol{\beta} + \varepsilon_t, \qquad (7.5)$$

where the error term ε_t is iid $(0, \sigma^2)$. Note that this model is equivalent to assuming that

$$y_t | y_{t-1} \sim \mathsf{D}[\rho y_{t-1} + \mathbf{x}_t'\boldsymbol{\beta},\ \sigma^2], \qquad (7.6)$$

(i.e., y_t conditional on y_{t-1} and $\mathbf{x}_t$ has distribution with mean $\rho y_{t-1} + \mathbf{x}_t'\boldsymbol{\beta}$ and variance σ^2). More generally, additional lags of y and $\mathbf{x}$ may appear as regressors. If only $\mathbf{x}_t$ and lags of $\mathbf{x}_t$ appear, the model is instead called a *distributed lag model.* If $\mathbf{x}_t$ alone appears as a regressor, the model is called a *static model.*

The two preceding approaches, autoregression and error serial correlation, can be combined to yield an autoregressive model with serially correlated error.

Applying least squares to (7.1), ignoring (7.2), yields consistent and asymptotically normal estimates of $\boldsymbol{\beta}$. Efficient estimation for these models is by NLS, or is by ML if a distribution is specified for ε_t. In the latter case, specification of a normal distribution for ε_t leads by change of variable techniques to the joint density for $y_1, \ldots, y_T$. The NLS estimator instead minimizes the sum of squared residuals, $\sum_{t=1}^{T} \varepsilon_t^2$. For example, the model (7.1)–(7.2) leads to ε_t defined implicitly in (7.3), which will lead to first-order conditions that are nonlinear in parameters. Since ε_t is homoskedastic and uncorrelated, inference for the NLS estimator is the same as in the non-time series case. Note that when u_t is serially correlated, it is $\sum_t \varepsilon_t^2$ rather than $\sum_t u_t^2$ that is minimized. Minimizing the latter would lead to estimates that are inefficient, and even inconsistent if lagged dependent variables appear as regressors. The ML (based on normally distributed errors) and NLS estimators are asymptotically equivalent, though they differ in small samples due to different treatment of the first observation y_1. For models with an MA component in the error, estimation is more complicated but is covered in standard time-series texts.

To deal with the joint complications of nonstationarity and co-movements in multiple time series that occur commonly in economic time series, the literature on linear models for continuous data has been extended to models with unit roots, where ρ in (7.2) or (7.5) takes the value $\rho = 1$, and to the related analysis of cointegrated time series. Then y_t is nonstationary due to a nonstationary stochastic trend, and the usual asymptotic normal theory for estimators no longer applies. For count regression any nonstationarity is accommodated by deterministic trends, in which case the usual asymptotic theory is still applicable. Unit root tests for count data are mentioned briefly at the end of section 7.6.1.

7.2.2 Extensions of Linear Models

There are many extensions of the basic linear models sketched earlier: two are now presented because they are also considered later in the count context.

The *state-space* or *time-varying parameter model* is a modification of (7.1) that introduces dependence through parameters that vary over time, rather than through the error term. An example is

$$\begin{aligned} y_t &= \mathbf{x}_t'\boldsymbol{\beta}_t + \varepsilon_t \\ \boldsymbol{\beta}_t - \bar{\boldsymbol{\beta}} &= \boldsymbol{\Gamma}(\boldsymbol{\beta}_{t-1} - \bar{\boldsymbol{\beta}}) + \boldsymbol{v}_t, \end{aligned} \tag{7.7}$$

where ε_t is iid $(0, \sigma^2)$, $\boldsymbol{\Gamma}$ is a $k \times k$ matrix, and $\boldsymbol{v}_t$ is a $k \times 1$ iid $(\mathbf{0}, \boldsymbol{\Sigma})$ error vector. If the roots of $\boldsymbol{\Gamma}$ lie inside the unit circle, this model is stationary, and the model is called the return to normality model. Harvey and Phillips (1982) show how to estimate the model by reexpressing it in state space form and using the Kalman filter. This model is also widely used in Bayesian analysis of time series, where it is called the dynamic linear model. A detailed treatment is given in West and Harrison (1997).

The *hidden Markov model* or *regime shift model* allows the parameters to differ according to which of a finite number of regimes is currently in effect. The unobserved regimes evolve over time according to a Markov chain – hence the term "hidden Markov models." These models were popularized in economics by Hamilton (1989), who considered a two-regime Markov trend model

$$\widetilde{y}_t = \alpha_1 d_{t1} + \alpha_2 d_{t2} + \widetilde{y}_{t-1}, \tag{7.8}$$

where $\widetilde{y}_t$ is the trend component of y_t, and d_{tj} are indicator variables for whether in regime j, where $j = 1$ or 2. The transitions between the two regimes are determined by realization c_t of the first-order Markov chain C_t with transition probabilities

$$\gamma_{ij} = \Pr[C_t = j | C_{t-1} = j], \qquad i, j = 1, 2, \tag{7.9}$$

where $\gamma_{j1} + \gamma_{j2} = 1$. Then

$$d_{tj} = \begin{cases} 1 & \text{if } c_t = j \\ 0 & \text{otherwise} \end{cases} \qquad j = 1, 2. \tag{7.10}$$

Parameters to be estimated are the intercepts α_1 and α_2, the transition probabilities γ_{11} and γ_{21}, and the parameters in the model for the trend component $\widetilde{y}_t$ of the actual data y_t. An even simpler example sets $\widetilde{y}_t = y_t$ and omits $\widetilde{y}_{t-1}$ from (7.8), in which case dynamics are introduced solely via the Markov chain determining the regime switches.

7.2.3 Count Models

There are many possible time series models for count data. Different models arise through different models of the dependency of y_t on past y, current and past $\mathbf{x}$, and the latent process or error process ε_t; through different models of the latent process; and through different extensions of basic models.

Before presenting the various count models in detail, it is helpful to provide a summary. For simplicity the role of regressors other than lagged dependent variables is suppressed.

1. Serially correlated error models or latent variable models let y_t depend on a static component and a serially correlated latent variable. This is an extension of the serially correlated error model (7.1)–(7.2).
2. Autoregressive models or Markov models specify the conditional distribution of y_t to be a count distribution such as Poisson or NB2, with a mean parameter that is a function of lagged values of y_t. This is an extension of the autoregressive model (7.5). Here the conditional distribution of y_t is specified.
3. Integer-valued ARMA (or INARMA) models specify y_t to be the sum of an integer whose value is determined by past y_t and an independent innovation. Appropriate distributional assumptions lead to a count marginal distribution of y_t such as Poisson or NB2. This is a generalization of the autoregressive model (7.5). The INARMA model specifies the marginal distribution of y_t.
4. State space models or time-varying parameter models specify the distribution of y_t to be a count distribution such as Poisson or NB2, with conditional mean or parameters of the conditional mean that depend on their values in previous periods. This is an extension of the linear state space model (7.7).
5. Hidden Markov (HM) models or regime shift models specify the distribution of y_t to be a count distribution such as Poisson or NB2, with parameters that vary according to which of a finite number of regimes is currently in effect. The unobserved regimes evolve over time according to a Markov chain. This is an extension of the linear hidden Markov model (7.8) and (7.9).
6. Dynamic ordered probit models are a time series extension of the ordered probit model introduced in Chapter 3.
7. Discrete ARMA (or DARMA) models introduce time dependency through a mixture process.

Attempts have been made to separate these models into classes of models, but there is no simple classification system that nests all models. Some authors follow Cox (1981) and refer to models as either *parameter-driven*, with dependence induced by a latent variable process, or *observation-driven*, with time series dependence introduced by specifying conditional moments or densities as explicit functions of past outcomes.

Others distinguish between *conditional* models where the moments or density are conditional on both $\mathbf{x}_t$ and past outcomes of y_t, and *marginal* models where conditioning is only on $\mathbf{x}_t$ and not on past outcomes of y_t. This is most useful for distinguishing between models 1 and 2.

Once a model is specified, ML estimation is generally not as straightforward as in the normal linear case. NLS estimation is usually possible but may be inefficient because the error term may be heteroskedastic and/or autocorrelated. Estimation is often by nonlinear feasible GLS or by GMM.

Criteria for choosing between various models include ease of estimation – models 1–3 are easiest – and similarity to standard time series models such as having a serial correlation structure similar to ARMA models – models 3 and 7 are most similar. One should also consider the appropriateness of models to count data typically encountered and the problem at hand. If interest lies in the role of regressor variables, a static model of $\mathsf{E}[y_t|\mathbf{x}_t]$ may be sufficient. For forecasting, a conditional model of $\mathsf{E}[y_t|\mathbf{x}_t, y_{t-1}, y_{t-2}, \ldots]$ may be more useful.

7.3 STATIC COUNT REGRESSION

Before studying in detail various time-series count models, we consider static regression – for example, Poisson regression of y_t on $\mathbf{x}_t$ – and some simple extensions. We present a method for valid statistical inference in the presence of serial correlation. Residual-based tests for serial correlation are also presented; these are easiest to implement when standardized residuals are used.

Sometimes a static regression may be sufficient. Several regression applications of time series of counts, cited later, find little or no serial correlation. Then there is no need to use the models presented in this chapter. This may seem surprising, but it should be recalled that a pure Poisson point process will generate a time series of independent counts.

7.3.1 Estimation

We consider estimation of a static regression model, with exponential conditional mean

$$\mathsf{E}[y_t|\mathbf{x}_t] = \exp(\mathbf{x}_t'\boldsymbol{\beta}). \tag{7.11}$$

We allow for the possibility that the dependent variable y_t, conditional on the static regressors $\mathbf{x}_t$, is serially correlated, in addition to the usual concern of heteroskedasticity.

The Poisson QMLE, $\hat{\boldsymbol{\beta}}_{\mathsf{P}}$, solves $\sum_{t=1}^{T}(y_t - \exp(\mathbf{x}_t'\boldsymbol{\beta}))\mathbf{x}_t = \mathbf{0}$. It remains consistent in the presence of autocorrelation, provided (7.11) still holds. More difficult is obtaining a consistent estimator of the variance matrix of these estimators, which is necessary for statistical inference. We assume that autocorrelation is present to lag l, and define

$$\omega_{tj} = \mathsf{E}[(y_t - \mu_t)(y_{t-j} - \mu_{t-j})|\mathbf{x}_1, \ldots, \mathbf{x}_T], \qquad j = 0, \ldots, l, \tag{7.12}$$

where $\mu_t = \exp(\mathbf{x}_t'\boldsymbol{\beta})$. The time-series results of Chapter 2.7.1 are applicable, with $g_t = (y_t - \mu_t)x_t$. Then $\hat{\boldsymbol{\beta}}_{\mathsf{P}}$ is asymptotically normal with mean $\boldsymbol{\beta}$ and variance matrix

$$\mathsf{V}[\hat{\boldsymbol{\beta}}_{\mathsf{P}}] = \left(\sum_{t=1}^{T} \mu_t \mathbf{x}_t \mathbf{x}_t'\right)^{-1} \mathbf{B}_{\mathsf{P}} \left(\sum_{t=1}^{T} \mu_t \mathbf{x}_t \mathbf{x}_t'\right)^{-1}, \tag{7.13}$$

where

$$\mathbf{B}_{\mathsf{P}} = \sum_{t=1}^{T} \omega_{t0} \mathbf{x}_t \mathbf{x}_t' + \sum_{j=1}^{l} \sum_{t=l}^{T} \omega_{tj} (\mathbf{x}_t \mathbf{x}_{t-j}' + \mathbf{x}_{t-j} \mathbf{x}_t'). \tag{7.14}$$

A robust sandwich estimate of $\mathsf{V}[\hat{\boldsymbol{\beta}}_{\mathsf{P}}]$, called a heteroskedasticity and autocorrelation consistent (HAC) estimate, replaces μ_t by $\hat{\mu}_t = \exp(x_t'\hat{\boldsymbol{\beta}})$ and ω_{tj} by $(y_t - \hat{\mu}_t)(y_{t-j} - \hat{\mu}_{t-j})$ and uses Newey-West weights defined in Chapter 2.7.1, so

$$\hat{\mathbf{B}}_{\mathsf{P}} = \sum_{t=1}^{T} \hat{u}_t^2 \mathbf{x}_t \mathbf{x}_t' + \sum_{j=1}^{l} \left(1 - \frac{j}{l+1}\right) \times \left(\sum_{t=j+1}^{T} \hat{u}_t \hat{u}_{t-j} \mathbf{x}_t \mathbf{x}_{t-j}' + \sum_{t=j+1}^{T} \hat{u}_{t-j} \hat{u}_t \mathbf{x}_{t-j} \mathbf{x}_t'\right), \tag{7.15}$$

where $\hat{u}_t = y_t - \hat{\mu}_t$. Some models presented later place more structure on variances and autocovariances, that is, on ω_{tj}. In particular, we expect ω_{tj} to be a function of μ_t and μ_{t-j}. Then consistent estimates $\hat{\omega}_{tj}$ of ω_{tj} may be used in (7.14). In these cases it is better, of course, to use alternative estimators to the Poisson QMLE that use models for ω_{tj} to obtain more efficient estimators.

The variance matrix estimate also applies to distributed lag models. Then interpret $\mathbf{x}_t$ as including lagged exogenous variables as well as contemporaneous exogenous variables. The results also extend to dynamic models with lagged dependent variables as regressors, provided enough lags are included to ensure that there is no serial correlation in y_t after controlling for regressors, so $\omega_{tj} = 0$ for $j \neq 0$.

The results do not apply if lagged dependent variables are regressors, and there is serial correlation in y_t after controlling for regressors. Then the Poisson QMLE or the NLS estimator is inconsistent, just as the OLS estimator is inconsistent in similar circumstances in the linear model.

7.3.2 Detecting Serial Correlation

Tests based on residuals from static regressions can indicate whether any time-series corrections are necessary or whether results from the previous chapters can still be used.

We first consider the use of a standardized residual z_t that has mean zero and constant variance, the natural choice being the Pearson residual. For the Poisson GLM or, equivalently, Poisson QMLE with NB1 variance function, $z_t = (y_t - \hat{\mu}_t)/\sqrt{\hat{\mu}_t}$. For more general variance functions $\omega_t = \mathsf{V}[y_t|\mathbf{x}_t]$, use $z_t = (y_t - \hat{\mu}_t)/\sqrt{\hat{\omega}_t}$.

Li (1991) formally obtained the asymptotic distribution of autocorrelations based on Pearson residuals in GLMs, including $z_t = (y_t - \hat{\mu}_t)/\sqrt{\hat{\mu}_t}$ for the Poisson. In this subsection we specialize to the distribution under the null hypothesis that there is no serial correlation.

Then asymptotically we can follow Box-Jenkins modeling as in the continuous case, using the autocorrelation function (ACF), which plots the estimated correlations $\hat{\rho}_k$ against k, where

$$\hat{\rho}_k = \frac{\sum_{t=k+1}^{T} z_t z_{t-k}}{\sum_{t=1}^{T} z_t^2}. \tag{7.16}$$

If no correlation is present $\hat{\rho}_k \simeq 0$, $k \neq 0$. Formal tests are based on the standard result that $\hat{\rho}_k \overset{a}{\sim} \mathsf{N}[0, \frac{1}{T}]$ under the null hypothesis that $\rho_j \equiv \mathsf{Cor}[z_t, z_{t-j}] = 0$, $j = 1, \ldots, k$. Then the test statistic

$$\mathsf{T}_k = \sqrt{T}\hat{\rho}_k. \tag{7.17}$$

is asymptotically $\mathsf{N}[0, 1]$. An overall test for serial correlation is the Box-Pierce portmanteau statistic, based on the sum of the first l squared sample correlation coefficients, $\mathsf{T}_{\mathsf{BP}} = T\sum_{k=1}^{l} \hat{\rho}_k^2$. More common is to use a variant, the Ljung-Box portmanteau statistic,

$$\mathsf{T}_{\mathsf{LB}} = T(T+2)\sum_{k=1}^{l} \frac{1}{T-k}\hat{\rho}_k^2. \tag{7.18}$$

Both T_{BP} and T_{LB} are asymptotically $\chi^2(l)$ under the null hypothesis that $\rho_k = 0$, $k = 1, \ldots, l$.

The preceding tests require that z_t is standardized with mean zero and constant variance. If z_t is not standardized, or z_t is standardized but one wants to guard against incorrect specification of the variance, the correct test statistic to use is

$$\mathsf{T}_k^* = \frac{\sum_{t=k+1}^{T} z_t z_{t-k}}{\sqrt{\sum_{t=k+1}^{T} z_t^2 z_{t-k}^2}}, \tag{7.19}$$

rather than (7.17). $\mathsf{T}_k^* \overset{a}{\sim} \mathsf{N}[0, 1]$ under the null hypothesis of no autocorrelation; see the derivations section. The statistic (7.19) is the sample analog of

$$\rho_{kt} = \mathrm{E}[z_t z_{t-k}]/\sqrt{\mathrm{E}[z_t^2]\mathrm{E}[z_{t-k}^2]},$$

whereas the statistic (7.16) used the simplification that $\mathsf{E}[z_t^2] = \mathsf{E}[z_{t-k}^2]$ given standardization. An overall $\chi^2(l)$ test, analogous to T_{BP}, is

$$\mathsf{T}_{\mathsf{BP}}^* = \sum_{k=1}^{l} (\mathsf{T}_k^*)^2. \tag{7.20}$$

An alternative method, for unstandardized z_t, is to regress z_t on z_{t-k} and use robust sandwich (or Eicker-White) heteroskedastic-consistent standard errors to perform individual t-tests that are asymptotically standard normal if there is no correlation. A joint Wald test of autocorrelation to lag l can be based on regression of z_t on x_t and $z_{t-1}, \ldots, z_{t-l}$, again using heteroskedastic-consistent standard errors. This residual-augmented test is an extension of similar well-established tests of serial correlation in the time-series linear regression literature; see Godfrey (1988). Only heteroskedasticity is controlled for, but this is sufficient because under the null hypothesis there is no need to control for serial correlation. However, the standard errors should additionally be adjusted because the Pearson residuals are not the true residuals, but rather their generated counterparts, and hence are subject to sampling error. The robust sandwich variance estimate does not allow for this added complication. Under the null hypothesis of no autocorrelation at any lag, one can use a standard bootstrap for a two-step estimator to obtain the estimator's variance matrix.

Brännäs and Johansson (1994) find good small-sample performance of serial correlation tests based on T_k for samples of size 40. Cameron and Trivedi (1993) present tests for the stronger condition of independence in count data and argue that it is better to base tests on orthogonal polynomials rather than simply powers of z_t.

Several time-series count data applications have found little evidence of serial correlation in residuals in models with exogenous contemporaneous or lagged regressors. Examples include the annual number of bank failures in the United States in 1947–81 (Davutyan, 1989), daily data on homicides in California (Grogger, 1990), and the daily number of deaths in Salt Lake County (Pope, Schwartz, and Ransom, 1992).

It is important to note that the previous tests are valid only in static models. Autocorrelation tests based on residuals from dynamic regressions that include lagged dependent variables require adjustment, just as they do for the linear regression model.

7.3.3 Trends and Seasonality

Time dependence can be modeled by extending the pure Poisson process of Chapter 1.1.2 to a *nonhomogeneous* (or nonstationary) *Poisson process*. For this process the rate parameter μ is replaced by $\mu(t)$, allowing the rate to vary with the time elapsed since the start of the process. Let $N(t, t+h)$ denote the number of events occurring in the interval $(t, t+h]$. For the nonhomogeneous Poisson process, $N(t, t+h)$ has Poisson distribution with mean

$$\mathsf{E}[N(t, t+h)] = \int_t^{t+h} \mu(s)ds. \tag{7.21}$$

As an example, suppose there is an exponential deterministic time trend in the rate parameter. When $\mu(t) = \lambda \exp(\alpha t)$ and $h = 1$, (7.21) yields

$$\mathsf{E}[N(t, t+1)] = \frac{\lambda(\exp(1) - 1)}{\alpha} \exp(\alpha t),$$

which introduces an exponential time trend to the Poisson mean. For regression analysis, this suggests the model

$$\mathsf{E}[y_t|\mathbf{x}_t] = \exp(\mathbf{x}_t'\boldsymbol{\beta} + \alpha t),$$

ignoring the additional complication introduced by time-varying regressors $\mathbf{x}_t$.

Seasonality can be modeled in ways similar to those used in linear models. For example, consider monthly data with annual seasonal effects. A set of 11 monthly seasonal indicator variables might be included as regressors. Using the usual exponential conditional mean, the seasonal effects will be multiplicative. More parsimonious may be to include a mix of sine and cosine terms, $\cos(2\pi jt/12)$ and $\sin(2\pi jt/12)$, $j = 1, \ldots, p$, for some chosen p.

An early and illuminating discussion of trends and seasonality in count data models is given by Cox and Lewis (1966).

7.3.4 Example: Strikes

We analyze the effect of the level of economic activity on strike frequency, using monthly U.S. data from January 1968 to December 1976. The dependent variable *STRIKES* is the number of contract strikes in U.S. manufacturing beginning each month. The one independent variable, *OUTPUT*, is a measure of the cyclical departure of aggregate production from its trend level. High values of *OUTPUT* indicate a boom and low levels of recession. A static model is analyzed here, whereas application of some dynamic models is reported in section 7.11.

The application comes from Kennan (1985) and Jaggia (1991), who analyzed the data using duration models, notably the Weibull, applied to the completed length of each strike that began during this period. Here we instead model the number of strikes beginning each month during the period. Time series methods for counts are likely to be needed, since Kennan found evidence of

Table 7.1. *Strikes: Variable definitions and summary statistics*

Variable	Definition	Mean	Standard deviation
STRIKES	Number of strikes that began each month	5.241	3.751
OUTPUT	Deviation of monthly industrial production from its trend level	−.004	.055

duration dependence. The data are from table 1 of Kennan (1985). For the five months where there were no strikes (*STRIKES* = 0), the data on *OUTPUT* were not given. We interpolated these data by averaging adjacent observations, giving values of 0.06356, −0.0743, 0.04591, −0.04998, and −0.06035 in, respectively, the months 1969(12), 1970(12), 1972(11), 1974(12), and 1975(9).

Summary statistics and variable definitions are given in Table 7.1. The sample mean number of strikes is low enough to warrant treating the data as count data. Sixty percent of the observations take values between 0 and 5, and the largest count is 18. There is considerable overdispersion in the raw data, with the sample variance 2.68 times the raw mean.

The data on *STRIKES* and *OUTPUT* are plotted in Figure 7.1, with different scalings for the two series given on, respectively, the left and right axes. It appears that strike activity increases with an increase in economic activity, though with a considerable lag during the middle of the sample period. The data on *STRIKES* are considerably more variable than the data on *OUTPUT*.

Estimates from static Poisson regression with exponential conditional mean are given in Table 7.2. The standard errors reported are robust sandwich

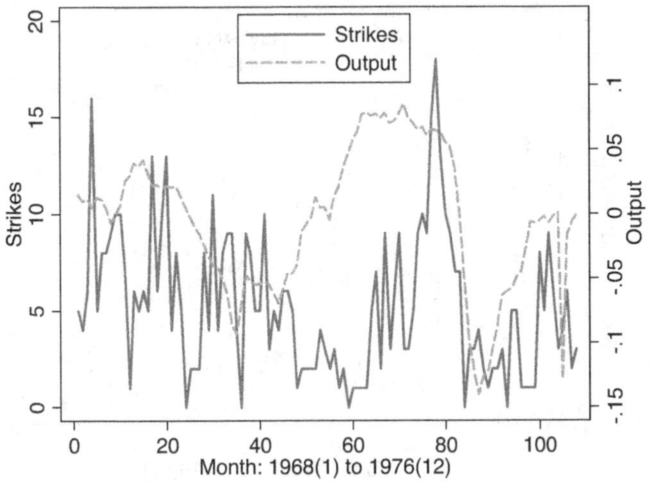

Figure 7.1. Strikes (left axis) and output (right axis) per month.

Table 7.2. *Strikes: Poisson QMLE with heteroskedastic and autocorrelation robust standard errors*

		Standard errors			
Variable	Coefficient	HAC0	HAC4	HAC8	HAC12
ONE	1.654	0.066	0.103	0.116	0.123
OUTPUT	3.134	1.184	1.921	2.222	2.402

Note: Standard errors correct for overdispersion and for autocorrelation of length 0, 4, 8, 12, and 16 months.

estimates based on (7.13) and (7.14) that control for overdispersion and for autocorrelation of, respectively, 0, 4, 8, and 12 months. The coefficient of OUTPUT indicates that as economic activity increases the number of strikes increases, with a one-standard-deviation change in OUTPUT leading to a 17% increase in the mean number of strikes ($0.055 \times 3.134 = 0.172$). The effect is statistically significant at 5% when autocorrelation is ignored ($t = 3.134/1.184 = 2.48$), but once correlation is controlled for the effect becomes statistically insignificant. For example, $t = 3.134/2.402 = 1.30$ if there is autocorrelation up to 12 months.

A plot of the predicted value of *STRIKES* from this static regression, $\exp(1.654 + 3.134 \times OUTPUT)$, is given in Figure 7.2. Clearly *OUTPUT* explains only a small part of the variation in *STRIKES*. This is also reflected in the low R-squareds. The Poisson deviance R^2 and Pearson R^2 defined in section 5.3.3 are, respectively, 0.049 and 0.024, and the squared correlation between

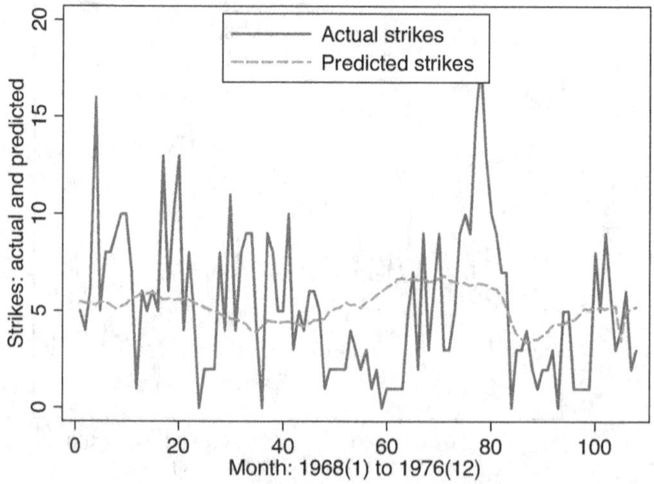

Figure 7.2. Actual and predicted strikes from a static regression model.

Table 7.3. *Strikes: Residuals autocorrelations and serial correlation tests*

	Autocorrelations			z Statistics for p	
Lag	y	r	p	Standardized	Unstandardized
1	0.49	0.47	0.44	4.60	4.07
2	0.43	0.40	0.38	3.91	3.96
3	0.37	0.34	0.32	3.36	3.40
4	0.23	0.20	0.20	2.10	2.21
5	0.13	0.11	0.11	1.13	1.22
6	0.04	0.03	0.03	0.31	0.33
7	0.01	0.01	0.01	0.15	0.17
8	0.02	0.04	0.04	0.40	0.44
9	0.05	0.09	0.08	0.85	1.06
10	0.01	0.06	0.05	0.47	0.55
11	0.06	0.10	0.10	1.04	1.17
12	0.00	0.03	0.03	0.31	0.38
BP				55.76	
LB				58.10	
BP*					53.47

Note: Autocorrelations up to lag 12 are based on three different residuals: the raw data (y) on STRIKES, and the raw residual (r) and the Pearson residual (p) from Poisson regression of STRIKES on OUTPUT. The standardized z statistics (T_k) for p require the variance to be a multiple of the mean. The unstandardized z statistics (T_k^*) for p do not require the variance to be a multiple of the mean. Both are N[0,1] asymptotically. BP is the Box-Pierce statistic, LB is the Ljung-Box statistic, and BP* is the statistic T_{BP}^*. All three are chi-squared distributed with 12 degrees of freedom.

y and $\hat{\mu}$ is 0.051. Considerable improvement is reported in section 7.11, when lagged dependent variables are included as regressors.

Measures of autocorrelation and tests for serial correlation are presented in Table 7.3. The first three columns give autocorrelation coefficients based on various residuals: (1) the raw data y on *STRIKES* or equivalently the residual $y - \bar{y}$ from Poisson regression of *STRIKES* on a constant; (2) the raw residual $r = y - \hat{\mu}$ from Poisson regression of *STRIKES* on a constant and *OUTPUT*; and (3) the Pearson residual $p = (y - \hat{\mu})/\sqrt{\hat{\mu}}$ from this regression. Since there is relatively little variability in μ around $\bar{y}$, see Figure 7.2, in this application we expect the autocorrelations to tell a similar story, even though those based on the Pearson residual are theoretically preferred.

The first three columns of Table 7.3 all indicate autocorrelation that dies out after five lags. The last two columns present two different tests for autocorrelation based on the Pearson residuals. The first test assumes that the variance is indeed a multiple of the mean, so the Pearson residual is standardized to have constant variance, and we can work with $\sqrt{T}\hat{\rho}_k$ in (7.17) and the associated Box-Pierce and Ljung-Box statistics. Autocorrelation is statistically significant at level 0.05 at the first four lags and is overall statistically significant

Table 7.4. *Strikes: Actual and fitted frequency distributions of Poisson and NB2 regressions*

	Count										
	0	1	2	3	4	5	6	7	8	9	10
Actual frequency	0.046	0.111	0.130	0.102	0.083	0.130	0.083	0.037	0.065	0.093	0.056
Fitted: Poisson	0.008	0.035	0.081	0.129	0.160	0.163	0.142	0.109	0.074	0.046	0.026
Fitted: NB2	0.049	0.094	0.119	0.125	0.118	0.105	0.088	0.072	0.057	0.044	0.034

Note: The fitted frequencies are the sample averages of the cell frequencies estimated from the fitted models.

since $\chi^2(12)$ at level 0.05 has the critical value 21.03. The second test no longer requires that the variance is a multiple of the mean and uses (7.19) and (7.20). The results are similar. Finally, OLS regression of the Pearson residual on *OUTPUT* and the first 12 lagged Pearson residuals yields an overall robust F-test statistic of 3.96, asymptotically equivalent to a $\chi^2(12)$ statistic of $12 \times 3.96 = 47.52$, similar to the portmanteau test statistics.

Although it is natural to test for serial correlation in a time-series model, overdispersion also remains a distinct possibility. Indeed serial correlation leads to overdispersion in the static model and will generally lead to a deficient fit of the model. The graphical comparison of the original time series and the fitted values from the Poisson regression conveys a clear impression of poor fit. Comparison of actual and fitted Poisson frequencies, given in Table 7.4, also confirms the impression of poor fit.

An NB2 regression fitted to the same data strongly rejects the null hypothesis of equidispersion and improves the fit of the model. However, although the discrepancies between fitted and actual distributions are smaller, even the NB2 model cannot generate the multimodal distribution of the raw data. Tests for autocorrelation following NB2 estimation should apply (7.17) and the associated Box-Pierce and Ljung-Box statistics to the Pearson residual $p = (y - \hat{\mu})/\sqrt{\hat{\mu} + \hat{\alpha}\hat{\mu}^2}$.

7.4 SERIALLY CORRELATED HETEROGENEITY MODELS

A commonly used extension of the static regression model in the linear case is to model the additive error as following an ARMA process, leading to more efficient FGLS estimation. For the Poisson model the analogous method introduces a latent multiplicative error that is serially correlated. The dynamics of y_t are then driven by the dynamics of the latent process, with resulting autocorrelations being a function of current and lagged values of $\mathsf{E}[y_t|\mathbf{x}_t]$.

7.4.1 Zeger Model

Zeger (1988) introduced serial correlation in y_t by multiplying the conditional mean in the static regression model of section 7.3 by a serially correlated multiplicative latent variable. This enables more efficient estimation than the

Poisson QMLE of section 7.3, assuming that the properties of the latent variable are correctly specified.

For counts the Poisson is used as a starting point, with variance equal to the mean. The *conditional distribution* of the dependent variable y_t is specified to be Poisson with mean $\lambda_t \varepsilon_t$, where

$$\lambda_t = \exp(\mathbf{x}_t' \boldsymbol{\beta}),$$

and $\varepsilon_t > 0$ is an unobserved latent variable. The first two moments of y_t conditional on both $\mathbf{x}_t$ and ε_t are

$$\begin{aligned} \mathsf{E}[y_t|\lambda_t, \varepsilon_t] &= \lambda_t \varepsilon_t \\ \mathsf{V}[y_t|\lambda_t, \varepsilon_t] &= \lambda_t \varepsilon_t. \end{aligned} \tag{7.22}$$

This set-up is similar to the mixture models in Chapter 4, except here ε_t is no longer assumed to be independently distributed across observations. Instead, the latent variable ε_t is assumed to follow a stationary process with the mean normalized to unity, variance σ_ε^2, covariances $\gamma_{k\varepsilon} \equiv \mathsf{Cov}[\varepsilon_t, \varepsilon_{t-k}]$, and autocorrelations

$$\begin{aligned} \rho_{k\varepsilon} &\equiv \mathsf{Cor}[\varepsilon_t, \varepsilon_{t-k}] \\ &= \mathsf{E}[(\varepsilon_t - 1)(\varepsilon_{t-k} - 1)] / \sqrt{\mathsf{E}[(\varepsilon_t - 1)^2]\mathsf{E}[(\varepsilon_{t-k} - 1)^2]} \\ &= \frac{\gamma_{k\varepsilon}}{\sigma_\varepsilon^2}, \qquad k = 1, 2, \ldots . \end{aligned} \tag{7.23}$$

It follows that the *marginal distribution* of y_t (i.e., marginal with respect to ε_t while still conditional on λ_t and hence $\mathbf{x}_t$), has first two moments

$$\begin{aligned} \mu_t &= \mathsf{E}[y_t|\lambda_t] = \lambda_t \\ \sigma_t^2 &= \mathsf{V}[y_t|\lambda_t] = \lambda_t + \sigma_\varepsilon^2 \lambda_t^2, \end{aligned} \tag{7.24}$$

inducing overdispersion of the NB2 form. The latter result uses

$$\begin{aligned} \mathsf{V}[y_t|\lambda_t] &= \mathsf{E}[\mathsf{V}[y_t|\lambda_t, \varepsilon_t]|\lambda_t] + \mathsf{V}[\mathsf{E}[y_t|\lambda_t, \varepsilon_t]|\lambda_t] \\ &= \mathsf{E}[\lambda_t \varepsilon_t|\lambda_t] + \mathsf{V}[\lambda_t \varepsilon_t|\lambda_t] \\ &= \lambda_t \mathsf{E}[\varepsilon_t] + \lambda_t^2 \mathsf{V}[\varepsilon_t]. \end{aligned}$$

The latent process also induces autocorrelation in y_t. The covariance between y_t and y_{t-k} can be shown to equal $\gamma_{k\varepsilon}\lambda_t\lambda_{t-k}$, implying that the (time-varying) autocorrelation function for y_t is

$$\rho_{ky_t} = \frac{\rho_{k\varepsilon}}{\sqrt{\left(1 + (\sigma_\varepsilon^2 \lambda_t^2)^{-1}\right)\left(1 + (\sigma_\varepsilon^2 \lambda_{t-k}^2)^{-1}\right)}}, \qquad k = 1, 2, \ldots . \tag{7.25}$$

We refer to this model as a *serially correlated error model* because of its obvious connection to the linear model (7.1)–(7.2). Other authors refer to it as a marginal model, because from (7.24) we model $\mu_t = \exp(\mathbf{x}_t'\boldsymbol{\beta})$, which does not condition on lagged y_t. This model is an extension of the cross-section Poisson model for overdispersion, one that simultaneously introduces overdispersion, via σ_ε^2, and serial dependence, via σ_ε^2 and $\rho_{k\varepsilon}$.

7.4.2 QML Estimation

An obvious starting point is the standard Poisson QMLE $\hat{\boldsymbol{\beta}}_{\mathsf{P}}$ obtained by Poisson regression of y_t on $\mathbf{x}_t$ with conditional mean $\exp(\mathbf{x}_t'\boldsymbol{\beta})$. This estimator ignores the complication of the latent variable ε_t. Just as this estimator was shown to be consistent under overdispersion, it remains consistent here.

Davis, Dunsmuir, and Wang (2000) prove the consistency and asymptotically normality of the Poisson QMLE under the assumption that $\varepsilon_t = \exp(\alpha_t)$, where $\{\alpha_t\}$ is a linear Gaussian process with mean $-\sigma_\alpha^2/2$ and variance σ_α^2, so ε_t has mean 1, and with bounded autocovariances. The regressors may include trend functions, that form triangular arrays, harmonic functions, and stationary processes; see also Brännäs and Johansson (1994). The asymptotic variance matrix is that given in (7.13) and (7.14), where in the current model $\omega_{tj} = \gamma_{j\varepsilon}\lambda_t\lambda_{t-j}$.

Davis et al. (2000) also consider tests of autocorrelation in this model. First, they note that estimation error in $\hat{\lambda}_t$ leads to biased estimates of σ_ε^2 and ρ_k in finite samples, and they propose improved bias-adjusted estimators. Second, they note that the usual Poisson model Pearson residual $p_t = (y_t - \lambda_t)/\sqrt{\lambda_t}$ has variance in the current model of $(1 + \sigma_\varepsilon^2\lambda_t)$, so the standard Ljung-Box test statistic defined in (7.18) cannot be used to test autocorrelation in y_t. They derive a valid modified Ljung-Box test statistic. Alternatively the test statistic $\mathsf{T}_{\mathsf{BP}}^*$ defined in (7.20) can still be used or (7.18) can be applied to $p_t = (y_t - \lambda_t)/\sqrt{\lambda_t + \sigma_\varepsilon^2\lambda_t}$, which has constant variance in the current model.

Third, Davis et al. (2000) note that the autocorrelations in y_t are smaller than those of ε_t. This is clear from (7.25) because $\rho_{ky_t} \to 0$ as $\sigma_\varepsilon^2 \to 0$ for given $\rho_{k\varepsilon}$. They also propose tests for autocorrelation of ε_t. It is quite possible to find little autocorrelation in the observed process y_t, even though there is considerable autocorrelation in the latent process ε_t. Application is to the asthma attacks data introduced in Chapter 1.3.4.

Davis and Wu (2009) provide an extension to the negative binomial (NB2) model, parameterized as

$$\begin{aligned}\mathsf{E}[y_t|\lambda_t, \varepsilon_t] &= r\lambda_t\varepsilon_t \\ \mathsf{V}[y_t|\lambda_t, \varepsilon_t] &= r\lambda_t\varepsilon_t(1 + r\lambda_t\varepsilon_t),\end{aligned}$$

where $\lambda_t = \exp(\mathbf{x}_t'\boldsymbol{\beta})$, and the serially correlated latent variable ε_t has mean one and autocovariances $\gamma_{k\varepsilon} \equiv \mathsf{Cov}[\varepsilon_t, \varepsilon_{t-k}]$. Then

$$\begin{aligned}\mathsf{E}[y_t|\lambda_t] &= \mu_t = r\lambda_t \\ \mathsf{V}[y_t|\lambda_t] &= \mu_t + \mu_t^2(1 + \gamma_{0\varepsilon}(1 + r))/r,\end{aligned}$$

and $\mathsf{Cov}[y_t, y_{t-k}|\lambda_t, \lambda_{t-k}] = \mu_t\mu_{t-k}\gamma_{k\varepsilon}$. They consider the QMLE of $\boldsymbol{\beta}$ that maximizes the NB2 log-likelihood with overdispersion parameter r specified, prove that it is consistent if $\{\varepsilon_t\}$ is a strongly mixing process, and obtain the asymptotic distribution.

7.4.3 GEE Estimation

The Poisson QMLE is inefficient when a serially correlated error is present. More efficient moment-based estimation uses knowledge of the specified model for the conditional mean, variance, and autocovariances of y_t. Zeger (1988) proposed estimation using the GEE estimator, presented in Chapter 2.4.6, adapted to serially correlated data.

The GEE estimator for $\boldsymbol{\beta}$ in the model of section 7.4.1 solves the first-order conditions

$$\mathbf{D}'\mathbf{V}^{-1}(\mathbf{y}-\boldsymbol{\lambda}) = \mathbf{0}, \tag{7.26}$$

where $\mathbf{D}$ is the $T \times k$ matrix with tj^{th} element $\partial\lambda_t/\partial\beta_j$, $\mathbf{V}^{-1}$ is a $T \times T$ weighting matrix defined later, $\mathbf{y}$ is the $T \times 1$ vector with t^{th} entry y_t, and $\boldsymbol{\lambda}$ is the $T \times 1$ vector with t^{th} entry $\lambda_t = \mathsf{E}[y_t|\mathbf{x}_t]$. Equation (7.26) is of the same form as the linear WLS estimator given in Chapter 2.4.1 (i.e., $\mathbf{X}'\mathbf{V}^{-1}(\mathbf{y}-\mathbf{X}'\boldsymbol{\beta}) = \mathbf{0}$), except that in moving to nonlinear models x_{tj} in $\mathbf{X}$ is replaced by $\partial\lambda_t/\partial\beta_j$ in $\mathbf{D}$ and the mean $\mathbf{X}'\boldsymbol{\beta}$ is replaced by $\boldsymbol{\lambda}$. Then by similar results to Chapter 2.4.1, $\hat{\boldsymbol{\beta}}_{\mathsf{GEE}}$ is asymptotically normal with mean $\boldsymbol{\beta}$ and variance

$$\mathsf{V}[\hat{\boldsymbol{\beta}}_{\mathsf{GEE}}] = (\mathbf{D}'\mathbf{V}^{-1}\mathbf{D})^{-1}\mathbf{D}'\mathbf{V}^{-1}\boldsymbol{\Omega}\mathbf{V}^{-1\prime}\mathbf{D}\,(\mathbf{D}'\mathbf{V}^{-1}\mathbf{D})^{-1}, \tag{7.27}$$

where $\boldsymbol{\Omega} = \boldsymbol{\Omega}(\boldsymbol{\beta}, \boldsymbol{\gamma}, \sigma^2) = \mathsf{V}[\mathbf{y}|\mathbf{X}]$ is the covariance matrix of $\mathbf{y}$, and $\boldsymbol{\gamma}$ are parameters of the autocorrelation function $\rho_\varepsilon(k)$.

The efficient GEE estimator sets $\mathbf{V}^{-1} = \hat{\boldsymbol{\Omega}}^{-1}$, where $\hat{\boldsymbol{\Omega}}$ is a consistent estimator of $\boldsymbol{\Omega} = \boldsymbol{\Omega}(\boldsymbol{\beta}, \boldsymbol{\gamma}, \sigma^2)$. This entails inversion of the $T \times T$ estimated covariance matrix of y_t, which poses problems for large T. Zeger instead proposed using a less efficient GEE estimator that specifies a working weighting matrix $\mathbf{V}^{-1}$ that is chosen to be reasonably close to $\boldsymbol{\Omega}^{-1}$.

Zeger (1988) applied this method to monthly U.S. data on polio cases, whereas Campbell (1994) applied this model to daily U.K. data on sudden infant death syndrome cases and the role of temperature. Brännäs and Johansson (1994) find that the Poisson QMLE yields quite similar estimates and efficiency to that using Zeger's method, once appropriate correction for serial correlation is made to the Poisson QMLE standard errors. Johansson (1995) presents Wald and LM type tests based on GMM estimation (see Newey and West, 1987b) for overdispersion and serial correlation in the Zeger model, and investigates their small-sample performance by a Monte Carlo study.

7.4.4 ML Estimation

Fully efficient ML estimation is possible given specification of a parametric model for the latent process.

In the cross-section case a Poisson-lognormal mixture model specifies $\ln(\varepsilon_t)$ to be normally distributed. A time series generalization has $\ln(\varepsilon_t)$ follow a Gaussian first-order autoregression. Defining $v_t = \ln(\varepsilon_t)$, we have

$$v_t = \delta v_{t-1} + \sigma u_t, \qquad u_t \sim \text{iid N}[0, 1]. \tag{7.28}$$

Here $\delta = 0$ yields the usual Poisson-lognormal mixture model, and the restrictions $\delta = 0$ and $\sigma = 0$ reduce the model to the Poisson case. This model is stationary if $|\delta| < 1$.

Jung et al. (2006) call the model of section 7.4.1 with ν_t as in (7.28) the stochastic autoregressive mean (SAM) model. Then $y_t|\mu_t, v_{t-1}, u_t \sim \text{P}[\exp(\mathbf{x}_t'\boldsymbol{\beta} + \delta v_{t-1} + \sigma u_t)]$. Let

$$f(y_t|\exp(\mathbf{x}_t'\boldsymbol{\beta} + \delta \nu_{t-1} + \sigma u_t))$$

denote this Poisson density. The marginal distribution $h(y_t|\mu_t, v_{t-1})$ can be obtained by integrating out u_t, so

$$h(y_t|\mu_t, v_{t-1}) = \int_{-\infty}^{\infty} f(y_t|\exp(\mathbf{x}_t'\boldsymbol{\beta} + \delta \nu_{t-1} + \sigma u_t)) \frac{1}{\sqrt{2\pi}} e^{-u^2/2} du_t.$$

There is no closed-form solution for $h(y_t|\mu_t, v_{t-1})$, but the expression can be approximated numerically by adapting simulated ML, presented in Chapter 4.2.4, to a time-series setting.

First, note that given an initial condition, $\nu_0 = \text{E}[\nu_t] = 0$, and random draws from the N[0, 1] distribution, say $\mathbf{u}^{(s)} = \{u_1^{(s)}, \ldots, u_T^{(s)}\}$, the sequence $\mathbf{v}^{(s)} = \{v_1^{(s)}, \ldots, v_T^{(s)}\}$ can be generated recursively for a given value of δ using (7.28). Therefore, the unobserved latent variables (u_t, ν_t) can be replaced by simulated values. Then

$$h(y_t|\, \mu_t, \nu_0, u_1^{(s)}, \ldots, u_t^{(s)}) \approx \frac{1}{S} \sum_{s=1}^{S} f(y_t|\exp(\mathbf{x}_t'\boldsymbol{\beta} + \delta v_{t-1}^{(s)} + \sigma u_t^{(s)}))]. \tag{7.29}$$

The simulated likelihood function can be built up using T terms like that above. If the heterogeneity term were observable, then the likelihood could be constructed directly. But absent such information, we use S (preferably a large integer) draws from a pseudorandom number generator. Simulated ML estimates are obtained by maximizing the log-likelihood function

$$\mathcal{L}(\boldsymbol{\beta}, \delta, \sigma) = \sum_{t=1}^{T} \ln \frac{1}{S} \sum_{s=1}^{S} f(y_t|\exp(\mathbf{x}_t'\boldsymbol{\beta} + \delta v_{t-1}^{(s)} + \sigma u_t^{(s)})), \tag{7.30}$$

with respect to $(\boldsymbol{\beta}, \delta, \sigma)$. Additional details of the properties of this method are given in Chapter 4.2.4. This approach can potentially be extended to nonnormal distributions of the latent variable as well as to higher order ARMA processes.

Other alternative simulation methods may also be applied to estimate the SAM model. Jung et al. (2006) apply efficient importance sampling in their analysis of the Sydney asthma counts data described in Chapter 1.3.4. The parametric formulations similar in spirit to that given earlier are also amenable to analysis by Bayesian Markov chain Monte Carlo (MCMC) methods as shown by Chan and Ledolter (1995).

7.5 AUTOREGRESSIVE MODELS

The most direct way to specify a time-series model is to include lagged values of the dependent count variable as regressors in a standard count regression model that may also include current and past values of exogenous variables as regressors. The models are then called autoregressive models or observation-driven models. Relative to the parameter-driven models of section 7.4, they are easier to estimate and to interpret and are more convenient for forecasting purposes. Given their nonlinearity, however, their time series properties are often difficult to establish.

A simple autoregressive count model adds y_{t-1} as a regressor, so the conditional mean is $\exp(x_t'\beta + \rho y_{t-1})$. However, this model is explosive for $\rho > 0$ since then $\rho y_{t-1} \geq 0$ because the count y is nonnegative. For example, in the pure time series case the conditional mean $\exp(\beta + \rho y_{t-1}) = \exp(\beta)\exp(\rho y_{t-1}) \simeq \exp(\beta)(1 + \rho y_{t-1})$, for small ρ, necessarily increases over time for $y > 0$. Simulations and discussion are given in Blundell, Griffith, and Windmeijer (2002), who call this model exponential feedback. A less explosive alternative is to add $\ln y_{t-1}$ as a regressor, but then adjustment is needed when $y_{t-1} = 0$.

The difficulty in incorporating lagged y has led to a wide range of autoregressive models. Unlike the linear case, there is currently no firmly established preferred model.

7.5.1 Zeger-Qaqish Model

Define $\mathbf{X}^{(t)} = (\mathbf{x}_t, \mathbf{x}_{t-1}, \ldots, \mathbf{x}_0)$ and $\mathbf{Y}^{(t-1)} = (y_{t-1}, y_{t-2}, \ldots, y_0)$. In principle the conditioning set for the dependent variable y_t is $(\mathbf{X}^{(t)}, \mathbf{Y}^{(t-1)})$, and we model

$$\mu_{t|t-1} = \mathsf{E}[y_t|\mathbf{X}^{(t)}, \mathbf{Y}^{(t-1)}].$$

Zeger and Qaqish (1988) considered GLMs with up to p lags of the dependent variable determining the conditional mean. They called their models *Markov models* of order p, because $y_{t-1}, \ldots, y_{t-p}$ are the only elements of the past history $Y^{(t-1)}$ that affect the conditional distribution of y_t. This terminology should not be confused with hidden Markov models presented in section 7.8.

Here we specialize to the Poisson with one lag of the dependent variable that appears multiplicatively, with conditional mean

$$\begin{aligned}\mu_{t|t-1} &= \exp(\mathbf{x}_t'\boldsymbol{\beta} + \rho \ln y_{t-1}^*) \qquad (7.31)\\ &= \exp(\mathbf{x}_t'\boldsymbol{\beta})(y_{t-1}^*)^{\rho},\end{aligned}$$

where y_{t-1}^* is a transformation of y_{t-1} that is strictly positive since otherwise $\mu_{t|t-1} = 0$ when $y_t = 0$. Possibilities for y_{t-1}^* include rescaling only the zero values of y_{t-1} to a constant c

$$y^*_{t-1} = \max(c, y_{t-1}), \qquad 0 < c < 1, \tag{7.32}$$

and translating all values of y_{t-1} by the same amount

$$y^*_{t-1} = y_{t-1} + c, \qquad c > 0. \tag{7.33}$$

The model (7.31) can be viewed as a multiplicative AR(1) model by comparison to the linear AR(1) model in (7.4).

An alternative to (7.31), proposed by Zeger and Qaqish (1988), is the model

$$\begin{aligned}\mu_{t|t-1} &= \exp(\mathbf{x}_t'\boldsymbol{\beta} + \rho(\ln y^*_{t-1} - \mathbf{x}_{t-1}'\boldsymbol{\beta})) \\ &= \exp(\mathbf{x}_t'\boldsymbol{\beta})\left(\frac{y^*_{t-1}}{\exp(\mathbf{x}_{t-1}'\boldsymbol{\beta})}\right)^{\rho}.\end{aligned} \tag{7.34}$$

This can be viewed as a multiplicative AR(1) *error* model with multiplicative "error" $y_t / \exp(\mathbf{x}_t'\boldsymbol{\beta})$, by comparison to the linear AR(1) error model that from (7.3) implies $\mu_{t|t-1} = \mathbf{x}_t'\boldsymbol{\beta} + \rho(y_{t-1} - \mathbf{x}_{t-1}'\boldsymbol{\beta})$.

Given the conditional mean specification in (7.31), the NLS estimator is consistent provided $y_t - \mu_{t|t-1}$ has zero mean and is serially uncorrelated, and heteroskedastic-robust standard errors lead to valid inference. As in the cross-section case, however, estimators based on a count model are likely to be more efficient.

Given a specified conditional density $f(y_t|\mathbf{x}_t, y_{t-1})$, such as the Poisson, the ML estimator maximizes

$$\mathcal{L}(\boldsymbol{\beta}, \rho) = \sum_{t=1}^{T} \ln f(y_t|\mathbf{x}_t, y_{t-1}). \tag{7.35}$$

Estimation theory is given in Wong (1986) and Fahrmeir and Kaufman (1987). Consistency requires that $f(y_t|\mathbf{X}^{(t)}, \mathbf{Y}^{(t-1)}) = f(y_t|\mathbf{x}_t, y_{t-1})$ so that $L = f(y_1, \ldots, y_T|\mathbf{x}_1, \ldots, \mathbf{x}_T) = \prod_{t=1}^{T} f(y_t|\mathbf{x}_t, y_{t-1})$ and that $\mu_{t|t-1}$ is correctly specified. To allow for overdispersion in the Poisson model, heteroskedastic-robust standard errors are used. Or overdispersion could be handled by instead specifying $f(y_t|\mathbf{x}_t, y_{t-1})$ to be the NB2 density.

More generally regressors should include sufficient lags of y_t, say p, to ensure that $f(y_t|\mathbf{X}^{(t)}, \mathbf{Y}^{(t-1)}) = f(y_t|\mathbf{X}^{(t)}, y_{t-1}, \ldots, y_{t-p})$. Then the Poisson QMLE is consistent, provided the functional form for $\mu_{t|t-1}$ is correctly specified, and inference is based on heteroskedastic-robust standard errors.

If the constant c in (7.32) or (7.33) is specified, often $c = 0.5$ in (7.32) and $c = 0.1$ in (7.33), then a standard Poisson program can be used. If instead c is an additional parameter to be estimated, standard Poisson software can still be used for $y^*_t = \max(c, y_{t-1})$ as in (7.32). Rewrite (7.34) as

$$\mu_{t|t-1} = \exp(\mathbf{x}_t'\boldsymbol{\beta} + \rho \ln y^{**}_{t-1} + (\rho \ln c)\, d_t) \tag{7.36}$$

where $y^{**}_{t-1} = y_{t-1}$ and $d_t = 0$ if $y_{t-1} > 0$, and $y^{**}_{t-1} = 1$ and $d_t = 1$ if $y_{t-1} = 0$. Poisson regression of y_t on $\mathbf{x}_t$, y^{**}_{t-1}, and d_t yields estimates of $\boldsymbol{\beta}$, ρ, and $\rho \ln c$.

Then use $c = \exp[(\rho \ln c)/\rho]$ to obtain an estimate of c. The autoregressive count model is attractive because it is relatively simple to implement.

Such autoregressive models have not been widely applied, especially for data where some counts are zero. Fahrmeier and Tutz (1994, chapter 6.1) gave an application to monthly U.S. data on the number of polio cases. Cameron and Leon (1993) applied the model to monthly U.S. data on the number of strikes and gave some limited simulations to investigate the properties of both estimators and the time-series process itself, since theoretical results on its serial correlation properties are not readily obtained. Leon and Tsai (1997) proposed and investigated by simulation quasi-likelihood analogs of LM, Wald, and LR tests, following the earlier study by Li (1991), who considered the LM test.

Zeger and Qaqish (1988) more generally considered GLMs. For GLMs other than Poisson the exponential function in (7.31) is replaced by the relevant canonical link function, and in most cases lags of y can be used because an adjustment for zero lagged value of y is not necessary. Zeger and Qaqish (1988) proposed QML estimation.

7.5.2 Alternative Conditional Mean Specifications

A weakness of the Zeger-Qaqish model in the special case of counts is that the adjustments for zero lagged values of y_t, such as (7.32) or (7.33), are ad hoc. Furthermore, such adjustments complicate evaluation of the impact of regressors on the change in the conditional mean. Papers subsequent to Zeger and Qaqish (1988) have proposed various alternative specifications for $\mu_{t|t-1} = \mathsf{E}[y_t|\mathbf{X}^{(t)}, \mathbf{Y}^{(t-1)}]$.

One alternative is to replace (7.31) by

$$\mu_{t|t-1} = \rho y_{t-1} + \exp(\mathbf{x}_t'\boldsymbol{\beta}). \tag{7.37}$$

This ensures a positive conditional mean if $\rho > 0$ and is not explosive. This can be interpreted as a partial adjustment model. Suppose μ_t is a target value of y_t and we adjust y_t according to $\Delta y_t = \lambda(y_t - y_{t-1})$. Then $y_t = (\frac{1}{1-\lambda})y_{t-1} - (\frac{1}{1-\lambda})\mu_t$, leading to a conditional mean of the form (7.37) with $(\frac{1}{1-\lambda})\mu_t = \exp(\mathbf{x}_t'\boldsymbol{\beta})$.

The specification (7.37) equals the conditional mean (7.50) of the Poisson INAR(1) model presented in section 7.6, although other features of the distribution such as the conditional variance (see [(7.44)]) will differ.

Another alternative, proposed by Shephard (1995), is to develop autoregressive models for a transformation of the dependent variable. For a GLM with log link function, corresponding to the Poisson, he proposes the model

$$z(y_t) = \mathbf{x}_t'\boldsymbol{\beta} + \sum_{j=1}^{p} \gamma_j z(y_{t-j}) + \sum_{k=1}^{q} \delta_k \frac{y_{t-k} - \mu_{t-k}}{\mu_{t-k}}, \tag{7.38}$$

where $z(y_t)$ is recursively generated given initial conditions.

Davis et al. (2003) propose the following specification for the conditional mean in an autoregressive model of order p:

$$\mu_{t|t-1} = \exp\left(\mathbf{x}_t'\boldsymbol{\beta} + \sum_{j=1}^{p} \gamma_j z_{t-j}^{(\lambda)}\right), \tag{7.39}$$

where $z_{t-j}^{(\lambda)} = (y_{t-j} - \mu_{t-j})/\mu_{t-j}^{\lambda}$, $0 < \lambda \le 1$. The parameter λ can be fixed or free; z_t is the Pearson residual if $\lambda = 1/2$. This specification augments the static conditional mean specification by adding lagged residual-like terms that control for potential serial correlation of an order up to q. The model is a variation on (7.34) with an alternative specification for the error term that does not have the explosiveness of exponential feedback. Interpretation of the slope coefficients $\boldsymbol{\beta}$ as semi-elasticities does not change because on average the residual terms equal zero and they only modify the intercept in the model. We can therefore interpret this specification as a model with a potentially serially correlated intercept. The models (7.38) and (7.39) are sometimes called examples of *generalized linear autoregressive moving average* (GLARMA) models.

Jung et al. (2006) suggest another dynamic specification that is in the spirit of Engle and Russell's (1998) autoregressive conditional duration model, and hence is called an *autoregressive conditional Poisson* (ACP) model. This model has the following mean specification:

$$\mu_{t|t-1} = \exp(\mathbf{x}_t'\boldsymbol{\beta})\varepsilon_t,$$

where

$$\varepsilon_t = \exp\left(\omega + \sum_{j=1}^{p} \alpha_j y_{t-j} + \sum_{k=1}^{q} \gamma_k \varepsilon_{t-k}\right).$$

This is an extension of the serial correlation model of section 7.4.1, which is the special case $\alpha_j = 0$, $j = 1, \ldots, p$. However, the inclusion of lagged y_t variables can also capture state dependence or persistence of the outcomes. Estimation of this model is discussed in Jung et al. (2006). They also compare the performance of several dynamic count models, using the Sydney asthma count data.

Fokianos, Rahbek, and Tjøstheim (2009) investigate the properties of a pure time-series model in which y_t is Poisson with a conditional mean that depends on both the conditional mean lagged and the dependent variable lagged, so

$$\mu_{t|t-1} = \beta + \gamma\mu_{t-1|t-2} + \alpha y_{t-1},$$

where $\beta > 0$, $\gamma > 0$, and $\alpha > 0$.

7.6 INTEGER-VALUED ARMA MODELS

INARMA models specify the realized value of y_t to be the sum of a count random variable, whose value depends on past outcomes, and the realization of an iid count random variable ε_t whose value does not depend on past outcomes. This model is similar to the linear AR model $y_t = \rho y_{t-1} + \varepsilon_t$, for example, though it explicitly models y_t as a count. Different choices of the distribution for ε_t lead to different marginal distributions for y_t, such as the Poisson. The model has the attraction of having the same serial correlation structure as linear ARMA models for continuous data.

INARMA models were independently proposed by McKenzie (1986) and Al-Osh and Alzaid (1987) for the pure time-series case and were extended to the regression case by Brännäs (1995a). They build on earlier work for continuous non-Gaussian time series, specifically exponential and gamma distributions (Lewis, 1985; Jacobs and Lewis, 1977).

7.6.1 Pure Time-Series Models

We begin with the pure time-series case before introducing other regressors $\mathbf{x}_t$ in the next subsection. Let $\mathbf{Y}^{(t-1)} = (y_{t-1}, y_{t-2}, \ldots, y_0)$. The simplest example is the INAR(1) process

$$y_t = \rho \circ y_{t-1} + \varepsilon_t, \qquad 0 \leq \rho < 1, \tag{7.40}$$

where ε_t is an iid latent count variable independent of $\mathbf{Y}^{(t-1)}$. The symbol $\circ$ denotes the *binomial thinning operator* of Stuetel and Van Harn (1979), whereby $\rho \circ y_{t-1}$ is the realized value of a binomial random variable with y_{t-1} trials and probability ρ of success on each trial. More formally $\rho \circ y = \sum_{j=1}^{y} u_j$, where u_j is a sequence of iid binary random variables that take value 1 with probability ρ and value 0 with probability $1 - \rho$. Thus each of the y components survives with probability ρ and dies with probability $1 - \rho$.

The model can be viewed as a birth (via ε_t) and death (via $\rho \circ y_{t-1}$) process or as a queueing model, stock model, or a branching process. The special feature is that both components are counts.

First consider the unconditional distribution of y. It can be shown that

$$\mu = \mathsf{E}[y] = \frac{\mathsf{E}[\varepsilon]}{1-\rho}, \tag{7.41}$$

and

$$\sigma^2 = \mathsf{V}[y] = \frac{\rho\mathsf{E}[\varepsilon] + \mathsf{V}[\varepsilon]}{1-\rho^2}; \tag{7.42}$$

see, for example, Brännäs (1995a) and exercise 7.3. Given a particular distribution for ε, the unconditional stationary distribution for y can be found by probability generating function techniques; see Steutel and Van Harn (1979) and McKenzie (1986). For example, y is Poisson if ε is Poisson.

For the conditional distribution, taking the conditional expectation of (7.40) yields the conditional mean,

$$\mu_{t|t-1} = \mathsf{E}[y_t|y_{t-1}] = \rho y_{t-1} + \mathsf{E}[\varepsilon_t], \tag{7.43}$$

a result similar to that for the Gaussian model. The conditional variance is

$$\sigma^2_{t|t-1} = \mathsf{V}[y_t|y_{t-1}] = \rho(1-\rho)y_{t-1} + \mathsf{V}[\varepsilon_t]. \tag{7.44}$$

The key step in obtaining (7.43) and (7.44) is to note that $\rho \circ y_{t-1}$, conditional on y_{t-1}, has mean ρy_{t-1} and variance $\rho(1-\rho)y_{t-1}$ using standard results on the binomial with y_{t-1} trials. It can be shown that the autocorrelation at lag k is ρ^k. Thus the INAR(1) model has the same autocorrelation function as the AR(1) model for continuous data. The conditional distribution of y_t given y_{t-1} is that of a Markov chain.

The *Poisson* INAR(1) model results from specifying the latent variable ε_t in (7.40) to be iid Poisson with parameter λ. Then y_t is unconditionally Poisson with parameter $\lambda/(1-\rho)$. Furthermore, in this case (y_t, y_{t-1}) is bivariate Poisson, defined in Chapter 8.4.1. The conditional moments using (7.43) and (7.44) are

$$\begin{aligned} \mu_{t|t-1} &= \rho y_{t-1} + \lambda \\ \sigma^2_{t|t-1} &= \rho(1-\rho)y_{t-1} + \lambda, \end{aligned} \tag{7.45}$$

so the Poisson INAR(1) model is conditionally underdispersed. The transition probabilities for the Markov chain conditional distribution are

$$\Pr[y_t|y_{t-1}] = \exp(-\lambda) \sum_{j=0}^{\min(y_t, y_{t-1})} \frac{\lambda^{y_t - j}}{(y_t - j)!} \binom{y_{t-1}}{j} \rho^j (1-\rho)^{y_{t-1}-j}. \tag{7.46}$$

Generalizations of Poisson INAR(1) to INAR(p) and INARMA(p, q) models, and to marginal distributions other than Poisson, are given in various papers by McKenzie and by Al-Osh and Alzaid. Papers by Al-Osh and Alzaid additionally considered estimation. In the INAR(p) process

$$y_t = \sum_{k=1}^{p} \rho_k \circ y_{t-k} + \varepsilon_t, \qquad 0 \leq \sum\nolimits_{k=1}^{p} \rho_k < 1.$$

McKenzie (1986) obtained an INARMA model with an unconditional negative binomial distribution for y_t by specifying ε_t to be iid negative binomial. McKenzie (1988) studied the Poisson INARMA model in detail. Al-Osh and Alzaid (1987) considered estimation of the Poisson INAR(1) model and detailed properties of INAR(1) and INAR(p) models in, respectively, Alzaid and Al-Osh (1988, 1990). Alzaid and Al-Osh (1993) obtained an INARMA model with an unconditional generalized Poisson distribution, see Chapter 4.11.4, which potentially permits underdispersion. This model specifies ε_t to be generalized Poisson and replaces the binomial thinning operator by quasi-binomial thinning. Whereas the INARMA models have the same autocorrelation function as linear ARMA models, the partial autocorrelation functions differ. Further models are given by Gauthier and Latour (1994), who define a generalized Steutel Van Harn operator. Still further generalization may be possible. In the AR(1) case, for example, essentially all that is needed is an operator that yields a discrete value for the first term in the right-hand side of (7.40). For example, there is no reason why the y_{t-1} trials need be independent, and one could, for example, use a correlated binomial model.

Initial research on INARMA models focused on stochastic properties, with less attention paid to estimation. The conditional least squares estimator is straightforward but inefficient. For the Poisson INAR(1) model, $\mathsf{E}[y_t|y_{t-1}] = \rho y_{t-1} + \lambda$. This allows simple estimation of ρ and λ by OLS regression of y_t on an intercept and y_{t-1}, a method proposed by Al-Osh and Alzaid (1987). The error in this regression is heteroskedastic, since $\mathsf{V}[y_t|y_{t-1}] = \rho(1-\rho)y_{t-1} + \lambda$, so care needs to be taken to obtain the correct variance covariance matrix and the estimator could potentially be quite inefficient. Du and Li (1989) more generally proposed estimation of the INAR(p) model based on the Yule-Walker equations. Brännäs (1994) proposed and investigated the use of GMM estimators for the Poisson INAR(1) model. GMM has the theoretical advantage of incorporating more of the moment restrictions, notably autocovariances, implied by the Poisson INAR(1) model. Savani and Zhigljavsky (2007) proposed estimation of the negative binomial INAR(1) by power estimation methods.

Exact ML estimators, as well as conditional ML estimators that condition on an initial value y_0, were proposed and investigated by Al-Osh and Alzaid (1987) and Ronning and Jung (1992) for the Poisson INAR(1) model. These estimators can be difficult to implement, especially for models other than the Poisson. Drost, Van Den Akker, and Werker (2008) propose a fully efficient two-step estimator for the INAR(p) model. Bu, McCabe, and Hadri (2008) derive the likelihood for the Poisson INAR(p) model, using a recursive formulation for conditional probabilities, and expressions for the score and information matrix. Enciso-Mora, Neal, and Rao (2008a) present MCMC methods to estimate INARMA(p, q) models and to select the orders p and q.

Drost, Van Den Akker, and Werker (2009a) consider nonparametric ML estimation for the INAR(p) model, where the distribution $G(\varepsilon)$ for ε_t is not specified. They prove consistency and root-n asymptotic normality for the estimators of both $\rho_1, \ldots, \rho_p$ and $G(\cdot)$. McCabe, Martin, and Harris (2011) consider efficient probability forecasts based on nonparametric ML estimation of the INAR(p) model.

Drost, Van Den Akker, and Werker (2009b) consider testing for unit roots in the INAR(1) model and estimation when the root is near to unity (with $\rho = 1 - c/T^2$).

7.6.2 Regression Models

The statistics literature has focused on the pure time-series case, a notable early exception was McKenzie (1985, p. 649), who briefly considered introduction of trends.

Brännäs (1995a) proposed a *Poisson* INAR(1) *regression model*, with regressors introduced into (7.40) through both the binomial thinning parameter ρ_t and the latent count variable ε_t. Then

$$y_t = \rho_t \circ y_{t-1} + \varepsilon_t, \tag{7.47}$$

where the latent variable ε_t in (7.47) is assumed to be Poisson distributed with mean

$$\lambda_t = \exp(\mathbf{x}_t'\boldsymbol{\beta}). \tag{7.48}$$

To ensure $0 < \rho_t < 1$, the logistic function is used:

$$\rho_t = \frac{1}{1+\exp(-\mathbf{z}_t'\boldsymbol{\gamma})}. \tag{7.49}$$

From (7.43) with ρ replaced by ρ_t, the conditional mean for this model is

$$\mu_{t|t-1} = \mathsf{E}[y_t|\mathbf{x}_t, \mathbf{z}_t, y_{t-1}] = \left(\frac{1}{1+\exp(-\mathbf{z}_t'\boldsymbol{\gamma})}\right) y_{t-1} + \exp(\mathbf{x}_t'\boldsymbol{\beta}). \tag{7.50}$$

A simpler specification sets $\mathbf{z}_t = 1$, so the parameter ρ is a constant. The conditional variance is

$$\sigma^2_{t|t-1} = \mathsf{V}[y_t|\mathbf{x}_t, \mathbf{z}_t, y_{t-1}] = \left(\frac{\exp(-\mathbf{z}_t'\boldsymbol{\gamma})}{[1+\exp(-\mathbf{z}_t'\boldsymbol{\gamma})]^2}\right) y_{t-1} + \exp(\mathbf{x}_t'\boldsymbol{\beta}), \tag{7.51}$$

from (7.44).

In Brännäs' application to annual Swedish data on the number of paper mills of a particular type, the parameters $\boldsymbol{\gamma}$ in the model for ρ_t are interpreted as representing the role of regressors in explaining the death of firms, whereas the parameters $\boldsymbol{\beta}$ of the Poisson density for ε_t represent the role of regressors in explaining the birth of firms.

Brännäs proposed NLS and GMM estimators for the INAR(1) model and additionally considered prediction. Using (7.50), the conditional NLS estimator minimizes with respect to $\boldsymbol{\beta}$ and $\boldsymbol{\gamma}$

$$S(\boldsymbol{\beta}, \boldsymbol{\gamma}) = \sum_{t=1}^{T} \left\{ y_t - \left(\frac{1}{1+\exp(-\mathbf{z}_t'\boldsymbol{\gamma})}\right) y_{t-1} - \exp(\mathbf{x}_t'\boldsymbol{\beta}) \right\}^2. \tag{7.52}$$

This estimator is relatively straightforward to implement given access to a statistical package that includes NLS estimation and computing robust standard errors that adjust for heteroskedasticity. An alternative more efficient estimator is conditional weighted LS, where the weighting function can be obtained from (7.51). Details of the GMM estimator, based on additional knowledge of the functional form for the autocovariances implied by the Poisson INAR(1) model, are given in Brännäs (1995a). In principle GMM should be more efficient than NLS, but Brännäs found little gain.

Brännäs and Hellström (2001) present a range of generalizations of the INAR(1) model, including allowing for correlation in the binomial thinning $\rho_t \circ y_{t-1}$ and a richer process for ε_t. They present conditional and unconditional first and second moments and GMM estimators based on these moments. Weiss (2008) provides further details on Poisson INAR(1) and INAR(q) regression models and finds that the models are closely related to each other. Enciso-Mora, Neal, and Rao (2008b) present MCMC methods to estimate an INAR(p) model with regressors using Bayesian MCMC methods with data augmentation.

7.7 STATE SPACE MODELS

The autoregressive and INAR models specify the conditional distribution of y_t to depend on a specified function of $(\mathbf{X}^{(t)}, \mathbf{Y}^{(t-1)})$. The state space model or time-varying parameters model instead specifies the conditional distribution of y_t to depend on stochastic parameters that evolve according to a specified distribution whose parameters are determined by $(\mathbf{X}^{(t)}, \mathbf{Y}^{(t-1)})$.

Analytical results can be obtained if only the mean parameter evolves over time, with density that is conjugate to the density of y_t. More generally the model has regression coefficients that evolve over time, in which case a normal distribution is typically chosen. Analytical results are then no longer attainable, so the models are estimated using computationally intensive methods.

These methods often cast the model in a Bayesian framework. Then interest lies in obtaining the posterior mean, which in the time-series case is used to generate forecasts that incorporate prior information. A standard reference is West and Harrison (1997, first edition 1990). Bayesian computational methods are also widely used in frequentist analyses where the focus is on parameter estimation, and it is not uncommon to see frequentist analyses take on a Bayesian flavor because of this use; see Chapter 12. Durbin and Koopman (2000) present both frequentist and Bayesian methods for state space models for non-Gaussian data such as count data. We focus on the frequentist interpretation here.

7.7.1 Conjugate Distributed Mean

West, Harrison, and Migon (1985) proposed Bayesian time-series models for regression models with prior density for the conditional mean chosen to be conjugate to an LEF density. These models are an extension of dynamic linear models, see West and Harrison (1997), to the GLM framework though this extension is not seamless. Interest lies in obtaining the posterior mean and forecasting, given specified values for the prior density parameters.

Harvey and Fernandes (1989) studied these models in a non-Bayesian framework and considered parameter estimation. The most tractable model for count data is a Poisson-gamma model. Assume y_t conditional on μ_t is $\mathsf{P}[\mu_t]$ distributed, so

$$f(y_t|\mu_t) = e^{-\mu_t}\mu_t/y_t!. \tag{7.53}$$

The mean parameter μ_t is modeled to evolve stochastically over time with distribution determined by past values of y_t. A convenient choice of distribution is the gamma

$$f(\mu_t|a_{t|t-1}, b_{t|t-1}) = \frac{e^{-b\mu_t}\mu_t^{a-1}}{\Gamma(a)b^{-a}}, \qquad a_{t|t-1} > 0,\ b_{t|t-1} > 0, \tag{7.54}$$

where a and b in (7.54) are evaluated at $a = a_{t|t-1} = \omega a_{t-1}$ and $b = b_{t|t-1} = \omega b_{t-1}$ and $0 < \omega \leq 1$. The conditional density of y_t given the observables $\mathbf{Y}^{(t-1)}$ is

$$f(y_t|\mathbf{Y}^{(t-1)}) = \int_0^\infty f(y_t|\mu_t) f(\mu_t|\mathbf{Y}^{(t-1)}) d\mu_t. \tag{7.55}$$

From Chapter 4.2.2, $f(y_t|\mathbf{Y}^{(t-1)})$ is the negative binomial with parameters $a_{t|t-1}$ and $b_{t|t-1}$. Estimation of ω and the parameters of μ_t is by ML, where the joint density of $\mathbf{Y}^{(t)}$ is the product of the conditional densities (7.55). The Kalman filter is used to recursively build $a_{t|t-1}$ and $b_{t|t-1}$.

Harvey and Fernandes (1989) apply this approach to count data on goals scored in soccer matches, purse snatchings in Chicago, and van driver fatalities. They also obtain tractable results for negative binomial with parameters evolving according to the beta distribution and for the binomial model. Singh and Roberts (1992) consider count data models. Harvey and Shephard (1993) consider the general GLM class. Brännäs and Johansson (1994) investigate small-sample performance of estimators. Johansson (1996) gives a substantive regression application to monthly Swedish data on traffic accident fatalities, which additionally uses the model of Zeger (1988) discussed in section 7.4.

7.7.2 Normally Distributed Parameters

A richer model with regressors is the Poisson regression model (7.53), with $\mu_t = \exp(\mathbf{x}_t'\boldsymbol{\beta}_t)$ where $\boldsymbol{\beta}_t$ evolves according to

$$\boldsymbol{\beta}_t = \mathbf{A}_t \boldsymbol{\beta}_{t-1} + \boldsymbol{\upsilon}_t, \tag{7.56}$$

where

$$\boldsymbol{\upsilon}_t \sim \mathsf{N}[\mathbf{0}, \boldsymbol{\Sigma}_t].$$

For this model there are no closed-form solutions analogous to (7.55). The development of numerical techniques for these models is an active area of research.

An example is Durbin and Koopman (1997), who model British data on van driver fatalities. They first follow Shephard and Pitt (1997) in developing an MCMC method to numerically evaluate the likelihood of the model. Durbin and Koopman (1997) then propose a faster procedure that calculates the likelihood for an approximating linear Gaussian model by Kalman filter techniques for linear models; then they compute the true likelihood as an adjustment to this approximating model. Durbin and Koopman (2000) survey computational methods for a class of models that includes this model. Frühwirth-Schnatter and Wagner (2006) propose use of auxiliary mixture sampling and data augmentation that enable the use of Gibbs sampling rather than the slower Metropolis-Hastings algorithm.

7.8 HIDDEN MARKOV MODELS

Finite mixture models presented in Chapter 4.8 model cross-section count data as coming from one of a finite number of classes or regimes. Hidden Markov time-series models similarly specify different parametric models in different regimes, but extend the cross-section case to modeling the unobserved regimes as evolving over time according to a Markov chain. Here we summarize results given in considerably more detail in MacDonald and Zucchini (1997).

For the model with m possible regimes, let C_t, $t = 1, \ldots, T$, denote a Markov chain on state space $\{1, 2, \ldots, m\}$. Thus $C_t = j$ if at time t we are in regime j. In the simplest case, considered here, C_t is an irreducible homogeneous Markov chain, with transition probabilities

$$\gamma_{ij} = \Pr[C_t = j | C_{t-1} = i], \qquad i, j = 1, \ldots, m, \tag{7.57}$$

that are time invariant. It is assumed that there exists a unique strictly positive stationary distribution

$$\delta_j = \Pr[C_t = j], \qquad j = 1, \ldots, m, \tag{7.58}$$

where the δ_j are a function of γ_{ij}.

The model is completed by specifying conditional densities $f(y_t | C_t = j)$, $j = 1, \ldots, m$.

7.8.1 Poisson Hidden Markov Model

For the Poisson hidden Markov model, it is assumed that the count data y_t in each regime are Poisson distributed and independent over t, with a mean parameter that varies with exogenous variables and the regime

$$\mu_{tj} = \exp(\mathbf{x}_t' \boldsymbol{\beta}_j), \qquad j = 1, \ldots, m. \tag{7.59}$$

The moments of y_t, unconditional on C_t, though still conditional on $\mathbf{x}_t$, can be shown to be

$$\mathsf{E}[y_t | \mathbf{x}_t] = \sum_{j=1}^{m} \delta_j \mu_{tj} \tag{7.60}$$

$$\mathsf{E}[y_t^2 | \mathbf{x}_t] = \sum_{j=1}^{m} \delta_j (\mu_{tj} + \mu_{tj}^2) \tag{7.61}$$

$$\mathsf{E}[y_t y_{t+k} | \mathbf{x}_t] = \sum_{i=1}^{m} \sum_{j=1}^{m} \delta_i \gamma_{ij}(k) \mu_{ti} \mu_{t+k,j}, \tag{7.62}$$

where $\gamma_{ij}(k) = \Pr[C_{t+k} = j | C_{t-1} = i]$ and $t = 1, \ldots, T$.

Equations (7.60) and (7.62) are just the weighted sums of the first two moments in each regime, with weights δ_j, and it can be shown that $\mathrm{V}[y_t | x_t] > \mathrm{E}[y_t | x_t]$, so the model induces overdispersion.

Equation (7.62) follows from the Markov chain. The autocorrelation function of y_t, which follows directly from (7.60) to (7.62), is a function of the Poisson parameters and the transition probabilities. In the case $m = 2$ and $\mu_{tj} = \mu_j$, $j = 1, 2$, $\mathsf{Cor}[y_t y_{t+k}] = a(\gamma_{11} + \gamma_{22} - 1)^k$, where a is defined in MacDonald and Zucchini (1997, p.71).

Applications to count data include the daily number of epileptic seizures by a particular patient, fitted by a two-state hidden Markov Poisson model with $\mu_{tj} = \mu_j$, and the weekly number of firearm homicides in Cape Town, fitted by a two-state hidden Markov Poisson model with $\mu_{tj} = \exp(\alpha_{1j} + \alpha_{2j}t + \alpha_{3j}t^2)$.

Count data can be directly modeled as a Markov chain, rather than via a hidden Markov chain. For counts there are potentially an infinite number of transition parameters, and additional structure is needed. An example of such structure is the Poisson $\mathsf{INAR}(1)$ model whose transition probabilities are given in (7.46).

7.8.2 ML Estimation

The parameters to be estimated are the regime-specific parameters $\boldsymbol{\beta}_j$, $j = 1, \ldots, m$, and the transition probabilities γ_{ij}. ML estimation, imposing the constraints that $\gamma_{ij} \geq 0$ and the constraints $\sum_{j \neq i} \gamma_{ij} \leq 1$, $i = 1, \ldots, m$, is presented in MacDonald and Zucchini (1997).

The joint density $f(y_t, C_t | \mathbf{Y}^{(t-1)})$ at time t is given by

$$f(y_t, C_t | \mu_1, \ldots, \mu_m) = \prod_{j=1}^{m} \left(\Pr[C_t = j | \mathbf{Y}^{(t-1)}] \times f(y_t | C_t = j) \right)^{d_{tj}}, \tag{7.63}$$

where the indicator variables $d_{tj} = \mathbf{1}[C_t = j]$, $j = 1, \ldots, m$. The marginal density is

$$f(y_t | \mathbf{Y}^{(t-1)}) = \sum_{j=1}^{m} \Pr[C_t = j | \mathbf{Y}^{(t-1)}] \times f(y_t | C_t = j). \tag{7.64}$$

Suppose the path of the Markov chain is known and is denoted as

$$\{C_1 = j_1, C_2 = j_2, \ldots, C_T = j_T\}. \tag{7.65}$$

Then the joint density of the sample can be written as

$$\begin{aligned} f(\mathbf{Y}^{(T)}, \mathbf{C}^{(T)}) &= f(y_1, \ldots, y_T, C_1 = j_1, \ldots, C_T = j_T) \\ &= \Pr[C_1 = j_1, \ldots, C_T = j_T] \times f(y_1, \ldots, y_T | C_1 = j_1, \ldots, C_T = j_T) \\ &= \Pr[C_1 = j_1] f(y_1 | C_1 = j_1) \prod_{t=2}^{T} \Pr[C_t = j_t | C_{t-1} = j_{t-1}) \\ &\quad \times \prod_{t=2}^{T} f(y_t | C_t = j_t) \\ &= \delta_{j_1} f(y_1 | C_1 = j) \prod_{t=2}^{T} \gamma_{j_t j_{t-1}} f(y_t | C_t = j_t). \end{aligned} \tag{7.66}$$

An alternative way of writing the complete-data likelihood is the following.

$$f(\mathbf{Y}^{(T)}, \mathbf{C}^{(T)}) = \prod_{j=1}^{m} [f(y_1|C_1 = j)\delta_j]^{d_{1j}} \left[\prod_{t=2}^{T}\prod_{k=1}^{m}\prod_{l=1}^{m} [\gamma_{kl} f(y_t|C_t = j)]^{d_{tkl}}\right], \tag{7.67}$$

where $d_{tkl} = \mathbf{1}[C_{t-1} = k, C_t = l], k, l = 1, \ldots, m$.

The corresponding marginal joint density is obtained by integrating out the unobserved Markov chain variables $C_1, \ldots, C_T$, so

$$f(\mathbf{Y}^{(T)}) = \sum_{j_1=1}^{m} \cdots \sum_{j_T=1}^{m} \left[\delta_{j_1} f(y_1|C_1 = j_1) \prod_{t=2}^{T} \gamma_{j_{(t-1)} j_t} f(y_t|C_t = j_t)\right]. \tag{7.68}$$

Written as in the preceding equation, the marginal likelihood is difficult to compute. However, following MacDonald and Zucchini (1997), it can be written using matrix-vector operations. Define the $m \times 1$ vector $\boldsymbol{\delta} = (\delta_1, \ldots, \delta_m)'$, the $m \times m$ diagonal matrix $\boldsymbol{\lambda}(y_t)$ with j^{th} diagonal entry $f(y_t|C_t = j)$, the $m \times m$ transition probabilities matrix $\boldsymbol{\Gamma}$ with ij^{th} entry γ_{ij}, and the $m \times 1$ vector $\mathbf{l} = (1, \ldots, 1)'$. Then the joint marginal density, and hence the marginal likelihood, can be written as

$$\mathsf{L}(\boldsymbol{\theta}) = \boldsymbol{\delta}'\boldsymbol{\lambda}(y_1) \prod_{t=2}^{T} \boldsymbol{\gamma}\boldsymbol{\lambda}(y_t)\mathbf{l}. \tag{7.69}$$

Alternatively one can work with the complete-data likelihood and estimate the parameters of the model using the EM algorithm. From equation (7.67)

$$\begin{aligned} \ln f(\mathbf{Y}^{(T)}, \mathbf{C}^{(T)}) &= \sum_{j=1}^{m} d_{1j} \ln \left(\delta_j f(y_1|C_1 = j)\right) \\ &\quad + \sum_{t=2}^{T}\sum_{k=1}^{m}\sum_{l=1}^{m} d_{tkl} \ln(\gamma_{kl} f(y_t|C_t = j)). \end{aligned} \tag{7.70}$$

This is linear in d_{1j} and d_{tkl} and we can proceed as in Chapter 4.8. The expected likelihood is obtained by replacing d_{1j} and d_{tkl} by their expected values given the current parameter estimated, the expected likelihood is then maximized with respect to model parameters, and so on. The appendix to Hyppolite and Trivedi (2012) provides considerable detail.

Yet another approach is to use Bayesian methods. Scott (2002) presents Bayesian methods for hidden Markov models, including the Poisson.

7.9 DYNAMIC ORDERED PROBIT MODEL

In Chapter 3.6.2 we considered the possibility of using a discrete ordered outcome model, specifically the ordered probit model, for count data. When modeling a time series of counts in which the support of the distribution is restricted to a small number of values, the dynamic ordered probit model is an attractive alternative to the preceding models. The reader is reminded that this framework interprets discrete counts as the outcomes of an underlying continuous latent process.

Let y_t denote the observed count that takes m distinct values and let y_t^* denote a latent variable. The two variables are related by the following observational rule:

$$y_t = \begin{cases} 0 & \text{if } -\infty < y_t^* \leq c_1 \\ 1 & \text{if } c_1 < y_t^* \leq c_2 \\ \vdots & \vdots \\ m & \text{if } c_m < y_t^* < \infty. \end{cases} \tag{7.71}$$

Here there are m intervals into which the latent variable y_t^* can fall, and the coefficients $c_1, \ldots, c_m$ are the unobserved boundary points ("cutoffs") of these intervals. The standard specification given in Chapter 3.6.2 specifies $y_t^* = \mathbf{x}_t'\boldsymbol{\beta} + \varepsilon_t$, $\varepsilon_t \sim \mathsf{N}[0, 1]$.

To capture dynamic dependence the specification of the latent variable can be extended as follows:

$$y_t^* = \mathbf{x}_t'\boldsymbol{\beta} + u_t + \varepsilon_t, \tag{7.72}$$

$$u_t = \rho u_{t-1} + \alpha y_{t-1}. \tag{7.73}$$

This specification can capture pure serial correlation ($\rho \neq 0, \alpha = 0$) or state dependence ($\alpha \neq 0, \rho = 0$). Then given $\varepsilon_t \sim \mathsf{N}[0, 1]$, we obtain

$$\Pr[y_t = j] = \Phi(c_{j+1} - \mathbf{x}_t'\boldsymbol{\beta} - u_t) - \Phi(c_j - \mathbf{x}_t'\boldsymbol{\beta} - u_t), \; j = 0, \ldots, m. \tag{7.74}$$

The likelihood can be formed and maximized (see Chapter 3.6.2). ML estimation of this model is relatively straightforward, subject to the caveat that u_t in (7.74) is an unobserved variable and has to be recursively generated using (7.73). Simulation-based methods have an obvious role here also. For an empirical illustration, see Jung et al. (2006).

7.10 DISCRETE ARMA MODELS

The first serious attempt to define a time-series count model with similar autocorrelation structure to ARMA models was by Jacobs and Lewis (1978a, 1978b, 1983). They defined the class of *discrete autoregressive moving average models* (DARMA) models, for which the realized value of y_t is a *mixture* of past values $\mathbf{Y}^{(t-1)}$ and the current realization of a latent variable ε_t.

The simplest example is the DARMA(1, 0) model with

$$y_t = u_t y_{t-1} + (1 - u_t)\varepsilon_t, \tag{7.75}$$

where u_t is a binary mixing random variable that takes value 1 with probability ρ and value 0 with probability $1 - \rho$, and different distributional assumptions can be made for the iid discrete latent random variable ε_t. This model implies

$$\begin{aligned} \Pr[y_t = y_{t-1}] &= \rho \\ \Pr[y_t = \varepsilon_t] &= 1 - \rho. \end{aligned} \tag{7.76}$$

Clearly for this model the autocorrelation at lag k is ρ^k, as in the AR(1) model, and only positive correlation is possible. Extensions can be made to DARMA(p, q) models with correlation structures equal to that of standard linear ARMA(p, q) models, though with greater restrictions on the permissible range of correlation structures.

A major restriction of the model is that for high serial correlation the data will be characterized by a series of runs of a single value. This might be appropriate for some data, for example on the number of firms in an industry where there is very little entry and exit over time. But most time-series count data exhibit more variability over time than that data. For this reason this class of models is rarely used, and we do not consider estimation or possible extension to the regression case. Instead INARMA models are preferred.

7.11 APPLICATIONS

We illustrate some of the preceding models, using two examples. The first example of strike frequency is a continuation of the example introduced in section 7.3.4. The second example uses high-frequency financial transactions data on the number of trades in a single stock in consecutive five-minute intervals.

7.11.1 Strikes Data Revisited

We begin with a variant of the autoregressive model of Zeger and Qaqish given by (7.31)–(7.33), with up to three lags of the dependent variable appearing as explanatory variables. A Poisson regression model is estimated with conditional mean

$$\begin{aligned} &\mathsf{E}[y_t|x_t, y_{t-1}, y_{t-2}, \ldots] \\ &\quad = \exp(\beta_1 + \beta_2 x_t + \rho_1 \ln y^*_{t-1} + \rho_2 \ln y^*_{t-2} + \rho_3 \ln y^*_{t-3}), \end{aligned} \tag{7.77}$$

where y_t denotes *STRIKES*, x_t denotes *OUTPUT*, and $y^*_t = \max(c, y_t)$, where c is a value between 0 and 1 that prevents potential problems in taking the natural logarithm when $y_t = 0$. This model can be estimated using a standard Poisson regression program, as explained in Section 7.5.

In the first three columns of Table 7.5, estimates for the model (7.77) are presented with 0, 1, or 3 lags (models ZQ0, ZQ1, and ZQ3), with c set to the value 0.5. The first column reproduces the static regression estimates given in Table 7.2. Introducing lagged dependent variables, the biggest gain comes from introducing just one lag. The autocorrelations of the Pearson residuals reduce substantially. The fit of the model improves substantially, with the deviance

Table 7.5. *Strikes: Zeger-Qaqish autoregressive model estimates and diagnostics*

	Model						
Variable	ZQ0	ZQ1	ZQ3	ZQ1(c)	B0	B1	B3
ONE	1.654	1.060	0.846	0.896	1.655	1.017	0.597
OUTPUT	3.134	2.330	2.187	2.345	2.993	3.480	5.107
ln y*(-1)		0.396	0.267	0.482			
ln y*(-2)			0.163				
ln y*(-3)			0.114				
c		0.5	0.5	0.553			
y(-1)						0.469	0.319
y(-2)							0.198
y(-3)							0.127
ACF lag 1	0.44	−0.10	0.02	−0.13	0.44	−0.11	−0.01
ACF lag 2	0.38	0.13	0.05	0.15	0.38	0.16	0.05
ACF lag 3	0.32	0.16	0.03	0.13	0.32	0.15	0.02
ACF lag 4	0.20	0.08	0.01	0.03	0.20	0.07	0.00
ACF lag 5	0.11	0.03	−0.01	0.01	0.11	0.03	−0.03
ACF lag 6	0.03	0.00	−0.06	0.00	0.03	0.00	−0.06
LB	55.5	6.7	0.8	6.4	55.8	7.1	1.0
Corr$^2[\hat{y}$,y]	0.052	0.278	0.330	0.293	0.052	0.264	0.320
Observations	108	107	105	107	108	107	105

Note: ZQ0 is the static Poisson regression; ZQ1 and ZQ3 are the Zeger-Qaqish autoregressive model defined in (7.77) with c=0.5; ZQ1(c) is the ZQ1 model with c estimated; B0 is the static Poisson regression estimated by NLS; B1 and B3 are the Brannas INAR model defined in (7.78), estimated by NLS; ACF lags 1–6, autocorrelations to lag 6 from the Pearson residuals from each model; LB is the Ljung-Box statistic with 10 degrees of freedom; and Corr^2 [$\hat{y}$,y] is the squared correlation between $\hat{y}$ and y.

R-squared increasing from 0.053 to 0.245. Further gains occur in introducing additional lags, but these gains are relatively small and have little impact on the coefficient of *OUTPUT*. In model ZQ3 $\hat{\rho}_3 = 1.115$ has a robust t-statistic of 1.57 and is statistically insignificant at level 0.05.

There is still overdispersion. For example, for model ZQ3 the Pearson statistic leads to $\hat{\phi} = 1.97$ for the NB1 variance function. Using heteroskedastic robust standard errors that control for this overdispersion, the coefficient of *OUTPUT* is statistically significant at 5% in models ZQ0 to ZQ3, with the t-statistic ranging from 2.15 to 2.65. Even controlling for past strike activity, there is an independent positive effect when output rises above trend.

The model ZQ1(c) in column 5 of Table 7.5 is the same as model ZQ1, except that the coefficient of c is estimated rather than being set at 0.5. An indicator variable is constructed, as detailed in section 7.5. Then $\hat{c} = 0.55$, close to the arbitrarily chosen value of 0.5, with a 95% confidence interval (−0.19, 1.30) that is reasonably tight given that only five observations actually

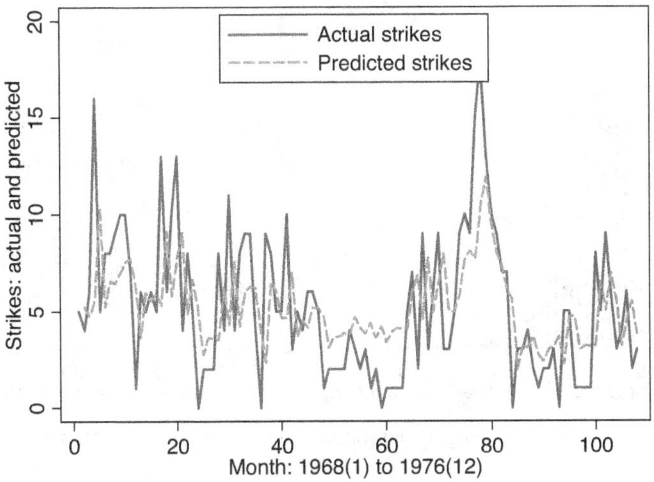

Figure 7.3. Strikes: Actual and predicted strikes from a dynamic regression model.

equal zero. Varying c in the range 0 to 1 makes little difference to estimates of other parameters and the residual ACF, and we use the midpoint 0.5.

The remaining columns present estimates for the Brännäs INAR model. For the INAR(3) model the conditional mean is

$$\mathsf{E}[y_t|x_t, y_{t-1}, y_{t-2}, \ldots] = \rho_1 y_{t-1} + \rho_2 y_{t-2} + \rho_3 y_{t-3} + \exp(\beta_1 + \beta_2 x_t). \tag{7.78}$$

Estimation is by NLS, which is not fully efficient, and the standard errors for $\hat{\beta}_2$ are around 15% higher than for the ZQ models. The estimated models B0 to B3 lead to quite similar results to those for the models ZQ0 to ZQ3, in terms of serial correlation in the residuals and fit of the model.

Neither model is straightforward to analyze for long-run impacts. In the long run, $y_t = y_{t-1} = y$. The B1 model estimates yield $y = [\exp(1.017 + 3.480x)]/0.531$. The ZQ1 model estimates yield $y = [\exp(1.060 + 2.330x)]^{.604}$.

The predictions of strikes from model B1, plotted in Figure 7.3, fit the data much better than the static regression model presented in Figure 7.2. ZQ1 model predictions are quite similar, with correlation of 0.978 between B1 and ZQ1 predictions.

A richer version of the INAR(1) model given in (7.50) models ρ_1 as varying over time with *OUTPUT*. Then

$$\mathsf{E}[y_t|x_t, y_{t-1}, y_{t-2}, \ldots] = \frac{1}{1 + \exp(-\gamma_1 - \gamma_2 x_t)} + \exp(\beta_1 + \beta_2 x_t).$$

From results not included in Table 7.5, the NLS estimate $\hat{\beta}_2$ becomes negative and statistically insignificant. The coefficient $\hat{\gamma}_2$ is borderline statistically

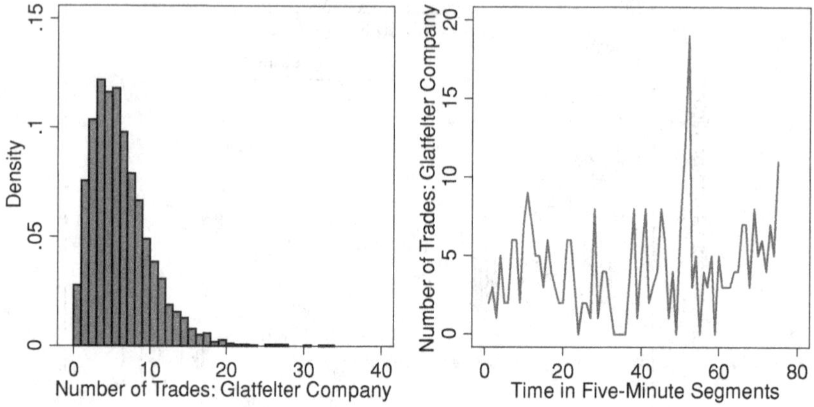

Figure 7.4. Histogram and time series of the number of trades.

significant ($t = 1.65$) and positive, implying that an increase in *OUTPUT* is associated with an increase in ρ_1 and hence the conditional number of strikes. The resulting estimate $\hat{\rho}_1$ ranges from 0.12 to 0.66 with an average of 0.41. The first six autocorrelations of the Pearson residuals from this model are, respectively, 0.00, 0.04, 0.02, 0.02 , −0.06, and −0.09; the Ljung-Box statistic is 1.6; and $\mathsf{Cor}^2[\hat{y}, y] = 0.283$. The model fits the autocorrelations as well as the BQ3 and Z3 models.

The application here illustrates a simple way to estimate dynamic count regression models. Richer models such as those in sections 7.7 and 7.8 might then be used. They involve a considerably higher level of complexity, which may require referring to the original sources.

7.11.2 Number of Stock Trades

This section provides a brief analysis of the number of trades in five-minute intervals between 9:45 a.m. and 4 p.m. on the New York Stock Exchange (NYSE) for the Glatfelter Company (NYSE ticker symbol: GLT) over 39 trading days in the first quarter of 2005 (January 3, 2005–February 18, 2005). The data are part of a larger dataset used by Jung, Liesenfeld, and Richard (2011).

Because there are 75 five-minute intervals in each trading day, the sample size is $T = 75 \times 39 = 2{,}925$. The number of trades ranges from 0 (3% of the time) to 34, with a mean of 5.66. The standard deviation is 3.90, so the data are overdispersed with a variance 2.7 times the mean. The left-hand panel of Figure 7.4 shows the histogram of the number of trades.

The right-hand panel of Figure 7.4 shows the time series of the number of trades on the first day of the sample. The first six autocorrelation coefficients of the data are, respectively, 0.33, 0.27, 0.23, 0.21 , 0.19, and 0.21. The autocorrelations after 15 lags are less than 0.10, except for a value of 0.13 at lag 75, which is the same segment on the preceding trading day. The first six partial autocorrelation coefficients are, respectively, 0.33, 0.19, 0.12, 0.08 , 0.06, and 0.09 and are not as persistent. (The asthma data of Chapter 1.3.4 also show considerable persistence in the autocorrelations.)

There are several well-established features of time-series analysis of high-frequency stock market data. The outcome variable is characterized by pronounced intraday seasonality, and usually there are no other available data on time-varying covariates measured at the same frequency. Hence the analysis usually concentrates on pure time-series features such as serial correlation and volatility properties.

There are several ways to proceed. The data may be analyzed for a single day at a time. All available data may be pooled into a long time series treated as a single sample. Or, a multiple time series approach may be taken with observations for different days treated as repeated samples from the same population. We take the second of these approaches, analyzing a single time series of 2,925 observations.

Table 7.6 presents results. The first column gives Poisson results for an intercept-only model. The second model follows Jung et al. (2011) by including four terms to capture the U-shaped pattern of intraday trading activity: $\mathbf{x}_t = (\cos(2\pi t/75), \sin(2\pi t/75), \cos(4\pi t/75), \sin(4\pi t/75))$. These four regressors are jointly highly statistically significant, with a $\chi^2(4)$ statistic of 188. The dominant term is the first, leading to a daily cycle that peaks at 7.1 shortly after the start of day and bottoms out at 4.3 shortly after the middle of the day. There is still considerable serial correlation in the residuals, however, with a Ljung-Box statistic based on the first 10 lags of the Pearson residual of 861.

The next three columns present various dynamic models that introduce a single lag – the Zeger-Qaqish model, the Brännäs INAR(1) model, and a model that includes as regressor the first lag of the Pearson residual from the static model. The three models yield similar results. The lagged variables are highly statistically significant, improve the model fit, and considerably reduce the autocorrelation, though the Ljung-Box statistics based on the first 10 lags of Pearson residuals are still around 200. The final column adds the first three lags of the static model Pearson residuals, leading to still further improvement.

Another way to analyze these data is to treat them as clustered samples, with 39 clusters that each have 75 observations. In that case panel data methods can be applied to analyze the time series. A small advantage is that the framework could then potentially control for intra-cluster correlation. However, in the present example where there are no covariates other than intercept and trigonometric terms, this approach does not produce any additional worthwhile information.

The more extensive study of Jung et al. (2011) analyzes time series for the number of trades in four distinct stocks. A multivariate framework is used, with dynamics introduced at the level of latent factors. It is assumed that y_{jt}, the number of trades of the jth stock at time t, is conditionally independently

Table 7.6. *Stock trades: Static and dynamic model estimates*

	Models for number of trades in GLT stock					
Variable	None	Static	ZQ1	B1	P1	P3
ONE	1.734	1.719	1.351	1.387	1.704	1.699
	(0.013)	(0.013)	(0.030)	(0.031)	(0.012)	(0.012)
X1		0.242	0.175	0.242	0.243	0.246
		(0.018)	(0.018)	(0.024)	(0.017)	(0.017)
X2		0.021	0.018	0.014	0.016	0.013
		(0.017)	(0.017)	(0.024)	(0.017)	(0.017)
X3		−0.020	−0.012	−0.014	−0.019	−0.018
		(0.018)	(0.017)	(0.024)	(0.017)	(0.017)
X4		0.032	0.027	0.028	0.030	0.031
		(0.017)	(0.017)	(0.023)	(0.017)	(0.017)
lag_1			0.240	0.283	0.105	0.081
			(0.018)	(0.022)	(0.007)	(0.008)
lag_2						0.051
lag_3						0.036
ACF lag 1	0.327	0.285	−0.013	−0.045	−0.031	0.009
ACF lag 2	0.274	0.230	0.131	0.125	0.129	−0.001
ACF lag 3	0.233	0.182	0.102	0.096	0.101	−0.025
ACF lag 4	0.207	0.157	0.088	0.083	0.089	0.036
ACF lag 5	0.189	0.142	0.070	0.067	0.072	0.038
ACF lag 6	0.207	0.166	0.108	0.111	0.118	0.086
LB	1305	861	204	197	218	53
$\text{Corr}^2[\hat{y},\text{y}]$	0.000	0.062	0.127	0.137	0.132	0.152
Observations	2,925	2,925	2,924	2,924	2,924	2,922

Note: Static is static Poisson regression; ZQ1 is the Zeger-Qaqish autoregressive model with $c = 0.5$; B1 is the Brannas INAR(1) model estimated by NLS; P1 and P3 are Poisson models with one lag and first three lags of the Pearson residual from the static model as regressors; ACF lags 1–6, autocorrelations to lag 6 from the Pearson residuals from each model; LB is the Ljung-Box statistic with 10 degrees of freedom; and $\text{Corr}^2\,[\hat{y}, \text{y}]$ is the squared correlation between $\hat{y}$ and y.

distributed as $\mathsf{P}[\mu_{tj}]$, $j = 1, \ldots, m$. The means μ_{tj} are latent variables that are determined by

$$\boldsymbol{\eta}(\boldsymbol{\mu}_t) = \boldsymbol{\mu} + \Gamma\mathbf{f}_t,$$

where $\boldsymbol{\mu} = (\mu_{1t}, \ldots, \mu_{mt})$, $\boldsymbol{\eta}(\cdot)$ is an m-dimensional link function, $\boldsymbol{\mu}$ denotes an $m \times 1$ vector of fixed intercepts, Γ is an $(m \times L)$ matrix of factor loadings, and $\mathbf{f}_t$ denotes L latent factors assumed to be orthogonal to each other. In the context of the previous example the factors can include a common market factor, industry factors, and stock-specific factors. Dynamics in the number of trades may arise from the presence of serial correlation in the factors. To complete the specification of the model a distributional assumption on $\mathbf{f}_t$ is added. Estimation of the model involves significant computational complications because numerical integration methods are used to handle latent factors.

Jung et al. (2011), for example, assume that the latent factors follow a first-order Gaussian autoregression and use efficient importance sampling for inference. Monte Carlo integration, using Halton sequences, is another computationally efficient procedure.

7.12 DERIVATIONS

7.12.1 Tests of Serial Correlation

We derive the section 7.3.2 results on serial correlation tests. Suppose z_t is distributed with mean 0 and is independently distributed, though is potentially heteroskedastic. Then $\sum_{t=1}^{T} z_t z_{t-k}$ has mean

$$\mathsf{E}\left[\sum_{t=1}^{T} z_t z_{t-k}\right] = 0,$$

since $\mathsf{E}[z_t z_{t-k}] = 0$, and variance

$$\mathsf{E}\left[\sum_{s=1}^{T}\sum_{t=1}^{T} z_s z_{s-k} z_t z_{t-k}\right] = \sum_{t=1}^{T} \mathsf{E}\left[z_t^2 z_{t-k}^2\right],$$

using $\mathsf{E}[z_t z_u z_v z_w] = \mathsf{E}[z_t]\mathsf{E}[z_u z_v z_w]$ by independence for $t \neq u, v, w$, and $\mathsf{E}[z_t] = 0$. By a law of large numbers

$$\left(\sum_{t=1}^{T} \mathsf{E}\left[z_t^2 z_{t-k}^2\right]\right)^{-1/2} \sum_{t=1}^{T} z_t z_{t-k} \xrightarrow{d} \mathsf{N}[0, 1], \tag{7.79}$$

which yields (7.19) for tests on unstandardized residuals.

Now specialize to the special case that z_t is scaled to have constant variance, in which case $\mathsf{E}[z_t^2]$ is constant. Then

$$\begin{aligned}\textstyle\sum_{t=1}^{T} \mathsf{E}\left[z_t^2 z_{t-k}^2\right] &= \textstyle\sum_{t=1}^{T} \mathsf{E}[z_t^2]\mathsf{E}[z_{t-k}^2] \\ &= \tfrac{1}{T}\left(\textstyle\sum_{t=1}^{T} \mathsf{E}[z_t^2]\right)\left(\textstyle\sum_{t=1}^{T} \mathsf{E}[z_{t-k}^2]\right) \\ &= \tfrac{1}{T}\left(\textstyle\sum_{t=1}^{T} \mathsf{E}[z_t^2]\right)\left(\textstyle\sum_{t=1}^{T} \mathsf{E}[z_t^2]\right),\end{aligned}$$

where the first equality uses independence, and the second and third equalities use constancy of $\mathsf{E}[z_t^2]$. It follows that $(\frac{1}{T}\sum_{t=1}^{T} z_t^2)^2$ is consistent for $\frac{1}{T}\sum_{t=1}^{T} \mathsf{E}[z_t^2 z_{t-k}^2]$, so

$$\left(\frac{1}{T}\sum_{t=1}^{T} z_t^2\right)^{-1} \sum_{t=1}^{T} z_t z_{t-k} \xrightarrow{d} \mathsf{N}[0, 1].$$

This implies $T\hat{\rho}_k \xrightarrow{d} \mathsf{N}[0, 1]$, as in (7.17), or equivalently the usual result that $\hat{\rho}_k \sim \mathsf{N}[0, \frac{1}{T}]$.

If $\mathsf{E}[z_t^2]$ is nonconstant this simplification is not possible. One instead uses $\frac{1}{T}\sum_{t=1}^{T} z_t^2 z_{t-k}^2$ as a consistent estimator of $\frac{1}{T}\sum_{t=1}^{T}\mathsf{E}[z_t^2 z_{t-k}^2]$, and the more general (7.19).

7.13 BIBLIOGRAPHIC NOTES

An early treatment is by Cox and Lewis (1966). McKenzie (1985) is a stimulating paper that raises, in the count context, many standard time-series issues, such as trends and seasonality, in the context of hydrological examples. Chapter 1 of MacDonald and Zucchini (1997) surveys many of the count time-series models presented in this chapter. Autoregressive models, serially correlated error models, and state space models for GLMs are covered in some detail in Fahrmeier and Tutz (1994, chapter 8). Jung et al. (2006) provide a survey and apply several different models to the asthma data introduced in Chapter 1.3.4.

Davis et al. (2000) extend the serially correlated error model of Zeger (1988). Davis et al. (2003) and Jung et al. (2006) propose richer versions of the autoregressive models advanced by Zeger and Qaqish (1988). The recent time series literature has focused on INARMA models. Papers covering forecasting, ML estimation, and nonparametric ML estimation include ones by Freeland and McCabe (2004), Bu, McCabe, and Hadri (2008), Enciso-Mora et al. (2008a), Drost et al. (2009a), and McCabe et al. (2011). This model has been less used in the regression context, a notable exception being Brännäs (1995a). Recent papers by Zhu and Joe (2006), Weiss (2008) and Enciso-Mora, Neal, and Rao (2008b) have also considered the properties of INARMA with regressors. State space models, both linear and nonlinear, are discussed in Harvey (1989) and West and Harrison (1997). Estimation of models with time-varying parameters is surveyed in Durbin and Koopman (2000); they also survey computational methods for a class of models that includes this model; see also Frühwirth-Schnatter and Wagner (2006). The hidden Markov model for count and binomial data is presented in detail by MacDonald and Zucchini (1997). Jung et al. (2011) propose a multivariate model in which components are Poisson distributed conditional on latent factors that have a joint multivariate distribution.

7.14 EXERCISES

7.1 The first-order conditions for the Poisson QMLE are $\sum_t g(y_t, \mathbf{x}_t, \boldsymbol{\beta}) = \mathbf{0}$, where $g(y_t, \mathbf{x}_t, \boldsymbol{\beta}) = (y_t - \exp(\mathbf{x}_t'\boldsymbol{\beta}))\mathbf{x}_t$. Apply the general result given in Chapter 2.8.1 to obtain (7.13) for y_t heteroskedastic and autocorrelated to lag l.

7.2 Consider OLS regression of y_t on scalar x_t without intercept in the cross-section case. The robust sandwich heteroskedastic-consistent estimate of the variance matrix of $\hat{\beta}$ is $[\sum_t x_t^2]^{-1}[\sum_t (y_t - x_t\hat{\beta})^2 x_t^2][\sum_t x_t^2]^{-1}$. Specialize this formula if there is no relationship between y_t and x_t (so $\beta = 0$), and obtain the resulting formula for $t = \hat{\beta}/\sqrt{\mathsf{V}[\hat{\beta}]}$. Apply this result to the case of regression of z_t on z_{t-k} and compare the formula to (7.19).

7.3 Suppose ε_t in (7.40) is iid. Then y_t is stationary with, say $\mathsf{E}[y_t] = \mu$ for all t and $\mathsf{V}[y_t] = \sigma^2$ for all t. Using the general result $\mathsf{E}[y] = \mathsf{E}_x[\mathsf{E}[y|x]]$ use (7.43) to obtain (7.41). Using the general result $\mathsf{V}[y] = \mathsf{E}_x[\mathsf{V}[y|x]] + \mathsf{V}_x[\mathsf{E}[y|x]]$ use (7.44) to obtain (7.42).

7.4 For the Poisson INAR(1) model give the objective function for the conditional weighted LS estimator discussed after (7.52).

7.5 Show that the binomial thinning operator in section 7.6 implies (i) $0 \circ y = 0$, (ii) $1 \circ y = y$, (iii) $\mathsf{E}[\alpha \circ y] = \alpha\mathsf{E}[y]$, (iv) $\mathsf{V}[\alpha \circ y] = \alpha^2\mathsf{V}[y] + \alpha(1 - \alpha)\mathsf{E}[y]$, and (v) for any $\gamma \in [0, 1]$, $\gamma \circ \alpha \circ y = (\gamma\alpha) \circ y$, by definition.

7.6 Given (7.22) and the assumptions on ε_t obtain (7.24) and (7.25).

CHAPTER 8

Multivariate Data

8.1 INTRODUCTION

In this chapter we consider regression models for an m-dimensional vector of jointly distributed and, in general, correlated random variables $\mathbf{y} = (y_1, y_2, \ldots, y_m)$, a subset of which are event counts. One special case of interest is that of m seemingly unrelated count regressions denoted as $\mathbf{y}|\mathbf{x} = (y_1|\mathbf{x}_1, y_2|\mathbf{x}_2, \ldots, y_m|\mathbf{x}_m)$, where $\mathbf{x} = (\mathbf{x}_1, \ldots, \mathbf{x}_m)$ are observed exogenous covariates and the counts are conditionally correlated. In econometric terminology this model is a multivariate reduced-form model in which multivariate dependence is not causal. Most of this chapter deals with such reduced-form dependence. Causal dependence, such as y_1 depending explicitly on y_2, is covered elsewhere, most notably in Chapter 10.

Depending on the multivariate model, ignoring multivariate dependence may or may not affect the consistency of the univariate model estimator. In either case, joint modeling of $y_1, \ldots, y_m$ leads to improved efficiency of estimation and the ability to make inferences about the dependence structure. A joint model can also support probability statements about the conditional distribution of a subset of variables, say $\mathbf{y}_1$, given realization of another subset, say $\mathbf{y}_2$.

Multivariate nonlinear, non-Gaussian models are used much less often than multivariate linear Gaussian models, and there is no model with the universality of the linear Gaussian model. Fully parametric approaches based on the joint distribution of non-Gaussian vector $\mathbf{y}$, given a set of covariates $\mathbf{x}$, are difficult to apply because analytically and computationally tractable expressions for such joint distributions are available for special cases only. Consequently, it is more convenient to analyze models that are of interest in specific situations.

Multivariate data appear in three contexts in this book. The first is basic cross-section data with more than one outcome variable, the main subject of this chapter. The second is longitudinal data with repeated measures over time on the same outcome variable, leading to time-series correlation structures detailed in Chapter 9. The third is multivariate cross-section data with endogeneity or feedback from one outcome to another; see Chapter 10. There are other forms of multivariate data, such as multivariate time series analogs of Gaussian vector autoregressions for several outcomes observed over time, that we do not cover.

Illustrative empirical examples are readily found. For example, a model of the frequency of entry and exit of firms to an industry is an example of a bivariate count process (Mayer and Chappell, 1992). A model of health care utilization using several measures of health care services (hospital admissions, days spent in hospitals, use of prescribed and nonprescribed medicines) is an example of jointly dependent counts. An example of a joint model of counts and binary or multinomial variables is a model of the frequency of recreational trips and discrete choice of recreational sites (Terza and Wilson, 1990; Hausman, Leonard and McFadden, 1995). Recent examples of joint modeling of counts are provided by Danaher (2007), who models the joint distribution of the number of page views across web sites and Miravete (2011), who models the number of tariff options offered by cellular firms in 305 cellular markets in the United States.

Application of multivariate count models is increasing, although to date practical experience has been restricted to some special computationally tractable cases. Section 8.2 discusses some general issues relevant to characterizing dependence. Section 8.3 deals with potential sources of dependence between counts. Section 8.4 considers parametric bivariate and multivariate models. Section 8.5 continues with the parametric approach and provides an introduction to copula-based multivariate models, including examples of count data applications. Section 8.6 deals with moment-based estimation of multivariate count models. Section 8.7 analyzes tests of dependence using an approach based on orthogonal polynomial series expansions. Section 8.8 considers mixed models with dependence between counts and other continuous or discrete variables. An empirical example of bivariate count regression is given in section 8.9.

8.2 CHARACTERIZING AND GENERATING DEPENDENCE

When multivariate data are available, it is natural to consider joint distributions that are informative about the nature of dependence. For continuous outcomes on $(-\infty, \infty)$, the multivariate normal distribution is the standard model and enables one to analyze linear dependence between variables through the correlation matrix.

For other types of data such as counts, the dependence may not depend so directly on the standard Pearson correlation measure, and in this section we consider alternative measures of dependence. We focus on the bivariate case of two random variables Y_1 and Y_2 with cdf's $F_1(\cdot)$ and $F_2(\cdot)$, respectively, and joint distribution function $F(\cdot)$. Definitions are given for the population version of the dependence measure and for continuous random variables. Some adaptation for the discrete case is discussed at the end of the section.

8.2.1 Desirable Properties of Dependence Measures

The random variables (Y_1, Y_2) are said to be dependent or associated if they are not statistically independent so that $F(Y_1, Y_2) \neq F_1(Y_1)F_2(Y_2)$. Let $\delta(Y_1, Y_2)$ denote a scalar measure of dependence in the bivariate case. Embrechts, McNeil, and Straumann (2002) list four desirable properties of this measure:

1. $\delta(Y_1, Y_2) = \delta(Y_2, Y_1)$ (symmetry)
2. $-1 \leq \delta(Y_1, Y_2) \leq +1$ (normalization)
3. $\delta(Y_1, Y_2) = 1 \Leftrightarrow (Y_1, Y_2)$ comonotonic; $\delta(Y_1, Y_2) = -1 \Leftrightarrow (Y_1, Y_2)$ countermonotonic
4. For a strictly monotonic transformation $T : \mathcal{R} \to \mathcal{R}$ of Y_1 :

$$\delta(T(Y_1), Y_2) = \begin{cases} \delta(Y_2, Y_1), & T \text{ increasing} \\ -\delta(Y_2, Y_1), & T \text{ decreasing.} \end{cases}$$

Cherubini, Luciano, and Vecchiato (2004) note that association can be measured using several alternative concepts. They examine four in particular – linear correlation, concordance, tail dependence, and positive quadrant dependence – and we consider these in turn.

Correlation and Dependence

By far the most familiar association (dependence) concept is Pearson's product-moment correlation coefficient between a pair of variables (Y_1, Y_2), defined as

$$\rho_{Y_1Y_2} = \frac{\mathsf{Cov}[Y_1, Y_2]}{\sigma_{Y_1}\sigma_{Y_2}},$$

where $\mathsf{Cov}[Y_1, Y_2] = \mathsf{E}[Y_1Y_2] - \mathsf{E}[Y_1]\mathsf{E}[Y_2]$ and $\sigma_{Y_1} > 0$ and $\sigma_{Y_2} > 0$ denote, respectively, the standard deviations of Y_1 and Y_2. This measure of association can be extended to the multivariate case $m \geq 3$, for which the covariance and correlation measures are symmetric positive definite matrices.

It is well known that (a) $\rho_{Y_1Y_2}$ is a measure of linear dependence, (b) $\rho_{Y_1Y_2}$ is symmetric, (c) the lower and upper bounds on the inequality $-1 \leq \rho_{Y_1Y_2} \leq 1$ measure perfect negative and positive linear dependence (a property referred to as normalization), and (d) the measure is invariant with respect to linear transformations of the variables. Furthermore, if the pair (Y_1, Y_2) follows a bivariate normal distribution, then the correlation is fully informative about their joint dependence, and $\rho_{Y_1Y_2} = 0$ implies and is implied by independence. In this case, the dependence structure is fully determined by the correlation, and zero correlation and independence are equivalent.

In the case of other multivariate distributions, such as the multivariate elliptical families that share some properties of the multivariate normal, the dependence structure is also fully determined by the correlation matrix; see Fang and Zhang (1990).

In general, however, zero correlation does not imply independence. For example, if $Y_1 \sim \mathsf{N}[0, 1]$ and $Y_2 = Y_1^2$, then $\mathsf{Cov}[Y_1, Y_2] = \mathsf{E}[Y_1^3] = 0$ by symmetry of the normal, but (Y_1, Y_2) are clearly dependent. Zero correlation only requires $\mathsf{Cov}[Y_1, Y_2] = 0$, whereas zero dependence requires $\mathsf{Cov}[\phi_1(Y_1), \phi_2(Y_2)] = 0$ for any functions $\phi_1(\cdot)$ and $\phi_2(\cdot)$. This represents a weakness of correlation as a measure of dependence.

A second limitation of correlation is that it is not defined for some heavy-tailed distributions whose second moment does not exist (e.g., some members of the stable class and Student's t distribution with degrees of freedom equal to 2 or 1). Many financial time series display the distributional property of heavy tails and nonexistence of higher moments; see, for example, Cont (2001). Boyer, Gibson, and Loretan (1999) find that correlation measures are not sufficiently informative in the presence of asymmetric dependence.

A third limitation of the correlation measure is that it is not invariant under strictly increasing nonlinear transformations. That is, $\rho[T(Y_1), T(Y_2)] \neq \rho_{Y_1 Y_2}$ for $T : \mathcal{R} \rightarrow \mathcal{R}$. Finally, attainable values of the correlation coefficient within the interval $[-1, +1]$ between a pair of variables depend on their respective marginal distributions $F_1(\cdot)$ and $F_2(\cdot)$.

These limitations motivate an alternative measure of dependence, rank correlation.

Rank Correlation and Concordance

Two well-established association measures based on rank correlation are described next.

Spearman's rank correlation ("Spearman's rho") is defined as

$$\rho_S(Y_1, Y_2) = \rho(F_1(Y_1), F_2(Y_2)). \tag{8.1}$$

This measure is the Pearson correlation between $F_1(Y_1)$ and $F_2(Y_2)$, which are integral transforms of Y_1 and Y_2. It is a measure of rank correlation, since a cdf ranks the value of a random variable on the (0, 1) interval.

Kendall's rank correlation ("Kendall's tau") is defined as

$$\begin{aligned} \rho_\tau(Y_1, Y_2) &= \Pr[(Y_1^{(1)} - Y_1^{(2)})(Y_2^{(1)} - Y_2^{(2)}) > 0] \\ &\quad - \Pr[(Y_1^{(1)} - Y_1^{(2)})(Y_2^{(1)} - Y_2^{(2)}) < 0], \end{aligned} \tag{8.2}$$

where the pairs $(Y_1^{(1)}, Y_2^{(1)})$ and $(Y_1^{(2)}, Y_2^{(2)})$ are two independent draws of (Y_1, Y_2) from the joint cdf $F(\cdot)$. The two pairs are concordant if the ranks of each component are in accord, so that both $Y_1^{(1)} > Y_1^{(2)}$ and $Y_2^{(1)} > Y_2^{(2)}$ or both $Y_1^{(1)} < Y_1^{(2)}$ and $Y_2^{(1)} < Y_2^{(2)}$. The first term on the right, $\Pr[(Y_1^{(1)} - Y_1^{(2)})(Y_2^{(1)} - Y_2^{(2)}) > 0]$, is therefore Pr[concordance], whereas the second is Pr[discordance]. Hence

$$\rho_\tau(Y_1, Y_2) = \Pr[\text{concordance}] - \Pr[\text{discordance}] \tag{8.3}$$

is a measure of the relative difference between the two.

Both $\rho_S(Y_1, Y_2)$ and $\rho_\tau(Y_1, Y_2)$ are measures of monotonic dependence between (Y_1, Y_2). Both measures are based on the concept of concordance, more loosely defined as large values of one random variable in the pair being associated with large values of the other.

Embrechts et al. (2002; theorem 3) show that both $\rho_S(Y_1, Y_2)$ and $\rho_\tau(Y_1, Y_2)$ have the properties of symmetry, normalization, and co- and countermonotonicity, and both assume the value zero under independence. Although the rank correlation measures have the property of invariance under monotonic transformations and can capture perfect dependence, they are not simple functions of moments and hence computation is more involved.

Tail Dependence

In some cases the concordance between extreme (tail) values of random variables is of interest. For example, one may be interested in the probability that stock indexes in two countries exceed (or fall below) given levels. This requires a dependence measure for upper and lower tails of the distribution. Such a dependence measure is essentially related to the conditional probability that one index exceeds some value given that another exceeds some value.

Specifically, lower tail dependence of Y_1 and Y_2 is defined as the limit of the probability that Y_1 is less than its vth quantile given that Y_2 is less than its vth quantile, as v goes to 0. Now

$$\begin{aligned}
\Pr[Y_1 < F_1^{-1}(v)|Y_2 < F_2^{-1}(v)] &= \Pr[Y_1 < F_1^{-1}(v), Y_2 < F_2^{-1}(v)]/v \\
&= \Pr[F_1(Y_1) < v, F_2(Y_2) < v]/v \\
&= \Pr[U_1 < v, U_2 < v]/v \\
&= C(v, v)/v,
\end{aligned} \tag{8.4}$$

where $U_1 = F_1(Y_1)$ and $U_2 = F_2(Y_2)$ are standard uniform random variables, and $C(u_1, u_2) = \Pr[U_1 < u_1, U_2 < u_2]$ is the copula function, detailed in section 8.5, associated with $F(y_1, y_2)$. This leads to the following measure of lower tail dependence:

$$\lambda_L = \lim_{v \to 0+} \frac{C(v, v)}{v}. \tag{8.5}$$

Because this measure is a function of the copula it is invariant under strictly increasing transformations.

Upper tail dependence of Y_1 and Y_2 is defined as the limit of the probability that Y_1 exceeds its vth quantile given that Y_2 exceeds its vth quantile – that is, $\Pr[Y_1 > F_1^{-1}(v)|Y_2 > F_2^{-1}(v)]$ – as v goes to 1. A similar argument to that in (8.4) yields for upper tail dependence

$$\lambda_U = \lim_{v \to 1^-} \frac{S(v, v)}{1 - v}, \tag{8.6}$$

where $S(u_1, u_2)$ denotes the joint survival function for standard uniform random variables U_1 and U_2. The measure λ_U is widely used in actuarial applications of extreme value theory to handle the probability that one event is extreme conditional on another extreme event.

For joint distributions with simple analytical expressions, the computation of λ_U can be straightforward, being a simple function of the dependence parameter; see section 8.5.2 for examples. The bivariate Gaussian copula has the property of asymptotic independence. According to Embrechts et al. (2002), this means that "regardless of how high a correlation we choose, if we go far enough into the tail, extreme events appear to occur independently in each margin." This is a great limitation, most notably in financial applications.

Positive Quadrant Dependence

Another measure of dependence is positive quadrant dependence (PQD). Two random variables Y_1, Y_2 are said to exhibit PQD if their copula is greater than their product (i.e., $C(u_1, u_2) > u_1 u_2$).

In terms of the associated cdf's of Y_1 and Y_2, PQD implies $F(y_1, y_2) \geq F_1(y_1)F_2(y_2)$ for all (y_1, y_2) in $\mathbb{R}^2$. Positive quadrant dependence implies nonnegative Pearson correlation and nonnegative rank correlation. But all these properties are implied by comonotonicity, which is the strongest type of positive dependence. Decreasing left-tail dependence and decreasing right-tail dependence imply PQD (Nelsen, 2006).

8.2.2 Dependence Measures for Discrete Data

Dependence measures for continuous data do not, in general, apply directly to discrete data. Concordance measures for bivariate discrete data are subject to constraints (Marshall, 1996; Denuit and Lambert, 2005).

Reconsider Kendall's tau defined in (8.2) in the context of discrete variables. Unlike the case of continuous random variables, the discrete case has to allow for ties (i.e., $\Pr[\text{tie}] = \Pr[Y_1^{(1)} = Y_1^{(2)} \text{ or } Y_2^{(1)} = Y_2^{(2)}]$). Some attractive properties of ρ_τ for continuous variables are consequently lost in the discrete case. Several modified versions of ρ_τ and other dependence measures exist that handle the ties in different ways; see Denuit and Lambert (2005) for discussion and additional references. When (Y_1, Y_2) are nonnegative integers

$$\Pr[\text{concordance}] + \Pr[\text{discordance}] + \Pr[\text{tie}] = 1.$$

Hence (8.3) becomes (Denuit and Lambert, 2005, p. 43)

$$\begin{aligned}\rho_\tau(Y_1, Y_2) &= 2\Pr[\text{concordance}] - 1 + \Pr[\text{tie}],\\ &= 4\Pr[Y_1^{(1)} < Y_1^{(2)}, Y_2^{(1)} < Y_2^{(2)}] - 1\\ &\quad + \Pr[Y_1^{(1)} = Y_1^{(2)} \text{ or } Y_2^{(1)} = Y_2^{(2)}]. \end{aligned} \tag{8.7}$$

This analysis shows that in the discrete case $\rho_\tau(Y_1, Y_2)$ is related to marginal distributions through the last term in (8.7), whereas in the continuous case it is determined only by the joint distribution – a result of the absence of ties in the continuous case. Furthermore it is reduced in magnitude by the presence of ties. When the number of distinct realized values of (X, Y) is small, there is likely to be a higher proportion of ties, and the attainable value of $\rho_\tau(Y_1, Y_2)$ will be smaller. Denuit and Lambert (2005) obtain an upper bound and show, for example, that in the bivariate case with identical $\mathsf{P}[\mu]$ marginal distributions, the upper bound for $\rho_\tau(Y_1, Y_2)$ increases monotonically with μ.

In econometric modeling it is common to include regressors. To accommodate this case, we should substitute $F(y|\mathbf{x})$ in place of $F(y)$. Loosely speaking we replace the original data on outcomes by residual-like quantities. Even if the original outcomes are discrete these residual-like variables generally are not. Potentially this change mitigates to some extent the loss of point identification of the dependence parameter and makes it easier to transfer the insights obtained from the analysis of continuous data.

8.3 SOURCES OF DEPENDENCE

Multivariate count models for cross-section data are essentially nonlinear multi-equation models whose statistical specification has motivation similar to that for the seemingly unrelated regression (SUR) linear model. Before going into specific and detailed model specification, we review some general considerations and mechanisms for generating dependence across equations.

One method models the jth component y_j, $j = 1, \ldots, m$, as generated by the sum of random variables $u_j + w$, where u_j has a standard distribution such as the Poisson and integrates out the common component w to obtain the joint distribution of $y_1, \ldots, y_m$. This method is used infrequently because it rarely leads to tractable models, and even when it does these models are very restrictive. A leading example is the bivariate Poisson presented in section 8.4.1.

Instead dependence is often modeled as arising due to exposure to one or more common shocks that act as unobserved regressors. The single common shock model postulates marginal distributions such as $F_j(y_j|\mathbf{x}_j, \nu)$, $j = 1, \ldots, m$, where ν denotes the common stochastic shock (assumed independent of $\mathbf{x}_j$) and one integrates out v to obtain $F(\mathbf{y}|\mathbf{x})$. The common shock ν is the unique source of dependence between outcomes and additionally imparts overdispersion in each marginal model. An example is the bivariate negative binomial presented in section 8.4.2. This single-factor model is restrictive, however, because the shock is simultaneously inducing overdispersion in the marginals and inducing dependence across marginals.

A more flexible model introduces multiple correlated shocks, so that overdispersion and dependence are modeled through different parameters. Suppose the marginal distributions are written as $F_j(y_j|\mathbf{x}_j, \nu_1, \ldots, \nu_m)$, where ν_j, $j = 1, \ldots, m$, denotes the set of m jointly distributed stochastic factors independent of $\mathbf{x}_j$. We refer to this as an m-factor model. These common shocks can be introduced as random intercepts; see section 8.4.3. Usually no tractable solution exists, but estimation is feasible using simulated ML or by reverting to a finite mixtures model.

Yet another approach specifies the marginal distributions $F_j(y_j|\mathbf{x}_j)$ to be a flexible model such as the negative binomial and then introduces dependence across the marginals using a copula function. This approach, detailed in section 8.5, leads to a tractable form for the joint distribution $F(y|\mathbf{x})$, so is estimable using standard ML techniques.

A final approach, detailed in section 8.6, specifies a model for the conditional means $\mathsf{E}[y_j|\mathbf{x}_j]$ and the conditional variances and covariances $\mathsf{Cov}[y_j, y_k|\mathbf{x}]$. Estimation is by feasible generalized least squares. This method is less parametric than the others because it is does not lead to specification of the joint distribution $F(y|\mathbf{x})$.

8.4 MULTIVARIATE COUNT MODELS

In the class of multivariate count models, the bivariate special case has attracted particular attention. Excellent summaries of results for bivariate count models are given in Kocherlakota and Kocherlakota (1993, pp. 87–158) and Johnson, Kotz, and Balakrishnan (1997, pp. 124–152). Bivariate models discussed there have closed-form joint distributions only in particular cases that are subject to some restrictions.

Poisson mixed models can be derived as a special case of the generalized linear mixed models. Diggle et al. (2002) consider a variety of generalized linear models for discrete response variables and discuss alternative methods to maximum likelihood estimation.

8.4.1 Bivariate Poisson

Unlike the case of the normal distribution, there is no unique multivariate Poisson. There are several ways in which a model with Poisson marginals can be derived. Often any distribution that leads to Poisson marginals is referred to as a multivariate Poisson.

A well-established technique for deriving multivariate (especially bivariate) count distributions is the method of mixtures and convolutions. The oldest and the most studied special case leads to a model widely called the bivariate Poisson model, notwithstanding the preceding comment that there is more than one bivariate Poisson model. This model is generated by sums of independent random counts with common components in the sums. It is also called the *trivariate reduction* technique (Kocherlakota and Kocherlakota, 1993).

Suppose count variables y_1 and y_2 are defined as

$$\begin{aligned} y_1 &= u + w \\ y_2 &= v + w, \end{aligned} \tag{8.8}$$

where u, v, and w are independently distributed as Poisson variables with parameters μ_1, μ_2, and μ_3, respectively, for $\mu_j > 0$, $j = 1, 2, 3$. Then the joint frequency distribution for the *bivariate Poisson*, derived in section 8.10, is given by

$$f(y_1 = r, y_2 = s) = \exp(\mu_1 + \mu_2 + \mu_3) \sum_{l=0}^{\min(r,s)} \frac{\mu_1^{r-l} \mu_2^{s-l} \mu_3^l}{(r-l)!(s-l)!l!}. \tag{8.9}$$

The corresponding marginals are Poisson, with mean $(\mu_1 + \mu_3)$ for y_1 and mean $(\mu_1 + \mu_3)$ for y_2. Note that this model has the limitation of not allowing for overdispersion.

The squared Pearson correlation between y_1 and y_2 is given by

$$\rho^2 = \frac{\mu_3^2}{(\mu_1 + \mu_3)(\mu_2 + \mu_3)} \tag{8.10}$$

(see Johnson and Kotz, 1969). Allowing for heterogeneity by allowing the parameters to vary across individuals implies that the correlation between events also varies across individuals. However, the maximum correlation between y_1 and y_2 is given by $\mu_3 / \left[\mu_3 + \min(\mu_1, \mu_2)\right]$. Gourieroux et al. (1984b) present a different derivation of the bivariate Poisson.

The use of the trivariate reduction technique leads to the following properties:

1. In general, in multivariate nonnormal distributions the correlation does not fully describe the dependence structure of variables. However, in this special case the correlation coefficient fully characterizes dependence, and no additional measures of dependence are needed.
2. The marginal distributions for y_1 and y_2 are both Poisson. Hence, with a correctly specified conditional mean function the marginal models can be estimated consistently, but not efficiently, by maximum likelihood. Joint estimation is more efficient even if the two mean functions depend on the same covariates.
3. This model only permits positive correlation between counts.

The log-likelihood for the model (8.9) with observations independent over i is

$$\mathcal{L}(\boldsymbol{\beta}, \mu_3, \boldsymbol{\gamma} | y_1, y_2, \mathbf{x}) = n\mu_3 - \sum_{i=1}^{n} (\mu_{1i} + \mu_{2i}) + \sum_{i=1}^{n} \ln S_i, \tag{8.11}$$

where

$$S_i = \sum_{l=0}^{\min(y_{1i}, y_{2i})} \frac{\mu_1^{y_{1i}-l} \mu_2^{y_{2i}-l} \mu_3^l}{(y_{1i}-l)!(y_{2i}-l)!l!}. \tag{8.12}$$

At this stage there are two ways to proceed. One approach parameterizes the sum $\mu_{1i} + \mu_{2i}$ as an (exponential) function of $\mathbf{x}_i$. A second approach parameterizes μ_{1i} and μ_{2i} individually in terms of the same or different covariates. The two approaches imply different specifications of the conditional means.

Taking the second approach, assume that $\mu_{3i} = \mu_3$, $\mu_{1i} = \exp(\mathbf{x}_i'\boldsymbol{\beta})$, and $\mu_{2i} = \exp(\mathbf{x}_i'\boldsymbol{\gamma})$. This leads to the log-likelihood

$$\mathcal{L}(\boldsymbol{\beta}, \mu_3, \boldsymbol{\gamma}) = n\mu_3 - \sum_{i=1}^{n} \left(\exp(\mathbf{x}_i'\boldsymbol{\beta}) + \exp(\mathbf{x}_i'\boldsymbol{\gamma})\right) + \sum_{i=1}^{n} \ln S_i, \quad (8.13)$$

where S_i is defined in (8.12).

King (1989a) calls this a model of seemingly unrelated Poisson regression by analogy with the well-known seemingly unrelated least squares model. Jung and Winkelmann (1993), in their application of the bivariate Poisson to the number of voluntary and involuntary job changes, assume a constant covariance and exponential mean parameterization for the marginal means so that $\mu_j + \mu_3 = \exp(\mathbf{x}_j'\boldsymbol{\beta}_j)$, $j = 1, 2$. This allows the two means to depend on separate or common sets of covariates.

The bivariate Poisson is a special case of the generalized exponential family (Jupp and Mardia, 1980; Kocherlakota and Kocherlakota, 1993):

$$\begin{aligned} g(y_1, y_2; \mu_1, \mu_2, \rho) = \exp\{&\mu_1' v(y_1) + v(y_1)'\rho w(y_2) + \mu_2' w(y_2) \\ &- c(\mu_1, \mu_2, \rho) + d_1(y_1) + d_2(y_2)\} \end{aligned} \quad (8.14)$$

where $v(\cdot)$, $w(\cdot)$, $d_1(\cdot)$, and $d_2(\cdot)$ are functions, and $c(\mu_1, \mu_2, 0) = c_1(\mu_1)c_2(\mu_2)$. Under independence $\rho = 0$, in which case the right-hand side is a product of two exponential families. In this family $\rho = 0$ is a necessary and sufficient condition for independence. The correlation coefficient is increasing in μ_3 and decreasing steeply in both μ_1 and μ_2.

8.4.2 Bivariate Negative Binomial

The bivariate Poisson can be generalized and extended to allow for unobserved heterogeneity and overdispersion in the respective marginal distributions, using mixtures and convolutions as in the univariate case presented in chapter 4.2.

Marshall and Olkin (1988, 1990) provide a quite general treatment of the multivariate case. Consider the bivariate distribution,

$$f(y_1, y_2|x_1, x_2) = \int_0^{\infty} f_1(y_1|x_1, \nu) f_2(y_2|x_2, \nu) g(\nu) d\nu, \quad (8.15)$$

where f_1, f_2, and g are univariate densities and ν may be interpreted as common unobserved heterogeneity affecting both counts. Multivariate distributions generated in this way have univariate marginals in the same family (Kocherlakota and Kocherlakota, 1993). Thus, a bivariate negative binomial mixture generated in this way will have univariate negative binomial mixtures. This approach suggests a way of specifying or justifying overdispersed and correlated count models, based on a suitable choice of $g(.)$, which is more general than in the example given earlier.

Marshall and Olkin (1990) generate a *bivariate negative binomial* distribution beginning with $f(y_1)$ and $f(y_2)$, which are Poisson with parameters $\mu_1\nu$

and $\mu_2 \nu$, respectively, while ν has a gamma distribution with shape parameter α^{-1} and scale parameter 1. Then

$$h(y_1, y_2|\alpha^{-1}) = \int_0^\infty \left[\frac{(\mu_1 \nu)^{y_1} e^{-\mu_1 \nu}}{y_1!}\right]\left[\frac{(\mu_2 \nu)^{y_2} e^{-\mu_2 \nu}}{y_2!}\right]\left[\frac{\nu^{\alpha^{-1}-1} e^{-\nu}}{\Gamma(\alpha^{-1})}\right] d\nu$$

$$= \frac{\Gamma(y_1 + y_2 + \alpha^{-1})}{y_1! y_2! \Gamma(\alpha^{-1})}\left[\frac{\mu_1}{\mu_1 + \mu_2 + 1}\right]^{y_1}\left[\frac{\mu_2}{\mu_1 + \mu_2 + 1}\right]^{y_2}$$

$$\times \left[\frac{1}{\mu_1 + \mu_2 + 1}\right]^{\alpha^{-1}}. \tag{8.16}$$

This mixture has a closed-form solution, a special result that arises because the model restricts the unobserved heterogeneity to be identical for both count variables. The marginal distributions are univariate negative binomial, with means $\alpha^{-1}\mu_j$ and variances $\mu_j + \alpha^{-1}\mu_j^2$, $j = 1, 2$. The rescaling of the mean, absorbed in the intercept in regression with exponential conditional mean, occurs because the parameterization of the exponential in (8.16) differs from that in Chapter 4.2.2 and implies $\mathsf{E}[\upsilon] = \mathsf{V}[\upsilon] = \alpha^{-1}$. The unconditional correlation between the two count variables,

$$\mathsf{Cor}[y_1, y_2] = \frac{\mu_1 \mu_2}{\sqrt{(\mu_1^2 + \alpha\mu_1)(\mu_2^2 + \alpha\mu_2)}}, \tag{8.17}$$

is restricted to be positive.

An alternative derivation of the model (8.16), based on specifying a probability generating function that is a natural extension of that for the negative binomial, is given in Johnson et al. (1997). The model (8.16) extends easily to the multivariate case and is referred to either as a *multivariate negative binomial* or a *negative multinomial model.* Nguyen, Gupta, Nguyen, and Wang (2007) provide further details on the negative multinomial.

Miles (2001) provides an application to individual consumer data on the number of purchases of bread and cookies in a one-week period, parameterizing $\mu_1 = \exp(\mathbf{x}_1'\boldsymbol{\beta})$ and $\mu_2 = \exp(\mathbf{x}_2'\boldsymbol{\beta}_2)$ and estimating by ML. Miles also estimates a trivariate negative binomial model for wheat bread, other breads, and cookies. A result of Lindeboom and van den Berg (1994) for bivariate survival models indicates that it is hazardous to estimate bivariate models in which mutual dependence between survival times arises purely from unobserved heterogeneity characterized as a univariate random variable. Miles (2001) finds that this is also the case for his count application, so it is desirable to have more flexible models of the correlation structure between counts.

8.4.3 More Flexible Bivariate Count Models

One convenient way to generate flexible dependence structures between counts is to begin by specifying latent factor models. Munkin and Trivedi (1999) propose a more flexible dependence structure using a correlated unobserved heterogeneity model. Suppose y_1 and y_2 are, respectively, $\mathsf{P}[\mu_1|\nu_1]$ and $\mathsf{P}[\mu_2|\nu_2]$ with means

$$\mathrm{E}[y_j|\mathbf{x}_j, \nu_j] = \mu_j = \exp(\mathbf{x}_j'\boldsymbol{\beta}_j + \lambda_j \nu_j), \quad j = 1, 2, \tag{8.18}$$

where ν_1 and ν_2 represent correlated latent factors or unobserved heterogeneity, λ_1 and λ_2 are factor loadings, and $\mathbf{x}_j$ includes an intercept so ν_j has mean zero. Dependence is induced if ν_1 and ν_2 are correlated with joint distribution $g(\nu_1, \nu_2)$. Integrating out (ν_1, ν_2), we obtain the joint distribution for y_1 and y_2:

$$f(y_1, y_2|\mathbf{x}_1, \mathbf{x}_2) = \int \int f_1(\mathbf{y}_1|\mathbf{x}_1, \nu_1) f_2(\mathbf{y}_2|\mathbf{x}_2, \nu_2) g(\nu_1, \nu_2) d\nu_1 d\nu_2. \tag{8.19}$$

Unlike the preceding examples, there is no choice of $g(\cdot)$ that leads to a closed-form solution. Instead (8.19) can be replaced by the simulation-based numerical approximation

$$\hat{f}(y_1, y_2|\mathbf{x}_1, \mathbf{x}_2) = \frac{1}{S} \sum_{s=1}^{S} f_1(y_1|\mathbf{x}_1, \nu_1^{(s)}) f_2(y_2|\mathbf{x}_2, \nu_2^{(s)}), \tag{8.20}$$

where $\nu_1^{(s)}$ and $\nu_2^{(s)}$ are draws from the specified distributions of unobserved heterogeneity. The simulated ML method estimates the unknown parameters using the log-likelihood based on such an approximation.

Munkin and Trivedi (1999) specify (ν_1, ν_2) to be bivariate normal distributed with zero means, variances σ_1^2 and σ_2^2, and correlation ρ, $-1 \leq \rho \leq 1$. They normalize $\lambda_1 = 1$ and $\lambda_2 = 1$ to ensure identification. It is more convenient to sample ν_1 and ν_2 from the standard normal, using

$$\begin{bmatrix} \nu_1 \\ \nu_2 \end{bmatrix} = \begin{bmatrix} \sigma_1 & 0 \\ \sigma_2 \rho & \sigma_2 \sqrt{1-\rho^2} \end{bmatrix} \begin{bmatrix} \varepsilon_1 \\ \varepsilon_2 \end{bmatrix},$$

where ε_1 and ε_2 are iid standard normal. Then the SMLE maximizes with respect to $(\boldsymbol{\beta}_1, \boldsymbol{\beta}_2, \sigma_1, \sigma_2, \rho)$

$$\sum_{i=1}^{n} \log \left\{ \frac{1}{S} \sum_{s=1}^{S} f_1 \left(y_{1i}|\mathbf{x}_{1i}, \sigma_1 \varepsilon_{1i}^{(s)} \right) f_2 \left(y_{2i}|\mathbf{x}_{2i}, \sigma_2 \rho \varepsilon_{1i}^{s} + \sigma_2 \sqrt{1-\rho^2} \varepsilon_{2i}^{(s)} \right) \right\}, \tag{8.21}$$

where $\varepsilon_{1i}^{(s)}$ and $\varepsilon_{2i}^{(s)}$ are draws from the standard normal and, for example, $f_1(\cdot)$ is the Poisson density with mean $\exp(\mathbf{x}_{1i}'\boldsymbol{\beta}_1 + \sigma_1 \varepsilon_{1i})$. The SML objective function converges to the log-likelihood likelihood function as $S \to \infty$, and the SMLE is consistent and asymptotically equivalent to the MLE estimator when $S, n \to \infty$ and $n/S \to 0$ or $\sqrt{n}/S \to 0$. The covariance matrix can be consistently estimated by the robust sandwich estimator. The model accommodates both overdispersion, through σ_1 and σ_2, and correlation, through ρ.

Munkin and Trivedi (1999) show that although SML estimation is feasible, it may not be computationally straightforward for all specified joint distributions of latent factors. They note that in their parameterization,

$$\mathsf{E}[y_{2i}|x_{2i}, \varepsilon_{1i}, \varepsilon_{2i}] = \exp\left(\mathbf{x}'_{2i}\boldsymbol{\beta}_2 + \sigma_2\rho\varepsilon_{1i} + \sigma_2\sqrt{1-\rho^2}\varepsilon_{2i}\right),$$

so the derivative

$$\frac{\partial \mathsf{E}[y_{2i}|x_{2i}, \varepsilon_{1i}, \varepsilon_{2i}]}{\partial \rho} = \exp\left(\mathbf{x}'_{2i}\boldsymbol{\beta}_2 + \sigma_2\rho\varepsilon_{1i} + \sigma_2\sqrt{1-\rho^2}\varepsilon_{2i}\right)\frac{-\sigma_2\rho\varepsilon_{2i}}{\sqrt{1-\rho^2}},$$

which is not bounded at $\rho = 1$. It follows that the derivative of the sum in equation (8.21) with respect to ρ is unbounded at $\rho = 1$, since the derivative is the product of two functions, one of which is bounded at $\rho = 1$ and the other is $\partial \mathsf{E}[y_{2i}|x_{2i}, \varepsilon_{1i}, \varepsilon_{2i}]/\partial \rho$, which is unbounded at $\rho = 1$. Similarly, the second derivative of the sum with respect to ρ is also unbounded. Therefore, for values of ρ in a neighborhood of 1, maximization of the SML objective is not feasible using standard gradient methods. See Munkin and Trivedi (1999) for a discussion of methods for handling this problem.

Several alternatives to SMLE have emerged that may offer computational convenience while still allowing for some flexibility in modeling the dependence structure. Alfo and Trovato (2004) propose finite mixture distributions for ν_1 and ν_2 in (8.18) and additionally allow for random slope parameters and the random intercepts captured by ν_1 and ν_2. However, these authors place additional structure on the correlation matrix of the random components of the slope coefficients, which generates patterns in the dependence structure. Chib and Winkelmann (2001) use Bayesian Monte Carlo Markov Chain methods; see Chapter 12.5.4. And copulas can be used to generate a joint distribution whose parameters can be estimated without simulation; see section 8.5.

8.4.4 Bivariate Hurdle

For univariate counts the hurdle model, presented in Chapter 4.5, is the most common departure from the Poisson or related models such as negative binomial or Poisson-lognormal. When we observe multivariate count outcomes in which there is substantial probability mass at zero frequency, there is motivation for a multivariate version of the hurdle model. The closely related zero-inflated Poisson (ZIP) model of Chapter 4.6 is also a candidate for a similar extension.

Hellstrom (2006) provides an empirical application of a bivariate hurdle model in which the outcome variables are the number of trips taken and the number of overnight stays. Li, Lu, Park, Kim, Brinkley, and Peterson (1999) present a multivariate ZIP model for several types of defects in items produced by an industrial process. In section 8.5.7 we present a bivariate hurdle model based on copulas.

8.4.5 Sarmanov Models

A general way to produce a bivariate distribution from given marginals was proposed by Sarmanov (1966). Given marginal densities $f_1(y_1)$ and $f_2(y_2)$, the joint density is specified as

$$f_{12}(y_1, y_2) = f_1(y_1)f_2(y_2) \times [1 + \omega\phi_1(y_1)\phi_2(y_2)], \tag{8.22}$$

where $\phi_j(y_j)$, $j = 1, 2$, are mixing functions that satisfy $\int \phi_j(t)f_j(t)dt = 0$, or $\sum_{x_k} \phi_j(x_k)f_j(x_k) = 0$ in the discrete case, so that the joint probabilities integrate or sum to one. To ensure that $f_{12}(y_1, y_2) \geq 0$ the mixing functions are determined in part by the marginals. M-L.T. Lee (1996) shows that for marginal distributions for nonnegative counts

$$\phi_j(y_j) = e^{-y_j} - \mathsf{E}[e^{-y_j}] = e^{-y_j} - \sum_{y_j=0}^{\infty} e^{-y_j}\phi_j(y_j). \tag{8.23}$$

When $\omega = 0$ the random variables y_1 and y_2 are independent,

Danaher (2007) specifies negative binomial marginals and applies a (truncated) multivariate extension of this model to individual-level data on the number of page views across a number of websites. He estimates models without regressors by a two-step method – separate estimation of the parameters of each density $f_j(y_j)$ is followed by estimation of the remaining parameters (just ω in the bivariate case). He considers a model without regressors. Famoye (2010) estimates this model by ML with application to joint modeling of doctor consultations and nondoctor consultations using data from Cameron et al. (1988).

Miravete (2011) specifies double Poisson marginals, better suited to his data on the number of tariff options (a measure of firm responsiveness to customer tastes) offered by cellular firms in 305 cellular markets in the United States. He estimates by ML a bivariate regression model for the incumbent (y_1) and entrant (y_2) firms. The coefficient ω is constrained to lie in bounds determined by the parameters and regressors of $f_1(y_1)$ and $f_2(y_2)$. If the bounds are reached statistical inference becomes nonstandard, and Miravete (2011) uses a subsampling bootstrap.

8.5 COPULA-BASED MODELS

For notational simplicity, in what follows we suppress the presence of covariates in the model. Thus, when we write the cdf $F_1(y_1)$, the reader may interpret this to mean the conditional cdf $F_1(y_1|\mathbf{x}_1)$. For simplicity we focus on the bivariate case.

For nonnormal multivariate data, joint distributions are often not readily available and harder to derive. This applies both to discrete and mixed discrete/continuous multivariate data. In contrast, marginal distributions for the component variables may be easier to specify. The copula approach allows one to combine given marginal distributions, using a specified functional form, to yield a joint distribution with dependence parameter(s) and a tractable functional form.

The generation of a parametric joint distribution by combining fixed marginals is based on Sklar's theorem (Sklar, 1973). A variety of joint distributions can be generated by specifying different functional forms for the copula. These functional forms may, however, impose restrictions on the form of dependence. Therefore care must be taken to specify a sufficiently flexible copula. Furthermore, the choice of marginal distributions also requires care because they too are in no sense known and, if poorly chosen, may not provide a good fit to the data.

8.5.1 Specification

Copula-based joint estimation is based on Sklar's theorem, which provides a method of generating joint distributions by combining marginal distributions using a copula.

An m-dimensional *copula* or linking function $C(\cdot)$ is any m-dimensional cumulative distribution function with all m univariate margins being standard uniform, so

$$C(u_1, \ldots, u_m|\theta) = \Pr[U_1 \leq u_1, \ldots, U_m \leq u_m], \tag{8.24}$$

where U_j are iid U[0, 1] and θ is a dependence parameter. Now if Y_j has continuous cdf $F_j(\cdot)$, the integral transform $F_j(Y_j)$ is uniformly distributed. It follows that for a continuous m-variate distribution function $F(y_1, \ldots, y_m)$ with univariate marginal distributions $F_1(y_1), \ldots, F_m(y_m)$, the joint cdf

$$\begin{aligned} F(y_1, \ldots, y_m) &= \Pr[Y_1 \leq y_1, \ldots, Y_m \leq y_m] \\ &= \Pr[F_1(Y_1) \leq F_1(y_1), \ldots, F_m(Y_m) \leq F_m(y_m)] \\ &= \Pr[U_1 < F_1(y_1), \ldots, U_m \leq F_m(y_m)] \\ &= C(F_1(y_1), \ldots, F_m(y_m)|\theta) \end{aligned} \tag{8.25}$$

where, by Sklar's theorem, $C(\cdot)$ is the unique copula function associated with the cdf $F(\cdot)$. An alternative way to express (8.25) is

$$\begin{aligned} F(y_1, \ldots, y_m) &= F(F_1^{-1}(u_1), \ldots, F_m^{-1}(u_1)) \\ &= C(u_1, \ldots, u_m|\theta), \end{aligned} \tag{8.26}$$

as $F_j(Y_j) = U_j$ so $Y_j = F_j^{-1}(U_j)$. Zero dependence implies that the joint distribution is the product of marginals. Sklar's theorem is extended by Patton (2006) to the case of conditional copulas that are relevant in the regression context.

The term "discrete copula" refers to a discrete multivariate distribution, usually with discrete marginals. The cdf in the discrete case is a step function with jumps at integer values, so the inverse $F_j^{-1}(\cdot)$ is not uniquely determined without imposing a convention. Usually the minimum of the interval is chosen.

Table 8.1. *Copulas: Five leading examples*

Copula type	Function $C(u_1, u_2))$	θ–domain	Kendall's τ
Clayton	$(u_1^{-\theta} + u_2^{-\theta} - 1)^{-1/\theta}$	$\theta \in (0, \infty)$	$\frac{\theta}{\theta+2}$
Survival Clayton	$((1-u_1)^{-\theta} + (1-u_2)^{-\theta} - 1)^{-1/\theta}$	$\theta \in (0, \infty)$	$\frac{\theta}{\theta+2}$
Frank	$\frac{1}{\theta}\ln\left(1 + e^{\theta u_1 - 1} e^{\theta u_2 - 1}/e^{\theta - 1}\right)$	$\theta \in (-\infty, \infty)$	$1 + \frac{4}{\theta}\left[\int_0^\theta \frac{t}{\theta(e^t - 1)} dt - 1\right]$
Gaussian	$\Phi_G(\Phi^{-1}(u_1)\Phi^{-1}(u_2)\vert\theta)$	$-1 < \theta < +1$	$\frac{2}{\pi}\arcsin(\theta)$
Gumbel	$\exp(-[(-\ln u_1)^\theta + (-\ln u_2)^\theta])^{1/\theta}$	$\theta \in [1, \infty)$	$(\theta - 1)/\theta$

Note: $\Phi_G[y_1, y_2|\theta]$ denotes the standardized bivariate normal cdf with correlation θ, and $\Phi(\cdot)$ denotes the standard normal cdf.

Sklar's theorem implies that copulas provide a "recipe" to derive joint distributions when only marginal distributions are specified. The approach is attractive because copulas (1) provide a fairly general approach to joint modeling of count data; (2) neatly separate the inference about marginal distribution from the inference on dependence; (3) represent a method for deriving joint distributions given fixed marginals such as Poisson and negative binomial; (4) in the bivariate case can be used to define nonparametric measures of dependence, which can capture asymmetric (tail) dependence, as well as correlation or linear association; and (5) are easier to estimate than multivariate latent factor models with unobserved heterogeneity. However, copulas and latent factor models are closely related; see Trivedi and Zimmer (2007) and Zimmer and Trivedi (2006).

8.5.2 Functional Forms of Copulas

Sklar's theorem states that there is a unique copula, but this is unknown a priori, just as in practice the cdfs for the marginal distributions are unknown. The literature offers a vast array of copula functional forms from which to choose (Nelsen, 2006; Joe, 1997). In practice one can experiment with a variety of copulas or choose one that is relatively unrestricted.

For concreteness Table 8.1 presents five functional forms for the copula: Clayton, Survival Clayton, Gaussian, Frank, and Gumbel copulas. Closed-form expressions exist for all the copulas, except for the Gaussian copula,

$$\Phi_G\left(\Phi^{-1}(u_1), \Phi^{-1}(u_2)|\theta\right)$$
$$= \int_{-\infty}^{\Phi^{-1}(u_1)} \int_{-\infty}^{\Phi^{-1}(u_2)} \frac{1}{2\pi(1-\theta^2)^{1/2}} \left\{\frac{-(s^2 - 2\theta st + t^2)}{2(1-\theta^2)}\right\} dsdt.$$

The Gaussian and Frank copulas permit both positive and negative dependence. Any observed dependence is symmetric, with no tail dependence allowed

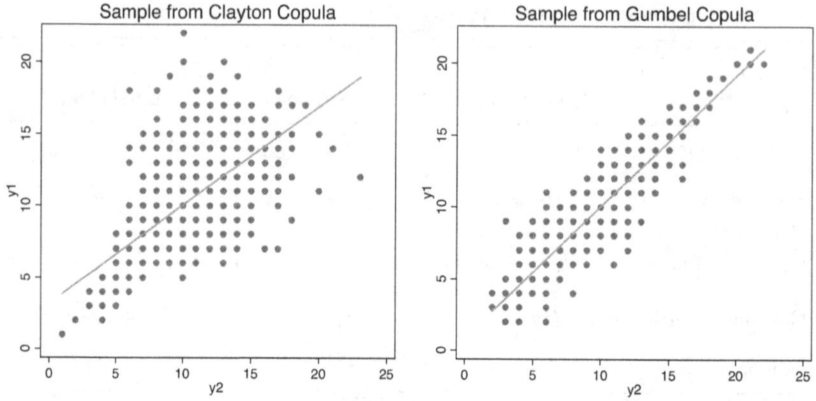

Figure 8.1. Samples from Clayton and Gumbel copulas with Poisson marginals.

in the Gaussian case and limited tail dependence in the Frank copula. By contrast, the Clayton, Survival Clayton, and Frank copulas allow for asymmetric tail dependence, but none of them allows for negative dependence. Clayton exhibits lower tail dependence, whereas Survival Clayton and Gumbel show upper tail dependence. By changing the functional form of the copula, many different dependence patterns between marginal distributions can be explored. Properties of these well-established functional forms are discussed in the literature (Cherubini, Luciano, and Vecchiato, 2004; Trivedi and Zimmer, 2007).

Figure 8.1 illustrates lower and upper tail dependence using samples generated using Monte Carlo draws from the Clayton and Gumbel copulas, where Y_1 and Y_2 are $\mathsf{P}[10]$ distributed and for both copulas $\theta = 2$ and hence Kendall's τ equals 0.5. The samples have different tail dependence properties. The Clayton copula diagram shows clustering of observations at low values, whereas observations are more dispersed at high values; the opposite description applies for the draws from the Gumbel copula. Upper tail dependence has been reported especially in returns on financial assets.

8.5.3 Interpretation of the Dependence Parameter

The dependence parameter θ can be difficult to interpret directly because it is generally not a Pearson correlation parameter that lies in the interval $(-1, 1)$. A common approach converts θ to a measure of concordance such as Kendall's tau, defined in (8.2). Converting θ to τ is straightforward for many copulas. The last column in Table 8.1 gives the conversion formula for the five copulas considered here, and Trivedi and Zimmer (2007) provide additional examples. When θ is the parameter of intrinsic interest, an alternative estimation approach first estimates τ and then converts back to θ (Genest and Neslehova, 2008).

8.5.4 Estimation

Copula modeling begins with specification of the marginal distributions and a copula. There are many possible choices of copula functional forms; see Nelsen (2006) for an especially extensive list. The resultant model can be estimated by a variety of methods, most notably joint ML of all parameters and two-step estimation. We provide a brief summary, with further details given in Trivedi and Zimmer (2007).

Equation (8.25) provides a fairly general approach to modeling complex joint distributions. By plugging the specified marginal distributions (F_1, F_2) into a copula function C, the right-hand side of equation (8.25) provides a parametric representation of the unknown joint cdf.

Estimation is then based on the pdf version of the copula. In the continuous outcomes case the pdf is $f(y_1, y_2) = \partial F(y_1, y_2)/\partial y_1 \partial y_2$. Using (8.25) and (8.26) yields

$$\begin{aligned} f(y_1, y_2) &\equiv c_{12}(F_1(y_1), F_2(y_2)|\theta) \\ &= \frac{\partial C(F_1(y_1), F_2(y_2)|\theta)}{\partial y_1 \partial y_2} \\ &= \frac{\partial C(u_1, u_2|\theta)}{\partial u_1 \partial u_2} \times \frac{\partial u_1}{\partial y_1}\frac{\partial u_2}{\partial y_2} \\ &= \left.\frac{\partial C(u_1, u_2|\theta)}{\partial u_1 \partial u_2}\right|_{u_1=F_1(y_1), u_2=F_2(y_2)} \times f_1(y_1) f_2(y_2), \end{aligned} \tag{8.27}$$

where $f_1(y_1)$ and $f_2(y_2)$ are the marginal densities corresponding to the cdfs $F_1(y_1)$ and $F_2(y_2)$, and we use $u_1 = F_1(y_1)$ so $\partial u_1/\partial y_1 = \partial F_1(y_1)/\partial y_1 = f_1(y)$ and similarly for $f_2(y_2)$.

In discrete settings, the analogous quantity is obtained by taking a finite difference. In the count data case where y_1 and y_2 take integer values,

$$\begin{aligned} f(y_1, y_2) &\equiv c_{12}(F_1(y_1), F_2(y_2)|\theta) \\ &= C(F_1(y_1), F_2(y_2)|\theta) - C(F_1(y_1 - 1), F_2(y_2)|\theta) \\ &\quad - C(F_1(y_1), F_2(y_2 - 1)|\theta) + C(F_1(y_1 - 1), F_2(y_2 - 1)|\theta). \end{aligned} \tag{8.28}$$

Taking the natural logarithm of (8.27) or (8.28) and summing over all observations give the log-likelihood function. Joint estimation means that the parameters of the marginals, suppressed above for notational simplicity, and the copula function are jointly estimated. Two-step or stepwise or sequential estimation means that first the parameters of $F_1(y_1)$ and $F_2(y_2)$ are estimated, and then θ is estimated by maximizing the log-likelihood based on $C(\hat{F}_1(y_1), \hat{F}_2(y_2)|\theta)$. Both steps may be parametrically executed, but the second step may also be nonparametric.

When joint estimation is used, one can choose the copula function that leads to the largest log-likelihood value in the fitted model. For comparison across models that specify different marginals one can use information criteria such as BIC.

8.5.5 Discrete Copulas

The presence of ties in count data can potentially reduce the identifiability of discrete copulas (Genest and Neslehova, 2008). In a bivariate context a tie occurs when pairs of observations have the same value. For count data this seems likely when the sample is dominated by zeros. However, the problem is mitigated when regressors are present and we are dealing with conditional marginals and conditional copulas.

If the marginal cdfs $F_j(\cdot)$, $j = 1, \ldots, m$, are continuous, then the corresponding copula in equation (8.25) is unique. If (F_1, F_2) are not all continuous, the joint distribution function can always be expressed as (8.25), although in such a case the copula lacks uniqueness (see Schweizer and Sklar, 1983, chapter 6). For applied work in the regression context that involves conditioning on covariates, this does not invalidate the application of copulas. As Genest and Neslehova (2008, p. 14) observe, "The fact that there exist (infinitely many) copulas for the same discrete joint distribution does not invalidate models of this sort."

Copulas have been used most often for continuous data, notably returns on multiple financial assets. Several discrete data examples are given later. Additionally, copulas can be used for a mixture of discrete and continuous data. Zimmer and Trivedi (2006) use a trivariate copula framework to develop a joint distribution of two counted outcomes and one binary treatment variable.

In the m-variate discrete case, the formula (8.28) generalizes to have 2^m terms, making estimation difficult. Smith and Khaled (2011) propose augmenting the likelihood with uniform latent variables and estimating using Bayesian MCMC methods. They apply their method to a 16-variate model for data on bike path use by hour ($m = 16$) within each of 565 days ($n = 565$).

8.5.6 Example: A Bivariate Negative Binomial Model

For standard count data that are overdispersed a natural copula-based bivariate model is one with negative binomial marginals. Trivedi and Zimmer (2007) model the individual use of prescription drugs (y_1) and nonprescription drugs (y_2), using data from Cameron et al. (1988). The data are characterized by low counts, with $y_1 > 3$ for 6% of observations and $y_2 > 3$ for 1% of the sample. There are many ties, especially at counts of 0 and 1. The data actually exhibit some negative dependence, suggesting that prescribed and nonprescribed drugs may be substitutes.

The authors estimate by ML models based on density (8.28), where the marginals are NB2, regressors are the same as those for the doctor visits example in Chapter 3.2.6, and a range of copula functions are employed. Copulas with weak tail dependence, such as the Frank copula, fit the data best. The

Clayton, Gumbel, and Gaussian copulas encountered convergence problems, perhaps due to the large number of ties.

8.5.7 Example: A Bivariate Hurdle Model

Bivariate hurdle models were introduced in section 8.4.4. Here we present a copula-based bivariate hurdle model for counts, based on the model of Deb, Trivedi, and Zimmer (2011), who considered continuous outcomes (health expenditures) rather than count outcomes. The model uses copulas twice – to model correlation when the hurdles are not crossed by either variable and to model correlation when the hurdles are crossed by both variables.

Consider two nonnegative integer-valued outcomes y_1 and y_2, each with a significant fraction of zeros. Let y_1^0 denote $y_1 = 0$, y_1^+ denote $y_1 > 0$, y_2^0 denote $y_2 = 0$, and y_2^+ denote $y_2 > 0$. Then there are four configurations of outcomes with the following probabilities:

$$\begin{aligned} y_1^0, y_2^0 &\longrightarrow F^0(y_1 = 0, y_2 = 0) \\ y_1^+, y_2^0 &\longrightarrow F^0(y_1 > 0, y_2 = 0) \times f_1^+(y_1|y_1 > 0) \\ y_1^0, y_2^+ &\longrightarrow F^0(y_1 = 0, y_2 > 0) \times f_2^+(y_2|y_2 > 0) \\ y_1^+, y_2^+ &\longrightarrow F^0(y_1 > 0, y_2 > 0) \times f_{12}^+(y_1, y_2|y_1 > 0, y_2 > 0), \end{aligned} \tag{8.29}$$

where F^0 is a bivariate distribution function defined over binary outcomes, f_1^+ and f_2^+ are univariate mass functions defined over positive, integer-valued outcomes, and f_{12}^+ is a bivariate joint mass function also defined over a pair of positive, integer-valued outcomes.

We first consider specification of $F^0(\cdot)$ for the bivariate hurdle part of the model. A natural choice for the marginal distributions is a probit model, so $F_j^0(\cdot) = \Phi_j(\cdot)$, $j = 1, 2$. A copula-based model for the joint probability distribution of y_1 and y_2 is

$$F^0(y_1 > 0, y_2 > 0) = C^0\left(\Phi_1(\cdot), \Phi_2(\cdot)|\theta^0\right) \tag{8.30}$$

where C^0 is a selected copula function with a given functional form, and θ^0 is a dependence parameter. Since, for example, $F^0(y_1 > 0, y_2 = 0) = F_1^0(y_1 > 0) - F^0(y_1 > 0, y_2 > 0)$ it follows from (8.29) that the other three discrete probabilities are

$$\begin{aligned} F^0(y_1 = 0, y_2 = 0) &= 1 - \Phi_1(\cdot) - \Phi_2(\cdot) + C^0\left(\Phi_1(\cdot), \Phi_2(\cdot)|\theta^0\right) \\ F^0(y_1 > 0, y_2 = 0) &= \Phi_1(\cdot) - C^0\left(\Phi_1(\cdot), \Phi_2(\cdot)|\theta^0\right) \\ F^0(y_1 = 0, y_2 > 0) &= \Phi_2(\cdot) - C^0\left(\Phi_1(\cdot), \Phi_2(\cdot)|\theta^0\right). \end{aligned} \tag{8.31}$$

The direct use of a copula is an alternative to a bivariate probit model or deriving the joint distribution from a pair of marginal distributions with common unobserved heterogeneity that is integrated out as in section 8.4.3, an approach also used in Hellstrom (2006). The copula approach avoids numerical integration, yet the dependence between y_1^0 and y_2^0 can be characterized in a flexible way.

Next consider specification of the densities $f^+(\cdot)$ for the bivariate positive outcomes part of the model. The univariate mass functions $f_1^+(\cdot)$ and $f_2^+(\cdot)$ may be, for example, zero-truncated Poisson or negative binomial. Using a copula, the zero-truncated joint distribution is

$$f_{12}^+(y_1, y_2|y_1 > 0, y_2 > 0) = c^+\left(F_1^+(\cdot), F_2^+(\cdot)|\theta^+\right) \times f_1^+(\cdot) \times f_2^+(\cdot), \tag{8.32}$$

where lower case $c^+(\cdot)$ represents the copula *density*, and $F_j^+(\cdot)$ is the cumulative distribution function (cdf) version of $f_j^+(\cdot)$, $j = 1, 2$.

The joint log-likelihood is based on the logarithm of the four component densities given in (8.29). Estimation is simplified by exploiting the convenient property of the hurdle model that the log-likelihood decomposes into two parts that can be maximized separately. Then introducing regressors $\mathbf{x}_i$ but suppressing the regression parameters for simplicity, $\ln L = \ln L_1 + \ln L_2$, where for the bivariate hurdle part

$$\begin{aligned} \ln L_1(\theta^0) &= \sum_{0,0} \ln F^0(y_{1i} = 0, y_{2i} = 0|\mathbf{x}_i, \theta^0) \\ &+ \sum_{+,0} \ln F^0(y_{1i} > 0, y_{2i} = 0|\mathbf{x}_i, \theta^0) \\ &+ \sum_{0,+} \ln F^0(y_{1i} = 0, y_{2i} > 0|\mathbf{x}_i, \theta^0) \\ &+ \sum_{+,+} \ln F^0(y_{1i} > 0, y_{2i} > 0|\mathbf{x}_i, \theta^0), \end{aligned} \tag{8.33}$$

and for the bivariate positive part

$$\begin{aligned} \ln L_2(\theta^+) = &\sum_{+,0} \ln f_1^+(y_{1i}|y_{1i} > 0|\mathbf{x}_i) + \sum_{0,+} \ln f_2^+(y_{2i}|y_{2i} > 0|\mathbf{x}_i) \\ &+ \sum_{+,+} \ln f_{12}^+(y_1, y_2|y_1 > 0, y_2 > 0|\mathbf{x}_i). \end{aligned} \tag{8.34}$$

8.5.8 Example: Distribution of Differences of Counts

The difference between two counts is an integer-valued variable with support at both negative and positive values. Karlis and Ntzoufras (2003) provide references to the early history of this topic as well as an application to sports data. The case of the difference between two independent Poisson variates with different means is considered by Skellam (1946). Karlis and Ntzoufras consider the correlated case, initially using the bivariate Poisson. This approach restricts the counts to be positively correlated, an assumption too strong for their data, so they use a variant of a zero-inflated Poisson that provides inflation when there are ties so that $y_1 = y_2$.

Cameron, Li, Trivedi, and Zimmer (2004) instead use a copula framework. In their application y_1 denotes self-reported doctor visits and y_2 denotes actual doctor visits as recorded administratively. They use a copula model for the joint distribution $F(y_1, y_2)$, with NB2 marginals. The distribution of the measurement error $y_1 - y_2$ is obtained by using the transformation $z = y_1 - y_2$, which allows the joint distribution to be equivalently expressed in terms of z and y_2 as

$$f(z, y_2) = c_{12}\left(F_1(z + y_2), F_2(y_2)|\theta\right), \tag{8.35}$$

where $c_{12}(\cdot)$ is defined in (8.28). The corresponding probability mass function of z is obtained by summing over all possible values of y_2, so

$$g(z) = \sum_{y_2=0}^{\infty} c_{12}\left(F_1(z + y_2), F_2(y_2)|\theta\right). \tag{8.36}$$

For any value of z, (8.36) gives the corresponding probability mass. The cdf of $g(z)$ is calculated by accumulating masses at each point z, $G(z) = \sum_{k=-\infty}^{z} g(k)$. Both $g(z)$ and $G(z)$ characterize the full distribution of z so that inference can be made regarding the difference between two count variables. Misreporting is found to be increasing in the number of actual visits, with results similar across the several copulas used.

This method can also be applied to any discrete or continuous variables when the marginal distribution of the components of the differences is parametrically specified.

8.6 MOMENT-BASED ESTIMATION

Moment-based estimators provide consistent estimators of the parameters in the models for the conditional means of $y_1, \ldots, y_m$ rather than fully parameterizing the distribution of $y_1, \ldots, y_m$ as in previous sections of this chapter. The approach is essentially the same as that for seemingly unrelated regressions (SUR) for the linear model. There are two complications. First, the conditional means are nonlinear, usually exponential models. Second, the SUR model presumes homoskedastic errors, whereas efficient estimation requires modeling the intrinsic heteroskedasticity of count data. Estimation is by feasible generalized nonlinear GLS, essentially a variant of GEE introduced in Chapter 2.4.6. Systems estimation can improve estimation efficiency, and tests of cross-equation restrictions are easy to implement.

8.6.1 Seemingly Unrelated Nonlinear Regressions

Consider a multivariate nonlinear heteroskedastic regression model of the form

$$\mathbf{y}_j = \boldsymbol{\mu}(\mathbf{X}_j \boldsymbol{\beta}_j) + \mathbf{u}_j,\ j = 1, 2, \ldots, m,$$

where $\mathbf{y}_j$ is an $n \times 1$ vector on the jth component of the regression, $\boldsymbol{\mu}(\mathbf{X}_j\boldsymbol{\beta}_j)$ is an $n \times 1$ vector with i^{th} entry $\exp(\mathbf{x}'_{ij}\boldsymbol{\beta}_j)$, and $\mathsf{E}[\mathbf{u}_j\mathbf{u}'_k] = \boldsymbol{\Sigma}(\mathbf{X}_j, \mathbf{X}_k)$. We assume independence across individuals $i = 1, \ldots, n$, but permit correlation over the j components for given individual i. Define the following stacked $nm \times 1$ vectors $\mathbf{y}$ and $\mathbf{u}$, $nm \times 1$ vector $\boldsymbol{\mu}(\boldsymbol{\beta})$, $km \times 1$ vector $\boldsymbol{\beta}$, and $nm \times nm$ matrix $\boldsymbol{\Sigma}(\mathbf{X}) = \mathsf{E}[\mathbf{uu}']$:

$$\mathbf{y} = \begin{bmatrix} \mathbf{y}_1 \\ \vdots \\ \mathbf{y}_m \end{bmatrix}, \quad \boldsymbol{\mu}(\boldsymbol{\beta}) = \begin{bmatrix} \boldsymbol{\mu}(\mathbf{X}_1\boldsymbol{\beta}_1) \\ \vdots \\ \boldsymbol{\mu}(\mathbf{X}_m\boldsymbol{\beta}_m) \end{bmatrix}, \quad \boldsymbol{\beta} = \begin{bmatrix} \boldsymbol{\beta}_1 \\ \vdots \\ \boldsymbol{\beta}_m \end{bmatrix}, \quad \mathbf{u} = \begin{bmatrix} \mathbf{u}_1 \\ \vdots \\ \mathbf{u}_m \end{bmatrix}.$$

$$\boldsymbol{\Sigma}(\mathbf{X}) = \begin{bmatrix} \boldsymbol{\Sigma}(\mathbf{X}_1) & \cdots & \boldsymbol{\Sigma}(\mathbf{X}_1, \mathbf{X}_m) \\ \vdots & \ddots & \vdots \\ \boldsymbol{\Sigma}(\mathbf{X}_1, \mathbf{X}_m) & \cdots & \boldsymbol{\Sigma}(\mathbf{X}_m) \end{bmatrix}.$$

In this case association between elements of $\mathbf{y}$ arises only through the covariance matrix $\boldsymbol{\Sigma}(\mathbf{X})$. If $\boldsymbol{\Sigma}(\mathbf{X})$ is block-diagonal, then simpler conditional modeling of each component of $\mathbf{y}$ separately is equivalent to joint modeling.

The feasible nonlinear GLS estimator, given a first-step consistent estimate $\hat{\boldsymbol{\Sigma}}(\mathbf{X})$ of $\boldsymbol{\Sigma}(\mathbf{X})$ based on single equation Poisson or NLS estimation, minimizes the quadratic form

$$Q(\boldsymbol{\beta}) = (\mathbf{y} - \boldsymbol{\mu}(\boldsymbol{\beta}))'\hat{\boldsymbol{\Sigma}}(\mathbf{X})^{-1}(\mathbf{y} - \boldsymbol{\mu}(\boldsymbol{\beta})). \tag{8.37}$$

This is a variant of GEE estimation presented in Chapter 2.4.6.

A key component for implementation is specification of a model for $\boldsymbol{\Sigma}(\mathbf{X})$. An obvious model for counts is that $\mathsf{Var}[y_{ij}|\mathbf{x}_{ij}] = \alpha_j \exp(\mathbf{x}'_{ij}\beta)$ and $\mathsf{Cov}[y_{ij}, y_{ik}|\mathbf{x}_{ij}, \mathbf{x}_{ik}] = \alpha_{jk} \exp(\mathbf{x}'_{ij}\boldsymbol{\beta})\exp(\mathbf{x}'_{ik}\boldsymbol{\beta})$. A simpler, less efficient estimator lets $\mathsf{Var}[y_{ij}|\mathbf{x}_{ij}]$ and $\mathsf{Cov}[y_{ij}, y_{ik}|\mathbf{x}_{ij}, \mathbf{x}_{ik}]$ be constants, in which case one can simply use a standard nonlinear SUR command. In either case standard errors can be obtained that are robust to misspecification of $\boldsymbol{\Sigma}(\mathbf{X})$, as in Chapter 2.4.6.

8.6.2 Semiparametric Estimation

An alternative is to nonparametrically estimate $\boldsymbol{\Sigma}(\mathbf{X})$ given a first-stage estimate of $\boldsymbol{\beta}$. In the context of a linear multiple regression with a linear conditional expectation $\mathsf{E}[y_i|\mathbf{x}_i] = \mathbf{x}'_i\boldsymbol{\beta}$, Robinson (1987b) gives conditions under which asymptotically efficient estimation is possible in a model in which the variance function $\sigma_i(\mathbf{x}_i)$ has an unknown form. His technique is to estimate the error variance function $\sigma_i^2(\mathbf{x}) = \mathsf{V}[y_i|\mathbf{x}] = \mathsf{E}[y_i^2|\mathbf{x}] - (\mathsf{E}[y_i|\mathbf{x}])^2$ by k-nearest neighbor (k-NN) nonparametric estimation of $\mathsf{E}[y_i|\mathbf{x}_i]$ and $\mathsf{E}[y_i^2|\mathbf{x}_i]$. Then estimate $\boldsymbol{\beta}$ by weighted least squares estimation with weights based on $\hat{\sigma}_i^2$ or, equivalently, OLS regression of $(y_i/\hat{\sigma}_i)$ on $(\mathbf{x}_i/\hat{\sigma}_i)$. The estimator is fully efficient with $\mathsf{V}[\widetilde{\boldsymbol{\beta}}] = \left[\sum_{i=1}^{n} \mathbf{x}_i\mathbf{x}_i'\hat{\sigma}_i^{-2}\right]^{-1}$. We refer to this technique as semiparametric generalized least squares (SPGLS). Chapter 11.6 provides additional details of the implementation of this approach to a single-equation count model.

Delgado (1992) extends SPGLS to multivariate nonlinear models, including the model considered here. The technique amounts to feasible nonlinear GLS estimation that minimizes the objective function in (8.37), based on a consistent estimate $\hat{\boldsymbol{\Sigma}}(\mathbf{X})$ obtained using the k-NN method given first-step nonlinear SUR estimates of $\boldsymbol{\beta}$. Practical experience with the multivariate SPGLS estimator is limited, but Delgado provides some simulation results. The efficiency of such estimators is a complex issue and beyond the scope of the present chapter (Newey, 1990b; Chamberlain, 1992a).

8.6.3 Bivariate Poisson with Heterogeneity

Gourieroux et al. (1984b, pp. 716–717) propose moment-based estimation based on an extension of the bivariate Poisson model of section 8.4.1 that introduces unobserved heterogeneity into the conditional means of the components of the bivariate Poisson; see also Gurmu and Elder (2000).

The two counts are determined by $y_1 = u + w$ and $y_2 = v + w$, as in (8.8), where u, v, and w are independent Poisson with means of, respectively, μ_1, μ_2, and μ_3. The departure is to let $\mu_j = \exp(\mathbf{x}'\boldsymbol{\beta}_j + \nu_j)$, $j = 1, 2, 3$. This introduces unobserved heterogeneity components ν_j that allow for overdispersion and also may be correlated. Specifically it is assumed that $\mathsf{E}[\exp(\nu_j)] = 1$, $\mathsf{V}[\exp(\nu_j)] = \sigma_{jj}$, and $\mathsf{Cov}[\exp(\nu_j), \exp(\nu_k)] = \sigma_{jk}$, $j, k = 1, 2, 3$. It follows that

$$\mathsf{E}[y_{1i}|\mathbf{x}_i] = \exp(\mathbf{x}_i'\boldsymbol{\beta}_1) + \exp(\mathbf{x}_i'\boldsymbol{\beta}_3)$$

$$\mathsf{E}[y_{2i}|\mathbf{x}_i] = \exp(\mathbf{x}_i'\boldsymbol{\beta}_2) + \exp(\mathbf{x}_i'\boldsymbol{\beta}_3).$$

The more complicated expression for $\mathsf{Var}[\mathbf{y}_i|\mathbf{x}_i]$, where $\mathbf{y}_i = (y_{1i}, y_{2i})$, is given in Gourieroux et al. (1984b, p. 716). Then the QGPML estimator of $(\boldsymbol{\beta}_1, \boldsymbol{\beta}_2, \boldsymbol{\beta}_3)$ may be obtained by minimizing $\sum_{i=1}^{n}(\mathbf{y}_i - \mathsf{E}[\mathbf{y}_i|\mathbf{x}_i])'\hat{\mathbf{V}}_i^{-1}(\mathbf{y}_i - \mathsf{E}[\mathbf{y}_i|\mathbf{x}_i])$, where $\hat{\mathbf{V}}_i$ is a consistent estimator of $\mathsf{Var}[\mathbf{y}_i|\mathbf{x}_i]$. For details of implementing this estimator, see Gourieroux et al. (1984b).

8.7 TESTING FOR DEPENDENCE

Before estimating joint models, it may be desirable to test the hypothesis of dependence. In principle one can apply standard Wald, LR, or LM tests to the parametric models presented in sections 8.5 and 8.6, but these models can be challenging to estimate and rely heavily on parametric assumptions. Instead we consider simpler score and conditional moment tests based on orthogonal polynomial series expansions of a baseline density. These expansions can also be a basis for estimation; see Chapter 11.

8.7.1 Definitions and Conditions

A useful (but not widely used) technique for generating and approximating multivariate discrete distributions is via a sequence of orthogonal or orthonormal polynomial expansions for the unknown joint density. For example, the bivariate density $f(y_1, y_2)$ may be approximated by a series expansion in which the terms are orthonormal polynomials of the univariate marginal densities $f(y_1)$ and $f(y_2)$. A required condition for the validity of the expansion is the existence of finite moments of all orders, denoted $\mu^{(k)}$, $k = 1, 2, \ldots$. Begin by considering an important definition in the univariate case.

Definition (Orthogonality). *An* orthogonal *polynomial of integer order j, denoted by $Q_j(y)$, $j = 1, \ldots, K$, has the property that*

$$\int Q_j(y)Q_k(y)f(y)dy = \delta_{jk}\sigma_{jj}, \tag{8.38}$$

where δ_{jk} is the Kronecker delta, which equals zero if $j \neq k$ and one otherwise, and σ_{jj} denotes the variance of $Q_j(y)$. That is,

$$\mathsf{E}_f\left[Q_j(y_i)Q_k(y_i)\right] = \delta_{jk}\sigma_{jj}. \tag{8.39}$$

An orthogonal polynomial obtained by a scale transformation of $Q_j(y)$ such that it has unit variance is referred to as an *orthonormal* polynomial of degree j. Thus $Q_j(y)/\sqrt{\sigma_{jj}}$ is an orthonormal polynomial; it has zero mean and unit variance. For convenience we use the notation $P_j(y)$ to denote a jth order orthonormal polynomial.

Let $\mathbf{\Delta}$ be a matrix whose ij^{th} element is $\mu^{(i+j-2)}$, for $i \geq 1$, $j \geq 1$. Then the necessary and sufficient condition for an arbitrary sequence $\{\mu^{(k)}\}$ to give rise to a sequence of orthogonal polynomials, unique up to an arbitrary constant, is that $\mathbf{\Delta}$ should be positive definite (Cramer, 1946). An orthonormal sequence $P_j(y)$ is complete if, for every function $R(y)$ with finite variance, $\mathsf{V}[R(y)] = \sum_{j=0}^{\infty} a_j^2$, where $a_j = \mathsf{E}[R(y)P_j(y)]$.

8.7.2 Univariate Expansion

A well-behaved or regular pdf has a series representation in terms of orthogonal polynomials with respect to that density (Cramer, 1946, chapter 12; Lancaster, 1969). Let $\{Q_j(y), j = 0, 1, 2, \ldots; Q_0(y) = 1\}$ be a sequence of orthogonal polynomials for $f(y)$. Let $H(y)$ denote the true but unknown distribution function and $h(y)$ denote a data density that satisfies regularity conditions (Lancaster, 1969). Then the following series expansion of $h(y)$ around a baseline density $f(y)$ is available:

$$h(y) = f(y)\left[1 + \sum_{j=1}^{\infty} a_j Q_j(y)\right]. \tag{8.40}$$

Multiplying both sides of (8.40) by $Q_j(y)$ and integrating shows that the coefficients in the expansion are defined by

$$a_j = \int Q_j(y)h(y)dy = \mathsf{E}_h\left[Q_j(y)\right], \tag{8.41}$$

which are identically zero if $h(y) = f(y)$. The terms in the series expansion reflect the divergence between the true but unknown pdf $h(y)$ and the assumed (baseline) pdf $f(y)$. A significant deviation implies that these coefficients are significantly different from zero. The orthogonal polynomials $Q_j(y)$ have zero mean

$$\mathsf{E}\left[Q_j(y)\right] = 0. \tag{8.42}$$

Furthermore, the variance of $Q_j(y)$, evaluated under $f(y)$, is given by $\mathsf{E}[Q_j^2(y)]$. A general procedure for deriving orthogonal polynomials is discussed in Cameron and Trivedi (1990b). For selected densities the orthogonal polynomials can also be derived using generating functions. These generating functions are known for the classical cases and for the Meixner class of distributions, which includes the normal, binomial, negative binomial, gamma, and Poisson densities (Cameron and Trivedi, 1993). For ease of later reference, we also note the following expressions for $Q_j(y),\ j = 0, 1, 2$:

$$Q_0(y) = 1$$
$$Q_1(y) = y - \mu$$
$$Q_2(y) = (y - \mu)^2 - (\mu_3/\mu_2)(y - \mu) - \mu_2,$$

where $\mu_k, k = 1, 2,$ and 3, denote, respectively, the first, second, and third central moments of y. Hence it is seen that the orthogonal polynomials are functions of the "raw" residuals $(y - \mu)$.

8.7.3 Multivariate Expansions

We first consider the bivariate case. Let $f(y_1, y_2)$ be a bivariate pdf of random variables y_1 and y_2 with marginal distributions $f_1(y)$ and $f_2(y)$ whose corresponding *orthogonal polynomial sequences* (OPS) are $Q_j(y)$ and $R_j(y)$, $j = 0, 1, 2, \ldots$. If $f(y_1, y_2)$ satisfies regularity conditions then the following expansion is formally valid:

$$f(y_1, y_2) = f_1(y_1)f_2(y_2)\left[1 + \sum_{j=1}^{\infty}\sum_{k=1}^{\infty}\rho_{jk}Q_j(y_1)R_k(y_2)\right] \tag{8.43}$$

where

$$\rho_{jk} = \mathsf{E}\left[Q_j(y_1)R_k(y_2)\right]$$
$$= \int\int Q_j(y_1)R_k(y_2)f(y_1, y_2)dy_1dy_2.$$

The derivation of this result is similar to that given earlier for the a_j coefficients in (8.40) (Lancaster, 1969, p. 97).

The general multivariate treatment has close parallels with the bivariate case. Consider r random variables $(y_1, \ldots, y_r)$ with joint density $f(y_1, \ldots, y_r)$ and r marginals $f_1(y_1), f_2(y_2), \ldots, f_r(y_r)$ whose respective OPSs are denoted by $Q^i_j(y)$, $s = 0, 1, \ldots, \infty$; $i = 1, 2, \ldots, r$. Under regularity conditions the joint pdf admits a series expansion of the same type as that given in equation (8.43); that is,

$$\begin{aligned} f(y_1, \ldots, y_r) &= f_1(y_1) \cdots f_r(y_r) \left[1 + \sum_{i<j}^{r} \sum_{j}^{r} \sum_{s}^{\infty} \sum_{t}^{\infty} \rho^{ij}_{st} Q^i_s(y_i) Q^j_t(y_j) \right. \\ &\quad \left. + \sum_{i<j}^{r} \sum_{j<k}^{r} \sum_{k}^{r} \sum_{s}^{\infty} \sum_{t}^{\infty} \sum_{o}^{\infty} \rho^{ijk}_{sto} Q^i_s(y_i) Q^j_t(y_j) Q^k_o(y_k) \right], \end{aligned} \tag{8.44}$$

where ρ^{ij}_{st} denotes the correlation coefficient between $Q^i_s(y_i)$ and $Q^j_t(y_j)$.

8.7.4 Tests of Independence

Series expansions can be used to generate and estimate approximations to unknown joint densities in a manner similar to the "semi-nonparametric" approach of Gallant and Tauchen (1989), adapted and applied to univariate count data by Cameron and Johansson (1997). A complication in estimation, not present in testing, is the need to ensure that the more general series expansion density is properly defined. This leads to replacing the term in square brackets in (8.40) by its square, requiring the introduction of a normalizing constant, and truncating the order of the polynomial at much less than infinity. Chapter 11.2 provides details of implementing this approach using nonorthogonal polynomial expansions. The objective is the estimation of flexible parametric forms. The approach has not yet been widely applied to multivariate count models.

In this section we present applications of series expansions only to tests of independence in a multivariate framework assuming given marginals. Our treatment of testing follows the developments in Cameron and Trivedi (1990b, 1993). The attraction of the approach is that it leads to score and conditional moment tests that are simpler to implement than Wald or LR tests. The tests are more general than tests of zero correlation, a necessary but not sufficient condition for independence.

Using the key idea that under independence the joint pdf factorizes into a product of marginals, Cameron and Trivedi (1993) developed score-type tests of independence based on a series expansion of the type given in (8.43). The leading term in the series is the product of the marginals; the remaining terms in the expansion are orthonormal polynomials of the univariate marginal densities.

The idea behind the test is to measure the significance of the higher order terms in the expansion using estimates of the marginal models only. The conditional moment test of independence consists of testing for zero correlation between all pairs of orthonormal polynomials. So the steps are as follows. First, specify the marginals and estimate their parameters. Then evaluate the orthogonal or orthonormal polynomials at the estimated parameter values. Finally, calculate the tests.

Given the marginals and corresponding orthogonal polynomials, the tests can be developed as follows. Using equation (8.43), the test of independence in the bivariate case requires us to test H_0: $\rho_{jk} = 0$ (all j, k). This onerous task may be simplified in one of two ways. The null may be tested against an alternative in which dependence is restricted to be a function of a small number of parameters, usually just one. Or we may approximate the bivariate distribution by a series expansion with a smaller number of terms and then derive a score (LM) test of the null hypothesis $H_0 : \rho_{jk} = 0$ (some j, k). For independence we require $\rho_{jk} = 0$ for all j and k. By testing only a subset of the restrictions, the hypothesis of approximate independence is tested. If $p = 2$, this is equivalent to the null hypothesis H_0: $\rho_{11} = \rho_{22} = \rho_{12} = \rho_{21} = 0$. For general p, the appropriate moment restriction is $\mathsf{E}_0[Q_j(y_1)R_k(y_2)] = 0$, $j, k = 1, 2, \ldots, p$, where E_0 denotes expectation under the null hypothesis of independence of y_1 and y_2.

The key moment restriction is

$$\mathsf{E}_0[S_{jk}(y, \mathbf{x}, \boldsymbol{\theta})] = 0,$$

where $S_{jk}(y, \mathbf{x}, \boldsymbol{\theta}) = Q_j(y_1|\mathbf{x}_1, \boldsymbol{\theta}_1)R_k(y_2|\mathbf{x}_2, \boldsymbol{\theta}_2)$. The conditioning operator is used to make explicit the presence of subsets of regressors $\mathbf{x}$ in the model. By independence of Q_j and R_k, and using the property that $\mathsf{E}_0[R_s(\cdot)] = \mathsf{E}_0[Q_t(\cdot)] = 0$, and conditional on $\mathbf{x}$,

$$\begin{aligned}\mathsf{V}_0[S_{jk}(\cdot)] &= \left(\mathsf{E}_0[R_j(\cdot)]\right)^2 \mathsf{V}_0[Q_k(\cdot)] + (\mathsf{E}_0[Q_k(\cdot)])^2\, \mathsf{V}_0[R_j(\cdot)] \\ &\quad + \mathsf{V}_0[Q_k(\cdot)]\mathsf{V}_0[R_j(\cdot)]\mathsf{V}_0[Q_k(\cdot)]\mathsf{V}_0[R_j(\cdot)].\end{aligned}$$

Assume initially that the parameters of the marginal distributions are known. By application of a central limit theorem for orthogonal polynomials $Q_{j,i}$ and $R_{k,i}$, which have zero means by construction, we can obtain the following test statistic for the null hypothesis of $H_0 : \rho_{jk} = 0$:

$$\mathsf{r}^2_{jk} = \left(\sum_{i=1}^{n} Q_{j,i}R_{k,i}\right)\left[\sum_{i=1}^{n} \left(Q_{j,i}R_{k,i}\right)^2\right]^{-1}\left(\sum_{i=1}^{n} Q_{j,i}R_{k,i}\right) \overset{a}{\sim} \chi^2(1). \tag{8.45}$$

Note that r^2_{jk} can be computed as n times the uncentered R^2_U (equals the proportion of the explained sum of squares) from the auxiliary regression of 1 on $Q_{j,i}R_{k,i}$.

For orthonormal polynomials, distinguished by an asterisk, we have the result that

$$\mathsf{E}_0\left(\sum_{i=1}^{n} Q_{j,i}^{*2}\right)\left(\sum_{i=1}^{n} R_{k,i}^{*2}\right) = n^{-1}\mathsf{E}_0\left(\sum_{i=1}^{n}(Q_{j,i}^{*}R_{k,i}^{*})^2\right)$$

by the properties of homoskedasticity and independence of $Q_{j,i}^{*}$ and $R_{k,i}^{*}$. A test of the null hypothesis of H_0: $\rho_{jk} = 0$ is

$$\mathsf{r}_{jk}^2 = n\left(\sum_{i=1}^{n} Q_{j,i}^{*}R_{k,i}^{*}\right)\left(\left(\sum_{i=1}^{n} Q_{j,i}^{*2}\right)\left(\sum_{i=1}^{n} R_{k,i}^{*2}\right)\right)^{-1}\left(\sum_{i=1}^{n} Q_{j,i}^{*}R_{k,i}^{*}\right)$$
$$\overset{a}{\sim} \chi^2(1). \tag{8.46}$$

These polynomials are functions of parameters $\boldsymbol{\theta}$. To implement the tests, they are evaluated at the maximum likelihood estimates. Consider the effect of substituting the estimated parameters $\hat{\boldsymbol{\theta}}_1$ for $\boldsymbol{\theta}_1$ and $\hat{\boldsymbol{\theta}}_2$ for $\boldsymbol{\theta}_2$ in the test statistics. Using the general theory of conditional moment tests given in Chapter 2.6.3 and specifically noting that the derivative condition (5.69),

$$\mathsf{E}_0[\nabla_{\boldsymbol{\theta}} S_{jk,i}(\mathbf{y}, \mathbf{x}, \hat{\boldsymbol{\theta}}|\mathbf{x})] = \mathbf{0},$$

is satisfied, it follows that the asymptotic distribution of the test statistics (8.45) and (8.46) is not affected by the replacement of $\boldsymbol{\theta}$ by $\hat{\boldsymbol{\theta}}$.

The application of these ideas to the multivariate case is potentially burdensome because it involves all unique combinations of the polynomials of all marginal distributions. If the dimension of $\mathbf{y}$ is large, it would seem sensible to exploit prior information on the structure of dependence in constructing a test. It is simpler to test for zero correlation between two subsets of $\mathbf{y}$, denoted $\mathbf{y}_1$ and $\mathbf{y}_2$, of dimensions r_1 and r_2, respectively, with the covariance matrix $\boldsymbol{\Sigma} = [\boldsymbol{\Sigma}_{jk}]$, $j, k = 1, 2$. Define the squared canonical correlation coefficient,

$$\begin{aligned}\rho_c^2 &= (\text{vec}\boldsymbol{\Sigma}_{21})'(\boldsymbol{\Sigma}_{11}^{-1} \otimes \boldsymbol{\Sigma}_{22}^{-1})(\text{vec}\boldsymbol{\Sigma}_{21})\\ &= \text{tr}(\boldsymbol{\Sigma}_{11}^{-1}\boldsymbol{\Sigma}_{12}\boldsymbol{\Sigma}_{22}^{-1}\boldsymbol{\Sigma}_{21}),\end{aligned} \tag{8.47}$$

which equals zero under the null hypothesis of independence of $\mathbf{y}_1$ and $\mathbf{y}_2$. Let r_c^2 denote the sample estimate of ρ_c^2. Then, analogous to the test in (8.46) we have the result that

$$n\mathsf{r}_c^2 \overset{d}{\to} \chi^2(r_1 r_2). \tag{8.48}$$

See Jupp and Mardia (1980) and Shiba and Tsurumi (1988) for related results.

In practice tests of independence may simply turn out to be misspecification tests of misspecified marginals. The tests may be significant because the baseline marginals are misspecified, not necessarily because the variables are dependent. However, rather remarkably, in the bivariate case if only one marginal is misspecified, the tests retain validity as tests of independence. See Cameron and Trivedi (1993, pp. 34–35), who also investigate properties of the tests in Monte Carlo experiments. The investigations of these tests for

Table 8.2. *Orthogonal polynomials: First and second order*

Density	$Q_1(y)$	$Q_2(y)$
Poisson	$y-\mu$	$(y-\mu)^2-y$
NB1	$y-\mu$	$(y-\mu)^2-(2\phi-1)(y-\mu)-\phi\mu$
NB2	$y-\mu$	$(y-\mu)^2-(1+2\alpha\mu)(y-\mu)-(1+\alpha\mu)\mu$

bivariate count regression models show that the tests have the correct size and high power if the marginals are correctly specified, but they overreject if the marginals are misspecified.

Table 8.2 shows first- and second-order polynomials for the specific cases of Poisson, NB1 (with $\mathsf{V}[y]=\phi\mu$), and NB2 (with $\mathsf{V}[y]=\mu+\alpha\mu^2$)).

8.8 MIXED MULTIVARIATE MODELS

We present several examples of multivariate models where only one of the variables is a count, whereas the others are multinomial or continuous variables.

8.8.1 Discrete Choice and Counts

In some studies the objective is to analyze several types of events that are mutually exclusive and collectively exhaustive. Such models involve two types of discrete outcomes. An example is the frequency of visits to alternative recreational sites. Typically one observes $(y_{ij}, \mathbf{x}_{ij};\ i=1,\ldots,n;\ j=1,\ldots,M)$, where i is the individual subscript and j is the destination subscript. The variable y measures trip frequency and $\mathbf{x}$ refers to covariates. In this section we consider a framework for modeling such data.

The starting point in such an analysis is to condition on the total number of events across all types, using the multinomial distribution,

$$f(\mathbf{y}|N)=N!\frac{\prod_{j=1}^{M}p_j^{y_j}}{\prod_{j=1}^{M}y_j!}, \tag{8.49}$$

where $\mathbf{y}=(y_1,\ldots,y_M)$, $y_j\in\{0,1,2,\ldots;\ j=1,\ldots,M\}$, where p_j denotes the probability of the jth type of event, so that $\sum_j p_j=1$ and $y_j$denotes the frequency of the jth event, and $N=\sum_j y_j$ denotes the total number of outcomes across the alternatives. The multinomial distribution is conditional on the total number of all types of events, N. This is a useful specification for analysis if N, the total number of outcomes, is given.

Next, suppose that the y_j are independent Poissons with means μ_j. Then the total number of outcomes N is Poisson distributed with parameter $\mu=\sum_{j=1}^{N}\mu_j$, and

$$\Pr\left[\sum_{j=1}^{M}y_j=N\right]=\frac{e^{-\mu}\mu^N}{N!}. \tag{8.50}$$

Note that then the probability of the jth type of event is $p_j = \mu_j/\mu$, so $\mu_j = \mu p_j$.

The unconditional distribution of $\mathbf{y}$ combines the conditional multinomial probability with the probability of the total number of events,

$$\Pr[\mathbf{y}] = \Pr\left[\sum y_j = N\right] \Pr[\mathbf{y}|N].$$

Using (8.49) and (8.50), Terza and Wilson (1990) therefore propose a mixed multinomial (logit) Poisson model with

$$f^{MP}(\mathbf{y}) = \left(\prod_{j=1}^{M} p_j^{y_j}\right)\left(\frac{e^{-\mu}\mu^{\left[\sum_{j=1}^{M} y_j\right]}}{\prod_{j=1}^{M} y_j!}\right), \tag{8.51}$$

with multinomial logit probabilities

$$p_j = \frac{\exp[\mathbf{x}_j'\boldsymbol{\beta}_j]}{\sum_{j=1}^{M}\exp[\mathbf{x}_j'\boldsymbol{\beta}_j]}.$$

The estimation of the mixed model by maximum likelihood is simplified on noting that the resultant log-likelihood is additive in the parameters $(\boldsymbol{\beta}_1, \ldots, \boldsymbol{\beta}_M)$ and μ. Hence, to maximize the likelihood one can *sequentially* estimate the sub log-likelihoods for the Poisson model for the total number of events, N, and for the choice probabilities of the multinomial model, although not necessarily in that order.

Terza and Wilson (1990) note that since $\mu_j = \mu p_j$ and $\mu = \sum_{j=1}^{N} \mu_j$, after some algebra $f^{MP}(\mathbf{y}) = \prod_{j=1}^{M}(e^{-\mu_j}\mu_j^{r_j}/r_j!)$, which is just the density of M independent Poissons with means μ_j, $j = 1, \ldots, M$. So the mixed multinomial-Poisson and an M individual Poisson model are equivalent. But the analysis remains valid if more generally the Poisson specification is replaced by one of the modified or generalized variants discussed in Chapter 4, and the multinomial logit is replaced by a nested multinomial logit. For example, Terza and Wilson (1990) use a mixed multinomial logit and Poisson hurdles specification.

Hausman, Leonard, and McFadden (1995) develop a joint model of choice of recreational sites and the number of recreational trips. They use panel data from a large-scale telephone survey of Alaskan residents. Although there are similarities with the approach of Terza and Wilson (1990), their model explicitly incorporates restrictions from utility theory. The model conforms to a two-stage budgeting process. First, a multinomial model is specified and estimated for explaining the choice of recreational sites in Alaska. Explanatory variables include the prices associated with the choice of sites. The estimates from this model are used to construct a price index for recreational trips. This price index subsequently becomes an explanatory variable in the count model for the total number of trips, which the authors specify as a fixed-effects Poisson (see Chapter 9). The two-step modeling approach is described as *utility consistent* in the sense that it is consistent with two-stage consumer budgeting. At step one the consuming unit allocates a utility-maximizing expenditure on the total number of trips. At the second stage this amount is optimally allocated across trips to alternative sites.

8.8.2 Counts and Continuous Variables

We consider two bivariate examples where one variable is a count and the other is a continuous variable, in one example a duration and in the other expenditures.

Cowling, Hutton, and Shaw (2006) present a joint model for (1) the number of epileptic seizures experienced by an individual in the six months before a randomized treatment intervention and (2) the post-treatment survival time to the first subsequent epileptic seizure. They specify a bivariate Marshall-Olkin type model as in (8.15) where the two marginal models are an exponential hazard model with gamma heterogeneity for the survival time and a Poisson with gamma heterogeneity for the number of prior seizures. This is essentially a one-factor model for heterogeneity. Because both exponential-gamma and Poisson-gamma mixtures have closed forms, no numerical integration is necessary. Treatment indicators are included in the set of covariates, with effective treatments expected to increase the duration between seizures and decrease the frequency.

Van Praag and Vermeulen (1993) analyze data on the number of events (shopping trips), denoted by y_1, and the vector of corresponding outcomes (recorded expenditures), denoted by $\mathbf{y}_2' = (y_{21}, \ldots, y_{2k})$ where k refers to the number of events, so $y_1 = k$. The objective is to formulate a joint probability model for y_1 and $\mathbf{y}_2$, and the authors specify the joint probability to be the product of the marginals, so

$$\Pr[y_1 = k]\Pr[\mathbf{y}_2' = (y_{21}, \ldots, y_{2k})] = f(y_1|\boldsymbol{\theta}_1, \mathbf{x})g(\mathbf{y}_2|\boldsymbol{\theta}_2, \mathbf{x}), \quad (8.52)$$

where $(\boldsymbol{\theta}_1, \boldsymbol{\theta}_2)$ are unknown parameters and $\mathbf{x}$ is a set of explanatory variables. This model assumes that any dependency is captured through the regressors $\mathbf{x}$, so that conditional on $\mathbf{x}$, the variables y_1 and $\mathbf{y}_2$ are stochastically independent. Under this assumption the joint log-likelihood $\mathcal{L}(\boldsymbol{\theta}_1, \boldsymbol{\theta}_2)$ factors into one component for count and another component for the amount, which may be estimated separately. The assumptions permit one to specify a flexible model for the counts – accounting, for example, for the presence of excess of zero counts due to taste differences in the population – and possible truncation of expenditures due to, for example, expenditures smaller than a specified amount, say $y_{2,\min}$, not being recorded. Van Praag and Vermeulen (1993) estimate a count-amount model for tobacco, bakery, and aggregate food expenditure in which the frequencies are modeled by a zero-inflated NB model and the amounts are modeled by a truncated normal.

The assumption of stochastic independence in (8.52) is convenient because it simplifies ML estimation, but in certain cases the assumption may be tenuous.

For example, the availability of bulk discounts in shopping may provide an incentive for larger but fewer transactions. This dependency might be captured by using bulk discount as an explanatory variable. More generally correlation could be introduced using some of the methods presented in this chapter, such as correlated unobserved heterogeneity or copulas.

8.9 EMPIRICAL EXAMPLE

To illustrate the use of bivariate count models in empirical work, we use the data set introduced in Chapter 6.3. Here we jointly model the number of emergency room visits (*EMR*) and number of hospitalizations (*HOSP*). A sample of 4,406 observations obtained from the National Medical Expenditure Survey conducted in 1987 and 1988 is used.

As expected, this pair of variables shows high raw positive correlation (0.47) because a stay in a hospital often follows an emergency room visit. The counted outcomes for *EMR* and *HOSP* are both dominated by zeros with the proportion of zeros exceeding 80%. The correlation is somewhat lower (0.39) if based only on the 454 positive-valued pairs of observations.

The count regressions use the same regressors as in Chapter 6.3. We begin with tests of independence based on univariate NB2 regressions and the first two orthogonal polynomials for the NB2 given in Table 8.2. Let Q_j and R_k denote, respectively, the jth and k^{th} orthogonal polynomials for *EMR* and *HOSP*. Then the test statistic (8.45) equals 88.45, 18.05, 0.01, and 3.93, respectively, for $(j, k) = (1, 1), (1, 2), (2, 1)$, and $(2, 2)$. The critical value is 3.84 for this $\chi^2(1)$ distributed test statistic, so independence is rejected at significance level 0.05 for all but the third test.

Table 8.3 provides estimates of two bivariate specifications discussed earlier: the bivariate negative binomial (BVNB) model of section 8.4.2 and the nonlinear seemingly unrelated regression (NLSUR) model assuming exponential conditional mean as in section 8.6.1. As previously indicated the BVNB model has a single parameter α to capture both overdispersion and correlation between unobserved heterogeneity, and it is estimated by ML. The NLSUR model does not explicitly accommodate either the integer-valued aspect of the outcomes or the presence of unobserved heterogeneity. The NLSUR model is estimated by two-step nonlinear GLS estimator in which NLS is applied to each equation separately and a covariance matrix of the residuals is constructed for the second-step estimation. The reported standard errors are based on the robust sandwich form in both cases.

Many of the health status variables are statistically significant, whereas many of the socioeconomic variables are statistically insignificance at level 0.05. The insurance variables *PRIVINS* and *MEDICAID* are statistically insignificant, except for NLSUR estimates for *HOSP*. A number of point estimates of the coefficients differ by more than 20%. However, in both cases the conditional correlation between the outcomes is significantly positive as expected. For NLSUR the correlation of 0.43 is close to the raw data correlation of 0.47.

Table 8.3. *ML estimates of bivariate negative binomial model and NLSUR estimates*

	ML (BVNB)				NLSUR			
	EMR		*HOSP*		*EMR*		*HOSP*	
Variable	Coefficient	Standard error	Coefficient	Standard error	Coefficient	Standard error	Coefficient	Standard error
CONS	−1.88	0.51	−2.82	0.49	−1.03	0.63	−1.95	0.60
EXCLHTH	−0.62	0.21	−0.70	0.19	−0.43	0.25	−0.65	0.21
POORHLTH	0.49	0.10	0.50	0.10	0.64	0.11	0.64	0.10
NUMCHRON	0.24	0.03	0.27	0.02	0.27	0.04	0.24	0.03
ADLDIFF	0.40	0.09	0.34	0.09	0.43	0.17	0.40	0.11
NOREAST	0.04	0.10	0.02	0.10	−0.12	0.17	−0.07	0.13
MIDWEST	0.03	0.10	0.16	0.10	−0.03	0.17	0.08	0.14
WEST	0.18	0.11	0.16	0.11	−0.08	0.18	−0.05	0.13
AGE	0.10	0.06	0.18	0.06	−0.15	0.09	−0.06	0.08
BLACK	0.17	0.12	0.10	0.11	0.28	0.21	0.13	0.14
MALE	0.11	0.09	0.21	0.08	0.08	0.16	0.05	0.12
MARRIED	−0.11	0.09	−0.02	0.09	−0.16	0.15	0.02	0.11
SCHOOL	−0.02	0.01	0.00	0.01	−0.00	0.02	0.01	0.02
FAMINC	0.00	0.01	0.00	0.01	0.01	0.02	0.01	0.01
EMPLOYED	0.19	0.13	0.05	0.13	0.32	0.20	0.10	0.15
PRIVINS	0.03	0.10	0.19	0.11	0.00	0.15	0.32	0.13
MEDICAID	0.15	0.13	0.18	0.14	0.05	0.20	0.26	0.14
α	2.17	0.13						
ρ					0.43			
$-\ln L$		5,223						

The BVNB standard errors are roughly 20% lower than those for NLSUR. From separate estimation of each outcome by NLS, Poisson, and NB2, not reported here, the efficiency gain comes mainly due to BVNB controlling for heteroskedasticity whereas NLSUR does not; in this example there is little efficiency gain due to joint modeling of the two outcomes. Overall, this comparison indicates that the computational convenience of NLSUR is an attractive starting point for multivariate count modeling, especially because it avoids numerical integration and strong distributional assumptions.

An improvement relative to the BVNB specification results from using a more flexible parametric specification such as the bivariate Poisson-Normal mixture. This additional flexibility comes at a computational cost as discussed in section 8.4.3. Munkin and Trivedi (1999) illustrate this and related estimators using the same data as analyzed here.

8.10 DERIVATIONS

Kocherlakota and Kocherlakota (1993) derive the bivariate Poisson distribution in several different ways. The method of trivariate reduction is one method that is commonly used. The joint pgf of y_1 and y_2 defined in (8.8) is

$$\begin{aligned} P(z_1, z_2) &= \mathsf{E}[z_1^{y_1} z_2^{y_2}] \\ &= \mathsf{E}[z_1^u z_2^v (z_1 z_2)^w] \\ &= \exp\left[\mu_1(z_1 - 1) + \mu_2(z_2 - 1) + \mu_3(z_1 z_2 - 1)\right] \\ &= \exp[(\mu_1 + \mu_3)(z_1 - 1) + (\mu_2 + \mu_3)(z_2 - 1) \\ &\quad + \mu_3(z_1 - 1)(z_2 - 1)]. \end{aligned} \tag{8.53}$$

The marginal pgf is

$$P(z) = \exp[(\mu_j + \mu_3)(z - 1)], \quad j = 1, 2, \tag{8.54}$$

whence the marginal distributions are $y_1 \sim \mathsf{P}[\mu_1 + \mu_3]$ and $y_2 \sim \mathsf{P}[\mu_2 + \mu_3]$. The condition for the independence of y_1 and y_2 is that the joint pgf is the product of the two marginals, which is true if $\mu_3 = 0$.

To derive the joint probability function expand (8.53) in powers of z_1 and z_2 as

$$P[z_1, z_2] = \exp\left[\mu_1 + \mu_2 + \mu_3\right] \sum_{l=0}^{\infty} \frac{\mu_1^l z_1^l}{l!} \sum_{j=0}^{\infty} \frac{\mu_2^j z_2^j}{j!} \sum_{k=0}^{\infty} \frac{\mu_3^k z_1^k z_2^k}{k!}, \tag{8.55}$$

which yields the joint frequency distribution as the coefficient of $z_1^r z_2^s$:

$$f(y_1 = r, y_2 = s) = \exp\left[\mu_1 + \mu_2 + \mu_3\right] \sum_{l=0}^{min(r,s)} \frac{\mu_1^{r-l}\mu_2^{s-l}\mu_3^{l}}{(r-l)!(s-l)!l!}. \tag{8.56}$$

The covariance between y_1 and y_2, using the independence of u, v, w, is given by

$$\begin{aligned} \mathsf{Cov}\left[y_1, y_2\right] &= \mathsf{Cov}\left[u + w, v + w\right] \\ &= \mathsf{V}\left[w\right] \\ &= \mu_3, \end{aligned}$$

and the correlation is given by

$$\mathsf{Cov}\left[y_1, y_2\right] / \sqrt{\mathsf{V}\left[y_1\right]\mathsf{V}\left[y_2\right]} = \mu_3 / \sqrt{(\mu_1 + \mu_3)(\mu_2 + \mu_3)}.$$

Jung and Winkelmann (1993, pp. 555–556) provide first and second derivatives of the log-likelihood.

If the method of trivariate reduction is used, zero correlation between any pair implies independence.

8.11 BIBLIOGRAPHIC NOTES

An introductory survey of multivariate extensions of GLMs is given in Fahrmeier and Tutz (1994); see especially their treatment of multivariate models with correlated responses. Formal statistical properties of bivariate discrete models are found in Kocherlakota and Kocherlakota (1993) and Johnson et al. (1997). Bivariate extensions of the zero-truncated Poisson distribution (without covariates) are studied in Hamdan (1972) and Patil, Patel, and Kovner (1977). Aitchison and Ho (1989) study a multivariate Poisson with log-normal heterogeneity. Lindeboom and van den Berg (1994) analyze the impact of heterogeneity on correlation between survival times in bivariate survival models; their results are suggestive of consequences to be expected in bivariate count models. Arnold and Strauss (1988, 1992) and Moschopoulos and Staniswalis (1994) have considered the problem of estimating the parameters of bivariate exponential family distributions with given conditionals. The econometric literature on the estimation of multivariate models under conditional moment restrictions is relevant if the less parametric nonlinear GLS or semiparametric GLS approach is followed. But only some of this discussion relates easily to count models; see Newey (1990a) and Chamberlain (1992a).

Bivariate count models have been widely used in applications. Jung and Winkelmann (1993) consider the number of voluntary and involuntary job changes as a bivariate Poisson process; Mayer and Chappell (1992) apply it to study determinants of entry and exit of firms. Good and Pirog-Good (1989) consider several bivariate count models for teenage delinquency and paternity, but without the regression component. King (1989a) presents a bivariate model of U.S. presidential vetoes of social welfare and defense policy legislation with a regression component. Duarte and Escario (2006) use a bivariate negative binomial to analyze alcohol abuse and truancy among the adolescent Spanish population. Meghir and Robin (1992) develop and estimate a joint model of the frequency of purchase and a consumer demand system for eight types of foodstuffs using French data on households that were observed to purchase all eight foodstuffs over the survey period. They show that consistent estimation of the demand system may require data on the frequency of purchase. They adopt a sequential approach in which a frequency of purchase equation is estimated by NLS and the ratio of the fitted mean to the actual frequency of purchase is used to weight all observed expenditures. A system of demand equations is fitted using these reweighted expenditures. Many more recent examples are given throughout this chapter.

As in the case of single-equation count regression, the zero inflation or excess zeros problem may also occur in multivariate contexts, but the issue gets relatively less attention. Both hurdle and zero-inflated bivariate variants have been developed. Hellström (2006) develops and applies a bivariate hurdle model to household tourism demand; estimation uses simulated ML. Li et al. (1999) develop a multivariate ZIP model but without a regression component. Wang (2003) proposes a bivariate variant of the zero-inflated negative binomial model and an estimation procedure based on the EM algorithm.

The copula framework has the attraction of enabling flexible models for the marginals. Trivedi and Zimmer (2007) consider several bivariate count models with dependence modeled via copulas.

CHAPTER 9

Longitudinal Data

9.1 INTRODUCTION

Longitudinal data or *panel data* are observations on a cross-section of individual units such as persons, households, firms, and regions that are observed over several time periods. The data structure is similar to that of multivariate data considered in Chapter 8. Analysis is simpler than for multivariate data because for each individual unit the same outcome variable is observed, rather than several different outcome variables. Yet analysis is also more complex because this same outcome variable is observed at different points in time, introducing times series data considerations presented in Chapter 7.

In this chapter we consider longitudinal data analysis if the dependent variable is a count variable. Remarkably, many count regression applications are to longitudinal data rather than simpler cross-section data. Econometrics examples include the number of patents awarded to each of many individual firms over several years, the number of accidents in each of several regions, and the number of days of absence for each of many persons over several years. A political science example is the number of protests in each of several different countries over many years. A biological and health science example is the number of occurrences of a specific health event, such as a seizure, for each of many patients in each of several time periods.

A key advantage of longitudinal data over cross-section data is that they permit more general types of individual heterogeneity. Excellent motivation was provided by Neyman (1965), who pointed out that panel data enable one to control for heterogeneity and thereby distinguish between true and apparent contagion. For example, consider estimating the impact of research and development expenditures on the number of patent applications by a firm, controlling for the individual firm-specific propensity to patent. For a single cross-section these controls can only depend on observed firm-specific attributes such as industry, and estimates may be inconsistent if there is additionally an unobserved component to the individual firm-specific propensity to patent. With longitudinal data one can additionally include a firm-specific term for the unobserved firm-specific propensity to patent.

The simplest longitudinal count data regression models are standard count models, with the addition of an individual specific term reflecting individual heterogeneity. In a *fixed effects* model this is a separate parameter for each individual. Creative estimation methods are needed if there are many individuals in the sample, because then there are many parameters. In a *random effects* model this individual specific term is instead drawn from a specified distribution. Then creativity is required either in choosing a distribution that leads to tractable analytical results or in obtaining estimates if the model is not tractable.

Asymptotic theory requires that the number of observations, here the number of individual units times the number of time periods, goes to infinity. We focus on the most common case of a *short panel* in which only a few time periods are observed and the number of cross-sectional units goes to infinity. We also consider briefly the case where the number of cross-sectional units is small but is observed for a large number of periods, as can be the case for cross-country studies. Then the earlier discussion for handling individual specific terms is mirrored in a similar discussion for time specific effects. It is important to realize that the distribution of estimators and the choice of estimator varies according to the type of sampling scheme.

In longitudinal data analysis the data are assumed to be independent over individual units for a given year but are permitted to be correlated over time for a given individual unit. In the simplest models this correlation over time is assumed to be adequately controlled for by individual-specific effects. In more general models *correlation* over time is additionally introduced in ways similar to those used in time series analysis. Finally, as in time series models, one can consider *dynamic models* or *transition models* that add a dynamic component to the regression function, allowing the dependent variable this year to depend on its own value in previous years.

A review of the standard linear models for longitudinal data, with fixed effects and random effects, is given in section 9.2, along with a statement of the analogous models for count data and a summary of statistical inference with data from short panels. Pooled or population-averaged estimators are presented in section 9.3. In section 9.4 fixed effects models for count data are presented, along with application to data on the number of patents awarded to each of 346 firms in each of the years 1975 through 1979. Random effects models are studied in section 9.5. In sections 9.4 and 9.5 both moment-based and ML estimators are detailed. A discussion of applications and of the relative merits of fixed effects and random effects approaches is given in section 9.6. Model specification tests are presented in section 9.7. Dynamic models, in which the regressors include lagged dependent variables, are studied in section 9.8. Estimation with endogenous regressors is presented in section 9.9, and more flexible count models for panel data are presented in section 9.10.

9.2 MODELS FOR LONGITUDINAL DATA

We review linear panel models, with emphasis on models that include fixed or random individual-specific effects. For count data models with a positive conditional mean, individual-specific effects enter multiplicatively rather than additively.

Throughout we consider statistical inference for *short panels*. Then the number of individuals $n \to \infty$ while the number of time periods T is finite, and data are assumed to be independent over i but possibly correlated over t for given i. In the standard case in econometrics the time interval is fixed and the data are equi-spaced through time. However, in most cases panel estimators can be adapted to unbalanced panels and to repeated events that are not necessarily equi-spaced through time. An example of such data is the number of epileptic seizures during a two-week period preceding each of four consecutive clinical visits; see Diggle, Heagerty, Liang, and Zeger (2002).

9.2.1 Linear Models and Linear Estimators

A quite general *linear model* for longitudinal data is

$$y_{it} = \alpha_{it} + \mathbf{x}_{it}'\boldsymbol{\beta}_{it} + u_{it}, \quad i = 1, \ldots, n, \quad t = 1, \ldots, T, \tag{9.1}$$

where y_{it} is a scalar dependent variable, $\mathbf{x}_{it}$ is a $k \times 1$ vector of independent variables that here excludes an intercept, and u_{it} is a scalar disturbance term. The subscript i indexes an individual person, firm, or country in a cross-section, and the subscript t indexes time. The distinguishing feature of longitudinal data models is that the intercept α_{it} and regressor coefficients $\boldsymbol{\beta}_{it}$ may differ across individuals and/or time. Such variation in coefficients reflects individual-specific and time-specific effects.

The model (9.1) is too general and is not estimable. Further restrictions need to be placed on the extent to which α_{it} and $\boldsymbol{\beta}_{it}$ vary with i and t, and on the behavior of the error u_{it}.

The *population-averaged* or *pooled linear model* restricts regression coefficients to be constant over individuals and time, so

$$y_{it} = \alpha + \mathbf{x}_{it}'\boldsymbol{\beta} + u_{it}. \tag{9.2}$$

This is just the usual linear model, without adjustment for the data being longitudinal data. The *pooled* OLS *estimator* is obtained by OLS regression of y_{it} on an intercept and $\mathbf{x}_{it}$. It is consistent if the true model is (9.2) and u_{it} is uncorrelated with $\mathbf{x}_{it}$. For short panels any correlation of the error over time for a given individual should be controlled for by computing *panel-robust standard errors* that are *cluster-robust standard errors* with clustering on the individual. More efficient feasible GLS estimation is possible given a model for the correlation in u_{it}. A common assumption is *equicorrelated errors*, with $\mathsf{Cor}[u_{it}, u_{is}] = \rho$ for $s \neq t$. Such estimators are called *pooled estimators* or *population-averaged estimators*.

The *between model* averages model (9.2) over time for each individual, so

$$\bar{y}_i = \alpha + \bar{\mathbf{x}}_i'\boldsymbol{\beta} + \bar{u}_i, \tag{9.3}$$

where $\bar{y}_i = \frac{1}{T}\sum_{t=1}^{T} y_{it}$ and $\bar{\mathbf{x}}_i = \frac{1}{T}\sum_{t=1}^{T} \mathbf{x}_{it}$. The *between estimator* is obtained by OLS regression of $\bar{y}_i$ on an intercept and $\bar{\mathbf{x}}_i$, and it uses only variation between individuals.

The most commonly used models for short panels are variants of the *one-way individual-specific effect linear model*,

$$y_{it} = \alpha_i + \mathbf{x}'_{it}\boldsymbol{\beta} + u_{it}. \tag{9.4}$$

This is the standard linear regression model, except that rather than one intercept α there are n individual-specific intercepts $\alpha_1, \ldots, \alpha_n$. Note that for short panels time-specific effects λ_t, $t = 1, \ldots, T$, are easily accommodated by including time dummies in the regressors $\mathbf{x}_{it}$. There are several variants of this model, leading to different estimators.

The *fixed effects linear model* treats α_i in (9.4) as a random variable that is potentially correlated with the regressors $\mathbf{x}_{it}$. The parameters $\boldsymbol{\beta}$ in this model are potentially inestimable, because the number of α_i terms goes to infinity as $n \rightarrow \infty$. Fortunately it is possible to eliminate the intercepts, enabling consistent estimation of $\boldsymbol{\beta}$, with two distinct methods most often used.

The *within model* or *mean-differenced model*,

$$y_{it} - \bar{y}_i = (\mathbf{x}_{it} - \bar{\mathbf{x}}_i)'\boldsymbol{\beta} + (u_{it} - \bar{u}_i), \tag{9.5}$$

is obtained by subtraction from model (9.4) of $\bar{y}_i = \alpha_i + \bar{\mathbf{x}}'_i\boldsymbol{\beta} + \bar{u}_i$, obtained by averaging (9.4) over time for individual i. The *within estimator* is obtained by OLS regression of $(y_{it} - \bar{y}_i)$ on $(\mathbf{x}_{it} - \bar{\mathbf{x}}_i)$, and it uses only variation over time within each individual. This estimator is also called the *fixed effects estimator* since if the α_i in (9.4) are viewed as parameters to be estimated, then joint estimation of $\alpha_1, \ldots, \alpha_n$ and $\boldsymbol{\beta}$ in (9.4) by OLS can be shown to yield this estimator for $\boldsymbol{\beta}$. The estimator is also a *conditional* ML *estimator*, since if u_{it} in (9.4) are assumed to be normally distributed, then maximizing with respect to $\boldsymbol{\beta}$ the conditional likelihood function given $\sum_{t=1}^{T} y_{it}$, $i = 1, \ldots, n$, yields the within estimator. A major limitation of the fixed effects model is that the coefficients of time-invariant regressors, say $\mathbf{z}_i$, are not identified since then $\mathbf{z}_i - \bar{\mathbf{z}}_i = \mathbf{0}$. The individual-specific fixed effects can be estimated by $\hat{\alpha}_i = \bar{y}_i - \bar{\mathbf{x}}'_i\hat{\boldsymbol{\beta}}_{\text{LFE}}$. For a short panel, $\hat{\alpha}_i$ is not consistent for α_i because only T observations are used in estimating each α_i.

The *first-difference model*

$$y_{it} - y_{it-1} = (\mathbf{x}_{it} - \mathbf{x}_{it-1})'\boldsymbol{\beta} + (u_{it} - u_{it-1}), \tag{9.6}$$

is obtained by subtraction from model (9.4) of the first lag of model (9.4). The *first-difference estimator* is obtained by OLS regression of $(y_{it} - y_{it-1})$ on $(\mathbf{x}_{it} - \mathbf{x}_{it-1})$, and it uses only variation over time within each individual.

The *random effects linear model* treats α_i in (9.4) as an iid random variable with mean α and variance σ_α^2, and it additionally assumes that u_{it} is iid with mean 0 and variance σ_u^2. The *random effects estimator* is the feasible GLS estimator in this model. It can be obtained by OLS estimation of $(y_{it} - \hat{\theta}_i\bar{y}_i)$ on an intercept and $(\mathbf{x}_{it} - \hat{\theta}_i\bar{\mathbf{x}}_i)$, where $\hat{\theta}_i$ is a consistent estimate of $\theta_i = 1 -$

$\sqrt{\sigma_u^2/[T_i\sigma_\alpha^2 + \sigma_u^2]}$. The model assumptions imply $\mathsf{Cor}[u_{it}, u_{is}] = \sigma_\alpha^2/(\sigma_\alpha^2 + \sigma_u^2)$ for $s \neq t$, so the random effects estimator is asymptotically equivalent to the population-averaged estimator with equicorrelated errors. The estimator is also asymptotically equivalent to the MLE under the additional assumption that α_i and u_{it} are normally distributed.

The random effects and fixed effects linear models are compared by, for example, Hsiao (2003, pp. 41–49). The models are conceptually different, with the fixed effects analysis being conditional on the effects for individuals in the sample; random effects is an unconditional or marginal analysis with respect to the population. A major practical difference is that the fixed effects analysis provides only estimates of time-varying regressors. Thus, for example, it does not allow estimation of an indicator variable for whether or not a patient in a clinical trial was taking the drug under investigation (rather than a placebo). Another major difference is that the random effects model assumption that individual effects are iid implies that individual effects are uncorrelated with the regressors. If, instead, unobserved individual effects are correlated with observed effects, the random effects estimator is inconsistent. The pooled and between estimators are also inconsistent in this case. Many econometrics studies in particular prefer fixed effects estimators (within or first-difference) because of this potential problem.

The *conditional correlated random effects linear* model is a variation of the random effects model that allows α_i in (9.4) to be correlated with the regressors $\mathbf{x}_{it}$ according to the model $\alpha_i = \pi + \overline{\mathbf{x}}_i'\boldsymbol{\lambda} + v_i$, where v_i is iid $[0, \sigma_v^2]$. It can be shown that random effects estimation with regressors $\mathbf{x}_{it}$ and $\overline{\mathbf{x}}_i$ is equivalent to within estimation.

The *random coefficients* model allows both intercept and slope coefficients to vary across individuals, so

$$y_{it} = \alpha_i + \mathbf{x}_{it}'\boldsymbol{\beta}_i + u_{it}, \tag{9.7}$$

where α_i is iid $[\alpha, \sigma_\alpha^2]$ and $\boldsymbol{\beta}_i$ is iid $[\boldsymbol{\beta}, \boldsymbol{\Sigma}_\beta]$. This model is a special case of a mixed linear model. The *random coefficients estimator* is the feasible GLS estimator in this model. In the linear case the other estimators remain consistent since this model implies $\mathsf{E}[y_{it}|\mathbf{x}_{it}] = \alpha + \mathbf{x}_{it}'\boldsymbol{\beta}$.

Dynamic models include lags of y_{it} as regressors. For example,

$$y_{it} = \alpha_i + \rho y_{it-1} + \mathbf{x}_{it}'\boldsymbol{\beta} + u_{it}.$$

If α_i is a fixed effect the within estimator is no longer consistent. Instead IV estimation based on the first-difference model, with appropriate lags of y_{it} used as instruments, yields consistent estimates.

9.2.2 Individual Effects in Count Models

The subsequent sections present a range of models and estimators for longitudinal count data, many with individual-specific effects. In the linear case the individual-specific effect enters additively, so $\mu_{it} = \alpha_i + \mathbf{x}_{it}'\boldsymbol{\beta}$. For count data the individual-specific effect instead enters multiplicatively.

For models with an exponential conditional mean, $\exp(\mathbf{x}_{it}'\boldsymbol{\beta})$, a model with *multiplicative individual-specific effect* α_i, specifies the conditional mean to be

$$\begin{aligned} \mathsf{E}[y_{it}|\mathbf{x}_{it}, \alpha_i] &= \mu_{it} \\ &= \alpha_i \lambda_{it} \\ &= \alpha_i \exp(\mathbf{x}_{it}'\boldsymbol{\beta}). \end{aligned} \tag{9.8}$$

Note that α used here refers to the individual effect and is not used in the same way as in previous chapters, where it was an overdispersion parameter.

In the *fixed effects* model the α_i are potentially correlated with the regressors $\mathbf{x}_{it}$. Like the linear model, however, consistent estimation of $\boldsymbol{\beta}$ is possible by eliminating α_i; see section 9.4. In the *random effects* model the α_i are instead iid random variables; see section 9.5.

A key departure from the linear model is that the individual-specific effects in (9.8) are *multiplicative*, rather than additive as in the linear model (9.4). Given the exponential form for λ_{it}, the multiplicative effects can still be interpreted as a shift in the intercept because

$$\mu_{it} = \alpha_i \exp(\mathbf{x}_{it}'\boldsymbol{\beta}) = \exp(\delta_i + \mathbf{x}_{it}'\boldsymbol{\beta}), \tag{9.9}$$

where $\delta_i = \ln \alpha_i$.

Note that this equality between multiplicative effects and intercept shift relies on an exponential conditional mean. The equality does not hold in some count data models, nor does it hold in noncount models such as binary models to which similar longitudinal methods might be applied. Suppose the starting point is a more general conditional mean function $g(\mathbf{x}_{it}'\boldsymbol{\beta})$. Then some models and estimation methods continue with multiplicative effects, so

$$\mathsf{E}[y_{it}|\mathbf{x}_{it}, \alpha_i] = \mu_{it} = \alpha_i g(\mathbf{x}_{it}'\boldsymbol{\beta}), \tag{9.10}$$

whereas other methods use a shift in the intercept,

$$\mathsf{E}[y_{it}|\mathbf{x}_{it}, \alpha_i] = \mu_{it} = g(\delta_i + \mathbf{x}_{it}'\boldsymbol{\beta}). \tag{9.11}$$

Models (9.10) and (9.11) differ if $g(x) \neq \exp(x)$.

The *marginal effect* given (9.8) is

$$\mathsf{ME}_j = \frac{\partial \mathsf{E}[y_{it}|\mathbf{x}_{it}, \alpha_i]}{\partial \mathbf{x}_{itj}} = \alpha_i \exp(\mathbf{x}_{it}'\boldsymbol{\beta})\beta_j = \beta_j \mathsf{E}[y_{it}|\mathbf{x}_{it}, \alpha_i]. \tag{9.12}$$

Unlike the linear FE model, ME_j depends on the unknown α_i, so that consistent estimation of $\boldsymbol{\beta}$ is not sufficient for computing the marginal effect. From the last equality, however, the slope coefficient β_j can still be interpreted as a semi-elasticity, giving the proportionate increase in $\mathsf{E}[y_{it}|\mathbf{x}_{it}, \alpha_i]$ associated with a one-unit change in x_{itj}.

We first present estimation for models without dynamics. In particular, for multiplicative effects models the regressors $\mathbf{x}_{it}$ are initially assumed to be *strictly exogenous*, so that

$$\mathsf{E}[y_{it}|\mathbf{x}_{i1},\ldots,\mathbf{x}_{iT},\alpha_i]=\alpha_i\lambda_{it}. \tag{9.13}$$

This is a stronger condition than $\mathsf{E}[y_{it}|\mathbf{x}_{it},\alpha_i]=\alpha_i\lambda_{it}$. Condition (9.13) is relaxed when time-series models are presented in section 9.8.

Some results require only moment conditions (9.8) or (9.13), so are not restricted to panel data that are count data. For example, they can be applied to right-skewed nonnegative continuous data such as expenditure data.

Parametric model results are most easily obtained for the Poisson. Extensions to the negative binomial do not always work, and when they do work they do so only for some methods for the NB1 model and in other cases for the NB2 model. It should be kept in mind, however, that a common reason for such extensions is to control for unobserved heterogeneity. Access to longitudinal data can control for heterogeneity through the individual-specific effect α_i, so the efficiency gains in going beyond Poisson models may not be as great as in the cross-section case.

9.2.3 Panel-Robust Inference

We present panel-robust inference for the generalized method of moments estimator, introduced in Chapter 2.5.2, for the case of short panels. The results nest both ML and moment-based estimation methods. Many panel count data estimators are moment-based, especially those for models that are dynamic or have endogenous regressors.

Collecting data over all time periods for a given individual, define

$$\mathbf{y}_i=\begin{bmatrix} y_{i1}\\ \vdots \\ y_{iT}\end{bmatrix},\quad \mathbf{X}_i=\begin{bmatrix} \mathbf{x}'_{i1}\\ \vdots \\ \mathbf{x}'_{iT}\end{bmatrix}.$$

Estimation is based on the population moment conditions

$$\mathsf{E}[\mathbf{h}_i(\mathbf{y}_i,\mathbf{X}_i,\boldsymbol{\theta}_0)]=\mathbf{0},\qquad i=1,\ldots,n, \tag{9.14}$$

where $\boldsymbol{\theta}_0$ is the value of the $q\times 1$ parameter vector $\boldsymbol{\theta}$ in the dgp, $\mathbf{h}_i(\cdot)$ is an $r\times 1$ vector, and $r\geq q$ so that the number of moment conditions potentially exceeds the number of parameters.

The GMM estimator $\hat{\boldsymbol{\theta}}_{\mathsf{GMM}}$ minimizes

$$\left[\sum_{i=1}^{n}\mathbf{h}_i(\mathbf{y}_i,\mathbf{X}_i,\boldsymbol{\theta})\right]'\mathbf{W}_n\left[\sum_{i=1}^{n}\mathbf{h}_i(\mathbf{y}_i,\mathbf{X}_i,\boldsymbol{\theta})\right], \tag{9.15}$$

where $\mathbf{W}_n\xrightarrow{p}\mathbf{W}$ is a possibly stochastic symmetric positive definite $r\times r$ weighting matrix.

Given essentially that (9.14) holds, $\hat{\boldsymbol{\theta}}_{\mathsf{GMM}}\xrightarrow{p}\boldsymbol{\theta}_0$ and

$$\sqrt{n}(\hat{\boldsymbol{\theta}}_{\mathsf{GMM}}-\boldsymbol{\theta}_0)\xrightarrow{d}\mathsf{N}[\mathbf{0},(\mathbf{H}'\mathbf{W}\mathbf{H})^{-1}\mathbf{H}'\mathbf{W}\mathbf{S}\mathbf{W}\mathbf{H}(\mathbf{H}'\mathbf{W}\mathbf{H})^{-1}], \tag{9.16}$$

where $\mathbf{W}$ is consistently estimated by $\mathbf{W}_n$, $\mathbf{H}$ is consistently estimated by

$$\hat{\mathbf{H}} = \frac{1}{n}\sum_{i=1}^{n} \left.\frac{\partial \mathbf{h}_i}{\partial \boldsymbol{\theta}'}\right|_{\hat{\boldsymbol{\theta}}} \tag{9.17}$$

and, assuming independence over i, $\mathbf{S}$ is consistently estimated by

$$\hat{\mathbf{S}} = \frac{1}{n}\sum_{i=1}^{n} \mathbf{h}_i(\hat{\boldsymbol{\theta}})\mathbf{h}_i(\hat{\boldsymbol{\theta}})'. \tag{9.18}$$

The variance matrix estimate based on (9.16) to (9.18) is a *panel-robust* or *cluster-robust* estimate. It can also be obtained by a *panel bootstrap* or *panel jackknife*, see Chapter 2.6.4; these resample over individuals rather than over individual-time pairs. More generally independence over i can be relaxed to independence over clusters such as villages, see section 2.7.1, provided the number of clusters is large.

In the just identified case ($r = q$) the GMM estimator simplifies to the estimating equations estimator, presented in Chapter 2.5.1. Using the current notation, $\hat{\boldsymbol{\theta}}_{\mathsf{EE}}$ that solves

$$\sum_{i=1}^{n} \mathbf{h}_i(\mathbf{y}_i, \mathbf{X}_i, \boldsymbol{\theta}) = \mathbf{0}, \tag{9.19}$$

is consistent and

$$\sqrt{n}(\hat{\boldsymbol{\theta}}_{\mathsf{EE}} - \boldsymbol{\theta}_0) \overset{d}{\to} \mathsf{N}[\mathbf{0}, \mathbf{H}^{-1}\mathbf{S}\mathbf{H}'^{-1}]. . \tag{9.20}$$

In the special case of the MLE, $\mathbf{h}_i(\mathbf{y}_i, \mathbf{X}_i, \boldsymbol{\theta}) = \partial \ln f(\mathbf{y}_i|\mathbf{X}_i, \boldsymbol{\theta})/\partial\boldsymbol{\theta}$, where $f(\mathbf{y}_i|\mathbf{X}_i, \boldsymbol{\theta})$ is the joint density for the ith observation $\mathbf{y}_i$.

As an example consider the linear fixed effects model and specify

$$\mathbf{h}_i(\mathbf{y}_i, \mathbf{X}_i, \boldsymbol{\beta}) = \sum_{t=1}^{T} (\mathbf{x}_{it} - \bar{\mathbf{x}}_i)\{(y_{it} - \bar{y}_i) - (\mathbf{x}_{it} - \bar{\mathbf{x}}_i)'\boldsymbol{\beta}\},$$

which has an expected value $\mathbf{0}$ given (9.4) and $\mathsf{E}[u_{it}|\mathbf{x}_{i1}, \ldots, \mathbf{x}_{iT}, \alpha_i] = 0$. Then the estimating equations (9.20) can be solved to yield

$$\hat{\boldsymbol{\beta}} = \left[\sum_{i=1}^{n}\sum_{t=1}^{T} (\mathbf{x}_{it} - \bar{\mathbf{x}}_i)(\mathbf{x}_{it} - \bar{\mathbf{x}}_i)'\right]^{-1} \sum_{i=1}^{n}\sum_{t=1}^{T} (\mathbf{x}_{it} - \bar{\mathbf{x}}_i)(y_{it} - \bar{y}_i), \tag{9.21}$$

which is just the within estimator. Since $\partial\mathbf{h}_i/\partial\boldsymbol{\theta}' = \sum_{t=1}^{T}(\mathbf{x}_{it} - \bar{\mathbf{x}}_i)(\mathbf{x}_{it} - \bar{\mathbf{x}}_i)'$, the variance matrix given in (9.20) can be consistently estimated by n times

$$\left[\sum_{i=1}^{n}\sum_{t=1}^{T} \tilde{\mathbf{x}}_{it}\tilde{\mathbf{x}}_{it}'\right]^{-1} \sum_{i=1}^{n}\sum_{t=1}^{T}\sum_{s=1}^{T} \tilde{u}_{it}\tilde{u}_{is}\tilde{\mathbf{x}}_{it}\tilde{\mathbf{x}}_{is}' \left[\sum_{i=1}^{n}\sum_{t=1}^{T} \tilde{\mathbf{x}}_{it}\tilde{\mathbf{x}}_{it}'\right]^{-1},$$

where $\tilde{\mathbf{x}}_{it} = \mathbf{x}_{it} - \bar{\mathbf{x}}_i$ and $\tilde{u}_{it} = \hat{u}_{it} - \bar{u}_i = (y_{it} - \bar{y}_i) - (\mathbf{x}_{it} - \bar{\mathbf{x}}_i)'\hat{\boldsymbol{\beta}}$.

9.3 POPULATION AVERAGED MODELS

Pooled estimators combine the between cross-section and within (time-series) variation in the data by simply regressing y_{it} on an intercept and $\mathbf{x}_{it}$. More efficient pooled estimators adjust for correlation over time for a given individual.

9.3.1 Pooled Poisson QMLE

The obvious pooled estimator for count data is the Poisson regression with exponential conditional mean. Defining $\mathbf{z}_{it}' = [1 \;\; \mathbf{x}_{it}']$ and $\boldsymbol{\gamma}' = [\delta \;\; \boldsymbol{\beta}']$, so $\exp(\mathbf{z}_{it}'\boldsymbol{\gamma}) = \exp(\delta + \mathbf{x}_{it}'\boldsymbol{\beta})$, the first-order conditions for the *pooled Poisson* QMLE are

$$\sum_{i=1}^{n}\sum_{t=1}^{T}(y_{it} - \exp(\mathbf{z}_{it}'\boldsymbol{\gamma}))\mathbf{z}_{it} = \mathbf{0}. \tag{9.22}$$

The estimator is consistent if

$$\mathsf{E}[y_{it}|\mathbf{x}_{it}] = \exp(\delta + \mathbf{x}_{it}'\boldsymbol{\beta}) = \alpha\exp(\mathbf{x}_{it}'\boldsymbol{\beta}), \tag{9.23}$$

where $\alpha = \exp(\delta)$.

For cross-section data with independent observations, robust inference is based on heteroskedastic-robust standard errors; see Chapter 3.2.2. For short panels where observations are independent over i but correlated over t for given i, inference is based on cluster-robust standard errors, see Chapter 2.7.1 or section 9.2.3, with clustering on the individual. Then the Poisson QMLE has the variance matrix estimate,

$$\left[\sum_{i=1}^{n}\sum_{t=1}^{T}\hat{\mu}_{it}\mathbf{z}_{it}\mathbf{z}_{it}'\right]^{-1}\sum_{i=1}^{n}\sum_{t=1}^{T}\sum_{s=1}^{T}\hat{u}_{it}\hat{u}_{is}\mathbf{z}_{it}\mathbf{z}_{is}'\left[\sum_{i=1}^{n}\sum_{t=1}^{T}\hat{\mu}_{it}\mathbf{z}_{it}\mathbf{z}_{it}'\right]^{-1}, \tag{9.24}$$

where $\hat{\mu}_{it} = \exp(\mathbf{z}_{it}'\hat{\boldsymbol{\gamma}})$, and $\hat{u}_{it} = y_{it} - \exp(\mathbf{z}_{it}'\hat{\boldsymbol{\gamma}})$. It is very important that this variance matrix estimate be used because the default matrix estimate is usually erroneously much smaller – it fails to control for correlation over time for a given individual and for any overdispersion.

The individual-specific multiplicative effects model (9.8) implies that

$$\begin{aligned}\mathsf{E}[y_{it}|\mathbf{x}_{it}] &= \mathsf{E}_{\alpha_i}[\mathsf{E}[y_{it}|\mathbf{x}_{it}, \alpha_i]] \\ &= \mathsf{E}_{\alpha_i}[\alpha_i\exp(\mathbf{x}_{it}'\boldsymbol{\beta})].\end{aligned}$$

In the random effects model with α_i independent of $\mathbf{x}_{it}$, it follows that

$$\mathsf{E}[y_{it}|\mathbf{x}_{it}] = \alpha\exp(\mathbf{x}_{it}'\boldsymbol{\beta}),$$

where $\alpha = \mathsf{E}_{\alpha_i}[\alpha_i]$, so that condition (9.23) is satisfied. More generally, the pooled estimator is called the *population-averaged estimator* in the statistics literature, since (9.23) is assumed to hold after averaging out any individual-specific effects. In the fixed effects model, however, no such simplification occurs due to the correlation of α_i with $\mathbf{x}_{it}$, condition (9.23) no longer holds, and the pooled Poisson QMLE is inconsistent. Alternative estimators, given in section 9.4, need to be used.

9.3.2 Generalized Estimating Equations Estimator

The pooled Poisson QMLE essentially bases estimation on the assumption that observations are independent over both i and t and then performs inference that relaxes the assumption of independence over t. More efficient estimation is possible by specifying a model for the correlation over time for a given individual and then estimating subject to this model.

This is an example of generalized estimating equations estimation. The results presented in Chapter 2.4.6 can be immediately applied, with subscript i replaced with subscript t. Let $\boldsymbol{\mu}_i(\boldsymbol{\gamma}) = [\mu_{i1} \cdots \mu_{iT}]'$ where $\mu_{it} = \exp(\mathbf{z}'_{it}\boldsymbol{\gamma})$, and let $\boldsymbol{\Sigma}_i$ be a working model for $\mathsf{V}[\mathbf{y}_i|\mathbf{X}_i]$ with ts^{th} entry $\mathsf{Cov}[y_{it}, y_{is}|\mathbf{X}_i]$. For example, if we assume data are equicorrelated with overdispersion of the NB1 form, then we use $\mathsf{Cor}[y_{it}, y_{is}|\mathbf{X}_i] = \rho$ for all $s \neq t$ and $\mathsf{Cov}[y_{it}, y_{is}|\mathbf{X}_i] = \rho\sigma_{it}\sigma_{is}$, where $\sigma^2_{it} = \phi\mu_{it}$. A second possibility is to assume correlation for the first m lags, so $\mathsf{Cor}[y_{it}, y_{i,t-k}|\mathbf{X}_i] = \rho_k$ where $\rho_k = 0$ for $|k| > m$.

From Chapter 2.4.6, the Poisson GEE estimator solves

$$\sum_{i=1}^{n} \frac{\partial \boldsymbol{\mu}'_i(\boldsymbol{\gamma})}{\partial \boldsymbol{\gamma}} \hat{\boldsymbol{\Sigma}}_i^{-1} (\mathbf{y}_i - \boldsymbol{\mu}_i(\boldsymbol{\gamma})) = \mathbf{0}, \tag{9.25}$$

where $\hat{\boldsymbol{\Sigma}}_i$ is obtained from initial first-stage estimation of $\boldsymbol{\beta}$ (by pooled Poisson QMLE) and any other parameters that determine $\boldsymbol{\Sigma}_i$. The formula for a cluster-robust estimate of the asymptotic variance matrix, one robust to misspecification of $\boldsymbol{\Sigma}_i$, is given in Chapter 2.4.6.

Zeger and Liang (1986) and Liang and Zeger (1986) call the GEE approach *marginal analysis*, because estimation is based on moments of the distribution marginal to the random effects. Zeger, Liang, and Albert (1988) consider mixed GLMs with random effects that may interact with regressors. They call the approach *population-averaged*, because the random effects are averaged out, and contrast this with subject-specific models that explicitly model the individual effects. These papers present estimating equations of the form (9.25) with little explicit discussion of the random effects or the precise form of the correlation and modified heteroskedasticity that they induce. More formal treatment of random effects using this approach is given in Thall and Vail (1990). Liang, Zeger, and Qaqish (1992) consider *generalized* GEE estimators that jointly estimate the regression and correlation parameters.

Brännäs and Johansson (1996) consider a more tightly specified model than Zeger and Liang (1986): the Poisson model with multiplicative random effects in which the random effects are also time varying; that is, α_{it} replaces α_i in (9.9). They consider estimation by GMM, which exploits more of the moment conditions implied by the model. They apply these estimation methods to data on the number of days absent from work for 895 Swedish workers in each of 11 years.

9.3.3 Pooled Parametric Models

For fully parametric models, pooled models assume that the marginal density for a single observation y_{it},

$$f(y_{it}|\mathbf{x}_{it}) = f(\alpha + \mathbf{x}_{it}'\boldsymbol{\beta}, \boldsymbol{\gamma}), \tag{9.26}$$

is correctly specified, regardless of the (unspecified) form of the joint density

$$f(y_{it}, \ldots, y_{iT}|\mathbf{x}_{i1}, \ldots, \mathbf{x}_{iT}, \alpha, \boldsymbol{\beta}, \boldsymbol{\gamma}).$$

This is a strong assumption.

The *pooled* MLE maximizes the log-likelihood formed assuming independence over i and t,

$$\mathcal{L}(\alpha, \boldsymbol{\beta}, \boldsymbol{\gamma}) = \sum_{i=1}^{n}\sum_{t=1}^{T} \ln f(\alpha + \mathbf{x}_{it}'\boldsymbol{\beta}, \boldsymbol{\gamma}), \tag{9.27}$$

where, for example, $f(y_{it}|\mathbf{x}_{it})$ may be the NB2 density or a hurdle model density. The pooled MLE is consistent if (9.26) holds. Inference should be based on a panel-robust or cluster-robust (with clustering on i) estimator of the covariance matrix,

$$\left[\sum_{i=1}^{n}\sum_{t=1}^{T} \frac{\partial \mathbf{s}_{it}'}{\partial \boldsymbol{\theta}}\bigg|_{\hat{\boldsymbol{\theta}}}\right]^{-1} \sum_{i=1}^{n}\sum_{t=1}^{T}\sum_{s=1}^{T} \mathbf{s}_{it}\mathbf{s}_{is}'\big|_{\hat{\boldsymbol{\theta}}} \left[\sum_{i=1}^{n}\sum_{t=1}^{T} \frac{\partial \mathbf{s}_{it}}{\partial \boldsymbol{\theta}'}\bigg|_{\hat{\boldsymbol{\theta}}}\right]^{-1}, \tag{9.28}$$

where $\mathbf{s}_{it} = \partial \ln f(\alpha + \mathbf{x}_{it}'\boldsymbol{\beta}, \boldsymbol{\gamma})/\partial\boldsymbol{\theta}$, and $\boldsymbol{\theta}' = [\alpha\ \boldsymbol{\beta}'\ \boldsymbol{\gamma}']$.

9.4 FIXED EFFECTS MODELS

The econometrics literature places great emphasis on the estimation of panel models with individual-specific fixed effects. In short panels it is a challenge to consistently estimate $\boldsymbol{\beta}$ because, given asymptotic theory that relies on $n \to \infty$ with T fixed, we need to first control for the infinite number of fixed effects $\alpha_1, \ldots, \alpha_n$. For the linear case this is possible, but for most nonlinear models it is not. For counts it is possible for the Poisson model and, more generally, for models with a conditional mean function that is multiplicative in the fixed effects. But it is not possible in many other commonly used count models, such as hurdle models and models with censoring or truncation.

There are three distinct leading approaches that, in a few special nonlinear fixed effects models, lead to consistent estimation of $\boldsymbol{\beta}$. First, moment-based estimators use a differencing transformation to eliminate the fixed effects. Second, both $\boldsymbol{\beta}$ and the fixed effects α_i may be jointly estimated by maximum likelihood. Third, conditional maximum likelihood estimation performs ML estimation of $\boldsymbol{\beta}$ conditional on sufficient statistics for the individual effects.

Remarkably, for the Poisson model with multiplicative fixed effects, all three approaches lead to the consistent estimation of $\boldsymbol{\beta}$. We present moment-based methods first, because they rely on fewer assumptions and in later sections are adapted to handle regressors that are endogenous or predetermined. For NB models the first approach is also applicable, and the third approach works for a restricted version of the NB1 model.

9.4.1 Moment-Based Methods

In the linear model (9.4) with additive fixed effects, there are several ways to transform the model to eliminate the fixed effects and hence obtain a moment-based consistent estimator of $\boldsymbol{\beta}$. Examples given in section 9.2.1 are mean differencing, see (9.5), and first differencing, see (9.6).

Related transformations are possible for models with multiplicative fixed effects and strictly exogenous regressors. From (9.13) we assume

$$\mathsf{E}[y_{it}|\mathbf{x}_{i1},\ldots,\mathbf{x}_{iT},\alpha_i] = \alpha_i\lambda_{it}.$$

Averaging over time for individual i, it follows that

$$\mathsf{E}[\bar{y}_i|\mathbf{x}_{i1},\ldots,\mathbf{x}_{iT},\alpha_i] = \alpha_i\bar{\lambda}_i,$$

where $\bar{y}_i = \frac{1}{T}\sum_t \lambda_{it}$ and $\bar{\lambda}_i = \frac{1}{T}\sum_t \lambda_{it}$. Subtracting

$$\mathsf{E}\left[(y_{it} - (\lambda_{it}/\bar{\lambda}_i)\bar{y}_i)\,|\mathbf{x}_{i1},\ldots,\mathbf{x}_{iT}\right] = 0, \tag{9.29}$$

and hence by the law of iterated expectations,

$$\mathsf{E}\left[\mathbf{x}_{it}\left(y_{it} - \frac{\lambda_{it}}{\bar{\lambda}_i}\bar{y}_i\right)\right] = \mathbf{0}. \tag{9.30}$$

The *Poisson fixed effects estimator* $\hat{\boldsymbol{\beta}}_{\mathsf{PFE}}$ is the method of moments estimator that solves the corresponding sample moment conditions

$$\sum_{i=1}^{n}\sum_{t=1}^{T}\mathbf{x}_{it}\left(y_{it} - \frac{\bar{y}_i}{\bar{\lambda}_i}\lambda_{it}\right) = \mathbf{0}. \tag{9.31}$$

The estimator is called the Poisson fixed effects estimator because (9.31) are the first-order conditions of both the Poisson fixed effects MLE and the Poisson fixed effects conditional MLE, derived in subsequent subsections. The coefficients $\boldsymbol{\beta}_{\mathbf{z}}$ say of time-invariant regressors $\mathbf{z}_i$ are not identified since $\mathbf{z}_i'\boldsymbol{\beta}_{\mathbf{z}}$ drops out of the ratio $\lambda_{it}/\bar{\lambda}_i$ in (9.31) given an exponential conditional mean. This mirrors the result for the linear model. From inspection of (9.31), observations with $y_{it} = 0$ for all T make no contribution to the estimation of $\boldsymbol{\beta}$.

The asymptotic variance of $\hat{\boldsymbol{\beta}}_{\mathsf{PFE}}$ can be obtained using the results of section 9.2.3. The first-order conditions (9.31) have first-derivative matrix with respect to $\boldsymbol{\beta}'$

$$\mathbf{A}_n = \sum_{i=1}^{n}\left[\sum_{t=1}^{T}\mathbf{x}_{it}\mathbf{x}_{it}'\frac{\bar{y}_i}{\bar{\lambda}_i}\lambda_{it} - \frac{1}{T}\sum_{t=1}^{T}\sum_{s=1}^{T}\mathbf{x}_{it}\mathbf{x}_{is}'\frac{\bar{y}_i}{\bar{\lambda}_i}\lambda_{it}\lambda_{is}\right], \tag{9.32}$$

for $\lambda_{it} = \exp(\mathbf{x}_{it}'\boldsymbol{\beta})$, while the outer product of these conditions on taking expectations and eliminating cross-products in i and $j \neq i$ due to independence is

$$\mathbf{B}_n = \sum_{i=1}^{n}\sum_{t=1}^{T}\sum_{s=1}^{T} \mathbf{x}_{it}\mathbf{x}_{is}' \left(y_{it} - \frac{\bar{y}_i}{\bar{\lambda}_i}\lambda_{it} \right) \left(y_{is} - \frac{\bar{y}_i}{\bar{\lambda}_i}\lambda_{is} \right)'. \qquad (9.33)$$

The panel-robust or cluster-robust variance matrix estimate is

$$\hat{\mathsf{V}}_{\mathsf{RS}}[\hat{\boldsymbol{\beta}}_{\mathsf{PFE}}] = \hat{\mathbf{A}}^{-1}\hat{\mathbf{B}}\hat{\mathbf{A}}^{-1}, \qquad (9.34)$$

where $\hat{\mathbf{A}}$ and $\hat{\mathbf{B}}$ are $\mathbf{A}_n$ and $\mathbf{B}_n$ evaluated at $\hat{\boldsymbol{\beta}}_{\mathsf{PFE}}$.

It should be clear that the essential condition for consistency of $\hat{\boldsymbol{\beta}}_{\mathsf{PFE}}$ is that (9.13) holds and λ_{it} is correctly specified. The variance matrix estimate in (9.34) requires only the additional assumption that $\lambda_{it} = \exp(\mathbf{x}_{it}'\boldsymbol{\beta})$ and is easily adapted to other models for λ_{it}. Furthermore, the results are not restricted to count data. For example, the Poisson fixed effects estimator may be a good estimator for nonnegative right-skewed continuous data such as that on expenditures, provided there are not too many zero observations in which case a two-part or Tobit model may be needed.

A quite general treatment is given by Wooldridge (1990c), who considers a multiplicative fixed effect for general specifications of $\lambda_{it} = g(\mathbf{x}_{it}'\boldsymbol{\beta})$. In addition to giving robust variance matrix estimates, he gives more efficient GMM estimators if the conditional mean is specified to be of form $\alpha_i\lambda_{it}$ with other moments not specified, and when additionally the variance is specified to be of the form $\psi_i\alpha_i\lambda_{it}$. Chamberlain (1992a) gives semiparametric efficiency bounds for models using only the specified first moment of form (9.9). Attainment of these bounds is theoretically possible but practically difficult, because it requires high-dimensional nonparametric regressions.

It is conceivable that a fixed effects type formulation may account for much of the overdispersion of counts. But there are other complications that generate overdispersion, such as excess zeros and fat tails. At present little is known about the performance of moment-based estimators when the dgp deviates significantly from Poisson-type behavior. Also, moment-based models do not exploit the integer-valued aspect of the dependent variable. Whether this leads to significant efficiency loss, and if so when, is a topic that deserves future investigation.

Models based on alternative moment conditions that enable estimation when regressors are predetermined or endogenous are presented in section 9.8.

9.4.2 Poisson Maximum Likelihood

The simplest parametric fixed effects model for count data is the *Poisson fixed effects* model where, conditional on λ_{it} and individual-specific effects α_i,

$$y_{it} \sim \mathsf{P}[\mu_{it} = \alpha_i\lambda_{it}], \qquad (9.35)$$

where λ_{it} is a specified function of $\mathbf{x}_{it}$ and $\boldsymbol{\beta}$, and at times we specialize to the exponential form (9.8).

ML estimation treats $\alpha_i, i = 1, \ldots, n$, as parameters to be estimated jointly with $\boldsymbol{\beta}$. Then in general $\boldsymbol{\beta}$ may be inconsistently estimated if T is small and $n \to \infty$, because as $n \to \infty$ the number of parameters, $n + \dim[\boldsymbol{\beta}]$, to be estimated goes to infinity, possibly negating the benefit of a larger sample size, nT. The individual fixed effects can be viewed as *incidental parameters*, because real interest lies in the slope coefficients. The α_i are estimated based on T observations for individual i, so are clearly inconsistently estimated unless $T \to \infty$. For some nonlinear fixed effects panel data models, this inconsistent estimation of the α_i leads to inconsistent parameter estimates of $\boldsymbol{\beta}$. A leading example is the logit model with fixed effects, with

$$\Pr[y_{it} = 1] = [\exp(\alpha_i + \mathbf{x}_{it}'\boldsymbol{\beta})]/[1 + \exp(\alpha_i + \mathbf{x}_{it}'\boldsymbol{\beta})].$$

Hsiao (2003, section 7.3.1) demonstrates the inconsistency of the MLE for $\boldsymbol{\beta}$ in this case, for fixed T and $n \to \infty$. This inconsistency can be large (e.g., $\text{plim}\,\hat{\boldsymbol{\beta}} = 2\boldsymbol{\beta}$ for $T = 2$ in this logit example), though disappears as $T \to \infty$, and Monte Carlo experiments for several different models by Greene (2004) suggest that for $T \geq 5$ the bias may not be too severe.

For the linear model there is no incidental parameters problem. This is also the case for the Poisson fixed effects model, as first shown independently by Lancaster (1997) and Blundell, Griffith, and Van Reenan (1999). For y_{it} iid $\mathsf{P}[\alpha_i\lambda_{it}]$, the joint density for the ith observation is

$$\begin{aligned}\Pr[y_{i1}, \ldots, y_{iT}|\alpha_i, \boldsymbol{\beta}] &= \prod_t \left[\exp(-\alpha_i\lambda_{it})(\alpha_i\lambda_{it})^{y_{it}}/y_{it}!\right] \qquad (9.36)\\ &= \exp\left(-\alpha_i \sum_t \lambda_{it}\right) \prod_t \alpha_i^{y_{it}} \prod_t \lambda_{it}^{y_{it}} \Big/ \prod_t y_{it}!\,.\end{aligned}$$

The corresponding log-density is

$$\begin{aligned}\ln \Pr[y_{i1}, \ldots, y_{iT}|\alpha_i, \boldsymbol{\beta}] &= -\alpha_i \sum_t \lambda_{it} + \ln \alpha_i \sum_t y_{it}\\ &\quad + \sum_t y_{it} \ln \lambda_{it} - \sum_t \ln y_{it}!.\end{aligned}$$

Differentiating with respect to α_i and setting to zero yields

$$\hat{\alpha}_i = \frac{\sum_t y_{it}}{\sum_t \lambda_{it}}. \qquad (9.37)$$

Substituting this back into (9.36), simplifying, and considering all n observations yields the concentrated likelihood function,

$$\begin{aligned}\mathsf{L}_{\text{conc}}(\boldsymbol{\beta}) &= \prod_i \left[\exp\left(-\sum_t y_{it}\right) \times \prod_t \left(\frac{\sum_t y_{it}}{\sum_t \lambda_{it}}\right)^{y_{it}} \prod_t \lambda_{it}^{y_{it}} / \prod_t y_{it}!\right] \qquad (9.38)\\ &\propto \prod_i \prod_t \left(\frac{\lambda_{it}}{\sum_s \lambda_{is}}\right)^{y_{it}},\end{aligned}$$

where the last expression drops terms not involving $\boldsymbol{\beta}$. This is the likelihood for n independent observations on a T-dimensional multinomial variable with cell probabilities

$$p_{it} = \frac{\lambda_{it}}{\sum_s \lambda_{is}} = \frac{\exp(\mathbf{x}'_{it}\boldsymbol{\beta})}{\sum_s \exp(\mathbf{x}'_{is}\boldsymbol{\beta})},$$

where the second equality specializes to an exponential conditional mean. It follows that for the Poisson fixed effects model there is no incidental parameters problem.

The MLE for $\boldsymbol{\beta}$, consistent for fixed T and $n \to \infty$, can be obtained by maximization of $\ln \mathsf{L}_{conc}(\boldsymbol{\beta})$ in (9.38). The first-order conditions for this estimator of $\boldsymbol{\beta}$, when $\lambda_{it} = \exp(\mathbf{x}'_{it}\boldsymbol{\beta})$, are exactly those given in (9.31) for the moment-based estimator. Estimates of α_i can be obtained from (9.37), given $\hat{\boldsymbol{\beta}}$, and are inconsistent unless $T \to \infty$.

Inference for the MLE of $\boldsymbol{\beta}$ should be based on the cluster-robust variance matrix given in (9.34). The default ML variance matrix estimate is $\hat{\mathbf{A}}_n^{-1}$, where $\mathbf{A}_n$ is defined in (9.32), and it can greatly understate the true variance due to the failure to control for overdispersion and any within-cluster correlation.

The consistency of the MLE for $\boldsymbol{\beta}$ despite the presence of incidental parameters is a special result that holds for the Poisson multiplicative fixed effects model and, for continuous data, the normal linear additive fixed effects model. It holds in few other models, if any.

The Poisson fixed effects estimator can be obtained by estimating a Poisson dummy variables model. The exponential mean specification (9.8) can be rewritten as $\exp(\sum_{j=1}^{n} \delta_j d_{jit} + \mathbf{x}'_{it}\boldsymbol{\beta})$, where d_{jit} is an indicator variable equal to one if the it^{th} observation is for individual j and zero otherwise. Thus we can use standard Poisson software to regress y_{it} on $d_{1it}, d_{2it}, \ldots, d_{nit}$ and $\mathbf{x}_{it}$. This is impractical, however, if n is so large that $(n + \dim(\boldsymbol{\beta}))$ exceeds software restrictions on the maximum number of regressors. It is computationally quicker to work with specialized software that solves (9.31) that is of smaller dimension $\dim(\boldsymbol{\beta})$.

9.4.3 Poisson Conditional Maximum Likelihood

The *conditional maximum likelihood* approach of Andersen (1970) performs inference conditional on the sufficient statistics for $\alpha_1, \ldots, \alpha_n$, which for LEF densities such as the Poisson are the individual-specific totals $T\bar{y}_i = \sum_{t=1}^{T} y_{it}$. Consistent estimation of $\boldsymbol{\beta}$ is possible if the conditional log-likelihood does not depend on the α_i.

Palmgren (1981) and Hausman, Hall, and Griliches (1984) show that this is the case for the Poisson fixed effects model. Furthermore the estimator then coincides with the estimators of sections 9.4.1 and 9.4.2. This conditional ML method provided the original derivation of the Poisson fixed effects estimator.

In section 9.11.1 it is shown that for y_{it} iid $\mathsf{P}[\mu_{it}]$, the conditional joint density for the ith observation is

$$\Pr\left[y_{i1},\ldots,y_{iT}\left|\sum_{t=1}^{T} y_{it}\right.\right] = \frac{(\sum_t y_{it})!}{\prod_t y_{it}!} \times \prod_t \left(\frac{\mu_{it}}{\sum_s \mu_{is}}\right)^{y_{it}}. \quad (9.39)$$

This is a multinomial distribution, with probabilities $p_{it} = \mu_{it}/\sum_s \mu_{is}$, $t = 1,\ldots,T$, which has already been used in Chapter 8.8.

Models with multiplicative effects set $\mu_{it} = \alpha_i \lambda_{it}$. This has the advantage that simplification occurs because α_i cancels in the ratio $\mu_{it}/\sum_s \mu_{is}$. Then (9.39) becomes

$$\Pr\left[y_{i1},\ldots,y_{iT}\left|\sum_{t=1}^{T} y_{it}\right.\right] = \frac{(\sum_t y_{it})!}{\prod_t y_{it}!} \times \prod_t \left(\frac{\lambda_{it}}{\sum_s \lambda_{is}}\right)^{y_{it}}. \quad (9.40)$$

Because $y_{i1},\ldots,y_{iT}|\sum_t y_{it}$ is multinomial distributed with probabilities $p_{i1},\ldots,p_{iT}$, where $p_{it} = \lambda_{it}/\sum_s \lambda_{is}$, it follows that y_{it} has mean $p_{it}\sum_s y_{is}$. Given (9.9) this implies that we are essentially estimating the fixed effects α_i by $\sum_s y_{is}/\sum_s \lambda_{is}$.

In the special case $\lambda_{it} = \exp(\mathbf{x}_{it}'\boldsymbol{\beta})$ this becomes

$$\Pr\left[y_{i1},\ldots,y_{iT}\left|\sum_{t=1}^{T} y_{it}\right.\right] = \frac{(\sum_t y_{it})!}{\prod_t y_{it}!} \times \prod_t \left(\frac{\exp(\mathbf{x}_{it}'\boldsymbol{\beta})}{\sum_s \exp(\mathbf{x}_{is}'\boldsymbol{\beta})}\right)^{y_{it}}. \quad (9.41)$$

The *conditional* MLE of the Poisson fixed effects model $\hat{\boldsymbol{\beta}}_{\mathsf{PFE}}$ therefore maximizes the conditional log-likelihood function,

$$\mathcal{L}_c(\boldsymbol{\beta}) = \sum_{i=1}^{n}\left[\ln\left(\sum_{t=1}^{T} y_{it}\right)! - \sum_{t=1}^{T}\ln(y_{it}!) + \sum_{t=1}^{T} y_{it}\ln\left(\frac{\exp(\mathbf{x}_{it}'\boldsymbol{\beta})}{\sum_{s=1}^{T}\exp(\mathbf{x}_{is}'\boldsymbol{\beta})}\right)\right]. \quad (9.42)$$

This log-likelihood function is similar to that of the multinomial logit model, except that y_{it} is not restricted to taking only values zero or one and to sum over t to unity. More importantly, it is proportional to the natural logarithm of $\mathsf{L}_{\mathsf{conc}}(\boldsymbol{\beta})$ defined in (9.38), and therefore here the conditional MLE equals the MLE.

Differentiation of (9.42) with respect to $\boldsymbol{\beta}$ yields first-order conditions for $\hat{\boldsymbol{\beta}}_{\mathsf{PFE}}$ that can be reexpressed as (9.31); see, for example, Blundell, Griffith, and Windmeijer (2000). The distribution of the resulting estimator can be obtained using standard ML results, but as previously explained the panel-robust estimate given in (9.34) should instead be used.

McCullagh and Nelder (1989, section 7.2) consider the conditional ML method in a quite general setting. Diggle et al. (2002) specialize to GLMs with canonical link function, see Chapter 2.4.4, in which case we again obtain the multinomial form (9.40). They also consider more general fixed effects in which the conditional mean function is of the form $g(\mathbf{x}'_{it}\boldsymbol{\beta} + \mathbf{d}'_{it}\boldsymbol{\alpha}_i)$ where $\mathbf{d}_{it}$ takes a finite number of values and $\boldsymbol{\alpha}_i$ is now a vector. Hsiao (2003) specializes to binary models and finds that the conditional maximum likelihood approach is tractable for the logit model but not the probit model; that is, the method is tractable for individual intercepts if the canonical link function is used.

9.4.4 Negative Binomial Models

Negative binomial fixed effects models specify the conditional mean function to satisfy (9.13), so the Poisson fixed effects estimator still yields a consistent estimator in short panels. A negative binomial fixed effects estimator has the attraction of potentially more efficient estimation, although the inclusion of individual effects will control for much of the overdispersion.

Hausman, Hall, and Griliches (1984) present a *negative binomial fixed effects* model. Then y_{it} is iid NB1 with parameters $\alpha_i\lambda_{it}$ and ϕ_i, where $\lambda_{it} = \exp(\mathbf{x}'_{it}\boldsymbol{\beta})$, so y_{it} has mean $\alpha_i\lambda_{it}/\phi_i$ and variance $(\alpha_i\lambda_{it}/\phi_i) \times (1 + \alpha_i/\phi_i)$. This negative binomial model is of the less common NB1 form, with the variance a multiple of the mean. The parameter α_i is the individual-specific fixed effect; the parameter ϕ_i is the negative binomial overdispersion parameter that is permitted to vary across individuals. An important restriction of this model is that α_i and ϕ_i can only be identified up to the ratio α_i/ϕ_i.

Estimation is by conditional MLE. Some considerable algebra yields the conditional joint density for the ith observation

$$\Pr\left[y_{i1}, \ldots, y_{iT} \left| \sum_{t=1}^{T} y_{it} \right. \right] = \left(\prod_t \frac{\Gamma(\lambda_{it} + y_{it})}{\Gamma(\lambda_{it})\Gamma(y_{it}+1)} \right) \times \frac{\Gamma\left(\sum_t \lambda_{it}\right)\Gamma\left(\sum_t y_{it} + 1\right)}{\Gamma\left(\sum_t \lambda_{it} + \sum_t y_{it}\right)}, \tag{9.43}$$

which involves $\boldsymbol{\beta}$ through λ_{it} but does not involve the individual effects α_i and ϕ_i. The distribution in (9.43) for integer λ_{it} is the negative hypergeometric distribution. The conditional log-likelihood function follows from this density, and the *negative binomial fixed effects* conditional ML estimator $\hat{\boldsymbol{\beta}}_{\text{NB1FE}}$ is obtained in the usual way.

Allison and Waterman (2002) argue that this model is not strictly speaking a fixed effects model. This is apparent both from setting $\alpha_i/\phi_i = \delta$, because then the conditional mean is simply $\delta\lambda_{it}$ and δ is absorbed in the intercept, and from the fact that the coefficients of time-invariant regressors are actually identified in this model. Allison and Waterman (2002), Greene (2007), and Guimarães (2008) suggest various alternatives, but they have an incidental parameters problem. The simplest approach is to jointly estimate $\boldsymbol{\beta}$, $\alpha_1, \ldots, \alpha_n$ and the overdispersion model in a standard NB2 model with a full set of individual-specific dummy variables. Greene (2004), for example, presents a practical

Table 9.1. *Patents awarded: Actual frequency distribution*

Count	0	1–5	6–10	11–20	21–50	51–100	101–200	201–515
Frequency	337	565	139	140	223	146	107	73
Relative frequency	0.195	0.327	0.080	0.081	0.129	0.084	0.062	0.042

computation method when n is very large so that there are many α_i to estimate. Simulations by Allison and Waterman (2002) and Greene (2004) suggest that the resultant incidental parameters bias in $\boldsymbol{\beta}$ may not be too large for moderately small T.

9.4.5 Example: Patents

Many longitudinal count data studies, beginning with Hausman, Hall, and Griliches (1984), consider the relationship between past research and development (R&D) expenditures and the number of patents y_{it} awarded to the ith firm in the tth year, using data in a short panel. Here we consider data used by Hall, Griliches, and Hausman (1986) on 346 firms for the five years 1975 through 1979.

The range of the dependent variable *PAT* is summarized in Table 9.1. The counts take a wide range of values with 20% of the counts being zero, whereas 20% exceed 50. The mean number of patents is 34.7 with variance $70.9^2 = 5{,}023$, so the data are highly overdispersed. The between standard deviation of 69.8 is much larger than the within standard deviation of 12.6. The five-year average number of patents per firm varies greatly, with lower quartile 1.4, median 5, and upper quartile of 33. Most of the overdispersion is across firms, with some patenting little and some a lot, rather than over time. It is possible that inclusion of individual effects can greatly reduce overdispersion.

The regressors of interest are $\ln R_0, \ldots, \ln R_{-5}$, the logarithm of current and up to five past years' research and development expenditures. Given the logarithmic transformation of R and the exponential conditional mean, the coefficient of $\ln R$ is an elasticity, so that the coefficients of $\ln R_0, \ldots, \ln R_{-5}$ should sum to unity if a doubling of R&D expenditures leads to a doubling of patents. To control for firm-specific effects, the estimated model includes two time-invariant regressors: ln*SIZE*, the logarithm of firm book value in 1972, which is a measure of firm size, and *DSCI*, an indicator variable equal to 1 if the firm is in the science sector. Time dummies are included to control for the decline in patents over time, from an average of 36.9 in 1975 to 32.1 in 1979.

Regression results are given in Table 9.2. The reported panel-robust standard errors correct for both clustering and the considerable overdispersion in the data. The first two estimates are from pooled regression. The pooled Poisson QML estimates treat the data as one long cross-section, with y_{it} having the conditional mean $\exp(\alpha + \mathbf{x}_{it}'\boldsymbol{\beta})$. The pooled Poisson GEE estimates, see section 9.3.2, are based on a working variance matrix that assumes equicorrelation and overdispersion that is a multiple of the mean. The panel-robust standard errors

Table 9.2. *Patents: Pooled and fixed effects estimates*

	Pooled Poisson QMLE		Pooled GEE Poisson		Poisson fixed effects		NB1 fixed effects CMLE	
Variable	Coefficient	Standard error	Coefficient	Standard error	Coefficient	Standard error	Coefficient	Standard error
$\ln R_0$	0.135	0.183	0.316	0.062	0.322	0.081	0.273	0.080
$\ln R_{-1}$	−0.053	0.106	−0.052	0.060	−0.087	0.071	−0.098	0.080
$\ln R_{-2}$	0.008	0.093	0.105	0.054	0.079	0.062	0.032	0.060
$\ln R_{-3}$	0.066	0.114	0.020	0.067	0.001	0.078	−0.020	0.072
$\ln R_{-4}$	0.090	0.093	0.023	0.054	−0.005	0.064	0.016	0.063
$\ln R_{-5}$	0.240	0.123	0.049	0.054	0.003	0.076	−0.010	0.066
ln *SIZE*	0.253	0.059	0.270	0.057			0.207	0.103
DSCI	0.454	0.167	0.440	0.175			0.018	0.326
Sum ln *R*	0.486	0.075	0.460	0.070	0.313	0.143	0.193	0.114

Note: All standard errors are panel-robust standard errors that cluster on the individual. All models include an intercept and four time dummies for the years 1976–1979.

for the time-varying regressors are smaller than those for the Poisson QMLE, so the expected efficiency gains occur here. The only statistically significant regressors at level 0.05 are $\ln R_0$, ln*SIZE*, *DSCI*, and the year dummies, not reported in the table, that for the first estimator take values of, respectively, −0.04, −0.05, −0.17, and −0.20. For the two pooled estimators the coefficients of current and lagged R&D expenditures, $\ln R_{-j}$, sum to 0.486 and 0.460, respectively. There is a statistically significant positive effect of R&D, but the elasticity is considerably less than 1 and statistically so, at conventional levels of significance.

One possible explanation is that this result is an artifact of the failure to control adequately for firm-specific effects. However, the sum of R&D coefficients for the Poisson fixed effects estimator is 0.313, which is even further away from one. The standard errors for this estimator that uses only within variation are substantially larger than those for the pooled estimators, and are 50%–100% larger than the default standard errors, not reported, that fail to control for any overdispersion and correlation over time remaining after the inclusion of fixed effects. Poisson coefficients of time-invariant regressors are not identified in the Poisson fixed effects model, so are missing in the table. The negative binomial fixed effects estimates are similar to those for the Poisson fixed effects, with a very slight improvement in precision. As discussed in section 9.4.4, the coefficients of time-invariant regressors are, surprisingly, estimable in this model.

The various longitudinal estimators imply that in the long run a doubling of R&D expenditures leads to much less than a doubling in the number of patents. Qualitatively similar results have been found with other data sets and estimators, leading to a large literature on alternative estimators that may lead to results closer to a priori beliefs.

9.5 RANDOM EFFECTS MODELS

The simplest random effects model for count data is the *Poisson random effects* model. This model is given by (9.8); that is, y_{it} conditional on α_i and λ_{it} is iid Poisson ($\mu_{it} = \alpha_i \lambda_{it}$), and λ_{it} is a function of $\mathbf{x}_{it}$ and parameters $\boldsymbol{\beta}$. But in a departure from the fixed effects model, the α_i are iid random variables.

One approach is to specify the density $f(\alpha_i)$ of α_i, and then integrate out α_i to obtain the joint density of $y_{i1}, \ldots, y_{iT}$ conditional on just $\lambda_{i1}, \ldots, \lambda_{iT}$. Then

$$\begin{aligned} \Pr[y_{i1}, \ldots, y_{iT}] &= \int_0^\infty \Pr[y_{i1}, \ldots, y_{iT} | \alpha_i] \, f(\alpha_i) d\alpha_i \\ &= \int_0^\infty \left[\prod_t \Pr[y_{it} | \alpha_i] \right] f(\alpha_i) d\alpha_i, \end{aligned} \tag{9.44}$$

where, for notational simplicity, dependence on $\lambda_{i1}, \ldots, \lambda_{iT}$ is suppressed as in the fixed effects case. This one-dimensional integral appears similar to those in Chapter 4, except that here there is only one draw of α_i for the T random variables $y_{i1}, \ldots, y_{iT}$, so that this integral does not equal the product $\prod_t \left[\int_0^\infty \Pr[y_{it} | \alpha_i] \, f(\alpha_i) d\alpha_i \right]$ of mixtures considered in Chapter 4.

Different distributions for α_i lead to different distributions for $y_{i1}, \ldots, y_{iT}$. Analytical results can be obtained as they would be obtained in a similar Bayesian setting, by choosing $f(\alpha_i)$ to be conjugate to $\prod_t \Pr[y_{it} | \alpha_i]$. Conjugate densities exist for Poisson and NB2.

These conjugate densities are not the normal. Nonetheless there is considerable interest in results if $f(\alpha_i)$ is the normal density, because if results can be obtained for scalar α_i then they can be extended to random effects in slope coefficients. A number of methods have been proposed.

Another solution if analytical results for the distribution are not available is to use the population-averaged model and estimator, presented in section 9.3.2. Fixed and random effects panel count models that model omitted factors via individual-specific effects have relatively greater credibility than the PA model. Recent developments have affected random effects panel models more than fixed effect models, in part because computational advances have made them more accessible.

9.5.1 Conjugate-Distributed Random Effects

The gamma density is conjugate to the Poisson. In the pure cross-section case a Poisson-gamma mixture leads to the negative binomial; see Chapter 4.2.2. A similar result is obtained in the longitudinal setting. In section 9.11.2 it is shown that for y_{it} iid $\mathsf{P}[\alpha_i \lambda_{it}]$, where α_i is iid gamma(δ, δ) so that $\mathsf{E}[\alpha_i] = 1$ and $\mathsf{V}[\alpha_i] = 1/\delta$, integration with respect to α_i leads to the joint density for the ith individual

$$\Pr[y_{i1},\ldots,y_{iT}] = \left[\prod_t \frac{\lambda_{it}{}^{y_{it}}}{y_{it}!}\right] \times \left(\frac{\delta}{\sum_t \lambda_{it}+\delta}\right)^{\delta} \tag{9.45}$$

$$\times \left(\sum_t \lambda_{it}+\delta\right)^{-\sum_t y_{it}} \frac{\Gamma\left(\sum_t y_{it}+\delta\right)}{\Gamma(\delta)}.$$

This is the density of the *Poisson random effects model* (with gamma-distributed random effects). For this distribution $\mathsf{E}[y_{it}] = \lambda_{it}$ and $\mathsf{V}[y_{it}] = \lambda_{it} + \lambda_{it}^2/\delta$ so that overdispersion is of the NB2 form.

Maximum likelihood estimation of $\boldsymbol{\beta}$ and δ is straightforward. For $\lambda_{it} = \exp(\mathbf{x}_{it}'\boldsymbol{\beta})$, where regressors $\mathbf{x}_{it}$ include an intercept, the first-order conditions for $\hat{\boldsymbol{\beta}}_{\mathsf{PRE}}$ can be expressed as

$$\sum_{i=1}^{n}\sum_{t=1}^{T} \mathbf{x}_{it}\left(y_{it} - \lambda_{it}\frac{\bar{y}_i + \delta/T}{\bar{\lambda}_i + \delta/T}\right) = \mathbf{0}; \tag{9.46}$$

see exercise 9.3. Thus this estimator, like the Poisson fixed effects estimator, can be interpreted as being based on a transformation of y_{it} to eliminate the individual effects, and consistency essentially requires correct specification of the conditional mean of y_{it}. As for NB2 in the cross-section case the information matrix is block-diagonal, and the first-order conditions for δ are quite complicated.

Hausman et al. (1984) propose this model and additionally consider the negative binomial case. Then y_{it} is iid NB2 with parameters $\alpha_i\lambda_{it}$ and ϕ_i, where $\lambda_{it} = \exp(\mathbf{x}_{it}'\boldsymbol{\beta})$, and hence y_{it} has mean $\alpha_i\lambda_{it}/\phi_i$ and variance $\left(\alpha_i\lambda_{it}/\phi_i\right) \times (1+\alpha_i/\phi_i)$. It is assumed that $(1+\ \alpha_i/\phi_i)^{-1}$ is a beta-distributed random variable with parameters (a, b).

Hausman et al. (1984) show after considerable algebra that the *negative binomial random effects* model (with beta distributed random effects) has joint density for the ith individual

$$\Pr[y_{i1},\ldots,y_{iT}] = \left(\prod_t \frac{\Gamma(\lambda_{it}+y_{it})!}{\Gamma(\lambda_{it})!\Gamma(y_{it}+1)!}\right) \tag{9.47}$$

$$\times \frac{\Gamma(a+b)\,\Gamma\left(a+\sum_t \lambda_{it}\right)\Gamma\left(b+\sum_t y_{it}\right)}{\Gamma(a)\,\Gamma(b)\,\Gamma\left(a+b+\sum_t \lambda_{it}+\sum_t y_{it}\right)}.$$

This is the basis for maximum likelihood estimation of $\boldsymbol{\beta}$, a, and b.

In the linear model the total sample variability is split into between-group variability and within-group variability, where variability is measured by sums of squares. Hausman et al. attempt a similar decomposition for count data models, where sample variability is measured by the log-likelihood function. For the Poisson model with gamma-distributed random effects, the log-likelihood of the iid $\mathsf{P}[\lambda_{it}]$ can be decomposed as the sum of the conditional (on $\sum_t y_{it}$) log-likelihood and the marginal (for $\sum_t y_{it}$) log-likelihood. The conditional log-likelihood is naturally interpreted as measuring within variation; the marginal log-likelihood can be interpreted as measuring between variation, although it depends on $\sum_t \lambda_{it}$, which depends on $\boldsymbol{\beta}$, rather than on $\bar{\mathbf{x}}_i$ alone. A similar decomposition for negative binomial is not as neat.

9.5.2 Gaussian Random Effects

An alternative random effects model is to allow the random effects to be normally distributed. In these models it is standard to assume an exponential conditional mean function. Thus for the Poisson model the data y_{it} are assumed to be iid $\mathsf{P}[\exp(\delta_i + \mathbf{x}_{it}'\boldsymbol{\beta})]$, where the random effect δ_i is iid $\mathsf{N}[0, \sigma_\delta^2]$ and regressors $\mathbf{x}_{it}$ here include an intercept. From (9.9) this model can be rewritten as $y_{it} \sim \mathsf{P}[\alpha_i \exp(\mathbf{x}_{it}'\boldsymbol{\beta})]$, where $\alpha_i = \exp \delta_i$; it is therefore model (9.44) where α_i is lognormally distributed with mean $e^{\sigma_\delta^2/2}$ and variance $e^{\sigma_\delta^2/2}(e^{\sigma_\delta^2/2} - 1)$.

There is no analytical expression for the unconditional density (9.44) in this case. One solution (Schall, 1991; McGilchrist, 1994) is to linearize the model and use linear model techniques. A better alternative is to directly compute the unconditional density by numerical integration, using Gaussian quadrature, or by using simulation methods (Fahrmeir and Tutz, 2001, chapter 7). An example, discussed further in Chapter 12.5.3, is given by Chib, Greenberg, and Winkelmann (1998), who use a Markov-chain Monte Carlo scheme to simulate. They apply their methods to epilepsy data from Diggle et al. (2002), patent data from Hall et al. (1986), and German work absence data.

9.5.3 Random Coefficients Models

The *random coefficients model* with an exponential conditional mean sets $\lambda_{it} = \exp(\delta_i + \mathbf{x}_{it}'\boldsymbol{\beta}_i)$, where both intercepts δ_i and slopes $\boldsymbol{\beta}_i$ are normally distributed, independent of the regressors $\mathbf{x}_{it}$ that here exclude an intercept. In practice only a subset of the slopes are treated as random, to speed up computation if numerical integration methods are used.

Given $\delta_i \sim \mathsf{N}[\delta, \sigma_\delta^2]$ and $\boldsymbol{\beta}_i \sim \mathsf{N}[\mathbf{0}, \boldsymbol{\Sigma}_\beta]$, we have $\delta_i + \mathbf{x}_{it}'\boldsymbol{\beta}_i \sim \mathsf{N}[\delta + \mathbf{x}_{it}'\boldsymbol{\beta}, \sigma_\delta^2 + \mathbf{x}_{it}'\boldsymbol{\Sigma}_\beta \mathbf{x}_{it}]$. It follows that

$$\mathsf{E}[y_{it}|\mathbf{x}_{it}] = \exp(\delta + \mathbf{x}_{it}'\boldsymbol{\beta}) \times \exp((\sigma_\delta^2 + \mathbf{x}_{it}'\boldsymbol{\Sigma}_\beta \mathbf{x}_{it})/2). \tag{9.48}$$

So, unlike for the linear model, the conditional mean for the random slopes model differs from that for the pooled and random effects models, making model comparison and interpretation more difficult.

This model falls in the class of generalized linear latent and mixed models; see Skrondal and Rabe-Hesketh (2004). The likelihood now involves multi-dimensional integrals, rather than the univariate integral in the case of only a random intercept. Numerical methods to approximate the MLE include

Gauss–Hermite quadrature, Laplace approximations, penalized quasilikelihood, and Markov chain Monte Carlo.

9.5.4 Conditionally Correlated Random Effects

The standard random effects panel model assumes that α_i and $\mathbf{x}_{it}$ are uncorrelated. Instead we can relax this and assume that they are conditionally correlated. This idea, originally developed in the context of a linear panel model by Mundlak (1978) and Chamberlain (1982), can be interpreted as intermediate between fixed and random effects. That is, if the correlation between α_i and the regressors can be controlled by adding some suitable sufficient statistic for the regressors, then the remaining unobserved heterogeneity can be treated as random and uncorrelated with the regressors. Although in principle we may introduce a subset of regressors, in practice it is more parsimonious to introduce time-averaged values of time-varying regressors.

The conditionally correlated random (CCR) effects model specifies that α_i in (9.8) can be modeled as

$$\alpha_i = \exp(\bar{\mathbf{x}}_i'\boldsymbol{\lambda} + \varepsilon_i), \tag{9.49}$$

where $\bar{\mathbf{x}}_i$ denotes the time average of the time-varying exogenous variables and ε_i may be interpreted as unobserved heterogeneity that is uncorrelated with the regressors. Substituting (9.49) into (9.8) yields

$$\mathsf{E}[y_{it}|\mathbf{x}_{i1}, \ldots, \mathbf{x}_{iT}, \alpha_i] = \exp(\mathbf{x}_{it}'\boldsymbol{\beta} + \bar{\mathbf{x}}_i'\boldsymbol{\lambda} + \varepsilon_i). \tag{9.50}$$

This can be estimated using standard random effects model software, with $\bar{\mathbf{x}}_i$ as an additional regressor. For example, if $\exp(\varepsilon_i)$ is gamma distributed then we can use the Poisson-gamma random effects model. Alternatively, pooled Poisson or GEE estimation can be performed, again with $\bar{\mathbf{x}}_i$ as an additional regressor. This formulation may also be used when dynamics are present in the model.

Because the CCR formulation is intermediate between the FE and RE models, it may serve as a useful substitute for not being able to deal with fixed effects in some specifications. For example, a panel version of the hurdle model with fixed effects is rarely used because the fixed effects cannot be easily eliminated. In such a case the CCR specification is feasible.

9.5.5 Example: Patents (Continued)

Table 9.3 provides estimates of several different random effects estimators. Panel-robust standard errors are obtained using a cluster bootstrap with clustering on the individual and 400 replications or using a cluster jackknife.

The Poisson-gamma and Poisson-normal random effects estimates and standard errors are very close to each other. The NB2 random effects estimates are estimated somewhat more precisely, and the log-likelihood is considerably higher. These random effects model estimates are closest to the Poisson GEE

Table 9.3. *Patents: Random effects estimates*

	Poisson RE (gamma effects)		Poisson RE (normal effects)		NB2 RE (beta effects)		Poisson CCRE (gamma effects)	
Variable	Coefficient	Standard error	Coefficient	Standard error	Coefficient	Standard error	Coefficient	Standard error
$\ln R_0$	0.404	0.081	0.415	0.078	0.350	0.072	0.322	0.090
$\ln R_{-1}$	−0.046	0.077	−0.040	0.074	−0.003	0.072	−0.087	0.076
$\ln R_{-2}$	0.108	0.064	0.112	0.065	0.105	0.058	0.079	0.065
$\ln R_{-3}$	0.030	0.084	0.035	0.089	0.016	0.077	0.000	0.082
$\ln R_{-4}$	0.011	0.067	0.013	0.067	0.036	0.059	−0.005	0.070
$\ln R_{-5}$	0.041	0.076	0.047	0.080	0.072	0.061	0.002	0.080
ln *SIZE*	0.292	0.077	0.292	0.083	0.162	0.054	0.062	0.060
DSCI	0.257	0.136	0.444	0.153	0.118	0.139	−0.049	0.121
Sum ln *R*	0.546	0.094	0.582	0.098	0.577	0.070	0.312	0.145
lnL	−5,234		−5,245		−4,949		−5,212	

Note: Poisson CCRE is conditionally correlated random effects. All standard errors are panel-robust standard errors that cluster on the individual. All models include four time dummies for the years 1976–1979, and random effects variance parameters that are not reported.

estimates in Table 9.2, though GEE has smaller standard errors in this application. The sum of current and lagged R&D is now higher, ranging from 0.55 to 0.58.

The final estimates given in Table 9.3 are for the Poisson conditionally correlated effects model. It additionally includes as regressors the individual averages over the five years of $\ln R_0, \ldots, \ln R_{-5}$, whose coefficients are omitted from the table. The estimates and standard errors are remarkably close to those for the Poisson fixed effects estimator given in Table 9.2.

9.6 DISCUSSION

Moment-based methods can more generally be used for all models with multiplicative individual effects as in (9.10). The fixed effects model can be generalized from linear models to count data models. The conditional maximum likelihood approach leads to tractable results for some count models – for example, for the Poisson (9.39) that simplifies to (9.40) – but not for all count data models. The random effects model can also be used in a wide range of settings. The maximum likelihood approach is generally computationally difficult, unless a model with conjugate density for the random effects, such as Poisson-gamma, is used.

9.6.1 Fixed versus Random Effects

The strengths and weaknesses of fixed effects versus random effects models in the linear case carry over to nonlinear models. For the linear model a considerable literature exists on the difference between fixed and random effects; see, in particular, Mundlak (1978) and a summary by Hsiao (2003). The random effects model is appropriate if the sample is drawn from a population and one wants to do inference on the population; the fixed effects model is appropriate if one wishes to confine oneself to explaining the sample. The random effects model more easily accommodates random slope parameters as well as random intercepts. For the fixed effects model, coefficients of time-invariant regressors are absorbed into the individual-specific effect α_i and are not identified. For the random effects model, coefficient estimates are inconsistent if the random effects are correlated with regressors. A test of correlation with regressors is presented in section 9.7.1.

We have focused on individual fixed effects in a short panel. Time-specific effects can additionally be included to form a two-way fixed effects error-component model. This can be estimated using conditional maximum likelihood as outlined previously, where the regressors $\mathbf{x}_{it}$ include time dummies. The results can clearly be modified to apply to a long panel with few individuals. Conditional maximum likelihood would then condition on $\sum_i y_{it}$, where, for example, y_{it} is iid $\mathsf{P}[\alpha_t \lambda_{it}]$.

9.6.2 Applications

References to count applications outside economics are given in Diggle et al. (2002) and Fahrmeir and Tutz (2001). Many of these applications use random effects models with Gaussian effects or use the generalized estimating equations approach of Liang and Zeger (1986). Here we focus on economics applications, which generally use the fixed effects models or random effects models with conjugate density for the random effects.

The paper by Hausman, Hall, and Griliches (1984) includes a substantial application to the number of patents for 121 U.S. firms observed from 1968 through 1975. This paper estimates Poisson and NB models with both fixed and random effects. Other studies using patent data are discussed in section 9.8.4.

Ruser (1991) studies the number of workdays lost at 2,788 manufacturing establishments from 1979 through 1984. He uses the NB fixed effects estimator and finds that the number of workdays lost increases as workers' compensation benefits increase, with most of the effect occurring in smaller establishments whose workers' compensation insurance premiums are less experience-rated.

Blonigen (1997) applies the NB2 random effects model to data on the number of Japanese acquisitions in the United States across 365 three-digit Standard Industry Classification industries from 1975 through 1992. The paper finds that if the U.S. dollar is weak relative to the Japanese yen, Japanese acquisitions increase in industries that are more likely to involve firm-specific assets, notably high R&D manufacturing sectors, which can generate a return in yen without involving a currency transaction.

In a novel application, Page (1995) applies the Poisson fixed effects model to data on the number of housing units shown by housing agents to each of two paired auditors, where the two auditors are as much as possible identical except that one auditor is from a minority group and the other is not. Specifically black/white pairs and Hispanic/Anglo pairs are considered. Here the subscript i refers to a specific auditor pair, $i = 1, \ldots, n$; subscript $t = 1, 2$ refers to whether the auditor is minority (say $t = 1$) or nonminority (say $t = 2$). A simple model without covariates is that $\mathsf{E}[y_{it}] = \alpha_i \exp(\beta d_{it})$, where $d_{it} = 1$ if minority and equals 0 otherwise. Then $\exp(\beta)$ equals the ratio of population-mean housing units shown to minority auditors to those shown to nonminority auditors, and $\exp(\beta) < 1$ indicates discrimination is present. Page shows that in this case the Poisson fixed effects conditional MLE has explicit solution $\exp(\hat{\beta}) = \bar{y}_1/\bar{y}_2$. For the data studied by Page (1995) $\exp(\hat{\beta})$ lies between 0.82 and 0.91, with robust standard errors using (9.34) of between 0.022 and 0.028. Thus discrimination is present. Further analysis includes regressors that might explain the aggregate difference in the number of housing units shown.

Van Duijn and Böckenholt (1995) analyze the number of spelling errors by 721 first-grade pupils on each of four dictation tests. They consider a Poisson-gamma mixture model that leads to a conditional multinomial distribution. It does not adequately model overdispersion, so they consider a finite mixtures version of this model, using the methods of Chapter 4.8. On the basis of chi-square goodness-of-fit tests, they prefer a model with two classes, essentially good spellers and poor spellers.

Pinquet (1997) uses estimates of individual effects from longitudinal models of the number and severity of insurance claims to determine "bonus-malus" coefficients used in experience-rated insurance. In addition to an application to an unbalanced panel of more than 100,000 policyholders, the paper gives considerable discussion of discrimination between true and apparent contagion. A range of models, including the random effects model of section 9.5, is considered.

9.7 SPECIFICATION TESTS

If fixed effects are present then random effects estimators are inconsistent. For those count models, notably Poisson and NB1, for which consistent estimation of $\boldsymbol{\beta}$ is possible in the presence of fixed effects, the need for fixed effects estimation can be tested using a Hausman test, presented in section 9.7.1.

If there is correlation over time for a given individual, then the estimators presented in sections 9.3 to 9.5 remain consistent if regressors are strictly exogenous, but there may be efficiency gains in modeling the correlation. Tests for correlation are given in section 9.7.2.

9.7.1 Fixed versus Random Effects

The random effects estimator assumes that α_i in (9.8) is iid distributed, which in particular implies that the random effects are uncorrelated with the regressors. Thus it is assumed that individual-specific unobservables are uncorrelated with individual-specific observables, which is a strong assumption. The fixed effects model makes no such assumption, and α_i could be determined by individual-specific time-invariant regressors.

If the random effects model is correctly specified, then both fixed and random effects estimators are consistent, whereas if the random effects are correlated with regressors the random effects estimator loses its consistency. The difference between the two estimators can therefore be used as the basis for a Hausman test, introduced in Chapter 5.6.6, based on

$$\mathsf{T_H} = (\hat{\boldsymbol{\beta}}_{\mathsf{RE}} - \widetilde{\boldsymbol{\beta}}_{\mathsf{FE}})' \left[\hat{\mathsf{V}}[\widetilde{\boldsymbol{\beta}}_{\mathsf{FE}} - \hat{\boldsymbol{\beta}}_{\mathsf{RE}}]\right]^{-1} (\hat{\boldsymbol{\beta}}_{\mathsf{RE}} - \widetilde{\boldsymbol{\beta}}_{\mathsf{FE}}). \tag{9.51}$$

If $\mathsf{T_H} < \chi^2_\alpha(\dim(\boldsymbol{\beta}))$ then at significance level α, we do not reject the null hypothesis that the individual-specific effects are uncorrelated with regressors. In that case there is no need for fixed effects estimation.

The challenge is to construct $\hat{\mathsf{V}}[\widetilde{\boldsymbol{\beta}}_{\mathsf{FE}} - \hat{\boldsymbol{\beta}}_{\mathsf{RE}}]$, a consistent estimate of the variance of $(\widetilde{\boldsymbol{\beta}}_{\mathsf{FE}} - \hat{\boldsymbol{\beta}}_{\mathsf{RE}})$. Under the assumption that $\hat{\boldsymbol{\beta}}_{\mathsf{RE}}$ is fully efficient, Hausman (1978) shows that we can use $\hat{\mathsf{V}}[\widetilde{\boldsymbol{\beta}}_{\mathsf{FE}}] - \hat{\mathsf{V}}[\hat{\boldsymbol{\beta}}_{\mathsf{RE}}]$, since then the variance of the difference is the difference of the variances. But modern econometric analysis emphasizes inference under weaker distributional assumptions than those needed for the MLE to be fully efficient. For example, full efficiency of the Poisson-gamma random effects estimator requires that the gamma-distributed α_i soaks up all the overdispersion and correlation in the data. In fact, for the Poisson-gamma example in Table 9.3, the reported panel-robust standard errors are 60% larger than default ML standard errors that impose the information matrix equality.

For short panels a panel bootstrap that resamples over individuals can be used. Then

$$\hat{\mathsf{V}}[\widetilde{\boldsymbol{\beta}}_{\mathsf{FE}} - \hat{\boldsymbol{\beta}}_{\mathsf{RE}}] = \frac{1}{B-1}\sum_{b=1}^{B} \left(\widetilde{\boldsymbol{\beta}}^{(b)}_{\mathsf{FE}} - \hat{\boldsymbol{\beta}}^{(b)}_{\mathsf{RE}}\right)\left(\widetilde{\boldsymbol{\beta}}^{(b)}_{\mathsf{FE}} - \hat{\boldsymbol{\beta}}^{(b)}_{\mathsf{RE}}\right)', \tag{9.52}$$

where $\widetilde{\boldsymbol{\beta}}^{(b)}_{\mathsf{FE}}$ and $\hat{\boldsymbol{\beta}}^{(b)}_{\mathsf{RE}}$ are the estimates obtained from the bth of B bootstrap replications. Note that resampling here is over i and not (i, t). Furthermore, care is needed in doing this bootstrap in the presence of fixed effects – if the same observation i appears twice in a bootstrap resample, then $\widetilde{\boldsymbol{\beta}}^{(b)}_{\mathsf{FE}}$ needs to treat the fixed effect α_i as being the same for both observations i. An alternative is a cluster leave-one-out-jackknife that does not have this complication. In practice for short panel count data a Hausman test is applied to a Poisson model, since there is then an estimator $\widetilde{\boldsymbol{\beta}}_{\mathsf{FE}}$ consistent for β, using the bootstrap variance matrix estimate in (9.52).

9.7.2 Tests for Serial Correlation

Tests for serial correlation are considered by Hausman et al. (1984). If individual effects are present, then models that ignore such effects will have residuals that are serially correlated. If this serial correlation disappears after controlling for individual effects, then time-series methods introduced in section 9.8 are not needed. We consider in turn tests for these two situations.

The natural model for the initial analysis of count longitudinal data is Poisson regression of y_{it} on $\mathbf{x}_{it}$, where independence is assumed over both i and t. Residuals from this regression are serially correlated if in fact individual effects α_i are present. Furthermore, the serial correlation between residuals from periods t and s is approximately constant in $(t - s)$, because it is induced by α_i which is constant over time. It is natural to base tests on standardized residuals such as the Pearson residual $\varepsilon_{it} = (y_{it} - \lambda_{it})/\sqrt{\lambda_{it}}$. Then we expect the correlation coefficient between ε_{it} and ε_{is}, estimated as $\sum_i \varepsilon_{it}\varepsilon_{is}/\sqrt{\sum_i \varepsilon_{it}^2}\sqrt{\sum_i \varepsilon_{is}^2}$, to equal zero, $t \neq s$, if individual effects are not present. In practice these correlations are often sufficiently large that a formal test is unnecessary.

If models with individual effects are estimated, the methods yield consistent estimates of $\boldsymbol{\beta}$ but not α_i. Thus residuals $y_{it} - \alpha_i\lambda_{it}$ cannot be readily computed and tested for the lack of serial correlation. For the fixed effects Poisson, $y_{i1}, \ldots, y_{iT} \mid \sum_t y_{it}$ is multinomial-distributed with probability $p_{it} = \lambda_{it}/\sum_s \lambda_{is}$. It follows that y_{it} has mean $p_{it}\sum_s y_{is}$ and variance $p_{it}(1 - p_{it})\sum_s y_{is}$, and the covariance between y_{it} and y_{is} is $-p_{it}p_{is}\sum_s y_{is}$. The residual $u_{it} = \left(y_{it} - p_{it}\sum_s y_{is}\right) / \sqrt{\sum_s y_{is}}$ therefore satisfies $\mathsf{E}[u_{it}^2] = (1 - p_{it})p_{it}$ and $\mathsf{E}[u_{it}u_{is}] = -p_{it}p_{is}$, $t \neq s$. Hausman, Hall, and Griliches (1984) propose a conditional moment test based on these moment conditions, where one of the residuals is dropped because predicted probabilities sum to one.

For dynamic longitudinal model applications discussed in section 9.8, it is necessary to implement tests of serial correlation to ensure estimator consistency. Blundell, Griffith, and Windmeijer (2002) adapt serial correlation tests proposed by Arellano and Bond (1991) for the linear model. Crepon and Duguet (1997a) and Brännäs and Johansson (1996) apply serial correlation tests in the GMM framework.

9.8 DYNAMIC LONGITUDINAL MODELS

The preceding models introduce time variation only through exogenous regressors that may vary over time. These regressors can include functions of time, such as a quadratic in time, and, for short panels, a time dummy for each period. Estimation of a semiparametric model for the conditional mean of form $\gamma(t)\exp(\mathbf{x}'\boldsymbol{\beta})$, with $\gamma(\cdot)$ unspecified, is presented in Wellner and Zhang (2007).

In this section we consider dynamic or transition longitudinal models that allow current realizations of the count y_{it} to depend on past realizations $y_{i,t-k}$, $k > 0$, where $y_{i,t-k}$ is the realization for individual i in period $t - k$.

An example with one lag is

$$\mathsf{E}[y_{it}|\mathbf{X}_i^{(t)}, \mathbf{Y}_i^{(t-1)}, \alpha_i] = \alpha_i\lambda_{it} = \alpha_i \exp(\rho y_{i,t-1} + \mathbf{x}_{it}'\boldsymbol{\beta}), \tag{9.53}$$

where $\mathbf{X}_i^{(t)} = (\mathbf{x}_{it}, \mathbf{x}_{i,t-1}, \ldots, \mathbf{x}_{i1})$ and $\mathbf{Y}_i^{(t-1)} = (y_{i,t-1}, y_{i,t-2}, \ldots, y_{i1})$. This is an autoregressive model like that studied in Chapter 7.5, except here the data are short panel data with asymptotics in $n \to \infty$ rather than $T \to \infty$, and an individual effect α_i may be present.

We consider in turn alternative functional specifications for how y_{it-1} enters λ_{it}, pooled models where $\alpha_i = \alpha$, fixed effects models for α_i, and random effects models for α_i. But first we review qualitatively similar results for the linear model.

9.8.1 Some Approaches

Consider the linear dynamic model with individual specific effect

$$y_{it} = \alpha_i + \rho y_{it,t-1} + \mathbf{x}'_{it}\boldsymbol{\beta} + u_{it}. \tag{9.54}$$

Then y_{it} may be autocorrelated over time through $\rho \neq 0$ (true state dependence), through α_i (unobserved heterogeneity), through autocorrelation in $\mathbf{x}_{it}$ (observed heterogeneity), or through autocorrelation in u_{it}.

In the simplest model $\alpha_i = \alpha$. Then OLS regression of y_{it} on an intercept, $y_{it,t-1}$ and $\mathbf{x}_{it}$ yields consistent parameter estimates, provided the error u_{it} is serially uncorrelated. If instead u_{it} is serially correlated then sufficient further lags of y_{it} need to be included in the regression to ensure u_{it} is serially uncorrelated. Alternatively if u_{it} is MA(q) then $y_{i,t-q-1}$, and later lags, can be used as instrument(s) for $y_{i,t-1}$.

When α_i is a fixed effect then the standard within estimator is inconsistent, as discussed, for example, in Nickell (1981), Hsiao (2003), and Baltagi (2008). In the simplest case with $y_{i,t-1}$ the only regressor, so $\boldsymbol{\beta} = \mathbf{0}$ in (9.54), the mean-differenced model (9.5) is

$$(y_{it} - \bar{y}_i) = \rho(y_{i,t-1} - \bar{y}_{i,-1}) + (u_{it} - \bar{u}_i),$$

where $\bar{y}_{i,-1} = \frac{1}{T-1}\sum_{t=2}^{T} y_{i,t-1}$. OLS estimation for finite T leads to an inconsistent estimate of ρ because the regressor $(y_{i,t-1} - \bar{y}_{i,-1})$ is correlated with $\bar{u}_i$ (to see this, lag the preceding equation by one period) and hence is correlated with the error term $(u_{it} - \bar{u}_i)$. This inconsistency disappears as $T \to \infty$, since then $\bar{u}_i$ is a small component of $(u_{it} - \bar{u}_i)$. For small T, however, there is a problem. Instead estimation can be based on the first-differenced model (9.6) that here is

$$(y_{it} - y_{i,t-1}) = \rho(y_{i,t-1} - y_{i,t-2}) + (u_{it} - u_{i,t-1}).$$

OLS estimation is again inconsistent. A consistent estimate of ρ can be obtained by instrumental variables methods, using for example $(y_{i,t-2} - y_{i,t-3})$ or $y_{i,t-2}$ as an instrument for $(y_{i,t-1} - y_{i,t-2})$, provided u_{it} is serially uncorrelated. If instead u_{it} is serially correlated then the method needs to be adjusted for the addition of sufficient further lags of y_{it} to ensure that u_{it} is serially uncorrelated, or for u_{it} being an MA(q) error. A considerable literature has developed on increasing the efficiency of such moment-based estimators and testing for error correlation; see especially Arellano and Bond (1991).

When α_i is a random effect, a distribution is specified for α_i, estimation is by ML, and consistency depends crucially on assumptions regarding starting values.

9.8.2 Alternative Specifications for Dynamic Models

As noted in Chapters 7.5 and 7.6 there are various ways that lagged dependent variables may be included in the conditional mean. Similar issues arise here in modeling

$$\mu_{it} = \mathsf{E}[y_{it}|\mathbf{X}_i^{(t)}, \mathbf{Y}_i^{(t-1)}, \alpha_i], \tag{9.55}$$

which for individual i conditions on current $\mathbf{x}_{it}$, past values of $\mathbf{x}_{it}$ and y_{it}, and the individual effect α_i. For simplicity we consider models where μ_{it} depends on just the first lag of y_{it}.

The obvious specification is an *exponential feedback model* (EFM) with

$$\mu_{it} = \alpha_i \exp(\rho y_{i,t-1} + \mathbf{x}_{it}'\boldsymbol{\beta}). \tag{9.56}$$

As noted at the beginning of Chapter 7.5, this model is explosive for $\rho > 0$ since then $\rho y_{it-1} \geq 0$ because the count y is nonnegative. Also, this specification introduces potentially sharp discontinuities that may result in a poor fit to the data, especially when the range of counts is very wide as in the patents data application of section 9.4.5. A less explosive model instead uses $\rho \ln y_{i,t-1}$ in (9.56), but then an ad hoc adjustment needs to be used when $y_{i,t-1} = 0$; see Chapter 7.5.1. The dynamic panel literature has emphasized alternative functional forms that we now present.

Blundell, Griffith, and Windmeijer (2002) propose a *linear feedback model* (LFM)

$$\mu_{it} = \rho y_{i,t-1} + \alpha_i \exp(\mathbf{x}_{it}'\boldsymbol{\beta}), \tag{9.57}$$

where the lagged value enters linearly. This formulation avoids awkward discontinuities, and the additive form is that of a Poisson INAR(1) model; see Chapter 7.6.

Crepon and Duguet (1997a) propose the alternative model

$$\mu_{it} = h(y_{i,t-1}, \boldsymbol{\gamma})\alpha_i \exp(\mathbf{x}_{it}'\boldsymbol{\beta}), \tag{9.58}$$

where the function $h(y_{it-1}, \boldsymbol{\gamma})$ parameterizes the dependence on lagged values of y_{it}. A simple example is $h(y_{i,t-1}, \gamma) = \exp(\gamma \mathbf{1}[y_{i,t-1} > 0])$, where $\mathbf{1}[\cdot]$ is the indicator function. More generally a set of dummies determined by ranges taken by y_{it-1} might be specified.

9.8.3 Pooled Dynamic Models

One approach is to ignore the panel nature of the data. Simply assume that all regression coefficients are the same across individuals, so that there are no individual-specific fixed or random effects. Then one can directly apply the time series methods presented in Chapter 7, even for small T provided $n \to \infty$. This approach is given in Diggle et al. (2002, chapter 10), who use autoregressive models that directly include $y_{i,t-k}$ as regressors. The discussion in section 9.8.2 on functional form is relevant here. Also Brännäs (1995a) briefly discusses a generalization of the INAR(1) time-series model to longitudinal data.

For the specifications given in section 9.8.2 a pooled or population-averaged approach considers estimation ignoring individual effects, so $\alpha_i = \alpha$. Under weak exogeneity of regressors, which requires that there is no serial correlation in $(y_{it} - \mu_{it})$, the models can be estimated by nonlinear least squares (or GEE) or by method of moments or GMM estimation based on the sample moment condition $\sum_i \sum_t \mathbf{z}_{it}(y_{it} - \mu_{it})$, where $\mathbf{z}_{it}$ can include $y_{i,t-1}$ and $\mathbf{x}_{it}$ and, if desired, additional lags in these variables.

Pooling may be adequate. In particular, lagged values of the dependent variable might be an excellent control for correlation between y_{it} and lagged y_{it}, so that there is no need to include individual-specific effects. For example, the firm-specific propensity to patent might be adequately controlled for simply by including patents last year as a regressor. As in the linear case, however, estimator consistency requires that sufficient lags in y_{it} are included to ensure there is no remaining serial correlation.

Currently the published literature does not provide detailed comparative information on the performance of the available estimators for dynamic panels. Their development is in early stages, and not surprisingly we are unaware of commercial software to handle such models. However, the relative ease of interpretation favors the EFM over the LFM assumption, and the explosiveness of EFM is not a problem in short panel applications with T small.

9.8.4 Fixed Effects Dynamic Models

The methods to eliminate fixed effects given in section 9.4 require that regressors are strictly exogenous, that is,

$$\mathsf{E}[y_{it}|\mathbf{X}_i^{(T)}] = \mathsf{E}[y_{it}|\mathbf{x}_{iT}, \ldots, \mathbf{x}_{i1}] = \alpha_i \lambda_{it}, \tag{9.59}$$

where usually $\lambda_{it} = \exp(\mathbf{x}_{it}'\boldsymbol{\beta})$. Defining $u_{it} = y_{it} - \alpha_i \lambda_{it}$, (9.59) implies $\mathsf{E}[u_{it}|\mathbf{x}_{iT}, \ldots, \mathbf{x}_{i1}] = 0$ and hence $\mathsf{E}[u_{it}|\mathbf{x}_{is}] = 0$ for all s.

This rules out *predetermined regressors* that may be correlated with past shocks, though are still uncorrelated with current and future shocks. To allow for predetermined regressors, we make the weaker assumption that regressors are *weakly exogenous*, or

$$\mathsf{E}[y_{it}|\mathbf{X}_i^{(t)}] = \mathsf{E}[y_{it}|\mathbf{x}_{it}, \ldots, \mathbf{x}_{i1}] = \alpha_i \lambda_{it}, \tag{9.60}$$

where now conditioning is only on current and past regressors. Then $\mathsf{E}[u_{it}|\mathbf{x}_{is}] = 0$ for $s \leq t$, so future shocks are indeed uncorrelated with current $\mathbf{x}$, but there is no restriction that $\mathsf{E}[u_{it}|\mathbf{x}_{is}] = 0$ for $s > t$.

For dynamic models in which lagged dependent variables also appear as regressors this condition becomes

$$\mathsf{E}[y_{it}|\mathbf{X}_i^{(t)}, \mathbf{Y}_i^{(t-1)}] = \mathsf{E}[y_{it}|\mathbf{x}_{it}, \ldots, \mathbf{x}_{i1}, y_{i,t-1}, \ldots, y_{i1}] = \alpha_i \lambda_{it}, \quad (9.61)$$

where conditioning is now also on past values of y_{it}. For the linear feedback model $\alpha_i \lambda_{it}$ in (9.61) is replaced by $\rho y_{i,t-1} + \alpha_i \lambda_{it}$.

If regressors are predetermined then the Poisson fixed effects estimator is inconsistent. This result is analogous to the inconsistency of the within- or mean-differenced fixed effects estimator in the linear model discussed in section 9.8.1. The problem is that quasidifferencing subtracts $(\lambda_{it}/\bar{\lambda}_i)\bar{y}_i$ from y_{it}, see (9.29), but $\bar{y}_i$ includes future values y_{is}, $s > t$. Condition (9.60) is not enough to ensure $\mathsf{E}[\frac{1}{T}\sum_{t=1}^{T} y_{it}|\mathbf{x}_{iT}, \ldots, \mathbf{x}_{i1}] = \alpha_i \bar{\lambda}_i$.

The solution is to instead use nonlinear instrumental variables or GMM estimation based on one of several possible variations of the first-differencing used for the dynamic linear model. Given (9.60), Chamberlain (1992b) proposes eliminating the fixed effects α_i by the transformation

$$q_{it}(\boldsymbol{\theta}) = \frac{\lambda_{i,t-1}}{\lambda_{it}} y_{it} - y_{i,t-1}, \quad (9.62)$$

where $\lambda_{it} = \lambda_{it}(\boldsymbol{\theta})$. Wooldridge (1997a) instead proposes eliminating the fixed effects using

$$q_{it}(\boldsymbol{\theta}) = \frac{y_{i,t-1}}{\lambda_{i,t-1}} - \frac{y_{it}}{\lambda_{it}}. \quad (9.63)$$

See section 9.11.3 for justification of these moment conditions.

For predetermined regressors it is shown in section 9.11 that there exist instruments $\mathbf{z}_{it}$ that satisfy

$$\mathsf{E}[q_{it}(\boldsymbol{\theta})|\mathbf{z}_{it}] = 0, \quad (9.64)$$

for $q_{it}(\boldsymbol{\theta})$ defined in (9.62) or (9.63), where $\mathbf{z}_{it}$ can be drawn from $\mathbf{x}_{i,t-1}, \mathbf{x}_{i,t-2}, \ldots$ In dynamic models with lags up to $y_{i,t-p}$ appearing as regressors, $\mathbf{z}_{it}$ can be drawn from $\mathbf{y}_{i,t-p-1}, \mathbf{y}_{i,t-p-2}, \ldots$. Often $p = 1$ and the available instruments are $y_{i,t-2}, y_{i,t-3}, \ldots$.

In the just-identified case in which there are as many instruments as parameters, the method of moments estimator solves

$$\sum_{i=1}^{n}\sum_{t=1}^{T} \mathbf{z}_{it} q_{it}(\boldsymbol{\theta}) = \mathbf{0}. \quad (9.65)$$

If there are more instruments $\mathbf{z}_{it}$ than regressors, such as through adding additional lags of regressors into the instrument set (potentially all the way back to $\mathbf{x}_{i1}$), one can consistently estimate $\boldsymbol{\theta}$ by the GMM estimator that minimizes

$$\left(\sum_{i=1}^{n}\sum_{t=1}^{T} \mathbf{z}_{it} q_{it}(\boldsymbol{\theta})\right)' \mathbf{W}_n \left(\sum_{i=1}^{n}\sum_{t=1}^{T} \mathbf{z}_{it} q_{it}(\boldsymbol{\theta})\right). \quad (9.66)$$

The GMM results of section 9.2.3 can be applied directly, with $\mathbf{h}_i(\boldsymbol{\theta}) = \sum_{t=1}^{T} \mathbf{z}_{it} q_{it}(\boldsymbol{\theta})$. One-step GMM typically uses $\mathbf{W}_n = (\mathbf{Z}'\mathbf{Z})^{-1}$, whereas more efficient two-step GMM uses $\mathbf{W}_n = \hat{\mathbf{S}}^{-1}$ where $\hat{\mathbf{S}}$ is defined in (9.18). Another possibility is continuously updated GMM.

Given two-step GMM, model adequacy can be tested using the overidentifying restrictions test given in Chapter 2.5.3. It is also important to test for serial correlation in $q_{it}(\boldsymbol{\theta})$ using $q_{it}(\hat{\boldsymbol{\theta}})$, because correct model specification requires that $\mathsf{Cor}[q_{it}(\boldsymbol{\theta}), q_{is}(\boldsymbol{\theta})] = 0$ for $|t - s| > 1$. These tests are analogous to those following Arellano-Bond estimation of the linear dynamic fixed effects model.

Windmeijer (2008) provides a good survey of GMM methods for the Poisson panel model, including models with regressors that are predetermined, lagged dependent variables, or endogenous. One issue is that two-step GMM estimated coefficients and standard errors can be biased in finite samples. Windmeijer (2008) proposes an extension of the variance matrix estimate of Windmeijer (2005) to nonlinear models and shows in a Monte Carlo exercise with a predetermined regressor that this approach leads to improved finite sample inference, as does the Newey and Windmeijer (2009) method applied to the continuous updating estimator variant of GMM.

Table 9.4 presents a summary of moment conditions and estimating equations for some of the leading panel count data estimators.

Blundell, Griffith, and Windmeijer (2002) propose an alternative transformation, the *mean scaling transformation*,

$$q_{it} = y_{it} - \frac{\overline{y}_{i0}}{\lambda_{i0}} \lambda_{it}, \tag{9.67}$$

where $\overline{y}_{i0}$ is the presample mean value of y_i and the instruments are $(\mathbf{x}_{it} - \mathbf{x}_{i0})$. This transformation leads to inconsistent estimation, but in a simulation this inconsistency is shown to be small, efficiency is considerably improved, and the estimator is especially useful if data on the dependent variable go back farther in time than data on the explanatory variables.

Blundell, Griffith, and Windmeijer (2002) model the U.S. patents data of Hall, Griliches, and Hausman (1986). Application to patents is of particular interest, because for several reasons there are few ways to measure innovation aside from patents and patents are intrinsically a count. R&D expenditures affect patents with a considerable lag, so there is potentially parsimony and elimination of multicollinearity in having patents depend on lagged patents, rather than having a long distributed lag in R&D expenditures. And, as noted in the example earlier, most studies using distributed lags on R&D expenditure find the R&D expenditure elasticity of patents to be much less than unity. Blundell and colleagues pay particular attention to the functional form for dynamics and the time-series implications of various functional forms. In their application up to two lags of patents and three lags of R&D expenditures appear as regressors. The estimates indicate long lags in the response of patents to R&D expenditures.

Table 9.4. *Moment conditions for selected panel count models*

Model	Moment and/or model specification	Estimating Equations
Pooled Poisson	$\mathsf{E}[y_{it}\|\mathbf{x}_{it}] = \mu_{it}$; $\mu_{it} = \exp(\mathbf{x}_{it}'\boldsymbol{\beta})$.	$\sum_i \sum_t \mathbf{x}_{it}\left(y_{it} - \mu_{it}\right) = \mathbf{0}$.
Population averaged	Same as pooled Poisson.	$\sum_i \sum_s \sum_t \hat{\gamma}^{st} \mu_{it}\mathbf{x}_{it}\left(y_{it} - \mu_{it}\right) = \mathbf{0}$; $\hat{\gamma}^{st}$ is $(s,t)^{th}$ element of $(\hat{\mathsf{V}}[\mathbf{y}_i\|\mathbf{x}_i])^{-1}$.
Poisson RE (gamma effects)	$\mathsf{E}[y_{it}\|\alpha_i, \mathbf{x}_{it}] = \alpha_i\lambda_{it}$; $\lambda_{it} = \exp(\mathbf{x}_{it}'\boldsymbol{\beta})$; $\alpha_i \sim \text{Gamma}[1, 1/\delta]$.	$\sum_i \sum_t \mathbf{x}_{it}\left(y_{it} - \lambda_{it}\frac{\bar{y}_i + \delta/T}{\bar{\lambda}_i + \delta/T}\right) = \mathbf{0}$; $\bar{y}_i = \frac{1}{T}\sum_t y_{it}$; $\overline{\mu}_i = \frac{1}{T}\sum_t \exp(\mathbf{x}_{it}'\boldsymbol{\beta})$.
Poisson FE	Same mean as Poisson RE $\text{Cor}[\alpha_i, \mathbf{x}_{it}] \neq \mathbf{0}$ possible.	$\sum_i \sum_t \mathbf{x}_{it}\left(y_{it} - \lambda_{it}\frac{\bar{y}_i}{\bar{\lambda}_i}\right) = \mathbf{0}$; $\overline{\lambda}_i = \frac{1}{T}\sum_t \exp(\mathbf{x}_{it}'\boldsymbol{\beta})$.
Chamberlain GMM	$\mathsf{E}\left[y_{it}\frac{\lambda_{it-1}}{\lambda_{it}} - y_{it-1}\,\|\mathbf{X}_i^{(t-1)}\right] = 0$; $y_{it} = \alpha_i\lambda_{it} + u_{it}$; $\mathsf{E}[u_{it}\|\mathbf{X}_i^{(t-1)}] = 0$.	$\sum_i \sum_t \mathbf{z}_{it}\left(y_{it}\frac{\lambda_{it-1}}{\lambda_{it}} - y_{it-1}\right) = \mathbf{0}$; $\mathbf{z}_{it}$ is subset of $\mathbf{X}_i^{(t-1)}$
Wooldridge GMM	$\mathsf{E}\left[\frac{y_{it}}{\lambda_{it}} - \frac{y_{it-1}}{\lambda_{it-1}}\,\|\mathbf{X}_i^{(t-1)}\right] = 0$; $y_{it} = \alpha_i\lambda_{it}u_{it}$; $\mathsf{E}[u_{it}\|\mathbf{X}_i^{(t-1)}] = 0$.	$\sum_i \sum_t \mathbf{z}_{it}\left(\frac{y_{it}}{\lambda_{it}} - \frac{y_{it-1}}{\lambda_{it-1}}\right) = \mathbf{0}$; $\mathbf{z}_{it}$ is subset of $\mathbf{X}_i^{(t-1)}$.
Dynamic FE (linear feedback)	$\mathsf{E}[y_{it}\|\alpha_i, y_{i,t-1}, \mathbf{x}_{it}]$ $= \rho y_{it-1} + \alpha_i\lambda_{it}$; $\lambda_{it} = \exp(\mathbf{x}_{it}'\boldsymbol{\beta})$	$\sum_i \sum_t \left(\mathbf{z}_{it}(y_{it} - \rho y_{it-1})\frac{\lambda_{it}}{\lambda_{it-1}} - (y_{it-1} - \rho y_{it-2})\right) = \mathbf{0}$; $\mathbf{z}_{it}$ is subset of $\mathbf{X}_i^{(t-1)}, \mathbf{Y}_i^{(t-2)}$.

Note: The Chamberlain and Wooldridge GMM estimators are for predetermined regressors and $X_i^{(t-1)} = (x_{it-1}, x_{it-2}, \ldots)$ and $Y_i^{(t-2)} = (y_{it-2}, y_{it-3}, \ldots)$. Estimating equations for GMM estimators and linear feedback are written for the just-identified case. For overidentified models the estimator minimizes a quadratic form in the expression given in the final column of the table.

Related studies by Blundell, Griffith, and Van Reenan (1995, 1999) model the number of 'technologically significant and commercially important' innovations commercialized by British firms. Dynamics are introduced more simply by including the lagged value of the knowledge stock, an exponentially weighted sum of past innovations.

Montalvo (1997) uses the Chamberlain (1992b) transformation to model the number of licensing agreements by individual Japanese firms and to analyze the Hall et al. (1986) data. Lagged dependent variables do not appear as regressors. Instead Montalvo argues that current R&D expenditures cannot be assumed to be strictly exogenous because patents depend on additional R&D expenditures for their full development. So there is still a need for quasidifferenced estimators.

Crepon and Duguet (1997a) apply GMM methods to French patents data. They also use a relatively simple functional form for dynamics. First, as regressor they use a measure of R&D capital. This capital measure is calculated as the weighted sum of current and past depreciated R&D expenditure and can be viewed as imposing constraints on R&D coefficients in a distributed lag model. Dynamics in patents are introduced by including indicator variables for whether y_{it-1} is in the ranges 1 to 5, 6 to 10, or 11 or more. Particular attention is paid to model specification testing and the impact of increasing the size of the instrument set $\mathbf{z}_{it}$ in (9.66).

In a more applied study, Cincera (1997) includes not only a distributed lag in firm R&D expenditures but also a distributed lag in R&D expenditures by other firms in the same sector to capture spillover effects. Application is to a panel of 181 manufacturing firms from six countries.

Hill, Rothchild, and Cameron (1998) model the monthly incidence of protests using data from 17 Western countries for 35 years. To control for overdispersion and dynamics, they use a negative binomial model with lagged y_{it} appearing as $\ln(y_{i,t-1} + c)$, where c is a constant whose role was explained in Chapter 7.5. Country-specific effects are additionally controlled for by inclusion of country-specific indicator variables. This poses no consistency problems in this example where $T \to \infty$, so the GMM methods of this section are not needed.

9.8.5 Random Effects Dynamic Models

The obvious dynamic version of a random effects model for counts is to use the same models as presented in section 9.5, with lagged y_{it} added as a regressor as in section 9.8.2. In a short panel, however, initial conditions play an important role in the evolution of the outcome, and we may expect the initial condition y_{i0} to be correlated with the individual-specific effect α_i. For a given individual the initial condition is a time-invariant factor that in the preceding fixed effects dynamic models is swept out along with other time-invariant variables. But in a random effects model it is important to control for the initial condition. Heckman (1981) and Wooldridge (2005) propose approaches for correlating α_i and y_{i0}.

Heckman (1981) writes the joint distribution of $y_{i0}, y_{i1}, \ldots, y_{iT}|\mathbf{x}_{it}$ as

$$\begin{aligned}\Pr\left[y_{i0,}y_{i1}, \ldots, y_{iT}|\mathbf{X}_i^{(T)}\right] \\ = \Pr[y_{i1}, \ldots, y_{iT}|\mathbf{X}_i^{(T)}, y_{i0}, \alpha_i] \Pr[y_{i0}|\mathbf{X}_i^{(T)}, \alpha_i] \Pr[\alpha_i|\mathbf{X}_i^{(T)}]. \qquad (9.68)\end{aligned}$$

To implement this requires specification of the functional forms $\Pr[y_{i0}|x_{it}, \alpha_i]$ and $\Pr[\alpha_i|x_{it}]$. Furthermore, numerical integration may be necessary if the convolution does not have a closed-form solution.

Wooldridge (2005) instead proposes a conditional approach based on the decomposition

$$\Pr\left[y_{i0}, y_{i1}, \ldots, y_{iT} | \mathbf{X}_i^{(T)}, y_{i0}\right] = \Pr[y_{i1}, \ldots, y_{iT} | \mathbf{X}_i^{(T)}, y_{i0}, \alpha_i] \Pr[\alpha_i | y_{i0}, \mathbf{X}_i^{(T)}], \tag{9.69}$$

which does away with having to specify the distribution of the initial condition y_{i0}. Under the assumption that the initial conditions are nonrandom, the standard random effects ML approach, with y_{i0} a regressor, identifies the parameters of interest. Wooldridge (2005) analyzes this model for a class of nonlinear dynamic panel models that includes the Poisson model.

Estimation of a dynamic panel model requires additional assumptions about the relationship between the initial observations y_0 and the α_i. This can be easily done using the CCR model (9.49), with y_{i0} added as an extra control. Then

$$\alpha_i = \exp(\delta_0 y_{i0} + \bar{\mathbf{x}}_i' \boldsymbol{\lambda} + \varepsilon_i), \tag{9.70}$$

where $\bar{\mathbf{x}}_i$ denotes the time average of the time-varying exogenous variables, and ε_i is an iid random variable. This specification jointly controls for the initial condition and makes the Mundlak correction. Conditional ML estimation treats the initial condition y_{i0} as given and can be implemented using the same software as for the regular random effects model. The alternative Heckman (1981) approach in (9.68) of taking the initial condition as random, specifying a distribution for y_{i0}, and then integrating out is computationally more demanding; see Stewart (2007).

9.8.6 Example: Patents (Continued)

Table 9.5 reports results for some dynamic count models. The lagged dependent variable PAT_{t-1} is included using the exponential feedback mechanism (9.56). Because this captures in part the longer term dynamic impact of past R&D expenditure, the specification is simplified by including only $\ln R$, $\ln R_{-1}$, and $\ln R_{-2}$. Throughout it is assumed that the models specify sufficient dynamics that the errors $y_{it} - \mu_{it}$ are serially uncorrelated.

The first two estimators are pooled estimators, and there is considerable efficiency gain in using GEE with equicorrelated errors. The third estimator is the usual random effects estimator, which here is a bit less efficient than GEE.

Table 9.6 presents methods that control for richer models of the individual-specific effects. A simple method is to use the correlated random effects approach of section 9.8.5, including as additional regressors *PAT* at initial period 0 and the individual averages over the five years of $\ln R_0, \ldots, \ln R_{-2}$, whose coefficients are omitted from the table. The regressor *PAT_INITIAL* is highly statistically significant.

Table 9.5. *Patents: Dynamic model pooled and random effects estimates*

	Pooled Poisson		Pooled GEE Poisson		Poisson RE (gamma effects)	
Variable	Coefficient	Standard error	Coefficient	Standard error	Coefficient	Standard error
PAT_{-1}	0.0034	0.0006	0.0019	0.0003	0.0013	0.0006
$\ln R_0$	0.359	0.144	0.379	0.066	0.446	0.080
$\ln R_{-1}$	−0.159	0.100	−0.080	0.074	−0.060	0.090
$\ln R_{-2}$	0.133	0.139	0.078	0.059	0.103	0.066
ln *SIZE*	0.183	0.046	0.222	0.045	0.300	0.055
DSCI	0.289	0.135	0.371	0.156	0.281	0.116
Sum ln *R*	0.333	0.062	0.376	0.053	0.490	0.066
lnL	−14717		–		−5188	

Note: All standard errors are panel-robust standard errors that cluster on the individual. All models include four time dummies for years 1976 to 1979, and random effects variance parameters that are not reported.

Alternatively estimation can be by GMM as in section 9.8.4. We use as instruments $\ln R_{-1}$, $\ln R_{-2}$, $\ln R_{-3}$, PAT_{-2}, PAT_{-3}, and PAT_{-4}, so there are two overidentifying restrictions. Because presample data are available in this example, only one year of data is lost, estimation uses four years of data, and just three year dummies are included. The reported estimates are two-step GMM estimates based on the Chamberlain transformation (9.62). The coefficient of PAT_{-1} is similar to that in the other models, whereas the coefficient on

Table 9.6. *Patents: Dynamic model correlated effects estimates*

	Poisson CCRE (gamma effects)		Poisson fixed effects GMM	
Variable	Coefficient	Standard error	Coefficient	Standard error
PAT_{-1}	0.0012	0.0006	−0.0001	0.0015
$\ln R_0$	0.341	0.087	0.300	0.800
$\ln R_{-1}$	−0.104	0.090	−0.068	0.110
$\ln R_{-2}$	0.047	0.067	0.132	0.078
ln *SIZE*	0.045	0.054	–	–
DSCI	−0.040	0.113	–	–
PAT_INITIAL	0.05	0.001	–	–
Sum ln *R*	0.284	0.096	0.365	0.796
lnL	−5157			

Note: Poisson CCRE is conditionally correlated random effects. All standard errors are panel-robust standard errors that cluster on the individual. All models include four time dummies for years 1976 to 1979, and random effects variance parameters that are not reported.

$\ln R_0$ is large and negative, though imprecisely estimated. The overidentifying restrictions test statistic is 4.13 ($N = 1{,}384$ times the GMM objective function minimized value of 0.00298) compared to a $\chi^2(2)$ critical value of 5.99 at level 0.05. The model is rejected at the 5% significance level, though not at the 10% level. A more complete analysis should also check that the residuals from the first-differenced model are not serially correlated beyond the first lag.

PAT_{-1} is statistically significant at level 0.05 in all cases except Poisson fixed effects GMM. The coefficient is slightly greater than zero, so the model is explosive though mildly so over a five-year period.

9.9 ENDOGENOUS REGRESSORS

In the linear case, estimation of models with endogenous regressors by IV and GMM methods is well established, provided valid instruments are available. For nonlinear models, such as count models, several distinct methods have been proposed in the case of cross-section data. They are presented in sections 10.2 to 10.4. Despite a growing theoretical literature on nonparametric estimation of simultaneous equations, the methods currently in use for counts rely greatly on functional form specification, in addition to exclusion restrictions, and are fragile to misspecification.

For panel data the same reliance on functional form specification arises, one that may be further complicated by the need to model correlation across time periods. At the same time, instruments are usually more readily available since lags of exogenous regressors, if excluded from the original model, can become available as instruments.

Here we simply consider moment-based nonlinear instrumental variables or GMM methods, because they are the most commonly used methods for panel count data with endogenous regressors. When there is no individual effect, a pooled approach directly applies the GMM estimators of Chapter 10.4, with inference then based on panel-robust standard errors that cluster on the individual. When there is an individual-specific fixed effect, the methods of section 9.8.4 can be adapted. The conditional mean of y_{it} is specified to be

$$\mu_{it} = \alpha_i \exp(\mathbf{x}_{it}'\boldsymbol{\beta}). \tag{9.71}$$

The problem is that one or more of the regressors in $\mathbf{x}_{it}$ is endogenous, correlated with the error, so that the standard Poisson fixed effects estimator is inconsistent. From Chapter 10.4 the moment conditions to base estimation on vary according to whether the error is defined to be additive or multiplicative. Windmeijer (2000) shows that, in the panel case, the individual-specific fixed effects α_i can be eliminated only if a multiplicative errors specification is assumed and if the Wooldridge transformation is used. Then nonlinear IV or GMM estimation is based on $q_{it}(\boldsymbol{\theta})$ defined in (9.63), where the instruments $\mathbf{z}_{it}$ can be drawn from $\mathbf{x}_{i,t-2}, \mathbf{x}_{i,t-3}, \ldots$.

9.10 MORE FLEXIBLE FUNCTIONAL FORMS FOR LONGITUDINAL DATA

The fully parametric models presented in the preceding sections on panel models have been panel versions of the Poisson or negative binomial. Elsewhere in this book, and especially in Chapters 4 and 11, a case has been made that more flexible functional forms, such as hurdle, zero-inflated, and finite mixtures models, may be more appropriate in a fully parametric framework.

In this section we present some panel versions of these models. The models are population-averaged or random effects models, with the exception of the fixed effects model of Deb and Trivedi (2013) that is also presented.

9.10.1 Finite Mixture Longitudinal Models

Finite mixtures models for cross-section data are presented in section 4.8.

Bago d'Uva (2005) applies a finite mixture version of the negative binomial model to panel data. Class membership is modeled as a multinomial logit model with probabilities π_{ij} (for individual i in class j) that are functions of time-invariant individual characteristics. Within each class the outcome of doctor visits is modeled using a pooled NB2 grouped-data model for panel data on doctor visits that are recorded in the ranges 0, 1–2, 3–5, 6–10, and more than 10 visits, with all coefficients varying with the latent class. As in the nonmixture case, the estimates are consistent if this pooled (or population-averaged) panel data parametric model is correctly specified, though estimates will be inconsistent if fixed effects are present.

Deb and Trivedi (2013) therefore propose a fixed effects Poisson version of the finite mixture model. They assume

$$y_{it}|\mathbf{x}_{it}, \boldsymbol{\lambda}_{jit}, \alpha_{ji} \sim \mathsf{P}[\alpha_{ji}\boldsymbol{\lambda}_{jit}], \; j = 1, 2, \ldots, C, \tag{9.72}$$

where j denotes the latent class, $\lambda_{jit} = \exp(\mathbf{x}_{it}'\boldsymbol{\beta}_j)$, π_j is the probability of being in the jth class, $\sum_{j=1}^{C} \pi_j = 1$, and here π_j does not vary across individuals or over time. For each latent class j the fixed effects can be eliminated as in section 9.4.2 to yield the concentrated likelihood function given in (9.38). Combining across latent classes, the full-data concentrated log-likelihood is

$$\mathsf{L}_{\mathsf{conc}}(\boldsymbol{\beta}, \pi_1, \ldots, \pi_1) \propto \prod_{i=1}^{n}\prod_{t=1}^{T}\sum_{j=1}^{C}\left[\pi_j\left(\frac{\lambda_{jit}}{\sum_{s=1}^{T}\lambda_{jis}}\right)^{y_{jit}}\right]^{d_{ji}}, \tag{9.73}$$

where $d_{ji} = 1$ if individual i is in the jth class and equals zero otherwise.

The log-likelihood can be maximized after adapting the EM algorithm given in Chapter 4.8. Given current estimates, the E-step estimates the unobserved d_{ji} by the current estimate $\hat{z}_{ji}$ of its expected value $\mathsf{E}\left[d_{ji}\right]$, where $\hat{z}_{ji}$ is the posterior probability of the classification of individual i into class j. The M-step of the EM procedure maximizes EL by maximizing with respect to $\boldsymbol{\beta}_j$ and $\pi_1, \ldots, \pi_C$

the log-likelihood function based on (9.73) with d_{ji} replaced by $\hat{z}_{ji}$. For cross-section data, likelihood maximization can use gradient methods based on the marginal likelihood formed using the mixture density $\sum_{j=1}^{C} \pi_j f_j(y_i|\mathbf{x}_i, \boldsymbol{\theta}_j)$, as an alternative to using the EM algorithm. In the panel case, however, use of the EM algorithm becomes essential. For additional details, see Deb and Trivedi (2011), who apply their model to panel data on doctor office visits from six waves of the Health and Retirement Study.

Böckenholt (1999) specifies a finite mixture dynamic panel model, a pooled version of the Poisson INAR(1) presented in Chapter 7.6. Estimation is by the EM algorithm. Application is to panel data on the frequency of weekly detergent purchases by households, with class membership determined by household demographic characteristics.

Applications of finite mixture models to cross-section data necessarily assume that the latent class probabilities are time invariant. With panel data this restriction can be relaxed, and in general the class membership probability π_{jit} may vary across individuals and over time. Using the random effect framework within each latent class, Hyppolite and Trivedi (2012) develop a suite of mixture models of increasing degrees of complexity and flexibility for use in a panel count data setting. The baseline model is a two-component mixture of Poisson distributions with latent classes that are time invariant. The first extension allows the mixing proportions to be smoothly varying continuous functions, such as logit or probit models, of time-varying covariates. The second extension accommodates at least two types of heterogeneity: heterogeneity with respect to class membership and heterogeneity with respect to within-class differences in individual characteristics by introducing additional random effects. A third extension adds time dependence by modeling the class indicator variable as a first-order Markov chain, as is done for the hidden Markov models described in Chapter 7.8. The fourth model introduces dynamics and makes the transition matrix covariate dependent. Hyppolite and Trivedi (2012) use dynamic versions of the mixture model to reanalyze the effect of 1996 health care reform in Germany, using the German Socio-Economic Panel data originally analyzed by Winkelmann (2004).

9.10.2 Hurdles and With-Zeroes Models

Several studies have proposed random effects or pooled finite mixtures versions of hurdle and zero-inflated models.

Olsen and Schafer (2001) estimate a random effects version of the two-part or hurdle model for panel data, where the outcome is continuous rather than a count once the hurdle is crossed. The hurdle probability is modeled as a logit model with random intercept and possibly slopes, the continuous response (or an appropriate transformation such as the natural logarithm) is modeled as a linear model with random intercepts and possibly slopes and a normally distributed error, and the random coefficients from the two parts are assumed to be joint normal and possibly correlated. The authors discuss the computational challenges and favor use of a Laplace approximation to evaluate the integral in the log-likelihood. Application is to longitudinal data on adolescent consumption of alcohol.

Min and Agresti (2005) also consider a panel hurdle model, but one for counts. The hurdle probability is a logit with random intercept, the count response is modeled as Poisson or NB2 with random intercept, and the two intercepts are assumed to be iid bivariate normal distributed. They also estimate a finite mixture version of this hurdle model, as well as zero-inflated Poisson models with random intercepts. Application is to the number of side-effect episodes following drug treatment.

Bago d'Uva (2006) applies a finite mixture version of the hurdle model for counts. Class membership is modeled with probabilities π_j that do not vary with individual characteristics. Within each class the outcome of doctor visits is modeled using the negative binomial hurdle model of Chapter 4.5. The estimates are consistent if the pooled (or population-averaged) panel model is correctly specified, but will be inconsistent if fixed effects are present. Application is to the annual number of outpatient visits using data from the Rand Health Insurance Experiment.

9.11 DERIVATIONS

9.11.1 Conditional Density for Poisson Fixed Effects

Consider the conditional joint density for observations in all time periods for a given individual, where for simplicity the individual subscript i is dropped. In general the density of $y_1, \ldots, y_T$ given $\sum_t y_t$ is

$$\begin{aligned}\Pr\left[y_1, \ldots, y_T \left|\textstyle\sum_t y_t\right.\right] &= \Pr\left[y_1, \ldots, y_T, \textstyle\sum_t y_t\right] / \Pr\left[\textstyle\sum_t y_t\right] \\ &= \Pr\left[y_1, \ldots, y_T\right] / \Pr\left[\textstyle\sum_t y_t\right],\end{aligned}$$

where the last equality arises because knowledge of $\sum_t y_t$ adds nothing given knowledge of $y_1, \ldots, y_T$.

Now specialize to y_t iid $\mathsf{P}[\mu_t]$. Then $\Pr[y_1, \ldots, y_T]$ is the product of T Poisson densities, and $\sum_t y_t$ is $\mathsf{P}[\sum_t \mu_t]$. It follows that

$$\begin{aligned}\Pr\left[y_1, \ldots, y_T \left|\textstyle\sum_t y_t\right.\right] &= \frac{\prod_t \left(\exp(-\mu_t)\mu_t^{y_t}/y_t!\right)}{\exp(-\sum_t \mu_t)\left(\sum_t \mu_t\right)^{\sum_t y_t} / \left(\sum_t y_t\right)!} \\ &= \frac{\exp(-\sum_t \mu_t)\prod_t \mu_t^{y_t} / \prod_t y_t!}{\exp(-\sum_t \mu_t)\prod_t \left(\sum_s \mu_s\right)^{y_t} / \left(\sum_t y_t\right)!} \\ &= \frac{\left(\sum_t y_t\right)!}{\prod_t y_t!} \times \prod_t \left(\frac{\mu_t}{\sum_s \mu_s}\right)^{y_t}.\end{aligned}$$

Introducing the subscript i yields (9.39) for $\Pr\left[y_{i1}, \ldots, y_{iT} \middle| \sum_t y_{it}\right]$.

9.11.2 Density for Poisson with Gamma Random Effects

Consider the joint density for observations in all time periods for a given individual, where for simplicity the individual subscript i is dropped. From (9.44) the joint density of $y_1, \ldots, y_T$ if $y_t|\alpha$ is $\mathsf{P}[\alpha\lambda_t]$ is

$$\begin{aligned}
\Pr[y_1, \ldots, y_T] &= \int_0^\infty \left[\prod_t \left(e^{-\alpha\lambda_t}(\alpha\lambda_t)^{y_t}/y_t!\right)\right] f(\alpha)d\alpha \\
&= \int_0^\infty \left[\prod_t \lambda_t^{y_t}/y_t!\right]\left(e^{-\alpha\sum_t\lambda_t} \cdot \alpha^{\sum_t y_t}\right) f(\alpha)d\alpha \\
&= \left[\prod_t \lambda_t^{y_t}/y_t!\right] \times \int_0^\infty \left(e^{-\alpha\sum_t\lambda_t} \cdot \alpha^{\sum_t y_t}\right) f(\alpha)d\alpha.
\end{aligned}$$

Now let $f(\alpha)$ be the gamma density with $\mathsf{E}[\alpha] = 1$, $\mathsf{V}[\alpha] = 1/\delta$. Similar algebra to that in Chapter 4.2.2 yields the Poisson random effects density given in (9.45).

9.11.3 Moment Conditions for Dynamic Fixed Effects

We consider a model for y_{it} with conditional mean $\alpha_i\lambda_{it}$ where $\lambda_{it} = \lambda_{it}(\mathbf{x}_{it})$ or, more generally, $\lambda_{it} = \lambda_{it}(\mathbf{x}_{it}, \mathbf{x}_{it-1}, \ldots)$, and α_i is potentially correlated with $\mathbf{x}_{it}$. Define the additive error

$$u_{it} = y_{it} - \alpha_i\lambda_{it}, \tag{9.74}$$

so $y_{it} = \alpha_i\lambda_{it} + u_{it}$. The predetermined regressor is uncorrelated with current and future shocks, so

$$\mathsf{E}[u_{it}|\mathbf{x}_{it}, \mathbf{x}_{it-1}, \ldots] = 0. \tag{9.75}$$

The Chamberlain transformation is

$$\begin{aligned}
y_{it}\tfrac{\lambda_{it-1}}{\lambda_{it}} - y_{it-1} &= (\alpha_i\lambda_{it} + u_{it})\tfrac{\lambda_{it-1}}{\lambda_{it}} - (\alpha_i\lambda_{it-1} + u_{it-1}) \\
&= \alpha_i\lambda_{it-1} + u_{it}\tfrac{\lambda_{it-1}}{\lambda_{it}} - \alpha_i\lambda_{it-1} + u_{it-1} \\
&= u_{it}\tfrac{\lambda_{it-1}}{\lambda_{it}} - u_{it-1}.
\end{aligned} \tag{9.76}$$

It follows that conditioning on past $\mathbf{x}$'s only,

$$\begin{aligned}
\mathsf{E}\left[u_{it}\tfrac{\lambda_{it-1}}{\lambda_{it}} - u_{it-1}|\mathbf{x}_{it-1}, \ldots\right] &= \mathsf{E}\left[u_{it}\tfrac{\lambda_{it-1}}{\lambda_{it}}|\mathbf{x}_{it-1}, \ldots\right] - \mathsf{E}[u_{it-1}|\mathbf{x}_{it-1}, \ldots] \\
&= \mathsf{E}_{\mathbf{x}_{it}}\left[\mathsf{E}\left[u_{it}\tfrac{\lambda_{it-1}}{\lambda_{it}}|\mathbf{x}_{it}, \mathbf{x}_{it-1}, \ldots\right]\right] - 0 \\
&= \mathsf{E}_{\mathbf{x}_{it}}\left[\mathsf{E}[u_{it}|\mathbf{x}_{it}]\tfrac{\lambda_{it-1}}{\lambda_{it}}|\mathbf{x}_{it-1}, \ldots\right] \\
&= 0,
\end{aligned} \tag{9.77}$$

where the first term in the second equality additionally conditions on $\mathbf{x}_{it}$ and then unconditions, the second term in the second equality is zero by (9.75), and the final equality uses $\mathsf{E}[u_{it}|\mathbf{x}_{it}] = 0$ given (9.75). Therefore

$$\mathsf{E}\left[y_{it}\frac{\lambda_{it-1}}{\lambda_{it}} - y_{it-1}|\mathbf{x}_{it-1},\ldots\right] = 0. \tag{9.78}$$

It follows that the Chamberlain transformation (9.62) satisfies the conditional moment condition (9.64) where the instruments $\mathbf{z}_{it}$ can be drawn from $\mathbf{x}_{i,t-1}, \mathbf{x}_{i,t-2}, \ldots$. In a dynamic model with, for example, $y_{i,t-1}$ included as a regressor, the conditioning set in (9.75) becomes $\mathbf{x}_{it}, \mathbf{x}_{it-1}, \ldots, \mathbf{y}_{it-1}, \mathbf{y}_{it-2}, \ldots$ and the instruments $\mathbf{z}_{it}$ can be drawn from $\mathbf{x}_{i,t-1}, \mathbf{x}_{i,t-2}, \ldots, \mathbf{y}_{it-2}, \mathbf{y}_{it-3}, \ldots$.

The Wooldridge transformation (9.62) divides all terms in (9.76), (9.77), and (9.78) by μ_{it}, and the method of proof used to obtain (9.78) still applies.

Windmeijer (2008) shows that the Chamberlain and Wooldridge transformations are still applicable with predetermined regressors if the error is instead defined to be a multiplicative error

$$u_{it} = \frac{y_{it}}{\alpha_i \lambda_{it}}, \tag{9.79}$$

so $y_{it} = \alpha_i \lambda_{it} u_{it}$. Then condition (9.75) becomes $\mathsf{E}[u_{it}|\mathbf{x}_{it}, \mathbf{x}_{it-1}, \ldots] = 1$.

9.12 BIBLIOGRAPHIC NOTES

Longitudinal data models fall in the class of multilevel models, surveyed by Goldstein (1995), who includes a brief treatment of Poisson. Standard references for linear models for longitudinal data include Hsiao (2003), Diggle et al. (2002), Baltagi (2008), and Wooldridge (2010). Diggle et al. (2002) and Fahrmeir and Tutz (2001) consider generalized linear models in detail. A useful reference for general nonlinear longitudinal data models is Mátyás and Sevestre (2008), including the Windmeijer chapter.

Remarkably, there are many different approaches to nonlinear models and many complications including serial correlation, dynamics, and unbalanced panels. The treatment in this book is comprehensive for models used in econometrics and covers many of the approaches used in other areas of statistics. Additional statistical references can be found in Diggle et al. (2002) and Fahrmeir and Tutz (2001). Lawless (1995) considers both duration and count models for longitudinal data for recurrent events. For dynamic models the GMM fixed effects approach is particularly promising. In addition to the count references given in section 9.8, it is useful to refer to earlier work on the linear model by Arellano and Bond (1991) and Keane and Runkle (1992). Recent research has extended panel count models to more flexible hurdle models and increasingly used finite mixture models; see section 9.10.

9.13 EXERCISES

9.1 Show that the Poisson fixed effects conditional MLE of $\boldsymbol{\beta}$ that maximizes the log-likelihood function given in (9.42) is the solution to the first-order conditions (9.31).

9.2 Find the first-order conditions for the negative binomial fixed effects conditional MLE of $\boldsymbol{\beta}$ that maximizes the log-likelihood function based on the density (9.43). (Hint: Use the gamma recursion as in Chapter 3.3.) Do these first-order conditions have a simple interpretation, like those for the Poisson fixed effect conditional MLE?

9.3 Verify that the first-order conditions for the Poisson random effects MLE for $\boldsymbol{\beta}$ can be expressed as (9.46).

9.4 Show that the Poisson fixed effects conditional MLE that solves (9.31) reduces to $\exp(\hat{\beta}) = \bar{y}_1/\bar{y}_2$ in the application by Page (1995) discussed in section 9.6.2.

CHAPTER 10

Endogenous Regressors and Selection

10.1 INTRODUCTION

Count regressions with endogenous regressors occur frequently. Ignoring the feedback from the response variable to the endogenous regressor, and simply conditioning the outcome on variables with which it is jointly determined, leads in general to inconsistent parameter estimates. The estimation procedure should instead allow for stochastic dependence between the response variable and endogenous regressors. In considering this issue the existing literature on simultaneous equation estimation in nonlinear models is of direct relevance (T. Amemiya, 1985).

The empirical example of Chapter 3 models doctor visits as depending in part on the individual's type of health insurance. In Chapter 3 the health insurance indicator variables were treated as exogenous, but health insurance is frequently a choice variable rather than exogenously assigned. A richer model is a simultaneous model with a count outcome depending on endogenous variable(s) that may be binary (two insurance plans), multinomial (more than two insurance plans), or simply continuous.

This chapter deals with several classes of models with endogenous regressors, tailored to the outcome of interest being a count. It discusses estimation and inference for both fully parametric full-information methods and less parametric limited-information methods. These approaches are based on a multiple equation model in which that for the count outcome is of central interest, but there is also an auxiliary model for the endogenous regressor, sometimes called the first-stage or reduced-form equation. Estimation methods differ according to the detail in which the reduced form is specified and exploited in estimation.

Models with endogenous regressors are closely related to models of *self-selection* in which both the selection into the process (a binary variable denoted y_2) and the count outcome y_1 conditional on selection (y_2) are partly determined by common or correlated unobserved factors. Such models may be distinguished from *selection bias* models in which the outcome variable is observed for a subset of the population that satisfies an observability condition and hence is not representative of the full population. This chapter considers both types of models; the simplest example of the latter is a standard truncation model presented in Chapter 4.3.

Standard methods may also need to be adjusted when samples do not represent the full population because the sampling scheme departs from simple random sampling. A leading example is endogenous sampling in which sampling is based on the value of the outcome. For example, those with low values of y are less likely to be sampled. Although the set-up of this problem is different from the models with endogenous regressors, it is also discussed in this chapter.

To keep the discussion manageable, our analysis is restricted to recursive models as defined in section 10.2. In section 10.3 parametric models based on selection on unobservables are given special attention. Section 10.4 covers less parametric approaches using instrumental variables (or GMM) estimators and control function estimators. An empirical example is given in section 10.5. Section 10.6 analyzes endogeneity in two-part models. The final section of this chapter covers the identification implications of different sampling frames including choice-based sampling and endogenous stratification.

10.2 ENDOGENEITY IN RECURSIVE MODELS

In this section we consider count models for outcome y_1 with discrete or continuous endogenous regressors y_2.

For linear models with additive errors, estimation of models with endogenous regressors by IV and GMM methods is well established, provided valid instruments are available. For nonlinear models, several distinct methods have been proposed. They vary with the strength of functional form assumptions and on whether or not error terms are additive (or separable). To minimize dependence on strong assumptions, a literature on semiparametric and nonparametric regression has developed. Blundell and Powell (2003) provide an early review that emphasizes the role of separable errors and contrasts IV and control function estimators. Recent references, not restricted to recursive systems, include Blundell and Matzkin (2010) and Matzkin (2010).

Count data have the added problem of discreteness. Chesher (2010) and Chesher (2005) consider conditions for point identification of single equation models for a discrete outcome using instrumental variable estimation. He shows that in general only set identification is possible. But some restrictions, such as a recursive model structure defined in section 10.2.1, make point identification possible. For the count regression models considered in this chapter we make restrictions on functional form and model structure that are sufficient to yield point identification.

10.2.1 Recursive Model

In our initial set-up y_1 depends on exogenous variables $\mathbf{x}_1$, y_2, and a random shock u_1; y_2 depends on exogenous variables $\mathbf{z}$ and a random shock u_2. Then

$$\begin{aligned} y_1 &= f_1(\mathbf{x}_1, y_2, u_1) \\ y_2 &= f_2(\mathbf{z}, u_2). \end{aligned} \tag{10.1}$$

It is assumed that $\mathbf{z}$ contains at least one variable not in $\mathbf{x}_1$ and that u_1 and u_2 are potentially correlated. We refer to (10.1) as a *recursive model* or *triangular model* because the model for y_2 specifies no feedback from y_1 to y_2. The y_1-equation is *structural* in the sense that the coefficient of y_2 in the y_1-equation is intended to have a causal interpretation. The y_2-equation is instead simply called a *first-stage* or *reduced-form* equation. Exogenous variables are referred to as either *instruments* or *instrumental variables.*

The complication in model (10.1) is that u_1 and u_2 are potentially correlated, which in turn means u_1 and y_2 are correlated. As a result the structural y_1-equation includes a regressor y_2 that is correlated with the error u_1, leading to inconsistent estimation of model parameters, suppressed in (10.1) for simplicity, if this correlation is ignored. If instead u_1 and u_2 are independent, there is no problem.

All model specifications we consider later have important restrictions.

1. $\mathsf{E}[y_1|\mathbf{x}_1, y_2, u_1]$ and $\mathsf{E}[y_2|\mathbf{z}, u_2]$ are smooth, continuous, and differentiable functions.
2. The unobserved shocks (u_1, u_2) are separable and independent of $\mathbf{x}_1$ and $\mathbf{z}$, and are jointly dependent.
3. The variable $\mathbf{z}$ includes an instrumental variable, excluded from $\mathbf{x}_1$, that accounts for nontrivial variation in y_2 and affects y_1 only through y_2.
4. The model is *recursive* or *triangular* in the sense that there is no feedback from y_1 to y_2.

Assumption 1 would be violated by a piecewise conditional mean function as in a two-part model. Assumption 2 would be violated if there were interaction effects between unobserved heterogeneity and exogenous variables. Assumption 3 would be violated if the instrumental variable accounts for negligible variation in the outcome variables.

These assumptions ensure *point* identification of the model parameters by the method of instrumental variables. We use the term "separable functional form" to refer to the case in which unobserved heterogeneity (u_1) enters the y_1 equation such that the marginal effects $\partial y_1/\partial x_j$ are independent of u_1. A specification in which unobserved heterogeneity affects regression coefficients directly violates the separability restriction. The assumption of a recursive structure avoids potentially incoherent models that are known to arise in models with binary outcomes; see, for example, Windmeijer and Santos Silva (1997).

10.2.2 Alternative Approaches

Consider the two-element vector $\mathbf{y} = (y_1 \; y_2)'$ with joint density $f(\mathbf{y}|\mathbf{w}, \boldsymbol{\theta})$, where analysis is conditional on $\mathbf{w}$. In the notation of model (10.1), $\mathbf{w} = (\mathbf{x}_1' \; \mathbf{z}')'$. For count applications, y_1 is a count, whereas y_2 may be either discrete or continuous. The case of binary-valued y_2, which often represents a policy variable (y_2 equals 1 if policy is on, and zero otherwise), has received special attention in the literature.

Under the assumption that the joint density factorizes as

$$f(\mathbf{y}|\mathbf{w}, \boldsymbol{\theta}) = g(y_1|\mathbf{w}, y_2, \boldsymbol{\theta}_1)h(y_2|\mathbf{w}, \boldsymbol{\theta}_2), \quad \boldsymbol{\theta} \in \boldsymbol{\Theta}, \tag{10.2}$$

where $\boldsymbol{\theta}_1$ and $\boldsymbol{\theta}_2$ are distinct, the standard result is that y_2 may be treated as *exogenous*. In that case a two-part model can be estimated; see section 10.6. If instead the density $h(y_2|\mathbf{w})$ depends additionally on $\boldsymbol{\theta}_1$, then y_2 is said to be *endogenous*. This is the focus of this chapter.

We consider estimation of both complete and incomplete models.

- *Complete models* are two-equation models, one each for y_1 and y_2. Then one approach to handle endogeneity is to jointly model (y_1, y_2) within a fully parametric or *full-information* setting that requires specifying the distributions $g(\cdot)$ and $h(\cdot)$ and estimating by maximum likelihood.
- *Incomplete models* are single equation models that specify only a model for y_1 as a function of y_2, while the variation in y_2 is left unrestricted. Then a *limited-information* approach is to specify one or two conditional moments of $g(y_1|\mathbf{w}, y_2, \boldsymbol{\theta}_1)$, assume the existence of instruments, and estimate by GMM or control function methods. Relative to the full-information approach, moment-based methods are computationally simpler and founded on fewer assumptions.

In either approach the major focus in applied work is on the identification and consistent estimation of the causal parameters attached to the endogenous variables.

10.3 SELECTION MODELS FOR COUNTS

We consider models in which endogeneity results from selection. These models consist of a count outcome (y_1) equation and a binary outcome (y_2) equation, with endogeneity arising because the errors in the two equations are correlated with each other. In some applications the errors may be uncorrelated, in which case the much simpler two-part model presented in section 10.6 may be used.

We present several selection models. In the selection bias model we only observe y_1 when $y_2 = 1$. In the dummy endogenous variable model we always observe y_1, but the value of y_1 depends directly on y_2. This is a recursive simultaneous equations model where the endogenous variable in the count equation is binary. In a richer switching equations model or Roy model, slope parameters in addition to intercepts in the y_1 equation vary according to whether $y_2 = 0$ or $y_2 = 1$. In even richer models y_2 may be multinomial rather than binary.

10.3.1 Selection Bias Model

Selection bias arises if some phenomenon of interest is not fully observed, unobservables determining the phenomenon of interest are correlated with unobservables in the selection mechanism, and this complication is ignored. A simple example is left truncation at zero, in which case the methods presented in Chapter 4.3 should be applied. Here we consider a richer model for the selection mechanism.

In a count model event counts may be observed only for a selected subpopulation. An example given by Greene (1994) considers the number of major derogatory reports for a sample of credit-card holders. Suppose this sample were used to make inferences about the probability of loan default assuming a credit card was issued to an individual with specified characteristics. Such inference would exhibit selectivity bias because it would be based only on major derogatory reports of individuals who have already been issued credit cards. The sample would not be random if the conditional default probability differs between the two subpopulations, and hence one subpopulation is underrepresented in the sample of existing card holders. To estimate a count model in the presence of selectivity bias, it is necessary to model both the process of issuing credit cards (selection) and the counts of major derogatory reports (outcome).

The classic sample selection model in the econometric literature is the Gronau-Heckman female labor supply model, where a substantial fraction of women choose not to have positive hours of work. The parametric version of this model, assuming bivariate normal errors, is well covered in standard texts such as Cameron and Trivedi (2005), Wooldridge (2010), and Greene (2011). Because it is based on the linear model, it does not cover three distinguishing features of count regression models: nonnegativity, discreteness of the dependent variable, and the frequently observed high incidence of zero observations.

Terza (1998) and Greene (1997) develop analogous full ML and two-step procedures for a sample selection model for count data. The model has one outcome equation for the count variably y_1, and one selection equation for the binary participation indicator y_2 that takes values 1 or 0. Data for both $y_2 = 1$ and $y_2 = 0$ are observed (e.g., whether or not a credit card is issued), but y_1 is only observed if $y_2 = 1$ (e.g., number of major derogatory reports given that a credit card is issued). Estimates of the impact of regressors on y_1 based on data on the cardholders alone ($y_2 = 1$) are biased and inconsistent due to selection bias that arises from using a truncated sample that excludes nonholders.

Fully Parametric Model

Selection is controlled for by specifying a count distribution for y_1 that depends on both covariates $\mathbf{x}$ and unobserved heterogeneity ε_1, where ε_1 and the error term in the model for the binary outcome y_2 are correlated bivariate normal. The count is assumed to have Poisson or negative binomial density $\Pr[y_{1i}|\mathbf{x}_i, y_{2i}, \varepsilon_{1i}] = \Pr[y_{1i}|\mathbf{x}_i, \varepsilon_{1i}]$ with conditional mean

$$\mathsf{E}[y_{1i}|\mathbf{x}_i, \varepsilon_{1i}] = \exp(\mathbf{x}_i'\boldsymbol{\beta}_1 + \varepsilon_{1i}), \tag{10.3}$$

while y_{2i} is a binary outcome from a standard latent variable model, so

$$y_{2i} = \begin{cases} 1 & \text{if } \mathbf{z}_i'\boldsymbol{\beta}_2 + \varepsilon_{2i} > 0 \\ 0 & \text{otherwise.} \end{cases} \tag{10.4}$$

The random variables ε_1 and ε_2 are assumed to have a bivariate normal distribution

$$\begin{bmatrix} \varepsilon_{1i} \\ \varepsilon_{2i} \end{bmatrix} \sim \mathsf{N}\left[\begin{bmatrix} 0 \\ 0 \end{bmatrix}, \begin{bmatrix} \sigma^2 & \rho\sigma \\ \rho\sigma & 1 \end{bmatrix}\right]. \tag{10.5}$$

Note that y_{1i} does not depend on y_{2i} once we condition on ε_{1i}. The problem is that we do not observe ε_{1i}.

This selection model is related to hurdle and zero-inflated models that set $\varepsilon_1 = 0$, so there is no error correlation. For the hurdle model the count outcome is 0 if $y_2 = 0$ and is restricted to be $y_1|y_1 > 0$ if $y_2 = 1$. For the zero-inflated model the count outcome is 0 if $y_2 = 0$ and is y_1 if $y_2 = 1$. By contrast, in the selection model we only observe count y_1 if $y_2 = 0$.

The joint distribution in (10.5) implies that the conditional distribution of ε_2 given ε_1 is

$$\varepsilon_{2i}|\varepsilon_{1i} \sim \mathsf{N}[(\rho/\sigma)\varepsilon_{1i}, (1-\rho^2)]. \tag{10.6}$$

This implies that

$$\begin{aligned} \Pr[y_{2i} = 1|\mathbf{z}_i, \varepsilon_{1i}] &= \Pr\left[\mathbf{z}_i'\boldsymbol{\beta}_2 + \varepsilon_{2i} > 0|\mathbf{z}_i, \varepsilon_{1i}\right] \\ &= \Pr\left[\varepsilon_{2i} > -\mathbf{z}_i'\boldsymbol{\beta}_2|\mathbf{z}_i, \varepsilon_{1i}\right] \\ &= \int_{-\mathbf{z}_i'\boldsymbol{\beta}_2}^{\infty} \frac{1}{\sqrt{2\pi(1-\rho^2)}} \exp\left[-\frac{1}{2(1-\rho^2)}\left(\varepsilon_{2i} - \frac{\rho}{\sigma}\varepsilon_{1i}\right)\right] d\varepsilon_{2i} \\ &= \Phi\left[\frac{1}{\sqrt{1-\rho^2}}\left(\mathbf{z}_i'\boldsymbol{\beta}_2 + \frac{\rho}{\sigma}\varepsilon_{1i}\right)\right], \end{aligned} \tag{10.7}$$

where $\Phi(\cdot)$ is the standard normal cdf.

The two contributions to the density, y_1 and y_2 both observed if $y_2 = 1$ and only y_2 observed if $y_2 = 0$, can be combined as

$$(\Pr[y_{1i}, y_{2i} = 1|\mathbf{x}_i, \mathbf{z}_i])^{y_{2i}}(\Pr[y_{2i} = 0|\mathbf{z}_i])^{1-y_{2i}}. \tag{10.8}$$

These probabilities are obtained by conditioning on the unobserved heterogeneity ε_1, in addition to the exogenous regressors $\mathbf{x}$ and $\mathbf{z}$, and then eliminating ε_1 by quadrature or by using simulation methods.

For the first term in (10.8), conditioning on unobserved heterogeneity ε_1 leads to simplification because y_1 and y_2 are independent conditional on ε_1. So

$$\begin{aligned}\Pr[y_{1i}, y_{2i} = 1|\mathbf{x}_i, \mathbf{z}_i, \varepsilon_{1i}] &= \Pr[y_{1i}|\mathbf{x}_i, \varepsilon_{1i}]\Pr[y_{2i} = 1|\mathbf{x}_i, \mathbf{z}_i, \varepsilon_{1i}] \\ &= \Pr[y_{1i}|\mathbf{x}_i, \varepsilon_{1i}]\Phi\left[\frac{1}{\sqrt{1-\rho^2}}\left(\mathbf{z}_i'\boldsymbol{\beta}_2 + \frac{\rho}{\sigma}\varepsilon_{1i}\right)\right].\end{aligned} \tag{10.9}$$

For the second term in (10.8), $\Pr[y_2 = 0] = 1 - \Pr[y_2 = 1]$, so using (10.7)

$$\Pr[y_{2i} = 0|\mathbf{z}_i, \varepsilon_{1i}] = \Phi\left[\frac{1}{\sqrt{1-\rho^2}}\left(-(\mathbf{z}_i'\boldsymbol{\beta}_2 + \frac{\rho}{\sigma}\varepsilon_{1i})\right)\right]. \tag{10.10}$$

The log-likelihood, given (10.8), is

$$\mathcal{L}(\boldsymbol{\beta}_1, \boldsymbol{\beta}_2, \rho, \sigma) = \sum_{i=1}^{n}\{y_{2i}\ln\Pr[y_{1i}, y_{2i} = 1|\mathbf{x}_i, \mathbf{z}_i] + (1-y_{2i})\ln\Pr[y_{2i} = 0|\mathbf{z}_i]\}. \tag{10.11}$$

Implementation requires eliminating ε_{1i} in (10.9) and (10.10). We present two ways to do so.

The first method is simulated maximum likelihood; see Chapter 4. It is convenient to transform from $\varepsilon_{1i} \sim \mathsf{N}[0, \sigma^2]$, which depends on the unknown σ^2, to $v_i = \varepsilon_{1i}/\sigma \sim \mathsf{N}[0, 1]$. Given pseudorandom draws $v_i^{(s)}$ from the $\mathsf{N}[0, 1]$ distribution, the simulated log-likelihood is

$$\begin{aligned}&\hat{\mathcal{L}}(\boldsymbol{\beta}_1, \boldsymbol{\beta}_2, \rho, \sigma) \\ &= \sum_{i=1}^{n}\frac{1}{S}\sum_{s=1}^{S}\left\{y_{2i}\ln\left(\Pr[y_{1i}|\mathbf{x}_i, \sigma v_i^{(s)}]\Phi\left[\frac{1}{\sqrt{1-\rho^2}}\left(\mathbf{z}_i'\boldsymbol{\beta}_2 + \rho v_i^{(s)}\right)\right]\right)\right. \\ &\qquad\left. + (1-y_{2i})\ln\Phi\left[\frac{1}{\sqrt{(1-\rho^2}}\left(-(\mathbf{z}_i'\boldsymbol{\beta}_2 + \rho v_i^{(s)})\right)\right]\right\},\end{aligned} \tag{10.12}$$

where $\Pr[y_{1i}|\mathbf{x}_i, \sigma v_i] = \Pr[y_{1i}|\mathbf{x}_i, \varepsilon_{1i}]$ may be the Poisson or negative binomial with conditional mean given in (10.3).

Second, one can numerically integrate using Gaussian quadrature, which from Chapter 4.2.4 approximates integrals of the form $\int_{-\infty}^{\infty} e^{-u^2}g(u)du$. Consider the contribution to the likelihood when $y_2 = 0$. Using (10.10) and $\varepsilon_1 \sim \mathsf{N}[0, \sigma^2]$, we have

$$\begin{aligned}&\Pr[y_{2i} = 0|\mathbf{z}_i] \\ &= \int_{-\infty}^{\infty}\Pr[y_{2i} = 0|\mathbf{z}_i, \varepsilon_{1i}]\,g(\varepsilon_{1i})d\varepsilon_{1i} \\ &= \int_{-\infty}^{\infty}\Phi\left[\frac{1}{\sqrt{(1-\rho^2)}}\left(-(\mathbf{z}_i'\boldsymbol{\beta}_2 + \frac{\rho}{\sigma}\varepsilon_{1i})\right)\right]\frac{1}{\sqrt{2\pi\sigma^2}}\exp\left(\frac{-\varepsilon_{1i}^2}{2\sigma^2}\right)d\varepsilon_{1i} \\ &= \int_{-\infty}^{\infty}\frac{1}{\sqrt{\pi}}\exp(-u_i^2)\Phi\left[-\left(\mathbf{z}_i'\frac{\boldsymbol{\beta}_2}{\sqrt{1-\rho^2}} + \frac{\sqrt{2}}{\sqrt{1-\rho^2}}u_i\right)\right]du_i \\ &\simeq \sum_{j=1}^{m} w_j\frac{1}{\sqrt{\pi}}\Phi\left[-\left(\mathbf{z}_i'\frac{\boldsymbol{\beta}_2}{\sqrt{1-\rho^2}} + \frac{\sqrt{2}}{\sqrt{1-\rho^2}}t_j\right)\right].\end{aligned} \tag{10.13}$$

where the third equality uses the change of variable $u_i = \varepsilon_{1i}/\sqrt{2\sigma^2}$, and the last equality uses the Gaussian quadrature weights w_j and evaluation points t_j for m evaluations. Similarly

$$\Pr[y_{1i}, y_{2i} = 1|\mathbf{x}_i, \mathbf{z}_i] \tag{10.14}$$
$$\simeq \sum_{j=1}^{m} w_j \frac{1}{\sqrt{\pi}} \Pr\left[y_{1i}|\mathbf{x}_i, \sqrt{2\sigma^2}t_j\right] \Phi\left[\left(\mathbf{z}_i' \frac{\boldsymbol{\beta}_2}{\sqrt{1-\rho^2}} + \frac{\sqrt{2}}{\sqrt{1-\rho^2}}t_j\right)\right].$$

The MLE maximizes (10.11) on substitution of the approximations (10.13) and (10.14).

An alternative fully parametric model for counts with selection has been proposed by Lee (1997). Counts are generated by a renewal process, see Chapter 4.10.3, with normal interarrival times that are correlated with the normal error in the latent variable model for y_2. ML estimation is by simulation methods.

Partially Parametric Model

A less parametric approach models the conditional mean of the observed y_1 given the selection rule. Then, suppressing conditioning on regressors,

$$\begin{aligned}
\mathsf{E}[y_{1i}|y_{2i}=1] &= \mathsf{E}_{\varepsilon_{1i}}[y_{1i}|y_{2i}=1, \varepsilon_{1i}] \\
&= \mathsf{E}_{\varepsilon_{1i}}[\exp(\mathbf{x}_i'\boldsymbol{\beta}_1 + \varepsilon_{1i})|\varepsilon_{2i} > -\mathbf{z}_i'\boldsymbol{\beta}_2] \\
&= \mathsf{E}[\exp(\mathbf{x}_i'\boldsymbol{\beta}_1 + \rho\sigma\varepsilon_{2i} + \xi_i)|\varepsilon_{2i} > -\mathbf{z}_i'\boldsymbol{\beta}_2] \\
&= \exp(\mathbf{x}_i'\boldsymbol{\beta}_1)\mathsf{E}[\exp(\xi_i)]\mathsf{E}[\exp(\rho\sigma\varepsilon_{2i})|\varepsilon_{2i} > \mathbf{z}_i'\boldsymbol{\beta}_2] \\
&= \exp(\mathbf{x}_i'\boldsymbol{\beta}_1)\exp(\tfrac{1}{2}\sigma^2(1-\rho^2))\exp(\tfrac{1}{2}\rho^2\sigma^2)\tfrac{\Phi(\rho\sigma+\mathbf{z}_i'\boldsymbol{\beta}_2)}{\Phi(\mathbf{z}_i'\boldsymbol{\beta}_2)} \\
&= \exp(\mathbf{x}_i'\boldsymbol{\beta}_1 + \tfrac{1}{2}\sigma^2)\tfrac{\Phi(\rho\sigma+\mathbf{z}_i'\boldsymbol{\beta}_2)}{\Phi(\mathbf{z}_i'\boldsymbol{\beta}_2)},
\end{aligned} \tag{10.15}$$

where the second equality uses (10.3) and the assumption that y_{1i} is assumed to be independent of y_{2i} conditional on ε_{1i} and (10.4); the third equality uses $\varepsilon_{1i} = \rho\sigma\varepsilon_{2i} + \xi_i$, where $\xi_i \sim \mathsf{N}[0, \sigma^2(1-\rho^2)]$ is independent of ε_{2i}, given the bivariate normality assumption (10.5); the fourth equality follows given the independence of ξ_i; the fifth equality uses the untruncated mean and the truncated mean of the lognormal (see, e.g., Terza, 1998, p. 134); and the last equality follows on simplification. From (10.15) it is clear that there is no simultaneity problem if $\rho = 0$, since then $\mathsf{E}[y_{1i}|y_{2i}=1] = \exp(\mathbf{x}_i'\boldsymbol{\beta}_1)\exp(\frac{1}{2}\sigma^2)$, the usual result for the Poisson-normal model. The selection adjustment in (10.15) differs from the inverse-Mills ratio term obtained for the linear model under normality.

The model can be estimated by nonlinear least squares regression based on (10.15), or a Heckman (1976) two-step type estimator where $\boldsymbol{\beta}_2$ in (10.15) is replaced by a first-step probit estimate $\hat{\boldsymbol{\beta}}_2$ before second-step NLS estimation. This moment-based approach requires normal errors and that (10.3) holds, but unlike the ML approach does not require specification of a parametric count distribution for y_1 such as the Poisson.

Greene (1997) applies this model to major derogatory reports given that the individual is a credit-card holder. Identification is secured though socio-economic characteristics such as the number of dependents that are assumed to determine whether the person holds a credit card, but not to directly determine the number of derogatory reports. He compares FIML estimates based on (10.11) with NLS estimates based on (10.15).

10.3.2 Single Endogenous Binary Regressor

Terza (1998) focuses on a closely related model that he notes can easily be adapted to the preceding sample selection model. In Terza (1998), the data on count y_1 and binary outcome y_2 are completely observed, but y_2 is a regressor in the equation for y_1. This model is a nonlinear simultaneous equations model that is a count version of the dummy endogenous variables model of Heckman (1978).

The count has density $\Pr[y_{1i}|\mathbf{x}_i, y_{2i}, \varepsilon_{1i}]$ where the conditional mean

$$\mathsf{E}[y_{1i}|\mathbf{x}_i, y_{2i}, \varepsilon_{1i}] = \exp(\mathbf{x}_i'\boldsymbol{\beta}_1 + \gamma_1 y_{2i} + \varepsilon_{1i}), \tag{10.16}$$

where $y_{2i} = \mathbf{1}[\mathbf{z}_i'\boldsymbol{\beta}_2 + \varepsilon_{2i} > 0]$ as in (10.4), and ε_1 and ε_2 are correlated bivariate normals as in (10.5).

The model is a recursive simultaneous equations system, see (10.1), with the complication that y_1 is a count and y_2 is binary. Terza shows that the joint density can be expressed as

$$\begin{aligned}\Pr[y_{1i}, y_{2i}|\mathbf{x}_i, \mathbf{z}_i] = \textstyle\int_{-\infty}^{\infty} &\Pr[y_{1i}|\mathbf{x}_i, y_{2i}, \varepsilon_{1i}]\,\{y_{2i}\Phi^*(\varepsilon_{1i}) \\ &+ (1 - y_{2i})(1 - \Phi^*(\varepsilon_{1i}))\}\,\frac{1}{\sqrt{2\pi\sigma^2}}\exp\left(\frac{-\varepsilon_{1i}^2}{2\sigma^2}\right)d\varepsilon_{1i},\end{aligned} \tag{10.17}$$

where $\Phi^*(\varepsilon_{1i}) = \Phi\left([\mathbf{z}_i'\beta_2 + (\rho/\sigma)\varepsilon_{1i}]/\sqrt{1-\rho^2}\right)$. This is the basis for ML estimation, with simulation methods or Gaussian quadrature used to eliminate the one-dimensional integral in (10.17).

Terza (1998) also obtains an analytical expression for

$$\begin{aligned}\mathsf{E}[y_{1i}|y_{2i}, \mathbf{x}_i, \mathbf{z}_i] &= y_{2i}\mathsf{E}[y_{1i}|y_{2i} = 1, \mathbf{x}_i, \mathbf{z}_i] + (1 - y_{2i})\mathsf{E}[y_{1i}|y_{2i} = 0, \mathbf{x}_i, \mathbf{z}_i] \\ &= \exp(\mathbf{x}_i'\boldsymbol{\beta}_1 + \tfrac{1}{2}\sigma^2)\left(y_{2i}\frac{\Phi(\rho\sigma + \mathbf{z}_i'\boldsymbol{\beta}_2)}{\Phi(\mathbf{z}_i'\boldsymbol{\beta}_2)} + (1 - y_{2i})\frac{1-\Phi(\rho\sigma + \mathbf{z}_i'\boldsymbol{\beta}_2)}{1-\Phi(\mathbf{z}_i'\boldsymbol{\beta}_2)}\right)\end{aligned} \tag{10.18}$$

and considers NLS estimation of (10.18) with $\mathbf{z}_i'\boldsymbol{\beta}_2$ replaced by $\mathbf{z}_i'\hat{\boldsymbol{\beta}}_2$, where $\hat{\boldsymbol{\beta}}_2$ is obtained by first-stage probit regression of y_{2i} on $\mathbf{z}_i$. Application is to the number of trips taken by members of a household in the preceding 24 hours, with the endogenous dummy variable being whether or not the household owns a motorized vehicle.

Weiss (1999) considers estimation of a recursive model with the reverse ordering to that of Terza, so that the binary variable y_2 depends on the count y_1. He considers richer models for the bivariate error distribution, based on Lee (1983). In his application complete data on both major derogatory reports and credit-card application success are available, and successful credit-card application depends on the endogenous past number of the major derogatory reports.

10.3.3 Several Mutually Exclusive Endogenous Binary Regressors

The model (10.16) can be viewed as one where the intercept for the count outcome y_1 can take one of two values, depending on a correlated binary outcome y_2. In the treatment effects terminology the binary treatment y_2 is endogenous, rather than randomly assigned, and the parametric model of Terza (1998) controls for this endogeneity.

More generally the treatment may take one of several values. Then the intercept for the count outcome y_1 may take one of several values, depending on a correlated multinomial outcome $\mathbf{d}$, a set of J indicator variables, denoted $d_j,\ j = 1, \ldots, J$. Deb and Trivedi (2006a) propose a model for this case, in an application where the utilization of health care services, a count y_1, depends on the individual's health insurance plan, represented by a set of J mutually exclusive indicator variables $d_1, \ldots, d_J$. In this model the endogenous treatment variables $d_1, \ldots, d_j$ are categorical, but the approach can be extended to the case of continuous variables.

The expected outcome equation for the counted outcomes y_{1i} is

$$\begin{aligned}\mathsf{E}[y_{1i}|\mathbf{d}_i, \mathbf{x}_i, \mathbf{l}_i] &= \exp(\mathbf{x}_i'\boldsymbol{\beta}_1 + \mathbf{d}_i'\boldsymbol{\gamma} + \mathbf{l}_i'\boldsymbol{\lambda}) \\ &= \exp(\mathbf{x}_i'\boldsymbol{\beta}_1 + \textstyle\sum_{j=1}^J \gamma_j d_{ij} + \sum_{j=1}^J \lambda_j l_{ij}),\end{aligned} \qquad (10.19)$$

where $\mathbf{x}_i$ is a set of exogenous covariates, $\boldsymbol{\gamma} = [\gamma_1 \cdots \gamma_J]'$, $\mathbf{l}_i = [l_{i1} \cdots l_{iJ}]'$, $\boldsymbol{\lambda} = [\lambda_1 \cdots \lambda_J]'$, and l_{ij} are latent or unobserved factors that are iid $\mathsf{N}[0, 1]$ with loading parameters λ_j. If the jth insurance plan is chosen, then $d_{ij} = 1$ and $\mathsf{E}[y_{1i}|d_{ij} = 1, \mathbf{x}_i, \mathbf{l}_i] = \exp(\mathbf{x}_i'\boldsymbol{\beta}_1 + \gamma_j d_{ij} + \varepsilon_{ij})$, where $\varepsilon_{ij} = \lambda_j l_{ij} \sim \mathsf{N}[0, \lambda_j^2]$, so the count model reduces to (10.16) for the case of a single binary treatment. The outcome y_{1i} is assumed to have a negative binomial distribution with mean (10.19) and dispersion parameter ψ, with density denoted $f(\mathbf{x}_i'\boldsymbol{\beta}_1 + \mathbf{d}_i'\boldsymbol{\gamma} + \mathbf{l}_i'\boldsymbol{\lambda}, \psi)$.

The distribution of the J dichotomous treatment variables is specified as a *mixed multinomial logit* (MMNL) model,

$$\Pr[d_{ij} = 1|\mathbf{z}_i, \mathbf{l}_i] = \frac{\exp(\mathbf{z}_i'\boldsymbol{\alpha}_j + \delta_j l_{ij})}{\sum_{k=1}^J \exp(\mathbf{z}_i'\boldsymbol{\alpha}_k + \delta_k l_{ik})}, \quad j = 1, \ldots, J, \qquad (10.20)$$

where $\mathbf{z}_i$ is a set of exogenous covariates, the latent factors l_{ij} are the same as those in (10.19) with loading parameters δ_j, and for identification we normalize $\boldsymbol{\alpha}_1 = \mathbf{0}$. Then $\Pr[\mathbf{d}_i|\mathbf{z}_i, \mathbf{l}_i] = \prod_{j=1}^{J}(\Pr[d_{ij} = 1|\mathbf{z}_i, \mathbf{l}_i])^{d_{ij}}$, which we denote $\mathbf{g}(\mathbf{d}_i|\mathbf{z}_i'\boldsymbol{\alpha}_1 + \delta_1 l_{i1}, \ldots, \mathbf{z}_i'\boldsymbol{\alpha}_J + \delta_J l_{iJ})$.

The self-selection problem arises from unobserved heterogeneity due to the common latent factors $l_{i1}, l_{i2}, \ldots, l_{iJ}$ that simultaneously affect treatment choice (insurance plan) and outcome (health care utilization). Policy interest lies in $\partial\mathsf{E}[y_1|\mathbf{d}, \mathbf{x}, \mathbf{z}]/\partial d_j$, the pure treatment effect of d_j on y_1. If endogeneity of $\mathbf{d}$ is ignored, then the identified effect combines the desired pure treatment effect and the selection effect. To obtain the pure treatment effect, we need to base inference on $f(y_1, \mathbf{d}|\mathbf{x}, \mathbf{z})$, with the joint density of the outcome and treatment variables conditional only on exogenous variables $\mathbf{x}$ and $\mathbf{z}$, because knowledge of $f(y_1, \mathbf{d}|\mathbf{x}, \mathbf{z})$ yields knowledge of the marginal density $f(y_1|\mathbf{d}, \mathbf{x}, \mathbf{z})$

The joint distribution of the outcome and treatment variables, conditional on exogenous regressors and the unobserved latent factors $\mathbf{l}_i$, is

$$\begin{aligned}\Pr[y_{1i}, \mathbf{d}_i|\mathbf{x}_i, \mathbf{z}_i, \mathbf{l}_i] &= f(y_{1i}|\mathbf{d}_i, \mathbf{x}_i, \mathbf{l}_i) \times \Pr(\mathbf{d}_i|\mathbf{z}_i, \mathbf{l}_i)\\ &= f(y_{1i}|\mathbf{x}_i'\boldsymbol{\beta}_1 + \mathbf{d}_i'\boldsymbol{\gamma} + \mathbf{l}_i'\boldsymbol{\lambda}, \psi) \qquad (10.21)\\ &\quad \times g(\mathbf{d}_i|\mathbf{z}_i'\boldsymbol{\alpha}_1 + \delta_1 l_{i1}, \ldots, \mathbf{z}_i'\boldsymbol{\alpha}_J + \delta_J l_{iJ}).\end{aligned}$$

The log-likelihood is based on $\Pr[y_i, \mathbf{d}_i|\mathbf{x}_i, \mathbf{z}_i]$, which requires eliminating $\mathbf{l}_i$ by quadrature or simulation-based methods, because there is no closed-form solution in integrating out $\mathbf{l}_i$. The maximum simulated likelihood estimator maximizes

$$\begin{aligned}\hat{\mathcal{L}}(\boldsymbol{\beta}_1, \boldsymbol{\gamma}, \boldsymbol{\lambda}, \psi, \boldsymbol{\alpha}, \boldsymbol{\delta}|\mathbf{y}_1, \mathbf{X}, \mathbf{Z}) &= \sum_{i=1}^{n}\frac{1}{S}\sum_{s=1}^{S}\Pr[y_{1i}, \mathbf{d}_i|\mathbf{x}_i, \mathbf{z}_i, \mathbf{l}_i^{(s)}]\\ &= \sum_{i=1}^{n}\frac{1}{S}\sum_{s=1}^{S}\ln\left(\left[f(y_{1i}|\mathbf{x}_i'\boldsymbol{\beta}_1 + \mathbf{d}_i'\boldsymbol{\gamma} + \tilde{\mathbf{l}}_i^{(s)\prime}\boldsymbol{\lambda}, \psi)\right.\right.\\ &\quad \left.\left.\times\, g(\mathbf{d}_i|\mathbf{z}_i'\boldsymbol{\alpha}_1 + \delta_1\tilde{l}_{i1}^{(s)}, \ldots, \mathbf{z}_i'\boldsymbol{\alpha}_J + \delta_J\tilde{l}_{iJ}^{(s)})\right]\right),\end{aligned} \qquad (10.22)$$

where $\tilde{\mathbf{l}}_i^{(s)}$ is the sth draw (from a total of S draws) of pseudorandom numbers from the distribution of $\tilde{\mathbf{l}}_i^{(s)}$. For calculating standard errors the usual approach is to use a numerical approximation to the robust sandwich estimator. For operational details, see Deb and Trivedi (2006a, 2006b).

For identification the scale of each choice equation should be normalized, and the covariances between choice equation errors should be fixed. Deb and Trivedi (2006a) set $\delta_j = 1\ \forall j$. Although the model is identified through nonlinearity even when $\mathbf{z} = \mathbf{x}$, for robust identification there should be some exclusion restrictions so that some variables in $\mathbf{z}$ (for insurance choice) are not included in $\mathbf{x}$ (for health service use), and these excluded variables should contribute nontrivially to explaining the variation in $\mathbf{d}$.

Deb and Trivedi (2006a) apply their model to data from the 1996 Medical Expenditure Panel Survey, with insurance choice collapsed to one of three mutually exclusive plans ($J = 3$). The parameter estimates from the maximization of (10.22) are used to compute the average treatment effects of a health maintenance organization (i.e., $\mathsf{E}[y_1|\mathbf{x}, \mathbf{z}, d_j = 1] - \mathsf{E}[y_1|\mathbf{x}, \mathbf{z}, d_j = 0]$), and associated standard errors are evaluated for a variety of hypothetical individuals, where $d_j = 1$ if the individual's health insurance plan is a health maintenance organization.

Fabbri and Monfardini (2011) extend this model to the case where there is more than one count outcome of interest. Then the multivariate counts are modeled by adding common latent factors in Poisson models for each outcome, see Chapter 8.4.3, a multivariate extension of (10.19). The dummy endogenous regressors are modeled as in (10.20). Their application is to Italian data, with two count outcomes – the number of visits to public specialist doctors and the number of visits to private specialist doctors. There are three types of health insurance: statutory insurance only, not-for-profit voluntary health insurance, and for-profit voluntary health insurance. Voluntary health insurance is both complementary with and supplementary to statutory insurance, with the for-profit variant providing greater coverage. Once the endogeneity of insurance choice is controlled for, the authors find that those with for-profit voluntary health insurance substitute away from public specialists to private specialist rather than increasing specialists visits.

10.3.4 Endogenously Switching Regressions

The model (10.16) of section 10.3.2 specifies the conditional mean of a count outcome to have an intercept that varies with the value taken by an endogenous binary regressor. A richer model allows the slope coefficients to also vary with the value of the endogenous binary regressor. In the linear case, $y_1 = \mathbf{x}_1'\boldsymbol{\beta}_1 + \varepsilon_1$, $y_2 = \mathbf{x}_2'\boldsymbol{\beta}_2 + \varepsilon_2$, $y = y_1$ if $\mathbf{z}'\boldsymbol{\alpha} + \varepsilon_3 > 0$ and $y = y_2$ if $\mathbf{z}'\boldsymbol{\alpha} + \varepsilon_3 < 0$, and the errors ε_1, ε_2, and ε_3 are correlated normal variates. This model is called the *Roy model* or an endogenous *switching regressions model.*

Deb, Munkin, and Trivedi (2006) propose a count version of this model. The count y comes from one of two models

$$y_i = \begin{cases} y_{1i} & \text{if } \mathbf{z}_i'\boldsymbol{\alpha} + \varepsilon_{3i} > 0 \\ y_{2i} & \text{if } \mathbf{z}_i'\boldsymbol{\alpha} + \varepsilon_{3i} < 0, \end{cases} \tag{10.23}$$

where

$$\begin{aligned} y_{1i} &\sim \mathsf{P}[\exp(\mathbf{x}_i'\boldsymbol{\beta}_1 + \pi_1\varepsilon_{3i} + \varepsilon_{1i})] \\ y_{2i} &\sim \mathsf{P}[\exp(\mathbf{x}_i'\boldsymbol{\beta}_2 + \pi_2\varepsilon_{3i} + \varepsilon_{2i})]. \end{aligned} \tag{10.24}$$

There are two regimes generating y_1 and y_2, but only one value is observed for each individual. The alternative state serves to generate a counterfactual. The assignment to the state is not random but is endogenous, with endogeneity modeled by assuming that the random disturbances have joint normal distribution

$$\begin{bmatrix} \varepsilon_{1i} \\ \varepsilon_{2i} \\ \varepsilon_{3i} \end{bmatrix} \sim \mathsf{N}\left[\begin{bmatrix} 0 \\ 0 \\ 0 \end{bmatrix}, \begin{bmatrix} \sigma_1^2 & 0 & 0 \\ 0 & \sigma_1^2 & 0 \\ 0 & 0 & 1 \end{bmatrix}\right], \tag{10.25}$$

where the variance of ε_3 is restricted to unity for identification.

Deb et al. (2006) use MCMC Bayesian methods to implement this model. Application is to the 1996–2002 Medical Expenditure Panel Surveys, with doctor visits as the count outcome and the model varying according to whether the individual has private health insurance.

10.4 MOMENT-BASED METHODS FOR ENDOGENOUS REGRESSORS

The methods of section 10.3 for controlling for endogenous regressors are quite specialized, being highly parametric and focused on examples where the endogenous regressors are binary. In the terminology of section 10.2.2, the methods are full-information approaches based on a complete model. In this section we consider less parametric methods that are extensions of instrumental variables and control function methods for the linear regression model.

It is important to note that the two-stage interpretation of linear 2SLS does not carry over to nonlinear models. Specifically, first-stage regression of an endogenous regressor on instruments followed by second-stage Poisson regression with the endogenous regressor replaced by its first-stage predicted value leads to inconsistent parameter estimates; see, for example, Windmeijer and Santos Silva (1997, p. 286). Instead, the methods presented in this section need to be used.

We consider cross-section data. If panel data are available then a simple solution to endogeneity is the Poisson fixed effects estimator, see Chapter 9.4, in the special case that the endogenous regressor is time varying but is correlated only with a time-invariant component of the error.

10.4.1 Nonlinear Instrumental Variables Estimation

In the linear regression model $y = \mathbf{x}'\boldsymbol{\beta} + u$, the OLS estimator is inconsistent if $\mathsf{E}[u|\mathbf{x}] \neq 0$. The instrumental variables approach assumes that there are instruments $\mathbf{z}$ such that $\mathsf{E}[u|\mathbf{z}] = 0$. Then $\mathsf{E}[\mathbf{z}u] = \mathsf{E}[\mathbf{z}(y - \mathbf{x}'\boldsymbol{\beta})] = \mathbf{0}$. If there are as many instruments as regressors, the instrumental variables estimator solves the associated sample moment conditions $\sum_{i=1}^{n} \mathbf{z}_i(y_i - \mathbf{x}_i'\boldsymbol{\beta}) = \mathbf{0}$. If there are more instruments than regressors, then estimation is by linear GMM that minimizes a quadratic form in $\sum_{i=1}^{n} \mathbf{z}_i(y_i - \mathbf{x}_i'\boldsymbol{\beta})$, a leading example being the 2SLS estimator.

Linear and nonlinear IV estimation are presented in further detail in Chapter 2.5.4. In this section we consider the special case of nonlinear IV estimation for models with exponential conditional means, the usual specification for Poisson and negative binomial count models. Then two distinct IV estimators are obtained, depending on whether the model is assumed to have an additive error (Grogger, 1990) or a multiplicative error (Mullahy, 1997b). Windmeijer and Santos Silva (1997) contrast these two different approaches and also consider optimal GMM estimation. We summarize the relevant results.

Additive Heterogeneity

We first consider estimation of an exponential regression with an additive error (*additive heterogeneity*) that is correlated with the regressors $\mathbf{x}$. Then

$$y_i = \exp(\mathbf{x}_i'\boldsymbol{\beta}) + u_i. \tag{10.26}$$

The Poisson QMLE is inconsistent if $\mathsf{E}[u_i|\mathbf{x}_i] \neq 0$, since then $\mathsf{E}[y_i|\mathbf{x}_i] \neq \exp(\mathbf{x}_i'\boldsymbol{\beta})$ and the essential condition for consistency, see Chapter 3.2.2, does not hold.

Instead assume the existence of r linearly independent instruments $\mathbf{z}_i$ that satisfy

$$\mathsf{E}[u_i|\mathbf{z}_i] = \mathsf{E}[y_i - \exp(\mathbf{x}_i'\boldsymbol{\beta})|\mathbf{z}_i] = 0. \tag{10.27}$$

Then $\mathsf{E}[\mathbf{z}_i(y_i - \exp(\mathbf{x}_i'\boldsymbol{\beta}))] = \mathbf{0}$ or

$$\mathsf{E}[\mathbf{Z}'(\mathbf{y} - \boldsymbol{\mu}(\boldsymbol{\beta}))] = \mathbf{0}, \tag{10.28}$$

where $\mathbf{Z}$ denotes the $n \times r$ matrix of instruments, and $\boldsymbol{\mu}(\boldsymbol{\beta})$ is the $n \times 1$ vector with i^{th} entry $\mu_i = \exp(\mathbf{x}_i'\boldsymbol{\beta})$.

If there are as many instruments as regressors ($r = k$), the nonlinear IV estimator solves the corresponding sample moment condition

$$\mathbf{Z}'(\mathbf{y} - \boldsymbol{\mu}(\boldsymbol{\beta})) = \mathbf{0}. \tag{10.29}$$

More generally, if there are more instruments than regressors ($r > k$), the nonlinear IV estimator $\hat{\boldsymbol{\beta}}_{\mathsf{NLIV}}$ minimizes the quadratic form

$$(\mathbf{y} - \boldsymbol{\mu}(\boldsymbol{\beta}))'\,\mathbf{Z}[\mathbf{Z}'\mathbf{Z}]^{-1}\mathbf{Z}'\,(\mathbf{y} - \boldsymbol{\mu}(\boldsymbol{\beta}))\,. \tag{10.30}$$

Using results in Chapter 2.5, $\hat{\boldsymbol{\beta}}_{\mathsf{NLIV}}$ is asymptotically normal distributed with the variance matrix estimated by

$$\begin{aligned}\hat{\mathsf{V}}[\hat{\boldsymbol{\beta}}_{\mathsf{NLIV}}] = {} & \left(\hat{\mathbf{G}}'\mathbf{Z}[\mathbf{Z}'\mathbf{Z}]^{-1}\mathbf{Z}'\hat{\mathbf{G}}\right)^{-1}\hat{\mathbf{G}}'\mathbf{Z}[\mathbf{Z}'\mathbf{Z}]^{-1}\hat{\mathbf{S}}[\mathbf{Z}'\mathbf{Z}]^{-1}\mathbf{Z}'\hat{\mathbf{G}} \\ & \times\left(\hat{\mathbf{G}}'\mathbf{Z}[\mathbf{Z}'\mathbf{Z}]^{-1}\mathbf{Z}'\hat{\mathbf{G}}\right)^{-1},\end{aligned} \tag{10.31}$$

where $\hat{\mathbf{G}} = \partial\boldsymbol{\mu}(\boldsymbol{\beta})/\partial\boldsymbol{\beta}'\big|_{\hat{\boldsymbol{\beta}}_{\mathsf{NLIV}}}$, and for heteroskedastic errors and the exponential conditional mean

$$\hat{\mathbf{S}} = \sum_{i=1}^{n} \hat{u}_i^2\mathbf{z}_i\mathbf{z}_i',$$

where $\hat{u}_i = y_i - \exp(\mathbf{x}_i'\hat{\boldsymbol{\beta}}_{\mathsf{NLIV}})$.

As noted in Chapter 2.5.4, the NLIV estimator presumes homoskedastic errors. When $r > k$, the more efficient two-step GMM estimator minimizes

$$(\mathbf{y} - \boldsymbol{\mu}(\boldsymbol{\beta}))' \mathbf{Z}\hat{\mathbf{S}}^{-1}\mathbf{Z}' (\mathbf{y} - \boldsymbol{\mu}(\boldsymbol{\beta})). \tag{10.32}$$

Then more simply

$$\hat{\mathsf{V}}\left[\hat{\boldsymbol{\beta}}^{\text{2step}}_{\mathsf{GMM}}\right] = \left(\hat{\mathbf{G}}'\mathbf{Z}\hat{\mathbf{S}}^{-1}\mathbf{Z}'\hat{\mathbf{G}}\right)^{-1}. \tag{10.33}$$

Alternatively one can assume a model for the heteroskedasticity. For example, if the conditional variance is a multiple of the conditional mean then $\hat{\mathbf{S}}$ in (10.32) can be replaced by $\hat{\mathbf{S}} = \sum_{i=1}^{n} \exp(\mathbf{x}_i'\hat{\boldsymbol{\beta}}_{\mathsf{NLIV}})\mathbf{z}_i\mathbf{z}_i'$.

Condition (10.27) implies not only that $\mathbf{z}_i$ is an instrument but also that any transformation of $\mathbf{z}_i$ may be an instrument. Optimal instrumental variables, denoted $\mathbf{z}_i^*$, are those that yield the smallest asymptotic variances. In the present case the optimal instrumental variable matrix, using the results in Chapter 2.5.3, is

$$\mathbf{Z}^* = \mathsf{E}\left[(\mathsf{V}[\mathbf{y}|\mathbf{X}])^{-1}\frac{\partial\boldsymbol{\mu}(\boldsymbol{\beta})}{\partial\boldsymbol{\beta}'}|\mathbf{Z}\right]. \tag{10.34}$$

This specializes to just $\mathbf{X}$ if $\mathbf{Z} = \mathbf{X}$ and $\mathsf{V}[\mathbf{y}|\mathbf{X}] = c\mathsf{Diag}[\mu_i]$, for some constant $c > 0$, so that in the case of no endogenous regressors the optimal IV estimator solves $\mathbf{X}'(\mathbf{y} - \boldsymbol{\mu}(\boldsymbol{\beta})) = \mathbf{0}$, yielding the Poisson QMLE. Given endogenous regressors, so $\mathbf{Z} \neq \mathbf{X}$, the quantity $\mathsf{V}[\mathbf{y}|\mathbf{X}]^{-1}\left[\partial\boldsymbol{\mu}/\partial\boldsymbol{\beta}'\right]$ is a complicated function of $\mathbf{X}$ and $\mathbf{Z}$, even if a model for $\mathsf{V}[\mathbf{y}|\mathbf{X}]$ is specified, so it is not possible to obtain $\mathbf{Z}^*$. If one accepts as a working hypothesis that $\mathsf{V}[\mathbf{y}|\mathbf{X}]$ is a scalar multiple of $\mathsf{Diag}[\mu_i]$, and $\mu_i = \exp(\mathbf{x}_i'\boldsymbol{\beta})$, then $\partial\boldsymbol{\mu}(\boldsymbol{\beta})/\partial\boldsymbol{\beta} = \mathsf{Diag}[\mu_i] \times \mathbf{X}$ so the optimal instruments are $\mathbf{Z}^* = \mathsf{E}[\mathbf{X}|\mathbf{Z}]$. Windmeijer and Santos Silva (1997) suggest that in practice one should use as instruments an estimate $\hat{\mathsf{E}}[\mathbf{X}|\mathbf{Z}]$ of $\mathsf{E}[\mathbf{X}|\mathbf{Z}]$, augmented by variables in $\mathbf{Z}$ that are not collinear with $\hat{\mathsf{E}}[\mathbf{X}|\mathbf{Z}]$.

Multiplicative Heterogeneity

NLIV estimation for the Poisson model based on additive heterogeneity has been proposed by Grogger (1990). Mullahy (1997a) instead views the more natural model to work with to be one with *multiplicative heterogeneity*, since it treats the error (or unobserved heterogeneity) symmetrically to the regressors.

Given an exponential conditional mean, a natural way to introduce an error term is

$$\begin{aligned} y_i &= \exp(\mathbf{x}_i'\boldsymbol{\beta} + u_i) \\ &= \exp(\mathbf{x}_i'\boldsymbol{\beta})\exp(u_i) \\ &= \exp(\mathbf{x}_i'\boldsymbol{\beta})v_i, \end{aligned} \tag{10.35}$$

so an additive error within $\exp(\cdot)$ implies a multiplicative error for $\exp(\mathbf{x}_i'\boldsymbol{\beta})$. In order for $\mathrm{E}[y_i|\mathbf{x}_i] = \exp(\mathbf{x}_i'\boldsymbol{\beta})$, it must be that $\mathrm{E}[v_i|\mathbf{x}_i] = 1$ or, equivalently,

$$\mathsf{E}\left[\frac{v_i}{\exp(\mathbf{x}_i'\boldsymbol{\beta})} - 1|\mathbf{x}_i\right] = 0. \tag{10.36}$$

Endogeneity of some of the regressors implies that (10.36) does not hold. Instead we assume that the existence of instruments $\mathbf{z}_i$ satisfies

$$\mathsf{E}\left[\frac{v_i}{\exp(\mathbf{x}_i'\boldsymbol{\beta})} - 1|\mathbf{z}_i\right] = 0. \tag{10.37}$$

Given (10.35), this condition can be reexpressed as

$$\mathsf{E}\left[\frac{y_i - \exp(\mathbf{x}_i'\boldsymbol{\beta})}{\exp(\mathbf{x}_i'\boldsymbol{\beta})}|\mathbf{z}_i\right] = \mathbf{0}. \tag{10.38}$$

This leads to IV estimation based on the moment condition

$$\mathsf{E}[\mathbf{Z}'\mathbf{M}^{-1}(\mathbf{y} - \boldsymbol{\mu}(\boldsymbol{\beta}))] = \mathbf{0}, \tag{10.39}$$

where $\mathbf{M} = \mathsf{Diag}[\exp(\mathbf{x}_i'\boldsymbol{\beta})]$.

If there are as many instruments as regressors ($r = k$), the nonlinear IV estimator solves the corresponding sample moment condition

$$\mathbf{Z}'\mathbf{M}^{-1}(\mathbf{y} - \boldsymbol{\mu}(\boldsymbol{\beta})) = \mathbf{0}. \tag{10.40}$$

More generally, if there are more instruments than regressors ($r > k$), the nonlinear IV estimator $\hat{\boldsymbol{\beta}}_{\mathsf{NLIV}}$ minimizes the quadratic form

$$(\mathbf{y} - \boldsymbol{\mu}(\boldsymbol{\beta}))'\,\mathbf{M}^{-1}\mathbf{Z}\mathbf{W}_n\mathbf{Z}'\mathbf{M}^{-1}\,(\mathbf{y} - \boldsymbol{\mu}(\boldsymbol{\beta}))\,, \tag{10.41}$$

where $\mathbf{W}_n = [\mathbf{Z}'\mathbf{Z}]^{-1}$ in the simplest case of NLIV. Given heteroskedasticity it is more efficient to use the two-step GMM estimator. For even more efficient estimation, the optimal instruments in the multiplicative case are given by

$$\mathbf{Z}^* = \mathsf{E}\left[\mathbf{M}(\mathsf{V}[\mathbf{y}|\mathbf{X}])^{-1}\mathbf{M}\frac{\partial \mathbf{v}(\boldsymbol{\beta})}{\partial \boldsymbol{\beta}'}|\mathbf{Z}\right], \tag{10.42}$$

where $\mathbf{v}(\boldsymbol{\beta}) = \mathbf{M}^{-1}(\mathbf{y} - \boldsymbol{\mu}(\boldsymbol{\beta}))$.

The essential conditional moment conditions are (10.27) for multiplicative heterogeneity and (10.38) for additive heterogeneity. Given endogeneity, instruments $\mathbf{z}_i$ cannot simultaneously satisfy both (10.27) and (10.38); see Windmeijer and Santos Silva (1997, p. 284). An important point is that instrumental variables that are orthogonal to a multiplicative error v are not in general orthogonal to an additive error u. The specification of the error term affects the objective function, the choice of instruments, and the estimates of $\boldsymbol{\beta}$.

An alternative interpretation of the two IV estimators, suggested by Joao Santos Silva, is that they correspond to two different weighting schemes. The first, based on $\mathsf{E}[\mathbf{z}_i(y_i - \exp(\mathbf{x}_i'\boldsymbol{\beta}))] = \mathbf{0}$, gives equal weight to each observation as does the Poisson, so it is appropriate when the variance is a multiple of

$\exp(\mathbf{x}_i'\boldsymbol{\beta})$. The second, based on $\mathsf{E}[(\mathbf{z}_i(y_i - \exp(\mathbf{x}_i'\boldsymbol{\beta})))/\exp(\mathbf{x}_i'\boldsymbol{\beta})] = \mathbf{0}$, downweights observations with large $\exp(\mathbf{x}_i'\boldsymbol{\beta})$ and is appropriate when the variance is proportional to the square of $\exp(\mathbf{x}_i'\boldsymbol{\beta})$. Again the instruments $\mathbf{z}_i$ cannot simultaneously satisfy both, except in the trivial case of exogenous $\mathbf{x}_i$.

Most of the discussion in this section is not tailored specifically to count data models but applies more generally to models with exponential conditional means. However, heteroskedasticity is a major feature of count data, so customization to count models with given variance functions may improve efficiency. For simplicity, the foregoing discussion uses a variance function without a nuisance parameter, but this extension is feasible. Consistent estimation of such a nuisance parameter can be based on a two-step procedure. At the first stage NLIV is applied without heteroskedasticity adjustment. At the second stage the nuisance parameter is estimated using methods analogous to those discussed in Chapter 3.2.

Finally, although we have emphasized single equation estimation, as pointed out in Chapter 9, the NLIV approach can be generalized to panel data models with endogenous or predetermined regressors (Blundell, Griffith, and Windmeijer, 2002).

10.4.2 Recursive Estimation Using Control Functions

Control function estimation is a two-step method, introduced by Heckman and Robb (1985). Our general presentation follows Navarro (2008). A recursive model

$$
\begin{aligned}
y_1 &= g(\mathbf{x}, y_2, u_1) \\
y_2 &= h(\mathbf{z}, u_2),
\end{aligned}
$$

is specified where $\mathbf{z}$ includes $\mathbf{x}$ plus at least one extra regressor, and u_1 and u_2 are correlated so y_2 is endogenous in the first equation. Interest lies in estimating a function $a(\mathbf{x}, y_2)$. In this model if u_1 is known then the model for y_1 can be worked with directly, without an endogeneity problem.

A *control function* defines a function of u_1 that depends on observables, and conditioning on this eliminates the endogeneity. Formally, a function $K(\mathbf{z}, y_2)$ is a control function if it allows recovery of $a(\mathbf{x}, y_2)$, u_1 is independent of y_2 conditional on a known function $\rho(\mathbf{x}, K(\mathbf{z}, y_2))$, and $K(\cdot)$ is identified. It is not always possible to find a control function, and when it does exist it may rely on strong distributional assumptions. A major topic in the semiparametric literature is establishing conditions under which the control function can be recovered nonparametrically. This approach leads to a generated regressor, so inference on estimates of parameters in the outcome equation should use a bootstrap or a comparable methodology.

For linear regression with one right-hand side endogenous variable, an example of a control function is the *residual* $\hat{u}_2$ from the reduced-form equation for that endogenous variable. The least squares estimate from regression of y_1 on

$\mathbf{x}$, y_2, and $\hat{u}_2$ is easily shown to be equivalent to the two-stage least squares estimator.

Consider the following recursive model for counts:

$$\begin{aligned} \mathsf{E}[y_{1i}|\mathbf{x}_i, y_{2i}, u_{1i}] &= \exp(\mathbf{x}_i'\boldsymbol{\beta}_1 + \gamma_1 y_{2i} + u_{1i}) \\ y_{2i} &= \mathbf{z}_i'\boldsymbol{\beta}_2 + u_{2i}, \end{aligned} \tag{10.43}$$

where $\mathbf{z}_i$ includes at least one variable not in $\mathbf{x}_i$. If the errors u_1 and u_2 are correlated, then y_2 and u_1 are correlated so Poisson regression of y_1 on $\mathbf{x}$ and y_2 yields inconsistent parameter estimates.

In this recursive model, however, the problem would disappear if the error u_1 was observed, since then one can do Poisson regression of y_1 on $\mathbf{x}$, y_2, and u_2. Wooldridge (1997b) assumes that

$$u_{1i} = \rho u_{2i} + \varepsilon_i, \tag{10.44}$$

where ε_i is independent of u_{2i}, which implies

$$\mathsf{E}[y_{1i}|\mathbf{x}_i, y_{2i}, u_{2i}] = \exp(\mathbf{x}_i'\boldsymbol{\beta}_1 + \gamma_1 y_{2i} + \rho u_{2i}). \tag{10.45}$$

Wooldridge therefore proposes Poisson regression of y_1 on $\mathbf{x}$, y_2 and $\hat{u}_2$, where $\hat{u}_{2i} = y_{2i} - \mathbf{z}_i'\hat{\boldsymbol{\beta}}_2$ and $\hat{\boldsymbol{\beta}}_2$ is the estimate from OLS estimation of y_2 on $\mathbf{z}$. This yields consistent estimates of $\boldsymbol{\beta}_1$ and γ_1 under assumptions (10.43) and (10.44), though statistical inference should be based on standard errors that control for the estimation error in $\hat{u}$. This can be done using a bootstrap for a two-step estimator. Wooldridge notes that under the null hypothesis of no endogeneity, $H_0 : \rho = 0$, there is no need to correct for two-step estimation – one can simply do a t-test based on $\hat{\rho}$ with the usual robust standard errors for Poisson regression.

Terza, Basu, and Rathouz (2008) call this method *two-stage residual inclusion* and provide many relevant references. More generally the method is called a *control function approach*, discussed at the start of this subsection, since the additional regressor $\hat{u}_2$ is included as a control for the endogeneity of y_2. In the linear model case the method coincides with the usual 2SLS estimator. For nonlinear models this method requires the stronger assumption that ε is statistically independent of u_2 and not just mean independent. The resulting estimator differs from the nonlinear IV estimator, and which approach, if any, works well in practice has not been established.

10.5 EXAMPLE: DOCTOR VISITS AND HEALTH INSURANCE

A key issue in health policy is the role of health insurance in the utilization of health services. In many countries purchasing basic health insurance, or supplemental insurance if there is universal basic health insurance, is an individual choice and is likely to be an endogenous regressor even after conditioning on other regressors. The goal is to estimate the pure causal or incentive effects of insurance, due to the reduction in the out-of-pocket costs of care, from the effects of self-selection due to individuals sorting themselves out across different types of insurance plans based on their health status and preferences.

10.5.1 Data Summary

The data are a cross-section sample from the U.S. Medical Expenditure Panel Survey for 2003. We model the annual number of doctor visits (*DOCVIS*) using a sample of the elderly Medicare population aged 65 and higher. These individuals all receive basic health insurance through Medicare, so we consider the role of supplemental health insurance obtained privately or publicly.

The covariates in regressions are age (*AGE*), squared age (*AGE2*), years of education (*EDUCYYR*), presence of activity limitation (*ACTLIM*), number of chronic conditions (*TOTCHR*), having private insurance that supplements Medicare (*PRIVATE*), and having public Medicaid insurance for low-income individuals that supplements Medicare (*MEDICAID*). The last two are indicator variables that take the value zero for the uninsured, which is a base category.

To simplify analysis, observations are dropped if the survey participant reports dual supplemental insurance coverage through simultaneously having private supplemental insurance and Medicaid coverage. There are 45 such observations. Additionally three observations with $DOCVIS > 70$ are dropped (for reasons explained in Chapter 6), yielding a final sample with 3,629 observations. The distribution over insurance categories is *PRIVATE*, 49.1%; *MEDICAID*, 15.6%; and *NOINSR* (no private supplemental insurance), 35.3%.

The sampled individuals are aged 65–90 years, and a considerable portion have an activity limitation or chronic condition. The sample mean of *DOCVIS* is 6.74, and the sample variance is 45.56 so there is great overdispersion. The distribution has a long right tail, with 22% of observations exceeding 10 though less than 1% exceed 40. The proportion of zeros is 10.9%, relatively low for this type of data, partly because the data pertain to the elderly population.

To control for endogeneity of supplemental insurance we use as instruments *INCOME* and *SSIRATIO*. The first is a measure of total household income, and the second is the ratio of Social Security income to total income. Jointly the two variables reflect the affordability of private insurance. High *INCOME* makes private insurance more accessible, whereas a high value of *SSIRATIO* indicates an income constraint and is expected to be negatively associated with the variable *PRIVATE*. For these to be valid instruments, we need to assume that for people aged 65–90 years doctor visits are not directly determined by *INCOME* or *SSIRATIO*.

10.5.2 Single Endogenous Regressor

The initial analysis treats only *PRIVATE* as potentially endogenous, with *MEDICAID* treated as exogenous. Table 10.1 presents the estimated coefficients of

Table 10.1. *Doctor visits: One endogenous regressor*

Regressor	Poisson	NB2	Poisson-CF	NB2-CF	GMM-add	GMM-mult
PRIVATE	0.170	0.185	0.410	0.596	0.446	0.678
	(0.036)	(0.037)	(0.238)	(0.258)	(0.278)	(0.251)
MEDICAID	0.082	0.085	0.199	0.284	0.248	0.289
	(0.048)	(0.048)	(0.120)	(0.128)	(0.177)	(0.103)
residual			−0.246	−0.419		
			(0.239)	(0.258)		
OIR statistic					1.392	0.940
OIR p-value					(0.238)	(0.332)
log-likelihood	−14,288	−10,405	−14,285	−10,403		

Note: Heteroskedasticity-robust standard errors are reported in parentheses; those for PoissonCF and NegbinCF are based on 400 bootstrap replications. Coefficients of exogenous regressors other than *MEDICAID* are not reported. Residual is the residual from the first-stage regression of PRIVATE on exogenous regressors and the instruments *INCOME* and *SSIRATIO*. OIR test is the overidentifying restrictions test.

PRIVATE and *MEDICAID*; coefficients of the other regressors *AGE*, *AGE2*, *EDUCYYR*, *ACTLIM*, and *TOTCHR* are not reported for brevity and because their coefficients do not vary greatly across models.

The first two columns estimate standard count models that treat *PRIVATE* as exogenous. The Poisson and negative binomial (NB2) estimates are quite similar, and the heteroskedastic-robust standard errors are similar, though the NB2 specification provides a much better fit as measured by the log-likelihood. Private supplementary insurance is highly statistically significant and is associated with close to a 20% increase in doctor visits.

The next two estimators use the control function approach of section 10.4.2 to allow for endogeneity of *PRIVATE*. At the first stage, *PRIVATE* is OLS regressed on *MEDICAID*, *AGE*, *AGE2*, *EDUCYYR*, *ACTLIM*, and *TOTCHR* and the two instruments *INCOME* or *SSIRATIO*. The second stage is a Poisson or NB2 regression of *DOCVIS* on the same regressors as in columns 1 and 2, including *PRIVATE*, plus the fitted residual from the first-stage regression. The coefficient of *PRIVATE* is now two to three times as large as under the exogeneity assumption, which suggests that private supplementary insurance generates 40–60% more doctor visits. The standard errors associated with this coefficient are more than five times as large as under exogeneity, so that for the Poisson the coefficient is barely statistically significant at 5% using a one-sided test. Such a loss of precision is typical when instruments are used in cross-section applications. Endogeneity of *PRIVATE* can be tested based on the coefficient of the first-stage residual. From Table 10.1, $t = -0.419/0.258 = 1.62$, so we fail to reject the null hypothesis of exogeneity using a two-sided test at level 0.05.

The final two estimators are two-step GMM estimators that use *INCOME* and *SSIRATIO* as instruments for *PRIVATE*. The first estimator is based on

Table 10.2. *Doctor visits: Two endogenous regressors*

Regressor	Poisson-CF	GMM-add	GMM-mult	MMNL-NB1	MMNL-NB2
PRIVATE	0.530	0.621	0.741	0.290	0.557
	(0.269)	(0.359)	(0.260)	(0.089)	(0.065)
MEDICAID	0.673	0.689	0.670	0.127	0.115
	(0.424)	(0.438)	(0.447)	(0.078)	(0.251)
residual1	−0.365				
	(0.269)				
residual2	−0.598				
	(0.433)				
$\lambda_{PRIVATE}$				−0.134	−0.449
				(0.100)	(0.059)
$\lambda_{MEDICAID}$				−0.026	0.023
				(0.082)	(0.318)
log-likelihood	–	–	–	−13,430	−13,483

Note: Heteroskedasticity-robust standard errors are reported in parentheses; those for Poisson-CF are based on 400 replications. Coefficients of exogenous regressors are not reported. Residuals 1 and 2 are, respectively, the residuals from the first-stage regression of *PRIVATE* and of *MEDICAID* on regressors and instruments *INCOME* and *SSIRATIO*.

additive heterogeneity, see (10.32), and yields results similar to the Poisson control function estimates, while the standard errors are somewhat larger. The second estimator is based on multiplicative heterogeneity, see (10.41), and yields a larger coefficient for *PRIVATE.* Since there are two instruments and one endogenous regressor, there is one overidentifying restriction yielding a $\chi^2(1)$ test statistic that for both GMM estimators has a p-value greater than 0.05, so we do not reject the null hypothesis that the overidentifying restrictions are valid.

10.5.3 Two Endogenous Regressors

We now consider models with two endogenous regressors – *PRIVATE* and *MEDICAID* – with the same regressors and the same instruments *INCOME* and *SSIRATIO* as in section 10.5.2. The preceding methods can again be applied.

The first column of Table 10.2 presents control function results, where two fitted residuals are added as additional regressors in the Poisson regression for *DOCVIS* – one from OLS regression of *PRIVATE* on exogenous variables and instruments, and one from a similar regression with *MEDICAID* the dependent variable.

The second and third columns present GMM estimates with additive and multiplicative heterogeneity. In this case the model is just-identified, with two endogenous variables and two instruments, so an overidentifying restrictions test is not possible.

Finally the more structural model (10.19) and (10.20) of section 10.3.3 is estimated by maximum simulated likelihood (Deb and Trivedi, 2006a, 2006b).

Unlike the preceding methods, the two endogenous regressors are explicitly treated as binary. This model has three endogenous insurance categories – *PRIVATE*, *MEDICAID*, and *NOINSR* – that are modeled using an MMNL specification, with *NOINSR* as the base category. The regressors for the MMNL part of the model for treatment selection are the exogenous variables and instruments used there. The outcome variable *DOCVIS* is modeled using both NB1 and NB2 specifications. The MSL estimates are obtained using 300 Halton sequence-based quasirandom draws. For evidence in support of the hypothesis of endogeneity (self-selection) of the insurance variables, the factor-loading coefficients λ_1 and λ_2 should be jointly statistically significant, which is the case here. According to the MMNL-NB1 results, supplementary private insurance leads to 30% more doctor visits, and Medicaid to 13% more visits, whereas for MMNL-NB1 supplementary private insurance has an even bigger effect. MMNL-NB1 is preferred here because the log-likelihood is much larger, with the same number of parameters.

Because this example is only intended to be illustrative, only a few covariates were included in the model even though many available socioeconomic variables are known to be significant drivers of doctor visits. Including such information should improve the precision of estimates.

10.6 SELECTION AND ENDOGENEITY IN TWO-PART MODELS

In the notation used in this chapter, the two-part or hurdle model of Chapter 4 specifies

$$y_{1i} = \begin{cases} y_{1i} > 0 & \text{if } y_{2i} = 1 \\ 0 & \text{if } y_{2i} = 0. \end{cases} \tag{10.46}$$

This is closely related to the selection models of section 10.3, but there are important differences. In particular, when the binary variable equals one, we observe only positive counts from the zero-truncated density $f(y_1|y_1 > 0,$ whereas in section 10.3.1, for example, we observed y_1 from the untruncated density $f(y_1)$. And in this benchmark model the two parts are assumed to be independent, conditional on regressors, and all regressors are assumed to be strictly exogenous.

In this section we briefly consider selection and endogeneity in the two-part model. Qualitatively similar analysis holds for the with-zeros model.

A selection version of the two-part model can be obtained by allowing the conditional means in the models for y_1 and y_2 to depend in part on errors ε_1 and ε_2. These errors represent correlated unobserved factors that affect both the probability of the binary outcome and the zero-truncated count outcome, so the two parts are connected via unobserved heterogeneity. This is a variant of the selection bias model of section 10.3.1. Greene (2007) gives relevant algebraic details for ML estimation.

A second variant of the two-part model further extends the selection version of the two-part model to allow for the presence of endogenous regressors in both parts. This can be done by appropriately adapting the model of section 10.3.3, by specializing (10.19)–(10.20) to the case of one dichotomous variable and one truncated count distribution, and using the latent factors to control for both selection and endogeneity. Identification of such a model will require restrictions on the joint covariance matrix of errors, but simulation-based estimation appears to be a promising alternative.

A third version of the two-part model allows for endogeneity only. This special case of the second variant makes the assumption that conditional on exogenous regressors and common endogenous regressor(s), the two parts are independent. This assumption is not easy to justify, especially if endogeneity is introduced via dependent latent factors. Estimation of a class of binary outcome models with endogenous regressors is well established in the literature and has been incorporated in several software packages. Both two-step and ML estimators have been developed for the case of a continuous endogenous regressor (see Newey, 1987) but are not applicable for the case of an endogenous discrete regressor.

10.7 ALTERNATIVE SAMPLING FRAMES

Departures from simple random sampling, except for exogenous sampling, which is defined later, can mean that standard estimation procedures need to be adjusted to obtain consistent parameter estimates. An example from Chapter 4 is adjustment of the standard Poisson or negative binomial estimators for zero-truncated counts. An example in section 10.3.1 is selection, where observation of the count outcome is determined by a correlated binary outcome.

In this section we provide a more general treatment of the consequences of endogenous sampling, where the probability of inclusion in the sample depends on the value taken by the count outcome. Then the resulting likelihood function is a weighted version of that obtained under random sampling. We also specialize this result to the case of count data from on-site samples, a particular form of choice-based sampling.

10.7.1 Simple Random Samples

As a benchmark for subsequent discussion, consider simple random samples for count data. These generally involve a nonnegative integer-valued count variable y and a set of covariates $\mathbf{x}$ whose joint distribution, denoted $f(y, \mathbf{x})$, can be factored as the product of the conditional and marginal distributions, thus

$$f(y, \mathbf{x}) = g(y|\mathbf{x}, \boldsymbol{\theta})h(\mathbf{x}). \qquad (10.47)$$

In the preceding chapters attention has been largely focused on $g(y|\mathbf{x}, \boldsymbol{\theta})$, modeling y given $\mathbf{x}$. *Simple random sampling* involves drawing $(y, \mathbf{x})$

combinations at random from the entire population. A variation of simple random sampling is *stratified random sampling.* It involves partitioning the population into strata defined in terms of $(y, \mathbf{x})$ and making random draws from each stratum. The number of draws from a stratum is some preselected fraction of the total survey sample size. We now consider how departures from simple random sampling arise and the consequent complications if the strata are not based on $\mathbf{x}$ alone.

10.7.2 Exogenous Sampling

Exogenous sampling from survey data occurs if the analyst segments the available sample into subsamples based only on a set of exogenous variables $\mathbf{x}$, but not on the response variable y, here the number of events. Perhaps it is more accurate to depict this type of sampling as exogenous subsampling because it is done by reference to an existing sample that has already been collected. Segmenting an existing sample by gender, health, or socioeconomic status is very commonplace. For example, in their study of hospitalizations in Germany, Geil et al. (1997) segment the data into two categories, those with and without chronic conditions. Classification by income categories is also common. Under exogenous sampling the probability distribution of the exogenous variables is independent of y and contains no information about the population parameters of interest, $\boldsymbol{\theta}$. Therefore, one may ignore the marginal distribution of the exogenous variables and simply base estimation on the conditional distribution $g(y|\mathbf{x}, \boldsymbol{\theta})$.

10.7.3 Endogenous or Choice-Based Sampling

Endogenous or choice-based sampling occurs if the probability of an individual being included in the sample depends on the value taken by the count outcome. The practical significance of this feature is that consistent estimation of $\boldsymbol{\theta}$ can no longer be carried out using the conditional population density $g(y|\mathbf{x})$ alone. The effect of the sampling scheme must also be taken into account.

In what follows f denotes the joint, g the conditional, and h the marginal densities. For further analysis we distinguish the sampling probability from the population probability by superscript s on $f(\cdot)$. In particular we assume endogenous sampling based on y according to the density $h^s(y)$. Then the joint sampling density is

$$f^s(y, \mathbf{x}) = g(\mathbf{x}|y)h^s(y),$$

which can be reexpressed using the relations

$$\begin{aligned} f(y, \mathbf{x}) &= g(y|\mathbf{x})h(\mathbf{x}), \\ &= g(\mathbf{x}|y)h(y), \end{aligned}$$

where the marginal distributions are $h(\mathbf{x}) = \sum_y f(y, \mathbf{x})$ and $h(y) = \int g(y|\mathbf{x})h(\mathbf{x})d\mathbf{x}$. Combining the above, and introducing dependence on the parameters $\boldsymbol{\theta}$ in $g(y|\mathbf{x}, \boldsymbol{\theta})$, we obtain

$$
\begin{aligned}
f^s(y, \mathbf{x}) &= \frac{f(y, \mathbf{x})h^s(y|\boldsymbol{\theta})}{h(y)} \\
&= \frac{g(y|\mathbf{x}, \boldsymbol{\theta})h(\mathbf{x})h^s(y|\boldsymbol{\theta})}{\int g(y|\mathbf{x}, \boldsymbol{\theta})h(\mathbf{x})d\mathbf{x}} \\
&= g(y|\mathbf{x}, \boldsymbol{\theta})\omega(y, \mathbf{x}, \boldsymbol{\theta}),
\end{aligned}
$$

where the second term is the weight

$$\omega(y, \mathbf{x}|\boldsymbol{\theta}) = \frac{h(\mathbf{x})h^s(y|\boldsymbol{\theta})}{\int g(y|\mathbf{x})h(\mathbf{x})d\mathbf{x}}.$$

The log-likelihood function based on $f^s(y, \mathbf{x})$ is

$$\mathcal{L}(\boldsymbol{\theta}\,|\,y, \mathbf{x}) = \sum_{i=1}^{n} \ln g(y_i|\mathbf{x}_i, \boldsymbol{\theta}) + \sum_{i=1}^{n} \ln \omega(y_i, \mathbf{x}_i|\boldsymbol{\theta}). \tag{10.48}$$

This can be interpreted as a weighted log-likelihood or a log-likelihood based on weighted probabilities, where the weights are ratios of sample and population probabilities and differ from unity (the case for simple random samples). If there exists prior information on the weights, then likelihood estimation based on weighted probabilities is straightforward. In the more usual situation in which such information is not available, estimation is difficult because the distribution of $\mathbf{x}$ is involved. The literature on the estimation problem in the context of discrete choice models is extensive (Manski and McFadden, 1981).

Standard conditional estimation considers the case when $f^s(y, \mathbf{x}) = f(y, \mathbf{x})$. Then $\omega(y, \mathbf{x}|\boldsymbol{\theta}) = h(\mathbf{x})$, which does not depend on $\boldsymbol{\theta}$, and maximizing $\mathcal{L}(\boldsymbol{\theta}\,|\,y, \mathbf{x})$ with respect to $\boldsymbol{\theta}$ is the same as just considering the first term in (10.48). If $f^s(y, \mathbf{x}) \neq f(y, \mathbf{x})$, however, analysis using standard conditional estimation that ignores the last term leads to inconsistent estimates of $\boldsymbol{\theta}$.

10.7.4 Counts with Endogenous Stratification

Count data are sometimes collected by *on-site sampling* of users. For example, on-site recreational or shopping mall surveys (Shaw, 1988; Englin and Shonkwiler, 1995; Okoruwa, Terza, and Nourse, 1988) may be carried out to study the determinants of frequency of use, often using travel-cost models. Then only those who use the facility at least once are included in the survey, and even among users the likelihood of being included in the sample depends on the frequency of use. This is an example of endogenous stratification or sampling because the selection of persons surveyed is based on a stratified sample, with random sampling within each stratum, the latter being defined by the number of events of interest.

Endogenous stratification has some similarities with choice-based sampling. As in that case, lower survey costs provide an important motivation for using stratified samples in preference to simple random samples. It requires a very large random sample to generate enough observations (information) about a relatively rare event, such as visiting a particular recreational site. Hence, it is deemed cheaper to collect an on-site sample. However, the problem is that of making inferences about the population from the sample. To do so, we need the relation between the sample frequency function and the population density function (T. Amemiya, 1985, pp. 319–338; Pudney, 1989, pp. 102–105).

To derive a density function suitable for analyzing on-site samples, the effect of stratification has to be controlled for. Shaw (1988) considers the estimation problem for on-site sampling. First, assume that in the absence of endogenous stratification the probability density of visits by individual i, given characteristics $\mathbf{x}^0$, is $g(y_i|\mathbf{x}^0)$. Suppose there are m sampled individuals with $\mathbf{x} = \mathbf{x}^0$. Then the probability that individual i is observed to make y^0 visits is

$$\begin{aligned}\Pr[y_i = y^0|\mathbf{x}^0] &= \Pr[\text{sampled value is } y^0 \text{ and sampled individual is } i]\\ &= \Pr[\text{sampled value is } y^0]\Pr[\text{sampled individual is } i]\\ &= g(y^0|\mathbf{x}^0)\frac{y_i}{y_1 + y_2 + \ldots + y_{i-1} + y_i + \ldots + y_m}.\end{aligned}$$

Next consider the probability of observing y^0 visits across all individuals, not just the ith individual, $\Pr[y = y^0|\mathbf{x}^0]$, denoted P_m. This is the weighted sum of probability of y^0 visits, where the weight of the ith individual, $i = 1, \ldots, m$, is $y^0/(y_1 + y_2 + \ldots + y_{i-1} + y^0 + y_{i+1} + \ldots + y_m)$. Then

$$\begin{aligned}P_m &= g(y^0|\mathbf{x}^0)\left(\frac{y^0}{y^0+y_2+\ldots+y_{i-1}+y_i+\ldots+y_m} + \ldots + \frac{y^0}{y_1+y_2+\ldots+y_{i-1}+y_i+\ldots+y^0}\right)\\ &= g(y^0|\mathbf{x}^0)\frac{1}{m}\left(\frac{y^0}{(y^0+y_2+\ldots+y_{i-1}+y_i+\ldots+y_m)/m} + \ldots\right.\\ &\quad \left. + \frac{y^0}{(y_1+y_2+\ldots+y_{i-1}+y_i+\ldots+y^0)/m}\right).\end{aligned}$$

If we let $m \to \infty$, the denominators inside the brackets above approaches the population mean value of y, given $\mathbf{x} = \mathbf{x}^0$,

$$\mathsf{E}[y^0|\mathbf{x}^0] = \sum_{y^0=1}^{\infty} y^0 g(y^0|\mathbf{x}^0).$$

Then

$$\lim_{m\to\infty} P_m = g(y^0|\mathbf{x}^0)\frac{1}{m}\left(\frac{y^0}{\sum_{y^0=1}^{\infty} y^0 g(y^0|\mathbf{x}^0)} + \ldots + \frac{y^0}{\sum_{y^0=1}^{\infty} y^0 g(y^0|\mathbf{x}^0)}\right)$$
$$= \frac{y^0 g(y^0|\mathbf{x}^0)}{\sum_{y^0=1}^{\infty} y^0 g(y^0|\mathbf{x}^0)}.$$

This argument and derivation show that the relation between the density, $g^s(y_i|\mathbf{x}_i)$, for the endogenously stratified sample, and the population density, $g(y_i|\mathbf{x}_i)$, is given by

$$\begin{aligned} f^s(y_i|\mathbf{x}_i) &= g(y_i|\mathbf{x}_i)\frac{y_i}{\sum_{y^0=1}^{\infty} y^0 g(y^0|\mathbf{x}_i)} \\ &= g(y_i|\mathbf{x}_i)\omega(y_i, \mu_i), \end{aligned} \tag{10.49}$$

where $\omega(y_i|\mu_i) = y_i/\mu_i$. The key result is (10.49). This expression specializes to the following if the population density is $\mathsf{P}[\mu_i]$:

$$f^s(y_i|\mathbf{x}_i) = \frac{e^{-\mu_i}\mu_i^{y_i-1}}{(y_i-1)!}, \tag{10.50}$$

where

$$\begin{aligned} \mathsf{E}[y_i|\mathbf{x}_i] &= \mu_i + 1 \\ \mathsf{V}[y_i|\mathbf{x}_i] &= \mu_i. \end{aligned} \tag{10.51}$$

Notice that the sample displays underdispersion even though the population shows equidispersion. Maximization of the likelihood based on (10.50) can be interpreted as maximizing a weighted likelihood. The case considered here is a special case of the more general discussion of endogenous sampling in the preceding section.

An interesting implication of the analysis is that there is a computationally simple way of maximizing this particular weighted likelihood. This is achieved by making the transformation $w_i = y_i - 1$ because the resulting sample space for w_i is the same as that for the regular Poisson likelihood for w_i. That is, applying the Poisson model to the original data with 1 subtracted from all y observations yields consistent estimates of the population mean parameter because

$$\frac{e^{-\mu_i}\mu_i^{w_i}}{w_i!} = \frac{e^{-\mu_i}\mu_i^{y_i-1}}{(y_i-1)!}, \qquad y_i = 1, 2, \ldots. \tag{10.52}$$

Hence, estimation can be implemented with the usually available software for Poisson maximum likelihood, whereas maximum likelihood estimation based on (10.50) requires additional (although not difficult) programming.

Although the support for the zero-truncated Poisson (section 4.5) and the choice-based Poisson is the same, the two distributions are different. Specifically, in the truncated Poisson case,

$$\mathsf{E}[y_i|y_i \geq 0] = \mu_i/(1 - e^{-\mu_i}) > \mathsf{V}[y_i|y_i \geq 0],$$

which implies underdispersion. Choice-based sampling also displays underdispersion, but it arises from a shift in the probability distribution.

The approach developed here can be extended for any parent population density by specializing (10.49). Englin and Shonkwiler (1995) substitute the weighted negative binomial in place of the weighted Poisson and show that the subtraction device shown earlier for the Poisson also works for the negative binomial. Santos Silva (1997) considers the implications of unobserved heterogeneity in the same model.

10.8 BIBLIOGRAPHIC NOTES

Terza (1998) presents the selection model in detail; see also Greene (2007). For parametric models of binary endogenous regressors see Greene (2009) and, for the multinomial case, Deb and Trivedi (2006a). A good reference for the different GMM estimators for endogenous regressors, and associated optimal instruments, is Windmeijer and Santos Silva (1997). The control function approach, which has a long history in the discrete choice models literature (see Heckman and Robb, 1985), was introduced to count models by Wooldridge (1997a); see also Terza et al. (2008). Manski and McFadden (1981) survey and discuss choice-based sampling in the context of discrete choice models. Imbens and Lancaster (1994) deal with endogenous sampling and give many useful references.

CHAPTER 11

Flexible Methods for Counts

11.1 INTRODUCTION

Much of this book has focused on Poisson and negative binomial regression. Chapter 4 introduced more flexible parametric models, most notably finite mixture models. Here we present additional flexible models. The focus is on the cross-section case, though some of the methods given here have potential extension to time series, multivariate, or longitudinal count data and to treatment of sample selection.

There are several types of flexible models. Flexible *parametric* models are fully parametrized models, but are less restrictive than simpler models such as the negative binomial. *Semiparametric* methods are partly parameterized models that entail, in part, an infinite-dimensional component. For example, this is the case if the conditional mean is specified to be of form $\exp(\mathbf{x}_i'\boldsymbol{\beta} + \lambda(z))$ where the function $\lambda(\cdot)$ is not specified and instead needs to be nonparametrically estimated. In some cases the infinite-dimensional component may not affect estimation of desired parameters. For example, this is the case if we specify the conditional mean as $\exp(\mathbf{x}_i'\boldsymbol{\beta})$, estimate by the usual Poisson estimator, and then do inference based on no additional assumptions about the model. *Nonparametric* methods do not entail a parametric component. For example, the conditional mean may be $g(\mathbf{x}_i)$ where the function $g(\cdot)$ is not specified. We focus on conditional density estimation and conditional mean estimation, but also consider conditional variance and conditional quantile estimation. We also consider potential efficiency gains to moment-based estimation when higher-order conditional moments are specified.

In section 11.2 we present a class of flexible functional forms for parametric distributions for count data based on polynomial series expansions. In section 11.3 we consider more flexible models for the conditional mean, focusing on the case in which a functional form is given for part but not all of the conditional mean function. In section 11.4 we focus on efficient estimation when a functional form is specified for the conditional mean but not for the conditional variance. In section 11.5 we present quantile regression methods that account for the discreteness and nonnegativity of count data.

Section 11.6 presents kernel-based nonparametric methods for conditional density estimation and conditional mean estimation. Section 11.7 deals with efficient moment-based estimation, an extension of quasilikelihood estimation that is cast here in the GMM framework. Section 11.8 presents an application that illustrates several of the flexible parametric, semiparametric and nonparametric methods presented in this chapter.

11.2 FLEXIBLE DISTRIBUTIONS USING SERIES EXPANSIONS

A range of parametric models for count data that can be viewed as adaptations of the basic Poisson or NB models to accommodate unobserved heterogeneity and/or adjustments for excess zeros that were presented in Chapter 4. In this section we present an alternative way to generate more flexible parametric models, by a series expansion of a baseline density such as Poisson or NB. This approach was used in the multivariate case in Chapter 8.7 to generate tests of independence. Here we focus on the univariate case.

11.2.1 Semi-Nonparametric Maximum Likelihood

Gallant and Nychka (1987) propose approximating the distribution of an iid m-dimensional random variable $\mathbf{y}$ using a squared power-series expansion around an initial choice of density or baseline density, say $f(\mathbf{y}|\boldsymbol{\lambda})$. Thus

$$h_p(\mathbf{y}|\boldsymbol{\lambda}, \mathbf{a}) = \frac{(\mathcal{P}_p(\mathbf{y}|\mathbf{a}))^2 f(\mathbf{y}|\boldsymbol{\lambda})}{\int (\mathcal{P}_p(\mathbf{z}|\mathbf{a}))^2 f(\mathbf{z})dz}, \tag{11.1}$$

where $\mathcal{P}_p(\mathbf{y}|\mathbf{a})$ is an m-variate p^{th} order polynomial, $\mathbf{a}$ is the vector of coefficients of the polynomial, and the term in the denominator is a normalizing constant. Squaring $\mathcal{P}_p(\mathbf{y}|\mathbf{a})$ has the advantage of ensuring that the density is positive. This is closely related to the series expansions in Chapter 8.7, where $\mathcal{P}_p(\mathbf{y}|\mathbf{a})$ was not squared and emphasis was placed on choosing $\mathcal{P}_p(\mathbf{y}|\mathbf{a})$ to be the orthogonal or orthonormal polynomials for the baseline density $f(\mathbf{y}|\boldsymbol{\lambda})$.

The estimators of $\boldsymbol{\lambda}$ and $\mathbf{a}$ maximize the log-likelihood $\sum_{i=1}^{n} \ln h_p(\mathbf{y}_i|\boldsymbol{\lambda}, \mathbf{a})$. Gallant and Nychka (1987) show that, under fairly general conditions, if the order p of the polynomial increases with sample size n then the estimator yields consistent estimates of the density. This result holds for a wide range of choices of baseline density. The estimator is called the *semi-nonparametric maximum likelihood estimator*, where the nonparametric component is that the order p is not fixed but grows with the sample size. This is analogous to not specifying the number of components in a finite mixture model; see Chapter 4.8.

This result provides a strong basis for using (11.1) to obtain a class of flexible distributions for any particular data. There are, however, several potential problems. First, the method may not be very parsimonious. Thus a poor initial choice of baseline density may require a fairly high-order polynomial, a potential problem even with large data sets of, say, 1,000 observations. Second, it may be difficult to obtain analytical expressions for the normalizing constant, especially in the multivariate case. Third, the normalizing constant usually leads to a highly nonlinear log-likelihood function with multiple local maxima.

Finally, Gallant and Nychka establish only consistency. As with similar nonparametric approaches, it is difficult to establish the asymptotic distribution. One solution is to bootstrap, although the asymptotic properties of this particular bootstrap do not appear to have been established.

A simpler approach is to select the order of the polynomial on the basis of the BIC of Schwarz (1978), defined in Chapter 5.4.1, because this gives a relatively large penalty for the lack of parsimony. Then inference could be based on the assumption of correct specification of the ultimately chosen model. In principle this can be formally tested on the basis of the information matrix test of White (1980).

Gallant and Tauchen (1989) and various coauthors in many studies applied models based on (11.1) to continuous finance data. There the baseline density is the multivariate normal with mean $\boldsymbol{\mu}$ and variance $\boldsymbol{\Sigma}$, with a particular transformation used so that the normalizing constant is simply obtained as a weighted sum of the first $2p$ moments of the univariate standard normal. They advocated selecting the order of the polynomial on the basis of the BIC of Schwarz (1978), defined in Chapter 5.4.1, which gives a relatively large penalty for lack of parsimony. Here we adapt these methods to count data.

A more practical approach, that taken here, selects a baseline density and a polynomial of order high enough to feel that the data is being well fit by the model, makes the stronger assumption that this model is the dgp and apply standard maximum likelihood results. This is similar to starting with the Poisson, rejecting this in favor of NB2 and then using usual ML standard errors of the NB2 model for inference.

11.2.2 General Results

We consider just a single dependent variable, presenting general results for the univariate case, before specializing to count data with a Poisson density as a baseline in the next subsection. Derivations are given in Cameron and Johansson (1997).

Consider a scalar random variable y with baseline density $f(y|\boldsymbol{\lambda})$, where $\boldsymbol{\lambda}$ is possibly a vector. The density based on a squared polynomial series expansion is

$$h_p(y|\boldsymbol{\lambda}, \mathbf{a}) = f(y|\boldsymbol{\lambda})\frac{P_p^2(y|\mathbf{a})}{\eta_p(\boldsymbol{\lambda}, \mathbf{a})}, \tag{11.2}$$

where $P_p(y|\mathbf{a})$ is a pth-order polynomial

$$P_p(y|\mathbf{a}) = \sum_{k=0}^{p} a_k y^k, \tag{11.3}$$

$\mathbf{a} = (a_0\ a_1, \ldots, a_p)'$ with the normalization $a_0 = 1$, and $\eta_p(\boldsymbol{\lambda}, \mathbf{a})$ is a normalizing constant term that ensures that the density $h_p(y|\boldsymbol{\lambda}, \mathbf{a})$ sums to unity. Squaring the polynomial ensures that the density is nonnegative. This is just the univariate version of (11.1). It can be shown that

$$\eta_p(\boldsymbol{\lambda}, \mathbf{a}) = \sum_{k=0}^{p} \sum_{l=0}^{p} a_k a_l m_{k+l}, \tag{11.4}$$

where $m_r \equiv m_r(\boldsymbol{\lambda})$ denotes the rth moment (not centered around the mean) of the baseline density $f(y|\boldsymbol{\lambda})$.

The moments of the random variable y with density $g_p(y|\boldsymbol{\lambda}, \mathbf{a})$ can be readily obtained from those of the baseline density $f(y|\boldsymbol{\lambda})$ as

$$\mathsf{E}[y^r] = \frac{\sum_{k=0}^{p} \sum_{l=0}^{p} a_k a_l m_{k+l+r}}{\eta_p(\boldsymbol{\lambda}, \mathbf{a})}. \tag{11.5}$$

The rth moment of y generally differ from the rth moment of the baseline density. In particular, the mean for the series expansion density $h_p(y|\boldsymbol{\lambda}, \mathbf{a})$ usually differs from that for the baseline density $f(y|\boldsymbol{\lambda})$.

We consider estimation based on a sample $\{(y_i, \mathbf{x}_i),\ i = 1, \ldots, n\}$ of independent observations. Then $y_i|\mathbf{x}_i$ has density $h_p(y_i|\boldsymbol{\lambda}_i, \mathbf{a}_i)$ where regressors can be introduced by letting $\boldsymbol{\lambda}_i$ or $\mathbf{a}_i$ be a specified function of $\mathbf{x}_i$ and the parameters that are to be estimated. As a simple example, suppose λ_i is a scalar determined by a known function of regressors $\mathbf{x}_i$ and an unknown parameter vector $\boldsymbol{\beta}$,

$$\lambda_i = \lambda(\mathbf{x}_i, \boldsymbol{\beta}), \tag{11.6}$$

and the polynomial coefficients $\mathbf{a}$ are unknown parameters that do not vary with regressors. The log-likelihood function is then

$$\mathcal{L}(\boldsymbol{\beta}, \mathbf{a}) = \sum_{i=1}^{n} \{\ln f(y_i|\lambda(\mathbf{x}_i, \boldsymbol{\beta})) + 2 \ln P_p(y_i|\mathbf{a}) - \ln \eta_p(\lambda(\mathbf{x}_i, \boldsymbol{\beta}), \mathbf{a})\}, \tag{11.7}$$

with first-order conditions that, given $\eta_p(\boldsymbol{\lambda}, \mathbf{a})$ in (11.4), can be reexpressed as

$$\frac{\partial \mathcal{L}}{\partial \boldsymbol{\beta}} = \sum_{i=1}^{n} \left\{ \frac{\partial \ln f(y_i|\lambda_i)}{\partial \lambda_i} - \frac{\sum_{k=0}^{p} \sum_{l=0}^{p} a_k a_l \partial m_{k+l,i}/\partial \lambda_i}{\sum_{k=0}^{p} \sum_{l=0}^{p} a_k a_l m_{k+l,i}} \right\} \frac{\partial \lambda_i}{\partial \boldsymbol{\beta}} = \mathbf{0}, \tag{11.8}$$

$$\frac{\partial \mathcal{L}}{\partial a_j} = \sum_{i=1}^{n} 2 \left\{ \frac{y^j}{\sum_{k=0}^{p} a_k y^k} - \frac{\sum_{k=0}^{p} a_k m_{k+j,i}}{\sum_{k=0}^{p} \sum_{l=0}^{p} a_k a_l m_{k+l,i}} \right\} = 0,$$
$$j = 1, \ldots, p. \tag{11.9}$$

Cameron and Johansson do not establish semiparametric consistency of this method. Instead, the density with chosen p is assumed to be correctly specified. Inference is based on the standard result that the MLE for $\boldsymbol{\beta}$ and $\mathbf{a}$ is asymptotically normal distributed with the variance matrix equal to the inverse of the information matrix, under the assumption that the data are generated by (11.2) and (11.6).

In principle this method is very easy to apply, provided analytical moments of the baseline density are easily obtained. The maximum likelihood first-order conditions simply involve derivatives of these moments with respect to λ, and even then some optimization routines do not require specification of first derivatives.

In practice there are two potential problems. First, the objective function is very nonlinear in parameters and there are multiple optima. This is discussed in the next subsection. Second, these series expansion densities may not be very parsimonious or may not fit the data very well. This method's usefulness can really only be established by applications.

11.2.3 Poisson Polynomial Model

The preceding framework can be applied to a wide range of baseline densities. Cameron and Johansson illustrated the use of this method if the baseline density is the Poisson, so $f(y|\mu) = e^{-\mu}\mu^y/y!$. The motivation was a goal of modeling underdispersed data, in which case the Poisson is a better starting point than negative binomial.

They called this model the PPp model, for *Poisson polynomial of order p*. Then $\lambda = \mu$, and the normalizing constant $\eta_p(\mu, \mathbf{a})$ defined in (11.4) and the moments $\mathsf{E}[y^r]$ defined in (11.5) are evaluated using the moments $m_r(\mu)$ of the Poisson, which can be obtained from the moment generating function using $m_r(\mu) = \partial^r \exp(-\mu + \mu e^t)/\partial t^r \mid_{t=0}$.

As an example the PP2 model is

$$h_2(y|\mu, \mathbf{a}) = \frac{e^{-\mu}\mu^y}{y!} \frac{(1 + a_1 y + a_2 y^2)^2}{\eta_2(\mathbf{a}, \mu)}, \tag{11.10}$$

where

$$\eta_2(\mathbf{a}, \mu) = 1 + 2a_1 m_1 + (a_1^2 + 2a_2)m_2 + 2a_1 a_2 m_3 + a_2^2 m_4. \tag{11.11}$$

Note that μ here refers to the mean of the baseline density. In fact the first two moments of y for the PP2 density are

$$\begin{aligned} \mathsf{E}[y] &= (m_1 + 2a_1 m_2 + (a_1^2 + 2a_2)m_3 + 2a_1 a_2 m_4 + a_2^2 m_5)/\eta_2(\mathbf{a}, \mu) \\ \mathsf{E}[y^2] &= (m_2 + 2a_1 m_3 + (a_1^2 + 2a_2)m_4 + 2a_1 a_2 m_5 + a_2^2 m_6)/\eta_2(\mathbf{a}, \mu). \end{aligned} \tag{11.12}$$

Estimation requires evaluation of (11.11) and hence the first four moments of the Poisson density; evaluation of the mean and variance from (11.12) requires the first six noncentral moments of the Poisson density. These moments are as follows:

$$\begin{aligned}
m_1 &= \mu \\
m_2 &= \mu + \mu^2 \\
m_3 &= \mu + 3\mu^2 + \mu^3 \\
m_4 &= \mu + 7\mu^2 + 6\mu^3 + \mu^4 \\
m_5 &= \mu + 15\mu^2 + 25\mu^3 + 10\mu^4 + \mu^5 \\
m_6 &= \mu + 31\mu^2 + 90\mu^3 + 65\mu^4 + 15\mu^5 + \mu^6.
\end{aligned}$$

The PPp model permits a wide range of models for count data, including multimodal densities and densities with either underdispersion or overdispersion.

For the PPp model if the baseline density has an exponential mean, so $\lambda_i = \exp(\mathbf{x}_i'\boldsymbol{\beta})$, the first-order condition (11.8) simplifies to

$$\frac{\partial \mathcal{L}_{\mathsf{PPp}}}{\partial \boldsymbol{\beta}} = \sum_{i=1}^{n} \left\{ y_i - \frac{\sum_{k=0}^{p} \sum_{l=0}^{p} a_k a_l m_{k+l+1,i}}{\sum_{k=0}^{p} \sum_{l=0}^{p} a_k a_l m_{k+l,i}} \right\} \mathbf{x}_i. \tag{11.13}$$

Using (11.5) with $r = 1$ and (11.4), (11.13) can be reexpressed as

$$\frac{\partial \mathcal{L}_{\mathsf{PPp}}}{\partial \boldsymbol{\beta}} = \sum_{i=1}^{n} (y_i - \mathsf{E}[y_i|\mathbf{x}_i])\, \mathbf{x}_i = \mathbf{0}. \tag{11.14}$$

Thus the residual is orthogonal to the regressors, and the residuals sum to zero if an intercept term is included in the model. The result (11.14) holds more generally if the baseline density is an LEF density with the conditional mean function corresponding to the canonical link function.

As is common for many nonlinear models, the likelihood function can have multiple optima. To increase the likelihood that a global maximum is obtained, Cameron and Johansson use fast simulated annealing (Szu and Hartley, 1987), a variation on simulated annealing (see Goffe, Ferrier, and Rogers, 1994), to obtain parameter estimates close to the global optima that are used as starting values for standard gradient methods. The advantage of simulated annealing techniques is that they permit movements that decrease the value of the objective function, so that one is not necessarily locked into moving to the local maxima closest to the starting values, though performance relies on user specification of a tuning parameter. Cameron and Johansson find that using a range of starting values improves considerably the success of the Newton-Raphson method in finding the global maximum, but it is better still to use the fast simulated annealing method.

For the underdispersed takeover bids data introduced in Chapter 5.2.5, Cameron and Johansson find that a PP1 model provides the best fit in terms of BIC and performs better than other models proposed for underdispersed data: the Katz, hurdle, and double Poisson.

11.2.4 Negative Binomial Polynomial Model

The PPp model permits a wide range of flexible models for count data, including multimodal densities and densities with either underdispersion or overdispersion. However, as previously mentioned there are potential limitations, including a lack of parsimony, inappropriate choice of the baseline density, and computationally challenging log-likelihood functions. For overdispersed data it is better to choose the negative binomial (NB2) as the baseline density. The resulting family of distributions are referred to by the acronym NBPp.

For the NBPp case the moments of the baseline density are as follows:

$$\begin{aligned}
m_1 &= \mu \\
m_2 &= \mu + (1+\alpha)\mu^2 \\
m_3 &= \mu + 3(1+\alpha)\mu^2 + (1+\alpha)(1+2\alpha)\mu^3 \\
m_4 &= \mu + 7(1+\alpha)\mu^2 + 6(1+\alpha)(1+2\alpha)\mu^3 + (1+\alpha)(1+2\alpha)(1+3\alpha)\mu^4 \\
m_5 &= \mu + 15(1+\alpha)\mu^2 + 25(1+\alpha)(1+2\alpha)\mu^3 \\
&\quad + 10(1+\alpha)(1+2\alpha)(1+3\alpha)\mu^4 + (1+\alpha)(1+2\alpha)(1+3\alpha)(1+4\alpha)\mu^5.
\end{aligned}$$

These moments were calculated in Guo and Trivedi (2002) by extending the calculations for the PPp model reported in Cameron and Johansson (1997).

A special case is the NBP1 model, which has log-likelihood

$$\begin{aligned}
\mathcal{L}(\boldsymbol{\beta}, \alpha, \mathbf{a}) = \sum_{i=1}^{n} \{ &\ln f(y_i|\mathbf{x}_i, \boldsymbol{\beta}, \alpha) + \ln((1+a_1 y_i)^2) \\
&- \ln(1 + 2a_1\mu_i + a_1^2(\mu_i + (1+\alpha)\mu_i^2))\},
\end{aligned} \tag{11.15}$$

where $\mu_i = \exp(\mathbf{x}_i'\boldsymbol{\beta})$ and $f(y_i|\mathbf{x}_i, \boldsymbol{\beta}, \alpha)$ is the NB2 density given in Chapter 3.3.1. The conditional mean is this model is

$$\mathrm{E}[y] = \frac{\mu + 2a[\mu + (1+\alpha)\mu^2] + a^2[\mu + 3(1+\alpha)\mu^2 + (1+\alpha)(1+2\alpha)\mu^3]}{1 + 2a\mu + a^2[\mu + (1+\alpha)\mu^2]},$$

so that marginal effects are complicated to compute.

11.2.5 Modified Power Series Distributions

The family of *modified power series distributions* (MPSDs) for nonnegative integer-valued random variables y is defined by the pdf

$$f(y|\lambda) = \frac{a(\lambda)^y b(y)}{c(\lambda)}, \tag{11.16}$$

where $c(\lambda)$ is a normalizing constant

$$c(\lambda) = \sum_{y \in I} a(\lambda)^y b(y), \tag{11.17}$$

$b(y) > 0$ depends only on y, and $a(\lambda)$ and $c(\lambda)$ are positive, finite, and differentiable functions of the parameter λ. If a distribution belongs to the MPSD class, then the truncated version of the same distribution is also an MPSD. The MPSD permits the range of y to be a subset, say T, of the set I of nonnegative integers, in which case the summation in (11.17) is for $y \in T$.

For the MPSD density, differentiating the identity $\sum_{y \in I} f(y|\lambda) = 1$ with respect to λ yields

$$\mathsf{E}[y] \equiv \mu = \frac{a(\lambda)}{a'(\lambda)} \frac{c'(\lambda)}{c(\lambda)} = \left(\frac{\partial \ln a(\lambda)}{\partial \lambda} \right)^{-1} \frac{\partial \ln c(\lambda)}{\partial \lambda}, \tag{11.18}$$

where $a'(\lambda) = \partial a(\lambda)/\partial \lambda$ and $c'(\lambda) = \partial c(\lambda)/\partial \lambda$. The variance is

$$\mathsf{V}[y] \equiv \mu_2 = \frac{a(\lambda)}{a'(\lambda)} \frac{\partial \mu}{\partial \lambda} = \left(\frac{\partial \ln a(\lambda)}{\partial \lambda} \right)^{-1} \frac{\partial \mu}{\partial \lambda}.$$

Higher order central moments, $\mu_r = \mathsf{E}[(y - \mu)^r]$, can be derived from the recurrence relation

$$\mu_r = \frac{a(\lambda)}{a'(\lambda)} \frac{d\lambda_{r-1}}{d\lambda} + r\mu_2\mu_{r-2}, \qquad r \geq 3, \tag{11.19}$$

where $\mu_{r-2} = 0$ for $r = 3$.

The MPSD family, proposed by Gupta (1974), is a generalization of the family of power series distributions (PSDs). The PSD is obtained by replacing $a(\lambda)$ in (11.16) by λ, in which case $c(\lambda) = \sum_{y \in I} b(y)\lambda^y$. This provides the motivation for the term *power series density*, because it is based on a power series expansion of the function $c(\lambda)$ with different choices of $c(\lambda)$ leading to different densities. An early reference for the PSD is Noack (1950). The generalized PSD family is obtained from the PSD if the support of y is restricted to a subset of the nonnegative integers. A generalization of the MPSD is the class of Lagrange probability distributions proposed by Consul and Shenton (1972). There is an extensive statistical literature on these families. Some discussion is given in Johnson, Kemp, and Kotz (2005), with more extensive discussion in various entries in Kotz and Johnston (1982–1989). These densities have rarely been applied in a regression setting.

For nonnegative integer-valued random variables, the MPSD is a generalization of the LEF. To see this, note that (11.16) can be reexpressed as $f(y|\lambda) = \exp\{-\ln c(\lambda) + \ln b(y) + \ln a(\lambda) y\}$. This is exactly the same functional form as the LEF defined in Chapter 2.4.2, except it is parametrized in terms of λ rather than the mean μ. However, it is not a mean parameterization of the density. The difference is that the LEF places strong restrictions on $a(\lambda)$ and $c(\lambda)$. Here the restrictions are not as strong. The MPSD therefore includes the Poisson and the NB2 (with overdispersion parameter specified). For the Poisson, $a(\lambda) = \lambda$, $b(y) = 1/y!$ and $c(\lambda) = e^{\lambda}$. For the NB2, $a(\lambda) = [1 - \lambda/(\lambda + \alpha^{-1})]^{-1/\alpha}$, $b(y) = \Gamma(\alpha^{-1} + y)/[\Gamma(y+1)\Gamma(\alpha^{-1})$, and $c(\lambda) = \lambda/(\lambda + \alpha^{-1})$. The MPSD also includes the logarithmic series distributions.

The MPSD family is potentially very flexible. One modeling strategy is to begin with a particular choice of $a(\lambda)$ and $b(y)$ and then progressively generalize $b(y)$, which also changes the normalizing constant $c(\lambda)$. To be specific, the function $b(y)$ in (11.16) can be modified so that it also depends on additional parameters to be estimated, say $\mathbf{a}$, and consequently the term $c(\lambda)$ will also depend on $\mathbf{a}$.

The PPp model presented in section 11.2.3 is an MPSD model. For example, the PP2 density (11.10) is (11.16) with $a(\lambda) = \lambda$, $b(y, \mathbf{a}) = (1 + a_1 y + a_2 y^2)^2/y!$, and $c(\lambda, \mathbf{a}) = \eta_2(\mathbf{a}, \lambda)/e^{-\lambda}$. This results from choosing the baseline density, here the Poisson, to be an LEF density. With other choices of baseline density $f(y|\lambda)$ in (11.2), models found by series expansion as in section 11.2.2 need not be related to MPSD models.

11.3 FLEXIBLE MODELS OF THE CONDITIONAL MEAN

The standard semiparametric methods that have been developed for conditional mean estimation, such as those for single-index models, can still be applied when the dependent variable is a count. But these methods generally assume iid errors, so they will be inefficient in count models that are intrinsically heteroskedastic. These methods have been adapted for more efficient estimation when the variance depends on a specified function of the mean.

11.3.1 Flexible Parametric Models

Before presenting semiparametric methods, we consider parametric models for the conditional mean that are more flexible than the exponential model for the conditional mean.

The standard extension of an exponential conditional mean, corresponding to a log link function, is to model the linear predictor as a Box-Cox modified power transformation of the mean. Thus

$$\eta_i = \mathbf{x}_i'\boldsymbol{\beta} = \begin{cases} \dfrac{\mu_i^\lambda - 1}{\lambda} & \text{if } \lambda \neq 0 \\ \ln \mu_i & \text{if } \lambda = 0. \end{cases} \tag{11.20}$$

It follows that the conditional mean

$$\mu_i = \mu(\mathbf{x}_i, \boldsymbol{\beta}, \lambda) = \begin{cases} (\lambda \mathbf{x}_i'\boldsymbol{\beta} + 1)^{1/\lambda} & \text{if } \lambda \neq 0 \\ \exp(\mathbf{x}_i'\boldsymbol{\beta}) & \text{if } \lambda = 0. \end{cases} \tag{11.21}$$

This model has the attraction of including the exponential conditional mean as a special case and still being of single-index form. To ensure that $\mu_i > 0$ we need $\lambda \mathbf{x}_i'\boldsymbol{\beta} + 1 > 0$, which in turn requires $\mathbf{x}_i'\boldsymbol{\beta} > (-1/\lambda)$ if $\lambda > 0$ and $\mathbf{x}_i'\boldsymbol{\beta} < (-1/\lambda)$ if $\lambda < 0$.

From Chapter 2.4 the GLM estimator for the Poisson model with variance $\phi\mu_i$ solves the first-order conditions $\sum_{i=1}^n \frac{1}{\mu_i}(y - \mu_i)\frac{\partial \mu_i}{\partial \theta} = \mathbf{0}$. Here

$\boldsymbol{\theta} = (\boldsymbol{\beta}' \ \lambda)'$, and some algebra yields for $\lambda \neq 0$, $\partial\mu/\partial\eta = \mu^{1-\lambda}$, $\partial\eta/\partial\lambda = (\mu/\lambda^2)(1 - \mu^{-\lambda} - \lambda \ln \mu)$, and $\partial\eta/\partial\boldsymbol{\beta} = \mathbf{x}$. It follows that $\boldsymbol{\beta}$ and λ solve the estimating equations

$$\sum_{i=1}^{n}(y - \mu_i)\mu_i^{-\lambda}\mathbf{x}_i = \mathbf{0}, \tag{11.22}$$

$$\sum_{i=1}^{n}(y - \mu_i)(1 - \mu_i^{-\lambda} - \lambda \ln \mu_i)/\lambda^2 = \mathbf{0}.$$

As is clear from the second equation, we need to ensure $\mu_i > 0$, which requires $\lambda\mathbf{x}_i'\boldsymbol{\beta} + 1 > 0$.

Once the model is estimated, interpretation is based on the marginal effect

$$\frac{\partial \mathsf{E}[y|\mathbf{x}]}{\partial x_j} = \mu_i^{1-\lambda}\beta_j. \tag{11.23}$$

The marginal effects are a positive multiple of the coefficients and equal $\mu_i\beta_j$ when $\lambda = 0$.

Pregibon (1980) proposed use of such transformations as a diagnostic for testing correct specification of the link function in generalized linear models. More generally he considered the two-parameter transformation $\eta = ((\mu + \alpha)^{\lambda} - 1)/\lambda$; McCullagh and Nelder (1989, pp. 375–378) provide a useful summary. Scallan, Gilchrist, and Green (1984) discuss in more detail estimation based on the estimating equations (11.22). Basu and Rathouz (2005) provide simulations and an application that also introduces a flexible parametric model for the variance; see section 11.4.1.

11.3.2 Single-Index Models

We consider single-index models with the conditional mean a scalar function of a linear combination of the regressors, with $\mathrm{E}[y|\mathbf{x}] = g(\mathbf{x}'\boldsymbol{\beta})$, where the scalar function $g(\cdot)$ is unspecified.

For an unknown function $g(\cdot)$ the single-index model $\boldsymbol{\beta}$ is only identified up to location and scale. To see this, note that for scalar v the function $g^*(a + bv)$ can always be expressed as $g(v)$, so the function $g^*(a + b\mathbf{x}'\boldsymbol{\beta})$ is equivalent to $g(\mathbf{x}'\boldsymbol{\beta})$. Common normalizations are to drop the intercept and restrict $\|\boldsymbol{\beta}\| = \sum_{j=1}^{k}\beta_j^2 = 1$. Additionally $g(\cdot)$ must be differentiable. In the simplest case all regressors are continuous. If instead some regressors are discrete, then at least one regressor must be continuous; see Ichimura (1993).

Several different estimators have been proposed that lead to a root-n consistent and asymptotically normal estimator of $\boldsymbol{\beta}$ and an estimator of the function $g(\cdot)$ that is consistent, though with a convergence rate less than root-n. These estimators include semiparametric least squares (Ichimura, 1993) and average derivative estimation (Härdle and Stoker, 1989). See, for example, Pagan and

Ullah (1999) and Li and Racine (2007). These estimators ignore the intrinsic heteroskedasticity of count data, so will be inefficient.

For generalized linear models with a specified variance function, Weisberg and Welsh (1994) propose a more efficient version of Ichimura's semiparametric least squares. We suppose

$$\begin{aligned} \mathrm{E}[y_i|\mathbf{x}_i] &= g(\mathbf{x}_i'\boldsymbol{\beta}) \\ \mathrm{V}[y_i|\mathbf{x}_i] &= \phi v(g(\mathbf{x}_i'\boldsymbol{\beta})), \end{aligned} \tag{11.24}$$

where the functional form for the mean function $g(\cdot)$ is not specified, but that for the variance function $v(\cdot)$ is specified. For counts usually $\phi v(\mu) = \phi\mu$ or $\phi v(\mu) = \mu + \alpha\mu^2$. If $g(\cdot)$ were known, then $\boldsymbol{\beta}$ solves

$$\sum_{i=1}^{n} \frac{(y - g(\mathbf{x}_i'\boldsymbol{\beta}))}{v(g(\mathbf{x}_i'\boldsymbol{\beta}))} g'(\mathbf{x}_i'\boldsymbol{\beta})\mathbf{x}_i = \mathbf{0}. \tag{11.25}$$

With $g(\cdot)$ unknown estimation follows an alternating procedure. Given an initial estimate $\hat{\boldsymbol{\beta}}$, for example from standard Poisson regression, estimate $\hat{g}(\cdot)$ by kernel regression of y_i on $\mathbf{x}_i'\hat{\boldsymbol{\beta}}$ and then, given $\hat{g}(\cdot)$ and $\hat{\boldsymbol{\beta}}$, estimate the first-derivative $\hat{g}'(\cdot)$ by kernel methods. Then reestimate $\boldsymbol{\beta}$ based on the equations (11.25) with the unknown functions $g(\cdot)$ and $g'(\cdot)$ replaced by estimates $\hat{g}(\cdot)$ and $\hat{g}'(\cdot)$, and so on.

Weisberg and Welsh (1994) show that the resulting estimator of $\boldsymbol{\beta}$ has the same asymptotic distribution as in the usual GLM case where $g(\cdot)$ is known, and that if a second-order kernel is used the estimate $\hat{g}(\cdot)$ converges to $g(\cdot)$ at the optimal convergence rate of $n^{2/5}$.

11.3.3 Generalized Partially Linear Models

For linear regression a common semiparametric model is the partially linear model with conditional mean $\mathsf{E}[y|\mathbf{x},\mathbf{z}] = \mathbf{x}'\boldsymbol{\beta} + \gamma(\mathbf{z})$, where the function $\gamma(\mathbf{z})$ is not specified. Several estimators for this model have been proposed. For example, Robinson (1988) proposes estimation of $\boldsymbol{\beta}$ by OLS regression of $y - \hat{\mathsf{E}}[y|\mathbf{z}]$ on $(\mathbf{x} - \hat{\mathsf{E}}[\mathbf{x}|\mathbf{z}])$, where $\hat{\mathsf{E}}[y|\mathbf{z}]$ and $\hat{\mathsf{E}}[\mathbf{x}|\mathbf{z}]$ are nonparametric kernel regression estimates of $\mathsf{E}[y|\mathbf{z}]$and $\mathsf{E}[\mathbf{x}|\mathbf{z}]$.

We now consider generalization to GLMs, where the baseline conditional mean function is a specified nonlinear function and a model for the intrinsic heteroskedasticity is also specified. Then

$$\begin{aligned} \mathsf{E}[y_i|\mathbf{x}_i,\mathbf{z}_i] &= \mu(\mathbf{x}_i'\boldsymbol{\beta} + \gamma(\mathbf{z}_i)) \\ \mathsf{V}[y_i|\mathbf{x}_i,\mathbf{z}_i] &= \phi\, v(\mu(\mathbf{x}_i'\boldsymbol{\beta} + \gamma(\mathbf{z}_i))), \end{aligned} \tag{11.26}$$

where the functions $\mu(\cdot)$ and $v(\cdot)$ are specified, and $\gamma(\cdot)$ is an unknown smooth function. For counts $\mu(\cdot)$ is typically an exponential function and $v(\mu) = \mu$. In the case of scalar z the traditional parametric approach to obtaining a flexible specification with respect to z is to let $\gamma(z)$ be a polynomial function of $\mathbf{z}$. Greater generality can be achieved by treating this component nonparametrically.

Severini and Staniswalis (1994) proposed the following approach. Let $\sum_{i=1}^{n} q(\mu(\mathbf{x}_i'\boldsymbol{\beta} + \gamma(\mathbf{z}_i)))$ denote the quasilikelihood function. If $\gamma(\mathbf{z}_i) = \mathbf{z}_i'\boldsymbol{\alpha}$, then the estimates $\boldsymbol{\alpha}$ and $\boldsymbol{\beta}$ solve

$$\sum_{i=1}^{n} \frac{\partial q(\mu(\mathbf{x}_i'\boldsymbol{\beta} + \mathbf{z}_i'\boldsymbol{\alpha}))}{\partial \boldsymbol{\alpha}} = \mathbf{0} \tag{11.27}$$

and

$$\sum_{i=1}^{n} \frac{\partial q(\mu(\mathbf{x}_i'\boldsymbol{\beta} + \mathbf{z}_i'\boldsymbol{\alpha}))}{\partial \boldsymbol{\beta}} = \mathbf{0}. \tag{11.28}$$

One way to solve these two equations is to iterate, first solving (11.27) for $\boldsymbol{\alpha}$ given an estimate of $\boldsymbol{\beta}$ and then solving (11.28) for $\boldsymbol{\beta}$ given an estimate of $\boldsymbol{\alpha}$, and so on.

Here the functional form $\gamma(\mathbf{z}_i)$ is unspecified. Then instead of solving (11.27) when $\boldsymbol{\beta}$ is fixed, the quasilikelihood estimate of $\eta = \gamma(\mathbf{z}, \boldsymbol{\beta})$ – that is, η evaluated at a particular value $\mathbf{z}$ – solves the equation

$$\sum_{j=1}^{n} K\left(\frac{\mathbf{z} - \mathbf{z}_j}{b}\right) \frac{\partial q\left(\mu(\mathbf{x}_j'\boldsymbol{\beta} + \eta)\right)}{\partial \eta} = \mathbf{0}, \tag{11.29}$$

where $K(\cdot)$ is a given kernel function $K(\cdot)$ with bandwidth parameter $b > 0$, and observables $\mathbf{z}$. The motivation for this estimator is that, given $\boldsymbol{\beta}$, the problem reduces to semiparametric estimation in a model with density $f(y|\gamma(\mathbf{z}))$ where $f(\cdot)$ is specified and $\gamma(\cdot)$ is not specified. In that case one can use the kernel-weighted log-likelihood estimator of Staniswalis (1989). Given $\hat{\eta}(\mathbf{z}_i, \boldsymbol{\beta})$, $i = 1, \ldots, n$ that solves (11.29), the maximum quasilikelihood estimates of $\boldsymbol{\beta}$ solve

$$\sum_{i=1}^{n} \frac{\partial q\left(\mu(\mathbf{x}_i'\boldsymbol{\beta} + \hat{\eta}(\mathbf{z}_i, \boldsymbol{\beta}))\right)}{\partial \boldsymbol{\beta}} = \mathbf{0}, \tag{11.30}$$

which is a parametric estimation problem. In general these equations are solved iteratively. The resulting estimator of $\boldsymbol{\beta}$ is root-n consistent and an asymptotically normal estimator of $\boldsymbol{\beta}$, while $\hat{\gamma}(\cdot)$ converges at rate $n^{1/4}$.

For the Poisson model the quasilikelihood is $\sum_{i=1}^{n} y_i(\mathbf{x}_i'\boldsymbol{\beta} + \lambda(\mathbf{z}_i)) - \exp(\mathbf{x}_i'\boldsymbol{\beta} + \lambda(\mathbf{z}_i))$. Then (11.29) can be solved explicitly to yield

$$\hat{\eta}(\mathbf{z}, \boldsymbol{\beta}) = \ln\left(\frac{\sum_{j=1}^{n} K\left(\frac{\mathbf{z}-\mathbf{z}_j}{b}\right) y_j}{\sum_{j=1}^{n} K\left(\frac{\mathbf{z}-\mathbf{z}_j}{b}\right) \exp(\mathbf{x}_j'\boldsymbol{\beta})}\right).$$

Then $\boldsymbol{\beta}$ is estimated by Poisson regression of y_i on $\mathbf{x}_i$ and $\hat{\eta}(\mathbf{z}_i, \boldsymbol{\beta})$, where the coefficient of $\hat{\eta}(\mathbf{z}_i, \boldsymbol{\beta})$ is constrained to be one.

In practice the user of this method needs to determine which variables constitute the $\mathbf{z}$ subset and which ones the $\mathbf{x}$ subset. Although such choices are context specific, given the constraint that $\gamma(\cdot)$ should be smooth, $\mathbf{z}$ should be continuous variable.

11.3.4 Generalized Additive Models

An alternative class of flexible models of the conditional mean, one embedded in the GLM framework, is the generalized additive model due to Hastie and Tibshirani (1990).

Then the linear component $\mathbf{x}'\boldsymbol{\beta}$ in the GLM class model is replaced by an additive model of the form $\sum_{j=2}^{k} f_j(x_j)$, where $f_j(\cdot)$ are nonparametric univariate functions, one for each covariate. Then

$$\mathsf{E}[y_i|\mathbf{x}_i] = g\left(\beta_1 + \sum_{j=2}^{k} f_j(x_{ji})\right), \tag{11.31}$$

where x_{ji} is the jth component of $\mathbf{x}_i$, the function $g(\cdot)$ is specified while the $f_j(\cdot)$ are not, and $\mathsf{V}[y_i|\mathbf{x}_i] = \phi\, v(\mathsf{E}[y_i|\mathbf{x}_i])$ for specified function $v(\cdot)$. For Poisson regression $g(\cdot) = \exp(\cdot)$ and $v(\mu) = \mu$.

Despite its additive nature, this model can accommodate interactions by defining, for example, $x_{3i} = x_{1i}x_{2i}$.

11.4 FLEXIBLE MODELS OF THE CONDITIONAL VARIANCE

The preceding section considered flexible models for the conditional mean, but restricted the conditional variance to be a specified function of the conditional mean, such as a multiple of the mean. The estimators remain consistent if the variance function is misspecified, and robust standard errors can be obtained, but there is a potential efficiency loss.

In this section we consider more flexible models for the variance. In some cases the model for the conditional mean is specified, whereas in other cases it is not.

11.4.1 Flexible Parametric Models

Basu and Rathouz (2005) consider models with flexible parameterization of the mean and the variance. For the conditional mean they specify

$$\mu_i = \mu(\mathbf{x}_i, \boldsymbol{\beta}, \lambda) = \begin{cases} (\lambda \mathbf{x}_i'\boldsymbol{\beta} + 1)^{1/\lambda} & \text{if } \lambda \neq 0 \\ \exp(\mathbf{x}_i'\boldsymbol{\beta}) & \text{if } \lambda = 0. \end{cases} \tag{11.32}$$

Then $\mu(\mathbf{x}_i, \boldsymbol{\beta}, \lambda) = (\lambda \mathbf{x}_i'\boldsymbol{\beta} + 1)^{1/\lambda}$ if $\lambda \neq 0$ and $\mu(\mathbf{x}_i, \boldsymbol{\beta}, \lambda) = \exp(\mathbf{x}_i'\boldsymbol{\beta})$ if $\lambda = 0$. This is the Box-Cox modified power transformation, defined in section 11.3.1. The conditional variance is modeled using either a power family

$$v_i = \mathsf{V}[y_i|\mathbf{x}_i] = \phi_1 \mu_i^{\phi_2} \tag{11.33}$$

or a quadratic variance function

$$v_i = \mathsf{V}[y_i|\mathbf{x}_i] = v(\mu_i) = \phi_1 \mu_i + \phi_2 \mu_i^2. \tag{11.34}$$

Estimation is by extended quasilikelihood. Defining $\boldsymbol{\theta} = (\boldsymbol{\beta}'\ \lambda)'$ and $\boldsymbol{\phi} = (\phi_1\ \phi_2)'$, the estimators $\hat{\boldsymbol{\theta}}$ and $\hat{\boldsymbol{\phi}}$ solve

$$\begin{aligned} \textstyle\sum_{i=1}^{n} \frac{(y_i-\mu_i)}{\upsilon_i}\frac{\partial \upsilon_i}{\partial \boldsymbol{\theta}} &= \mathbf{0} \\ \textstyle\sum_{i=1}^{n} \frac{(y_i-\mu_i)^2-\upsilon_i}{\upsilon_i^2}\frac{\partial \upsilon_i}{\partial \boldsymbol{\phi}} &= \mathbf{0}. \end{aligned} \tag{11.35}$$

Basu and Rathouz (2005) report in detail results from a range of simulations from parametric models, including for a gamma-distributed dgp, which they note yields very similar results to the Poisson and negative binomial. They provide a substantial application to hospital expenditure data that are not well modeled by a gamma model and, if modeled as log-expenditures, need strong distributional assumptions to transform back to enable predictions of the level of expenditures.

11.4.2 Mixture Models for Heterogeneity

Given a baseline Poisson model, richer models for the conditional variance can be obtained by introducing unobserved heterogeneity. Letting $f(y_i|\mathbf{x}_i, \nu_i)$ denote the Poisson density with conditional mean $\mu_i\nu_i = \exp(\mathbf{x}_i'\boldsymbol{\beta})\nu_i$, where ν_i has density $g(\nu_i)$, leads to the density

$$h(y_i|\mathbf{x}_i) = \int f(y_i|\mathbf{x}_i, \nu_i)g(\nu_i)d\nu_i, \tag{11.36}$$

introduced in Chapter 4. Unlike Chapter 4, the distribution of the unobserved heterogeneity component is treated as unknown. In this section we consider flexible models for the mixture density $g(\nu_i)$, leading to more flexible models for $h(y_i|\mathbf{x}_i)$.

One method, already detailed in Chapter 4.8, is to specify a finite mixtures model

$$h(y_i|\mathbf{x}_i, \boldsymbol{\beta}, \boldsymbol{\pi}) = \sum_{j=1}^{C} \pi_j f(y_i|\mathbf{x}_i, \boldsymbol{\beta}, \nu_j), \tag{11.37}$$

where only the intercept varies across models. For the Poisson with an exponential mean, this representation of heterogeneity may be interpreted as a random-intercept model in which the intercept is $(\beta_0 + \nu_j)$ with probability π_j. It is detailed in Chapter 4.8. The subpopulation with each intercept is treated as a "type," and the number of types, C, is estimated from the data along with $\boldsymbol{\pi}$ and $\boldsymbol{\beta}$. The method has both a "nonparametric" component, because it avoids distributional assumptions on ν, and a parametric component, the density $f(y|\mathbf{x}, \boldsymbol{\beta}, \nu)$. It is standard terminology in the statistics literature to call the estimator an SPMLE if C is taken as given and maximum likelihood estimation is done for the unknown parameters $(\boldsymbol{\beta}, \pi_j)$.

A second method, which we present in considerable detail, uses a continuous mixture model with a flexible parametric distribution for the random heterogeneity component ν. If y conditional on μ and ν is P$[\mu\nu]$-distributed, the mixture density (11.36), suppressing the subscript i, is

$$\begin{aligned} h(y|\mu) &= \int \frac{e^{-\mu\nu}(\nu\mu)^y}{y!} g(\nu)d\nu \\ &= \frac{\mu^y}{y!}\int \nu^y e^{-\mu\nu} g(\nu)d\nu. \end{aligned} \tag{11.38}$$

Gurmu, Rilstone, and Stern (1999) proposed approximating the unknown mixture density $g(\nu)$ by $g_p^*(\nu)$, a squared p^{th}-order orthonormal polynomial expansion defined later, and analytically calculating the integral in (11.38) with respect to $g_p^*(\nu)$ rather than $g(\nu)$. This model, which they call semiparametric MLE, will approximate any density $g(\nu)$ as $p \to \infty$, and it includes the Poisson and NB as special cases.

Specifically, the approximating density $g_p^*(\nu)$ is

$$g_p^*(\nu|\lambda,\gamma) = f(\nu|\lambda,\gamma)\frac{P_p^2(\nu|\lambda,\gamma,\mathbf{a})}{\eta_p(\lambda,\gamma,\mathbf{a})}, \tag{11.39}$$

where the baseline density $f(\nu|\lambda,\gamma)$ is the two-parameter gamma density

$$f(\nu|\lambda,\gamma) = \frac{\nu^{\gamma-1}\lambda^\gamma}{\Gamma(\gamma)}e^{-\lambda\nu}, \tag{11.40}$$

and the polynomial $P_p(\nu|\alpha,\gamma,\mathbf{a})$ is the pth-order orthonormal (see Chapter 8.7.1) generalized Laguerre polynomial

$$P_p(\nu|\lambda,\gamma,\mathbf{a}) = \sum_{k=0}^{p} a_k\eta_k^{-1/2}Q_k(\nu). \tag{11.41}$$

In (11.41) the kth-order orthogonal generalized Laguerre polynomial is

$$Q_k(\nu) = \sum_{l=0}^{k}\binom{k}{l}\frac{\Gamma(k+\gamma)}{\Gamma(l+\gamma)\Gamma(k+1)}\lambda^l(-\nu)^l, \tag{11.42}$$

and orthonormalization is achieved by premultiplying $Q_k(\nu)$ in (11.41) by $\eta_k^{-1/2}$, with

$$\eta_k = \frac{\Gamma(k+\gamma)}{\Gamma(\gamma)\Gamma(k+1)}. \tag{11.43}$$

Orthonormalization leads to the normalizing constant in (11.39) being simply

$$\eta_p(\lambda,\gamma,\mathbf{a}) = \sum_{k=0}^{p} a_k^2. \tag{11.44}$$

Clearly, evaluating $h(y) = \int \nu^y e^{-\mu\nu} g_p^*(\nu|\lambda,\gamma)d\nu$ is not straightforward. Gurmu et al. observe, however, that the random variable ν with density $g(\nu)$ has the moment generating function $\mathsf{M}_\nu(t) = \int e^{t\nu}g(\nu)d\nu$, with y^{th}-order derivative $\mathsf{M}_\nu^{(y)}(t) = \partial^y \mathsf{M}_\nu(t)/\partial t^y = \int \nu^y e^{t\nu} g(\nu) d\nu$. Thus (11.38) can be re-expressed as

$$h(y|\mu) = [\mu^y/y!]\mathsf{M}_\nu^{(y)}(-\mu). \tag{11.45}$$

Gurmu et al. obtain the analytical expression for $\mathsf{M}^*_{\nu,p}(t)$, the moment generating function for $g^*_p(\nu|\lambda,\gamma)$ defined by (11.39) through (11.44), and its yth-order derivative $\mathsf{M}^{*(y)}_{\nu,p}(t) = \partial^y \mathsf{M}^*_{\nu,p}(t)/\partial t^y$. Evaluation at $t = -\mu$ yields

$$\mathsf{M}^{*(y)}_{\nu,p}(-\mu) = \left(1+\frac{\mu}{\lambda}\right)^{-\gamma} \frac{\Gamma(\gamma)}{(\lambda+\mu)^y} \left(\sum\nolimits_{k=0}^{p} a_k^2\right)^{-1} \sum\nolimits_{k=0}^{p} \sum\nolimits_{l=0}^{p} a_k a_l \left(\eta_k \eta_l\right)^{1/2}$$
$$\times \sum\nolimits_{r=0}^{k} \sum\nolimits_{s=0}^{l} \binom{k}{r}\binom{l}{s} \frac{\Gamma(\gamma+r+s+y)}{\Gamma(\gamma+r)\Gamma(\gamma+s)} \left(-1-\frac{\mu}{\lambda}\right)^{-(r+s)}. \tag{11.46}$$

Premultiplication by $[\mu^y/y!]$ yields at last the approximating density (11.38) for the count variable.

The log-likelihood function for a sample of size n is

$$\ln \mathcal{L}(\boldsymbol{\beta}, \gamma, \lambda, \mathbf{a}) = \sum_{i=1}^{n} \left\{ y_i \ln \mu_i - \ln(y_i!) + \ln\left(\mathsf{M}^{*(y_i)}_{\nu,p}(-\mu_i)\right)\right\}, \tag{11.47}$$

where in practice $\mu_i = \exp(\mathbf{x}_i'\boldsymbol{\beta})$. The MLE is obtained in the usual way. Identification requires that $\mathsf{E}[\nu_i] = \mathsf{M}^{*(1)}_{\nu,p}(0) = 1$ and $a_0 = 1$. Gurmu et al. (1999) provide a formal statement and proof of the semiparametric consistency of this procedure, meaning that $g^*_p(\nu)$ can approximate any density $g(\nu)$ as $p \to \infty$ and $n \to \infty$, but to date its asymptotic distribution as $p \to \infty$ has not been established.

This type of mixture representation generalizes the treatment of heterogeneity, but does not alter the specification of the conditional mean in any way. One advantage of using Laguerre polynomial expansion is that the leading term is the gamma density, which has been used widely as a mixing distribution in count and duration literature. Thus, if higher terms in the expansion vanish and $\gamma = \lambda$, we obtain the popular NB2 model of the earlier chapters, with $\gamma = \lambda = \alpha^{-1}$. Furthermore, if $\gamma^{-1} = \lambda^{-1} \to 0$ and $a_j = 0$ for $j \geq 1$, the Poisson model is obtained. Unlike semiparametric methods based on discrete mixtures, such as that in the preceding subsection, the series expansion method provides smooth estimation of the distribution of unobserved heterogeneity.

This form of heterogeneity representation is computationally demanding. The complexity of the last term in the log-likelihood hinders an analytical study of the likelihood. As with other series-based likelihoods there remains a possibility of multiple maxima. Gurmu et al. (1999) treat p as unknown and compute the standard errors by computer-intensive bootstrap methods. Alternatively, if the estimated value of p is treated as correct, one can use the outer product of numerically evaluated gradients of (11.47). As in other studies, information criteria are used to select the number of terms p in the expansion. In principal one can test for a global maximum using the test of Gan and Jiang (1999).

Gurmu et al. (1999) show that this method can be extended to Poisson models with censoring, truncation, and excess zeros and ones. They apply the model to censored data on the number of shopping trips to a shopping mall in a month taken by 828 shoppers, where 7.9% of the observations are right-censored due to the highest category being recorded as 3 or more and the data are somewhat overdispersed with a sample variance that is 2.1 times the sample mean. They prefer a model with $p = 2$, meaning two terms more than the NB2. Application by Gurmu and Trivedi (1996) to uncensored data is presented in section 11.7. In Gurmu (1997) the method is extended and applied to the hurdle model, leading to a model that nests the hurdle specifications considered in Chapter 4.7. Although appealing in principle, a potential problem with the hurdle specification is overfitting due to estimation of two parts of the model.

11.4.3 Semiparametric Nonlinear Weighted Least Squares

The methods of the previous subsection model the conditional distribution of the count. Now consider modeling only the conditional mean, when the conditional variance is completely unspecified.

Then the conditional variance is estimated nonparametrically. In ideal circumstances one can obtain estimates of the conditional mean parameters that are as efficient when the variance function is unspecified as they are when it is specified. Then the estimation method is called *adaptive*. Note that it is more ambitious than methods presented in Chapter 3 for an unknown variance function. With those methods, if the variance function was not specified or at least not assumed to be correctly specified, the goal was to obtain consistent parameter estimates and valid standard errors. Efficient estimation was not a goal.

For the linear model, with $\mathsf{E}[y_i|\mathbf{x}_i]$ Robinson introduced a semiparametric WLS estimator for the linear regression model in which the weights, the inverse of the square root of the error variance σ_i^2, are consistently estimated from residuals $\hat{u}_i = y_i - \mathbf{x}_i'\hat{\boldsymbol{\beta}}$ generated by the first-stage regression of y on $\mathbf{x}$. If there were many observations on y_i for a given value of $\mathbf{x}_i$ we could let $\hat{\sigma}_i^2$ simply be the average of $\hat{u}_i^2$ for those observations. More generally a local average is used, and the estimated variances $\hat{\sigma}_i^2$ are obtained by nonparametric regression of $\hat{u}_i^2$ on $\mathbf{x}_i$, where the method of k nearest neighbors (see, for example, Altman, 1992), is used rather than kernel methods. Here k can be viewed as a smoothing parameter, similar to the bandwidth in kernel methods, which determines the number of observations that are used in estimating each σ_i^2. A technical requirement is that the degree of smoothness should increase with the sample size, albeit at a slower rate. The properties of the estimator depend on the choice of the number of nearest neighbors. Values of $k = n^{1/2}$ and

$k = n^{3/5}$ have been used in empirical work; better is to use data-driven cross-validation methods. The resulting WLS estimator is shown to be adaptive. Formally this means it attains the semiparametric efficiency bound among estimators, given the specification of the conditional mean.

Delgado and Kniesner (1997) adapt the Robinson approach to heteroskedasticity of unknown functional form when the conditional mean function is exponential rather than linear. Following Robinson, given an initial root-n consistent estimator $\widetilde{\boldsymbol{\beta}}$, such as the NLS estimator or the Poisson MLE, they estimate the conditional variances σ_i^2 by the local averages

$$\hat{\sigma}_i^2 = \sum_{j=1}^{n} (y_j - \exp(\mathbf{x}_j'\widetilde{\boldsymbol{\beta}}))^2 w_{ij}, \tag{11.48}$$

where the w_{ij} are k nearest neighbor weights that equal $1/k$ for the k observations $\mathbf{x}_j$ closest to $\mathbf{x}_i$ according to a squared distance metric defined in the appendix to their paper, and equal 0 otherwise.

In general for y_i independent with mean $\mu_i(\mathbf{x}_i, \boldsymbol{\beta})$ and known variance σ_i^2, the nonlinear GLS estimator solves $\sum_{i=1}^{n}(y_i - \mu_i)(\partial\mu_i/\partial\boldsymbol{\beta})/\sigma_i^2 = 0$, and here $\partial\mu_i/\partial\boldsymbol{\beta} = \mu_i\mathbf{x}_i$. Thus the semiparametric WLS estimator $\hat{\boldsymbol{\beta}}_{\mathsf{SP}}$ solves the first-order conditions

$$\sum_{i=1}^{n} (y_i - \exp(\mathbf{x}_i'\hat{\boldsymbol{\beta}}_{\mathsf{SP}}))\exp(\mathbf{x}_i'\hat{\boldsymbol{\beta}}_{\mathsf{SP}})\mathbf{x}_i\hat{\sigma}_i^{-2} = \mathbf{0}. \tag{11.49}$$

Under regularity conditions, $\hat{\boldsymbol{\beta}}_{\mathsf{SP}}$ is root-n consistent and asymptotically normal with the variance matrix

$$\mathsf{V}[\hat{\boldsymbol{\beta}}_{\mathsf{SP}}] = \left(\sum_{i=1}^{n} \mathbf{x}_i\mathbf{x}_i'\hat{\mu}_i^2\hat{\sigma}_i^{-2}\right)^{-1}. \tag{11.50}$$

In theory a robust sandwich variant of (11.50) is not needed. But in finite samples one may want to guard against a poor choice of the number of neighbors and use

$$\mathsf{V}_{\mathsf{RS}}[\hat{\boldsymbol{\beta}}_{\mathsf{SP}}] = \left(\sum \mathbf{x}_i\mathbf{x}_i'\hat{\mu}_i^2\hat{\sigma}_i^{-2}\right)^{-1} \sum \mathbf{x}_i\mathbf{x}_i'\hat{u}_i^2\hat{\mu}_i^2\hat{\sigma}_i^{-4} \left(\sum \mathbf{x}_i\mathbf{x}_i'\hat{\mu}_i^2\hat{\sigma}_i^{-2}\right)^{-1},$$

where $\hat{u}_i = y_i - \exp(\mathbf{x}_i'\hat{\boldsymbol{\beta}}_{\mathsf{SP}})$ is the raw residual.

Delgado and Kniesner (1997) apply their method to a study of the factors determining the number of spells of absence of a week or less for London bus drivers. Their results appear to be more sensitive to the specification of the conditional mean than the estimator for the conditional variance. Indeed, their results suggest that the changes resulting from the use of the nonparametric variance estimation compared with the NB assumption are not large, even though the latter makes a very strong assumption that variances depend on the mean. A possible reason for this is that in practice the NB variance specification is a good approximation to heteroskedasticity of unknown form.

Delgado (1992) extended Robinson's approach to the estimation of a multivariate (multi-equation) nonlinear regression. This estimator is a systems GLS estimator based on k nearest neighbor estimates of the conditional variance matrices. The approach is an attractive alternative to likelihood-based methods in those cases in which the likelihood involves awkward integrals that cannot be simplified analytically. Multivariate count regressions with unrestricted patterns of dependence, and mixed multivariate models with continuous and discrete variables, fall into this category.

11.4.4 Semiparametric Estimation for GLMs

Chiou and Müller (1999) propose a semiparametric WLS estimator for GLMs with specified conditional moment function and variance $\sigma_i^2 = v(\mu_i)$, where the function $v(\cdot)$ is unspecified. It makes the stronger assumption that the variance is a function of the mean, as is typical for GLMs. It also has the advantage of reducing the nonparametric component of the problem to one dimension, rather than $\dim[\mathbf{x}_i]$ for Delgado and Kniesner (1997), so the estimator may perform better in finite samples.

In the special case of an exponential conditional mean, the estimator of Chiou and Müller also solves (11.49). The difference is that the conditional variances $\hat{\sigma}_i^2$ are obtained by kernel or local polynomial regression of $(y - \widetilde{\mu})^2$ on $\widetilde{\mu}$ where $\widetilde{\mu} = \exp(\mathbf{x}'\widetilde{\boldsymbol{\beta}})$ and $\widetilde{\boldsymbol{\beta}}$ is a consistent first-step estimator of $\boldsymbol{\beta}$.

11.4.5 Semiparametric Estimation of Mean and Variance

Chiou and Müller (1998) consider efficient estimation within the class of GLM estimators when, as in (11.24),

$$\begin{aligned} \mathrm{E}[y_i|\mathbf{x}_i] &= g(\mathbf{x}_i'\boldsymbol{\beta}) \\ \mathrm{V}[y_i|\mathbf{x}_i] &= v(g(\mathbf{x}_i'\boldsymbol{\beta})), \end{aligned} \tag{11.51}$$

but now the functional forms for neither the mean function $g(\cdot)$ nor the variance function $v(\cdot)$ are specified.

The estimator is similar to that of Weisberg and Welsh (1994) detailed in section 11.3.2, with an additional step for nonparametric estimation of $v(\cdot)$. First, for given estimate $\hat{\boldsymbol{\beta}}$ estimate $g(\cdot)$ by local linear regression of y on $\mathbf{x}_i'\hat{\boldsymbol{\beta}}$, and estimate $g'(\cdot)$ by local quadratic regression. Second, given $\hat{\mu}_i = \hat{g}(\mathbf{x}_i'\hat{\boldsymbol{\beta}})$, $i = 1, \ldots, n$, estimate $v(\cdot)$ by local linear regression of $(y - \hat{\mu})^2$ on $\hat{g}(\cdot)$. Third, given $\hat{\mu}_i = \hat{g}(\mathbf{x}_i'\hat{\boldsymbol{\beta}})$ and $\hat{\sigma}_i^2 = \hat{v}(\hat{g}(\mathbf{x}_i'\hat{\boldsymbol{\beta}}))$, $i = 1, \ldots, n$, estimate $\hat{\boldsymbol{\beta}}$ based on the equations (11.25) with the unknown functions $g(\cdot)$, $g'(\cdot)$, and $v(\cdot)$ replaced by estimates $\hat{g}(\cdot)$, $\hat{g}'(\cdot)$, and $\hat{v}(\cdot)$. This cycle is repeated until it converges. Chiou and Müller (1998) provide data-driven bandwidth selection based on the deviance and on Pearson's chi-squared statistic.

Chiou and Müller (2005) provide a generalization to clustered data or longitudinal data. Then y_{ij} is independent over i, y_{ij} has conditional mean and variance as in (11.51), while the conditional covariance within cluster j,

$$\mathrm{Cov}[y_{ij}, y_{i'j}|\mathbf{x}_{ij}, \mathbf{x}_{i'j}] = h(g(\mathbf{x}_{ij}'\boldsymbol{\beta}), g(\mathbf{x}_{i'j}'\boldsymbol{\beta})),$$

for univariate function $h(\cdot)$ that is unspecified. Then an extra step is added to nonparametrically estimate $h(\cdot)$, and the estimate $\hat{\boldsymbol{\beta}}$ is based on generalized estimating equations, see Chapter 2.4.6, rather than (11.25).

11.5 QUANTILE REGRESSION FOR COUNTS

We have focused on models for the conditional mean of y and for the conditional density of y and how they change as key regressors change. Interest may also lie in the impact of regressors on the conditional quantiles of y, such as the median, quartiles, and deciles. In principle these can be obtained from the conditional density, but doing so requires correct specification of the conditional density. Quantile regression instead directly estimates the conditional quantile, analogous to least squares regression to estimate the conditional mean.

Quantile regression (QR) methods for continuous unbounded variables are well established; see Koenker (2005) for a thorough treatment of properties of QR. Here we consider an extension to count regression that accounts for both the discreteness and nonnegativity of count data. The extension presented here requires additional assumptions that can be avoided by taking the fully nonparametric approach of Li and Racine (2008).

11.5.1 QR for Linear Regression

Let $F(y) = \Pr[Y \leq y]$ define the cumulative distribution function. The qth quantile of y, $q \in (0, 1)$, is that value of y that splits the data into proportions q below and $1 - q$ above. For example, the median y_{med} is such that $F(y_{med}) = 0.5$, so $y_{med} = F^{-1}(0.5)$. The median can be shown to minimize $\sum_i |y_i - y_{med}|$, and more generally the qth quantile α_q minimizes $\sum_{i:y_i \geq \alpha_q}^{n} q|y_i - \alpha_q| + \sum_{i:y_i < \alpha_q}^{n} (1-q)|y_i - \alpha_q|$.

Conditional quantile regression extends this to the regression case by supposing that the qth quantile equals $\mathbf{x}'\boldsymbol{\beta}_q$. Although this is a linear combination of the regressors, it need not necessarily be the case that the conditional mean itself is linear in regressors. The conditional quantile estimator of $\boldsymbol{\beta}_q$ minimizes

$$Q(\boldsymbol{\beta}_q) = \sum_{i:y_i \geq \mathbf{x}_i'\boldsymbol{\beta}_q}^{n} q|y_i - \mathbf{x}_i'\boldsymbol{\beta}_q| + \sum_{i:y_i < \mathbf{x}_i'\boldsymbol{\beta}_q}^{n} (1-q)|y_i - \mathbf{x}_i'\boldsymbol{\beta}_q|, \quad 0 < q < 1. \tag{11.52}$$

A major attraction of QR is that it potentially allows for response heterogeneity at different conditional quantiles of the variables of interest. There are some similarities between the QR and finite mixture (FM) approaches.

As in the case of FM, QR facilitates a richer interpretation of the data. But FM is fully parametric, whereas QR is consistent under weaker stochastic assumptions. And conditional quantiles are equivariant to monotone transformations, provided the functional form for the conditional quantile is correctly specified.

11.5.2 QR for Count Regression

We begin by noting that, whereas parametric models for counts can be used to recover the conditional quantiles, the standard models are too restrictive. Suppose y is distributed with Poisson density $f(y|\mu)$, where μ is modeled by a single-index model, so $\mu = g(\mathbf{x}'\boldsymbol{\beta})$. Then some algebra yields $\partial f(y|\mu)/\partial x_j = \beta_j f(y|\mu)(y-\mu)/\mu$. Winkelmann (2006) points out that this result implies that the marginal effect of a change in the regressor on $\Pr[Y = y|\mu]$ obeys a single-crossing property; in the case of $\beta_j > 0$ the marginal effect is negative for $y < \mu$, and positive for $y > \mu$. Then the marginal effect is necessarily negative at low conditional quantiles and positive at high quantiles. Winkelmann shows that this is also the case for the NB2 model.

This gives added motivation for directly modeling the conditional quantiles. Since the conditional quantiles for counts are positive, an exponential functional form is chosen, rather than the usual linear form. More substantively the discreteness of counts needs to accommodated. Machado and Santos Silva (2005) propose a method to do so, and Miranda (2006, 2008) and Winkelmann (2006) provide interesting applications to fertility and doctor visits data, respectively.

For counts the quantiles are not unique since the cdf is discontinuous with discrete jumps between flat sections. By convention the lower boundary of the interval defines the quantile in such a case. The lack of continuity is not in itself a serious violation for count data that have a large dispersion and relatively few tied values. But for many economics data sets, there can be a high proportion of zeros and numerous cases of tied values. Then some modification is necessary to allow the application of QR. This modification becomes less reasonable the less dispersed are the data. For example, if 95% of count observations are zeros then application of QR cannot be recommended.

The key step in the quantile count regression (QCR) model of Machado and Santos Silva (2005) involves replacing the discrete count outcome y with a continuous variable $z = h(y)$, where $h(\cdot)$ is a smooth continuous transformation. The particular continuous transformation used is

$$z_i = y_i + u_i, \tag{11.53}$$

where $u \sim \mathcal{U}[0, 1]$ is a pseudorandom draw from the uniform distribution on $(0, 1)$. This step, called "jittering," eliminates the discontinuities in the cdf. We can study the conditional quantiles of z, but it is the quantiles of y that are of interest. We next show how they are related.

The conditional quantile for $Q_q(z|\mathbf{x})$ is specified to be

$$Q_q(z|\mathbf{x}) = q + \exp(\mathbf{x}'\boldsymbol{\beta}_q). \tag{11.54}$$

As already noted, the exponential functional form is chosen, while the additional term q appears in the equation because $Q_q(z|\mathbf{x})$ is bounded from below by q, due to the uniform jittering operation. To be able to estimate a quantile model in the usual linear form $\mathbf{x}'\boldsymbol{\beta}$, a log transformation is applied so that $\ln(z - q)$ is modeled, with the adjustment that if $z - q < 0$ then we use $\ln(\varepsilon)$ where ε is a small positive number. The transformation is justified by the equivariance property of the quantiles, as well as the property that quantiles above the censoring point are not affected by censoring from below.

Postestimation the z-quantiles $Q_q(z|\mathbf{x})$ are transformed back to y-quantiles $Q_q(y|\mathbf{x})$. Because y is discrete the ceiling function is used, with

$$Q_q(y|\mathbf{x}) = \lceil Q_q(z|\mathbf{x}) - 1 \rceil, \tag{11.55}$$

where the symbol $\lceil r \rceil$ in the right-hand side of (11.55) denotes the smallest integer greater than or equal to r. This means that $Q_q(y|\mathbf{x}) = [Q_q(z|\mathbf{x})] - 1$ if $Q_q(z|\mathbf{x})$ is integer-valued and equals $[Q_q(z|\mathbf{x})]$ otherwise.

Jittering adds additional variation to the data. To reduce the effect of noise due to jittering, the model is estimated multiple times using independent draws from the $\mathcal{U}(0, 1)$ distribution, and the multiple estimated coefficients and confidence interval endpoints are averaged. Hence the estimates of the quantiles of y counts are based on

$$\hat{Q}_q(y|\mathbf{x}) = \lceil Q_q(z|\mathbf{x}) - 1 \rceil = \lceil q + \exp(\mathbf{x}'\overline{\hat{\boldsymbol{\beta}}}_q) - 1 \rceil, \tag{11.56}$$

where $\overline{\hat{\boldsymbol{\beta}}}_q$ denotes the average over the jittered replications. A number of variance estimators for QR have been proposed in the literature. Machado and Santos Silva (2005) discuss the related estimators for the count version of the same.

Once $\hat{Q}_q(y|\mathbf{x})$ is available, the impact of small changes in a selected covariate (under the assumption of all else being equal) can be calculated as the difference between $\hat{Q}_q(y|\mathbf{x}^p) - \hat{Q}_q(y|\mathbf{x}^c)$ where the superscripts p and c denote perturbed and control values.

The specific issue of how to choose the quantiles is discussed by Winkelmann (2006); the usual practice is to select a few values such as q equal to 0.25, 0.50, and 0.75. This practice has to be modified if there are many zeros in the sample. For example, if there are 35% zeros in a sample, then q must be greater than 0.35.

11.5.3 QR for Panel Counts

The preceding analysis applies to cross-section data. It extends to balanced and unbalanced panel data if the panel is treated as a pooled cross-section; see Winkelmann (2006) for an empirical example.

For panel data, however, the fixed and random effects framework is also widely used. For the linear model there has been some attempt to extend QR to panel data with fixed effects, but this literature is at a formative stage. Unlike the linear regression case, eliminating the individual-specific fixed effects parameter in the QR context is not straightforward. And the typical formulation of random effects based on parametric distributional assumption is not in the spirit of the semiparametric methodology of QR.

Therefore, an alternative strategy such as that of Abrevaya and Dahl (2006) is potentially attractive. They use the conditionally correlated random effects model presented in Chapter 9.5.4 as a way to obtain an estimator of the individual-specific effect. Then quantile regression is implemented to the model with estimated individual-specific effects as additional regressors.

11.6 NONPARAMETRIC METHODS

The analysis to date has presented models that are semiparametric or flexible parametric. Here we briefly review fully nonparametric methods. For concreteness we present only kernel methods, presented in detail in Li and Racine (2007), though for any given nonparametric setting there can be a range of alternative estimators such as nearest neighbors, smoothing splines, series, and sieve estimators.

11.6.1 Nonparametric Conditional Density Estimation

For univariate continuous random variable Y, the kernel estimate of the density at the evaluation point y is

$$\hat{f}(y) = \frac{1}{nh}\sum_{i=1}^{n} K\left(\frac{y_i - y}{h}\right), \tag{11.57}$$

where the function $K(\cdot)$ is a kernel function with properties that include at least $K(z) = K(-z)$, $\int K(z)dz = 1$, and $\int z^2 K(z)dz = \kappa > 0$. A commonly used kernel is the Gaussian kernel with $K(z) = (1/\sqrt{2\pi})\exp(-z^2/2)$. The choice of kernel function $K(\cdot)$ is less important than the choice of the bandwidth h that theoretically should approach zero at an appropriate rate as $n \to \infty$.

For multivariate continuous random variable $\mathbf{Y}$ of dimension m, the product kernel estimate of the density at the evaluation point $\mathbf{y} = (y_1, \ldots, y_m)$ is

$$\begin{aligned} \hat{f}(\mathbf{y}) &= \frac{1}{n(h_1 \times \cdots h_m)}\sum_{i=1}^{n}\prod_{j=1}^{m} K\left(\frac{y_{ji} - y_j}{h}\right) \\ &= \frac{1}{n(h_1 \times \cdots h_m)}\sum_{i=1}^{n}\mathbf{K}(\mathbf{y}_i - \mathbf{y}, \mathbf{h}), \end{aligned} \tag{11.58}$$

where the second equality defines the m-dimensional kernel $\mathbf{K}(\cdot)$. In the multivariate case the bandwidth $\mathbf{h} = (h_1, \ldots, h)$ is usually chosen by leave-one-out cross-validation, presented later for the regression case.

In regression applications, interest lies in the conditional density $f(y|\mathbf{x}) = f(y, \mathbf{x})/f(\mathbf{x})$. The kernel estimate of the conditional density of y given regressors $\mathbf{x}$ is

$$\hat{f}(y|\mathbf{x}) = \frac{\hat{f}(y, \mathbf{x})}{\hat{f}(\mathbf{x})}, \tag{11.59}$$

where $\hat{f}(y, \mathbf{x})$ and $\hat{f}(\mathbf{x})$ are obtained using (11.58) to estimate the joint data density $f(y, \mathbf{x})$ and the marginal density of the regressors $f(\mathbf{x})$.

Given conditional density estimation, the conditional cdf and conditional quantiles can also be estimated. The nonparametric conditional quantiles, due to Li and Racine (2008), provide an alternative to the count conditional quantiles in section 11.5 that needed to introduce jittering to handle the discreteness of count data.

11.6.2 Nonparametric Conditional Mean Estimation

For univariate continuous random variable Y and continuous multivariate random variable $\mathbf{X}$, the kernel estimate of the conditional mean of Y given $\mathbf{X} = \mathbf{x}$ is

$$\hat{g}(\mathbf{x}) = \frac{\sum_{i=1}^{n} y_i \mathbf{K}(\mathbf{x}_i - \mathbf{x}, \mathbf{h})}{\sum_{i=1}^{n} \mathbf{K}(\mathbf{x}_i - \mathbf{x}, \mathbf{h})}, \tag{11.60}$$

where $\mathbf{K}(\cdot)$ is a product kernel as in (11.58). The kernel regression estimate $\hat{g}(\mathbf{x})$ in (11.60) is also called the local constant estimator because it is equivalent to $\hat{a}(\mathbf{x})$ that minimizes $\sum_{i=1}^{n}(y_i - a)^2\mathbf{K}(\mathbf{x}_i - \mathbf{x}, \mathbf{h})$.

In addition to the conditional mean, interest often lies in the marginal effects $g'(\mathbf{x}) = \partial g'(\mathbf{x})/\partial \mathbf{x}$. One way to estimate these marginal effects is to form the local linear estimator that minimizes over a and $\mathbf{b}$

$$\sum_{i=1}^{n}(y_i - a - (\mathbf{x}_i - \mathbf{x})'\mathbf{b})^2\mathbf{K}(\mathbf{x}_i - \mathbf{x}, \mathbf{h}), \tag{11.61}$$

in which case $\hat{g}'(\mathbf{x}) = \hat{\mathbf{b}}(\mathbf{x})$. and $\hat{g}(\mathbf{x}) = \hat{a}(\mathbf{x})$

The bandwidth $\mathbf{h}$ can be chosen by leave-one-out cross-validation. In the case of local linear estimation the leave-one-out estimator when observation i is dropped is $(\hat{a}_i, \hat{\mathbf{b}}_i)$ that minimizes

$$\sum_{j=1, j\neq i}^{n} (y_i - a - (\mathbf{x}_j - \mathbf{x}_i)'\mathbf{b})^2\mathbf{K}\left(\mathbf{x}_j - \mathbf{x}_i, \mathbf{h}\right). \tag{11.62}$$

For m regressors, the bandwidth $\mathbf{h} = (h_1, \ldots, h_m)$ is chosen to minimize the cross-validation measure

$$\text{CV}(\mathbf{h}) = \frac{1}{n}\sum_{i=1}^{n}(y_i - \hat{a}_i)^2 M(\mathbf{x}_i), \tag{11.63}$$

where $M(\cdot)$ is a suitable weight function that trims outlying values. Variations include generalized cross-validation and expected Kullback-Liebler cross-validation (based on AIC for the nonparametric regression model).

11.6.3 Extensions

The preceding assumes continuous random variables. When random variables are discrete we can always use frequency methods, replacing the kernel weighting function $K((y_i - y)/h)$ by the indicator function $\mathbf{1}[y_i = y]$. But in practice this requires a large sample size and discrete random variables that take only a few distinct values.

Li and Racine, in a series of papers including Hall, Racine, and Li (2004) that are summarized in Li and Racine (2007), propose use of alternative weighting functions that lead to smoother estimation, thereby reducing estimator variance at the expense of introducing some bias as in the continuous case, and that enable use of cross-validation for bandwidth selection.

For scalar probability mass function estimation with discrete random variable Y that takes c distinct values, the kernel function $K((y_i - y)/h)$ in (11.57), for example, is replaced by the weight function

$$K_d(y_i, y, \lambda) = \begin{cases} 1 - \lambda & \text{if } y_i = y \\ \lambda/(1-c) & \text{if } y_i \neq y, \end{cases} \tag{11.64}$$

where $\lambda = 0$ yields the frequency estimate and $\lambda = 1$ corresponds to a uniform weight. For nonparametric regression with discrete regressor X, one can more simply replace the kernel $K((x_i - x)/h)$ with $K_d(x_i, x, \lambda) = 1$ if $x_i = x$ and $K_d(x_i, x, \lambda) = \lambda$ if $x_i \neq x$. Although these weights no longer sum to one, in the regression case this poses no problem because cancellation occurs since the weight function appears in both the numerator and denominator of (11.60).

When discrete data are ordered, nearby observations can be exploited in estimation, as in the continuous case. Then the kernel $K((y_i - y)/h)$ is replaced with

$$K_{ord}(y_i, y, \lambda) = \frac{c!}{j!(c-j)!}\lambda^j(1-\lambda)^{c-j} \quad \text{if } |y_i - y| = j, \tag{11.65}$$

where y takes the ordered values $\{0, 1, \ldots, c-1\}$. If the discrete data take a large number of values, as can be the case for count data, then this will yield similar results to the continuous case and it can be simpler to use the usual kernel methods.

11.7 EFFICIENT MOMENT-BASED ESTIMATION

In Chapter 2 and in sections 11.3 and 11.4 we considered efficient moment-based methods for efficient estimation of parameters of the model for the conditional mean based on models for the conditional mean and variance. Here we consider more efficient moment-based estimation that may arise if higher order moments than the second are additionally specified. We summarize results in the GLM literature before demonstrating that they follow naturally from a GMM formulation.

11.7.1 Estimating Equations and Quasilikelihood

The extended QL approach is a refinement of the moment-based approach discussed in earlier chapters. The version given here owes much to Crowder (1987) and Godambe and Thompson (1989). In this approach the central focus of estimation is on an estimating equation, whose solution defines an estimator, rather than an objective function that is maximized or minimized. See Chapter 2.5.1 for the general estimating equation approach. The equations are analogous to the score equations in maximum likelihood theory, leading to the terminology *extended* QL.

We consider models in which functional forms $\mu_i = \mu(\mathbf{x}_i, \boldsymbol{\theta})$ and $\sigma_i^2 = \omega(\mathbf{x}_i, \boldsymbol{\theta})$ are specified for the conditional mean and variance of the scalar dependent variable y_i. Crowder (1987) proposed the following general estimating equation

$$\sum_{i=1}^{n} \{\mathbf{a}(\mathbf{x}_i, \boldsymbol{\theta})(y_i - \mu_i) + \mathbf{b}(\mathbf{x}_i, \boldsymbol{\theta})\{(y_i - \mu_i)^2 - \sigma_i^2\}\} = \mathbf{0}, \tag{11.66}$$

where $\mathbf{a}(\mathbf{x}_i, \boldsymbol{\theta})$ and $\mathbf{b}(\mathbf{x}_i, \boldsymbol{\theta})$ are $q \times 1$ nonstochastic functions of $\boldsymbol{\theta}$, the $q \times 1$ unknown parameter vector to be estimated. Typically, $\boldsymbol{\theta} = (\boldsymbol{\beta}' \ \alpha)'$, where $\boldsymbol{\beta}$ are the parameters in the mean function, and the parameter α appears, in addition to $\boldsymbol{\beta}$, in the variance function. This class includes unweighted least squares where $\mathbf{a}(\mathbf{x}_i, \boldsymbol{\theta}) = \partial \mu_i / \partial \boldsymbol{\theta}$, and $\mathbf{b}(\mathbf{x}_i, \boldsymbol{\theta}) = \mathbf{0}$, and QL estimation in which case $\mathbf{a}(\mathbf{x}_i, \boldsymbol{\theta}) = (1/\sigma_i^2)\, \partial \mu_i / \partial \boldsymbol{\theta}$ and $\mathbf{b}(\mathbf{x}_i, \boldsymbol{\theta}) = \mathbf{0}$.

Setting $\mathbf{b}(\mathbf{x}_i, \boldsymbol{\theta}) \neq \mathbf{0}$ in (11.66) yields estimating equations that are quadratic in $(y_i - \mu_i)$. These *quadratic estimating equations* (QEEs) are a potential refinement to the QL approach. QL estimation is an appropriate approach when the variance specification is doubtful, whereas the quadratic approach is better if the variance specification is more certain. Cubic and higher order terms may be added if there is more information about higher moments, but the practical usefulness of such extensions is uncertain.

If $\mu_i = \mu(\mathbf{x}_i, \boldsymbol{\beta})$ and $\sigma_i^2 = \omega(\mathbf{x}_i, \boldsymbol{\beta}, \alpha)$ are correctly specified, and $\hat{\boldsymbol{\theta}}_{\text{QEE}}$ is the solution to the QEE, then from the results of Crowder (1987) it is known that the estimator is consistent and asymptotically normal with variance $\mathsf{V}[\hat{\boldsymbol{\theta}}_{\mathsf{QEE}}] = \mathbf{A}_n^{-1}\mathbf{B}_n\mathbf{A}_n^{-1\prime}$, where

$$\mathbf{A}_n = -\sum_{i=1}^{n}\left\{\mathbf{a}_i\frac{\partial\mu_i}{\partial\boldsymbol{\theta}'} + 2\sigma_i\mathbf{b}_i\frac{\partial\sigma_i}{\partial\boldsymbol{\theta}'}\right\}, \tag{11.67}$$

$$\mathbf{B}_n = \sum_{i=1}^{n}\sigma_i^2\left\{\mathbf{a}_i\mathbf{a}_i' + \sigma_i\gamma_{1i}(\mathbf{a}_i\mathbf{b}_i' + \mathbf{b}_i\mathbf{a}_i') + \sigma_i^2\left(\gamma_{2i}+2\right)\left(\mathbf{b}_i\mathbf{b}_i'\right)\right\}, \tag{11.68}$$

where $\mathbf{a}_i = \mathbf{a}(\mathbf{x}_i, \boldsymbol{\theta})$ and $\mathbf{b}_i = \mathbf{b}(\mathbf{x}_i, \boldsymbol{\theta})$ are $(k+1)\times 1$ vectors and γ_{1i} and γ_{2i} denote the skewness and kurtosis coefficients. The sandwich form of the variance matrix is used. Consistent estimation of $\boldsymbol{\theta}$ requires correct specification of the first two moments of y, whereas $\mathsf{V}[\hat{\boldsymbol{\theta}}_{\mathsf{QEE}}]$ depends on the first four moments. If $\mathbf{b}_i = \mathbf{0}$, the asymptotic covariance matrix does not depend on skewness or kurtosis parameters. This is consistent with earlier results for QL estimation.

As already noted, minimization of the first term only in the objective function corresponds to QL estimation. The second term is identically zero in some cases, such as the LEF density for which the variance function fully characterizes the distribution. In such a case, specification of higher order moments is redundant. In other cases, however, efficiency gains result from the inclusion of a correctly specified second term. An example of this follows.

Dean and Lawless (1989b) and Dean (1991), following Firth (1987), Crowder (1987), and Godambe and Thompson (1989), have discussed the estimation of a mixed Poisson model using the extended QL approach. It employs the following specification of the first four moments of the NB2 model:

$$\begin{aligned}
\mathsf{E}[y_i|\mathbf{x}_i] &= \mu_i = \mu_i(\mathbf{x}_i, \boldsymbol{\beta}) \qquad (11.69)\\
\mathsf{E}[(y_i-\mu_i)^2|\mathbf{x}_i] &= \sigma_i^2 = \mu_i(1+\alpha\mu_i)\\
\mathsf{E}[(y_i-\mu_i)^3|\mathbf{x}_i] &= \sigma_i^3\gamma_{1i} = \sigma_i^2(1+2\alpha\mu_i)\\
\mathsf{E}[(y_i-\mu_i)^4|\mathbf{x}_i] &= \sigma_i^4\gamma_{2i} = \sigma_i^2 + 6\alpha\sigma_i^4 + 3\sigma_i^4,
\end{aligned}$$

where $\alpha \geq 0$, and γ_{1i} and γ_{2i} denote skewness and kurtosis coefficients.

Suppose one assumes that the first four moments are known, but one does not wish to use the negative binomial distribution. The optimal quadratic estimating equations for estimation can be shown to be

$$\sum_{i=1}^{n}\frac{(y_i-\mu_i)}{\sigma_i^2}\frac{\partial\mu_i}{\partial\boldsymbol{\beta}} = \mathbf{0}, \tag{11.70}$$

$$\sum_{i=1}^{n}\left\{\frac{(y_i-\mu_i)^2-\sigma_i^2}{(1+\alpha\mu_i)^2} - \frac{(y_i-\mu_i)(1+2\alpha\mu_i)}{(1+\alpha\mu_i)^2}\right\} = \mathbf{0}. \tag{11.71}$$

Given correct specification of the first four moments, these equations yield the most efficient estimator for $(\boldsymbol{\beta}, \alpha)$. Given α, the solution of the first equation yields the QL estimate of $\boldsymbol{\beta}$. This is the special case of (11.66) with

$$\mathbf{a}(\mathbf{x}_i, \boldsymbol{\theta}) = \begin{bmatrix} (1/\sigma_i^2)(\partial\mu_i/\partial\boldsymbol{\beta}) \\ -(1+2\alpha\mu_i)/(1+\alpha\mu_i)^2 \end{bmatrix},$$

$$\mathbf{b}(\mathbf{x}_i, \boldsymbol{\theta}) = \begin{bmatrix} \mathbf{0} \\ 1/(1+\alpha\mu_i)^2 \end{bmatrix}.$$

The limit distribution can be obtained using (11.67) and (11.68).

Other estimators for α, given $\boldsymbol{\beta}$, have been suggested in the literature; for example, moment estimators that assume $\gamma_{1i} = \gamma_{2i} = 0$ and hence are less efficient than the previous one if the higher moment assumptions given earlier are correct. Dean and Lawless (1989b) have evaluated the resulting loss of efficiency in a simulation context. Dean (1991) shows that the asymptotic variance of $\boldsymbol{\beta}$ is unaffected by the choice of the estimating equation for α. An application to the mixed P-IG regression is in Dean, Lawless, and Willmot (1989).

11.7.2 Generalized Method of Moments

The literature generally does not motivate well the QEE estimator (11.66) and optimal cases such as (11.70) and (11.71). Results on optimal GMM provide a simple way to obtain the optimal formulation of the QEE.

From Chapter 2.5.3, the optimal GMM estimator for general conditional moment restriction $\mathsf{E}[\boldsymbol{\rho}(y_i, \mathbf{x}_i, \boldsymbol{\theta})|\mathbf{x}_i] = \mathbf{0}$, where $(y_i, \mathbf{x}_i)$ is iid, is the solution to the system of equations $\sum_{i=1}^{n} \mathbf{h}_i^*(y_i, \mathbf{x}_i, \boldsymbol{\theta}) = \mathbf{0}$, where

$$\mathbf{h}_i^*(y_i, \mathbf{x}_i, \boldsymbol{\theta}) = \mathsf{E}\left[\frac{\partial\boldsymbol{\rho}(y_i, \mathbf{x}_i, \boldsymbol{\theta})'}{\partial\boldsymbol{\theta}}\middle|\mathbf{x}_i\right] \left\{\mathsf{E}\left[\boldsymbol{\rho}(y_i, \mathbf{x}_i, \boldsymbol{\theta})\boldsymbol{\rho}(y_i, \mathbf{x}_i, \boldsymbol{\theta})'|\mathbf{x}_i\right]\right\}^{-1} \boldsymbol{\rho}(y_i, \mathbf{x}_i, \boldsymbol{\theta}). \tag{11.72}$$

We evaluate (11.72) for estimation based on the first two moments of the NB2 model:

$$\mu_i = \mu_i(\mathbf{x}_i, \boldsymbol{\beta}) \qquad \sigma_i^2 = \mu_i(1+\alpha\mu_i), \tag{11.73}$$

in which case $\boldsymbol{\beta}$ and α satisfy $\mathsf{E}[\boldsymbol{\rho}(y_i, \mathbf{x}_i, \boldsymbol{\theta})] = \mathbf{0}$, where

$$\boldsymbol{\rho}(y_i, \mathbf{x}_i, \boldsymbol{\theta}) = \begin{bmatrix} \rho_1(y_i, \mathbf{x}_i, \boldsymbol{\theta}) \\ \rho_2(y_i, \mathbf{x}_i, \boldsymbol{\theta}) \end{bmatrix} = \begin{bmatrix} y_i - \mu_i \\ (y_i - \mu_i)^2 - \sigma_i^2 \end{bmatrix}, \tag{11.74}$$

and $\boldsymbol{\theta} = (\boldsymbol{\beta}' \; \alpha)'$. Note that the conditional mean function is not restricted to be exponential.

For notational simplicity drop the subscript i and the conditioning on $\mathbf{x}_i$ in the expectation. The first term in the right-hand side of (11.72) is

$$\mathsf{E}\begin{bmatrix} \frac{\partial \rho_1}{\partial \boldsymbol{\beta}} & \frac{\partial \rho_2}{\partial \boldsymbol{\beta}} \\ \frac{\partial \rho_1}{\partial \alpha} & \frac{\partial \rho_2}{\partial \alpha} \end{bmatrix} = \mathsf{E}\begin{bmatrix} -\frac{\partial \mu}{\partial \boldsymbol{\beta}} & \{-2(y-\mu)-1-2\alpha\mu\}\frac{\partial \mu}{\partial \boldsymbol{\beta}} \\ 0 & -\mu^2 \end{bmatrix}$$

$$= \begin{bmatrix} -\frac{\partial \mu}{\partial \boldsymbol{\beta}} & -(1+2\alpha\mu)\frac{\partial \mu}{\partial \boldsymbol{\beta}} \\ 0 & -\mu^2 \end{bmatrix}, \qquad (11.75)$$

and the second term in the right-hand side of (11.72) is

$$\left\{\mathsf{E}\begin{bmatrix} \rho_1^2 & \rho_1\rho_2 \\ \rho_1\rho_2 & \rho_2^2 \end{bmatrix}\right\}^{-1}$$

$$= \left\{\mathsf{E}\begin{bmatrix} (y-\mu)^2 & (y-\mu)\{(y-\mu)^2-\sigma^2\} \\ (y-\mu)\{(y-\mu)^2-\sigma^2\} & \{(y-\mu)^2-\sigma^2\}^2 \end{bmatrix}\right\}^{-1}$$

$$= \left\{\begin{bmatrix} \sigma^2 & \sigma^3\gamma_1 \\ \sigma^3\gamma_1 & \sigma^4\gamma_2-\sigma^4 \end{bmatrix}\right\}^{-1}$$

$$= \frac{1}{\sigma^6(\gamma_2-1-\gamma_1^2)}\begin{bmatrix} \sigma^4\gamma_2-\sigma^4 & -\sigma^3\gamma_1 \\ -\sigma^3\gamma_1 & \sigma^2 \end{bmatrix}, \qquad (11.76)$$

using $\mathsf{E}[(y-\mu)^3|\mathbf{x}] = \sigma^3\gamma_1$ and $\mathsf{E}[(y-\mu)^4|\mathbf{x}] = \sigma^4\gamma_2$, where γ_1 and γ_2 denote skewness and kurtosis coefficients. Substituting (11.74) through (11.76) into (11.72) yields

$$\begin{bmatrix} \mathbf{h}_1^*(y_i,\mathbf{x}_i,\boldsymbol{\theta}) \\ h_2^*(y_i,\mathbf{x}_i,\boldsymbol{\theta}) \end{bmatrix}$$

$$= \begin{bmatrix} \frac{(\sigma_i^4\gamma_{2i}-\sigma_i^4)-(1+2\alpha\mu_i)\sigma_i^3\gamma_{1i}}{\sigma_i^6(\gamma_{2i}-1-\gamma_{1i}^2)}\frac{\partial \mu_i}{\partial \boldsymbol{\beta}} & \frac{\sigma_i^3\gamma_{1i}-(1+2\alpha\mu_i)\sigma_i^2}{\sigma^6(\gamma_2-1-\gamma_{1i}^2)}\frac{\partial \mu_i}{\partial \boldsymbol{\beta}} \\ \frac{\sigma_i^3\gamma_{1i}\mu_i^2}{\sigma_i^6(\gamma_{2i}-1-\gamma_{1i}^2)} & \frac{-\mu_i^2\sigma_i^2}{\sigma_i^6(\gamma_{2i}-1-\gamma_{1i}^2)} \end{bmatrix}$$

$$\times \begin{bmatrix} y-\mu_i \\ (y-\mu_i)^2-\sigma_i^2 \end{bmatrix} \qquad (11.77)$$

The optimal GMM estimator of $\boldsymbol{\beta}$ and α solves $\sum_{i=1}^n \mathbf{h}^*(y_i,\mathbf{x}_i,\boldsymbol{\theta}) = 0$ defined in (11.72). Its distribution can be obtained by applying the general results given in section 2.5.1.

Comparing (11.77) with (11.66), the optimal GMM estimator in this case is of the QEE form (11.66), with $\mathbf{a}(\mathbf{x}_i, \boldsymbol{\theta})$ and $\mathbf{b}(\mathbf{x}_i, \boldsymbol{\theta})$, respectively, equal to the first and second columns of the first matrix in the right-hand side (11.77), which is a $(k+1) \times 2$ matrix.

Now specialize to the case in which the skewness and kurtosis parameters γ_{1i} and γ_{2i} are the functions given in (11.69). Then, in section 11.9 it is shown that

$$
\begin{aligned}
(\sigma_i^4\gamma_{2i} - \sigma_i^4) - (1 + 2\alpha\mu_i)\sigma_i^3\gamma_{1i} &= \sigma_i^4(\gamma_{2i} - 1 - \gamma_{1i}^2) \qquad (11.78)\\
\sigma_i^3\gamma_{1i} - (1 + 2\alpha\mu_i)\sigma_i^2 &= 0\\
\sigma_i^6(\gamma_{2i} - 1 - \gamma_{1i}^2) &= 2(1+\alpha)\sigma_i^2.
\end{aligned}
$$

Using these results, (11.77) reduces to

$$
\begin{bmatrix} h_1^*(y_i, \mathbf{x}_i, \boldsymbol{\theta}) \\ h_2^*(y_i, \mathbf{x}_i, \boldsymbol{\theta}) \end{bmatrix} = \begin{bmatrix} \dfrac{1}{\sigma_i^2}\dfrac{\partial \mu_i}{\partial \boldsymbol{\beta}} & 0 \\ \dfrac{(1+2\alpha\mu_i)\mu_i^2}{2(1+\alpha)\sigma_i^4} & \dfrac{\mu_i^2}{2(1+\alpha)\sigma_i^4} \end{bmatrix} \begin{bmatrix} y - \mu_i \\ (y - \mu_i)^2 - \sigma_i^2 \end{bmatrix}. \qquad (11.79)
$$

For σ_i^2 defined in (11.73), $\mu_i^2/\sigma_i^4 = 1/(1+\alpha\mu_i)^2$. Thus (11.79) yields the optimal QEE estimator defined in (11.70) and (11.71), on premultiplication of (11.71) by the constant $2(1+\alpha)$.

The optimal GMM estimator requires specification of $\mathsf{E}[\boldsymbol{\rho}_i\boldsymbol{\rho}_i'|\mathbf{x}_i]$, which in this example requires specification of the first four conditional moments of y_i. Newey (1993) has proposed a semiparametric method of estimation, which replaces these elements by nonparametric estimates. For example, an estimate of the (1, 1) element, $\mathsf{E}[(y_i - \mu_i)^2|\mathbf{x}_i]$, may be formed from a kernel or series regression of $(y_i - \mu_i)^2$ on an intercept, $\hat{\mu}_i$ and $\hat{\mu}_i^2$, where the latter are consistent estimators. A similar treatment may be applied to the other two elements.

Unfortunately, even after determining that $\boldsymbol{\theta}$ is identified under GMM, this semiparametric method may run into practical difficulties. First, if the individually estimated elements are combined, there is no guarantee that the resulting estimate of $\mathsf{E}[\boldsymbol{\rho}_i\boldsymbol{\rho}_i'|\mathbf{x}_i]$ will be positive definite as required. Second, even if the procedure produces a positive definite estimate, the estimate may be highly variable. Finally, in small samples the resulting estimator may be biased, perhaps badly so as indicated by several studies of the GMM method (Smith, 1997). The econometric literature includes several studies in which use of a constant weighting matrix rather than the optimal matrix $(\mathsf{E}[\boldsymbol{\rho}_i\boldsymbol{\rho}_i'|\mathbf{x}_i])^{-1}$ produced better estimates.

11.8 ANALYSIS OF PATENT COUNTS

We consider the same cross-section data example as in Chapter 4.8.8, based on data from Hausman et al. (1984) and Hall et al. (1986). We use 1979 data

Table 11.1. *Patents: Poisson, NB2, FMNB2(2), and NB2P1 models with 1979 data*

			FMNB2(2)		
Variable	Poisson	NB2	Component 1	Component 2	NB2P1
Intercept	−0.340	−0.811	−1.274	1.516	−0.935
	(0.190)	(0.180)	(0.161)	(0.191)	(0.199)
ln R&D	0.475	0.816	0.918	0.352	0.611
	(0.064)	(0.083)	(0.063)	(0.090)	(0.066)
ln SIZE	0.271	0.115	0.065	0.277	0.110
	(0.055)	(0.069)	(0.057)	(0.062)	(0.050)
DSCI	0.467	−0.090	0.103	−0.573	0.006
	(0.155)	(0.138)	(0.142)	(0.346)	(0.107)
α (NB2)		0.885	0.651	0.062	1.507
		(0.100)	(0.080)	(0.047)	(0.236)
π			0.929		
			(0.022)		
a					0.082
					(0.026)
Number of parameters	4	5	11		6
Log-likelihood	−3,366	−1,128	−1,106		−1,116
AIC	6,740	2,265	2,234		2,245
BIC	6,755	2,284	2,277		2,269

Note: Standard errors are given in parentheses.

on the number of patents (PAT) awarded to 346 firms. From Table 4.2 the data are very widely dispersed and very overdispersed. The three regressors are the logarithm of total research over the current and preceding five years (ln $R\&D$), the logarithm of firm book value in 1972, which is a measure of firm size (ln $SIZE$), and a binary indicator variable equal to one if the firm is in the science sector ($DSCI$). Given the logarithmic transformation and the exponential conditional mean, the coefficient of ln $R\&D$ is an elasticity, so that the coefficient equals unity if there are constant returns to scale.

11.8.1 Parametric Models

Table 11.1 presents ML estimates for four parametric models. The first three also appear in Table 4.3 and have already been discussed in Chapter 4.8.8. The negative binomial model (NB2) fits the distribution of the data much better than the Poisson, with log-likelihood increasing from −3,366 to −1,128. The coefficient of ln $R\&D$ varies across the two models – 0.82 for the NB2 model compared to 0.49 for the Poisson – and lies between zero and one. The FMNB2(2) model, a two-component finite mixture of NB2, has improved fit, with log-likelihood increasing by 22 compared to the NB2 model. Using information criteria presented in Chapter 5.4.1 that penalize for model size, the

Table 11.2. *Patents: Actual versus nonparametric prediction*

	Predicted								
Actual	0	1	2–5	6–20	21–50	51–100	101–200	201–300	301–515
0	73	3	0	0	0	0	0	0	0
1	34	7	0	0	0	0	0	0	0
2–5	51	21	1	0	1	0	0	0	0
6–20	11	30	3	6	0	1	0	0	0
21–50	0	12	11	9	10	1	0	0	0
51–100	0	5	2	7	8	8	0	0	0
101–200	0	0	0	2	5	7	5	0	0
201–300	0	0	0	1	1	1	3	2	0
301–515	0	0	0	0	1	1	0	0	2

FMNB2(2) model is preferred to the more parsimonious NB2 model on the grounds of a lower AIC and BIC.

The final column of Table 11.1 presents a first-order squared polynomial model based on the NB2 density, denoted NB2P1. The slope coefficients can no longer be interpreted as semi-elasticities because the conditional mean is no longer $\exp(\mathbf{x}'\boldsymbol{\beta})$; see section 11.2.4. The model is an improvement on NB2 using either AIC or BIC, and it is preferred to the FMNB2(2) model using BIC though not AIC.

11.8.2 Nonparametric Conditional Density Estimate

We also performed nonparametric kernel estimation of the conditional density of PAT given ln $R\&D$, ln $SIZE$, and $DSCI$. Then $\hat{f}(y|\mathbf{x}) = \hat{f}(y, \mathbf{x})/\hat{f}(\mathbf{x})$ is obtained as the ratio of a four-dimensional kernel density estimate to a three-dimensional kernel density estimate, where PAT is treated as ordered discrete data with the weighting function given in (11.65), ln $SIZE$ and ln $R\&D$ are treated as continuous with a second-order Gaussian kernel of fixed bandwidth, and $DSCI$ is an unordered binary discrete variable with the weighting function given in (11.64). The bandwidth is chosen using expected Kullback-Liebler cross-validation. The resulting log-likelihood equals -1080, which is substantially larger than -1106 for FMNB2(2).

Table 11.2 is a classification table that compares the actual count y_i to the predicted count $\hat{y}_i$, where $\hat{y}_i = k$ if the conditional density estimate $\hat{f}(y|\mathbf{x}_i)$ is maximized when $y = k$. Since y and $\hat{y}$ take many values, Table 11.2 groups them into ranges. The nonparametric estimates predict zeros well and underpredict intermediate and larger counts. By contrast the NB2 model, in results not presented here, does similarly well in predicting zeros and ones, but underpredicts intermediate and larger counts much more. For example, for the 15 observations with PAT between 101 and 200 the NB2 model predicts that only 1 count exceeds 50, whereas the semiparametric model predicts that 12 counts exceed 50. And the NB2 predicts only three counts in excess of 43 (predicted

Table 11.3. *Patents: Summary of various fitted means*

Predicted mean	Mean	Standard Deviation	Minimum	25%	75%	Maximum	Fit $\text{Cor}[y,\hat{y}]^2$
PAT (y)	32.1	66.4	0.0	1.0	31.0	515.0	–
Poisson	32.1	64.4	0.4	3.8	35.5	669.0	0.71
NB2	41.6	130.0	0.1	2.3	30.7	1532.7	0.47
NP	32.5	58.5	−0.2	1.6	35.9	514.5	0.82
INDEX	31.5	55.8	0.0	2.2	35.9	483.7	0.77
PL	30.6	49.6	−18.4	2.4	38.5	323.5	0.70

Note: PAT, actual value; NB2, NB2 model; NP, nonparametric model; INDEX, single-index model; PL, partial linear model.

values of 124, 136, and 176 for PAT taking values of, respectively, 515, 183, and 319).

11.8.3 Nonparametric and Semiparametric Conditional Mean Estimation

We additionally estimate several semiparametric and nonparametric models of the conditional mean of PAT given the three regressors ln $R\&D$, ln $SIZE$, and $DSCI$.

The first model, denoted NP, estimates a fully nonparametric regression using local linear regression; see section 11.6.2. A second-order Gaussian kernel is used for ln $R\&D$ and ln $SIZE$, and the weight function in (11.64) is used for $DSCI$. The second model, denoted INDEX, is a single-index model estimated using the semiparametric least squares method of Ichimura (1993) that jointly estimates the bandwidth and coefficients using leave-one-out nonlinear least squares; see section 11.3.2. The third model, denoted PL, is a partially linear model with conditional mean $\mathbf{x}_1'\boldsymbol{\beta} + g(x_2)$ where ln $SIZE$ and $DSCI$ are the components of $\mathbf{x}_1$ and x_2 is ln $R\&D$. Estimation uses the method of Robinson; see section 11.3.3. This model is not expected to do as well as the first two models since for count data it is better to use the generalized partial linear model with conditional mean $\exp(\mathbf{x}_1'\boldsymbol{\beta} + g(x_2))$.

Table 11.3 presents descriptive statistics for the predicted values $\hat{y}$ of the conditional mean from the three nonparametric and semiparametric models, as well as for the Poisson and NB2 models. The NB2 model does particularly poorly, with the lowest squared correlation (of 0.47) between the actual and fitted values. Fitting the entire distribution using an NB2 model in this data example leads to poorer fit of the mean, as is also evident from the average fitted mean of 41.6 being substantially higher than the sample mean of 32.1.

The partial linear model fits the data as well as the Poisson in terms of $\text{Cor}[y, \hat{y}]^2$, though it does predict negative values. As already noted for counts, a generalized partial linear model should be used.

Table 11.4. *Patents: Correlations of various fitted means*

Predicted mean	y	Poisson	NB2	NP	INDEX	PL
PAT (y)	1.00	0.84	0.69	0.90	0.88	0.83
Poisson	0.84	1.00	0.92	0.94	0.96	0.93
NB2	0.69	0.92	1.00	0.77	0.79	0.83
NP	0.90	0.94	0.77	1.00	0.98	0.94
INDEX	0.88	0.96	0.79	0.98	1.00	0.95
PL	0.83	0.93	0.83	0.94	0.95	1.00

Note: y, actual value; NB2, NB2 model; NP, nonparametric model; INDEX, single-index model; PL, partial linear model.

The single-index model $g(\mathbf{x}'\boldsymbol{\beta})$ has estimated function $\hat{g}(\cdot)$, not plotted here, that has an exponential shape. Furthermore the estimated coefficients are, respectively, 0.678, 0.556, and 1.0, where the coefficient of $\ln SIZE$ is normalized to one. By contrast, the Poisson parameter estimates of 0.475, 0.271, and 0.467 from Table 11.1 become, on similar normalization, 1.072, 0.580, and 1.0, which are similar.

The nonparametric model leads to fitted values that are fairly similar to those for the index models. The nonparametric model is preferred because $\mathsf{Cor}[y, \hat{y}]^2$ is 0.82 compared to 0.77 for the single-index model.

Table 11.4 presents correlations for the fitted values. The single-index fitted values are highly correlated with those for the nonparametric model, suggesting that the single-index model may be a good model for these data.

Figure 11.1 presents plots of the fitted values from all these models except for the Poisson against the actual number of patents. These plots also suggest that the best fitting models are nonparametric and single-index semiparametric regression.

11.9 DERIVATIONS

11.9.1 Optimal GMM for NB2 First Four Moments

For notational simplicity we drop the subscript i in (11.77). From (11.69), $\gamma_1\sigma^3 = \sigma^2(1 + 2\alpha\mu)$ and $\gamma_2\sigma^4 = \sigma^2 + 6\alpha\sigma^4 + 3\sigma^4$. Then

$$\begin{aligned}(\sigma^4\gamma_2 - \sigma^4) - (1 + 2\alpha\mu)\gamma_1\sigma^3 &= \sigma^4\gamma_2 - \sigma^4 - (\sigma\gamma_1)\gamma_1\sigma^3 \\ &= \sigma^4\gamma_2 - \sigma^4 - \sigma^4\gamma_1^2 \\ &= \sigma^4\{\gamma_2 - 1 - \gamma_1^2\},\end{aligned}$$

$$\begin{aligned}\sigma^3\gamma_1 - (1 + 2\alpha\mu)\sigma^2 &= \sigma^2(1 + 2\alpha\mu) - (1 + 2\alpha\mu)\sigma^2 \\ &= 0,\end{aligned}$$

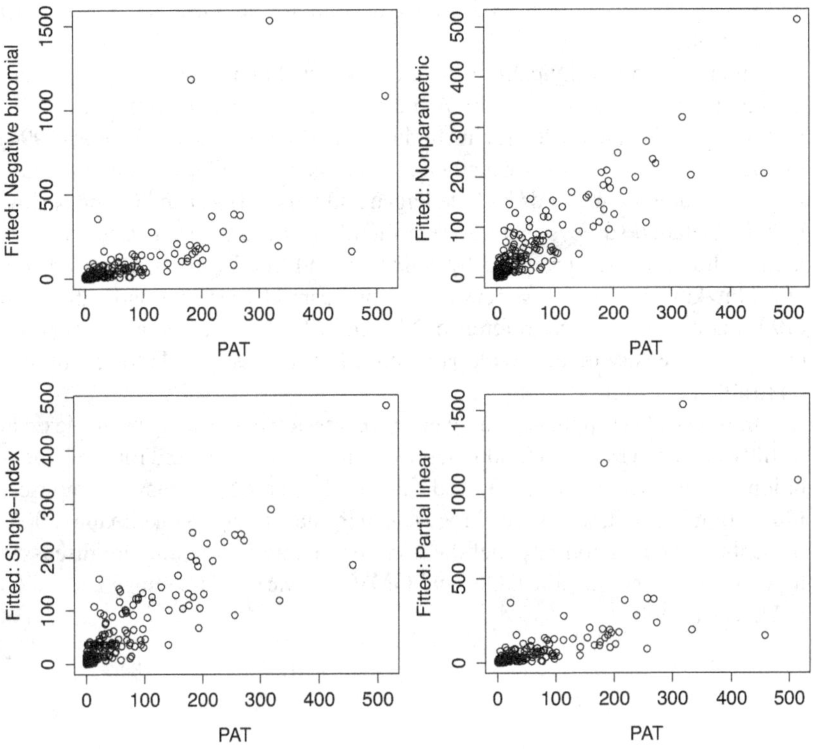

Figure 11.1. Fitted values from four models plotted against actual value.

and

$$\begin{aligned}\sigma^6(\gamma_2 - 1 - \gamma_1^2) &= \sigma^2(\sigma^2 + 6\alpha\sigma^4 + 3\sigma^4) - \sigma^6 - \sigma^4(1 + 2\alpha\mu)^2 \\ &= \sigma^4\{1 + 6\alpha\sigma^2 + 2\sigma^2 - (1 + 2\alpha\mu)^2\} \\ &= \sigma^4\{1+6\alpha\mu(1+\alpha\mu)+2\mu(1+\alpha\mu) - (1+4\alpha\mu+4\alpha^2\mu^2)\} \\ &= \sigma^4\{2(1 + \alpha)\mu(1 + \alpha\mu)\} \\ &= 2(1 + \alpha)\sigma^6.\end{aligned}$$

This yields (11.78).

11.10 BIBLIOGRAPHIC NOTES

The original articles by Cameron and Johansson (1999) and Gurmu et al. (1999) provide additional details on the computational aspects of their respective estimators based on squared polynomial series expansions. Johnson, Kemp, and Kotz (2005) provide an informative account of the power series family of distributions.

A relatively nontechnical introduction to kernel and nearest-neighbor nonparametric regression is given by Altman (1992), applied aspects of nonparametric regression are dealt with in Härdle (1990), and Härdle and Linton (1994) provide a survey of regression methods. Econometrics books on semiparametric and nonparametric methods include Pagan and Ullah (1999) and Li and Racine (2007). Detailed surveys that between them cover much of the recent econometrics literature are provided by Ichimura and Todd (2007), Chen (2007), and Matzkin (2007). For generalized additive models Hastie and Tibshirani (1990) is an authoritative reference. McLeod (2011) contrasts nonparametric conditional density models with parametric latent class models for count data on health utilization.

Miranda (2008) applies quantile count regression to the analysis of Mexican fertility data, and Miranda (2006) describes an add-on command for implementation in Stata. Cameron and Trivedi (2011, chapter 7.5) discuss an empirical illustration in detail, with special focus on marginal effects. Winkelmann (2006) presents an application to panel data on health care utilization and discusses how to choose the quantiles. Optimal GMM in a variety of settings is covered in Newey (1993).

CHAPTER 12

Bayesian Methods for Counts

12.1 INTRODUCTION

Bayesian methods provide a quite different way to view statistical inference and model selection and to incorporate prior information on model parameters. These methods have become increasingly popular over the past 20 years due to methodological advances, notably Markov chain Monte Carlo methods, and increased computational power.

Some applied studies in econometrics are fully Bayesian. Others merely use Bayesian methods as a tool to enable statistical inference in the classical *frequentist* maximum likelihood framework for likelihood-based models that are difficult to estimate using other methods such as simulated maximum likelihood.

Section 12.2 presents the basics of Bayesian analysis. Section 12.3 presents some results for Poisson models. Section 12.4 covers Markov chain Monte Carlo methods that are now the common way to implement Bayesian analysis when analytically tractable results cannot be obtained, and it provides an illustrative example. Section 12.5 summarizes Bayesian models for various types of count data. Section 12.6 concludes with a more complicated illustrative example, a count version of the Roy model that allows for endogenous selection.

12.2 BAYESIAN APPROACH

The Bayesian approach treats the parameters $\boldsymbol{\theta}$ as unknown random variables, with inference on $\boldsymbol{\theta}$ to be based both on the data $\mathbf{y}$ and on prior beliefs about $\boldsymbol{\theta}$. The data and prior beliefs are combined to form the posterior density of $\boldsymbol{\theta}$ given $\mathbf{y}$, and Bayesian inference is based on this posterior. This section presents a brief summary, with further details provided in subsequent sections.

12.2.1 Posterior Density

The *posterior density* of $\mathbf{y}$ given $\boldsymbol{\theta}$, denoted $p(\boldsymbol{\theta}|\mathbf{y})$, combines information from the current data $\mathbf{y}$ with *data density* $f(\mathbf{y}|\boldsymbol{\theta})$, and prior beliefs about $\boldsymbol{\theta}$, captured by a *prior density* $\pi(\boldsymbol{\theta})$. From Bayes theorem, $p(\boldsymbol{\theta}|\mathbf{y}) = f(\boldsymbol{\theta}, \mathbf{y})/f(\mathbf{y})$. Also by reexpressing Bayes theorem, $f(\boldsymbol{\theta}, \mathbf{y}) = f(\mathbf{y}|\boldsymbol{\theta}) \times \pi(\boldsymbol{\theta})$. It follows that $p(\boldsymbol{\theta}|\mathbf{y}) = f(\mathbf{y}|\boldsymbol{\theta}) \times \pi(\boldsymbol{\theta})/f(\mathbf{y})$.

In regression analysis we additionally condition on exogenous variables $\mathbf{X}$. Then we begin with the conditional density $f(\mathbf{y}|\boldsymbol{\theta}, \mathbf{X})$, yielding the following result for the posterior density:

$$p(\boldsymbol{\theta}|\mathbf{y}, \mathbf{X}) = \frac{f(\mathbf{y}|\boldsymbol{\theta}, \mathbf{X}) \times \pi(\boldsymbol{\theta})}{f(\mathbf{y}|\mathbf{X})}, \tag{12.1}$$

where $f(\mathbf{y}|\mathbf{X}) = \int f(\mathbf{y}|\boldsymbol{\theta}, \mathbf{X})\pi(\boldsymbol{\theta})d\boldsymbol{\theta}$ is a normalizing constant that ensures that the posterior density integrates to one. The density $f(\mathbf{y}|\boldsymbol{\theta}, \mathbf{X})$ is called the likelihood because it gives the probability or likelihood of observing the data $\mathbf{y}$ given regressors $\mathbf{X}$ and parameters $\boldsymbol{\theta}$. Frequentist analysis views $f(\mathbf{y}|\boldsymbol{\theta}, \mathbf{X})$ as a function of $\boldsymbol{\theta}$ and the MLE is that value of $\boldsymbol{\theta}$ that maximizes $f(\mathbf{y}|\boldsymbol{\theta}, \mathbf{X})$. The denominator density $f(\mathbf{y}|\mathbf{X})$ is called the *marginal likelihood* (with respect to $\boldsymbol{\theta}$). For basic count regression a Poisson or negative binomial model for $f(\mathbf{y}|\boldsymbol{\theta}, \mathbf{X})$ would be specified. We discuss later various specifications of the prior $\pi(\boldsymbol{\theta})$.

Rescaling the numerator in the right-hand side of (12.1) has no effect on the posterior $p(\boldsymbol{\theta}|\mathbf{y}, \mathbf{X})$ because the denominator in (12.1) changes by the same amount, providing that the rescaling is not a function of $\boldsymbol{\theta}$. The result (12.1) is often more simply written as

$$p(\boldsymbol{\theta}|\mathbf{y}, \mathbf{X}) \propto f(\mathbf{y}|\boldsymbol{\theta}, \mathbf{X}) \times \pi(\boldsymbol{\theta}), \tag{12.2}$$

where the symbol $\propto$ means "is proportional to." Even more simply we have

$$\text{posterior} \propto \text{ likelihood} \times \text{ prior}. \tag{12.3}$$

Although $f(\mathbf{y}|\mathbf{X})$ can be ignored in obtaining the key ingredient of Bayesian analysis, the posterior, some subsequent inference such as prediction and model selection requires computation of $f(\mathbf{y}|\mathbf{X})$.

12.2.2 Prior Density Specification

In some cases there is useful a priori knowledge about the likely values for $\boldsymbol{\theta}$. Then an *informative prior* is specified, one that has an impact on the posterior distribution. The functional form for the prior density of $\boldsymbol{\theta}$ is often chosen on grounds of analytical tractability or computational convenience. For iid data completely tractable results arise when the sample density is in the exponential family and the prior is a so-called natural conjugate prior; an example is given in section 12.3.1. For regression with count data such tractable results are unavailable, but some choices of functional form for the prior are more amenable to numerical computation of the posterior than others.

In most econometric applications with many regressors and parameters, there is little a priori knowledge about $\boldsymbol{\theta}$. Then an *uninformative prior* for $\boldsymbol{\theta}$ is specified, one that has little impact on the posterior. There are several ways to proceed.

The obvious approach is to use a proper prior for $\boldsymbol{\theta}$ that sets the variance of $\boldsymbol{\theta}$ to be very large. Such a prior is called a *diffuse prior* or *vague prior* or *flat prior* or *weak prior*. A second simpler approach is to use a *uniform prior* with $\pi(\boldsymbol{\theta}) = c$, for specified $c > 0$, making all values of $\boldsymbol{\theta}$ equally likely. This is usually an *improper prior*, since $\int c d\boldsymbol{\theta} = \infty$ for unbounded $\boldsymbol{\theta}$ so the density is improper. Nonetheless the resulting posterior is usually a proper density because constants cancel in the numerator and denominator of (12.1). A third approach is to use *Jeffrey's prior*, which is proportional to the square root of the determinant of $\partial^2 \ln f(\mathbf{y}|\boldsymbol{\theta}, \mathbf{X})/\partial\boldsymbol{\theta}\partial\boldsymbol{\theta}'$. This improper prior has the advantage of being invariant to reparameterization of $\boldsymbol{\theta}$, unlike a proper prior or the uniform prior. For example, a uniform prior $\pi(\theta) = c$ for $\theta > 0$ corresponds to the nonuniform prior $\pi^*(\gamma) = ce^{\gamma}$ for $\gamma = \ln\theta$.

An improper prior usually poses no problem for the posterior because constants cancel in the numerator and denominator of (12.1), and the methods detailed later permit estimation of the posterior density without knowing the marginal likelihood, the denominator in (12.1). But some subsequent analyses, notably Bayesian prediction, model selection, and hypothesis testing, do require computation of the marginal likelihood.

The prior is set without regard to the sample size, so as the sample size grows large the likelihood component of the posterior becomes dominant and the influence of the prior eventually becomes negligible.

12.2.3 Posterior Density Computation

In count applications there is usually no analytical solution for the posterior, especially once regressors are introduced. In that case a desired moment of the posterior distribution, say $\mathsf{E}[m(\boldsymbol{\theta})|\mathbf{y}] = \int m(\boldsymbol{\theta})p(\boldsymbol{\theta}|\mathbf{y})d\boldsymbol{\theta}$, can be computed using Monte Carlo methods for estimating an integral. Early work emphasized importance sampling methods; see, for example, Cameron and Trivedi (2005, p.444).

Usually, however, Monte Carlo methods presented in section 12.4 are used to instead obtain S draws $\boldsymbol{\theta}^{(1)}, \ldots, \boldsymbol{\theta}^{(S)}$ from the posterior $p(\boldsymbol{\theta}|\mathbf{y})$. Then moments such as the posterior mean and variance can be simply computed as $\hat{\mathsf{E}}[m(\boldsymbol{\theta})|\mathbf{y}] = \frac{1}{S}\sum_{s=1}^{S} m(\boldsymbol{\theta}^{(s)})$. Furthermore all interesting moments of functions of parameters *and* data can be estimated, given many draws from the posterior. This is convenient not only for coefficients but also treatment effects, especially when the quantiles of treatment effects are of interest. The hard work in Bayesian analysis is obtaining the draws. Once the draws are obtained, Bayesian analysis is simpler than frequentist analysis. The frequentist instead computes desired quantities evaluated at the MLE $\hat{\boldsymbol{\theta}}$ and then needs to use asymptotic theory and the delta method or a bootstrap to control for the sampling variability in $\hat{\boldsymbol{\theta}}$.

As the sample size gets large the likelihood component of the likelihood becomes dominant. Furthermore, the posterior density can be shown to be asymptotically normal, with the mean being the usual MLE $\hat{\boldsymbol{\theta}}$ based on the specified likelihood $f(\mathbf{y}|\boldsymbol{\theta}, \mathbf{X})$, and the variance matrix equal to minus the inverse of the information matrix $\partial^2 \ln f(\mathbf{y}|\boldsymbol{\theta}, \mathbf{X})/\partial\boldsymbol{\theta}\partial\boldsymbol{\theta}'$ evaluated at $\hat{\boldsymbol{\theta}}$. Thus asymptotically it is possible to provide a frequentist interpretation to Bayesian results, viewing the posterior mean as the MLE and the posterior standard deviation as the standard error of the MLE.

12.2.4 Bayesian Inference

Bayesian analysis bases inference on the posterior distribution of $\boldsymbol{\theta}$, obtained through either analytical or MCMC methods. In particular, a 95% *Bayesian credible interval* for θ is simply the 2.5 percentile to 97.5 percentile of the posterior distribution. Note that we are working directly with the posterior distribution. Furthermore, the Bayesian credible interval, also called a *Bayesian confidence interval,* is simpler to interpret than a classical confidence interval. If a 95% Bayesian credible interval is $(1, 4)$ then θ lies between 1 and 4 with posterior probability 0.95. By contrast, a 95% frequentist confidence interval of $(1, 4)$ is interpreted as meaning that if we were to have many different samples then 95% of similarly constructed intervals will include the single true unknown value of θ.

There are many different ways to construct a 95% interval for θ. Bayesians more often use the 95% *highest posterior density* (HPD) credible interval, which is the interval (a, b) such that the posterior probability of θ lying in a and b is 0.95 and $p(a|y) = p(b|y)$. For an asymmetric posterior, the usual case, the HPD credible interval is not the same as the interval from the 2.5 to the 97.5 percentiles. For vector parameter $\boldsymbol{\theta}$ this interval generalizes to the *HPD credible region.*

Posterior moments of θ are also informative, notably the posterior mean or median and the posterior standard deviation. If a loss function is specified then we can integrate out over this loss function using the posterior distribution to obtain a single optimal Bayesian estimate of $\boldsymbol{\theta}$. In particular, under quadratic loss the posterior mean is the best single estimate of $\boldsymbol{\theta}$.

Bayesian comparison of two models is based on the *posterior odds ratio,* which is the ratio of the marginal likelihoods of the two models, called the *Bayes factor*, weighted by the ratio of the prior probabilities of the two models. It includes as a special case a test of $\boldsymbol{\theta} = \boldsymbol{\theta}_1$ against $\boldsymbol{\theta} = \boldsymbol{\theta}_2$. The challenge is that it requires computation of the marginal likelihood, the denominator in (12.1), whereas evaluating the posterior does not.

In many applied studies the posterior is instead used as a means to enable classical frequentist interpretation of data generated by a process with a true value $\boldsymbol{\theta}_0$. An uninformative prior is specified, so the posterior is essentially the likelihood function. The posterior mean is treated as the MLE of $\boldsymbol{\theta}_0$, and the posterior variance is treated as the estimated variance of the MLE.

12.3 POISSON REGRESSION

For iid data the Poisson-gamma model is an example where an analytical expression for the posterior exists. Once regressors are introduced, however, there is no analytical expression.

12.3.1 Poisson-Gamma Model

We begin with no regressors, so the counts y_i, $i = 1, \ldots, n$, are Poisson distributed with mean μ, so

$$\begin{aligned} f(\mathbf{y}|\mu) &= \prod_{i=1}^{n} e^{-\mu}\mu^{y_i}/y_i! \qquad (12.4)\\ &= \exp\left\{-n\mu + n\bar{y}\ln\mu - \sum_{i=1}^{n}\ln y_i!\right\},\\ &\propto \exp\{-n\mu + n\bar{y}\ln\mu\}, \end{aligned}$$

where the final term can be ignored because it does not involve μ and will be captured by the normalizing constant in the posterior.

We specify an informative gamma prior for μ with mean a/b and variance a/b^2, where a and b are constants that are also specified, so

$$\begin{aligned} \pi(\mu) &= e^{-b\mu}\mu^{a-1}b^a/\Gamma(a) \qquad (12.5)\\ &= \exp\{-b\mu + (a-1)\ln\mu + a\ln b - \ln\Gamma(a)\},\\ &\propto \exp\{-b\mu + (a-1)\ln\mu\} \end{aligned}$$

where again multiplicative terms not involving μ will be captured by the normalizing constant in the posterior.

The resulting posterior $p(\mu|\mathbf{y}) \propto f(\mathbf{y}|\mu)\pi(\mu)$, yields

$$\begin{aligned} p(\mu|\mathbf{y}) &\propto \exp\{-n\mu + n\bar{y}\ln\mu\} \times \exp\{-b\mu + (a-1)\ln\mu\} \qquad (12.6)\\ &\propto \exp\{-(n+b)\mu + (n\bar{y}+a-1)\ln\mu\}. \end{aligned}$$

This is the kernel of a gamma distribution, so it follows that the posterior distribution for μ is a gamma with parameters $(n+b)$ and $(n\bar{y}+a)$. The posterior mean of μ is $(n\bar{y}+a)/(n+b) = \frac{n}{n+b}\bar{y} + \frac{na}{b(n+b)}b$, which is a weighted average of the sample mean $\bar{y}$ and the prior mean b. The posterior variance is $(n\bar{y}+a)/(n+b)^2$. Note that as $n \to \infty$ the posterior mean and variance go to $\bar{y}$ and $\bar{y}/n$, confirming the earlier statement that the prior has no influence as sample size $n \to \infty$.

It is interesting to compare the posterior to the original sample density $f(\mathbf{y}|\mu)$ viewed as a likelihood. The posterior essentially adds b observations on y that take value a. A similar augmented sample interpretation can be given in other exponential family examples where a natural conjugate prior exists.

Finally it can shown that for a single observation the marginal density $f(y_i) = \int p(\mu|y_i)d\mu$ is a negative binomial, so the prior can be viewed as introducing overdispersion to the Poisson.

As an example suppose we have a sample of size $n = 10$ with $\bar{y} = 3$, and our prior for μ is gamma with mean 4 and variance 4, so $a = 4$ and $b = 1$. Then the prior essentially adds one more observation with value 4. The posterior is gamma with posterior mean $(30+4)/(10+1) = 3.09$ and posterior variance $(30+4)/(10+1)^2 = 0.281$. A 95% credible region for μ is $(2.14, 4.12)$. By comparison the MLE is $\bar{y}$ with variance $\bar{y}/n$, so a 95% asymptotic confidence interval for μ is $3 \pm 1.96\sqrt{0.3} = (1.93, 3.07)$. An exact 95% confidence interval for the MLE is (2.0, 4.1), using $n\bar{y} = \sum_{i=1}^{n} y_i \sim \mathsf{P}[30]$ since the sum of iid Poissons is Poisson distributed.

12.3.2 Poisson Model

In the regression case $y_i|\mathbf{x}_i$ is Poisson distributed with mean $\mu_i = \exp(\mathbf{x}_i'\boldsymbol{\beta})$. Given prior density $\pi(\boldsymbol{\beta})$, the posterior is

$$p(\boldsymbol{\beta}|\mathbf{y}, \mathbf{X}) \propto \exp\left(\sum_{i=1}^{n}[-\exp(\mathbf{x}_i'\boldsymbol{\beta}) + y_i\mathbf{x}_i'\boldsymbol{\beta}]\right) \times \pi(\boldsymbol{\beta}). \tag{12.7}$$

Ideally we would have a prior for $\boldsymbol{\beta}$ that leads to a tractable posterior density $p(\boldsymbol{\beta}|\mathbf{y}, \mathbf{X})$, but there is no vector generalization of the preceding gamma prior that yields this result. Furthermore we should explicitly model the overdispersion that is usually present in count data.

The usual approach is to adopt a hierarchical modeling approach. Then $y_i|\mathbf{x}_i \sim \mathsf{P}[\mu_i]$ where μ_i themselves have an assumed distribution with parameters that priors are placed on.

One possibility is a Poisson-gamma mixture with $\mu_i = \upsilon_i \exp(\mathbf{x}_i'\boldsymbol{\beta})$, where υ_i is gamma with mean 1 and variance α. This leads to a negative binomial model; see Chapter 4.2. Then a prior is specified for $\boldsymbol{\beta}$ and α, with no closed-form solution for the posterior $p(\boldsymbol{\beta}, \alpha|\mathbf{y}, \mathbf{X})$.

More often a Poisson-lognormal mixture is considered with $\mu_i = \exp(\mathbf{x}_i'\boldsymbol{\beta} + \varepsilon_i)$ where $\varepsilon_i \sim \mathsf{N}[0, \sigma^2]$. Then a normal prior for $\boldsymbol{\beta}$ is specified, so $\boldsymbol{\beta} \sim \mathsf{N}[\underline{\boldsymbol{\beta}}, \underline{\mathbf{V}}]$ where, for example, $\underline{\boldsymbol{\beta}} = \mathbf{0}$ and $\underline{\mathbf{V}} = c\mathbf{I}$ for specified c, and an inverse-gamma prior is specified for the dispersion parameter σ^2. There is again no closed-form solution for $p(\boldsymbol{\beta}, \sigma^2|\mathbf{y}, \mathbf{X})$. However, this model is more amenable than the Poisson-gamma model to the computational methods summarized in the next section. We return to this particular model in section 12.5.1.

12.4 MARKOV CHAIN MONTE CARLO METHODS

Markov chain Monte Carlo (MCMC) methods provide a way to make (correlated) draws from the posterior, even if an analytical expression for the posterior is unavailable.

12.4.1 Overview

MCMC methods make sequential random draws $\boldsymbol{\theta}^{(s)}$, $s = 1, 2, \ldots$, that converge to draws from the posterior $p(\boldsymbol{\theta}|\cdot)$ as s become large. The process is a *Markov chain* because draws $\boldsymbol{\theta}^{(s)}$ depend in part on $\boldsymbol{\theta}^{(s-1)}$, but not earlier draws $\boldsymbol{\theta}^{(s-2)}, \boldsymbol{\theta}^{(s-3)}, \ldots$ once we condition on $\boldsymbol{\theta}^{(s-1)}$. Furthermore, the Markov chain is set up to be a stationary chain that converges to a stationary marginal distribution that is the desired posterior $p(\boldsymbol{\theta}|\mathbf{y})$. It is important that only draws from the stationary distribution are used, so initial "burn-in" draws from the chain are discarded and only the subsequent draws are treated as draws from the posterior.

The posterior draws are necessarily correlated. This causes no problem in estimating features of the posterior, such as the posterior mean and Bayesian credible intervals, except that these features will be more noisily estimated the more highly correlated the draws. In such cases more draws need to be made from the chain.

12.4.2 Gibbs Sampler

The *Gibbs sampler* is applicable to problems where it is difficult to make draws from the posterior $p(\boldsymbol{\theta})$, but the parameter vector can be partitioned in such a way that the full conditional posteriors are known.

For example, if we partition $p(\boldsymbol{\theta}) = p(\boldsymbol{\theta}_1, \boldsymbol{\theta}_2)$ then we need to know $p(\boldsymbol{\theta}_1|\boldsymbol{\theta}_2)$ and $p(\boldsymbol{\theta}_2|\boldsymbol{\theta}_1)$. The Gibbs sampler then makes alternating draws from $p(\boldsymbol{\theta}_1|\boldsymbol{\theta}_2)$ and $p(\boldsymbol{\theta}_2|\boldsymbol{\theta}_1)$. Thus given $\boldsymbol{\theta}_2^{(s-1)}$ we draw $\boldsymbol{\theta}_1^{(s)}$ from $p(\boldsymbol{\theta}_1|\boldsymbol{\theta}_2^{(s-1)})$, then $\boldsymbol{\theta}_2^{(s)}$ from $p(\boldsymbol{\theta}_2|\boldsymbol{\theta}_1^{(s)})$, then $\boldsymbol{\theta}_1^{(s+1)}$ from $p(\boldsymbol{\theta}_1|\boldsymbol{\theta}_2^{(s)})$, and so on. Even though

$$\begin{aligned} p(\boldsymbol{\theta}_1, \boldsymbol{\theta}_2) &= p(\boldsymbol{\theta}_1|\boldsymbol{\theta}_2) \times p(\boldsymbol{\theta}_2) \\ &\neq p(\boldsymbol{\theta}_1|\boldsymbol{\theta}_2) \times p(\boldsymbol{\theta}_2|\boldsymbol{\theta}_1), \end{aligned}$$

this method can be shown to converge to draws from the posterior $p(\boldsymbol{\theta}_1, \boldsymbol{\theta}_2)$.

The method generalizes to additional partitioning. In the general case $\boldsymbol{\theta}$ is partitioned into d blocks, so $p(\boldsymbol{\theta}) = p(\boldsymbol{\theta}_1, \boldsymbol{\theta}_2, \ldots, \boldsymbol{\theta}_d)$ and sequential draws are made from the d full conditional posteriors $p(\boldsymbol{\theta}_1|\boldsymbol{\theta}_2, \ldots, \boldsymbol{\theta}_d)$, $p(\boldsymbol{\theta}_2|\boldsymbol{\theta}_1, \boldsymbol{\theta}_3, \ldots, \boldsymbol{\theta}_d), \ldots, p(\boldsymbol{\theta}_d|\boldsymbol{\theta}_1, \ldots, \boldsymbol{\theta}_{d-1})$. The Gibbs sampler is a Markov chain that converges to a stationary distribution that is the desired full posterior $p(\boldsymbol{\theta}_1, \boldsymbol{\theta}_2, \ldots, \boldsymbol{\theta}_d)$; see Gelfand and Smith (1990).

The attraction of the Gibbs sampler is that it is generally faster than other MCMC methods, provided parameters in the different blocks are not too highly correlated. Bayesian models and priors are specified to exploit analytical results as much as possible, so that it is easy to make draws from all of the full conditional posteriors.

In some cases, however, it may be difficult to make draws from one (or more) of the conditional posteriors (e.g., from $p(\boldsymbol{\theta}_d|\boldsymbol{\theta}_1, \ldots, \boldsymbol{\theta}_{d-1})$). For such conditional posteriors the Metropolis-Hastings (MH) algorithm may be used to make draws: the method is called MH within Gibbs. The posterior should be partitioned so that correlated parameters are in the same MH block; see Chib (2001).

We begin with the Metropolis algorithm before turning to the more general MH algorithm.

12.4.3 Metropolis Algorithm

Consider the simplest case of desired draws from $p(\boldsymbol{\theta}) = p(\boldsymbol{\theta}|\mathbf{y})$, where there is no partitioning of the parameter vector.

The *Metropolis algorithm* makes a candidate draw $\boldsymbol{\theta}^*$ from the candidate distribution $g(\boldsymbol{\theta}|\boldsymbol{\theta}^{(s-1)})$, where $g(\cdot)$ must be symmetric, meaning $g(\boldsymbol{\theta}^a|\boldsymbol{\theta}^b) = g(\boldsymbol{\theta}^b|\boldsymbol{\theta}^a)$. The draw $\boldsymbol{\theta}^*$ is sometimes accepted as the new value $\boldsymbol{\theta}^{(s)}$, while sometimes we continue with the preceding value $\boldsymbol{\theta}^{(s-1)}$. The acceptance rule is

$$\boldsymbol{\theta}^{(s)} = \begin{cases} \boldsymbol{\theta}^* & \text{if } u < p(\boldsymbol{\theta}^*)/p(\boldsymbol{\theta}^{(s-1)}) \\ \boldsymbol{\theta}^{(s-1)} & \text{otherwise,} \end{cases} \tag{12.8}$$

where u is a draw from the uniform on $[0, 1]$. The algorithm is similar to the simple accept/reject method for making draws from $p(\boldsymbol{\theta})$, except the accept/reject method requires the stronger condition that there is a constant k such that $g(\boldsymbol{\theta}) > kp(\boldsymbol{\theta})$ for all $\boldsymbol{\theta}$.

The Metropolis algorithm has the attraction of requiring only computation of the ratio $p(\boldsymbol{\theta}^*)/p(\boldsymbol{\theta}^{(s-1)})$. Since the denominator in (12.1) does not depend on $\boldsymbol{\theta}$ it cancels out in the ratio, so $\boldsymbol{\theta}^*$ is accepted if $u < [f(\mathbf{y}|\boldsymbol{\theta}^*)\pi(\boldsymbol{\theta}^*)]/[f(\mathbf{y}|\boldsymbol{\theta}^{(s-1)})\pi(\boldsymbol{\theta}^{(s-1)})]$. Draws from the posterior can be made without requiring evaluation of the normalizing constant $f(\mathbf{y})$.

Ideally the candidate density is specified so that we are not at either extreme of accepting no draws or accepting all draws. The *random walk Metropolis algorithm* uses candidate density $\boldsymbol{\theta}^{(s)}|\boldsymbol{\theta}^{(s-1)} \sim \mathsf{N}[\boldsymbol{\theta}^{(s-1)}, \mathbf{V}]$ for specified $\mathbf{V}$. Then

$$g(\boldsymbol{\theta}^{(s)}|\boldsymbol{\theta}^{(s-1)}) = (2\pi)^{-k/2}|\mathbf{V}|^{-1/2}\exp(-\frac{1}{2}(\boldsymbol{\theta}^{(s)} - \boldsymbol{\theta}^{(s-1)})'\mathbf{V}^{-1/2}(\boldsymbol{\theta}^{(s)} - \boldsymbol{\theta}^{(s-1)}).$$

This clearly equals $g(\boldsymbol{\theta}^{(s-1)}|\boldsymbol{\theta}^{(s)})$, so the candidate density is symmetric and we can use the Metropolis algorithm.

12.4.4 Metropolis Algorithm for Poisson Regression with Uniform Prior

As an illustration of MCMC methods we apply the Metropolis algorithm to $y_i|\mathbf{x}_i \sim \mathsf{P}[\mu_i = \exp(\mathbf{x}_i'\boldsymbol{\beta})]$ with uniform prior $\pi(\boldsymbol{\beta}) = c$. More realistic examples that explicitly incorporate overdispersion and use a normal prior for $\boldsymbol{\beta}$ are given in section 12.5.

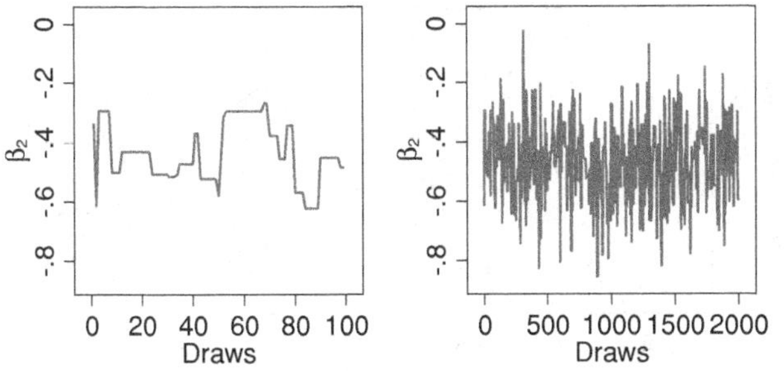

Figure 12.1. The first 100 retained draws (first panel) and the first 2,000 retained draws (second panel) of β_2 from the Metropolis algorithm.

From (12.7) with $\pi(\boldsymbol{\beta}) = c$ the posterior density is

$$p(\boldsymbol{\beta}|\mathbf{y}, \mathbf{X}) \propto \exp\left(\sum_{i=1}^{n}[-\exp(\mathbf{x}_i'\boldsymbol{\beta}) + y_i\mathbf{x}_i'\boldsymbol{\beta}]\right).$$

The acceptance rule (12.8), $u < p(\boldsymbol{\beta}^*)/p(\boldsymbol{\beta}^{(s-1)})$, is equivalently $\ln u < \ln(p(\boldsymbol{\beta}^*) - p(\boldsymbol{\beta}^{(s-1)})$, or here

$$\ln u < \sum_{i=1}^{n}[-\exp(\mathbf{x}_i'\boldsymbol{\beta}^*) + y_i\mathbf{x}_i'\boldsymbol{\beta}^*] - [-\exp(\mathbf{x}_i'\boldsymbol{\beta}^{(s-1)}) + y_i\mathbf{x}_i'\boldsymbol{\beta}^{(s-1)}].$$

We apply this to a simulated data set of size $n = 100$ of a count y_i and single regressor x_i. The regression model is $y_i|x_i \sim \mathsf{P}(\beta_1 + \beta_2 x_i)$, and the Metropolis algorithm uses draw $\boldsymbol{\beta}^* = \boldsymbol{\beta}^{(s-1)} + \mathbf{v}$ where $\mathbf{v}$ is drawn from $\mathsf{N}[\mathbf{0}, 0.25\mathbf{I}_2]$ and $c = 0.25$ was chosen after some trial and error. The first 10,000 Metropolis draws were discarded (burn-in) and the next 10,000 draws were kept.

Figure 12.1 presents the first 100 retained draws and the first 2,000 retained draws of the slope parameter β_2. The flat sections on the first panel are cases in which the new draw is not accepted. The second panel suggests that the chain has converged because it displays no upward or downward trend and it draws quite a range of values of β_2. In this particular example 26% of the draws are accepted. The draws are correlated, but this correlation disappears reasonably quickly. For the draws of β_2 the first five autocorrelations are, respectively, 0.80, 0.64, 0.51, 0.40, and 0.32.

Figure 12.2 plots the posterior density of β_2 estimated using the 10,000 retained draws from the Metropolis algorithm. The posterior will be normal asymptotically. Here the posterior is quite close to that of the normal, even though the sample size is only 100.

Table 12.1 provides some summary statistics for the posterior density of β_2. The posterior mean of β_2 is -0.464 and the posterior standard deviation is 0.121. A 95% percent Bayesian credible for interval β_2 is $(-0.706, -0.228)$. The results are similar to those for the MLE of the Poisson model, also presented in Table 12.1. This is expected because an uninformative prior was used.

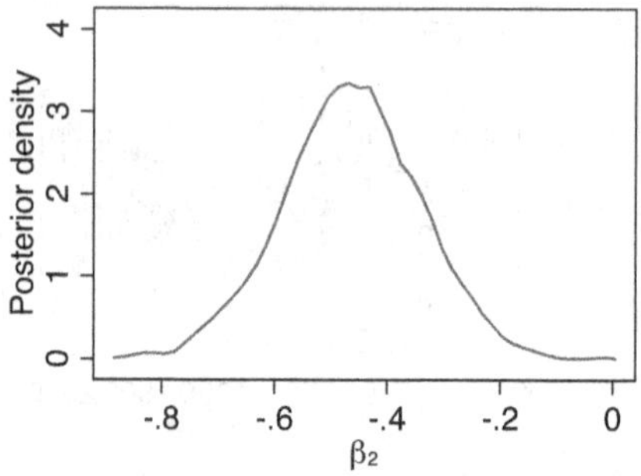

Figure 12.2. Estimate of posterior density of β_2 from 10,000 retained draws of β_2 from the Metropolis algorithm.

Using the results in Table 12.1, the posterior mean estimate $\hat{\mathsf{E}}[\beta_2|\mathbf{y}, \mathbf{X}] = \frac{1}{10000}\sum_{s=1}^{n}\beta_2^{(s)}$ has a standard error of $0.121/\sqrt{10{,}000} = 0.00121$ assuming that the MCMC draws are uncorrelated. In fact they are correlated, and controlling for this correlation using Newey-West standard errors gives a substantially larger standard error of 0.00336. The ratio $(0.00336/0.00121)^2 = 7.71$ is called the inefficiency factor, because it gives the efficiency loss in computation of the posterior mean of β_2 from these correlated draws compared to independent draws, and its inverse 0.13 is called the relative numerical efficiency. Despite this loss of efficiency, the posterior mean of −0.464 is still quite precisely estimated.

As a simple check of convergence a second chain was run with different seed and different starting values for β_1 and β_2. The results were similar, with a posterior mean for β_2 of −0.457, posterior standard deviation of 0.119, and an acceptance rate of 26%. In practice more detailed checks for convergence should be implemented.

12.4.5 Metropolis-Hastings Algorithm

The *Metropolis-Hastings algorithm* is a generalization of the Metropolis algorithm that does not require the candidate distribution $g(\boldsymbol{\theta}^{(s)}|\boldsymbol{\theta}^{(s-1)})$ to be

Table 12.1. *Metropolis Poisson example: Posterior summary and ML estimates*

	Bayesian posterior results			Maximum likelihood estimation results		
Variable	Posterior Mean	Posterior standard deviation	95% credible interval	Coefficient	Standard error	95% confidence interval
ONE	−0.135	0.110	(−0.356, 0.172)	−0.126	0.109	(−0.340, 0.089)
x	−0.464	0.121	(−0.706, −0.228)	−0.457	0.122	(−0.696, −0.219)

symmetric. Then the acceptance rule (12.8) becomes

$$\theta^{(s)} = \begin{cases} \theta^* & \text{if } u < \frac{p(\theta^*)\times g(\theta^{(s-1)}|\theta^*)}{p(\theta^{(s-1)})\times g(\theta^*|\theta^{(s-1)})} \\ \theta^{(s-1)} & \text{otherwise.} \end{cases} \tag{12.9}$$

The Gibbs sampler can be shown to be a special case of the MH algorithm.

A leading example is the *independence chain Metropolis-Hastings algorithm*, which uses $g(\theta^{(s)})$ not depending on $\theta^{(s-1)}$, but ensures that $g(\cdot)$ is a good approximation to $p(\cdot)$. Specifically, given the posterior $p(\theta|\mathbf{y},\mathbf{X})$ the usual ML computational methods are applied to obtain the MLE $\widetilde{\theta}$ and variance estimate $\widetilde{V}[\widetilde{\theta}] = -\left[\partial^2 \ln p(\theta|\mathbf{y},\mathbf{X})/\partial\theta\partial\theta'|_{\widetilde{\theta}}\right]^{-1}$. The candidate distribution $g(\theta^{(s)})$ is then the multivariate normal or, better still, the multivariate T that has fatter tails, with mean $\widetilde{\theta}$ and variance $\hat{V}[\widetilde{\theta}]$. Computational time can be saved by computing $\widetilde{\theta}$ using just the first few rounds of a Newton-Raphson algorithm rather than iterating to convergence.

For the independence chain the MH criteria simplifies to

$$\theta^{(s)} = \begin{cases} \theta^* & \text{if } u < \frac{p(\theta^*)\times g(\theta^{(s-1)})}{p(\theta^{(s-1)})\times g(\theta^*)} \\ \theta^{(s-1)} & \text{otherwise,} \end{cases} \tag{12.10}$$

where $g(\theta) \propto \exp(-\frac{1}{2}(\theta-\widetilde{\theta})'\widetilde{V}[\widetilde{\theta}]^{-1}(\theta-\widetilde{\theta}))$ for the normal independence chain.

For the preceding Poisson example with uniform prior and a normal independence chain the acceptance rule is

$$\begin{aligned}\ln u < {} & \textstyle\sum_{i=1}^N [-\exp(\mathbf{x}_i'\beta^*) + y_i\mathbf{x}_i'\beta^*] - [-\exp(\mathbf{x}_i'\beta^{(s-1)}) + y_i\mathbf{x}_i'\beta^{(s-1)}] \\ & -\tfrac{1}{2}(\beta^{(s-1)}-\widetilde{\beta})'\widetilde{V}[\widetilde{\theta}]^{-1}(\beta^{(s-1)}-\widetilde{\beta}) + \tfrac{1}{2}(\beta^*-\widetilde{\beta})'\widetilde{V}[\widetilde{\theta}]^{-1}(\beta^*-\widetilde{\beta})).\end{aligned}$$

In practice a more complex count model is specified, see section 12.5, and the MH algorithm is applied within Gibbs sampling. Consider the kth block and define θ_{-k} to include all components of θ other than θ_k, so $\theta = (\theta_k, \theta_{-k})$. Then in (12.9) the target posterior distribution $p(\theta^*)$ becomes $p(\theta_k^*, \theta_{-k})$, and the candidate distribution $g(\theta^*|\theta^{(s-1)})$ becomes $g_k(\theta^*|\theta^{(s-1)}, \theta_{-k})$ where $g_k(\cdot)$ may depend on the current value of θ_{-k} from the remaining blocks.

12.4.6 Data Augmentation for Latent Variable Models

Several leading nonlinear regression models are based on underlying latent (or hidden) variables. Leading examples are censored and selection models, such as the Tobit model; discrete choice models, such as the multinomial probit that are based on the additive random utility model; and random effects or random coefficients models.

In such cases Bayesian analysis would be simpler if the latent variables were known. This is especially so in the common case of normally distributed latent variables, because then standard Bayesian results for linear regression under normality can be applied. Let $\mathbf{y}$ denote the vector of observed variables and $\mathbf{y}^*$ the vector of latent variables, where the components are related by $y_i = g(y_i^*)$. Then rather than obtain the posterior $p(\boldsymbol{\theta}|\mathbf{y}) \propto f(\mathbf{y}|\boldsymbol{\theta})\pi(\boldsymbol{\theta})$, the latent variables $\mathbf{y}^*$ are introduced as additional parameters and analysis is based on the *augmented posterior*

$$p(\boldsymbol{\theta}, \mathbf{y}^*|\mathbf{y}) \propto f(\mathbf{y}|\boldsymbol{\theta}, \mathbf{y}^*)\pi(\boldsymbol{\theta}). \tag{12.11}$$

MCMC draws of $\boldsymbol{\theta}$ from this augmented posterior $p(\boldsymbol{\theta}, \mathbf{y}^*|\mathbf{y})$ can also be used as the draws from its marginal $p(\boldsymbol{\theta}|\mathbf{y})$, the original desired posterior.

The advantage of this procedure is computational because it enables use of Gibbs sampling, alternating between draws from the two conditional posteriors $p(\boldsymbol{\theta}|\mathbf{y}^*, \mathbf{y})$ and $p(\mathbf{y}^*|\mathbf{y}, \boldsymbol{\theta})$ of the full posterior given in (12.11). The conditional posterior $p(\boldsymbol{\theta}|\mathbf{y}^*, \mathbf{y})$ simplifies to $p(\boldsymbol{\theta}|\mathbf{y}^*)$, since $y_i = g(y_i^*)$, and it may be much simpler to draw from $p(\boldsymbol{\theta}|\mathbf{y}^*)$ than from the original $p(\boldsymbol{\theta}|\mathbf{y})$. The conditional posterior $p(\mathbf{y}^*|\boldsymbol{\theta}, \mathbf{y})$ is used to generate $\mathbf{y}^*$ given $\mathbf{y}$ and $\boldsymbol{\theta}$. The procedure is called *data augmentation*, because the observed data $\mathbf{y}$ are augmented by generated data $\mathbf{y}^*$, though mechanically it is the parameter vector $\boldsymbol{\theta}$ that has been augmented.

As an example, consider a latent variable formulation of the probit model. Then we observe $y_i = 1$ (or $y_i = 0$) if the latent variable $y_i^* = \mathbf{x}_i'\boldsymbol{\beta} + \varepsilon_i > 0$ (or is < 0), where $\varepsilon_i \sim \mathsf{N}[0, 1]$. It follows that $\Pr[y_i = 1|\mathbf{x}_i] = \Pr[y_i^* > 0|\mathbf{x}_i] = \Phi(\mathbf{x}_i'\boldsymbol{\beta})$, the probit model, where $\Phi(\cdot)$ is the standard normal cdf. Rather than draw from the posterior $p(\boldsymbol{\beta}|\mathbf{y}, \mathbf{X})$ we draw from the augmented posterior $p(\boldsymbol{\beta}, \mathbf{y}^*|\mathbf{y}, \mathbf{X})$. Here $\mathbf{y}^*|\boldsymbol{\beta}, \mathbf{X} \sim \mathsf{N}[\mathbf{X}\boldsymbol{\beta}, \mathbf{I}]$ and we can just use standard Bayesian results for the linear regression model under normality (with the simplification of a unit variance). With a uniform prior on $\boldsymbol{\beta}$ the conditional posterior $p(\boldsymbol{\beta}|\mathbf{y}^*, \mathbf{X})$ is $\mathsf{N}[(\mathbf{X}'\mathbf{X})^{-1}\mathbf{X}'\mathbf{y}^*, (\mathbf{X}'\mathbf{X})^{-1}]$, whereas with a normal prior the posterior is that in section 12.5.1 with $\mathbf{y}$ replaced by $\mathbf{y}^*$. To draw $\mathbf{y}^*$ from $p(\mathbf{y}^*|\boldsymbol{\beta}, \mathbf{y}, \mathbf{X})$ we note that $p(y_i^*|\boldsymbol{\beta}, y_i, \mathbf{x}_i)$ are independent truncated normals. So if $y_i = 1$ draw from $\mathsf{N}[\mathbf{x}_i'\boldsymbol{\beta}, 1]$ left truncated at 0, and if $y_i = 0$ draw from $\mathsf{N}[\mathbf{x}_i'\boldsymbol{\beta}, 1]$ right truncated at 0.

12.5 COUNT MODELS

We present several models for count data for several common types of data. These models are most often based on the Poisson with lognormal heterogeneity to control for overdispersion, normal priors for regression parameters, and inverse-gamma or Wishart priors for variance parameters. The Poisson-lognormal has computational advantage over the Poisson-gamma (the negative binomial) because it lends itself well to data augmentation.

12.5.1 Key Results for Linear Regression under Normality

The subsequent analysis uses the following analytical results for linear regression under normality with a normal-gamma prior.

Consider independent data $y_i, i = 1, \ldots, N$, from the homoskedastic linear model under normality,

$$\mathbf{y} \sim \mathsf{N}[\mathbf{X}\boldsymbol{\beta}, \sigma^2\mathbf{I}]. \tag{12.12}$$

Given independent priors for $\boldsymbol{\beta}$ and σ^2, the posterior is

$$p(\boldsymbol{\beta}, \sigma^2|\mathbf{y}, \mathbf{X}) \propto f(\mathbf{y}|\boldsymbol{\beta}, \sigma^2, \mathbf{X})\pi(\boldsymbol{\beta})\pi(\sigma^2). \tag{12.13}$$

The following choices for the priors lead to tractable posteriors for $\boldsymbol{\beta}$ and σ^2.

First, if the prior density for $\boldsymbol{\beta}$ is specified to be $\mathsf{N}[\underline{\boldsymbol{\beta}}, \underline{\mathbf{V}}]$, then the posterior density $p(\boldsymbol{\beta}|\sigma^2, \mathbf{y}, \mathbf{X})$ is $\mathsf{N}[\overline{\boldsymbol{\beta}}, \overline{\mathbf{V}}]$, where

$$\begin{aligned}\overline{\mathbf{V}} &= [\underline{\mathbf{V}}^{-1} + (\sigma^2(\mathbf{X}'\mathbf{X})^{-1})^{-1}]^{-1} \\ \overline{\boldsymbol{\beta}} &= \overline{\mathbf{V}}[\underline{\mathbf{V}}^{-1}\underline{\boldsymbol{\beta}} + (\sigma^2(\mathbf{X}'\mathbf{X})^{-1})^{-1}\hat{\boldsymbol{\beta}}_{\mathrm{OLS}})] \\ &= \overline{\mathbf{V}}[\underline{\mathbf{V}}^{-1}\underline{\boldsymbol{\beta}} + \sigma^{-2}\mathbf{X}'\mathbf{y}].\end{aligned} \tag{12.14}$$

The posterior is normal with precision $\overline{\mathbf{V}}^{-1}$, which is the sum of the prior precision and the sample precision, and posterior mean, which is a weighted sum of the prior mean and the sample mean (the OLS estimator). Common notation in the Bayesian literature is to use an underscore for the parameters of a prior density and an overscore for parameters of the posterior.

Second, if the prior density for σ^{-2} is specified to be the Gamma $[\frac{a}{2}, (\frac{c}{2})^{-1}]$, then the posterior $p(\sigma^{-2}|\boldsymbol{\beta}, \mathbf{y}, \mathbf{X})$ is

$$\text{Gamma}\left[\frac{a+n}{2}, \left(\frac{c}{2} + \sum\nolimits_{i=1}^{n} \frac{(y_i - \mathbf{x}_i'\boldsymbol{\beta})^2}{2}\right)^{-1}\right]. \tag{12.15}$$

When more generally $\text{Var}[\mathbf{y}|\mathbf{X}] = \boldsymbol{\Sigma}$, a Wishart prior for $\boldsymbol{\Sigma}^{-1}$ is specified, leading to a Wishart posterior for $\boldsymbol{\Sigma}^{-1}|\boldsymbol{\beta}, \mathbf{y}, \mathbf{X}$.

12.5.2 Cross-Section Data

The Poisson lognormal model specifies

$$\begin{aligned} y_i|\mathbf{x}_i &\sim \mathsf{P}[\mu_i = \exp(\gamma_i)] \\ \gamma_i &= \mathbf{x}_i'\boldsymbol{\beta} + \varepsilon_i. \\ \varepsilon_i &\sim \mathsf{N}[0, \sigma^2]. \end{aligned} \tag{12.16}$$

This is a lognormal model for heterogeneity since equivalently the Poisson mean is $\mu_i = \upsilon_i \exp(\mathbf{x}_i'\boldsymbol{\beta})$, where the heterogeneity term $\upsilon_i = \exp(\varepsilon_i)$ is lognormal with mean 1.

Bayesian analysis is based on the posterior $p(\boldsymbol{\beta}, \sigma^2|\mathbf{y}, \mathbf{X})$. This would be straightforward if we knew γ_i, since then we would have a standard linear normal regression model $\gamma_i = \mathbf{x}_i'\boldsymbol{\beta} + \varepsilon_i$ and could assume a normal-gamma prior and use the results of section 12.5.1. So data augmentation is employed to include the γ_i as parameters and make draws from the augmented posterior $p(\boldsymbol{\beta}, \sigma^2, \boldsymbol{\gamma}|\mathbf{y}, \mathbf{X})$ where $\boldsymbol{\gamma} = (\gamma_1, \ldots, \gamma_n)$.

The full conditional posterior $p(\boldsymbol{\gamma}|\boldsymbol{\beta}, \sigma^2, \mathbf{y}, \mathbf{X}_i)$ is the product of the likelihood, here $\mathsf{P}[\exp(\gamma_i)]$ for the ith observation, and the "prior" for γ_i which is $\mathsf{N}[\mathbf{x}_i'\boldsymbol{\beta}, \sigma^2]$ since $\gamma_i = \mathbf{x}_i'\boldsymbol{\beta} + \varepsilon_i$ with $\varepsilon_i \sim \mathsf{N}[0, \sigma^2]$. Thus for the ith observation the conditional posterior is

$$p(\gamma_i|\boldsymbol{\beta}, \sigma^2, y_i, \mathbf{x}_i) \propto \exp\{y_i\gamma_i - \exp(\gamma_i)\} \times \exp\{-0.5\sigma^{-2}(\gamma_i - \mathbf{x}_i'\boldsymbol{\beta})^2\}.$$

This is not tractable so we use the MH algorithm to make draws. For example we could use the normal with mean $\widetilde{\gamma}_i$, the maximum with respect to γ_i of $p(\gamma_i|\sigma^2, y_i, \mathbf{x}_i)$, and variance $-\partial^2 p(\gamma_i|\sigma^2, y_i, \mathbf{x}_i)/\partial^2\gamma_i$ evaluated at $\widetilde{\gamma}_i$.

We then make draws from the other full conditionals, $p(\boldsymbol{\beta}|\sigma^2, \boldsymbol{\gamma}, \mathbf{y}, \mathbf{X})$ and $p(\sigma^2|\boldsymbol{\beta}, \boldsymbol{\gamma}, \mathbf{y}, \mathbf{X})$. If the prior for $\boldsymbol{\beta}$ is $\mathsf{N}[\underline{\boldsymbol{\beta}}, \underline{\mathbf{V}}]$ then the conditional posterior $p(\boldsymbol{\beta}|\sigma^2, \boldsymbol{\gamma}, \mathbf{y}, \mathbf{X}) = p(\boldsymbol{\beta}|\sigma^2, \boldsymbol{\gamma}, \mathbf{X})$ is normal, with the posterior mean and variance that given in (12.14) with $\mathbf{y}$ replaced by $\boldsymbol{\gamma}$. If the prior for σ^{-2} is Gamma $[\frac{a}{2}, (\frac{c}{2})^{-1}]$ then the conditional posterior $p(\sigma^2|\boldsymbol{\beta}, \boldsymbol{\gamma}, \mathbf{y}, \mathbf{X}) = p(\sigma^2|\boldsymbol{\beta}, \boldsymbol{\gamma}, \mathbf{X})$ is gamma as in (12.15) with y_i replaced by γ_i.

Note that this data augmentation approach relies crucially on ε_i in (12.16) being normally distributed so that γ_i is normal and Bayesian results for the linear normal model can be used. With other distributions for heterogeneity, analysis would be much more complicated. With gamma heterogeneity it is actually easier to work directly with the negative binomial model with density $f(\mathbf{y}, \mathbf{X}|\boldsymbol{\beta}, \alpha)$, where α is the overdispersion parameter. Then calculate the posterior $p(\boldsymbol{\beta}, \alpha|\mathbf{y}, \mathbf{X}) \propto f(\mathbf{y}, \mathbf{X}|\boldsymbol{\beta}, \alpha) \times \pi(\boldsymbol{\beta}) \times \pi(\alpha)$ using the MH algorithm.

12.5.3 Panel Data

For a random effects model a similar approach to the preceding can be taken. Chib, Greenberg, and Winkelmann (1998) specify for individual i at time t the model

$$\begin{aligned} y_{it}|\mathbf{x}_{it} &\sim \mathsf{P}[\exp(\gamma_{it})] \\ \gamma_{it} &= \mathbf{x}_{it}'\boldsymbol{\beta} + \mathbf{w}_{it}'\boldsymbol{\alpha}_i \\ \boldsymbol{\alpha}_i &\sim \mathsf{N}[\boldsymbol{\eta}, \boldsymbol{\Sigma}]. \end{aligned} \tag{12.17}$$

The standard random effects model is the specialization $\mathbf{w}_{it}'\boldsymbol{\alpha}_i = \alpha_i$. In the more general case here Chib et al. (1998) argue that $\mathbf{x}_{it}$ and $\mathbf{w}_{it}$ should share no common variables to avoid identification and computational problems.

The structure of the model is similar to that in the previous subsection. Data augmentation is used to add γ_{it} as parameters, leading to the augmented posterior $p(\boldsymbol{\beta}, \boldsymbol{\eta}, \boldsymbol{\Sigma}, \boldsymbol{\gamma}|\mathbf{y}, \mathbf{X})$. A Gibbs sampler is used where draws from $p(\boldsymbol{\gamma}|\boldsymbol{\beta}, \boldsymbol{\eta}, \boldsymbol{\Sigma}, \mathbf{y}, \mathbf{X})$ use the MH algorithm, whereas draws from the other full conditionals $p(\boldsymbol{\beta}|\boldsymbol{\eta}, \boldsymbol{\Sigma}, \boldsymbol{\gamma}, \mathbf{y}, \mathbf{X})$, $p(\boldsymbol{\eta}|\boldsymbol{\beta}, \boldsymbol{\Sigma}, \boldsymbol{\gamma}, \mathbf{y}, \mathbf{X})$, and $p(\boldsymbol{\Sigma}|\boldsymbol{\beta}, \boldsymbol{\eta}, \boldsymbol{\gamma}, \mathbf{y}, \mathbf{X})$ are straightforward, assuming independent normal priors for $\boldsymbol{\beta}$ and $\boldsymbol{\eta}$ and a Wishart prior for $\boldsymbol{\Sigma}^{-1}$.

The conjugate-distributed random effects model of Chapter 9.5 instead specifies $y_{it}|\mathbf{x}_{it} \sim \mathsf{P}[\upsilon_i \exp(\mathbf{x}'_{it}\boldsymbol{\beta})]$ where υ_i is gamma distributed with parameter δ. This is equivalent to model (12.17) with $\mathbf{w}'_{it}\boldsymbol{\alpha}_i = \alpha_i$ and $\upsilon_i = e^{\alpha_i}$. Then a negative binomial model is obtained, and the MH algorithm applied with, say, a normal prior for $\boldsymbol{\beta}$ and a gamma prior for $\boldsymbol{\delta}$.

If again $y_{it}|\mathbf{x}_{it} \sim \mathsf{P}[\upsilon_i \exp(\mathbf{x}'_{it}\boldsymbol{\beta})]$ but a uniform prior is assumed for $\alpha_i = \ln \upsilon_i$, so $\pi(\upsilon_i) \propto 1/\alpha_i$, and $\boldsymbol{\beta}$ also has a uniform prior, then Lancaster (2004, p.298) shows that on integrating out α_i the posterior is

$$p(\boldsymbol{\beta}|\mathbf{y}, \mathbf{X}) \propto \prod\nolimits_{i=1}^{n} \prod\nolimits_{t=1}^{T} \left(\frac{\exp(\mathbf{x}'_{it}\boldsymbol{\beta})}{\sum_t \exp(\mathbf{x}'_{it}\boldsymbol{\beta}} \right)^{y_{it}},$$

which is the conditional likelihood for the Poisson fixed effects model given in Chapter 9.4.

12.5.4 Additional Count Models

From section 12.5.2, for the Poisson-lognormal model with data augmentation the key ingredient is applying known Bayesian results for linear regression under normality to $\gamma_i = \mathbf{x}'_i\boldsymbol{\beta} + \varepsilon_i$. Complications such as time-series or spatial error correlation can be added by specifying richer models for the normal error ε_i, and again applying known Bayesian results for the linear regression model with these complications.

Multivariate correlated count data can be analyzed using a variant of the Chib et al. (1998) panel data model. Chib and Winkelmann (2001) detailed an MCMC algorithm for the jth count variable for individual in the model

$$\begin{aligned} y_{ij}|\mathbf{x}_{ij} &\sim \mathsf{P}[\exp(\gamma_{ij})] \\ \gamma_{ij} &= \mathbf{x}'_{ij}\boldsymbol{\beta} + \alpha_{ij} \\ \boldsymbol{\alpha}_i &\sim \mathsf{N}[\mathbf{0}, \boldsymbol{\Sigma}]. \end{aligned} \tag{12.18}$$

where $\boldsymbol{\alpha}_i = (\alpha_{i1}, \ldots, \alpha_{iJ})$.

The cross-section model in section 12.5.2 can be adapted to a spatially correlated error model by instead specifying $\gamma_i = \mathbf{x}'_i\boldsymbol{\beta} + u_i$ in (12.16) and letting $\mathbf{u} = \lambda\mathbf{W}\mathbf{u} + \boldsymbol{\varepsilon}$, where $\mathbf{W}$ is specified and $\mathbf{u} = (u_1, \ldots, u_n)$ and $\boldsymbol{\varepsilon} = (\varepsilon_1, \ldots, \varepsilon_n)$. As an alternative, Lee and Bell (2009) present a spatial negative binomial model.

For time series we could instead specify an ARMA model for u_i in $\gamma_i = \mathbf{x}'_i\boldsymbol{\beta} + u_i$, though from Chapter 7 this is only one of many possible models for time series of counts. Jung, Liesenfeld, and J.-F. Richard (2011) propose a dynamic factor model for time-series data on multivariate counts and propose estimation using the efficient importance sampling Monte Carlo procedure. They apply the method to the number of stocks traded with daily data observed at five-minute intervals.

12.6 ROY MODEL FOR COUNTS

Suppose an outcome of interest takes one of two possible values, depending on an initial decision. For example, an individual's use of health care services will vary according to whether he or she chooses health insurance plan A or plan B. We observe decision

$$d_i = \begin{cases} 1 & \text{if } z_i^* = \mathbf{w}_i'\boldsymbol{\alpha} + u_i > 0 \\ 0 & \text{if } z_i^* = \mathbf{w}_i'\boldsymbol{\alpha} + u_i \leq 0, \end{cases} \tag{12.19}$$

where z_i^* is an unobserved latent variable and $\mathbf{w}_i$ are observed regressors. The outcome is then

$$y_i = \begin{cases} y_{1i} & \text{if } d_i = 1 \\ y_{2i} & \text{if } d_i = 0, \end{cases} \tag{12.20}$$

where regression models are specified for y_{1i} and y_{2i}.

Interest lies in the causative effect of insurance plan choice (d_i) on health care use (y_i), but insurance plans are self-selected rather than randomly assigned. The Roy model allows for selection not only on observables, by including regressors as controls, but also on unobservables through correlation of error terms in the models for z_i^*, y_{1i}, and y_{2i}.

The standard *Roy model*, presented in microeconometrics texts, assumes linear normal models for z_i^*, y_{1i}, and y_{2i}. This is suitable for continuously measured outcomes such as health expenditure or log health expenditure. In many cases, however, health care use is instead measured as a count, such as the number of doctor visits. Then y_{1i} and y_{2i} should be explicitly modeled as counts, and the likelihood becomes much less tractable.

Deb, Munkin, and Trivedi (2006) proposed a count version of the Roy model that can be implemented using Bayesian methods. They specify

$$\begin{aligned} y_{1i}|\mathbf{x}_i &\sim \mathsf{P}[\exp(\gamma_{1i})] \\ \gamma_{1i} &= \mathbf{x}_i'\boldsymbol{\beta}_1 + \pi_1 u_i + \varepsilon_{1i} \\ &= \mathbf{x}_i'\boldsymbol{\beta}_1 + \pi_1(z_i^* - \mathbf{w}_i'\boldsymbol{\alpha}) + \varepsilon_{1i}, \end{aligned} \tag{12.21}$$

and similarly

$$\begin{aligned} y_{2i}|\mathbf{x}_i &\sim \mathsf{P}[\exp(\gamma_{2i})] \\ \gamma_{2i} &= \mathbf{x}_i'\boldsymbol{\beta}_2 + \pi_2 u_i + \varepsilon_{2i} \\ &= \mathbf{x}_i'\boldsymbol{\beta}_2 + \pi_2(z_i^* - \mathbf{w}_i'\boldsymbol{\alpha}) + \varepsilon_{2i}. \end{aligned} \tag{12.22}$$

Model identification requires that the regressor $\mathbf{w}_i$ in the decision equation (12.19) includes at least one regressor that does not appear in $\mathbf{x}_i$. The errors ε_{1i} and ε_{2i} are assumed to be independent normals with means 0 and variances, respectively, σ_1^2 and σ_2^2. The error u_i is specified to be standard normal, leading to a probit model for the decision. If $\pi_1 = 0$ and $\pi_2 = 0$ then the errors in the models for y_{1i} and y_{2i} are uncorrelated with the decision equation error u_i, and there is no selection on unobservables. There is then no endogeneity, and insurance choice depends only on exogenous regressors. Otherwise there is endogeneity, and the more complicated model analyzed here is needed.

The posterior for the model is $p(\boldsymbol{\alpha}, \boldsymbol{\beta}_1, \boldsymbol{\beta}_2, \pi_1, \pi_2, \sigma_1^2, \sigma_2^2 | \mathbf{d}, \mathbf{y}_1, \mathbf{y}_2, \mathbf{W}, \mathbf{X})$ where, for example, $\mathbf{d} = (d_1, \ldots, d_n)$. Analysis would be much simpler if the latent variables z_i^*, γ_{1i}, and γ_{2i} were observed. So data augmentation is used to make draws from the augmented posterior $p(\boldsymbol{\alpha}, \boldsymbol{\beta}_1, \boldsymbol{\beta}_2, \pi_1, \pi_2, \sigma_1^2, \sigma_2^2, \mathbf{z}^*, \boldsymbol{\gamma}_1, \boldsymbol{\gamma}_2 | \mathbf{d}, \mathbf{y}_1, \mathbf{y}_2, \mathbf{W}, \mathbf{X})$ where, for example, $\boldsymbol{\gamma}_1 = (\gamma_{11}^*, \ldots, \gamma_{1n}^*)$. This can be done using a Gibbs sampler that sequentially draws from five conditional posteriors for (1) $\boldsymbol{\alpha}$; (2) $(\boldsymbol{\beta}_1, \boldsymbol{\beta}_2, \pi_1, \pi_2)$; (3) (σ_1^2, σ_2^2); (4) $\mathbf{z}^*$; and (5) $(\boldsymbol{\gamma}_1, \boldsymbol{\gamma}_2)$.

Details are presented in the appendix to Deb et al. (2006). Here we present a brief summary. First, for p-dimensional $\boldsymbol{\alpha}$ the prior is assumed to be $\mathsf{N}[\mathbf{0}, 10\mathbf{I}_p]$. The full conditional posterior is $p(\boldsymbol{\alpha} | \boldsymbol{\beta}_1, \boldsymbol{\beta}_2, \pi_1, \pi_2, \sigma_1^2, \sigma_2^2, \mathbf{z}^*, \boldsymbol{\gamma}_1, \boldsymbol{\gamma}_2, \mathbf{d}, \mathbf{y}_1, \mathbf{y}_2, \mathbf{W}, \mathbf{X})$. This posterior is normal because $\boldsymbol{\alpha}$ appears in linear normal models for z_i^*, γ_{1i} and γ_{2i}. Second, for $(\boldsymbol{\beta}_1, \pi_1)$ we have simply regression of γ_{1i} on $\mathbf{x}_i$ and $(z_i^* - \mathbf{w}_i'\boldsymbol{\alpha})$ with normal errors. Independent normal priors are specified that are $\mathsf{N}[\mathbf{0}, 10\mathbf{I}_k]$ for k-dimensional $\boldsymbol{\beta}_1$ and either $\mathsf{N}[\mathbf{0}, \frac{1}{2}]$ or $\mathsf{N}[\mathbf{0}, \frac{1}{4}]$ for scalar π_1. The full conditional posterior reduces to $p(\boldsymbol{\beta}_1, \pi_1 | \boldsymbol{\gamma}_1, \mathbf{z}^*, \mathbf{W}, \mathbf{X})$ and is normal. Similarly the full conditional posterior for $(\boldsymbol{\beta}_2, \pi_2)$ is normal. Third, σ_1^2 appears as the error variance in regression of γ_{1i} on $\mathbf{x}_i$ and $(z_i^* - \mathbf{w}_i'\boldsymbol{\alpha})$ with normal errors. An independent gamma prior for the precision, with $\sigma_1^{-2} \sim \text{Gamma}[0.5, 0.2]$, yielding a gamma posterior. Similarly for σ_2^2. Fourth, for the latent variable $z_i^* = \mathbf{w}_i'\boldsymbol{\alpha} + u_i$ it is necessary to condition on ε_{1i} and ε_{2i}, since the full conditional posterior conditions on y_{1i} and y_{2i}. This more complicated conditional distribution is nonetheless normal, leading to draws from the truncated normal such that $z_i^* > 0$ if $d_i = 1$ and $z_i^* \leq 0$ if $d_i = 0$. Finally the full conditionals for $\boldsymbol{\gamma}_{1i}$ and $\boldsymbol{\gamma}_{2i}$ are complicated, and so the MH algorithm needs to be used to make draws. Note that the model was set up so that there was an analytical form for all but one of the conditional posteriors in the Gibbs sampler, speeding up computation.

Deb et al. (2006) applied this method to annual doctor visits (y) made by people aged 25–64 years employed in the private sector in the United States in 1996–2002. The decision variable d takes value 1 if individuals have private health insurance and value 0 if they have no health insurance. Interest lies in how the number of doctor visits varies with insurance. In the raw data there is a big effect, with the 82% of the sample with private insurance having on average 4.28 doctor visits, whereas the 18% without insurance have on average 1.74 doctor visits. The model seeks to determine whether there is still an effect after controlling for observables such as age, education, income, gender, and health status and additionally for unobservables (endogeneity) through the error u that influences both the insurance decision and doctor visits. The key identification condition made to control for endogeneity is that the size of the firm a person works at determines insurance choice, because private insurance is typically obtained through the employer and larger employers are more likely to offer this benefit, but it does not directly determine doctor visits.

A Markov chain was run with the first 1,000 draws discarded and the next 19,000 draws retained for analysis. Table 2 in Deb et al. (2006) reports the posterior means and standard deviations of $\boldsymbol{\beta}_1$, $\boldsymbol{\beta}_2$, and $\boldsymbol{\alpha}$. It shows that socioeconomic and health status variables are both strong determinants of doctor visits, whereas health insurance choice is more strongly associated with socioeconomic characteristics than with health status.

From Table 3 of Deb et al. (2006) the error covariance parameter π_1 has posterior mean -0.096 and standard deviation 0.034, suggesting that there is selection on unobservables. For π_2 the corresponding numbers are -0.096 and 0.115. A more formal test is based on the Bayes factor that compares models with and without the constraint $\pi_1 = \pi_2 = 0$ imposed. This finds strong evidence against the constraint, so it is necessary to control for endogeneity of the insurance decision.

For those with health insurance the expected number of doctor visits is

$$\begin{aligned}
\mathsf{E}[y_1|\mathbf{x}] &= \mathsf{E}[\exp(\mathbf{x}'\boldsymbol{\beta}_1 + \pi_1 u + \varepsilon)|\mathbf{x}] \\
&= \mathsf{E}[\exp(\pi_1 u + \varepsilon) \times \exp(\mathbf{x}'\boldsymbol{\beta}_1)|\mathbf{x}] \\
&= \exp(0.5(\pi_1^2 + \sigma_1^2)) \times \exp(\mathbf{x}'\boldsymbol{\beta}_1) \\
&= \exp(\mathbf{x}'\boldsymbol{\beta}_1 + 0.5(\pi_1^2 u + \sigma_1^2)),
\end{aligned}$$

where the third equality uses $\pi_1 u + \varepsilon \sim \mathsf{N}[0, \pi_1^2 + \sigma_1^2]$ and $\mathsf{E}[\exp(x)] = \exp(0.5\sigma^2)$ for $x \sim \mathsf{N}[0, \sigma^2]$. A similar result holds for $\mathsf{E}[y_2|\mathbf{x}]$. It follows that the average treatment effect (ATE) of private insurance is

$$\mathsf{E}[y_1 - y_2|\mathbf{x}] = \exp(\mathbf{x}'\boldsymbol{\beta}_1 + 0.5(\pi_1^2 u + \sigma_1^2)) - \exp(\mathbf{x}'\boldsymbol{\beta}_2 + 0.5(\pi_2^2 u + \sigma_2^2)),$$

where $u = z^* - \mathbf{x}'\boldsymbol{\beta}$. The ATE is evaluated at a representative value of $\mathbf{x}$ for each of the 19,000 posterior draws of $\boldsymbol{\beta}_1$, $\boldsymbol{\beta}_2$, $\boldsymbol{\alpha}$, π_1, π_2, σ_1^2, σ_2^2, and $\mathbf{z}^*$. This yields a posterior mean ATE of 0.734 doctor visits (with posterior standard deviation 0.226) for a person in very good or excellent health, whereas the effect is larger for a person in fair or poor health. The study also reports estimates of the average treatment effect on the treated, as well as a local average treatment effect as the instrument (firm size) is varied. These estimates are similar to the average treatment effect.

12.6.1 Finite Mixtures and Extensions

Chapters 4 and 6 analyzed and applied finite mixture models to model heterogeneity in count models from a classical perspective. Analogous Bayesian treatment of this topic can also be found in the literature. A limitation of the classical approach is the need to specify a fixed number of components. Recent advances in Bayesian analysis of finite mixtures offer methods for overcoming these limitations. Of particular interest is the Dirichlet Process Mixture (DPM) model in which the data and specification of the prior jointly determine the number of components. If the prior of the mixing weights of FMM is assumed to be a Dirichlet Process prior, then Neal (2000) shows that the DPM is the infinite counterpart of FMM when the number of components goes to infinity. (The Dirichlet Process prior, sometimes described as a prior defined over the space of distributions, plays a central role in Bayesian nonparametric econometrics; see Ferguson (1973)). Hence the DPM is an alternative to the FMM, especially in the absence of good prior information on the number of components in the data. It is an "automatic" mechanism for selecting the appropriate number of components, thereby avoiding the computation of Bayes factors for determining the number of components, as in the case of FMM.

12.7 BIBLIOGRAPHIC NOTES

Much of the research on Bayesian methods appears in the statistics literature. Robert and Casella (1999) present MCMC methods, and a standard Bayesian reference for applied statisticians is Gelman et al. (2003). Econometrics books devoted to Bayesian methods include Geweke (2003), Koop (2003), Lancaster (2004), and Koop, Poirier, and Tobias (2007). Cameron and Trivedi (2005) provide a chapter-length treatment. Chib (2001) and Geweke and Keane (2001) provide detailed accounts of MCMC theory and methods for many econometrics models. Chernozhukov and Hong (2003) extend MCMC computational methods for likelihood-based models to models based on more general objective functions such as GMM. Computer software for Bayesian analysis is discussed in the appendix.

CHAPTER 13

Measurement Errors

13.1 INTRODUCTION

The benchmark measurement error model is the bivariate linear errors-in-variables (EIV) regression model with additive measurement errors in both the dependent variable and the regressor variable. The measurement errors are assumed to be classical, meaning that they are uncorrelated with the true value and have mean zero. Then the OLS estimator is inconsistent, with a bias toward zero. The measurement error is often large enough for this bias to be substantial; see, for example, Bound, Brown, and Mathiowetz (2001).

For nonlinear models the attenuation result does not necessarily hold, but measurement error still leads to inconsistency because the identified parameter is not the parameter of interest in the model free of measurement error. The essential problem lies in the correlation between the observed regressor variable and the measurement error. This leads to loss of identification and distorted inferences about the role of the covariate. A key objective of analysis is to establish an identification strategy for the parameter of interest.

There are important differences between nonlinear and linear measurement errors models. It is more difficult in nonlinear models to correct for classical measurement error in the regressors. Furthermore, in nonlinear models it may be more natural to allow measurement errors to be nonclassical and nonadditive. And although in linear models classical measurement error in the dependent variable is innocuous because it just contributes to the equation error as additive noise, in nonlinear models the presence of even classical measurement error in the dependent variable leads to loss of identification of model parameters. For count data, mechanisms that lead to measurement error in the dependent variable are plausible on a priori grounds and in some cases they also indicate the direction of measurement errors. For example, certain types of crime, accidents, and absences at the workplace may be typically underreported because of random failures in the mechanism for recording those events. A recurring theme in nonlinear measurement error models is that general results are relatively few and the results are often dependent on functional form and distributional assumptions.

This chapter considers estimation and inference in nonlinear count data models in the presence of measurement errors in regressors, as in the canonical linear EIV model. But we also analyze at some length measurement errors due to random variation in exposure time and errors due to the underreporting and misclassification of events. Our main focus remains on cross-section models. We also consider cross-section regression when there are replicated data available for the regressor measured with error. Panel data provide repeated observations on the same individual that may create additional opportunities for applying different identification strategies. This case differs from replicated data because both regressors and measurement errors may vary over time and is not pursued here.

Section 13.2 concentrates on measurement errors in the regressors under classical assumptions, ignoring any measurement error in counts. In some applications the regressors include a measure of exposure, such as miles driven in a model of auto accidents. Measurement error in exposure is given separate treatment in section 13.3. Section 13.4 studies measurement errors in the observed count, the dependent variable, including the practically important case of misclassified counts. Section 13.5 analyzes the specific case of underreported counts that arises in many commonly occurring situations. Extension to models with overreporting and underreporting is given in section 13.6. A simulation demonstrating the performance of some approximate corrections for measurement errors in regressors in a Poisson model is presented in section 13.7.

13.2 MEASUREMENT ERRORS IN REGRESSORS

Most methods to control for measurement error in the linear model do not carry over to the nonlinear model. The various methods developed for nonlinear models vary with the strength of distributional assumptions. The statistics literature focuses on methods for GLMs, whereas the recent econometrics literature has focused on nonparametric regression models. We focus on classical measurement error and its effect on estimation of regression parameters, although measurement error also affects marginal effects, prediction, and statistical inference.

13.2.1 Measurement Error Processes

Consider regression with dependent variable y_i, true regressors $\mathbf{x}_i^*$ and observed regressors $\mathbf{x}_i$. Interest lies in the relationship between y_i and $\mathbf{x}_i^*$, but $\mathbf{x}_i$ rather than $\mathbf{x}_i^*$ is observed.

The *classical* (or additive) *measurement error model* specifies

$$\mathbf{x}_i = \mathbf{x}_i^* + \mathbf{v}_i, \tag{13.1}$$

where $\mathsf{E}[\mathbf{v}_i|\mathbf{x}_i^*] = \mathbf{0}$, so $\mathbf{v}_i$ is mean independent of $\mathbf{x}_i^*$. Then $\mathsf{E}[\mathbf{x}_i|\mathbf{x}_i^*] = \mathbf{x}_i^*$, so $\mathbf{x}_i^*$ is measured correctly on average. An example is response y depends on true height x^*, which is mismeasured as x, but on average is correctly measured. Usually $\mathbf{v}_i$ is assumed to be iid $[\mathbf{0}, \mathbf{\Sigma}_{\mathbf{vv}}]$, but some studies relax this assumption to permit heteroskedasticity in $\mathbf{v}_i$.

An additional assumption usually made is that the measurement error is *nondifferential* with respect to y. In that case the distribution of $\mathbf{x}|y, \mathbf{x}^*$ is assumed to equal that of $\mathbf{x}|\mathbf{x}^*$. Equivalently the distribution of $y|\mathbf{x}, \mathbf{x}^*$ equals that of $y|\mathbf{x}^*$, and $\mathbf{x}$ is called a surrogate for $\mathbf{x}^*$. And equivalently y and $\mathbf{x}$ are independent conditional on $\mathbf{x}^*$.

A quite different model of measurement error is *Berkson error*. This specifies

$$\mathbf{x}_i^* = \mathbf{x}_i + \mathbf{u}_i, \tag{13.2}$$

where $\mathsf{E}[\mathbf{u}_i|\mathbf{x}_i] = \mathbf{0}$. An example is an experiment where the experimenter sets dosage nominally at level x, but there is error and the actual dosage, the determinant of the response y, is x^*. The consequences of Berkson error are similar to those for a proxy regressor variable and are more benign than those for classical measurement error. The literature focuses on classical measurement error, which is the relevant model for survey data, for example.

In some cases both $\mathbf{x}^*$ and $\mathbf{v}$ in the classical measurement error model are assumed to be normal. Suppose $\mathbf{x}^* \sim \mathsf{N}[\boldsymbol{\mu}_{\mathbf{x}}, \mathbf{\Sigma}_{\mathbf{xx}}]$ and $\mathbf{x}|\mathbf{x}^* \sim \mathsf{N}[\mathbf{x}^*, \mathbf{\Sigma}_{\mathbf{vv}}]$. Then $\mathbf{x}^*|\mathbf{x}$ is normally distributed with a mean that is linear in $\mathbf{x}$. This simplifies analysis because it is a linear generalization of the Berkson error model that specifies $\mathbf{x}^* = \mathbf{a} + \mathbf{B}\mathbf{x} + \mathbf{u}$.

Measurement error need not be classical. An obvious extension of (13.1) is linear measurement error with $\mathbf{x} = \mathbf{c} + \mathbf{D}\mathbf{x}^* + \mathbf{v}$. For categorical variables classical measurement error is necessarily nonclassical, and measurement error is called *classification error*.

13.2.2 Linear Errors in Variables Model

The canonical bivariate linear *errors in variables* (EIV) regression model specifies

$$\begin{aligned} y &= \beta x^* + u \\ x &= x^* + v, \end{aligned} \tag{13.3}$$

where $\mathsf{V}[x^*] = \sigma_{x^*}^2$, $\mathsf{E}[v|x^*] = 0$, $\mathsf{E}[v^2|x^*] = \sigma_v^2$, $\mathsf{E}[u|x^*] = 0$, $\mathsf{E}[u^2|x^*] = \sigma_u^2$, and $\mathsf{E}[vu] = 0$. Here x^* is the true unobserved value and x is its measured counterpart. The measurement error v is mean independent of x^* with constant variance measurement error.

The OLS regression of y on x identifies the parameter $\beta\sigma_{x^*}^2/(\sigma_v^2 + \sigma_{x^*}^2)$, since $\mathsf{V}[x] = \sigma_{x^*}^2 + \sigma_v^2$ and $\mathsf{Cov}[x, y] = \beta\sigma_{x^*}^2$, rather than the desired β. The bias of the inconsistent OLS estimator is toward zero, so the true impact of changing x^* on y is underestimated. This *attenuation bias* is increasing in the noise-to-signal ratio $\sigma_v^2/\sigma_{x^*}^2$. For example, the OLS estimator converges to only 0.5β if the noise is the same size as the signal.

Additional information is needed to obtain a consistent estimator for β. A range of additional sources of information and estimation methods are used. A natural approach is to use an instrumental variables estimator, since measurement error leads to endogeneity of the regressor x as it is correlated with the error term. To see this, note that (13.3) implies $y = \beta x + (u + \beta v)$, so the regressor $x = x^* + v$ is correlated with the composite error $(u + \beta v)$. Assume we have a valid instrument z correlated with x, but uncorrelated with both v and u. Then the instrumental variables estimator is consistent since $\mathsf{Cov}[z, y] = \mathsf{Cov}[z, \beta x]$, so $\mathsf{Cov}[z, y]/\mathsf{Cov}[z, x] = \beta$. The IV approach yields point identification without appealing to distributional knowledge about the measurement error v. An ideal instrument is a *replicate*, a second independent measurement of x^*.

Knowledge of key features of the measurement error process also enables consistent estimation. In this simple example knowledge of the measurement error variance σ_v^2 is sufficient. For measurements from a physical device σ_v^2 may be known. For standard tests such as those of intelligence the reliability ratio $(1 + \sigma_v^2/\sigma_{x^*}^2)^{-1}$ may be known. More often replicated data may be used. For example, if we have independent measures x_1 and x_2 of x^* with $x_j = x^* + v_j$ and $v_j \sim [0, \sigma_v^2]$, $j = 1, 2$, then $\sigma_v^2 = \mathsf{V}[x_1 - x_2]/2$.

In the case that x^* is skewed, ruling out normality, higher order moment conditions can be used to generate instruments (Lewbel, 1997) or to generate additional moment conditions that can be used in minimum distance or GMM estimation (Cragg, 1997).

13.2.3 Consequences for Nonlinear Models

When nonlinearity is introduced, it is difficult to obtain results on the effect of even classical measurement error on the functional form for the conditional mean and parameter estimates.

Consider a general nonlinear model

$$\mathsf{E}[y|\mathbf{x}^*] = m(\mathbf{x}^*, \boldsymbol{\beta}), \tag{13.4}$$

where $\mathbf{x}^*$ is a vector of covariates. The conditional mean given observed data is

$$\begin{aligned} \mathsf{E}[y|\mathbf{x}] &= \mathsf{E}[\mathsf{E}[y|\mathbf{x}^*, \mathbf{x}]|\mathbf{x}] \\ &= \mathsf{E}[\mathsf{E}[y|\mathbf{x}^*]|\mathbf{x}] \\ &= \textstyle\int m(\mathbf{x}^*, \boldsymbol{\beta}) f(\mathbf{x}^*|\mathbf{x}) d\mathbf{x}^*, \end{aligned} \tag{13.5}$$

where the second equality is satisfied under the mean independence assumption that $\mathrm{E}[y|\mathbf{x}^*, \mathbf{x}] = \mathrm{E}[y|\mathbf{x}^*]$. In general $\mathsf{E}[y|\mathbf{x}] \neq m(\mathbf{x}, \boldsymbol{\beta})$ and varies with the density $f(\mathbf{x}^*|\mathbf{x})$. Note that the classical measurement error model specifies $f(\mathbf{x}|\mathbf{x}^*)$ rather than $f(\mathbf{x}^*|\mathbf{x})$.

Chesher (1991) uses small variance approximations to determine the impact of small amounts of classical measurement error in the regressors in nonlinear models. For simplicity consider the bivariate case, with (y, x^*) having density $f(y, x^*) = g(y|x^*)h(x^*)$. We observe $x = x^* + v$ where the measurement error $v \sim [0, \sigma_v^2]$ independent of y and x^*. Chesher presents approximations that ignore terms of $o(\sigma_v^2)$. Then the mismeasurement x is more dispersed than x^*, and to order $O(\sigma_v^2)$ the effect of v on x is the same as if v was normally distributed. In some cases the usual attenuation result holds. But if $\mathsf{E}[y|x^*]$ is nonlinear then results depend on the curvature of $\mathsf{E}[y|x^*]$. At values of x^* for which $\mathrm{E}[y|x^*]$ is convex (concave), measurement error leads $\mathsf{E}[y|x]$ to be higher (lower) than $\mathsf{E}[y|x^*]$. Chesher obtains the same results in the more general case of multivariate dependent and regressor variables $(\mathbf{y}, \mathbf{x}^*)$.

Guo and Li (2002) consider models with underlying equidispersion so that $\mathsf{V}[y|\mathbf{x}^*] = \mathsf{E}[y|\mathbf{x}^*]$, the Poisson being a special case. They show that $\mathsf{V}[y|\mathbf{x}] < \mathsf{E}[y|\mathbf{x}]$ if there is classical measurement error, so measurement error induces underdispersion.

Analysis focuses on estimation of $\boldsymbol{\beta}$, though for nonlinear models the consequent predictions and marginal effects are of ultimate interest.

Suppose the true model is $\mathsf{E}[y|\mathbf{x}^*] = \exp(\mathbf{x}^{*\prime}\boldsymbol{\beta})$, and interest lies in $\mathsf{ME}_j = \partial\mathsf{E}[y|\mathbf{x}^*]/\partial x_j^* = \exp(\mathbf{x}^{*\prime}\boldsymbol{\beta})\beta_j$. It follows immediately that β_j retains its interpretation as a semi-elasticity, so any inconsistency in estimating $\boldsymbol{\beta}$ due to measurement error leads to the same inconsistency in the semi-elasticities.

Poisson regression of y on $\mathbf{x}$ yields the estimate $\widehat{\mathsf{ME}}_j = \exp(\mathbf{x}'\hat{\boldsymbol{\beta}})\hat{\beta}_j$, which varies with $\mathbf{x}$. The average ME does not, however, since from Chapter 3.5.1 it is necessarily the case that $\mathsf{AME}_j = \hat{\beta}_j\bar{y}$ for the Poisson QMLE in a model that includes an intercept. It follows that if Poisson regression of y on mismeasured $\mathbf{x}$ yields $\hat{\beta}_j = 0.8\beta_j$, say, then $\widehat{\mathsf{AME}}_j = 0.8\mathsf{AME}_j$. For the Poisson QMLE, inconsistent estimation of $\boldsymbol{\beta}$ due to measurement error is transmitted directly to inconsistent estimation of in-sample average marginal effects.

13.2.4 Instrumental Variable Estimation

Y. Amemiya (1985) has shown that for nonlinear (in regressors) regression models with classical measurement errors, the usual IV methods do not lead to consistent estimation. This result contrasts with the consistency of the IV estimator for the linear EIV model.

To see this, consider a model with exponential conditional mean in the scalar regressor case, so

$$y = \exp(\alpha + \beta x^*) + \varepsilon. \tag{13.6}$$

The observed regressor is $x = x^* + v$, where v is a measurement error. A Taylor series expansion of $\exp(\alpha + \beta x^*) = \exp(\alpha + \beta(x - v))$ around x yields

$$y = \exp(\alpha + \beta x) + \left\{ \sum_{j=1}^{\infty} (-v)^j \beta^j \exp(\alpha + \beta x) + \varepsilon \right\}. \tag{13.7}$$

IV estimation is based on an instrument z that is correlated with $\exp(\alpha + \beta x)$ while satisfying the moment condition $\mathsf{E}[z(y - \exp(\alpha + \beta x))] = 0$. The latter condition implies that z must be uncorrelated with the term in braces in (13.7), which in turn requires that z be uncorrelated with ε and powers of v since z does need to be correlated with $\exp(\alpha + \beta x)$. Clearly, mean independence between the instrument z and the errors ε and v is not enough to ensure consistent estimation; a much more stringent condition is required. Amemiya notes that consistent estimation is possible if the measurement error variance goes to zero faster than sample size $n \to \infty$, but this is unlikely in practice.

More generally, Amemiya shows that the instrumental variable does not yield consistent estimates for nonlinear errors-in-variables models because the error term involves both measurement error and observed error-contaminated variables. It is not possible to find an instrumental variable that is highly correlated with the observed variable but is uncorrelated with the residual term. Furthermore, from a practical viewpoint, it is not easy to verify the validity of an instrumental variable in estimation because of limited information about the latent variable x^* and measurement error.

When a replicate is available, consistent estimation by methods other than IV is possible in some models; see section 13.2.6. In the case where the instrument is not a replicate, Newey (2001) adds assumptions on the instrument, most notably that $x^* = \mathbf{z}'\boldsymbol{\delta} + \eta$, where the model error, measurement error, and η are conditionally independent. Hausman, Newey, and Powell (1995) also propose a consistent estimator in the model $y = m(\mathbf{x}, \boldsymbol{\beta}) + \varepsilon$, where the instrument need not necessarily be a replicate. And Carroll and Stefanski (1994) propose an *approximate* IV method that is presented in the next subsection.

13.2.5 Approximate Estimators

Approximate methods are convenient methods that are acknowledged to yield parameter estimates that are inconsistent in nonlinear models, but are expected to do considerably better than a naive model that ignores measurement error. They may be applied to a parametric or semiparametric estimator. These methods do provide consistent estimators in the linear case, so they may continue to do well in single-index models with conditional mean $\mathsf{E}[y|\mathbf{x}^*] = m(\mathbf{x}^{*\prime}\boldsymbol{\beta})$ where the function $m(\cdot)$ is not too curved.

Carroll et al. (2006) present these methods in considerable detail, with emphasis on application to GLMs and especially the logistic model. For simplicity we present results for the case where all regressors are measured with classical measurement error as in (13.1), but the methods are easily adapted to the more usual case where only a subset (often one regressor) is measured with error. For brevity we do not present the associated formulas for the estimated variance matrix of parameter estimates that correct for first-stage estimation. These are often estimated using an appropriate bootstrap. The methods require replicate data, or an estimate of $\mathsf{V}[\mathbf{v}]$, or an instrument in the case of approximate IV.

Replication means repeated measurement of the same underlying true value. In natural sciences where measurement takes place in controlled settings replicated measures arise naturally. In social sciences where measures are derived from survey responses, repeated measures are less often available.

Regression Calibration

The *regression calibration* approach for desired nonlinear regression of y on $\mathbf{x}^*$ instead regresses y on $\hat{\mathbf{x}}$, where $\hat{\mathbf{x}}$ is a prediction of $\mathbf{x}^*$. The approach presented here is due to Carroll and Stefanski (1990) and Gleser (1990).

Suppose for each observation we have K replicates $\mathbf{x}_1, \ldots, \mathbf{x}_K$ of $\mathbf{x}^*$. Then under squared error loss, the best linear prediction of $\mathbf{x}^*$ given $\bar{\mathbf{x}} = K^{-1} \sum_{j=1}^{K} \mathbf{x}_j$ is

$$\mathsf{E}[\mathbf{x}^*|\bar{\mathbf{x}}] = \mathsf{E}[\mathbf{x}^*] + \mathsf{Cov}[\mathbf{x}^*, \bar{\mathbf{x}}]\mathsf{V}[\bar{\mathbf{x}}]^{-1}(\bar{\mathbf{x}} - \mathsf{E}[\bar{\mathbf{x}}]).$$

Now $\bar{\mathbf{x}} = \mathbf{x}^* + \bar{\mathbf{v}}$, since $\mathbf{x}_j = \mathbf{x}^* + \mathbf{v}_j$. Given classical measurement error $\mathsf{E}[\mathbf{x}^*] = \mathsf{E}[\bar{\mathbf{x}}]$, $\mathsf{Cov}[\mathbf{x}^*, \bar{\mathbf{x}}] = \mathsf{V}[\mathbf{x}^*]$, and $\mathsf{V}[\bar{\mathbf{x}}] = \mathsf{V}[\mathbf{x}^*] + K^{-1}\mathsf{V}[\mathbf{v}]$. For the ith observation we use

$$\hat{\mathbf{x}}_i = \bar{\bar{\mathbf{x}}} + \hat{\mathsf{V}}[\mathbf{x}^*] \left(\hat{\mathsf{V}}[\mathbf{x}^*] + K^{-1}\hat{\mathsf{V}}[\mathbf{v}] \right)^{-1} (\bar{\mathbf{x}}_i - \bar{\bar{\mathbf{x}}}), \tag{13.8}$$

where $\bar{\bar{\mathbf{x}}} = n^{-1} \sum_{i=1}^{n} \bar{\mathbf{x}}_i$, $\bar{\mathbf{x}}_i = K^{-1} \sum_{j=1}^{K} \mathbf{x}_{ji}$, $\hat{\mathsf{V}}[\mathbf{x}^*] = \hat{\mathsf{V}}[\bar{\mathbf{x}}] - K^{-1}\hat{\mathsf{V}}[\mathbf{u}]$, $\hat{\mathsf{V}}[\bar{\mathbf{x}}] = \frac{K}{N(K-1)} \sum_{i=1}^{n} (\bar{\mathbf{x}}_i - \bar{\bar{\mathbf{x}}})(\bar{\mathbf{x}}_i - \bar{\bar{\mathbf{x}}})'$, and a key component is the use of the replicate data to estimate the measurement error variance matrix

$$\hat{\mathsf{V}}[\mathbf{v}] = \frac{1}{K-1} \sum_{i=1}^{n} \sum_{j=1}^{K} (\mathbf{x}_{ji} - \bar{\mathbf{x}}_i)(\mathbf{x}_{ji} - \bar{\mathbf{x}}_i)'. \tag{13.9}$$

These formulas can be adjusted to allow for the number of replicates K to vary with i. Alternatively $\mathsf{V}[\mathbf{v}]$ can be estimated or specified using external data or knowledge.

The method is only approximate in nonlinear models. Carroll et al. (2006) present some refinements for highly nonlinear problems.

Simulation Extrapolation

The *simulation extrapolation* (SIMEX) method, proposed by Cook and Stefanski (1994), assumes the measurement error is normally distributed, makes draws (simulates) from this distribution, and extrapolates to the case of no measurement error.

The naive regression results are obtained from regression of y on the regressor $\mathbf{x} = \mathbf{x}^* + \mathbf{v}$. If we could add to the observed $\mathbf{x}$ an amount $\lambda\mathbf{v}$ and reestimate, then we would have estimated a model with regressor $\mathbf{x}^* + \mathbf{v} + \lambda\mathbf{v} = \mathbf{x}^* + (1+\lambda)\mathbf{v}$. Doing so for a range of values of $\lambda > 0$ gives a range of estimates $\hat{\boldsymbol{\beta}}(\lambda)$. The case of no measurement error is the case $\lambda = -1$, so the desired estimate is obtained by estimating $\hat{\boldsymbol{\beta}}(\lambda)$ for, say, $\lambda = 0.5, 1.0, \ldots, 2.5$ and interpolating to the case $\lambda = -1$.

The measurement error $\mathbf{v}$ is not observed. To implement the method the measurement error is assumed to be normally distributed, and draws are made from $\mathsf{N}[\mathbf{0}, \hat{\mathsf{V}}[\mathbf{v}]]$ where, given replicate data, $\hat{\mathsf{V}}[\mathbf{v}]$ can be computed using (13.9).

Approximate IV Estimation

As noted in section 13.2.4, it is not possible to consistently estimate a nonlinear EIV model using an IV estimator. Nonetheless it is possible that IV methods provide a reasonable approximation.

For single-index models with $\mathsf{E}[y|\mathbf{x}^*] = m(\mathbf{x}^{*\prime}\boldsymbol{\beta})$, the obvious estimator is nonlinear IV or nonlinear 2SLS of y on $\mathbf{x}$ with instrument $\mathbf{z}$. In the just-identified case this solves the sample analog of the moment condition

$$\mathsf{E}[\mathbf{z}(y - m(\mathbf{x}'\boldsymbol{\beta})] = \mathbf{0}. \tag{13.10}$$

This estimator is not consistent, as explained in section 13.2.4, but may be better than simple nonlinear regression of y on $\mathbf{x}$.

Carroll and Stefanski (1994) proposed an alternative two-step estimator that they call an *approximate* IV estimator. It performs linear regression of $\mathbf{x}$ on $\mathbf{z}$ to yield predictor $\hat{\mathbf{x}}$ and, at the second stage, estimates the original model with $\mathbf{x}^*$ replaced by $\hat{\mathbf{x}}$. This method is similar in spirit to the regression calibration approach, but has the advantage that the instrument need not be a replicated observation of $\mathbf{x}^*$. Again this method does not yield a consistent estimator, but it may be a reasonable approximation method. For details see, in particular, Stefanski and Buzas (1995), who consider the logit case.

13.2.6 Consistent Estimators

Naive estimators that ignore measurement error are inconsistent because the resulting estimating equations have a nonzero probability limit. For example, in the Poisson case while $\text{plim}\, n^{-1}\sum_i \mathbf{x}_i^*(y_i - \exp(\mathbf{x}_i^{*\prime}\boldsymbol{\beta})) = \mathbf{0}$, inconsistency arises because $\text{plim}\, n^{-1}\sum_i \mathbf{x}_i(y_i - \exp(\mathbf{x}_i'\boldsymbol{\beta})) \neq \mathbf{0}$ if, for example, classical measurement error is present in $\mathbf{x}$.

We present two methods, developed for GLMs with classical measurement error, that use an adjusted estimating equation or an adjusted objective function that leads to consistent parameter estimates given knowledge of the measurement error distribution.

Conditional Scores

Consider regression of y on $\mathbf{x}^*$ with parameter $\boldsymbol{\beta}$ given a GLM density with canonical parameterization as defined in Chapter 2.4.4 that permits overdispersion, and with observed regressor $\mathbf{x} = \mathbf{x}^* + \mathbf{v}$ where $\mathbf{v} \sim \mathsf{N}[\mathbf{0}, \boldsymbol{\Sigma}_{\mathbf{vv}}]$ and $\boldsymbol{\Sigma}_{\mathbf{vv}}$ is known. Let the joint density of the data, conditional on $\mathbf{x}^*$, be denoted $f(y, \mathbf{x}|\mathbf{x}^*)$. Then Stefanski and Carroll (1987) show that a sufficient statistic for $\mathbf{x}^*$ is $\boldsymbol{\Delta} = \mathbf{x} + y\boldsymbol{\Sigma}_{\mathbf{vv}}\boldsymbol{\beta}$, yielding a conditional density $f(y|\boldsymbol{\Delta}, \mathbf{x})$ that does not depend on $\mathbf{x}^*$. Furthermore this density is also a GLM density, though with a reparameterized mean.

This leads to an estimator proposed by Stefanski and Carroll (1987) called a *conditional score* estimator due to the conditioning on $\boldsymbol{\Delta}$. For Poisson the conditional score estimator $\hat{\boldsymbol{\beta}}$ solves

$$\sum_{i=1}^{n}(y_i - m(\boldsymbol{\Delta}_i, \boldsymbol{\beta}))\mathbf{x}_i = \mathbf{0}, \tag{13.11}$$

where $\boldsymbol{\Delta}_i = \mathbf{x}_i + y_i \boldsymbol{\Sigma}_{\mathbf{vv}}\boldsymbol{\beta}$ and $m(\boldsymbol{\Delta}_i, \boldsymbol{\beta})$ is the mean $\mathsf{E}[y_i|\boldsymbol{\Delta}_i, \boldsymbol{\beta}]$, where y_i has probability mass function,

$$\Pr[y_i = k|\boldsymbol{\Delta}_i, \boldsymbol{\beta}] = \frac{\exp(k\boldsymbol{\Delta}_i'\boldsymbol{\beta} - .5\boldsymbol{\beta}'\Sigma_{\mathbf{vv}}\boldsymbol{\beta})/k!}{\sum_{j=0}^{\infty}\exp(j\boldsymbol{\Delta}_i'\boldsymbol{\beta} - .5\boldsymbol{\beta}'\boldsymbol{\Sigma}_{\mathbf{vv}}\boldsymbol{\beta})/j!}.$$

Analysis is complicated because of summation of an infinite series. By contrast a closed-form solution exists for the logit model. If $\boldsymbol{\Sigma}_{\mathbf{vv}}$ is estimated using replicate data, see (13.9), the variance matrix of $\hat{\boldsymbol{\beta}}$ can be obtained using bootstrap methods.

Corrected Scores

Stack y_i into vector $\mathbf{y}$, and $\mathbf{x}_i$ and $\mathbf{x}_i^*$ into matrices $\mathbf{X}$ and $\mathbf{X}^*$. The desired log-likelihood is $\mathcal{L}(\boldsymbol{\beta}|\mathbf{y}, \mathbf{X}^*)$, but $\mathbf{X}^*$ is not observed. The naive log-likelihood $\mathcal{L}(\boldsymbol{\beta}|\mathbf{y}, \mathbf{X})$ leads to inconsistent estimation. Instead Stefanski (1989) and Nakamura (1990) proposed a corrected log-likelihood that depends on the observed $\mathbf{y}, \mathbf{X}$ but whose expected value conditional on the unobserved $\mathbf{X}^*$ equals $\mathcal{L}(\boldsymbol{\beta}|\mathbf{y}, \mathbf{X}^*)$. Specifically, we use $\mathcal{L}^+(\boldsymbol{\beta}|\mathbf{y}, \mathbf{X})$ that has the property that $\mathsf{E}_\mathbf{X}[\mathcal{L}^+(\boldsymbol{\beta}|\mathbf{y}, \mathbf{X})|\mathbf{y}, \mathbf{X}^*] = \mathcal{L}(\boldsymbol{\beta}|\mathbf{y}, \mathbf{X}^*)$.

For Poisson with $\mathbf{X}^*$ known the log-likelihood is

$$\mathcal{L}(\boldsymbol{\beta}|\mathbf{y}, \mathbf{X}^*) = \sum_{i=1}^{n}\{-\exp(\mathbf{x}_i^{*\prime}\boldsymbol{\beta}) + y_i\mathbf{x}_i^{*\prime}\boldsymbol{\beta} - \ln y_i!\}.$$

The corrected log-likelihood assuming $\mathbf{x} = \mathbf{x}^* + \mathbf{v}$ and $\mathbf{v} \sim \mathsf{N}[\mathbf{0}, \boldsymbol{\Sigma}_{\mathbf{vv}}]$ is

$$\mathcal{L}^+(\boldsymbol{\beta}|\mathbf{y}, \mathbf{X}) = \sum_{i=1}^{n}\{-\exp(\mathbf{x}_i'\boldsymbol{\beta} - .5\boldsymbol{\beta}'\Sigma_{\mathbf{vv}}\boldsymbol{\beta}) + y_i\mathbf{x}_i'\boldsymbol{\beta} - \ln y_i!\}. \tag{13.12}$$

Consider the first term in the sum. Since $\mathbf{x} = \mathbf{x}^* + \mathbf{v}$, $\mathsf{E}_\mathbf{x}[\exp(\mathbf{x}'\boldsymbol{\beta})|\mathbf{x}^*] = \mathsf{E}_\mathbf{v}[\exp(\mathbf{x}^{*\prime}\boldsymbol{\beta})\exp(\mathbf{v}'\boldsymbol{\beta})|\mathbf{x}^*] = \exp(\mathbf{x}^{*\prime}\boldsymbol{\beta})\exp(.5\boldsymbol{\beta}'\boldsymbol{\Sigma}_{\mathbf{vv}}\boldsymbol{\beta})$. The last step uses $\mathbf{v}'\boldsymbol{\beta} \sim \mathsf{N}[\mathbf{0}, \boldsymbol{\beta}'\boldsymbol{\Sigma}_{\mathbf{vv}}\boldsymbol{\beta}]$ so $\mathsf{E}_\mathbf{v}[\exp(\mathbf{v}'\boldsymbol{\beta})] = \exp(.5\boldsymbol{\beta}'\boldsymbol{\Sigma}_{\mathbf{vv}}\boldsymbol{\beta})$. It follows that $\mathsf{E}_\mathbf{x}[\exp(\mathbf{x}'\boldsymbol{\beta} - .5\boldsymbol{\beta}'\boldsymbol{\Sigma}_{\mathbf{vv}}\boldsymbol{\beta})|\mathbf{x}^*] = \exp(\mathbf{x}_i^{*\prime}\boldsymbol{\beta})$ as desired. For the second term $\mathsf{E}_\mathbf{x}[y\mathbf{x}'\boldsymbol{\beta}|y, \mathbf{x}^*] = \mathsf{E}_\mathbf{v}[y(\mathbf{x}^{*\prime}\boldsymbol{\beta} + \mathbf{v}'\boldsymbol{\beta})|y, \mathbf{x}^*] = y\mathbf{x}'\boldsymbol{\beta}$.

The corresponding first-order conditions are called the *corrected score*. For Poisson the corrected score estimator solves

$$\sum_{i=1}^{n}\{y_i\mathbf{x}_i - \exp(\mathbf{x}_i'\boldsymbol{\beta} - .5\boldsymbol{\beta}'\boldsymbol{\Sigma}_{\mathbf{vv}}\boldsymbol{\beta})(\mathbf{x}_i - \boldsymbol{\Sigma}_{\mathbf{vv}}\boldsymbol{\beta})\} = \mathbf{0}. \tag{13.13}$$

Buonaccorsi (1996) relaxes the normality assumption for $\mathbf{v}$ and proposes an approximation called a *modified estimating equation* with first-order conditions

$$\sum_{i=1}^{n}\{y_i\mathbf{x}_i - (1 - .5\boldsymbol{\beta}'\boldsymbol{\Sigma}_{\mathbf{vv}}\boldsymbol{\beta})\exp(\mathbf{x}_i'\boldsymbol{\beta})\mathbf{x}_i + \exp(\mathbf{x}_i'\boldsymbol{\beta})\boldsymbol{\Sigma}_{\mathbf{vv}}\boldsymbol{\beta}\} = \mathbf{0}. \tag{13.14}$$

Guo and Li (2001) also relax the assumption of normally distributed measurement error. Then the exact corrected log-likelihood is

$$\mathcal{L}^{++}(\boldsymbol{\beta}|\mathbf{y},\mathbf{X}) = \sum_{i=1}^{n}\{-\mathsf{E}_{\mathbf{x}_i^*}[\exp(\mathbf{x}_i^{*\prime}\boldsymbol{\beta})] + y_i\mathbf{x}_i'\boldsymbol{\beta} - \ln y_i!\}, \tag{13.15}$$

and the corrected score estimator solves

$$\sum_{i=1}^{n}\{y_i\mathbf{x}_i - \mathsf{E}_{\mathbf{x}_i^*}[\exp(\mathbf{x}_i^{*\prime}\boldsymbol{\beta})\mathbf{x}_i^*]\} = \mathbf{0}. \tag{13.16}$$

For multivariate $\mathbf{x}^*$, the computation of $\mathsf{E}[\exp(\mathbf{x}^{*\prime}\boldsymbol{\beta})]$ is not straightforward, even if the distribution of $\mathbf{x}^*$ is known, because multiple integrals are involved. Simulation-based methods (Li, 2000) provide one possible approach to this problem.

13.2.7 Structural Methods

The preceding approaches did not specify a distribution of the regressors; only a distribution for the measurement error is assumed. They are called *functional methods*. By contrast *structural methods* specify a distribution for the regressors.

Hsiao (1989) considers the nonlinear model with $\mathsf{E}[y|\mathbf{x}^*] = m(\mathbf{x}^*;\boldsymbol{\beta})$ and assumes the classical measurement error model where both $\mathbf{x}^*$ and $\mathbf{v}$ are normally distributed. It follows that $\mathbf{x}^*|\mathbf{x}$ is normally distributed with parameters $\boldsymbol{\delta}$, say, and one can obtain

$$\mathsf{E}[y|\mathbf{x}] = \int m(\mathbf{x}^*;\boldsymbol{\beta})f(\mathbf{x}^*|\mathbf{x},\boldsymbol{\delta})d\mathbf{x}^* = g(\mathbf{x};\boldsymbol{\beta},\boldsymbol{\delta}). \tag{13.17}$$

Hsiao considers identification issues for the NLS estimator that minimizes

$$\sum_{i=1}^{n}(y_i - g(\mathbf{x}_i;\boldsymbol{\beta},\boldsymbol{\delta}))^2 \tag{13.18}$$

with respect to $\boldsymbol{\beta}$ and $\boldsymbol{\delta}$, and for a two-step estimator that first estimates $\boldsymbol{\delta}$ and then minimizes $\sum_i(y_i - g(\mathbf{x};\boldsymbol{\beta},\hat{\boldsymbol{\delta}}))^2$.

Shklyar and Schneeweiss (2005) specialize to the Poisson. Then rather than NLS estimation it may be more efficient to use the Poisson quasi-MLE with mean function $g(\mathbf{x}; \boldsymbol{\beta}, \boldsymbol{\delta})$. The authors consider the case where $\boldsymbol{\delta}$ is known and minimize $\sum_i \{y_i \ln g(\mathbf{x}_i; \boldsymbol{\beta}, \boldsymbol{\delta}) - g(\mathbf{x}_i; \boldsymbol{\beta}, \boldsymbol{\delta})\}$ with respect to $\boldsymbol{\beta}$. As expected, this estimator is more efficient than the corrected score estimator, which does not use knowledge of the distribution of $\mathbf{x}^*$.

Li (2000) considers a more general structural model where the distributions of $\mathbf{x}$ and $\mathbf{v}$ are not necessarily normal. Then simulation methods are needed because $g(\mathbf{x}; \boldsymbol{\beta}, \boldsymbol{\delta})$ is nontractable. He proposes a simulated minimum distance estimator.

More efficient estimation bases estimation on the density $f(y|\mathbf{x})$ rather than on the conditional mean $\mathsf{E}[y|\mathbf{x}]$. But the density is nontractable. Instead estimation could be by simulated maximum likelihood or by Bayesian methods with a diffuse prior. For normally distributed $\mathbf{x}^*$ and $\mathbf{v}$, Guo and Li (2001) use SML to estimate a negative binomial model for physician outpatient visits with the regressor income mismeasured. Jordan, Brubacher, Tsugane, Tsubono, Gey, and Moser (1997) use Bayesian methods for a Poisson model for the number of deaths, with the regressor plasma lycophene levels mismeasured.

With replicate data or data on an instrument, a fully structural model can be cast as a special case of a generalized linear latent and mixed model (GLAMM). Suppose y is Poisson distributed with a single mismeasured regressor x^* and other correctly measured regressor $\mathbf{w}$, so $\mu = \exp(\beta x^* + \mathbf{w}'\boldsymbol{\gamma})$. Let x_1 and x_2 be replicates of $\mathbf{x}$, with $x_j = x^* + u_j$, $u_j \sim \mathsf{N}[0, \sigma^2]$, $j = 1, 2$, so we have classical measurement error that is normally distributed. More generally x_2 can be an instrument satisfying $x_2 = \lambda x^* + u_2$, $u_2 \sim \mathsf{N}[0, \sigma_2^2]$. And assume that the true regressor is normally distributed, with $x^* = \mathbf{w}'\boldsymbol{\delta} + \varepsilon$, where $\varepsilon \sim \mathsf{N}[0, \tau^2]$. Then ML estimation is possible using a statistical package for GLAMMs.

13.2.8 Nonparametric Methods

More flexible methods, including nonparametric and semiparametric methods, use the following result for data observed with classical measurement error.

Let

$$\mathbf{z} = \mathbf{z}^* + \mathbf{v}, \tag{13.19}$$

where here $\mathbf{z}$ may be regressor variables or dependent and regressor variables. Let $\phi_{\mathbf{v}}(\mathbf{v}) = \mathrm{E}[e^{i\mathbf{t}'\mathbf{v}}]$ denote the characteristic function of $\mathbf{v}$, with the characteristic functions $\phi_{\mathbf{z}}(\mathbf{z})$ and $\phi_{\mathbf{z}^*}(\mathbf{z}^*)$ similarly defined. Given independent measurement error, the characteristic function of $\mathbf{z}$ is $\phi_{\mathbf{z}}(\mathbf{z}) = \phi_{\mathbf{z}^*}(\mathbf{z}^*)\phi_{\mathbf{v}}(\mathbf{v})$, so the characteristic function of $\mathbf{z}^*$ can be estimated using

$$\hat{\phi}_{\mathbf{z}^*}(\mathbf{z}^*) = \hat{\phi}_{\mathbf{z}}(\mathbf{z})/\hat{\phi}_{\mathbf{v}}(\mathbf{v}). \tag{13.20}$$

This in turn can be used to estimate the desired density $\hat{f}_{\mathbf{z}^*}(\mathbf{z}^*)$ of the true variable $\mathbf{z}^*$ using a truncated inverse Fourier transform. This approach is referred to as identification via *deconvolution*. Clearly $\hat{\phi}_{\mathbf{z}}(\mathbf{z})$ can be estimated from the observed data. Estimation of $\hat{\phi}_{\mathbf{v}}(\mathbf{v})$ requires additional information, which may be specification of a distributional model for $\mathbf{v}$ or replicated data.

Chen, Hong, and Nekipelov (2011) provide a detailed survey for nonlinear models of this active area of theoretical research. We summarize a few key papers that are applicable to a count model with an exponential conditional mean or, in some cases, a nonparametric conditional mean.

Li (2002) applies this approach to the nonlinear model $\mathsf{E}[y|\mathbf{x}^*] = m(\mathbf{x}^*, \boldsymbol{\beta})$ with $m(\cdot)$ specified, classical measurement error in $\mathbf{x}^*$, and two independent measurements of $\mathbf{x}^*$. The deconvolution method can be used to obtain the estimated conditional distribution $\hat{f}_{\mathbf{x}^*|\mathbf{x}}(\mathbf{x}^*|\mathbf{x})$, and given this one can estimate using the structural method of Hsiao (1989) as in section 13.2.7.

Hong and Tamer (2003) show that Li's model can be estimated even without replicate data, if the measurement error $\mathbf{v}$ is Laplace distributed.

Schennach (2007) generalizes the model of Newey (2001), see section 13.2.4, to a model with nonparametric regression function and nonparametric relationship between $\mathbf{x}^*$ and $\mathbf{z}$.

Ben-Moshe (2011) shows that estimators based on empirical first-order derivatives of characteristic functions can be used to derive the distribution of $\mathbf{x}^*$ under weaker conditions than previously realized. In particular the measurement error model may have more unobservables than equations, allowing for dependent unobservables.

13.2.9 Nonclassical Measurement Error

Nonclassical measurement error can arise in several ways. A common case is when a regressor is binary. We have $x = x^* + v$, where x is 0 or 1 and x^* is 0 or 1. Then $v = x - x^*$ is ≥ 0 if $x^* = 0$ and $v \leq 0$ if $x^* = 1$. It follows that the measurement error v is negatively correlated with the true value x^*. Most studies focus on estimation in the linear regression model. For nonlinear models Mahajan (2005) gives a quite general treatment that permits nonparametric identification given an instrument-like variable that satisfies stronger conditions than for standard IV estimation.

This binary regressor example is an example of a *misclassified regressor*. More generally categorical and count data may be misclassified. In the next sections we consider in detail count regression when the dependent variable is misclassified.

13.3 MEASUREMENT ERRORS IN EXPOSURE

From Chapter 1.1 the general Poisson density is

$$f(y_i|\mu_i t_i) = \frac{e^{-\mu_i t_i}\left(\mu_i t_i\right)^{y_i}}{y_i!}, \qquad y_i = 0, 1, \ldots. \tag{13.21}$$

Compactly, we refer to this density as $\mathsf{P}[\mu_i t_i]$. Here t_i denotes *exposure*, defined as the interval during which the subject is at risk of experiencing the event, and μ_i denotes *intensity*, the mean number of event occurrences in an exposure interval of unit length. Note that in most of the book exposure does not appear because we have (implicitly) assumed that exposure intervals are of the same length for all individuals (e.g., the past two weeks), normalized $t_i = 1$ for all i, and assumed no measurement error.

The preceding section considered measurement errors in regressors that determine the intensity rate μ. For example, in modeling the number of insurance claims for auto accidents, the desired regressor lifetime driving experience may be mismeasured when using number of years that a driving license has been held. Now we consider measurement error in the variable t that measures exposure. Continuing the auto insurance claims example, an error arises if it is assumed that all cases in the sample were covered by insurance for the entire period under study, but in fact some members were only covered for some part of the period.

A separate treatment of measurement error in exposure is warranted for several reasons. First, exposure t determines the mean μt multiplicatively. Second, there is no associated coefficient because the coefficient of t is constrained to be unity. Third, it is usually assumed that the measurement error is a multiplicative form of Berkson error; see section 13.2.1 for the additive form. Specifically, $y \sim \mathsf{P}[\mu t^*]$ with $t^* = tv$ where v is distributed independently of t with mean 1. For example, the interview may ask about events in the past year, so t is set at one year, but due to recall error the true time period t^* is not exactly one year. By contrast a multiplicative form of classical measurement error specifies $t = t^* v$.

Exposure need not necessarily measure only time, because other variables such as population, distance, and so forth may also be relevant. Measurement errors from exposures in the general case have been considered by Brännäs (1995b) and Alcañiz (1996).

In analyzing the consequences of measurement errors, one may either focus on the properties of a particular estimator, such as inconsistency, or study the effect on the entire distribution of the observed random variable. The first approach is widely used in the analysis of linear errors-in-variables models. The main technique in the latter case involves considering a small-variance local Taylor expansion around the true density (Cox, 1983) and studying the properties of the resulting approximate density.

In relation to exposures, three separate cases are considered. The first is the case of known but unequal exposures, which can be handled in a relatively straightforward manner. The second situation is that in which the exposure variable is not directly observed, but there is information on observable factors that affect it. The third case is one in which the exposure period is not observed and is incorrectly assumed to be the same for all subjects, giving rise to measurement errors. A potential advantage of specifying exposures explicitly is to permit separate identification of factors that affect exposure rather than the intensity parameter.

13.3.1 Correctly Observed Exposure

Using the exponential mean specification we have

$$\begin{aligned}\mu t &= \exp(\mathbf{x}'\boldsymbol{\beta})t \\ &= \exp(\mathbf{x}'\boldsymbol{\beta} + \ln t) \\ &= \exp(\mathbf{x}'\boldsymbol{\beta} + \beta_{k+1} \ln t),\end{aligned} \tag{13.22}$$

where $\beta_{k+1} = 1$. Estimation for standard models such as Poisson and negative binomial can proceed via constrained maximum likelihood, with the coefficient of $\ln t_i$ being restricted to unity. The constraint can be imposed directly by substitution into the likelihood, and the constrained coefficient is called an offset coefficient.

Another alternative is to estimate β_{k+1} freely. A Wald test of the null hypothesis that $\beta_{k+1} = 1$ can be used as a model specification test.

13.3.2 Multiplicative Berkson Error in Exposure

In this section we show that measurement error in exposure in the Poisson model is qualitatively analogous to measurement error in the dependent variable in the linear regression. A consequence is to inflate the variance through overdispersion. Under appropriate assumptions we find strong parallels between measurement errors and unobserved heterogeneity: both lead to overdispersion while the Poisson MLE remains consistent. There are also strong parallels in the algebraic analysis of the two problems.

Assume that the regression model for intensity μ_i is correctly specified and with no measurement error (e.g., $\mu_i = \exp(\mathbf{x}_i'\boldsymbol{\beta})$). But exposure t_i^* is measured with error that is uncorrelated with the explanatory variables $\mathbf{x}_i$. The Poisson model (13.21) yields for typical observation

$$f(y|\mu, t^*) = \frac{e^{-\mu t^*}(\mu t^*)^y}{y!}. \tag{13.23}$$

In considering the impact of measurement error on the properties of the estimator, notice that the model set-up is exactly analogous to the case of multiplicative heterogeneity considered in earlier chapters. The variable t^* replaces the term ν.

Consequently, in the presence of random measurement errors in t^*, the mixed Poisson model emerges. In particular, if $t^* = t\upsilon$, t is normalized to unity so $t^* = \upsilon$, υ is gamma distributed with mean 1, and the resulting marginal distribution of y is the NB2; see Chapter 4.2.2.

To handle the mixture problem quite generally we follow the approach of Gurmu, Rilstone, and Stern (1995) given in Chapter 11.4.2. Let t^* have density

$g(t^*)$; then the density of y conditional on μ only is

$$
\begin{aligned}
h(y|\mu) &= \int \frac{e^{-\mu t^*}(\mu t^*)^y}{y!} g(t^*)dt^* \\
&= \frac{\mu^y}{y!} \int (t^*)^y e^{-\mu t^*} g(t^*)dt^* \\
&= \frac{\mu^y}{y!} \mathsf{M}_{t^*}^{(y)}(-\mu), \qquad (13.24)
\end{aligned}
$$

where $\mathsf{M}_{t^*}(-\mu) = \mathsf{E}_{t^*}\left[e^{-\mu t^*}\right]$ denotes the moment generating function for t,

$$
\mathsf{M}_t^{(y)}(-\mu) = \mathsf{E}_{t^*}\left[(t^*)^y e^{-\mu t^*}\right] \qquad (13.25)
$$

is the yth order derivative of $\mathsf{M}_{t^*}(-\mu)$, and E_t denotes expectation with respect to the distribution of t^*. Essentially (13.24) gives the mixed Poisson density which can be analyzed after choosing a suitable $g(t^*)$ density and then doing the necessary algebra to derive the term $\mathsf{M}_{t^*}^{(y)}(\mu)$. A flexible specification of $g(t^*)$ can be obtained using a series expansion, though this generates formidable algebraic and computational detail as shown in Chapter 11.4.2.

Continuing with the assumption that the intensity function μ is correctly specified, the assumption of a multiplicative measurement error $t^* = tv$ with a specified parametric distribution for v leads to a mixed (overdispersed) Poisson model. Interestingly, this result implies that overdispersion in a count model may reflect measurement error in exposures. Under standard assumptions, such as a gamma distribution for v that leads to an NB2 distribution for y, the random measurement errors in exposures do not affect consistency of the Poisson MLE.

13.3.3 Small Measurement Error in Exposure

Furthermore, if the variance of the exposure measurement error is $O(n^{-1/2})$ – (a case of "modest overdispersion" in Cox's (1983) terminology) – then the estimator is also asymptotically efficient. A heuristic demonstration of this point follows from a second-order Taylor expansion of (13.21) around $(t_i - 1)$. In general, however, the limit distribution of the estimator is not affected by the measurement error.

Suppose observed exposure is of unit length with $t = 1$, so the measurement error is $v = t^* - t = t^* - 1$. Furthermore, assume the measurement error has finite variance and is uncorrelated with $(y - \mu)$:

$$
\begin{aligned}
\mathsf{E}[t^* - 1] &= 0 \\
\mathsf{E}[(t^* - 1)(y - \mu)] &= 0 \qquad (13.26) \\
\mathsf{E}\left[(t^* - 1)^2\right] &= \sigma_v^2.
\end{aligned}
$$

The true model is $\mathsf{P}[\mu t^*]$ but we instead set exposure to unit length and estimate $\mathsf{P}[\mu]$. A second-order Taylor expansion of $\mathsf{P}[\mu t^*]$ around $\mathsf{P}[\mu]$ yields

$$\frac{e^{-\mu t^*}(\mu t^*)^y}{y!} = \mathsf{P}[\mu_i][1 + (y-\mu)(t^*-1) + \tfrac{1}{2}\left((y-\mu)^2 - y\right)(t^*-1)^2 + O\left(t^*-1\right)^3].$$

Then taking expectation with respect to t^*, and using (13.26), yields as density for y given μ

$$h(y|\mu) \approx \mathsf{P}[\mu]\left[1 + \frac{1}{2}\left((y-\mu)^2 - y\right)\sigma_v^2\right]/a(\mu, \sigma_v^2), \tag{13.27}$$

where $a(\mu, \sigma_v^2)$ is a normalizing constant.

From (13.27) it can be seen that, under the assumption that the measurement error variance σ_v^2 is $O(n^{-1/2})$, $h(y|\mu) \approx \mathsf{P}[\mu]$ so neglect of overdispersion is not asymptotically a serious misspecification. We can also interpret (13.27) to mean that the use of mixed Poisson models may be justified by the presence of particular types of measurement errors.

Suppose instead that $(y-\mu)$ and $(t-1)$ are correlated, which is the case if measurement error in exposure is correlated with the regressor determining μ, with

$$\mathsf{E}\left[(y-\mu)\left(t^*-1\right)\right] \equiv \sigma_{\mathbf{x}v}.$$

Then (13.27) implies

$$h(y|\mu) \approx \mathsf{P}[\mu]\left[1 + \sigma_{\mathbf{xv}} + \frac{1}{2}\left((y-\mu)^2 - y\right)\sigma_{t^*}^2\right]/a(\mu, \sigma_v^2, \sigma_{\mathbf{xv}}).$$

Unlike the previous case, the moments of this distribution additionally depend on the distribution of the measurement error through the covariance $\sigma_{\mathbf{x}v}$. Consequently, because the first moment of the distribution is no longer μ, estimators based on the Poisson mean specification will be inconsistent. In general, the entire distribution of y, not just the variance, is affected by this type of measurement error.

Heuristically, the inclusion in the conditional mean function of variables correlated with exposure measurement error should reduce the extent of misspecification. This provides a motivation for using proxy variables for exposure.

13.3.4 Proxy Variables for Exposure

In some cases the exposure is more realistically specified as a function of a set of observables. For example, Dionne, Desjardins, Labergue-Nadeau, and Maag (1995) estimate the effect of different medical conditions on truck drivers' distribution of accidents, including exposure factors measured by hours behind the wheel, kilometers driven, and other qualitative factors. The exponential conditional mean function is specified as a function of variables that affect the intensity of the process ($\mathbf{x}_1$) and those that affect the exposure ($\mathbf{x}_2$). Then

$$\begin{aligned} \mu t^*|\mathbf{x}_1, \mathbf{x}_2 &= \exp(\mathbf{x}_1'\boldsymbol{\beta}_1 + \ln t^*) \\ &= \exp(\mathbf{x}_1'\boldsymbol{\beta}_1 + \mathbf{x}_2'\boldsymbol{\beta}_2), \end{aligned} \tag{13.28}$$

which does not require constrained estimation. Usually the variables determining exposure will determine t^* multiplicatively and hence determine $\ln t^*$ additively in logarithms. For example, $\mathbf{x}_2$ may include the natural logarithm of kilometers driven.

A random error may also be included in the conditional mean. For example, if

$$t^* = \exp(\mathbf{x}_2'\boldsymbol{\beta}_2)u,$$

where u denotes an iid error with unit mean, then one will get results similar to those in subsection 13.3.2.

13.3.5 Berkson Error in Regressors

Poisson models use an exponential specification for the conditional mean. Then an additive Berkson measurement error model and unobserved heterogeneity models can be shown to be algebraically similar. Different implications can be generated, however, by changes in those assumptions.

The additive Berkson error model specifies $\mathbf{x}^* = \mathbf{x} + \mathbf{v}$; see section 13.2.1. Then for the model $y|\mu \sim \mathsf{P}[\mu]$, where $\mu = \exp(\mathbf{x}^{*\prime}\boldsymbol{\beta})$. The conditional mean μ becomes

$$\begin{aligned} \mathsf{E}[\mathbf{y}|\mathbf{x}, \mathbf{v}] &= \exp\left[(\mathbf{x} + \mathbf{v})'\boldsymbol{\beta}\right] \\ &= \exp(\mathbf{x}'\boldsymbol{\beta}) \exp(\mathbf{v}'\boldsymbol{\beta}) \\ &= \exp(\mathbf{x}\boldsymbol{\beta})w, \end{aligned} \tag{13.29}$$

where $w = \exp(\mathbf{v}'\boldsymbol{\beta})$.

This formulation has an obvious parallel with unobserved heterogeneity. Consider the exponential mean model with multiplicative heterogeneity. Then $y|\mu, \nu \sim \mathsf{P}[\mu\nu]$, where

$$\mu\nu = \exp(\mathbf{x}'\boldsymbol{\beta})\nu. \tag{13.30}$$

Clearly one can interpret the unobserved heterogeneity term in (13.30) as due to Berkson measurement error.

Unobserved heterogeneity may also be interpreted as reflecting omitted regressors $\mathbf{z}$, as in

$$\begin{aligned} \mu|\mathbf{x}, \mathbf{z} &= \exp(\mathbf{x}'\boldsymbol{\beta} + \mathbf{z}'\boldsymbol{\gamma}) \\ &= \exp(\mathbf{x}'\boldsymbol{\beta}) \exp(\mathbf{z}'\boldsymbol{\gamma}) \\ &= \exp(\mathbf{x}'\boldsymbol{\beta})u, \end{aligned} \tag{13.31}$$

where $u = \exp(\mathbf{z}'\boldsymbol{\gamma})$, which is again algebraically similar to the preceding measurement error and heterogeneity models.

These similarities can be exploited by reinterpreting the analysis of section 13.3.3. For example, parallel to (13.26), we may specify $\mathsf{E}[w-1]=0$, $\mathsf{E}\big[(w-1)^2\big]=\sigma_w^2$, and $\mathsf{E}[(w-1)(y-\mu)]=\sigma_{\mathbf{x}w}$. Furthermore, parallel to (13.27), we can derive

$$h(y|\mu)\approx \mathsf{P}[\mu]\left[1+\sigma_{\mathbf{x}w}+\frac{1}{2}\left[(y-\mu)^2-y\right]\sigma_w^2\right]/a(\mu,\sigma_w^2).$$

So the effects of additive Berkson measurement error are qualitatively similar to those due to omitted heterogeneity, errors in exposure, or omitted regressors from the conditional mean. If the measurement errors are uncorrelated with the regressors $\mathbf{x}$, then $\mathsf{E}[\mathbf{v}|\mathbf{x}]=\mathsf{E}[\mathbf{v}]$ and w has unit mean and finite variance. Then the consequences of measurement error are essentially the same as those due to unobserved heterogeneity, namely overdispersion and loss of efficiency of the QML estimator.

An alternative assumption allows for possible correlation between $\mathbf{x}$ and $\mathbf{v}$. Mullahy (1997a) has argued by reference to the omitted variable case that nonzero correlation is more realistic. That is, one or more of the omitted regressors are likely to be correlated with the included variables. Then the standard QMLE is also inconsistent.

13.4 MEASUREMENT ERRORS IN COUNTS

Measurement error in the dependent count variable leads to inconsistency of standard count data estimators. The classical measurement error model for the dependent variable is inappropriate since it does not ensure that the observed y is nonnegative or that it is a count. A variety of parametric models have been proposed corresponding to different mechanisms for the measurement error. These models are presented in this section and the two subsequent sections.

13.4.1 Additive Measurement Errors in Counts

Suppose that error-contaminated counts, denoted by y, are nonnegative and integer valued. Let y^* denote the true unobserved count and ε the additive measurement error. That is,

$$y=y^*+\varepsilon, \tag{13.32}$$

where y^* and ε are nonnegative integers.

One simple model specifies $y^*|\mu\sim\mathsf{P}[\mu]$ and $\varepsilon|\gamma\sim\mathsf{P}[\gamma]$, with y^* and ε independently distributed. This implies that

$$y|\mu,\gamma\sim\mathsf{P}\left[\mu+\gamma\right].$$

So the measurement error leads to a larger mean and variance relative to the distribution of y^*. Because the model has a nonnegative measurement error it is useful only for characterizing count inflation.

More generally we might consider the case where ε in (13.32) is integer valued, but ε can be negative. This assumption permits undercounts as well as overcounts. But the nonnegativity restriction on y implies parallel restrictions on the distribution of ε. It is implausible to expect such restrictions to hold if it is assumed that y^* and ε are independently distributed. A joint distribution for y^* and ε in (13.32) that admits correlation may be more appropriate.

Binomial thinning is another example of a mechanism for generating measurement errors in counts. Then we observe

$$y = \pi \circ y^*, \tag{13.33}$$

where $\circ$ denotes the binomial thinning operator given in Chapter 7.6.1 and the observed count $\pi \circ y^*$ is the realization of y^* binomial trials, each with probability of success π, $\pi > 0$. This is necessarily a model of undercounts.

A variation is to suppose

$$y = \pi \circ y^* + \varepsilon, \tag{13.34}$$

where ε is nonnegative integer distributed. Under the strong assumption that y and y^* have the same distribution we can interpret the observed y as being the sum of two counts. The first is due to an undercount $\pi \circ y^*$ of the true count. The second is an independent overcount of ε. If ε is Poisson distributed, then y is Poisson distributed with mean $\mathsf{E}[\varepsilon]/(1 - \pi)$.

13.4.2 Misclassified Counts and Validation Data

Counted events may be categorized by types. Classification into types may itself be subject to error, leading to an incorrect total number of events in each category. If validation data are available, then one can correct for the misclassification.

Whittemore and Gong (1991) consider an application to estimating the mortality rate from cervical cancer using cross-country grouped data on deaths and population at risk in different age groups. They assume that no deaths from other forms of cancer are misdiagnosed as cervical cancer, but some deaths due to cervical cancer are misdiagnosed as due to other cancers. "The process of classifying has perfect specificity, but imperfect sensitivity denoted by π, and the sensitivity π may vary with covariates" (Whittemore and Gong, 1991, p. 83). The concern is that π_i varies across countries, and differences in age-adjusted mortality rates for cancer across countries may merely reflect differences in misdiagnosing cervical cancer.

In this example the counts are underreported, with a specific underreporting mechanism. Validation data are available to correct for this underreporting. Specifically suppose from a separate source that w_i known deaths due to cervical cancer are independently examined by a panel of doctors, and let w_{1i} and w_{2i} denote the number of correct and incorrect classifications of cervical cancer, respectively, with $w_{1i} + w_{2i} = w_i$, $i = 1, \ldots, n$.

Assume (i) $w_{1i} \sim \mathsf{Bin}[\pi_i, w_{1i} + w_{2i}]$; (ii) disease occurs in each population i as a Poisson process with $y_i^* \sim \mathsf{P}[\mu_i^*]$; and (iii) distinct populations are independent. In this application $\mu_i^* = \lambda_i^* L_i$, where λ_i^* is the true disease rate per person-year and the exposure variable L_i denotes person-years at risk. Then the Poisson assumption about the actual occurrence of the disease implies that observed disease counts y_i, subject to misclassification error, are mutually independent Poisson variates.

Specifically for the observed (with classification error) counts of cervical cancer,

$$\begin{aligned} y_i &\sim \mathsf{P}[\mu_i] \\ \mu_i &= \pi_i \mu_i^* = \pi_i \lambda_i^* L_i. \end{aligned}$$

The density for the ith observation is then the product of Poisson and binomial likelihoods,

$$f(y_i, w_{1i}, w_{2i} | \mu_i, \pi_i) = e^{-\mu_i} {\mu_i}^{y_i} \pi_i^{w_{1i}} (1 - \pi_i)^{w_{2i}} / y_i!,$$

and the log-likelihood is

$$\begin{aligned} \mathcal{L}(\boldsymbol{\lambda}^*, \boldsymbol{\pi}) = \sum_{i=1}^{n} \{ & y_i \ln \pi_i + y_i \ln \lambda_i^* - \pi_i \lambda_i^* L_i + w_{1i} \ln \pi_i \\ & + w_{2i} \ln (1 - \pi_i) \}. \end{aligned} \tag{13.35}$$

A natural parametrization of the model is an exponential model $\lambda_i^* = \exp(\mathbf{x}_i' \boldsymbol{\beta})$ for the Poisson rate parameter and a logit model $\pi_i = \Lambda(\mathbf{z}_i' \boldsymbol{\delta}) = \exp(\mathbf{z}_i' \boldsymbol{\delta}) / \left[1 + \exp(\mathbf{z}_i' \boldsymbol{\delta})\right]$ for the probability of correct classification.

Whittemore and Gong (1991, pp. 90–91) extend this model to one of classification of K different mutually exclusive diseases where each disease may be under- or overreported, subject to the restriction that $\sum_{j=1}^{K} \pi_{jk} = 1$, where π_{jk} is the probability that true disease k is classified as disease j. Then a multinomial logit model offers a suitable parameterization of the recording process. The approach is closely related to that for log-linear models for categorical data with misclassification errors. There are some similarities between this approach and that presented in section 13.5 for underrecorded counts, especially in relation to modeling the misclassification probability in terms of observed covariates. However, here it is assumed that validation data are available and that overreporting is possible.

13.4.3 Outlying Counts

In the data set on recreational trips used in Chapter 6 there was a suspicion of measurement errors arising from the curious "clustering" in the number of self-reported boating trips if that number was 10 or higher; see Table 6.9. Some clustering at 10 and 15 trips and the presence of responses in "anchoring" categories like 20, 25, 30, 40, and 50 raised a suspicion that the responses may be affected by recall errors.

To avoid distorted inferences from such data, robust estimation based on discarding some proportion of the data is sometimes recommended. Christmann (1994, 1996) shows that the least median of weighted squares estimator for the Poisson regression model has a high breakdown point and other desirable properties.

13.5 UNDERREPORTED COUNTS

In this section we consider how one might model event counts if there is reason to believe that some events that have occurred might not be recorded. It is easy to see that if events y^* are distributed with mean μ, and each event has a constant probability π of being observed, then the observed event count y has mean $\mu = \pi\mu$. So a count model fitted to observed data yields information about the product $\pi\mu$ that is clearly a downward-biased estimate of μ, the true mean of the event process.

This section considers refinements of this basic model. The analysis relies on distributional assumptions because validation data are presumed to be unavailable. It relies on prior information or theory that sharply distinguishes between the variables that affect the event occurrence μ and those that affect the recording probability π. And it depends on the functional forms chosen for $\mu(\cdot)$ and $\pi(\cdot)$. Discussion of identification is given in Papadapoulos and Santos Silva (2008).

13.5.1 Mechanism and Examples

The term "underrecording" refers to that feature of the method of data collection that causes the observed (recorded) count to understate on average the actual number of events. Parametric estimation and inference are to be based on the recorded counts of the event.

Suppose events are generated by a pure count process, but for each event, a Bernoulli process determines whether the event is recorded. The recording process and actual event occurrence may be independent or dependent. In either case, the recorded events are shown to follow a mixed binary process. A regression model can be developed by parameterizing the moments of the mixed process. The main idea is to combine a model of underrecording with a model of the count process.

Examples of underrecorded counts may be found in many fields. They include the frequency of absenteeism in workplaces (Barmby, Orme, and Treble, 1991; Johansson and Palme, 1996), the reporting of industrial injuries (Ruser, 1991), the number of violations of safety regulations in nuclear power plants (Feinstein, 1989), the frequency of criminal victimization (Fienberg, 1981; Schneider, 1981; Yannaros, 1993), needlestick injuries in hospital (Watermann, Jankowski, and Madan, 1994), and earthquakes and cyclones (Solow, 1993), to mention only a few.

Let y^* denote the true number of events. Suppose the probability that an event is recorded, conditional on occurrence, is $\pi, 0 \leq \pi \leq 1$. Assume that the recording and occurrence mechanisms are independent, an assumption relaxed later. Then the mean number of recorded events, denoted y, is

$$\mathsf{E}[y|\mu, \pi] = \mu\pi,$$

where $\mathsf{E}[y^*] = \mu$ is the true mean and π is the average recording probability.

This leads to a simple nonlinear regression model based on parameterizing the μ and π components. Letting $\mu = \mu(\mathbf{x}, \boldsymbol{\beta})$ and $\pi = \pi(\mathbf{z}, \boldsymbol{\gamma})$, where $\mu(\cdot)$ and $\pi(\cdot)$ are known functions, the approach leads to a nonlinear regression based on

$$\mathsf{E}[y|\mathbf{x}, \mathbf{z}] = \mu(\mathbf{x}, \boldsymbol{\beta})\pi(\mathbf{z}, \boldsymbol{\gamma}). \tag{13.36}$$

Natural choices are the exponential conditional mean for $\mu(\cdot)$ and a logit or probit model for $\pi(\cdot)$.

In the subsequent discussion, we take (13.36) as the basic equation of interest. Extensions considered include allowing dependence between event occurrence and the recording process, and modeling the distribution of y rather than just the conditional mean.

13.5.2 Dependence between Events and Recording

Consider a bivariate binomial random variable (Y, R), where $Y = 1$ denotes the single occurrence of an event of interest in some specified time interval, and $R = 1$ denotes the recording of that event, but the recording process is not directly observable. The following table establishes the notation for the joint probabilities associated with the four possible outcomes, as well as the associated marginal probabilities:

	$R = 1$	$R = 0$	
$Y = 1$	π_{11}	π_{10}	π
$Y = 0$	π_{01}	π_{00}	$1 - \pi$
	π'	$1 - \pi'$	

Recording and occurrence may be dependent. For example, if an event is an action of an informed individual, the probability that the event occurs may depend on whether the event will be recorded. Thus the commission of the crime is likely to be a function of the probability of detection. Similarly, unauthorized absences from work will depend on the probability of the absence being observed (recorded) by the employer.

We now consider the joint distribution of (Y, R) expressed in terms of the marginals. Interest lies especially in $\Pr[R = 1|Y = 1] = \pi_{11}/\pi$. We wish to re-parameterize this in terms of the marginal probabilities π and π' and the correlation $\rho = \mathsf{Cor}[Y, R]$. From the table it follows that π_{11}, π_{01}, π_{10}, and

π_{00} equal, respectively, $\pi_{11}, \pi' - \pi_{11}, \pi - \pi_{11}$, and $1 - \pi' - \pi + \pi_{11}$. Some considerable algebra yields $\mathsf{Cov}[Y, R] = \pi_{11} - \pi\pi'$. It follows that

$$\rho = \mathsf{Cor}[Y, R] = \frac{\pi_{11} - \pi\pi'}{\sqrt{\pi(1-\pi)\pi'(1-\pi')}}. \tag{13.37}$$

Under independence of Y and R, $\pi_{11} = \pi\pi'$ so (13.37) yields $\rho = 0$ as expected.

The conditional probability that the event is recorded, given that it has occurred, is

$$\begin{aligned} \pi^+ &= \Pr[R = 1|Y = 1] \\ &= \pi_{11}/\pi \\ &= \pi'\left[1 + \rho\frac{1-\pi}{\sqrt{\pi\,(1-\pi)}}\frac{1-\pi'}{\sqrt{\pi'\,(1-\pi')}}\right] \\ &= \pi'(1 + \rho C), \end{aligned} \tag{13.38}$$

where the third equality follows from (13.37) and the final equality defines $C \equiv (\pi\pi')^{-1/2}(1-\pi)^{1/2}\left(1-\pi'\right)^{1/2}$.

In the special case of independence $\rho = 0$, so $\pi^+ = \pi'$ and the recording probability does not depend on the event probability π.

One can choose to either model marginal probabilities (π, π') using a standard binary outcome model such as the logit or probit or to model the joint distribution of (Y, R) using a standard bivariate probit-type model or a copula-based joint distribution derived from specified marginals as in Chapter 8.5. The copula-based approach is also relevant when the objective is to model the joint distribution of the indicator of event occurrence and the number of recorded events.

From a count data perspective, however, the preceding discussion provides a background for the joint modeling of the number of recorded events and of the indicator variable for recording of the events. If the true number of events is generated by a Poisson-type process with the rate parameter μ, then the parameters of interest are μ and π' as well as the dependence ρ.

Distribution of Recorded Events

In the case of more than one recorded event for the ith individual we assume a Poisson process for the events, so there is no serial correlation in the event process. Similarly we assume there is no serial correlation in the recording process. Then, given the conditional probability of recording an event, the distribution of recorded events, y_i, follows the Poisson distribution by the following lemma derived in section 13.8.

Lemma. *If π^+ is the probability that an event is recorded and the number of events are distributed as $\mathsf{P}[\mu]$, then the number of recorded events is distributed as $\mathsf{P}[\mu\pi^+]$.*

The mean and variance ($\mu\pi^+$) of the recorded events are smaller than the mean and variance (μ) of the actual events. For given probability π' of recording the event, (13.38) implies that the understatement of the Poisson mean is greater if the correlation is negative. Underrecording can be interpreted as a source of excess zeros because it causes the distribution to shift left, reducing both the mean and the mode and leading to an excess of zeros relative to the parent distribution. This effect, which is similar to the statistical phenomenon of binomial thinning, appears to be a common feature of all underrecorded count models considered here. If the probability of underrecording for a given individual varies from event to event, the analysis given previously should be interpreted in terms of an average recording probability for an event (Feller, 1968, p. 282).

The basic result extends in a straightforward manner to the negative binomial model. The model can be motivated as follows. Let the number of recorded events be distributed according to $\mathsf{P}[\mu\pi^+\eta]$ where η is a random unobserved heterogeneity term distributed across individuals independently of $\mu\pi^+$. If η is gamma distributed with unit mean and variance α, then unconditionally the number of events follows the $\mathsf{NB2}[\mu\pi^+, \alpha]$ distribution with mean $\mu\pi^+$ and variance $\mu\pi^+(1 + \alpha\mu\pi^+)$.

Lemma. *If π^+ is the probability that an event is recorded and the number of events are distributed as $\mathsf{NB2}[\mu, \alpha]$, then the number of recorded events is distributed as $\mathsf{NB2}[\mu\pi^+, \alpha]$.*

An extension to the hurdle model along similar lines is also possible.

Note that in general if the event process has mean μ and each event occurrence is independently recorded with probability π^+, then the number of recorded events has mean $\mu\pi^+$.

13.5.3 Underrecorded Count Regressions under Independence

Independence, so $\pi^+ = \pi'$, may be a reasonable assumption in cases in which the observed events (actions) do not adapt to the recording mechanism; for example, see Solow (1993). To proceed to parametric regression models, additional functional form assumptions are required.

Suppose y_i^* has mean μ_i that is parameterized as $g(\mathbf{x}_i, \boldsymbol{\beta})$, where usually $g(\mathbf{x}_i, \boldsymbol{\beta}) = \exp(\mathbf{x}_i'\boldsymbol{\beta})$. And suppose the recording process R has probability $\pi_i' = F(\mathbf{z}_i, \boldsymbol{\delta})$ where for the logit model, for example, $F(\mathbf{z}_i, \boldsymbol{\delta}) = \Lambda(\mathbf{z}_i'\boldsymbol{\delta})$ for $\Lambda(z) = e^{z/(1+e^z)}$. Ideally $\mathbf{z}_i$ includes observable traits of the recording mechanism and differs from $\mathbf{x}_i$, though even then problems can arise as detailed later.

The mean $\mu_i^+ = \mu_i\pi_i'$ for the underrecorded count process is then parameterized as

$$\mathsf{E}[y_i|\mathbf{x}_i, \mathbf{z}_i] = g(\mathbf{x}_i, \boldsymbol{\beta})F(\mathbf{z}_i, \boldsymbol{\delta}). \tag{13.39}$$

The conditional mean has a *double-index* structure, and the parameters $\boldsymbol{\beta}$ and $\boldsymbol{\delta}$ in the model (13.39) can be estimated by NLS given specification of $g(\cdot)$ and $F(\cdot)$. When $g(\mathbf{x}_i, \boldsymbol{\beta}) = \exp(\mathbf{x}_i'\boldsymbol{\beta})$ and $F(\cdot)$ is the logistic cdf, then (13.39) becomes

$$\mu_i^+ \equiv \mathsf{E}[y_i|\mathbf{x}_i, \mathbf{z}_i] = \exp(\mathbf{x}_i'\boldsymbol{\beta})\Lambda(\mathbf{z}_i'\boldsymbol{\delta}). \tag{13.40}$$

Note that if π_i' is constant for all observations, then it cannot be identified. Then $F(\mathbf{z}_i, \boldsymbol{\delta}) = \pi'$ so (13.40) becomes $\mathsf{E}[y_i|\mathbf{x}_i, \mathbf{z}_i] = \exp(\beta_0 + \ldots + x_{ik}\beta_k)\pi'$; therefore π' enters the model through the intercept term as $\widetilde{\beta}_0 = (\beta_0 + \ln \pi')$. Hence both β_0 and π' cannot be individually identified from an intercept estimate $\widehat{\widetilde{\beta}}_0$. Under an alternative functional form for $g(\mathbf{x}_i, \boldsymbol{\beta})$ this problem may not arise.

Even when π_i' varies across observations, problems can arise. For model (13.40) we have

$$\begin{aligned}\mu_i^+ &= \exp(\mathbf{x}_i'\boldsymbol{\beta})\frac{\exp(\mathbf{z}_i'\boldsymbol{\delta})}{1+\exp(\mathbf{z}_i'\boldsymbol{\delta})} \qquad (13.41)\\ &= \exp(\mathbf{x}_i'\boldsymbol{\beta} + \mathbf{z}_i'\boldsymbol{\delta})\frac{\exp(-\mathbf{z}_i'\boldsymbol{\delta})}{1+\exp(-\mathbf{z}_i'\boldsymbol{\delta})},\end{aligned}$$

so there are two Poisson-logit models that lead to the same conditional mean function. In the case of scalar regressor with no intercept and $x_i = z_i$ this leads to two possible estimates: (i) (β, γ) and (ii) $(\beta + \gamma, -\gamma)$. Restrictions on $(\boldsymbol{\beta}, \boldsymbol{\delta})$ are required to choose between two observationally equivalent models. Papadapoulos and Santos Silva (2008) suggest identification via an a priori sign restriction on one of the parameters or the use of nonsample information, such as an exclusion restriction.

More efficient estimation is possible using ML or QML estimation. Using the first lemma we obtain the Poisson QMLE from regression of y_i on $\mathbf{x}_i$ and $\mathbf{z}_i$ where the Poisson mean has the nonstandard form μ_i^+ given in (13.40). Mukhopadhyay and Trivedi (1995) analyzed ML, QML, and moment-based estimators. In the case of the NB2 model, the log-likelihood function given independent observations is

$$\begin{aligned}\mathcal{L}(\boldsymbol{\beta}, \boldsymbol{\delta}, \alpha) = \sum_{i=1}^{n} [&\ln\Gamma(y_i + \alpha^{-1}) + \ln y! - \ln\Gamma(\alpha^{-1}) + \alpha^{-1}\ln(\alpha^{-1})\\ &- \alpha^{-1}\ln(\alpha^{-1} + \mu_i^+) + y_i \ln \mu_i^+ - y_i \ln(\alpha^{-1} + \mu_i^+)].\end{aligned} \tag{13.42}$$

An alternative to the double-index model presented here is a single-index model in which μ_i^+ is written in the form $g(\mathbf{x}_i, \mathbf{z}_i)$, where $g(\cdot)$ is treated as a known or unknown function. That is, no attempt might be made to distinguish between the two separate components of μ_i^+. Distinguishing between the double- and single-index models in small samples may be especially difficult, since this distinction relies heavily on functional forms.

Example: Safety Violations

An empirical example of the Poisson-binomial model analyzed earlier is Feinstein (1989). He used panel data from more than 1,000 inspections of 17 U.S. commercial nuclear reactors by the Nuclear Regulatory Commission over three years to study factors determining the rate of safety violations. The dependent variable is the number of safety violations cited. He reported a finding that economic incentives had a small impact on noncompliance, whereas technological and operating characteristics of the plant had a larger impact. The model he considers has an additional complication. A sampled plant may or may not comply with regulations, with probabilities $1 - f_1(\mathbf{x}_i'\boldsymbol{\beta}_1)$ and $f_1(\mathbf{x}_i'\boldsymbol{\beta}_1)$, respectively. Then a violation by a noncompliant plant may or may not be detected, with probabilities $f_2(\mathbf{z}_i'\boldsymbol{\beta}_2)$ and $1 - f_2(\mathbf{z}_i'\boldsymbol{\beta}_2)$, respectively. Reported nonviolations come from the set of undetected violators and from genuine nonviolators. These formulations lead to a likelihood function for violation and detection,

$$\mathcal{L}(\boldsymbol{\beta}_1, \boldsymbol{\beta}_2|\mathbf{x}, \mathbf{z}) = \sum_{i \in V} \left[\ln f_1(\mathbf{x}_i'\boldsymbol{\beta}_1) f_2(\mathbf{z}_i'\boldsymbol{\beta}_2)\right] + \sum_{i \in V^c} \left[\left(1 - \ln f_1(\mathbf{x}_{1i}'\boldsymbol{\beta}_1)\right) f_2(\mathbf{z}_i'\boldsymbol{\beta}_2)\right],$$

where V is the set of detected violators and V^c is its complement. In Feinstein's model the mean number of detected violations is the product of detection probability and the mean number of violations, $f_2(\mathbf{z}_i'\boldsymbol{\beta}_2)\mu_i$. Feinstein calls his estimator the *detection control estimator.*

13.5.4 Underreported Count Regressions under Dependence

Now suppose the observed counts refer to behavioral responses that incorporate knowledge of, or adaptation to, the recording process. Then $\pi^+ \neq \pi'$ but instead from (13.38) $\pi^+ = \pi'(1 + \rho C)$, where

$$C = \sqrt{\frac{(1-\pi)(1-\pi')}{\pi\pi'}} > 0.$$

It follows that $\mathsf{E}[y]$ is now

$$\mu^+ = \mu\pi'(1 + \rho C). \tag{13.43}$$

The second term ρC measures the effect of the departure from independence of the recording mechanism. This suggests that the functions μ, π', and C should be parameterized in such a way as to identify all parameters of interest. Doing so is difficult because from section 13.5.2 ρ and C depend on probabilities in the recording mechanism (R, Y) other than just $\pi' = \Pr[R = 1]$. In particular they depend on $\pi = \Pr[Y = 1]$ or, in the more general case, on $\mu = \mathrm{E}[y^*]$. Hence the ρC term depends on both $\mathbf{x}$ variables, through π or μ, and on the $\mathbf{z}$ variables, through π'. If we substitute known functional forms for all unknown parameters, the resulting expression for μ^+ may not be tractable. This provides some motivation for a functional form such as

$$\mu^+ = \mu\pi' + g^*(\mathbf{x}, \mathbf{z}), \tag{13.44}$$

where $g^*(\cdot)$ is an unknown function that can be handled by nonparametric methods; see Chapter 11.

To improve tractability, the use of an ad hoc "approximation" may simplify the expression for μ^+. For example, let

$$\mu_i^+ = g(\mathbf{x}_i\boldsymbol{\beta})F(\mathbf{z}_i, \boldsymbol{\delta}) + h(\mathbf{x}_i, \mathbf{z}_i, \boldsymbol{\theta}),$$

where the function $h(\cdot)$ may be specified to mimic the properties of the term ρC. Incorrectly ignoring the second term in (13.43) produces inconsistent estimates.

13.6 UNDERREPORTED AND OVERREREPORTED COUNTS

The approach of section 13.5 can be adapted to accommodate overreporting as well as underreporting.

13.6.1 Models

Zero inflation may reflect systematic underrecording in which either all event counts are understated or the occurrence of a single event is not recorded. Hence the with-zeros model introduced in Chapter 4.6 may be suitable for modeling underrecording. Recall that it can also be used for modeling inflation of other counts, such as a single event. Accordingly, the zero-inflation model has been used when zeros only are overreported, and a zero-deflated distribution has been used when the zeros are underrecorded (Cohen, 1960; Johnson, Kemp, and Kotz, 2005).

The model

$$\begin{aligned}\Pr[y = 0] &= \varphi + (1 - \varphi)e^{-\mu} \\ \Pr[y = r] &= (1 - \varphi)e^{-\mu}\mu^r/r!, \qquad r = 1, 2, \ldots,\end{aligned}$$

accommodates excess zeros relative to the Poisson (the case $\varphi = 0$) if $0 < \varphi < 1$. Furthermore, $\mathsf{E}[y] = \mu(1 - \varphi)$ is consistent with underreporting since $\mathsf{E}[y] = \mu$ in the Poisson case.

Overreporting of specific values, such as a count of two, leads to inflation at those values relative to the true frequency distribution. Some authors have proposed models that allow for an excess of certain positive values relative to a benchmark model such as the Poisson. From Chapter 6.5 fertility data have an excess of counts of both zero and two. To allow for this in their study of

Swedish fertility, Melkersson and Rooth (2000) formulate the following model:

$$\begin{aligned}
\Pr[y=0|\mathbf{x}] &= \varphi_1 + qh(y|\mathbf{x})\\
\Pr[y=2|\mathbf{x}] &= \varphi_2 + qh(y|\mathbf{x})\\
\Pr[y=r|\mathbf{x}] &= (1-\varphi_1-\varphi_2)h(y|\mathbf{x}), \qquad r=1,3,4,\ldots,
\end{aligned}$$

where $h(y|\mathbf{x})$ is the parent distribution, such as Poisson or NB2. Additional flexibility may result from parameterizing φ_1 and φ_2. This example shows how this model can potentially capture both underreporting of values other than zero and two. However, this example does require the assumption that the excess probability at particular values is due to measurement errors and nothing else, and such an assumption may be unwarranted.

A richer model is proposed by Li, Trivedi, and Guo (2003). A simple way to allow for over-recording or underrecording while maintaining a count distribution is to specify a count distribution for $y|y^*$ where y denotes the observed counts and a count distribution is also specified for the true (unobserved) counts y^*. To allow for underreporting of true zeroes, the distribution of $y|y^*$ must be such that the distribution of $y|y^*=0$ is not degenerate. Li et al. propose the following model

$$\begin{aligned}
y^* &\sim \mathsf{NB2}[\exp(\mathbf{x}'\boldsymbol{\beta})]\\
y|y^*=0 &\sim \mathsf{P}[\exp(\mathbf{z}'\boldsymbol{\delta})]\\
y|y^*>0 &\sim \mathsf{P}[\exp(\mathbf{z}'\boldsymbol{\gamma})y^*],
\end{aligned}$$

where regressors $\mathbf{x}$ determine the true counts and regressors $\mathbf{z}$ determine the recording process. The resulting density for $f(y|\mathbf{x},\mathbf{z},\boldsymbol{\beta},\boldsymbol{\delta},\boldsymbol{\gamma})$ involves an infinite sum over y^*. Rather than truncate this sum, Li et al. estimate the model by simulated ML.

They apply this model to National Crime Victims Survey data on the number of thefts experienced by students at school. For their data 88% of observed counts are 0 and about 1% exceed 4. The regressors $\mathbf{z}$ for the recording process are a subset of the regressors $\mathbf{x}$ for the true number of counts. A comparison of observed counts y with 500 replications from the fitted model of the true counts y^* finds that 83% of counts were reported accurately (88% of counts were 0), 12% were overreported, and 4% were underreported.

13.6.2 Empirical Evidence

Some evidence from the U.S. National Center for Health Statistics (1967) for chronic medical conditions suggests that self-reported medical records may understate the degree of health care usage. Hospitalizations have also been found to be underreported in the United States.

McCallum, Raymond, and McGilchrist (1995) carry out an interesting study using Australian data in which a comparison is made between the self-reported number of doctor visits over a three-month period and the same usage as measured by the (presumably more accurate) records of the Health Insurance

Commission. In a sample of around 500 individuals, they found that self-reported estimates were overreported relative to the Health Insurance Commission estimates for small numbers of visits (one to four), but generally underreported for higher usage. McCallum et al. analyze the relation between the reporting error and the characteristics of the respondents and conclude with a cautionary note about the use of self-reported data. Cameron et al. (2004) analyze the McCallum et al. data set, using a bivariate conditional copula for self-reported and directly observed number of doctor visits. Such a bivariate model can be used to study the conditional distribution of measurement errors in cases where two measures of the same outcome are available and to provide insights into the factors that affect these measurement errors.

13.7 SIMULATION EXAMPLE: POISSON WITH MISMEASURED REGRESSOR

We investigate the consequences of classical measurement error in a single regressor for Poisson regression, and the performance of several estimation methods given in section 13.2. These methods are intended to reduce estimator bias though are still inconsistent. We also present consistent ML estimates.

The model is a Poisson regression of y on x_1 and x_2. The regressor x_1 is measured with error with two measures of the true value available: x_1, and a replicate denoted z_1 which can be used as a replicate or instrument. The regressor x_2 is measured without error:

$$\begin{aligned} y_i &\sim \mathsf{P}[\beta_0 + \beta_1 \times x_{1i}^* + \beta_2 \times x_{2i}] \\ x_{1i} &= x_{1i}^* + v_{1i} \\ z_{1i} &= x_{1i}^* + v_{2i} \\ (\beta_0, \beta_1, \beta_2) &= (0, 1, 1). \end{aligned}$$

The random variables x_1^*, x_2, v_1, and v_2 are all independent draws with means 0 and standard deviation 0.4. It follows that there is considerable measurement error with a signal-to-noise ratio of one. The sample size is 10,000.

In the first simulation all draws are from the normal distribution (see Table 13.1). The count y has mean 1.157 and standard deviation 1.281. Most counts are low, with 70% equaling 0 or 1, and the highest count is 10. Poisson regression of y on x_1^* and x_2, given in the first column, yields estimates close to the dgp values of 0, 1, and 1, as expected. Poisson regression of y on x_1 and x_2, given in the second column, yields $\hat{\beta}_1 = 0.523$. This is essentially the linear regression result that a noise-to-signal ratio of 1 leads to $\hat{\beta}_1 = 0.5\beta_1$. By contrast, $\hat{\beta}_2 = 1.020$ so that there is no impact on the coefficient of x_2, a consequence of independence of x_2 and x_1 in this simulation design. Note also that the standard error of $\hat{\beta}_1$ has decreased from 0.024 to 0.016.

The next four estimators seek to correct the large inconsistency of $\hat{\beta}_1$, though all the methods are approximate methods that theoretically do not

Table 13.1. *Simulation exercise: Poisson regression with one mismeasured regressor*

	Estimates given normally distributed variates						
Variable	True	Naive	RCal	SIMEX	NL2SLS	IVapprox	MLE
ONE	−0.005	0.029	0.022	0.006	−0.072	0.039	−0.003
	(0.011)	(0.010)	(0.011)	(0.010)	(0.015)	(0.011)	(0.011)
*X1 or X1**	0.990	0.523	0.999	0.858	0.961	0.958	0.990
	(0.024)	(0.016)	(0.031)	(0.024)	(0.036)	(0.035)	(0.029)
X2	1.007	1.020	1.013	1.013	1.025	1.026	1.012
	(0.023)	(0.023)	(0.024)	(0.024)	(0.027)	(0.026)	(0.025)
	Estimates given rescaled chi-squared variates						
	True	Naive	RCal	SIMEX	NL2SLS	IVapprox	
ONE	0.010	0.033	0.015	0.008	−0.108	0.032	−0.018
	(0.010)	(0.010)	(0.011)	(0.011)	(0.018)	(0.013)	(0.012)
*X1 or X1**	0.975	0.692	1.254	0.956	0.979	1.356	1.214
	(0.010)	(0.010)	(0.025)	(0.015)	(0.023)	(0.047)	(0.021)
X2	0.987	1.000	0.980	0.984	0.993	0.976	0.970
	(0.010)	(0.010)	(0.017)	(0.018)	(0.025)	(0.023)	(0.016)

Note: Top panel is baed on simulated data with normally distributed variates. Bottom panel is based on rescaled chi-squared variates. Robust standard errors are given in parentheses.

yield consistent estimates; see section 13.2.5. The regression calibration and simulation extrapolation using the replicate z_1 are implemented using Stata programs due to Hardin, Schmiediche, and Carroll (2003a, 2003b) with 400 bootstrap replications. The fifth column uses the standard NL2SLS estimator based on the condition $\mathsf{E}\left[\mathbf{z}(y - \exp(\mathbf{x}'\boldsymbol{\beta}))\right] = \mathbf{0}$, where z_1 is an instrument for x_1 and z_2 is an instrument for itself. The sixth column uses the approximate IV method that obtains a predictor $\hat{x}_1$ and estimates a Poisson regression of y on $\hat{x}_1$ and x_2. All four methods do well. The only estimates that are more than two standard errors from the dgp values are $\hat{\beta}_1 = .858$ for SIMEX and the intercept for NL2SLS and approximate IV. For all but SIMEX there is a loss in precision of estimation of β_1 while β_2 is as precisely estimated as before. The final column is the exact MLE based on the structural model detailed at the end of section 13.2.7. The estimates are obtained using a Stata program for GLAMMs due to Rabe-Hesketh, Skrondal, and Pickles (2003). The estimates are consistent since the simulation is based on normally distributed measurement error and true regressor.

In the second simulation all draws are from the highly skewed adjusted chi-squared distribution. For example, $x_1^* = (0.4/\sqrt{2})(w - 1)$ where w is a draw from $\chi^2(1)$. All variables are more skewed and leptokurtic, though the variance is unchanged. The count y has a higher mean of 1.324 and a standard deviation of 2.807. Most counts are low, with 70% equaling 0 or 1, but there is a long tail with 1% of counts exceeding 8. This dispersion leads to improved estimator precision with the standard errors in the first column falling by 60%. The naive estimator is again inconsistent, with $\hat{\beta}_1 = 0.692$, but with less attenuation bias than in the normally distributed case. Unlike for the first design, the standard error of $\hat{\beta}_1$ is the same for the naive estimator. The SIMEX and NL2SLS methods correct for the measurement error, with $\hat{\beta}_1$ equal to, respectively, 0.956 and 0.979. By contrast, the regression calibration and approximate IV methods yield $\hat{\beta}_1$ equal to, respectively, 1.254 and 1.356. For these methods the bias is roughly as large as for the naive estimator, though now the true parameter value is overestimated. The ML estimator of 1.214 given in the last column is also inconsistent, a consequence of measurement error and the true regressor no longer being normally distributed.

13.8 DERIVATIONS

We derive the lemmas in section 13.5. Let Y^* be the number of events and let $R_1, R_2, \ldots, R_{Y^*}$ be a sequence of Y^* independent Bernoulli trials, in which each R_j takes 1 if the event is recorded (with probability π^+) and 0 otherwise (with probability $1 - \pi^+$). The number of recorded events Y can be written as $Y = R_1 + R_2 + \ldots + R_{Y^*}$.

If Y^* is $\mathsf{P}[\mu]$ distributed then the resulting distribution of Y is a compound Poisson distribution, or a binomial distribution stopped by Poisson distribution. This distribution can be derived using the pgf. Let $\varphi(t)$ be the pgf of the Poisson; then

$$\varphi(t) = \exp(-\mu + \mu t) \quad \text{for any real } t.$$

Let $\xi(t)$ be the pgf of the Bernoulli trial; then

$$\xi(t) = (1 - \pi^+) + \pi^+ t \quad \text{for any real } t.$$

Then the pgf of Y is

$$\begin{aligned}\varphi(\xi(t)) &= \exp(-\mu + \mu\xi(t)) \\ &= \exp(-(\mu\pi^+) + (\mu\pi^+)t) \quad \text{for any real } t.\end{aligned}$$

So Y is Poisson distributed with parameter $\mu\pi^+$.

The proof of the second lemma is similar to the first lemma. The pgf of the negative binomial distribution is

$$\varphi(t) = (1 + \gamma - \gamma t)^{-\alpha^{-1}} \quad \text{for any real } t,$$

where $\gamma = \alpha\mu$. Then the pgf of Y is

$$\begin{aligned}\varphi(\xi(t)) &= (1 + \gamma - \gamma\xi(t))^{-\alpha^{-1}} \\ &= (1 + (\gamma\pi^+) - (\gamma\pi^+)t)^{-\alpha^{-1}} \quad \text{for any real } t.\end{aligned}$$

So Y follows a negative binomial distribution with parameters $\mu\pi^+$ and α.

13.9 BIBLIOGRAPHIC NOTES

Carroll et al. (2006) and Buonaccorsi (2010) provide up-to-date accounts of recent research in measurement error in nonlinear models with an emphasis on the GLM literature, including some results for the Poisson model. Some of the estimators in these references are implemented using Stata code; see Newton and Cox (2003) and associated references. The econometrics literature has focused on identification in nonlinear models under weak parametric assumptions; see the survey by Chen et al. (2011).

A general discussion of the effects of measurement error on the distribution of the response variable, in leading to a possible attenuation bias in the LEF and LEFN classes of models, is given in Chesher (1991). Jordan et al. (1997) estimate a Poisson regression with overdispersion and normally distributed errors in variables for mortality data in a Bayesian framework using Monte Carlo techniques. Guo and Li (2001) estimate a negative binomial model for physician outpatient visits with normally distributed errors in variables using simulated ML.

The proxy variable approach to modeling exposures is illustrated in Dionne et al. (1995). Alcañiz (1996) examines computational algorithms for restricted estimation of Poisson and NB2 models with errors in exposure.

Because counted outcomes are affected by misclassification of the occurrence of the event of interest, the literature on the effects of misclassification of binary outcomes is also relevant to this chapter. Chen (1979) analyzes a log-linear model with misclassified data. Hausman, Abrevaya, and Scott-Morton (1998) and Whittemore and Gong (1991) provide further references.

The section on underrecorded counts borrows from Mukhopadhyay and Trivedi (1995). They also develop a score test for underrecording in a negative binomial model. The representation of the bivariate binomial that they use has been studied by Eagleson (1964, 1969). Eagleson's work follows the earlier contributions by Lancaster (1958, 1963). Cameron and Trivedi (1993) review some earlier contributions.

13.10 EXERCISES

13.1 Consider the following sequential estimator: (1) Estimate the Poisson-logistic model given in section 13.5.3 by maximum likelihood. Let $\hat{\mu}^+$ denote the MLE. (2) Regress the quantity $(y - \hat{\mu}^+)^2 - \hat{\mu}^+$ on $(\hat{\mu}^+)^2$. (3) Substitute the estimate of α, $\widetilde{\alpha}$, into the "score" equations obtained from (13.42) and solve the equations,

$$s(\boldsymbol{\theta}(\widetilde{\alpha})) = \sum_{1}^{n} \left(\frac{y_i - \mu_i^+}{1 + \alpha\mu_i^+} \right) \left[\begin{array}{c} x_i \\ (1 - \Lambda(\mathbf{z}_i'\delta))\mathbf{z}_i \end{array} \right] = \mathbf{0}.$$

Interpret this sequential estimator as a QGPML estimator. Compare the properties of the estimator with one in which α is estimated using the moment equation

$$\sum_{i=1}^{n}\left[\frac{(y_i-\hat{\mu}_i^{+})^2}{\hat{\mu}_i^{+}(1+\alpha\hat{\mu}_i^{+})}-\frac{n-k-m}{n}\right]=0.$$

13.2 Consider the recreational trips data from Chapter 6. It was suggested there that the data may be subject to recall or measurement errors. Assuming that these recall errors are potentially concentrated among high counts, reestimate the Poisson and NB2 models after sequentially deleting counts greater than (a) 15, (b) 20, and (c) 25. Which parameters from the estimated model would you expect to be most sensitive to such deletion? Does your expectation match the observed outcome?

13.3 Using the set-up in section 13.5.2, and assuming independence between event occurrence and event recording, show that the result of the first lemma also extends to the hurdle count model.

APPENDIX A

Notation and Acronyms

ACF	autocorrelation function
AIC	Akaike information criterion
AME	average marginal effect
ARMA	autoregressive moving average
BHHH	Berndt-Hall-Hall-Hausman algorithm
BIC	Bayes information criterion
BPoiss	binary Poisson
Boot	bootstrap
BVNB	bivariate negative binomial
BVP	bivariate Poisson
CAIC	consistent Akaike information criterion
CB	correlated binomial
CCRE	conditionally correlated random effects
cdf	cumulative distribution function
CFMNB	slope-constrained finite mixture of negative binomials
CFMP	slope-constrained finite mixture of Poissons
CM	conditional moment (function or test)
CML	conditional maximum likelihood
$\mathsf{COV}_{\mathsf{ML}}$	off-diagonal block of the (Fisher) information matrix
CSFE	cluster-specific fixed effects
CSRE	cluster-specific random effects
CV	coefficient of variation
DARMA	discrete ARMA
dgp	data-generating process
Diag	diagonal
DPM	Dirichlet Process Mixture
E	mathematical expectation
$\mathcal{EL}$	expected value of log-likelihood function
E[y|x]	conditional mean of y given $\mathbf{x}$
EIV	errors-in-variables model
Elast	elasticity
ELEF	extended linear exponential family

EM	expectation-maximization (algorithm)
EW	Eicker-White
f(y|x)	conditional density of y given $\mathbf{x}$
FE	fixed effects
FM	finite mixture
FMM	finite mixture model
FMNB-C	C-component finite mixture of negative binomial
FMP-C	C-component finite mixture of Poisson
GAM	generalized additive model
GEC(k)	generalized event count model
GEE	generalized estimating equations
GLLAMM	generalized linear latent and mixed model
GLM	generalized linear model
GLS	generalized least squares
GMM	generalized method of moments
HMM	hidden Markov model
HPD	highest posterior density
IC	information criteria
IG	inverse gamma
iid	independently identically distributed
IM	information matrix
INAR	integer-valued autoregressive
INARMA	integer ARMA
$\mathsf{L}[\tau, \mathsf{z}]$	Laplace transform
L	likelihood function
$\mathcal{L}$	log-likelihood function
LEF	linear exponential family
LEFN	linear exponential family with nuisance parameter
LM	Lagrange multiplier
LR	likelihood ratio
mgf	moment generating function
$\mathsf{m}(\mathsf{s};\pi_1, \ldots, \pi_n)$	multinomial distribution
M(t)	moment generating function
MCMC	Markov chain Monte Carlo
MEM	marginal effect at the mean
MER	marginal effect at a representative value
MH	Metropolis-Hastings
ML	maximum likelihood (method)
MLE	maximum likelihood estimator
MLH	maximum likelihood Hessian
MLOP	maximum likelihood outer product
MPSD	modified power series distributions
$\mathsf{N}[\mu, \sigma^2]$	normal distribution with mean μ and variance σ^2
N[0, 1]	standard normal distribution
NB	negative binomial

NB1	NB distribution with linear variance function
NB2	NB distribution with quadratic variance function
NBFE	negative binomial fixed effects model
NB1FE	negative binomial 1 fixed effects model
NBH	negative binomial hurdle
NBRE	negative binomial random effects model
NBPp	polynomial negative binomial of order p
NLIV	nonlinear instrumental variable
NLIV2	sequential two-step NLIV
NLS	nonlinear least squares
NLSUR	nonlinear seemingly unrelated regressions
NPML	nonparametric maximum likelihood
N(s,s+t)	number of events observed in interval (s,s+t)
OLS	ordinary (linear) least squares
OP	outer product
OPG	outer product of gradient (score)
OrdProb	ordered probit
P[μ]	Poisson distribution with mean μ
pdf	probability density function
PFE	Poisson fixed effects
pgf	probability generating function
PGLM	pseudo generalized linear model
PGP	Poisson-gamma polynomial
PH	Poisson hurdle
P-IG	Poisson inverse gamma mixture
plim	probability limit
pmf	probability mass function
PML	pseudo-maximum likelihood
PPp	polynomial Poisson (model) of order p
PQGPML	Poisson-based quasi-generalized PML
PRE	Poisson random effects
PSD	power series distribution
QEE	quadratic estimating equations
QGPML	quasigeneralized PML
QGPMLE	quasigeneralized PMLE
QL	quasilikelihood
QVF	quadratic variance function
RE	random effects
RS	robust sandwich
SC	Schwarz Bayesian information criterion
SML	simulated maximum likelihood
SMM	simulated method of moments
SNPML	semi-nonparametric maximum likelihood
SPML	semiparametric maximum likelihood
T_{CM}	conditional moment test

T_{GoF}	chi-square goodness-of-fit test
T_H	Hausman test
T_{LM}	Lagrange multiplier (score) test statistic
T_{LR}	likelihood ratio test statistic
T_W	Wald test statistic
T_Z	standard normal test statistic
V	variance
V[y|x]	conditional variance of y given $\mathbf{x}$
V_{ML}	diagonal block of the Fisher information matrix
WLS	weighted least squares
WZ	with (excess) zeros
ZINB	zero-inflated Negative binomial
ZIP	zero-inflated Poisson

Functions, Distributions, and Moments

For convenience we list in this appendix expressions and moment properties of several univariate distributions that have been used in this book, most notably the Poisson and negative binomial. But first we define the gamma function, a component of these distributions.

B.1 GAMMA FUNCTION

Definition: *The gamma function, denoted by* $\Gamma(a)$ *is defined by*

$$\Gamma(a) = \int_0^\infty e^{-t} t^{a-1} dt, \qquad a > 0.$$

Properties of the gamma function include the following:

1. $\Gamma(a) = (a-1)\Gamma(a-1)$
2. $\Gamma(a) = (a-1)!$ if a is a positive integer
3. $\Gamma(0) = \infty, \quad \Gamma(\frac{1}{2}) = \sqrt{\pi}$
4. $\Gamma(na) = (2\pi)^{(1-n)/2}(n)^{na-1/2}\prod_{k=0}^{n-1}\Gamma(a+\frac{k}{n})$, where n is a positive integer.

Definition: *The incomplete gamma function, denoted by* $\gamma(a,x)$*, is defined by*

$$\gamma(a,x) = \int_0^x e^{-t} t^{a-1} dt; \qquad a > 0, x > 0.$$

The ratio $\gamma(a,x)/\Gamma(a)$ is known as the *incomplete gamma function* ratio or the gamma cdf.

The derivative of the logarithm of the gamma function is the *digamma function*

$$\frac{d \ln \Gamma(a)}{da} = \psi(a).$$

The digamma function obeys the recurrence relation

$$\psi(a+1) = \psi(a) + 1/a,$$

and the jth derivative $\psi^{(j)}(a+1)$ with respect to a obeys the recurrence relation

$$\psi^{(j)}(a+1) = (-1)^j j! a^{-j-1} + \psi^{(j)}(a).$$

B.2 SOME DISTRIBUTIONS

B.2.1 Poisson

The Poisson density for the count random variable y is

$$f(y) = \frac{e^{-\mu}\mu^y}{y!}; \quad y = 0, 1, \ldots; \quad \mu > 0.$$

Then $y \sim \mathsf{P}[\mu]$ with mean μ, variance μ, mode $\lceil \mu \rceil - 1$, moment generating function (mgf) $exp\left[\mu(e^t - 1)\right]$, and probability generating function (pgf) $exp\left[\mu(t-1)\right]$.

B.2.2 Logarithmic Series

The logarithmic series density for the positive valued count random variable y is

$$f(y) = \alpha\frac{\theta^y}{y}; \quad y = 1, 2, \ldots, \quad 0 < \theta < 1, \quad \alpha = -\left[\log(1-\theta)\right]^{-1}.$$

Then y has mean $\alpha\theta/(1-\theta)$, variance $\alpha\theta(1-\alpha\theta)/(1-\theta)^2$, mgf $\log\left(1-\theta e^t\right)/\log(1-\theta)$, and pgf $\log(1-\theta t)/\log(1-\theta)$.

B.2.3 Negative Binomial

The negative binomial density for the count random variable y with parameters α and P is

$$f(y) = \binom{\alpha+y-1}{\alpha-1}\left(\frac{P}{1+P}\right)^y\left(\frac{1}{1+P}\right)^\alpha; \quad y = 0, 1, \ldots; \quad P, \alpha > 0.$$

Then y has mean αP, variance $\alpha P(1+P)$, and pgf $(1+P-Pt)^{-\alpha}$.

Different authors use different parameterizations of the negative binomial (Johnson, Kemp, and Kotz, 2005). In this book we reparameterize in terms of the mean $\mu = \alpha P$, which leads to

$$f(y) = \binom{\alpha+y-1}{\alpha-1}\left(\frac{\mu}{\alpha+\mu}\right)^y\left(\frac{\alpha}{\alpha+\mu}\right)^\alpha,$$

which has mean μ, variance $\mu(1+\mu/\alpha)$, and pgf $(1+\mu/\alpha-\mu t/\alpha)^{-\alpha}$.

The special case of the *geometric distribution* is obtained by setting $\alpha = 1$. *Pascal distribution* is obtained by setting α equal to an integer.

Yet another commonly used parameterization of the negative binomial is

$$f(y) = \binom{\alpha + y - 1}{\alpha - 1} p^y (1-p)^\alpha; \quad y = 0, 1, \ldots; \quad 0 < p < 1,\ \alpha > 0,$$

in which case y has mean $\alpha p/(1-p)$ and variance $\alpha p/(1-p)^2$; see Chapter 4.2.3.

B.2.4 Gamma

The gamma density for the positive continuous random variable y is

$$f(y) = \frac{e^{-y\phi} y^{\alpha-1} \phi^\alpha}{\Gamma(\alpha)}; \quad y > 0; \quad \phi > 0,\ \alpha > 0.$$

Then y has mean α/ϕ, variance α/ϕ^2, and mgf $exp\,[\phi/(\phi-1)^\alpha]$. Then α is called the shape parameter and β is called the inverse scale parameter.

An alternative parameterization of the gamma replaces β with $1/\theta$. Then y has mean $\alpha\theta$ and variance $\alpha\theta^2$, and β is called the scale parameter.

B.2.5 Lognormal

The lognormal density for the positive continuous random variable y is

$$f(y) = \frac{1}{\sigma y \sqrt{2\pi}} \exp\left[-\frac{1}{2\sigma^2} (\ln y - \xi)^2 \right]; \quad y > 0;$$
$$-\infty < \xi < \infty,\ \sigma > 0,\ \alpha > 0.$$

Then y has mean $\exp\left[\xi + (1/2\sigma^2)\right]$ and variance $\exp\left(2\xi + \sigma^2\right)\left[\exp\left(\sigma^2\right) - 1\right]$.

B.2.6 Inverse Gaussian

The inverse Gaussian density for the positive continuous random variable y is

$$f(y) = \sqrt{\frac{\theta}{2\pi y^3}} \exp\left[-\frac{\theta(y-\mu)^2}{2\mu^2 y} \right]; \quad y > 0; \quad \theta > 0,\ \mu > 0.$$

Then y has mean μ and variance μ^3/θ.

B.3 MOMENTS OF TRUNCATED POISSON

The kth order factorial moments of truncated Poisson can be derived conveniently in terms of the mean of the regular Poisson μ, the mean of the truncated Poisson γ, the adjustment factor δ, and the truncation point $r-1$. Both γ and δ are defined in Chapter 4.3. Let $\mathsf{E}\left[y^{(k)}\right]$ represent the kth descending factorial moment of y, and $y^{(k)} = \prod_{z=0}^{k-1}(y-z)$, $k \geq 1$. For the left-truncated Poisson model the factorial moments can be derived using

$$\mathsf{E}\left[y^{(k)}\right] = \mu^{k-1} + \delta \sum_{j=0}^{k-2} \mu^j \pi_j,$$

where

$$\pi_j = \prod_{i=0}^{k-2-j} (r - 1 - i).$$

Hence the first four factorial moments of the left truncated Poisson are

$$\begin{aligned}
\mathsf{E}\left[y^{(1)}\right] &= \gamma \\
\mathsf{E}\left[y^{(2)}\right] &= \mu\gamma + \delta(r - 1) \\
\mathsf{E}\left[y^{(3)}\right] &= \mu^2\gamma + \delta\left[(r - 1)(r - 2) + (r - 1)\mu\right] \\
\mathsf{E}\left[y^{(4)}\right] &= \mu^3\gamma + \delta\begin{bmatrix} (r - 1)(r - 2)(r - 3) \\ \quad + (r - 1)(r - 2)\mu + (r - 1)\mu^2 \end{bmatrix}.
\end{aligned}$$

Given the factorial moments, the uncentered and central moments can be obtained easily using the standard relationships between the three types of moments. For details, see Gurmu and Trivedi (1992).

APPENDIX C

Software

Most regression packages support estimation of cross-section Poisson and negative binomial regression models. Packages that claim to have a significant component for count models most likely cover cross-section truncated, zero-inflated, and hurdle models, and they may cover the standard panel count models. Packages that instead only include counts within a GLM module do not cover truncated, zero-inflated, and hurdle models and only cover panel counts to the extent that they have GEE and generalized linear mixed model modules. They do not cover the Poisson fixed effects model.

LIMDEP 9.0 covers a wide range of cross-section count models, including censored, truncated, hurdle, zero-inflated, latent heterogeneity, and latent class, as well as panel fixed and random effects Poisson and negative binomial.

Stata 12.0 covers a similar range of models. Additionally it covers GLM and GEE estimators, some mixed models, and a wide range of mixed models with user-written add-on GLLAMM.

SAS 9.3 covers a similar range of models to Stata through procedures COUNT, TRCOUNT (a recent addition that covers zero-truncated and panel), GENMOD, GEE, and NLMIXED. Procedure COPULA covers copula-based models.

EViews 7.2 covers cross-section Poisson, NB, and GLM.

TSP 5.1 covers cross-section Poisson and NB. It also provides a relatively simple syntax for ML and GMM estimation of user-provided objective functions, including panel count models.

SPSS 20 covers cross-section Poisson and NB and has procedures GENLIN, GEE, and GLMM for GLMs and their multi-equation extensions.

R 2.14 covers cross-section GLMs with function glm. The R package pscl includes hurdle and with-zeroes models, the package geepack covers GEE, and a number of R packages cover nonlinear mixed models. The R package np (Hayfield and Racine, 2008) provides kernel-based estimation of many nonparametric and semiparametric models and was used in Chapter 11.

Gauss package Gaussx covers Poisson regression and package Discrete Choice covers cross-section truncated, censored, and zero-inflated Poisson and NB. The Gauss program ExpEnd (Windmeijer, 2002) enables GMM estimation of fixed effects panel count models with regressors that are endogenous and/or lagged dependent variables.

For models where off-the-shelf software is not available, one needs to provide at least the likelihood function, for ML estimation, or the moment conditions and weighting matrix, for GMM estimation. In principle this can be done using many regression packages, using matrix programming languages such as R, MATLAB, GAUSS, SAS/IML, or Stata's Mata. In practice numerical problems can be encountered when models are quite nonlinear.

Software for simulation-based estimation is less readily available. The BUGS package, an acronym for Bayesian inference. Using the Gibbs Sampler, implements MCMC methods including Gibbs and Metropolis-Hastings. Lancaster (2004) includes a summary of BUGS and illustrative examples. WinBUGS is a Windows version of BUGS. BUGS can be run directly, run from R, and run from a number of packages including Matlab, SAS and Stata. For Stata see Thompson, Palmer, and Moreno (2006). For more advanced models most researchers write more customizable code in a matrix programming language such as R or Matlab. Bayesian MCMC methods are often coded in R.

All data sets and programs, written in Stata, that cover virtually all the analysis in this second edition, are available from the authors' respective web sites.

References

Abrevaya, J., and C.M. Dahl (2008), "The Effects of Birth Inputs on Birthweight: Evidence from Quantile Estimation on Panel Data," *Journal of Business and Economic Statistics*, 26, 379–397.

Aitchison, J., and C.H. Ho (1989), "The Multivariate Poisson-Log Normal Distribution," *Biometrika*, 76, 643.

Aitken, M., and D.B. Rubin (1985), "Estimation and Hypothesis Testing in Finite Mixture Models," *Journal of the Royal Statistical Society* B, 47, 67–75.

Akaike, H. (1973), "Information Theory and an Extension of the Maximum Likelihood Principle," in B.N. Petrov and F. Csaki, eds., *Second International Symposium on Information Theory*, 267–281, Budapest, Akademiai Kaido.

Alcaniz, M. (1996), *Modelos de Poisson generalizados con una variable de exposicion al riesgo*, Ph.D. dissertation (in Spanish), University of Barcelona, Spain.

Alfo, M., and G. Trovato (2004), "Semiparametric Approaches Based on Random Coefficient Framework," *Econometrics Journal*, 7, 426–454.

Allison, P.D., and R.P. Waterman (2002), "Fixed–Effects Negative Binomial Regression Models," *Sociological Methodology*, 32, 247–265.

Al-Osh, M.A., and A.A. Alzaid (1987), "First Order Integer Valued Autoregressive (INAR(1)) Process," *Journal of Time Series Analysis*, 8, 261–275.

Altman, N.S. (1992), "An Introduction to Kernel and Nearest-Neighbor Nonparametric Regression," *American Statistician*, 46, 175–185.

Alzaid, A.A., and M.A. Al-Osh (1988), "First-Order Integer-Valued Autoregressive (INAR(1)) Process: Distributional and Regression Properties," *Statistica Neerlandica*, 42, 53–61.

Alzaid, A.A., and M.A. Al-Osh (1990), "An Integer-Valued pth-Order Autoregressive Structure (INAR(p)) Process," *Journal of Applied Probability*, 27, 314–324.

Alzaid, A.A., and M.A. Al-Osh (1993), "Some Autoregressive Moving Average Processes with Generalized Poisson Distributions," *Annals of the Institute of Mathematical Statistics*, 45, 223–232.

Amemiya, T. (1974), "The Nonlinear Two-Stage Least Squares Estimator," *Journal of Econometrics*, 2, 105–110.

Amemiya, T. (1984), "Tobit Models: A Survey," *Journal of Econometrics*, 24, 3–61.

Amemiya, T. (1985), *Advanced Econometrics*, Cambridge, MA, Harvard University Press.

Amemiya, Y. (1985), "Instrumental Variable Estimator for the Nonlinear Error in Variables Model," *Journal of Econometrics*, 28, 273-289.

Andersen, E.B. (1970), "Asymptotic Properties of Conditional Maximum Likelihood Estimators," *Journal of the Royal Statistical Society* B, 32, 283–301.

Andersen, P.K., O. Borgan, R.D. Gill, and N. Keiding (1993), *Statistical Models Based on Counting Processes*, New York, Springer-Verlag.

Andrews, D.W.K. (1988a), "Chi-Square Diagnostic Tests for Econometric Models: Introduction and Applications," *Journal of Econometrics*, 37, 135–156.

Andrews, D.W.K. (1988b), "Chi-Square Diagnostic Tests for Econometric Models: Theory," *Econometrica*, 56, 1419–1453.

Arellano, M. (1987), "Computing Robust Standard Errors for Within-groups Estimators," *Oxford Bulletin of Economics and Statistics*, 49(4), 431–434.

Arellano, M., and S. Bond (1991), "Some Tests of Specification for Panel Data: Monte Carlo Evidence and an Application to Employment Equations," *Review of Economic Studies*, 58, 277–298.

Arnold, B.C., and D.J. Strauss (1988), "Bivariate Distributions with Exponential Conditionals," *Journal of the American Statistical Association*, 83, 522–527.

Arnold, B.C., and D.J. Strauss (1992), "Bivariate Distributions with Conditionals in Prescribed Exponential Families," *Journal of the Royal Statistical Society* B, 53, 365–375.

Bago d'Uva, T. (2005), "Latent Class Models for Use of Primary Care: Evidence from a British Panel," *Health Economics*, 14, 873–892.

Bago d'Uva, T. (2006), "Latent Class Models for Utilisation of Health Care," *Health Economics*, 15, 329–343.

Baltagi, B.H. (2008), *Econometric Analysis of Panel Data*, edition 4, Chichester, John Wiley and Sons.

Barlow, R.E., and F. Proschan (1965), *Mathematical Theory of Reliability*, New York, John Wiley.

Barmby, T.A., C. Orme, and J.G. Treble (1991), "Worker Absenteeism: An Analysis Using Micro Data," *Economic Journal*, 101, 214–229.

Basu, A., and P.J. Rathouz (2005), "Estimating Marginal and Incremental Effects on Health Outcomes Using Flexible Link and Variance Function Models," *Biostatistics*, 6, 93–109.

Ben-Moshe, D. (2011), "Identification and Estimation in Models with Dependent Multidimensional Unobservables," Working Paper, UCLA Department of Economics.

Berndt, E., B. Hall, R. Hall, and J. Hausman (1974), "Estimation and Inference in Nonlinear Structual Models," *Annals of Economic and Social Measurement*, 3/4, 653–665.

Besag, J. (1974), "Spatial Interaction and the Statistical Analysis of Lattice Systems," *Journal of the Royal Statistical Society* B, 36, 192–225.

Bishop, Y.M.M., S.E. Feinberg, and P.W. Holland (1975), *Discrete Multivariate Analysis: Theory and Practice*, Cambridge, MIT Press.

Blonigen, B. (1997), "Firm-Specific Assets and the Link between Exchange Rates and Foreign Direct Investment," *American Economic Review*, 87, 447–465.

Blundell, R., R. Griffith, and J. Van Reenan (1995), "Dynamic Count Models of Technological Innovation," *Economic Journal*, 105, 333–344.

Blundell, R., R. Griffith, and J. Van Reenan (1999), "Market Share, Market Value and Innovation in a Panel of British Manufacturing Firms," *Review of Economic Studies*, 66(3), 529–554.

Blundell, R., R. Griffith, and F. Windmeijer (2002), "Individual Effects and Dynamics in Count Data Models," *Journal of Econometrics*, 108(1), 113–131.

Blundell, R., and R.L. Matzkin (2010), "Conditions for the Existence of Control Functions in Nonseparable Simultaneous Equations Models," Working Paper, UCLA Department of Economics.

Blundell, R., and J. Powell (2003), "Endogeneity in Nonparametric and Semiparametric Regression Models," in M. Dewatripont, L.P. Hansen, and S.J. Turnovsky, eds., *Advances in Economics and Econonometrics: Theory and Applications*, Eighth World Congress, Volume II, 312–357, Cambridge, Cambridge University Press.

Böckenholt, U. (1999), "Mixed INAR(1) Poisson Regression Models: Analyzing Heterogeneity and Serial Dependencies in Longitudinal Count Data," *Journal of Econometrics*, 89, 317–338.

Böhning, D. (1995), "A Review of Reliable Maximum Likelihood Algorithms for Semiparametric Mixture Models," *Journal of Statistical Planning and Inference*, 47, 5–28.

Böhning, D., E. Dietz, R. Schaub, P. Schlattmann, and B.G. Lindsay (1994), "The Distribution of the Likelihood Ratio for Mixtures of Densities from the One-Parameter Exponential Family," *Annals of the Institute of Statistical Mathematics*, 46, 373–388.

Böhning, D., and R. Kuhnert (2006), "Equivalence of Truncated Count Mixture Distributions and Mixtures of Truncated Count Distributions," *Biometrics*, 62, 1207–1215.

Bortkiewicz, L. von (1898), *Das Gesetz de Kleinen Zahlen*, Leipzig, Teubner.

Boswell, M.T., and G.P. Patil (1970), "Chance Mechanisms Generating the Negative Binomial Distributions," in G.P. Patil, ed., *Random Counts in Models and Structures*, volume 1, 3–22, University Park, Pennsylvania State University Press.

Bound, J., C. Brown, and N. Mathiowetz (2001), "Measurement Error in Survey Data," in J.J. Heckman and E.E. Leamer, eds., *Handbook of Econometrics*, volume 5, 3705–3843, Amsterdam, North-Holland.

Boyer, B.H., M.S. Gibson, and M. Loretan (1999), "Pitfalls in Tests of Changes in Correlations," International Finance Discussion Papers Nr. 597. Board of Governors of The Federal Reserve System.

Brännäs, K. (1992), "Limited Dependent Poisson Regression," *The Statistician*, 41, 413–423.

Brännäs, K. (1994), "Estimation and Testing in Integer Valued AR(1) Models," Umea Economic Studies No. 355, University of Umea.

Brännäs, K. (1995a), "Explanatory Variables in the AR(1) Model," Umea Economic Studies No. 381, University of Umea.

Brännäs, K. (1995b), "Prediction and Control for a Time-Series Count Data Model," *International Journal of Forecasting*, 11, 263–270.

Brännäs, K., and J. Hellström (2001), "Generalized Integer-Valued Autoregression," *Econometric Reviews*, 20, 425–443.

Brännäs, K., and P. Johansson (1994), "Time Series Count Regression," *Communications in Statistics: Theory and Methods*, 23, 2907–2925.

Brännäs, K., and P. Johansson (1996), "Panel Data Regression for Counts," *Statistical Papers*, 37, 191–213.

Brännäs, K., and G. Rosenqvist (1994), "Semiparametric Estimation of Heterogeneous Count Data Models," *European Journal of Operations Research*, 76, 247–258.

Breslow, N. (1990), "Tests of Hypotheses in Overdispersed Poisson Regression and Other Quasi-Likelihood Models," *Journal of the American Statistical Association*, 85, 565–571.

Breusch, T.S., and A.R. Pagan (1979), "A Simple Test for Heteroscedasticity," *Econometrica*, 47, 1287–1294.

Bryk, A.S., and S.W. Raudenbush (2002), *Hierarchical Linear Models*, edition 2, Newberry Park, CA, Sage Publications.

Bu, R., B. McCabe, and K. Hadri (2008) "Maximum Likelihood Estimation of Higher-Order Integer-Valued Autoregressive Processes," *Journal of Time Series Analysis*, 29, 973–994.

Buonaccorsi, J.P. (1996), "A Modified Estimating Equation Approach to Correcting for Measurement Error in Regression," *Biometrika*, 83, 433–440.

Buonaccorsi, J.P. (2010), *Measurement Error: Models, Methods, and Applications*, Boca Raton, FL, Chapman and Hall / CRC.

Burguette, J., A.R. Gallant, and G. Souza (1982), "On the Unification of the Asymptotic Theory of Nonlinear Econometric Methods," *Econometric Reviews*, 1, 151–190.

Buzas, J.S., and L.A. Stefanski (1996), "Instrumental Variables Estimation in Generalized Linear Measurement Error Models," *Journal of the American Statistical Association*, 91, 999–1006.

Cameron, A.C. (1991), "Regression Based Tests of Heteroskedasticity in Models Where the Variance Depends on the Mean," Economics Working Paper Series No. 379, University of California, Davis.

Cameron, A.C., J. Gelbach, and D.L. Miller (2011), "Robust Inference with Multi-Way Clustering," *Journal of Business and Economic Statistics*, 29, 238–249.

Cameron, A.C., and P. Johansson (1997), "Count Data Regressions Using Series Expansions with Applications," *Journal of Applied Econometrics*, 12, 203–224.

Cameron, A.C., and L. Leon (1993), "Markov Regression Models for Time Series Data," presented at Western Economic Association Meetings, Lake Tahoe.

Cameron, A.C., T. Li, P.K. Trivedi, and D. Zimmer (2004), "Modeling the Differences in Counted Outcomes Using Bivariate Copula Models: With Application to Mismeasured Counts," *Econometrics Journal*, 7, 566–584.

Cameron, A.C., and D.L. Miller (2011), "Robust Inference with Clustered Data," in A. Ullah and D.E. Giles, eds., *Handbook of Empirical Economics and Finance*, 1–28, Boca Raton, FL, CRC Press.

Cameron, A.C., and P.K. Trivedi (1985), "Regression-Based Tests for Overdispersion," Econometric Workshop Technical Report No. 9, Stanford University.

Cameron, A.C., and P.K. Trivedi (1986), "Econometric Models Based on Count Data: Comparisons and Applications of Some Estimators," *Journal of Applied Econometrics*, 1, 29–53.

Cameron, A.C., and P.K. Trivedi (1990a), "Regression-Based Tests for Overdispersion in the Poisson Model," *Journal of Econometrics*, 46, 347–364.

Cameron, A.C., and P.K. Trivedi (1990b), "Conditional Moment Tests and Orthogonal Polynomials," Indiana University, Pre-print.

Cameron, A.C., and P.K. Trivedi (1990c), "Conditional Moment Tests with Explicit Alternatives," Economics Working Paper Series No. 366, University of California, Davis.

Cameron, A.C., and P.K. Trivedi (1990d), "The Information Matrix Test and Its Implied Alternative Hypotheses," Economics Working Paper Series No. 372, University of California, Davis.

Cameron, A.C., and P.K. Trivedi (1993), "Tests of Independence in Parametric Models: With Applications and Illustrations," *Journal of Business and Economic Statistics*, 11, 29–43.

Cameron, A.C., and P.K. Trivedi (1996), "Count Data Models for Financial Data," in G.S. Maddala and C.R. Rao, eds., *Handbook of Statistics, Volume 14: Statistical Methods in Finance*, Amsterdam, North-Holland.

Cameron, A.C., and P.K. Trivedi (2005), *Microeconometrics: Methods and Applications*, Cambridge, Cambridge University Press.

Cameron, A.C., and P.K. Trivedi (2011), *Microeconometrics Using Stata*, Revised Edition, College Station, TX, Stata Press.

Cameron, A.C., P.K. Trivedi, F. Milne, and J. Piggott (1988), "A Microeconometric Model of the Demand for Health Care and Health Insurance in Australia," *Review of Economic Studies*, 55, 85–106.

Cameron, A.C., and F.A.G. Windmeijer (1996), "R-Squared Measures for Count Data Regression Models with Applications to Health Care Utilization," *Journal of Business and Economic Statistics*, 14, 209–220.

Cameron, A.C., and F.A.G. Windmeijer (1997), "An R-Squared Measure of Goodness of Fit for Some Common Nonlinear Regression Models," *Journal of Econometrics*, 77, 329–342.

Campbell, M.J. (1994), "Time Series Regression for Counts: An Investigation into the Relationship between Sudden Infant Death Syndrome and Environmental Temperature," *Journal of the Royal Statistical Society* A, 157, 191–208.

Carroll, R.J., D. Ruppert, L.A. Stefanski, and C.M. Crainiceanu (2006), *Measurement Error in Nonlinear Models: A Modern Perspective*, edition 2, London, Chapman and Hall/CRC.

Carroll, R.J., and L.A. Stefanski (1990), "Approximate Quasilikelihood Estimation in Models with Surrogate Predictors," *Journal of the American Statistical Association*, 85, 652–663.

Carroll, R.J., and L.A. Stefanski (1994), "Measurement Error, Instrumental Variables and Corrections for Attenuation with Applications to Meta-Analyses," *Statistics in Medicine*, 13, 1265–1282.

Chamberlain, G. (1982), "Multivariate Regression Models for Panel Data," *Journal of Econometrics*, 18, 5–46.

Chamberlain, G. (1987), "Asymptotic Efficiency in Estimation with Conditional Moment Restrictions," *Journal of Econometrics*, 34, 305–334.

Chamberlain, G. (1992a), "Efficiency Bounds for Semiparametric Regression," *Econometrica*, 60, 567–596.

Chamberlain, G. (1992b), "Comment: Sequential Moment Restrictions in Panel Data," *Journal of Business and Economic Statistics*, 10, 20–26.

Chan, K.S., and J. Ledolter (1995), "Monte Carlo EM Estimation for Time Series Models Involving Counts," *Journal of the American Statistical Association*, 90, 242–252.

Chang, F.-R., and P.K. Trivedi (2003), "Economics of Self-Medication: Theory and Evidence," *Health Economics* 12, 721–739.

Chen, T.T. (1979), "Log-Linear Models for Categorical Data with Misclassification and Double Sampling," *Journal of the American Statistical Association*, 74, 481–488.

Chen, X. (2007), "Large Sample Sieve Estimation of Semiparametric Models," in E.E. Leamer, eds., *Handbook of Econometrics*, volume 6B, 5549–5632, Amsterdam, North-Holland.

Chen, X., H. Hong, and D. Nekipelov (2011), "Nonlinear Models of Measurement Error," *Journal of Economic Literature*, 49(4), 901–937.

Chernozhukov, V., and H. Hong (2003), "An MCMC Approach to Classical Estimation," *Journal of Econometrics*, 115, 293–346.

Cherubini, U., E. Luciano, and W. Vecchiato (2004), *Copula Methods in Finance*, New York, John Wiley Sons.

Chesher, A.D. (1984), "Testing for Neglected Heterogeneity," *Econometrica*, 52, 865–872.

Chesher, A.D. (1991), "The Effect of Measurement Error," *Biometrika*, 78, 451–462.

Chesher, A.D. (2005), "Nonparametric Identification under Discrete Variation," *Econometrica*, 73, 1525–1550.

Chesher, A.D. (2010), "Instrumental Variable Models for Discrete Outcomes," *Econometrica*, 78, 575–602.

Chesher, A.D., and M. Irish (1987), "Residual Analysis in the Grouped and Censored Normal Linear Model," *Journal of Econometrics*, 34, 33–62.

Chib, S. (2001), "Markov Chain Monte Carlo Methods: Computation and Inference," in J.J. Heckman and E. Leamer, eds., *Handbook of Econometrics*, volume 5, 3569–3649, Amsterdam, North-Holland.

Chib, S., E. Greenberg, and R. Winkelmann (1998), "Posterior Simulation and Bayes Factors in Panel Count Data Models," *Journal of Econometrics*, 86, 33–54.

Chib, S., and R. Winkelmann (2001), "Markov Chain Monte Carlo Analysis of Correlated Count Data," *Journal of Business and Economic Statistics*, 19, 428–435.

Chiou, J.-M., and H.-G. Muller (1998), "Quasi-Likelihood Regression with Unknown Link and Variance Functions," *Journal of the American Statistical Association*, 93, 1376–1387.

Chiou, J.-M., and H.-G. Muller (2005), "Estimated Estimating Equations: Semiparametric Inference for Clustered and Longitudinal Data," *Journal of the Royal Statistical Society* B, 67, 531–553.

Christmann, A. (1994), "Least Median of Weighted Squares in Logistic Regression with Large Strata," *Biometrika*, 81, 413–417.

Christmann, A. (1996), "On Outliers and Robust Estimation in Regression Models with Discrete Response Variables," paper presented at Workshop on Statistical Modelling of Discrete Data and Structures, Sonderforschungbereich 386, October 1996.

Cincera, M. (1997), "Patents, R&D and Technological Spillovers at the Firm Level: Some Evidence from Econometric Count Models for Panel Data," *Journal of Applied Econometrics*, 12, 265–280.

Cochran, W.G. (1940), "The Analysis of Variance When Experimental Errors Follow the Poisson or Binomial Law," *Annals of Mathematical Statistics*, 11, 335–347.

Cohen, A.C. (1960), "Estimation in a Truncated Poisson Distribution When Zeros and Ones Are Missing," *Journal of the American Statistical Association*, 55, 342–348.

Collings, B.J., and B.H. Margolin (1985), "Testing Goodness of Fit for the Poisson Assumption When Observations Are Not Identically Distributed," *Journal of the American Statistical Association*, 80, 411–418.

Conley, T.G. (1999), "GMM Estimation with Cross Sectional Dependence," *Journal of Econometrics*, 92, 1–45.

Conniffe, D.C. (1990), "Testing Hypotheses with Estimated Scores," *Biometrika*, 77, 97–106.

Consul, P.C. (1989), *Generalized Poisson Distributions: Properties and Applications*, New York, Marcel Dekker.

Consul, P.C., and G.C. Jain (1973), "A Generalization of the Poisson Distribution," *Technometrics*, 15, 791–799.

Consul, P.C., and L.R. Shenton (1972), "Use of Lagrange Expansion for Generating Discrete Generalized Probability Distributions," *SIAM Journal of Applied Mathematics*, 23, 239–248.

Cont, R. (2001), "Empirical Properties of Asset Returns: Stylized Facts and Statistical Issues," *Quantitative Finance*, 1, 223–236.

Cook, J.R., and L.A. Stefanski (1994), "Simulation-Extrapolation Estimation in Parametric Measurement Error Models," *Journal of the American Statistical Association*, 89, 1314–1328.

Cowling, B.J., J.L. Hutton, and J.E.H. Shaw (2006), "Joint Modelling of Event Counts and Survival Times," *Journal of the Royal Statistical Society* C, 55, 31–39

Cox, D.R. (1955), "Some Statistical Models Related with Series of Events," *Journal of the Royal Statistical Society* B, 17, 129–64.

Cox, D.R. (1961), "Tests of Separate Families of Hypotheses," *Proceedings of the Fourth Berkeley Symposium on Mathematical Statistics and Probability*, 1, 105–123.

Cox, D.R. (1962a), "Further Results on Tests of Separate Families of Hypotheses," *Journal of the Royal Statistical Society* B, 24, 406–424.

Cox, D.R. (1962b), *Renewal Theory*, London, Methuen & Co.

Cox, D.R. (1981), "Statistical Analysis of Time Series: Some Recent Developments," *Scandinavian Journal of Statistics*, 8, 93–115.

Cox, D.R. (1983), "Some Remarks on Overdispersion," *Biometrika*, 70, 269–274.

Cox, D.R., and P.A.W. Lewis (1966), *The Statistical Analysis of Series of Events*, London, Methuen & Co.

Cox, D.R., and E.J. Snell (1968), "A General Definition of Residuals (with discussion)," *Journal of the Royal Statistical Society* B, 30, 248–275.

Cragg, J.C. (1971), "Some Statistical Models for Limited Dependent Variables with Application to the Demand for Durable Goods," *Econometrica*, 39, 829–844.

Cragg, J.C. (1997), "Using Higher Moments to Estimate the Simple Errors-in-Variables Model," *Rand Journal of Economics*, 28, S71–S91.

Cramer, H. (1946), *Mathematical Methods of Statistics*, Princeton, NJ, Princeton University Press.

Creel, M.D., and J.B. Loomis (1990), "Theoretical and Empirical Advantages of Truncated Count Data Estimators for Analysis of Deer Hunting in California," *Journal of Agricultural Economics*, 72, 434–441.

Crepon, B., and E. Duguet (1997a), "Estimating the Innovation function from Patent Numbers: GMM on Count Data," *Journal of Applied Econometrics*, 12, 243–264.

Crepon, B., and E. Duguet (1997b), "Research and Development, Competition and Innovation: Pseudo-Maximum Likelihood and Simulated Maximum Likelihood Method Applied to Count Data Models with Heterogeneity," *Journal of Econometrics*, 79, 355–378.

Crowder, M.J. (1976), "Maximum Likelihood Estimation for Dependent Observations," *Journal of the Royal Statistical Society* B, 38, 45–53.

Crowder, M.J. (1987), "On Linear and Quadratic Estimating Functions," *Biometrika*, 74, 591–597.

Daley, D.J., and D. Vere-Jones (1988), *An Introduction to the Theory of Point Processes*, New York, Springer-Verlag.

Danaher, P.J. (2007), "Modeling Page Views across Multiple Websites with an Application to Internet Reach and Frequency Prediction," *Marketing Science*, 26, 422–437.

Davidson, R., and J.G. MacKinnon (1984), "Several Tests for Model Specification in the Presence of Alternative Hypotheses," *Econometrica*, 49, 781–793.

Davidson, R., and J.G. MacKinnon (1993, 2004), *Estimation and Inference in Econometrics*, Oxford, Oxford University Press.

Davis, R.A., W.T.M. Dunsmuir, and S.B. Streett (2003), "Observation-Driven Models for Poisson Counts," *Biometrika*, 90, 777–790.

Davis, R.A., W.T.M. Dunsmuir, and Y. Wang (1999), "Modelling Time Series of Count Data," in S. Ghosh, ed., *Asymptotics, Nonparametrics, and Time Series*, 63–114. New York, Marcel Dekker.

Davis, R.A., and R. Wu (2009), "A Negative Binomial Model for Time Series of Counts," *Biometrika*, 96, 735–749.

Davison, A.C., and A. Gigli (1989), "Deviance Residuals and Normal Scores Plots," *Biometrika*, 76, 211–221.

Davison, A.C., and E.J. Snell (1991), "Residuals and Diagnostics," in D.V. Hinkley, N. Reid, and E.J. Snell, eds., *Statistical Theory and Modelling: In Honor of Sir David Cox, FRS*, 83–106, London, Chapman and Hall.

Davutyan, N. (1989), "Bank Failures as Poisson Variates," *Economic Letters*, 29, 333–338.

Dean, C. (1991), "Estimating Equations for Mixed Poisson Models," in V. P. Godambe, ed., *Estimating Functions*, Oxford, Oxford University Press.

Dean, C. (1992), "Testing for Overdispersion in Poisson and Binomial Regression Models," *Journal of the American Statistical Association*, 87, 451–457.

Dean, C.B. (1993), "A Robust Property of Pseudo Likelihood Estimation for Count Data," *Journal of Statistical Planning and Inference*, 35, 309.

Dean, C.B. (1994), "Modified Pseudo-Likelihood Estimator of the Overdispersion Parameter in Poisson Mixture Models," *Journal of Applied Statistics*, 1994, 21, 523.

Dean, C.B., and R. Balshaw (1997), "Efficiency Lost by Analyzing Counts Rather than Event Times in Poisson and Overdispersed Poisson Regression Models," *Journal of the American Statistical Society*, 92, 1387–1398

Dean, C.B., D.M. Eaves, and C.J. Martinez (1995), "A Comment on the Use of Empirical Covariance Matrices in the Analysis of Count Data," *Journal of Statistical Planning and Inference*, 48, 197–206.

Dean, C., and F. Lawless (1989a), "Tests for Detecting Overdispersion in Poisson Regression Models," *Journal of the American Statistical Association*, 84, 467–472.

Dean, C., and J.F. Lawless (1989b), "Comment on Godambe and Thompson (1989)," *Journal of Statistical Planning and Inference*, 22, 155–158.

Dean, C., J.F. Lawless, and G.E. Willmot (1989), "A Mixed Poisson-Inverse Gaussian Regression Model," *Canadian Journal of Statistics*, 17, 171–182.

Deb, P., M.K. Munkin, and P.K. Trivedi (2006), "Private Insurance, Selection, and Health Care Use: A Bayesian Analysis of a Roy-Type Model," *Journal of Business and Economic Statistics*, 24, 403–415.

Deb, P., and P.K. Trivedi (1997), "Demand for Medical Care by the Elderly: A Finite Mixture Approach," *Journal of Applied Econometrics*, 12, 313–326.

Deb, P., and P.K. Trivedi (2002), "The Structure of Demand for Health Care: Latent Class versus Two-part Models," *Journal of Health Economics*, 21, 601–625.

Deb, P., and P.K. Trivedi. (2006a), "Specification and Simulated Likelihood Estimation of a Non-Normal Treatment-Outcome Model with Selection: Application to Health Care Utilization," *Econometrics Journal*, 9, 307–331.

Deb, P., and P.K. Trivedi (2006b), "Maximum Simulated Likelihood Estimation of a Negative-Binomial Regression Model with Multinomial Endogenous Treatment," *Stata Journal*, 6, 246–255.

Deb, P., and P.K. Trivedi (2013), "Finite Mixture for Panels with Fixed Effects," Forthcoming *Journal of Econometric Methods*, UK.

Deb, P., P.K. Trivedi, and D.M. Zimmer (2011), "Cost-Offsets of Prescription Drug Expenditures: Data Analysis via a Copula-based Bivariate Dynamic Hurdle Model," manuscript, Indiana University.

Delgado, M.A. (1992), "Semiparametric Generalized Least Squares in the Multivariate Nonlinear Regression Model," *Econometric Theory*, 8, 203–222.

Delgado, M.A., and T.J. Kniesner (1997), "Count Data Models with Variance of Unknown Form: An Application to a Hedonic Model of Worker Absenteeism," *Review of Economics and Statistics*, 79, 41–49.

Demidenko, E. (2007), "Poisson Regression for Clustered Data," *International Statistical Review*, 75, 96–113.

Denuit, M., and P. Lambert (2005), "Constraints on Concordance Measures in Bivariate Discrete Data," *Journal of Multivariate Analysis*, 93, 40–57.

Diggle, P.J., P. Heagerty, K.-Y. Liang, and S.L. Zeger (2002), *Analysis of Longitudinal Data*, edition 2, Oxford, Oxford University Press.

Dionne, G., D. Desjardins, C. Labergue-Nadeau, and U. Maag (1995), "Medical Conditions, Risk Exposure, and Truck Drivers' Accidents: An Analysis with Count Data Regression Models," *Accident Analysis and Prevention*, 27, 295–305.

Dionne, G., and C. Vanasse (1992), "Automobile Insurance Rate Making in the Presence of Asymmetrical Information," *Journal of Applied Econometrics*, 7, 149–165.

Dobbie, M.J., and A.H. Welsh (2001), "Models for Zero-inflated Count Data Using the Neyman Type A Distribution," *Statistical Modelling*, 1, 65–80.

Driscoll, J.C., and A.C. Kraay (1998), "Covariance Matrix Estimation with Spatially Dependent Data," *Review of Economics and Statistics*, 80, 549–560.

Drost, F.C., R. Van Den Akker, and B.J.M. Werker (2008), "Local Asymptotic Normality and Efficient Estimation for INAR(p) models," *Journal of Time Series Analysis*, 29, 783–801.

Drost, F.C., R. Van Den Akker, and B.J.M. Werker (2009a), "Efficient Estimation of Autoregression Parameters and Innovation Distributions for Semiparametric Integer-Valued AR(p) Models," *Journal of the Royal Statistical Society* B, 71, 467–485.

Drost, F.C., R. Van Den Akker, and B.J.M. Werker (2009b), "The Asymptotic Structure of Nearly Unstable Non-Negative Integer-Valued AR(1) Models," *Bernoulli*, 15, 297–324.

Du, J.-G., and Y. Li (1991), "The Integer Valued Autoregressive (INAR(p)) Model," *Journal of Time Series Analysis*, 12, 129–142.

Duan, N. (1983), "Smearing Estimate: A Nonparametric Retransformation Method," *Journal of the American Statistical Association*, 78, 605–610.

Duarte, R. and J.J. Escario (2006), "Alcohol Abuse and Truancy among Spanish Adolescents: A Count-Data Approach," *Economics of Education Review*, 25, 179–187.

Durbin, J. (1954), "Errors in Variables," *Review of the International Statistical Institute*, 22, 23–32.

Durbin, J., and S.J. Koopman (1997), "Monte Carlo Maximum Likelihood Estimation for Non-Gaussian State Space Models," *Biometrika*, 84, 669–684.

Durbin, J., and S.J. Koopman (2000), "Time Series Analysis of Non-Gaussian Observations Based on State Space Models from Both Classical and Bayesian Perspectives," *Journal of the Royal Statistical Society* B, 62, 3–56.

Eagleson, G.K. (1964), "Polynomial Expansions of Bivariate Distributions," *Annals of Mathematical Statistics*, 35, 1208–1215.

Eagleson, G.K. (1969), "A Characterization Theorem for Positive Definite Sequences on the Krawtchouk Polynomials," *Australian Journal of Statistics*, 11, 29–38.

Econometric Software Inc. (1991), *LIMDEP User's Manual and Reference Guide: Version 6.0*, New York.

Efron, B. (1979), "Bootstrapping Methods: Another Look at the Jackknife," *Annals of Statistics*, 7, 1–26.

Efron, B. (1986), "Double Exponential Families and Their Use in Generalized Linear Regressions," *Journal of the American Statistical Association*, 81, 709–721.

Efron, B., and R.J. Tibshirani (1993), *An Introduction to the Bootstrap*, London, Chapman and Hall.

Eggenberger, F., and G. Polya (1923), "Über die Statistik verketteter Vorgänge," *Zeitschrift für Angerwandte Mathematik and Mechanik*, 1, 279–289.

Eicker, F. (1967), "Limit Theorems for Regressions with Unequal and Dependent Errors," in L. LeCam and J. Neyman eds., *Proceedings of the Fifth Berkeley Symposium on Mathematical Statistics and Probability*, 59–82, Berkeley, University of California Press.

Embrechts, P., McNeil, A., and D. Straumann (2002), "Correlation and Dependence in Risk Management: Properties and Pitfalls," in M.A.H. Dempster, ed., *Risk Management: Value at Risk and Beyond*, 176–223, Cambridge, Cambridge University Press.

Enciso-Mora, V., P. Neal, and T.S. Rao (2008a), "Efficient Order Selection Algorithms for Integer-Valued ARMA Processes," *Journal of Time Series Analysis*, 30, 1–18.

Enciso-Mora, V., P. Neal, and T.S. Rao (2008b), "Integer Valued AR Processes with Explanatory Variables," Research Report No. 20, School of Mathematics, University of Manchester.

Engle, R.F., and J.R. Russell (1998), "Autoregressive Conditional Duration: A New Model for Irregularly Spaced Transaction Data," *Econometrica*, 66(5), 1127–1162.

Englin, J., and J.S. Shonkwiler (1995), "Estimating Social Welfare Using Count Data Models: An Application to Long-Run Recreation Demand under Conditions of Endogenous Stratification and Truncation," *Review of Economics and Statistics*, 77, 104–112.

Everitt, B.S., and D.J. Hand (1981), *Finite Mixture Distributions*, London, Chapman and Hall.

Fabbri, D., and C. Monfardini (2011), "Opt Out or Top Up? Voluntary Healthcare Insurance and the Public versus Private Substitution," IZA Working Paper DP-5952.

Fahrmeier, L., and H. Kaufman (1987), "Regression Models for Non-Stationary Categorical Time Series," *Journal of Time Series Analysis*, 8, 147–160.

Fahrmeier, L., and G.T. Tutz (1994) *Multivariate Statistical Modeling Based on Generalized Linear Models*, New York, Springer-Verlag.

Famoye, F. (2010), "On the Bivariate Negative Binomial Regression Model," *Journal of Applied Statistics*, 37, 969–981.

Fan, J. (1992), "Design-Adaptive Nonparametric Regression," *Journal of the American Statistical Association*, 87, 998–1004.

Fang, K.-T., and Y.-T. Zhang (1990), *Generalized Multivariate Analysis*, Berlin, Springer-Verlag.

Feinstein, J.S. (1989), "The Safety Regulation of U.S. Nuclear Power Plants: Violations, Inspections, and Abnormal Occurrences," *Journal of Political Economy*, 97, 115–154.

Feinstein, J.S. (1990), "Detection Controlled Estimation," *Journal of Law and Economics*, 33, 233–276.

Feller, W. (1943), "On a General Class of 'Contagious' Distributions," *Annals of Mathematical Statistics*, 14, 389–400.

Feller, W. (1968), *An Introduction to Probability Theory and Its Applications*, volume 1, edition 3, New York, John Wiley Sons.

Feller, W. (1971), *An Introduction to Probability Theory*, volume 2, New York, John Wiley Sons.

Fé., E. (2012), "Estimating Production Frontiers and Efficiency When Output is a Discretely Distributed Economic Bad," *Journal of Productivity Analysis*, forthcoming.

Feng, Z.D., and C.E. McCulloch (1996), "Using Bootstrap Likelihood Ratios in Finite Mixture Models," *Journal of the Royal Statistical Society* B, 609–617.

Ferguson, T.S. (1973), "A Bayesian Analysis of Some Nonparametric Problems," *The Annals of Statistics*, 1(2), 209–230.

Fienberg, S.E. (1981), "Deciding What and Whom to Count," in R.G. Lehnan and W.G. Skogan, eds., *The NCS Working Papers*, volume 1, 59–60, Washington, DC, U.S. Department of Justice.

Firth, D. (1987), "On the Efficiency of Quasi-likelihood Estimation," *Biometrika*, 74, 223–245.

Firth, D. (1991), "Generalized Linear Models," in D.V. Hinkley, N. Reid, and E.J. Snell, eds., *Statistical Theory and Modelling: In Honor of Sir David Cox, FRS*, 55–82, London, Chapman and Hall.

Fleming, T.R., and D.P. Harrington (2005), *Counting Process and Survival Analysis*, New York, John Wiley Sons.

Fokianos, F., A. Rahbek, and D. Tjøstheim (2009), "Poisson Autoregression," *Journal of the American Statistical Association*, 104, 1430–1439.

Freeland, R.K., and B.P.M. McCabe (2004), "Forecasting Discrete Valued Low Count Time Series," *International Journal of Forecasting*, 20, 427–434.

Frühwirth-Schnatter, S. (2001), "Markov Chain Monte Carlo Estimation of Classical and Dynamic Switching and Mixture Models," *Journal of the American Statistical Association*, 96, 194–209.

Frühwirth-Schnatter, S. (2006), *Finite Mixture and Markov Switching Models*, New York, Springer-Verlag

Frühwirth-Schnatter, S., and H. Wagner (2006), "Auxiliary Mixture Sampling for Parameter-Driven Models of Time Series of Counts with Applications to State Space Modelling," *Biometrika*, 93, 827–841.

Gallant, A.R., and D.W. Nychka (1987), "Seminonparametric Maximum Likelihood Estimation," *Econometrica*, 55, 363–390.

Gallant, A.R., and G. Tauchen (1989), "Seminonparametric Estimation of Conditionally Constrained Heterogeneous Processes: Asset Pricing Applications," *Econometrica*, 57, 1091–1120.

Gan, L., and J. Jiang (1999), "A Test of Global Maximum," *Journal of American Statistical Association*, 94, 847–854.

Gauthier, G., and A. Latour (1994), "Convergence forte des estimateurs des paramètres d'un processus GENAR(p)," *À paraître dans Ann. Sc. Math. Quèbec* (with extended English abstract).

Geil, P., A. Melion, R. Rotte, and K.F. Zimmermann (1997), "Economic Incentives and Hospitalization," *Journal of Applied Econometrics*, 12, 295–311.

Gelfand, A.E., and S.R. Dalal (1990), "A Note on Overdispersed Exponential Families," *Biometrika*, 77, 55–64.

Gelfand, A.E., and A.F.M. Smith (1990), "Sampling-Based Approaches to Calculating Marginal Densities," *Journal of the American Statistical Association*, 85, 398–409.

Gelman, A., J.B. Carlin, H.S. Stern, and D.B. Rubin (2003), *Bayesian Data Analysis*, edition 2, Boca Raton, FL, Chapman and Hall/CRC Press.

Genest, C., and J. Neslehova, (2008), "A Primer on Copulas for Count Data," *ASTIN Bulletin*, 37, 475–515.

Geweke, J. (2003), *Contemporary Bayesian Econometrics and Statistics*, New York, John Wiley Sons.

Geweke, J., and M. Keane (2001), "Computationally Intensive Methods for Integration in Econometrics," in J.J. Heckman and E. Leamer, eds., *Handbook of Econometrics*, volume 5, 3463–3568, Amsterdam, North-Holland.

Gleser, L.J. (1990), "Improvements of the Naive Approach to Estimation in Non-linear Errors-in-Variables Regression Models," in P.J. Brown and W.A. Fuller, eds., *Statistical Analysis of Measurement Error Models and Application*, 99–114. Providence, RI, American Mathematics Society.

Godambe, V.P., and M.E. Thompson (1989), "An Extension of Quasi-Likelihood Estimation," *Journal of Statistical Planning and Inference*, 22, 137–152.

Godfrey, L.G. (1988), *Misspecification Tests in Econometrics: The Lagrange Multiplier Principle and Other Approaches*, Cambridge, Cambridge University Press.

Goffe, W.L., G.D. Ferrier, and J. Rogers (1994), "Global Optimization of Statistical Functions with Simulated Annealing," *Journal of Econometrics*, 60, 65–99.

Goldstein, H. (2010), *Multilevel Statistical Models*, edition 4, Chichester, Wiley.

Good, D., and M.A. Pirog-Good (1989), "Models for Bivariate Count Data with an Application to Teenage Delinquency and Paternity," *Sociological Methods and Research*, 17, 409–431.

Gourieroux, C., and A. Monfort (1991), "Simulation Based Inference in Models with Heterogeneity," *Annales D'Economie Et De Statistique*, 20–21, 70–107.

Gourieroux, C., and A. Monfort (1995), *Statistics and Econometrics Models*, volumes 1 and 2, translated by Q. Vuong, Cambridge, Cambridge University Press. (Originally published as *Statistique et modèles économetriques*, 1989).

Gourieroux, C., and A. Monfort (1997), *Simulation Based Econometric Methods*, Oxford, Oxford University Press.

Gourieroux, C., A. Monfort, E. Renault, and A. Trognon (1987a), "Generalized Residuals," *Journal of Econometrics*, 34, 5–32.

Gourieroux, C., A. Monfort, E. Renault, and A. Trognon (1987b), "Simulated Residuals," *Journal of Econometrics*, 34, 201–252.

Gourieroux, C., A. Monfort, and A. Trognon (1984a), "Pseudo Maximum Likelihood Methods: Theory," *Econometrica*, 52, 681–700.

Gourieroux, C., A. Monfort, and A. Trognon (1984b), "Pseudo Maximum Likelihood Methods: Applications to Poisson Models," *Econometrica*, 52, 701–720.

Gourieroux, C., and M. Visser (1993), "A Count Data Model with Unobserved Heterogeneity," *Journal of Econometrics*, 79, 247–268.

Greene, W.H. (1994), "Accounting for Excess of Zeros and Sample Selection in Poisson and Negative Binomial Regression Models," Discussion Paper EC-94-10, Department of Economics, New York University.

Greene, W.H. (1997), "FIML Estimation of Sample Selection Models for Count Data," Discussion Paper EC-97-02, Department of Economics, Stern School of Business, New York University.

Greene, W. (2004), "The Behaviour of the Maximum Likelihood Estimator of Limited Dependent Variable Models in the Presence of Fixed Effects," *Econometrics Journal*, 7, 98–119.

Greene, W.H. (2007), "Functional Form and Heterogeneity in Models for Count Data," *Foundations and Trends in Econometrics*, 1, 113–218.

Greene, W.H. (2009), "Models for Count Data with Endogenous Participation," *Empirical Economics*, 36,133–173.

Greene, W.H. (2011), *Econometric Analysis*, edition 7, Upper Saddle River, NJ, Prentice Hall.

Greenwood, M., and G.U. Yule (1920), "An Inquiry into the Nature of Frequency Distributions of Multiple Happenings, with Particular Reference to the Occurrence of Multiple Attacks of Disease or Repeated Accidents," *Journal of Royal Statistical Society* A, 83, 255–279.

Griffith, D.A., and R. Haining (2006), "Beyond Mule Kicks: The Poisson Distribution in Geographical Analysis," *Geographical Analysis*, 38, 123–139.

Grogger, J. (1990), "The Deterrent Effect of Capital Punishment: An Analysis of Daily Homicide Counts," *Journal of the American Statistical Association*, 85, 295–303.

Grogger, J.T., and R.T. Carson (1991), "Models for Truncated Counts," *Journal of Applied Econometrics*, 6, 225–238.

Guimarães, P. (2008), "The Fixed Effects Negative Binomial Model Revisited," *Economics Letters*, 99, 63–66.

Guo, J., and T. Li (2001), "Simulation-Based Estimation of the Structural Errors-in-Variables Negative Binomial Regression Model with an Application," *Annals of Economics and Finance*, 2, 101–122.

Guo, J.Q., and T. Li (2002), "Poisson Regression Models with Errors-in-Variables," *Journal of Statistical Planning and Inference*, 104, 391–401.

Guo, J.Q., and P.K. Trivedi (2002) "Flexible Parametric Models for Long-Tailed Patent Count Distributions," *Oxford Bulletin of Economics and Statistics*, 64, 63–82.

Gupta, C.R. (1974), "Modified Power Series Distribution and Some of Its Applications," *Sankhya*, Series B, 36, 288–298.

Gurmu, S. (1991), "Tests for Detecting Overdispersion in the Positive Poisson Regression Model," *Journal of Business and Economic Statistics*, 9, 215–222.

Gurmu, S. (1993), "Testing for Overdispersion in Censored Poisson Regression Models," Discussion Paper 255, Jefferson Center for Political Economy, University of Virginia.

Gurmu, S. (1997), "Semiparametric Estimation of Hurdle Regression Models With an Application to Medicaid Utilization," *Journal of Applied Econometrics*, 12, 225–242.

Gurmu, S., and J. Elder (2000), "Generalized Bivariate Count Data Regression Models," *Economics Letters*, 68, 31–36.

Gurmu, S., P. Rilstone, and S. Stern (1995), "Nonparametric Hazard Rate Estimation, Discussion Paper," University of Virginia.

Gurmu, S., P. Rilstone, and S. Stern (1999), "Semiparametric Estimation of Count Regression Models," *Journal of Econometrics*, 88, 123–150.

Gurmu, S., and P.K. Trivedi (1992), "Overdispersion Tests for Truncated Poisson Regression Models," *Journal of Econometrics*, 54, 347–370.

Gurmu, S., and P.K. Trivedi (1993), "Variable Augmentation Specification Tests in the Exponential Family," *Econometric Theory*, 9, 94–113.

Gurmu, S., and P.K. Trivedi (1994), "Recent Developments in Models of Event Counts: A Survey," Discussion Paper No. 261, Thomas Jefferson Center, University of Virginia, Charlottesville.

Gurmu, S., and P.K. Trivedi (1996), "Excess Zeros in Count Models for Recreational Trips," *Journal of Business and Economic Statistics*, 14, 469–477.

Haab, T.C., and K.E. McConnell (1996), "Count Data Models and the Problem of Zeros in Reaction Demand Analysis," *American Journal of Agricultural Economics*, 78, 89–98.

Haight, F.A. (1967), *Handbook of the Poisson Distribution*, New York, John Wiley & Sons.

Hall, A. (1987), "The Information Matrix Test in the Linear Model," *Review of Economic Studies*, 54, 257–265.

Hall, B.H., Z. Griliches, and J.A. Hausman (1986), "Patents and R and D: Is There a Lag?," *International Economic Review*, 27, 265–283.

Hall, P., J. Racine, Q. Li (2004), "Cross-Validation and the Estimation of Conditional Density Functions," *Journal of the American Statistical Association*, 99, 1015–1026.

Hamdan, M.A. (1972), "Estimation in the Truncated Bivariate Poisson Distribution," *Technometrics*, 14, 37–45.

Hamilton, J.D. (1989), "A New Approach to the Economic Analysis of Nonstationary Time Series and the Business Cycle," *Econometrica*, 57, 357–384.

Hamilton, J.D. (1994), *Time Series Analysis*, Princeton, Princeton University Press.

Hannan, M.T., and J. Freeman (1987), "The Ecology of Organizational Founding: American Labor Unions, 1836–1985," *American Journal of Sociology*, 92, 910–943.

Hansen, L.P. (1982), "Large Sample Properties of Generalized Method of Moments Estimators," *Econometrica*, 50, 129–1054.

Hansen, L.P., J. Heaton, and A. Yaron (1996), "Finite Sample Properties of Some Alternative GMM Estimators," *Journal of Business and Economic Statistics*, 14(3), 262–280.

Hardin, J.W., H. Schmiediche, and R.J. Carroll (2003a), "The Regression-Calibration Method for Fitting Generalized Linear Models with Additive Measurement Error," *Stata Journal*, 3, 361–372.

Hardin, J.W., H. Schmiediche, and R.J. Carroll (2003b), "The Simulation Extrapolation Method for Fitting Generalized Linear Models with Additive Measurement Error," *Stata Journal*, 3, 373–385.

Härdle, W. (1990), *Applied Nonparametric Methods*, Cambridge, Cambridge University Press.

Härdle, W., and O. Linton, (1994), "Applied Nonparametric Methods," in R.F. Engle and D. McFadden, eds., *Handbook of Econometrics*, volume 4, Amsterdam, North-Holland.

Härdle, W., and T.M. Stoker (1989), "Investigating Smooth Multiple Regression by the Method of Average Derivatives," *Journal of the American Statistical Association*, 84, 986–995.

Harvey, A.C. (1989), *Forecasting, Structural Time Series Models and the Kalman Filter*, Cambridge, Cambridge University Press.

Harvey, A.C., and C. Fernandes (1989), "Time Series Models for Count or Qualitative Observations (with Discussion)," *Journal of Business and Economic Statistics*, 7, 407–417.

Harvey, A.C., and G.D.A. Phillips (1982), "The Estimation of Regression Models with Time-Varying Parameters," in M. Deistler, E. Furst, and G. Schwodiauer, eds., *Games, Economic Dynamics,and Time-Series Analysis*, Vienna, Physica-Verlag, 306–321.

Harvey, A.C., and N. Shephard (1993), "Stuctural Time Series Models," in G.S. Maddala, C.R. Rao, and H.D. Vinod, eds., *Handbook of Statistics, Volume 11: Econometrics*, Amsterdam, North-Holland.

Hastie, T.J., and R.J. Tibshirani (1990), *Generalized Additive Models*, New York, Chapman and Hall.

Hausman, J.A. (1978), "Specification Tests in Econometrics," *Econometrica*, 46, 1251–1271.

Hausman, J.A., J. Abrevaya, and F.M. Scott-Morton (1998), "Misclassification of the Dependent Variable in a Discrete-Response Setting," *Journal of Econometrics*, 87, 239–269.

Hausman, J.A., B.H. Hall, and Z. Griliches (1984), "Econometric Models for Count Data with an Application to the Patents–R and D Relationship," *Econometrica*, 52, 909–938.

Hausman, J.A., G.K. Leonard, and D. McFadden (1995), "A Utility-Consistent, Combined Discrete Choice and Count Data Model: Assessing Recreational Use Losses due to Natural Resource Damage," *Journal of Public Economics*, 56, 1–30.

Hausman, J.A., A.W. Lo, and A.C. MacKinlay (1992), "An Ordered Probit Analysis of Transaction Stock Prices," *Journal of Financial Economics*, 31, 319–379.

Hausman, J.A., W.K. Newey, and J.L. Powell (1995), "Nonlinear Errors in Variables: Estimation of Some Engel Curves," *Journal of Ecoometrics*, 65, 205–233.

Hayashi, F. (2000), *Econometrics*, Princeton, NJ, Princeton University Press.

Hayfield, T., and J. Racine (2008), "Nonparametric Econometrics: The np Package," *Journal of Statistical Software*, 27(5), 1–32.

Heckman, J.J. (1976), "The Common Structure of Statistical Models of Truncation, Sample Selection, and Limited Dependent Variables and a Simple Estimator for Such Models," *Annals of Economics and Social Measurement*, 5, 475–492.

Heckman, J.J. (1978), "Dummy Endogenous Variables in a Simultaneous Equations System," *Econometrica*, 46, 931–960.

Heckman, J.J. (1981), "The Incidental Parameters Problem and the Problem of Initial Conditions in Estimating a Discrete Time-Discrete Data Stochastic Process," in C. Manski and D. McFadden, eds., *Structural Analysis of Discrete Data with Econometric Applications*, 179–197, Cambridge, MIT Press.

Heckman, J.J. (1984), "The χ^2 Goodness of Fit test for Models with Parameters Estimated from Micro Data," *Econometrica*, 52, 1543–1547.

Heckman, J.J., and R. Robb (1985), "Alternative Methods for Evaluating the Impact of Interventions," in J.J. Heckman and B. Singer, eds., *Longitudinal Analysis of Labor Market Data*, 156–245, New York, Cambridge University Press.

Heckman, J., and B. Singer (1984), "A Method of Minimizing the Impact of Distributional Assumptions in Econometric Models for Duration Data," *Econometrica*, 52, 271–320.

Hellstrom, J. (2006), "A Bivariate Count Data Model for Household Tourism Demand," *Journal of Applied Econometrics*, 21, 213–226.

Hendry, D.F. (1995), *Dynamic Econometrics*, Oxford, Oxford University Press.

Hill, S., Rothchild, D., and A.C. Cameron, (1998), "Tactical Information and the Diffusion of Protests," in D.A. Lake and D. Rothchild, eds., *Ethnic Fears and Global Engagement: The International Spread and Management of Ethnic Conflict*, 61–88, Princeton, NJ, Princeton University Press.

Hinde, J. (1982), "Compound Poisson Regression Models," in R. Gilchrist, ed., *GLIM 82: Proceedings of the International Conference on Generalised Linear Models*, 109–121, New York, Springer-Verlag.

Holgate, P. (1970), "The Modality of Some Compound Poisson Distributions," *Biometrika*, 57, 666–667.

Holly, A. (1982), "A Remark on Hausman's Specification Test," *Econometrica*, 49, 749–759.

Holly, A. (1987), "Specification Tests: An Overview," in T.F. Bewley ed., *Advances in Economics and Econometrics: Theory and Applications*, volume 1, Fifth World Congress, 59–97, Cambridge, Cambridge University Press.

Honda, Y. (1988), "A Size Correction to the Lagrange Multiplier Test for Heteroskedasticity," *Journal of Econometrics*, 38, 375–386.

Hong, H., and E. Tamer (2003), "A Simple Estimator for Nonlinear Error in Variable Models," *Journal of Econometrics*, 117, 1–19.

Horowitz, J.L. (1997), "Bootstrap Methods in Econometrics: Theory and Numerical Performance," in D. M. Kreps and K. F. Wallis, eds., *Advances in Economics and Econometrics: Theory and Applications*, volume 3, Seventh World Congress, Tokyo. Cambridge, Cambridge University Press.

Horowitz, J.L. (2001), "The Bootstrap," *Handbook of Econometrics, Volume 5*, 3160–3228, Editors J.J. Heckman and E. Leamer, Amsterdam: Elsevier.

Hsiao, C. (1989), "Consistent Estimation for Some Nonlinear Errors-in-Variables Models," *Journal of Econometrics*, 41, 159–185.

Hsiao, C. (2003), *Analysis of Panel Data*, edition 2, Cambridge, Cambridge University Press.

Huber, P.J. (1967), "The Behavior of Maximum Likelihood Estimates under Nonstandard Conditions," in L. LeCam and J. Neyman, eds., *Proceedings of the Fifth Berkeley Symposium on Mathematical Statistics and Probability*, 221–234, Berkeley, University of California Press.

Hyppolite, J., and P.K. Trivedi (2012), "Alternative Approaches for Econometric Analysis of Panel Count Data Using Dynamic Latent Class Models (with Application to Doctor Visits Data)," *Health Economics*, 21, Supplement 1, 101–128.

Ichimura, H. (1993), "Semiparametric Least Squares (SLS) and Weighted SLS Estimation of Single-Index Models," *Journal of Econometrics*, 58, 71–120.

Ichimura, H., and P. Todd (2007), "Implementing Nonparametric and Semiparametric Estimators," *Handbook of Econometrics* volume 6, Part B, 5364–5468. Amsterdam, North-Holland.

Imbens, G.W. (2002), "Generalized Method of Moments and Empirical Likelihood," *Journal of Business and Economic Statistics*, 20, 493–506.

Imbens, G.W., and T. Lancaster (1994), "Combining Micro and Macro Data in Microeconometric Models," *Review of Economic Studies*, 61, 655–680.

Jacobs, P.A., and P.A.W. Lewis (1977), "A Mixed Autoregressive Moving Average Exponential Sequence and Point Process," *Advances in Applied Probability*, 9, 87–104.

Jacobs, P.A., and P.A.W. Lewis (1978a), "Discrete Time Series Generated by Mixtures I: Correlational and Runs Properties," *Journal of the Royal Statistical Society* B , 40, 94–105.

Jacobs, P.A., and P.A.W. Lewis (1978b), "Discrete Time Series Generated by Mixtures II: Asymptotic Properties," *Journal of the Royal Statistical Society* B , 40, 222–228.

Jacobs, P.A., and P.A.W. Lewis (1983), "Stationary Discrete Autoregressive Moving Average Time Series Generated by Mixtures," *Journal of Time Series Analysis*, 4, 19–36.

Jaggia, S. (1991), "Specification Tests Based on the Generalized Gamma Model of Duration," *Journal of Applied Econometrics*, 6, 169–180.

Jaggia, S., and S. Thosar (1993), "Multiple Bids as a Consequence of Target Management Resistance: A Count Data Approach," *Review of Quantitative Finance and Accounting*, 447–457.

Jeffries, N.O. (2003), "A Note on 'Testing the Number of Components in a Normal Mixture,'" *Biometrika*, 90, 991–994.

Jewel, N. (1982), "Mixtures of Exponential Distribtions," *Annals of Statistics*, 10, 479–484.

Joe, H. (1997), *Multivariate Models and Dependence Concepts*, London, Chapman and Hall.

Johansson, P (1995), "Tests for Serial Correlation and Overdispersion in a Count Data Regression Model," *Journal of Statistical Computation and Simulation*, 53, 153–164.

Johansson, P. (1996), "Speed Limitation and Motorway Casualties – A Time Series Count Data Regression Approach," *Accident Analysis and Prevention*, 28, 73–87.

Johansson, P., and M. Palme (1996), "Do Economics Incentives Affect Work Absence – Empirical Evidence Using Swedish Micro Data," *Journal of Public Economics*, 59, 195–218.

Johnson, N.L., A.W. Kemp, and S. Kotz (2005), *Univariate Discrete Distributions*, edition 3, New York, John Wiley.

Johnson, N.L., and S. Kotz (1969), *Discrete Distributions*, Boston, Houghton Mifflin.

Johnson, N.L., S. Kotz, and N. Balakrishnan (1997), *Continuous Univariate Distributions*, edition 2, New York, John Wiley Sons.

Jordan, P., D. Brubacher, S. Tsugane, Y. Tsubono, K.F. Gey, and U. Moser (1997), "Modeling Mortality Data from a Multi-Center Study in Japan by Means of Poisson Regression with Errors in Variables," *International Journal of Epidemiology*, 26, 501–507.

Jorgensen, B. (1987), "Exponential Dispersion Models (with discussion)," *Journal of the Royal Statistical Society* B, 49, 127–162.

Jorgensen, B. (1997), *The Theory of Dispersion Models*, London, Chapman and Hall.

Jorgenson, D.W. (1961), "Multiple Regression Analysis of a Poisson Process," *Journal of the American Statistical Association*, 56, 235–245.

Jung, R.C., M. Kukuk, and R. Liesenfeld (2006), "Time Series of Count Data: Modeling, Estimation and Diagnostics," *Computational Statistics & Data Analysis*, 51, 2350–2364

Jung, R.C., R. Liesenfeld, and J.-F. Richard (2011), "Dynamic Factor Models for Multivariate Count Data: An Application to Stock-Market Trading Activity," *Journal of Business and Economic Statistics*, 29, 73–86.

Jung, C.J., and R. Winkelmann (1993), "Two Apects of Labor Mobility: A Bivariate Poisson Regression Approach," *Empirical Economics*, 18, 543–556.

Jupp, P.E., and K.V. Mardia (1980), "A General Correlation Coefficient for Directional Data and Related Regression Problems," *Biometrika*, 67, 163–173.

Kaiser, M., and N. Cressie (1997), "Modeling Poisson Variables with Positive Spatial Dependence," *Statistics and Probability Letters* 35, 423–432.

Kalbfleisch, J., and R. Prentice (1980, 2002), *The Statistical Analysis of Failure Time Data*, 1st and 2nd editions, New York, John Wiley Sons.

Karlis, D., and I. Ntzoufras (2003), "Analysis of Sports Data Using Bivariate Poisson Models," *Journal of the Royal Statistical Society* D, 52, 381–393.

Karlis, D., and E. Xekalaki, (1998), "Minimum Hellinger Distance Estimation for Poisson Mixtures," *Computational Statistics and Data Analysis* 29, 81–103.

Katz, L. (1963), "Unified Treatment of a Broad Class of Discrete Probability Distributions," in G. P. Patil, ed., *Classical and Contagious Discrete Distributions*, Calcutta, Statistical Publishing Society.

Keane, M.P., and D.E. Runkle (1992), "On the Estimation of Panel-Data Models with Serial Correlation When Instruments Are Not Strictly Exogenous," *Journal of Business and Economic Statistics*, 10, 1–9.

Kennan, J. (1985), "The Duration of Contract Strikes in U.S. Manufacturing," *Journal of Econometrics*, 28, 5–28.

Kianifard, F., and P.P. Gallo (1995), "Poisson Regression Analysis in Clinical Research," *Journal of Biopharmaceutical Statistics*, 5, 115–129.

Kim, C-J., and C.R. Nelson (1999), *State-Space Models with Regime Switching*, Cambridge, MA, MIT Press.

King, G. (1987a), "Presidential Appointments to the Supreme Court. Adding Systematic Explanation to Probabilistic Description," *American Political Quarterly*, 15, 373–386.

King, G. (1987b), "Event Count Models for International Relations: Generalizations and Applications," *International Studies Quarterly*, 33, 123–147.

King, G. (1989a), "A Seemingly Unrelated Poisson Regression Model," *Sociological Methods and Research*, 17, 235–255.

King, G. (1989b), "Variance Specification in Event Count Models: From Restrictive Assumptions to a Generalized Estimator," *American Journal of Political Science*, 33, 762–784.

Kingman, J.F.C. (1993), *Poisson Processes*, Oxford, Oxford University Press.

Kocherlakota, S., and K. Kocherlakota (1993), *Bivariate Discrete Distributions*, New York, Marcel Dekker.

Koenker, R. (1982), "A Note on Studentizing a Test for Heteroskedasticity," *Journal of Econometrics*, 17, 107–112.

Koenker, R. (2005), *Quantile Regression*, Vol. 38, New York, Cambridge University Press.

Koop, G. (2003), *Bayesian Econometrics*, New York, John Wiley Sons.

Koop, G., D.J. Poirier, and J.L. Tobias (2007), *Bayesian Econometric Methods*, New York, Cambridge University Press.

Kotz, S., and N.L. Johnson (1982–89), *Encyclopedia of Statistical Sciences*, New York, John Wiley.

Laird, N. (1978), "Nonparametric Maximum Likelihood Estimation of a Mixing Distribution," *Journal of the American Statistical Association*, 73, 805–811.

Lambert, D. (1992), "Zero-Inflated Poisson Regression with an Application to Defects in Manufacturing," *Technometrics*, 34, 1–14.

Lancaster, H.O. (1958), "The Structure of Bivariate Distributions," *Annals of Mathematical Statistics*, 29, 719–736.

Lancaster, H.O. (1963), "Correlations and Canonical Forms of Bivariate Distributions," *Annals of Mathematical Statistics*, 34, 532–538.

Lancaster, H.O. (1969), *The Chi-Squared Distribution*, New York, John Wiley Sons.

Lancaster, T. (1990), *The Econometric Analysis of Transitional Data*, New York, Cambridge University Press.

Lancaster, T. (1997), "Orthogonal Parameters and Panel Data," Working Paper No. 1997–32, Department of Economics, Brown University.

Lancaster, T. (2004), *Introduction to Modern Bayesian Econometrics*, New York, John Wiley Sons.

Landwehr, J.M., D. Pregibon, and A.C. Shoemaker (1984), "Graphical Methods for Assessing Logistic Regression Models (with Discussion)," *Journal of the American Statistical Association*, 79, 61–83.

Lawless, J.F. (1987), "Negative Binomial and Mixed Poisson Regressions," *Canadian Journal of Statistics*, 15, 209–225.

Lawless, J.F. (1995), "The Analysis of Recurrent Events for Multiple Subjects," *Applied Statistics – Journal of the Royal Statistical Society* C, 44, 487–498.

Lee, K.L., and D.R. Bell (2009), "A Spatial Negative Binomial Regression of Individual-Level Count Data with Regional and Person-Specific Covariates," unpublished manuscript, Wharton School.

Lee, L.-F. (1983), "Generalized Econometric Models with Selectivity," *Econometrica*, 51, 507–512.

Lee, L.-F. (1986), "Specification Test for Poisson Regression Models," *International Economic Review*, 27, 689–706.

Lee, L.-F. (1997), "Specification and Estimation of Count Data Regression and Sample Selection Models – A Counting Process and Waiting Time Approach," presented at the North American Summer Meetings of the Econometric Society, Pasadena, June 1997.

Lee, M-L.T. (1996), "Properties and Applications of the Sarmanov Family of Bivariate Distributions," *Communication in Statististics, Part A – Theory and Methods*, 25, 1207–1222.

Lee, Y., and J.A. Nelder (1996), "Hierarchical Generalized Linear Models," *Journal of the Royal Statistical Society Series* B, 58, 619–678.

Leon, L.F., and C.-L. Tsai (1997), "The Assessment of Model Adequacy for Markov Regression Time Series Models," *Biometrics*, 54, 1165–1175.

Leroux, B.G. (1992), "Consistent Estimation of a Mixing Distribution," *Annals of Statistics*, 20, 3, 1350–1360.

Lewbel, A. (1997), "Constructing Instruments for Regressions with Measurement Error When no Additional Data Are Available, with an Application to Patents and R&D," *Econometrica*, 65, 1201–1213.

Lewis, P.A.W. (1985), "Some Simple Models for Continuous Variate Time Series," *Water Resources Bulletin*, 21, 635–644.

Li, C.-S., J.-C. Lu, J. Park, K. Kim, P.A. Brinkley, and J.P. Peterson (1999), "Multivariate Zero-Inflated Poisson Models and Their Applications," *Technometrics*, 41, 29–38.

Li, Q., and J. Racine (2007), *Nonparametric Econometrics*, Princeton, NJ, Princeton University Press.

Li, Q., and J. Racine (2008), "Nonparametric Estimation of Conditional CDF and Quantile Functions With Mixed Categorical and Continuous Data," *Journal of Business and Economic Statistics*, 26, 423–434.

Li, T. (2000), "Estimation of Nonlinear Errors-in-Variables Models: A Simulated Minimum Distance Estimator," *Statistics and Probability Letters*, 47, 243–248.

Li, T. (2002), "Robust and Consistent Estimation of Nonlinear Errors-in-Variables Models," *Journal of Econometrics*, 110, 1–26.

Li, T., P.K. Trivedi, and J. Guo (2003), "Modeling Response Bias in Count: A Structural Approach with an Application to the National Crime Victimization Survey Data," *Sociological Methods and Research*, 31, 514–545.

Li, W.K. (1991), "Testing Model Adequacy for Some Markov Regression Models for Time Series," *Biometrika*, 78, 83–89.

Liang, K.-Y., and S. Zeger (1986), "Longitudinal Data Analysis Using Generalized Linear Models," *Biometrika*, 73, 13–22.

Liang, K.-Y., S. Zeger, and B. Qaqish (1992), "Multivariate Regression Analyses for Categorical Data," *Journal of the Royal Statistical Society Series B (Methodological)*, 54(1), 3–40.

Lindeboom, M., and G.J. van den Berg (1994), "Heterogeneity in Models for Bivariate Survival: The Importance of Mixing Distribution," *Journal of the Royal Statistical Society*, B, 56, 49–60.

Lindsay, B.G. (1983), "The Geometry of Mixture Likelihoods," *The Annals of Statistics*, 11(1), 86–94.

Lindsay, B.G. (1995), *Mixture Models: Theory, Geometry and Applications*, NSF-CBMS Regional Conference Series in Probability and Statistics, volume 5, IMS-ASA.

Lindsay, B.G., and K. Roeder (1992), "Residual Diagnostics in the Mixture Model," *Journal of the American Statistical Association*, 87, 785–795.

Lindsay, B.G., and M. Stewart (2008), "Mixture Models," in S.N. Durlauf and L.E. Blume, eds., *The New Palgrave Dictionary of Economics*, edition 2, Palgrave Macmillan.

Lo, Y., N.R. Mendell, and D.R. Rubin (2001), "Testing the Number of Components in a Normal Mixture," *Biometrika*, 88, 767–778.

Long, J.S. (1997), *Regression Models for Categorical and Limited Dependent Variables*, Thousand Oaks, CA, Sage Publications.

Lourenco, O.D., and P.L. Ferreira (2005), "Utilization of Public Health Centres in Portugal: Effect of Time Costs and Other Determinants. Finite Mixture Models Applied to Truncated Samples," *Health Economics*, 14, 939–953.

Lu, Z., Y.V. Hui, and A.H. Lee (2003), "Minimum Hellinger Distance Estimation for Finite Mixtures of Poisson Regression Models and Its Applications," *Biometrics*, 59 1016–1026.

Lu, Z., and G.E. Mizon (1996), "The Encompassing Principle and Hypothesis Testing," *Econometric Theory*, 12, 845–858.

Luceño, A. (1995), "A Family of Partially Correlated Poisson Models for Overdispersion," *Computational Statistics and Data Analysis*, 20, 511–520.

Lunn, D.J., A. Thomas, N. Best, and D. Spiegelhalter (2000), "WinBUGS – a Bayesian Modelling Framework: Concepts, Structure, and Extensibility," *Statistics and Computing*, 10, 325–337.

MacDonald, I.L., and W. Zucchini (1997), *Hidden Markov and Other Models for Discrete-Valued Time Series*, London, Chapman and Hall.

Machado, J.A.F., and J.M.C. Santos Silva (2005), "Quantiles for Counts," *Journal of the American Statistical Association*, 100(472), 1226–1237.

Maddala, G.S. (1983), *Limited Dependent and Qualitative Variables in Econometrics*, Cambridge, Cambridge University Press.

Mahajan, A. (2005), "Identification and Estimation of Regression Models with Misclassification," *Econometrica*, 74, 631–665.

Manski, C.F., and D. McFadden (1981), eds., *Structural Analysis of Discrete Data with Econometric Applications*, Cambridge, MA, MIT Press.

Marshall, A. (1996), "Copulas, Marginals, and Joint Distributions," in L. Ruschendorf, B. Schweizer, and M.D. Taylor, eds., *Distributions with Fixed Marginals and Related Topics*, 213–222, Hayward, CA, Institute of Mathematic Statistics.

Marshall, A.W., and I. Olkin (1988), "Families of Multivariate Distributions," *Journal of the American Statistical Association*, 83, 834–841.

Marshall, A.W., and I. Olkin (1990), "Multivariate Distributions Generated from Mixtures of Convolution and Product Families," in H.W. Block, A.R. Sampson, and T.H. Savits, eds., *Topics in Statistical Dependence*, IMS Lecture Notes-Monograph Series, volume 16, 371–393, Hayward, CA, Institute of Mathematical Statistics.

Mátyás, L., and P. Sevestre (2008), *The Econometrics of Panel Data: Fundamentals and Recent Developments in Theory and Practice*, edition 3, New York, Springer.

Matzkin, R. (2007), "Nonparametric Identification," in E.E. Leamer, eds., *Handbook of Econometrics*, volume 68, 5307–5368, Amsterdam, North-Holland.

Matzkin, R.L. (2010), "Estimation of Nonparametric Models with Simultaneity," Working Paper, UCLA Department of Economics.

Mayer, W.J., and W.F. Chappell (1992), "Determinants of Entry and Exit: An Application of Bivariate Compounded Poisson Distribution to U. S. Industries, 1972–1977" *Southern Economic Journal*, 58, 770–778.

McCabe, B.P.M., G.M. Martin, and D. Harris (2011), "Efficient Probabilistic Forecasts for Counts," *Journal of the Royal Statistical Society* B, 73, 253–272.

McCallum, J., C. Raymond, and C. McGilchrist (1995), "How Accurate Are Self Reports of Doctor Visits? A Comparison Using Australian Health Insurance Commission Records," pre-print.

McCullagh, P. (1983), "Quasi-likelihood Functions," *Annals of Statistics*, 11, 59–67.

McCullagh, P. (1986), "The Conditional Distribution of Goodness-of-Fit Statistics for Discrete Data," *Journal of the American Statistical Association*, 81, 104–107.

McCullagh, P., and J.A. Nelder (1983, 1989), *Generalized Linear Models*, editions 1 and 2, London, Chapman and Hall.

McFadden, D., and P.A. Ruud (1994), "Estimation by Simulation," *Review of Economics and Statistics*, 76, 591–608.

McGilchrist, C.A. (1994), "Estimation in Generalized Mixed Models, *Journal of the Royal Statistical Society* B, 56, 61–69.

McKenzie, E. (1985), "Some Simple Models for Discrete Variate Time Series," *Water Resources Bulletin*, 21, 645–650.

McKenzie, E. (1986), "Autoregressive Moving-Average Processes with Negative Binomial and Geometric Marginal Distributions," *Advances in Applied Probability*, 18, 679–705.

McKenzie, E. (1988), "Some ARMA Models for Dependent Sequences of Poisson Counts," *Advances in Applied Probability*, 22, 822–835.

McLachlan, G.J., and K.E. Basford, (1988), *Mixture Models: Inference and Application to Clustering*, New York, Marcel Dekker.

McLachlan, G., and D. Peel. (2000), *Finite Mixture Models*, New York, John Wiley Sons.

McLeod, L. (2011), "A Nonparametric vs. Latent Class Model of General Practitioner Utilization," *Journal of Health Economics*, 20, 1261–1279.

McShane, B.M., M. Adrian, E.T. Bradlow, and P.S. Fader, (2008), "Count Models Based on Weibull Interarrival Times," *Journal of Business and Economic Statistics*, 26, 369–378.

Meghir, C., and J.-M. Robin (1992), "Frequency of Purchase and Estimation of Demand Equations," *Journal of Econometrics*, 53, 53–86.

Melkersson, M., and D. Rooth (2000) "Modeling Female Fertility Using Inflated Count Data Models," *Journal of Population Economics*, 13, 189–203.

Merkle, L., and K.F. Zimmermann (1992), "The Demographics of Labor Turnover: A Comparison of Ordinal Probit and Censored Count Data Models," *Reserches Economiques de Louvain*, 58, 283–307.

Miles, D. (2001), "Joint Purchasing Decisions: A Multivariate Negative Binomial Approach," *Applied Economics*, 33, 937–946.

Min, Y., and A. Agresti (2005), "Random Effect Models for Repeated Measures of Zero-Inflated Count Data," *Statistical Modelling*, 5, 1–19.

Miranda, A. (2006), "QCOUNT: Stata Program to Fit Quantile Regression Models for Count Data," Statistical Software Components Paper S456714, Boston College Department of Economics.

Miranda, A. (2008), "Planned Fertility and Family Background: A Quantile Regression for Counts Analysis," *Journal of Population Economics*, 21, 67–81.

Miranda, A. (2010), "A Double-Hurdle Count Model for Completed Fertility Data from the Developing World," DoQSS Working Papers 10-01, Department of Quantitative Social Science, Institute of Education, University of London.

Miravete, E.J. (2011), "Testing for Complementarities among Countable Strategies," manuscript, Department of Economics, University of Texas.

Mizon, G.E., and J.-F. Richard (1986), "The Encompassing Principle and Its Application to Testing Non-Nested Hypotheses," *Econometrica*, 54, 657–678.

Montalvo, J.G. (1997), "GMM Estimation of Count-Panel-Data Models with Fixed Effects and Predetermined Instruments," *Journal of Business and Economic Statistics*, 15, 82–89.

Moore, D.F. (1986), "Asymptotic Properties of Moment Estimators for Overdispersed Counts and Proportions," *Biometrika*, 73, 583–588.

Moran, P.A.P. (1971), "Maximum Likelihood Estimation in Non-Standard Conditions," *Proceedings of the Cambridge Philosophical Society*, 70, 441–450.

Morrison, D.G., and D.C. Schmittlein (1988), "Generalizing the NBD Model for Customer Purchases: What Are the Implications and Is It Worth the Effort?," *Journal of Economics and Business Statistics*, 6, 145–166.

Moschopoulos, P., and J.G. Staniswalis (1994), "Estimation Given Conditional from an Exponential Family," *American Statistician*, 48, 271–275.

Mukhopadhyay, K., and P.K. Trivedi (1995), "Regression Models for Under-reported Counts," Pre-print.

Mullahy, J. (1986), "Specification and Testing of Some Modified Count Data Models," *Journal of Econometrics*, 33, 341–365.

Mullahy, J. (1997a), "Heterogeneity, Excess Zeros and the Structure of Count Data Models," *Journal of Applied Econometrics*, 12, 337–350.

Mullahy, J. (1997b), "Instrumental Variable Estimation of Poisson Regression Models: Application to Models of Cigarette Smoking Behavior," *Review of Economics and Statistics*, 586–593.

Mundlak, Y. (1978), "On the Pooling of Time Series and Cross Section Data," *Econometrica*, 46, 69–85.

Munkin, M., and P.K. Trivedi (1999) "Simulated Maximum Likelihood Estimation of Multivariate Mixed-Poisson Regression Models, with Application," *Econometrics Journal*, 2, 29–48.

Murphy, K., and R. Topel (1985), "Estimation and Inference in Two Step Econometric Models," *Journal of Business and Economic Statistics*, 3, 370–379.

Nagin, D.S., and K.C. Land (1993), "Age, Criminal Careers and Population Heterogeneity: Specification and Estimation of a Nonparametric, Mixed Regression Model," *Criminology*, 31, 327–362.

Nakamura, T. (1990), "Corrected Score Function for Errors-in-Variables Models: Methodology and Application to Generalized Linear Models," *Biometrika*, 77, 127–137.

Navarro, S. (2008), "Control Functions," In S.N. Durlauf and L.E. Blume, eds., *The New Palgrave Dictionary of Economics*, edition 2, London, Palgrave Macmillan.

Neal, R.M. (2000), "Markov Chain Sampling Methods for Dirichlet Process Mixture Models," *Journal of Computational and Graphical Statistics*, 9(2), 249–265.

Nelder, J.A., and D. Pregibon (1987), "An Extended Quasi-likeihood Function," *Biometrika*, 74, 221–232.

Nelder, J.A., and R.W.M. Wedderburn (1972), "Generalized Linear Models," *Journal of the Royal Statistical Society* A, 135, 370–384.

Nelsen, R.B. (2006), *An Introduction to Copulas*, edition 2, New York, Springer.

Newell, D.J. (1965), "Unusual Frequency Distributions," *Biometrics*, 21, 159–168.

Newey, W.K. (1984), "A Methods of Moments Interpretation of Sequential Estimators," *Economics Letters*, 14, 201–206.

Newey, W.K. (1985), "Maximum Likelihood Specification Testing and Conditional Moment Tests," *Econometrica*, 53, 1047–1070.

Newey, W.K. (1987), "Efficient Estimation of Limited Dependent Variable Models with Endogenous Explanatory Variables," *Journal of Econometrics*, 36, 231–250.

Newey, W.K. (1990a), "Efficient Instrumental Variable Estimation of Nonlinear Models," *Econometrica*, 58, 809–838.

Newey, W.K. (1990b), "Semiparametric Efficiency Bounds," *Journal of Applied Econometrics*, 5, 99–136.

Newey, W.K. (1993), "Efficient Estimation of Models with Conditional Moment Restrictions," in G.S. Maddala, C.R. Rao, and H.D. Vinod, eds., *Handbook of Statistics, Volume 11: Econometrics*, 419–454, Amsterdam, North-Holland.

Newey, W.K. (2001), "Flexible Simulated Moment Estimation of Nonlinear Errors in Variables Models," *Review of Economics and Statistics*, 83, 616–627.

Newey, W.K., and D. McFadden (1994), "Large Sample Estimation and Hypothesis Testing," in R.F. Engle and D. McFadden, eds., *Handbook of Econometrics*, volume 4, Amsterdam, North-Holland.

Newey, W.K., and K.D. West (1987a), "A Simple, Positive Semi-Definite, Hetereroscedasticity and Autocorrelation Consistent Covariance Matrix," *Econometrica*, 55, 703–708.

Newey, W.K., and K.D. West (1987b), "Hypothesis Testing with Efficient Methods of Moments Estimators," *International Economic Review*, 28, 777–787.

Newey, W.K., and F. Windmeijer (2009), "GMM with Many Weak Moment Conditions," *Econometrica*, 77, 687–719.

Newhouse, J.P. (1993), *Free For All? Lessons from the Rand Health Insurance Experiment*. Cambridge, MA, Harvard University Press.

Newton, H.J., and N.J. Cox (2003), "A Special Issue of the Stata Journal," *Stata Journal*, 3, 327.

Neyman, J. (1939), "On a New Class of Contagious Distributions Applicable in Entomology and Bacteriology," *Annals of Mathematical Statistics*, 10, 35–57.

Neyman, J. (1965), "Certain Chance Mechanisms Involving Discrete Distributions," in G.N. Patil, ed., *Classical and Contagious Discrete Distributions*, 1–14, Calcutta, India, Statistical Publishing House.

Nguyen, T.T., A.K. Gupta, D.M. Nguyen, and Y. Wang (2007), "Characterizations of Negative Multinomial Distributions Based on Conditional Distributions," *Metrika*, 66, 315–322.

Nickell, S. (1981), "Biases in Dynamic Models with Fixed Effects," *Econometrica*, 49, 1399–1416.

Noack, A. (1950), "A Class of Random Variables with Discrete Distributions," *Annals of Mathematical Statistics*, 21, 127–132.

Nylund, K.L., T. Asparouhov, and B.O. Muthén (2007), "Deciding on the Number of Classes in Latent Class Analysis and Growth Mixture Modeling: A Monte Carlo Simulation Study," *Structural Equation Modeling: A Multidisciplinary Journal*, 14, 535–569.

Ogaki, M. (1993), "Generalized Method of Moments: Econometric Applications," in G.C. Maddala, C.R. Rao, and H.D. Vinod, eds., *Handbook of Statistics, Volume 11: Econometrics*, Amsterdam, North-Holland, 455–488.

Okoruwa, A.A., J.V. Terza, and H.O. Nourse (1988), "Estimating Patronization Shares for Urban Retail Centers: An Extension of the Poisson Gravity Model," *Journal of Urban Economics*, 24, 241–259.

Olsen, M.K., and J.L. Schafer (2001), "A Two-Part Random-Effects Model for Semicontinuous Longitudinal Data," *Journal of the American Statistical Association*, 96, 730–745.

Ord, J., and Whitmore, G. (1986), "The Poisson-Inverse Gaussian Distribution as a Model for Species Abundance," *Communications in Statistics A: Theory and Methods*, 15, 853–876.

Owen, A.B. (1988), "Empirical Likelihood Ratio Confidence Intervals for a Single Functional," *Biometrika*, 75, 237–249.

Ozuna, T., and I. Gomaz (1995), "Specification and Testing of Count Data Recreation Demand Functions," *Empirical Economics*, 20, 543–550.

Pagan, A.R. (1986), "Two Stage and Related Estimators and Their Applications," *Review of Economic Studies*, 53, 517–38.

Pagan, A.R., and A. Ullah (1999), *Nonparametric Econometrics*, Cambridge, UK, Cambridge University Press.

Pagan, A.R., and F. Vella (1989), "Diagnostic Tests for Models Based on Individual Data: A Survey," *Journal of Applied Econometrics*, 4, S29–S59.

Page, M. (1995), "Racial and Ethnic Discrimination in Urban Housing Markets – Evidence from a Recent Audit Survey," *Journal of Urban Economics*, 38, 183–206.

Palmgren, J. (1981), "The Fisher Information Matrix for Log-Linear Models Arguing Conditionally in the Observed Explanatory Variables," *Biometrika*, 68, 563–566.

Papadapoulos, G., and J.M.C. Santos Silva (2008), "Identification Issues in Models of Underreported Counts," University of Essex Discussion Paper 657.

Patil, G.P. (1970), editor, *Random Counts in Models and Structures*, volumes 1–3, University Park, Pennsylvania State University Press.

Patil, S.A., D.I. Patel, and J.L. Kovner (1977), "On Bivariate Truncated Poisson Distribution," *Journal of Statistical Computation and Simulation*, 6, 49–66.

Patton, A. (2006), "Modelling Asymmetric Exchange Rate Dependence," *International Economic Review*, 47, 527–556.

Pesaran, M.H. (1987), "Global and Partial Non-Nested Hypotheses and Asymptotic Local Power," *Econometric Theory*, 3, 677–694.

Pierce, D. (1982), "The Asymptotic Effect of Substituting Estimators for Parameters in Certain Types of Statistics," *Annals of Statistics*, 10, 475–478.

Pierce, D.A., and D.W. Schafer (1986), "Residuals in Generalized Linear Models," *Journal of the American Statistical Association*, 81, 977–986.

Pinquet, J. (1997), "Experience Rating through Heterogeneous Models," Working Paper No. 9725, THEMA, Universite Paris X-Nanterre.

Pohlmeier, W., and V. Ulrich (1995), "An Econometric Model of the Two-Part Decisionmaking Process in the Demand for Health Care," *Journal of Human Resources*, 30, 339–361.

Poisson, S.-D. (1837), *Recherches sur la probabilité des jugements en matière criminelle et en matière civile*, Paris, Bachelier.

Pope, A., J. Schwartz, and M. Ransom (1992), "Daily Mortality and PM10 Pollution in Utah Valley," *Archives of Environmental Health*, 47, 211–217.

Pregibon, D. (1980), "Goodness of Link Tests for Generalized Linear Models, *Journal of the Royal Statistical Society* C *(Applied Statistics)*, 29, 15–23.

Pregibon, D. (1981), "Logistic Regression Diagnostics," *Annals of Statistics*, 9, 705–724.

Press, W.H., S.A. Teukolsky, W.T. Vetterling, and B.P. Flannery (2008), *Numerical Recipes: The Art of Scientific Computing*, edition 3, Cambridge, Cambridge University Press.

Pudney, S. (1989), *Modelling Individual Choice: The Econometrics of Corners, Kinks and Holes*, New York, Basil Blackwell.

Qin, J., and J. Lawless (1994), "Empirical Likelihood and General Estimating Equations," *Annals of Statistics*, 22, 300–325.

Rabe-Hesketh, S., A. Skrondal, and A. Pickles (2003), "Maximum Likelihood Estimation of Generalized Linear Models with Covariate Measurement Error," *Stata Journal*, 3, 386–411.

Ramaswamy, V., E.N. Anderson, and W.S. DeSarbo (1994), "A Diasaggregate Negative Binomial Regression Procedure for Count Data Analysis," *Management Science*, 40, 405–417.

Ridout, M.S., and P. Besbeas (2004). "An Empirical Model of Underdispersed Count Data," *Statistical Modeling*, 4(1), 77–89.

Robert, C.P. (1996), "Mixtures of Distributions: Inference and Estimation," in W.R. Gilks, S. Richardson, and D.J. Spiegelhalter (eds.), *Markov Chain Monte Carlo in Practice*, 441–464, London, Chapman and Hall.

Robert, C.P., and G. Casella (1996), *Monte Carlo Methods*, New York: Springer-Verlag.

Robinson, P.M. (1987a), "Semiparametric Econometrics: A Survey," *Journal of Applied Econometrics*, 2, 35–52.

Robinson, P.M. (1987b), "Asymptotically Efficient Estimation in the Presence of Heteroskedasticity of Unknown Form," *Econometrica*, 55, 875–891.

Robinson, P.M. (1988), "Root-N Consistent Semiparametric Regression," *Econometrica*, 56, 931–954.

Ronning, G., and R.C. Jung (1992), "Estimation of a First Order Autoregressive Process with Poisson Marginals for Count Data," in L. Fahrmeir et al., eds., *Advances in GLIM and Statistical Modelling*, 188–194, New York, Springer-Verlag.

Rose, N. (1990), "Profitability and Product Quality: Economic Determinants of Airline Safety Performance," *Journal of Political Economy*, 98, 944–964.

Ross, S.M. (1996), *Stochastic Processes*, edition 2, New York, John Wiley Sons.

Ruser, J.W. (1991), "Workers' Compensation and Occupational Injuries and Illnesses," *Journal of Labor Economics*, 9, 325–350.

Santos Silva, J.M.C. (1997), "Unobservables in Count Data Models for On-site Samples," *Economics Letters* 54(3), 217–220.

Santos Silva, J., and F. Covas (2000), "A Modified Hurdle Model for Completed Fertility," *Journal of Population Economics*, 13, 173–188.

Santos Silva, J.M.C., and S. Tenreyro (2006), "The Log of Gravity," *The Review of Economics and Statistics*, 88, 641–658.

Santos Silva, J.M.C., and F. Windmeijer (2001), "Two-part Multiple Spell Models for Healthcare Demand," *Journal of Econometrics*, 104, 67–89.

Sargan, J.D. (1958), "The Estimation of Economic Relationships using Instrumental Variables," *Econometrica*, 26, 393–415.

Sarmanov, O.V. (1966), "Generalized Normal Correlations and Two-Dimensional Fréchet classes," *Soviet Mathematics – Doklady*, 168, 596–599.

Savani, V., and A.A. Zhigljavsky (2007), "Efficient Parameter Estimation for Independent and INAR(1) Negative Binomial Samples," *Metrika*, 65, 207–225.

Scallan, A., R. Gilchrist, and M. Green (1984), "Fitting Parametric Link Functions in Generalised Linear Models," *Computational Statistics and Data Analysis*, 2, 37–49.

Schall, R. (1991), "Estimation in Generalized Linear Models with Random Effects," *Biometrika*, 78, 719–727.

Schennach, S. (2007), "Instrumental Variable Estimation of Nonlinear Errors-in-Variables Models," *Econometrica*, 75, 201–239.

Schmidt, P., and A. Witte (1989), "Predicting Criminal Recidivism Using Split-Population Survival Time Models," *Journal of Econometrics*, 40, 141–159.

Schneider, A.L. (1981), "Differences between Survey and Police Information about Crime," in R.G. Lehnan and W.G. Skogan, eds., *The NCS Working Papers*, volume 1, Washington, DC, U.S. Department of Justice, 33–34.

Schwartz, E.S., and W.N. Torous (1993), "Mortgage Prepayments and Default Decisions: A Poisson Regression Approach," *AREUEA Journal: Journal of the American Real Estate Association*, 21, 431–449.

Schwarz, G. (1978), "Estimating the Dimension of a Model," *Annals of Statistics*, 6, 461–464.

Schweizer, B., and A. Sklar (1983), *Probabilistic Metric Spaces*, New York, North Holland.

Scott, S.L. (2002), "Bayesian Methods for Hidden Markov Models: Recursive Computing in the 21st Century," *Journal of the American Statistical Association*, 97, 337–351.

Sellar, C., J.R. Stoll, and J.P. Chavas (1985), "Validation of Empirical Measures of Welfare Change: A Comparison of Nonmarket Techniques," *Land Economics*, 61, 156–175.

Severini, T.A., and J.G. Staniswalis (1994), "Quasi-Likelihood Estimation in Semiparametric Models," *Journal of American Statistical Association*, 89, 501–511.

Shaban, S.A. (1988), "Poisson-Lognormal Distributions," in E.L. Crow and K. Shimizu, eds., *Lognormal Distributions*, 195–210, New York, Marcel Dekker.

Shaked, M. (1980), "On Mixtures from Exponential Families," *Journal of the Royal Statistical Society* B, 42, 415–433.

Shaw, D. (1988), "On-Site Samples' Regression Problems of Non-Negative Integers, Truncation and Endogenous Stratification," *Journal of Econometrics*, 37, 211–223.

Shephard, N. (1995), "Generalized Linear Autoregression," Nuffield College, Oxford University, pre-print.

Shephard, N., and M.K. Pitt (1997), "Likelihood Analysis of Non-Gaussian measurement Time Series," *Biometrika*, 84, 653–667.

Shiba, T., and H. Tsurumi (1988), "Bayesian and Non-Bayesian Tests of Independence in Seemingly Unrelated Regressions," *International Economic Review*, 29, 377–395.

Shklyar, S., and H. Schneeweiss (2005), "A Comparison of Asymptotic Covariance Matrices of Three Consistent Estimators in the Poisson Regression Model with Measurment Errors," *Journal of Multivariate Analysis*, 94, 250–270.

Sichel, H.S. (1971), "On a Family of Discrete Distributions Particularly Suited to Represent Long-tailed Frequency Data," in *Proceedings of the Third Symposium on Mathematical Statistics*. N.F. Laubscher (ed.), S.A.C.S.I.R (Pretoria), 51–97.

Sichel, H.S. (1974), "On a Distribution Representing Sentence-Length in Written Prose," *Journal of the Royal Statistical Society* A, 137, 25–34.

Sichel, H.S. (1975), "On a Distribution Law for Word Frequencies," *Journal of the American Statistical Association*, 70, 542–547.

Simar, L. (1976), "Maximum Likelihood Estimation of a Compound Poisson Process," *Annals of Statistics*, 4, 1200–1209.

Singh, A.C., and G.R. Roberts (1992), "State Space Modelling of Cross-Classified Time Series of Counts," *International Statistical Review*, 60, 321–335.

Skellam, J.G. (1946), "The Frequency Distribution of the Difference between Two Poisson Variates Belonging to Different Populations," *Journal of the Royal Statistical Society* A, 109, 296.

Sklar, A. (1973), "Random Variables, Joint Distributions, and Copulas," *Kybernetica*, 9, 449–460.

Skrondal, A., and S. Rabe-Hesketh (2004), *Generalized Latent Variable Modeling: Multilevel, Longitudinal, and Structural Equation Models*, London, Chapman and Hall.

Smith, M.S., and M.A. Khaled (2010), "Estimation of Copula Modles with Discrete Margins," manuscript, Melbourne Business School.

Smith, R.J. (1997), "Alternative Semi-Parametric Likelihood Approaches to Generalised Method of Moment Estimation," *Economic Journal*, 107, 503–519.

Smyth, G.K. (1989), "Generalized Linear Models with Varying Dispersion," *Journal of Royal Statistical Society* B, 51, 47–60.

Solow, A.R. (1993), "Estimating Record Inclusion Probability," *American Statistician*, 47, 206–209.

Staniswalis, J. (1989), "The Kernel Estimate of a Regression Function in Likelihood-Based Models," *Journal of the American Statistical Association*, 84, 426–431.

Stefanski, L.A. (1989), "Unbiased Estimation of a Nonlinear Function of a Normal Mean with the Application to Measurement Error Models," *Communications in Statistics*, Series A, 18, 4335–4358.

Stefanski, L.A., and J.S. Buzas (1995), "Instrumental Variable Estimation in Binary Regression Measurement Error Models," *Journal of the American Statistical Association*, 90, 541–550.

Stefanski, L.A., and R.J. Carroll (1987), "Conditional Scores and Optimal Scores in Generalized Linear Measurement Error Models," *Biometrika*, 74, 703–716.

Steutel, F.W., and K. Van Harn (1979), "Discrete Analogues of Self-Decomposability and Stability," *Annals of Probability*, 7, 893–899.

Stewart, M. (2007), "The Interrelated Dynamics of Unemployment and Low-Wage Employment," *Journal of Applied Econometrics*, 22, 511–531.

Stroud, A.H., and D. Secrest (1966), *Gaussian Quadrature Formulas*, Englewood Cliffs, NJ, Prentice Hall.

Stukel, T. (1988), "Generalized Logistic Models," *Journal of The American Statistical Association*, 83, 426–431.

Szu, H., and R. Hartley (1987), "Fast Simulated Annealing," *Physics Letters A*, 122, 157–162.

Tauchen, G. (1985), "Diagnostic Testing and Evaluation of Maximum Likelihood Models," *Journal of Econometrics*, 30, 415–443.

Taylor, H.M., and S. Karlin (1998), *An Introduction to Stochastic Modelling*, third edition, San Diego, CA, Academic Press.

Teicher, H. (1961), "The Identifiability of Mixtures," *Annals of Mathematical Statistics*, 32, 244–248.

Terza, J.V. (1985), "A Tobit-Type Estimator for the Censored Poisson Regression Model," *Economics Letters*, 18, 361–365.

Terza, J.V. (1998), "Estimating Count Data Models with Endogenous Switching: Sample Selection and Endogenous Switching Effects," *Journal of Econometrics*, 84, 129–139.

Terza, J.V., A. Basu, and P.J. Rathouz (2008), "Two-Stage Residual Inclusion Estimation: Addressing Endogeneity in Health Econometric Modeling," *Journal of Health Economics*, 27, 531–543.

Terza, J.V., and P.W. Wilson (1990), "Analyzing Frequencies of Several Types of Events: A Mixed Multinomial-Poisson Approach," *Review of Economics and Statistics*, 72, 108–115.

Thall, P.F., and S.C. Vail (1990), "Some Covariance Models for Longitudinal Count Data with Overdispersion," *Biometrika*, 46, 657–671.

Thompson, J., T. Palmer, and S. Moreno (2006), "Bayesian Analysis in Stata with WinBUGS," *The Stata Journal*, 6, 530–549.

Titterington, D.M., A.F. Smith, and U.E. Makov (1985), *Statistical Analysis of Finite Mixture Distributions*, Chichester, John Wiley.

Trivedi, P.K., and M. Munkin (2010), "Recent Developments in Cross Section and Panel Count Models," in A. Ullah and D. Giles, eds., *Handbook of Empirical Economics and Finance*, 87–131, Boca Rotan, FL, Francis and Taylor.

Trivedi, P.K., and D.M. Zimmer (2007), "Copula Modeling: An Introduction for Practitioners," *Foundations and Trends in Econometrics*, 1, 1–110.

Tweedie, M.C.K. (1957), "Statistical Properties of Inverse Gaussian Distributions: I," *Annals of Mathematical Statistics*, 28, 362–377.

U.S. National Center for Health Statistics (1965), *Health Interview Responses Compared with Medical Records*, Vital and Health Statistics Series 2, No. 7, Public Health Service, Washington, DC, U.S. Government Printing Office.

U.S. National Center for Health Statistics (1967), *Interview Data on Chronic Conditions Compared with Information Derived from Medical Records*, Vital and Health

Statistics Series 2, No. 23, Public Health Service, Washington, DC, U.S. Government Printing Office.

Van Duijn, M.A.J., and U. Böckenholt (1995), "Mixture Models for the Analysis of Repeated Count Data," *Applied Statistics – Journal of the Royal Statistical Society* C, 44, 473–485.

Van Praag, B.M.S., and E.M. Vermeulen (1993), "A Count-Amount Model With Endogenous Recording of Observations," *Journal of Applied Econometrics*, 8, 383–395.

Vose, D. (2008), *Risk Analysis: A Quantitative Guide*, edition 3, New York, John Wiley Sons.

Vuong, Q.H. (1989), "Likelihood Ratio Tests for Model Selection and Non-Nested Hypotheses," *Econometrica*, 57, 307–333.

Wang, K., K.K.W. Yau, and A.H. Lee (2002), "A Hierarchical Poisson Mixture Regression Model to Analyze Maternity Length of Hospital Stay," *Statistics in Medicine*, 21, 3639–3654.

Wang, P. (2003), "A Bivariate Zero-Inflated Negative Binomial Regression Model for Count Data with Excess Zeros," *Economics Letters*, 78, 373–378.

Watermann, J., R. Jankowski, and I. Madan (1994), "Under-Reporting of Needlestick Injuries by Medical Students," *Journal of Hospital Infection*, 26, 149.

Wedderburn, R.W.M. (1974), "Quasi-Likelihood Functions, Generalized Linear Models, and the Gauss-Newton Method," *Biometrika*, 61, 439–447.

Wedel, M., U. Bockenholt, and W.A. Kamakura (2003), "Factor Models for Multivariate Count Data," *Journal of Multivariate Analysis*, 87, 356–369.

Wedel, W., W.S. DeSarbo, J.R. Bult, and V. Ramaswamy (1993), "A Latent Class Poisson Regression Model for Heterogeneous Count Data," *Journal of Applied Econometrics*, 8, 397–411.

Weisberg, S., and A.H. Welsh (1994), "Adapting for the Missing Link," *Annals of Statistics*, 22, 1674–1700.

Weiss, A.A. (1999), "A Simultaneous Binary Choice/Count Model with an Application to Credit Card Approvals," in R.F. Engle and H. White, eds., *Cointegration, Causality, and Forecasting: A Festschrift in Honour of Clive W.J. Granger*, ch. 18, Oxford, Oxford University Press.

Weiss, C.H. (2008), "Serial Dependence and Regression of Poisson INARMA Models," *Journal of Statistical Planning and Inference*, 138, 2975–2990.

Wellner, J.A., and Y. Zhang (2007), "Two Likelihood-Based Semiparametric Estimation Methods for Panel Count Data with Covariates," *Annals of Statistics*, 35, 2106–2142.

West, M., and J. Harrison (1997), *Bayesian Forecasting and Dynamic Models*, edition 2, New York, Springer.

West, M., P.J. Harrison, and H.S. Migon (1985), "Dynamic Generalized Linear Models and Bayesian Forecasting (with Discussion)," *Journal of the American Statistical Association*, 80, 73–97.

White, H. (1980), "A Heteroskedasticity-Consistent Covariance Matrix Estimator and a Direct Test for Heteroskedasticity," *Econometrica*, 46, 817–838.

White, H. (1982), "Maximum Likelihood Estimation of Misspecified Models, *Econometrica*, 50, 1–25.

White, H. (1994), *Estimation, Inference, and Specification Analysis*, Cambridge, UK, Cambridge University Press.

Whittemore, A.S., and G. Gong (1991), "Poisson Regression with Misclassified Counts, Application to Cervical Cancer Mortality Rates," *Applied Statistics*, 40, 81–93.

Willmot, G.E. (1987), "The Poisson-Inverse Gaussian Distribution as an Alternative to the Negative Binomial," *Scandinavian Actuarial Journal*, 198, 113–127.

Williams, D.A. (1987), "Generalized Linear Model Diagnostics Using the Deviance and Single Case Deletions," *Applied Statistics*, 36, 181–191.

Windmeijer, F. (2000), "Moment Conditions for Fixed Effects Count Data Models with Endogenous Regressors," *Economics Letters*, 68, 21–24.

Windmeijer, F. (2002), "ExpEnd, A Gauss Programme for Non-Linear GMM Estimation of Exponential Models with Endogenous Regressors for Cross Section and Panel Data," CEMMAP Working Paper No. CWP 14/02.

Windmeijer F. (2005), "A Finite Sample Correction for the Variance of Linear Efficient Two-Step GMM estimators," *Journal of Econometrics*, 126, 25–517.

Windmeijer, F. (2008), "GMM for Panel Count Data Models," *Advanced Studies in Theoretical and Applied Econometrics*, 46, 603–624.

Windmeijer, F.A.G., and J.M.C. Santos Silva (1997), "Endogeneity in Count Data Models; an Application to Demand for Health Care," *Journal of Applied Econometrics*, 12, 281–294.

Winkelmann, R. (1995), "Duration Dependence and Dispersion in Count-Data Models," *Journal of Business and Economic Statistics*, 13, 467–474.

Winkelmann, R. (2004), "Health Care Reform and the Number of Doctor Visits – An Econometric Analysis," *Journal of Applied Econometrics*, 19, 455–472.

Winkelmann, R. (2006), "Reforming Health Care: Evidence from Quantile Regressions for Counts," *Journal of Health Economics*, 25, 131–145.

Winkelmann, R. (2008), *Econometric Analysis of Count Data*, edition 5, Berlin, Germany, Springer-Verlag.

Winkelmann, R., and K.F. Zimmermann (1991), "A New Approach for Modeling Economic Count Data," *Economics Letters*, 37, 139–143.

Winkelmann, R., and K.F. Zimmermann (1995), "Recent Developments in Count Data Modelling: Theory and Application," *Journal of Economic Surveys*, 9, 1–24.

Wong, W.H. (1986), "Theory of Partial Likelihood," *Annals of Statistics*, 14, 88–123.

Wooldridge, J.M. (1990a), "A Unified Approach to Robust, Regression-Based Specification Tests," *Econometric Theory*, 6, 17–43.

Wooldridge, J.M. (1990b), "An Encompassing Approach to Conditional Mean Tests with Applications to Testing Nonlinear Hypotheses," *Journal of Econometrics*, 45, 331–350.

Wooldridge, J.M. (1990c), "Distribution-Free Estimation of Some Nonlinear Panel Data Models," Working Paper No. 564, Department of Economics, Massachusetts Institute of Technology.

Wooldridge, J.M. (1991a), "On the Application of Robust, Regression-Based Diagnostics to Models of Conditional Means and Conditional Variances," *Journal of Econometrics*, 47, 5–46.

Wooldridge, J.M. (1991b), "Specification Testing and Quasi-Maximum-Likelihood Estimation," *Journal of Econometrics*, 48, 29–55.

Wooldridge, J.M. (1997a), "Multiplicative Panel Data Models without the Strict Exogeneity Assumption," *Econometric Theory*, 13, 667–679.

Wooldridge, J.M. (1997b), "Quasi-Likelihood Methods for Count Data," in M.H. Pesaran and P. Schmidt, eds., *Handbook of Applied Econometrics, Volume II: Microeconometrics*, Malden, MA, Blackwell Publishers.

Wooldridge, J.M. (2005), "Simple Solutions to the Initial Conditions Problem in Dynamic, Nonlinear Panel Data Models with Unobserved Heterogeneity," *Journal of Applied Econometrics*, 20, 39–54.

Wooldridge, J.M. (2010), *Econometric Analysis of Cross Section and Panel Data*, edition 2, Cambridge, MA, MIT Press.

Wu, D. (1973), "Alternative Tests of Independence between Stochastic Regressors and Disturbances," *Econometrica*, 41, 733–750.

Xiang, L., K.K.W. Yau, Y. Van Hui, and A.H. Lee (2008), "Minimum Hellinger Distance Estimation for k-Component Poisson Mixture with Random Effects," *Biometrics* 64, 508–518.

Yannaros, N. (1993), "Analyzing Incomplete Count Data," *The Statistician*, 42, 181–187.

Yule, G.U. (1944), *The Statistical Study of Literary Vocabulary*, Cambridge: Cambridge University Press.

Zeger, S.L. (1988), "A Regression Model for Time Series of Counts," *Biometrika*, 75, 621–629.

Zeger, S.L., and K.-Y. Liang (1986), "Longitudinal Data Analysis for Discrete and Continuous Outcomes," *Biometrics*, 42, 121–130.

Zeger, S.L., K.-Y. Liang, and P.S. Albert (1988), "Models for Longitudinal Data: A Generalized Estimating Equation Approach," *Biometrics*, 44, 1049–1060.

Zeger, S.L., and Qaqish B. (1988), "Markov Regression Models for Time Series: A Quasi-Likelihood Approach," *Biometrics*, 44, 1019–1031.

Zhu, R., and H. Joe (2006), "Modeling Count Data Time Series with Markov Processes Based on Binomial Thinning," *Journal of Time Series Analysis*, 27, 725–738.

Ziegler, A. (2011), *Generalized Estimating Equations*, New York, Springer.

Ziegler, A., C. Kastner, and M. Blettner (1998), "The Generalized Estimating Equations: An Annotated Bibliography," *Biometrical Journal*, 40, 115–139.

Zimmer, D.M., and P.K. Trivedi (2006), "Using Trivariate Copulas to Model Sample Selection and Treatment Effects," *Journal of Business and Economic Statistics*, 24(1), 63–76.

Author Index

Subject Index

Other titles in the series *(continued from page iii)*

Eric Ghysels, Norman R. Swanson, and Mark Watson, Editors, *Essays in econometrics: Collected papers of Clive W. J. Granger (Volume I)*, 9780521772976, 9780521774963

Eric Ghysels, Norman R. Swanson, and Mark Watson, Editors, *Essays in econometrics: Collected papers of Clive W. J. Granger (Volume II)*, 9780521792073, 9780521796491

Cheng Hsiao, *Analysis of panel data, second edition*, 9780521818551, 9780521522717

Mathias Dewatripont, Lars Peter Hansen, and Stephen J. Turnovsky, Editors, *Advances in economics and econometrics – Eighth World Congress (Volume I)*, 9780521818728, 9780521524117

Mathias Dewatripont, Lars Peter Hansen, and Stephen J. Turnovsky, Editors, *Advances in economics and econometrics – Eighth World Congress (Volume II)*, 9780521818735, 9780521524124

Mathias Dewatripont, Lars Peter Hansen, and Stephen J. Turnovsky, Editors, *Advances in economics and econometrics – Eighth World Congress (Volume III)*, 9780521818742, 9780521524131

Roger Koenker, *Quantile regression*, 9780521845731, 9780521608275

Charles Blackorby, Walter Bossert, and David Donaldson, *Population issues in social choice theory, welfare economics, and ethics*, 9780521825511, 9780521532587

John E. Roemer, *Democracy, education, and equality*, 9780521846653, 9780521609135

Richard Blundell, Whitney K. Newey, and Thorsten Persson, Editors, *Advances in economics and econometrics – Ninth World Congress (Volume I)*, 9780521871525, 9780521692083

Richard Blundell, Whitney K. Newey, and Thorsten Persson, Editors, *Advances in economics and econometrics – Ninth World Congress (Volume II)*, 9780521871532, 9780521692090

Richard Blundell, Whitney K. Newey, and Thorsten Persson, Editors, *Advances in economics and econometrics – Ninth World Congress (Volume III)*, 9780521871549, 9780521692106

Fernando Vega-Redondo, *Complex social networks*, 9780521857406, 9780521674096

Itzhak Gilboa, *Theory of decision under uncertainty*, 9780521517324, 9780521741231

Krislert Samphantharak and Robert M. Townsend, *Households as corporate firms: An analysis of household finance using integrated household surveys and corporate financial accounting*, 9780521195829, 9780521124164

Rakesh Vohra, *Mechanism design: A linear programming approach*, 9781107004368, 9780521179461

Daron Acemoglu, Manuel Arellano, and Eddie Dekel, Editors, *Advances in economics and econometrics – Tenth World Congress (Volume I)*, 9781107016040, 9781107638105

Daron Acemoglu, Manuel Arellano, and Eddie Dekel, Editors, *Advances in economics and econometrics – Tenth World Congress (Volume II)*, 9781107016057, 9781107674165

Daron Acemoglu, Manuel Arellano, and Eddie Dekel, Editors, *Advances in economics and econometrics – Tenth World Congress (Volume III)*, 9781107016064, 9781107627314

Andrew Harvey, *Dynamic models for volatility and heavy tails: With applications to financial and economic time series*, 9781107034723/ 9781107630024